Hopf algebras have important connections to quantum theory, Lie algebras, knot and braid theory, operator algebras, and other areas of physics and mathematics. They have been intensely studied in the last decade; in particular, the solution of a number of conjectures of Kaplansky from the 1970s has led to progress on the classification of semisimple Hopf algebras and on the structure of pointed Hopf algebras. There has been much progress also on actions and coactions of Hopf algebras and on Hopf Galois extensions. Many new methods have been used for these results: modular and braided categories, representation theory, algebraic geometry, and Lie methods such as Cartan matrices.

The contributors to this volume of expository papers were participants in the Hopf Algebras Workshop held at MSRI as part of the 1999–2000 Year on Noncommutative Algebra. Together the papers give a clear picture of the current trends in this active field, with a focus on what is likely to be important in future research.

Among the topics covered are results toward the classification of finite-dimensional Hopf algebras (semisimple and non-semisimple), as well as what is known about the extension theory of Hopf algebras. Some papers consider Hopf versions of classical topics, such as the Brauer group, while others are closer to recent work in quantum groups. The book also explores the connections and applications of Hopf algebras to other fields.

Mathematical Sciences Research Institute
Publications

43

New Directions in Hopf Algebras

Mathematical Sciences Research Institute Publications

Volumes 1–4 and 6–27 are published by Springer-Verlag

New Directions in Hopf Algebras

Edited by

Susan Montgomery
University of Southern California

Hans-Jürgen Schneider
Universität München

The Mathematical Sciences Research Institute wishes to acknowledge support by the National Science Foundation. This material is based upon work supported by NSF Grants 9701755 and 9810361.

CAMBRIDGE UNIVERSITY PRESS
Cambridge, New York, Melbourne, Madrid, Cape Town, Singapore,
São Paulo, Delhi, Dubai, Tokyo

Cambridge University Press
The Edinburgh Building, Cambridge CB2 8RU, UK

Published in the United States of America by Cambridge University Press, New York

www.cambridge.org
Information on this title: www.cambridge.org/9780521124317

© Mathematical Sciences Research Institute 2002

First published 2002
This digitally printed version 2009

A catalogue record for this publication is available from the British Library

ISBN 978-0-521-81512-3 Hardback
ISBN 978-0-521-12431-7 Paperback

Contents

Preface

This collection of expository papers highlights progress and new directions in Hopf algebras.

Most of the contributors were participants in the Hopf Algebras Workshop held at MSRI in late 1999, although the papers are not necessarily tied to lectures given at the workshop. The workshop was very timely, as much progress has been made recently within Hopf algebras itself (for example, some long-standing conjectures of Kaplansky have been solved) as well as in studying Hopf algebras that have arisen in other areas, such as mathematical physics and topology.

The first two papers discuss progress on classifying certain classes of Hopf algebras.

In the paper by Andruskiewitsch and Schneider, pointed Hopf algebras are studied in terms of their infinitesimal braiding. Important examples of pointed Hopf algebras are group algebras and the quantum groups coming from Lie theory, that is, $U_q(\mathfrak{g})$, $\mathfrak{g}$ a semisimple Lie algebra, introduced by Drinfel'd and Jimbo, and Lusztig's finite-dimensional Frobenius kernels $\mathfrak{u}(\mathfrak{g})$ where q is a root of unity. The classification of all finite-dimensional pointed Hopf algebras with abelian group of group-like elements in characteristic zero seems to be in reach today. In this classification, Hopf algebras that are closely related to the Frobenius kernels play the main role.

In the second paper, Gelaki gives a survey on what is known on finite-dimensional triangular Hopf algebras. In a sense these Hopf algebras are close to group algebras. Over the complex numbers, triangular semisimple, and more generally triangular Hopf algebras such that the tensor product of simple representations is semisimple, are Drinfeld twists of group algebras of finite groups or supergroups. Thus these Hopf algebras are completely classified by means of group-theretical data. A main ingredient in these results is Deligne's theorem on Tannakian categories.

Thus twisting is an important method of describing new Hopf algebras that in general are neither commutative nor cocommutative. Another important idea is to study extensions. The papers by Masuoka and Schauenburg explain the modern theory of extensions of Hopf algebras and (co)quasi Hopf algebras.

Here, recent versions and generalizations of an old exact sequence of G. I. Kac for Hopf extensions of group algebras from 1963 are crucial. Masuoka defined this sequence for universal enveloping algebras of Lie algebras. He showed how this sequnce can be used to compute many concrete examples of extensions in the group and Lie algebra case.

In Schauenburg's paper, a very general formulation and a completely new interpretation of Kac's sequence are given. The basic tools are reconstruction theorems, which allow one to construct Hopf algebras or (co)quasi Hopf algebras from various monoidal categories.

Letzter generalizes the classical representation theory of symmetric pairs, consisting of a semisimple complex Lie algebra and its fixed elements under an involution, to the quantum case. Since $U_q(\mathfrak{g})$ does not contain enough sub Hopf algebras, the theory has to deal with left coideal subalgebras instead of sub Hopf algebras. For general Hopf algebras, left and right coideal subalgebras are the natural objects to describe quotients and invariants.

Takeuchi's "short course on quantum matrices" starts with an elegant treatment of the 2×2 quantum matrices and their interpretation in the theory of knot invariants. Among many other aspects of quantum matrices for GL_n and SL_n, Takeuchi's theory of q-representations of quantum groups is discussed.

Radford's paper is a survey on his recent work with L. Kauffmann on invariants of knots and links. Motivated by the applications of quasitriangular and coquasitriangular Hopf algebras to topology, axioms for abstract quantum algebras and coalgebras are given. These new algebras and coalgebras are used in a very direct way to obtain topological invariants.

In another direction, the paper by Nikshych and Vainerman also goes beyond Hopf algebras in the strict sense. They study quantum groupoids, also called weak Hopf algebras. The groupoid algebra of a groupoid is a quantum groupoid, and a Hopf algebra if the groupoid is a group. One motivation comes from operator algebras in connection with depth 2 von Neumann subfactors. The theory is developed from scratch, and applications to topology and operator algebras are given.

The paper by van Oystaeyen and Zhang is an exposition of recent work on the Brauer group of a Hopf algebra. This new invariant of a Hopf algebra H is defined by introducing the notion of H-Azumaya algebras in the braided category of Yetter-Drinfeld modules over H. Thus Hopf algebra theory is related to the classical theory of the Brauer group of a commutative ring and of the Brauer-Long group of a graded algebra.

Susan Montgomery
Hans-Jürgen Schneider

New Directions in Hopf Algebras
MSRI Publications
Volume 43, 2002

Pointed Hopf Algebras

NICOLÁS ANDRUSKIEWITSCH AND HANS-JÜRGEN SCHNEIDER

ABSTRACT. This is a survey on pointed Hopf algebras over algebraically closed fields of characteristic 0. We propose to classify pointed Hopf algebras A by first determining the graded Hopf algebra gr A associated to the coradical filtration of A. The A_0-coinvariants elements form a braided Hopf algebra R in the category of Yetter–Drinfeld modules over the coradical $A_0 = \Bbbk\Gamma$, Γ the group of group-like elements of A, and gr $A \simeq R\#A_0$. We call the braiding of the primitive elements of R the infinitesimal braiding of A. If this braiding is of Cartan type [AS2], then it is often possible to determine R, to show that R is generated as an algebra by its primitive elements and finally to compute all deformations or liftings, that is pointed Hopf algebras such that gr $A \simeq R\#\Bbbk\Gamma$. In the last chapter, as a concrete illustration of the method, we describe explicitly all finite-dimensional pointed Hopf algebras A with abelian group of group-likes $G(A)$ and infinitesimal braiding of type A_n (up to some exceptional cases). In other words, we compute all the liftings of type A_n; this result is our main new contribution in this paper.

CONTENTS

2000 *Mathematics Subject Classification.* Primary: 17B37; Secondary: 16W30.

Key words and phrases. Pointed Hopf algebras, finite quantum groups.

This work was partially supported by ANPCyT, Agencia Córdoba Ciencia, CONICET, DAAD, the Graduiertenkolleg of the Math. Institut (Universität München) and Secyt (UNC).

Introduction

A Hopf algebra A over a field $\Bbbk$ is called *pointed* [Sw], [M1], if all its simple left or right comodules are one-dimensional. The coradical A_0 of A is the sum of all its simple subcoalgebras. Thus A is pointed if and only if A_0 is a group algebra.

We will always assume that the field $\Bbbk$ is algebraically closed of characteristic 0 (although several results of the paper hold over arbitrary fields).

It is easy to see that A is pointed if it is generated as an algebra by group-like and skew-primitive elements. In particular, group algebras, universal enveloping algebras of Lie algebras and the q-deformations of the universal enveloping algebras of semisimple Lie algebras are all pointed.

An essential tool in the study of pointed Hopf algebras is the *coradical filtration*

$$A_0 \subset A_1 \subset \cdots \subset A, \quad \bigcup_{n \geq 0} A_n = A$$

of A. It is dual to the filtration of an algebra by the powers of the Jacobson radical. For pointed Hopf algebras it is a Hopf algebra filtration, and the associated graded Hopf algebra $\operatorname{gr} A$ has a Hopf algebra projection onto $A_0 = \Bbbk\Gamma$, $\Gamma = G(A)$ the group of all group-like elements of A. By a theorem of Radford [Ra], $\operatorname{gr} A$ is a biproduct

$$\operatorname{gr} A \cong R\#\Bbbk\Gamma,$$

where R is a graded braided Hopf algebra in the category of left Yetter–Drinfeld modules over $\Bbbk\Gamma$ [AS2].

This decomposition is an analog of the theorem of Cartier–Kostant–Milnor–Moore on the semidirect product decomposition of a cocommutative Hopf algebra into an infinitesimal and a group algebra part.

The vector space $V = P(R)$ of the primitive elements of R is a Yetter–Drinfeld submodule. We call its braiding

$$c : V \otimes V \to V \otimes V$$

the **infinitesimal braiding** of A. The infinitesimal braiding is the key to the structure of pointed Hopf algebras.

The subalgebra $\mathfrak{B}(V)$ of R generated by V is a braided Hopf subalgebra. As an algebra and coalgebra, $\mathfrak{B}(V)$ only depends on the infinitesimal braiding of V. In his thesis [N] published in 1978, Nichols studied Hopf algebras of the form $\mathfrak{B}(V)\#\Bbbk\Gamma$ under the name of bialgebras of type one. We call $\mathfrak{B}(V)$ the *Nichols algebra* of V. These Hopf algebras were found independently later by Woronowicz [Wo] and other authors.

Important examples of Nichols algebras come from quantum groups [Dr1]. If $\mathfrak{g}$ is a semisimple Lie algebra, $U_q^{\geq 0}(\mathfrak{g})$, q not a root of unity, and the finite-dimensional Frobenius–Lusztig kernels $u_q^{\geq 0}(\mathfrak{g})$, q a root of unity of order N, are both of the form $\mathfrak{B}(V)\#\Bbbk\Gamma$ with $\Gamma = \mathbb{Z}^\theta$ resp. $(\mathbb{Z}/(N))^\theta, \theta \geq 1$. ([L3], [Ro1], [Sbg], and [L2], [Ro1], [Mu]) (assuming some technical conditions on N).

In general, the classification problem of pointed Hopf algebras has three parts:

(1) Structure of the Nichols algebras $\mathfrak{B}(V)$.
(2) The lifting problem: Determine the structure of all pointed Hopf algebras A with $G(A) = \Gamma$ such that $\operatorname{gr} A \cong \mathfrak{B}(V)\#\Bbbk\Gamma$.
(3) Generation in degree one: Decide which Hopf algebras A are generated by group-like and skew-primitive elements, that is $\operatorname{gr} A$ is generated in degree one.

We conjecture that all finite-dimensional pointed Hopf algebras over an algebraically closed field of characteristic 0 are indeed generated by group-like and skew-primitive elements.

In this paper, we describe the steps of this program in detail and explain the positive results obtained so far in this direction. It is not our intention to give a complete survey on all aspects of pointed Hopf algebras.

We will mainly report on recent progress in the classification of pointed Hopf algebras with *abelian* group of group-like elements.

If the group Γ is abelian, and V is a finite-dimensional Yetter–Drinfeld module, then the braiding is given by a family of non-zero scalars $q_{ij} \in \Bbbk, 1 \leq i \leq \theta$, in the form

$$c(x_i \otimes x_j) = q_{ij}x_j \otimes x_i, \text{ where } x_1, \ldots, x_\theta \text{ is a basis of } V.$$

Moreover there are elements $g_1, \ldots, g_\theta \in \Gamma$, and characters $\chi_1, \ldots, \chi_\theta \in \widehat{\Gamma}$ such that $q_{ij} = \chi_j(g_i)$. The group acts on x_i via the character χ_i, and x_i is a g_i-homogeneous element with respect to the coaction of Γ. We introduced braidings of Cartan type [AS2] where

$$q_{ij}q_{ji} = q_{ii}^{a_{ij}}, 1 \leq i, j \leq \theta, \text{ and } (a_{ij}) \text{ is a generalized Cartan matrix.}$$

If (a_{ij}) is a Cartan matrix of finite type, then the algebras $\mathfrak{B}(V)$ can be understood as twisting of the Frobenius–Lusztig kernels $\mathfrak{u}^{\geq 0}(\mathfrak{g})$, $\mathfrak{g}$ a semisimple Lie algebra.

By deforming the quantum Serre relations for simple roots which lie in two different connected components of the Dynkin diagram, we define finite-dimensional pointed Hopf algebras $\mathfrak{u}(\mathcal{D})$ in terms of a "linking datum $\mathcal{D}$ of finite Cartan type" [AS4]. They generalize the Frobenius–Lusztig kernels $\mathfrak{u}(\mathfrak{g})$ and are liftings of $\mathfrak{B}(V)\#\Bbbk\Gamma$.

In some cases linking data of finite Cartan type are general enough to obtain complete classification results.

For example, if $\Gamma = (\mathbb{Z}/(p))^s$, p a prime > 17 and $s \geq 1$, we have determined the structure of all finite-dimensional pointed Hopf algebras A with $G(A) \simeq \Gamma$. They are all of the form $\mathfrak{u}(\mathcal{D})$ [AS4].

Similar data allow a classification of infinite-dimensional pointed Hopf algebras A with abelian group $G(A)$, without zero divisors, with finite Gelfand–Kirillov dimension and semisimple action of $G(A)$ on A, in the case when the infinitesimal braiding is "positive" [AS5].

But the general case is more involved. We also have to deform the root vector relations of the $u(\mathfrak{g})'s$.

The structure of pointed Hopf algebras A with *non-abelian* group $G(A)$ is largely unknown. One basic open problem is to decide which finite groups appear as groups of group-like elements of finite-dimensional pointed Hopf algebras which are link-indecomposable in the sense of [M2]. In our formulation, this problem is the main part of the following question: given a finite group Γ, determine all Yetter–Drinfeld modules V over $k\Gamma$ such that $\mathfrak{B}(V)$ is finite-dimensional. On the one hand, there are a number of severe constraints on V [Gñ3]. See also the exposition in [A, 5.3.10]. On the other hand, it is very hard to prove the finiteness of the dimension, and in fact this has been done only for a few examples [MiS], [FK], [FP] which are again related to root systems. The examples over the symmetric groups in [FK] were introduced to describe the cohomology ring of the flag variety. At this stage, the main difficulty is to decide when certain Nichols algebras over non-abelian groups, for example the symmetric groups $\mathbb{S}_n$, are finite-dimensional.

The last chapter provides a concrete illustration of the theory explained in this paper. We describe explicitly all finite-dimensional pointed Hopf algebras with abelian group $G(A)$ and infinitesimal braiding of type A_n (up to some exceptional cases). The main results in this chapter are new, and complete proofs are given. The only cases which were known before are the easy case A_1 [AS1], and A_2 [AS3].

The new relations concern the root vectors $e_{i,j}$, $1 \leq i < j \leq n+1$. The relations $e_{i,j}^N = 0$ in $\mathfrak{u}_q^{\geq 0}(sl_{n+1})$, q a root of unity of order N, are replaced by

$$e_{i,j}^N = u_{i,j} \text{ for a family } u_{i,j} \in k\Gamma, 1 \leq i < j \leq n+1,$$

depending on a family of free parameters in k. See Theorem 6.25 for details.

Lifting of type B_2 was treated in [BDR].

To study the relations between a filtered object and its associated graded object is a basic technique in modern algebra. We would like to stress that finite-dimensional pointed Hopf algebras enjoy a remarkable rigidity; it is seldom the case that one is able to describe precisely all the liftings of a graded object, as in this context.

Acknowledgements. We would like to thank Jacques Alev, Matías Graña and Eric Müller for suggestions and stimulating discussions.

Conventions. As said above, our ground field k is algebraically closed field of characteristic 0. Throughout, "Hopf algebra" means "Hopf algebra with bijective antipode". Δ, $\mathcal{S}$, ε, denote respectively the comultiplication, the antipode, the counit of a Hopf algebra.

We denote by $\tau : V \otimes W \to W \otimes V$ the usual transposition, that is $\tau(v \otimes w) = w \otimes v$.

We use Sweedler's notation for the comultiplication and coaction; but, to avoid confusions, we use the following variant for the comultiplication of a braided Hopf algebra R: $\Delta_R(r) = r^{(1)} \otimes r^{(2)}$.

1. Braided Hopf Algebras

1.1. Braided categories. Braided Hopf algebras play a central rôle in this paper. Although we have tried to minimize the use of categorical language, we briefly and informally recall the notion of a braided category which is the appropriate setting for braided Hopf algebras.

Braided categories were introduced in [JS]. We refer to [Ka, Ch. XI, Ch. XIII] for a detailed exposition. There is a hierarchy of categories with a tensor product functor:

(a) A *monoidal* or *tensor* category is a collection $(\mathcal{C}, \otimes, a, \mathbb{I}, l, r)$, where

- $\mathcal{C}$ is a category and $\otimes : \mathcal{C} \times \mathcal{C} \to \mathcal{C}$ is a functor,
- $\mathbb{I}$ is an object of $\mathcal{C}$, and
- $a_{V,W,U} : V \otimes (W \otimes U) \to (V \otimes W) \otimes U$, $l_V : V \to V \otimes \mathbb{I}$, $r_V : V \to \mathbb{I} \otimes V$, V, W, U objects in $\mathcal{C}$, are natural isomorphisms;

such that the so-called "pentagon" and "triangle" axioms are satisfied, see [Ka, Ch. XI, (2.6) and (2.9)]. These axioms essentially express that the tensor product of a finite number of objects is well-defined, regardless of the place where parentheses are inserted; and that $\mathbb{I}$ is a unit for the tensor product.

(b) A *braided (tensor)* category is a collection $(\mathcal{C}, \otimes, a, \mathbb{I}, l, r, c)$, where

- $(\mathcal{C}, \otimes, a, \mathbb{I}, l, r)$ is a monoidal category and
- $c_{V,W} : V \otimes W \to W \otimes V$, V, W objects in $\mathcal{C}$, is a natural isomorphism;

such that the so-called "hexagon" axioms are satisfied, see [Ka, Ch. XIII, (1.3) and (1.4)]. A very important consequence of the axioms of a braided category is the following equality for any objects V, W, U:

$$(c_{V,W} \otimes \mathrm{id}_U)(\mathrm{id}_V \otimes c_{U,W})(c_{U,V} \otimes \mathrm{id}_W) = (\mathrm{id}_W \otimes c_{U,V})(c_{U,W} \otimes \mathrm{id}_V)(\mathrm{id}_U \otimes c_{V,W}),$$
$$(1\text{--}1)$$

see [Ka, Ch. XIII, (1.8)]. For simplicity we have omitted the associativity morphisms.

(c) A *symmetric* category is a braided category where $c_{V,W}c_{W,V} = \mathrm{id}_{W \otimes V}$ for all objects V, W. Symmetric categories have been studied since the pioneering work of Mac Lane.

(d) A left dual of an object V of a monoidal category, is a triple $(V^*, \mathrm{ev}_V, b_V)$, where V^* is another object and $\mathrm{ev}_V : V^* \otimes V \to \mathbb{I}$, $b_V : \mathbb{I} \to V \otimes V^*$ are morphisms such that the compositions

$$V \longrightarrow \mathbb{I} \otimes V \xrightarrow{b_V \otimes \mathrm{id}_V} V \otimes V^* \otimes V \xrightarrow{\mathrm{id}_V \otimes \mathrm{ev}_V} V \otimes \mathbb{I} \longrightarrow V$$

and

$$V^* \longrightarrow V^* \otimes \mathbb{I} \xrightarrow{\mathrm{id}_{V^*} \otimes b_V} V^* \otimes V \otimes V^* \xrightarrow{\mathrm{ev}_V \otimes \mathrm{id}_{V^*}} \mathbb{I} \otimes V^* \longrightarrow V^*$$

are, respectively, the identity of V and V^*. A braided category is *rigid* if any object V admits a left dual [Ka, Ch. XIV, Def. 2.1].

1.2. Braided vector spaces and Yetter–Drinfeld modules. We begin with the fundamental

DEFINITION 1.1. Let V be a vector space and $c : V \otimes V \to V \otimes V$ a linear isomorphism. Then (V, c) is called a *braided vector space*, if c is a solution of the *braid equation*, that is

$$(c \otimes \mathrm{id})(\mathrm{id} \otimes c)(c \otimes \mathrm{id}) = (\mathrm{id} \otimes c)(c \otimes \mathrm{id})(\mathrm{id} \otimes c). \tag{1-2}$$

It is well-known that the braid equation is equivalent to the *quantum Yang–Baxter equation*:

$$R_{12}R_{13}R_{23} = R_{23}R_{13}R_{12}. \tag{1-3}$$

Here we use the standard notation: $R_{13} : V \otimes V \otimes V \to V \otimes V \otimes V$ is the map given by $\sum_j r_j \otimes \mathrm{id} \otimes r^j$, if $R = \sum_j r_j \otimes r^j$. Similarly for R_{12}, R_{23}.

The equivalence between solutions of (1–2) and solutions of (1–3) is given by the equality $c = \tau \circ R$. For this reason, some authors call (1–2) the quantum Yang–Baxter equation.

An easy and for this paper important example is given by a family of non-zero scalars $q_{ij} \in \Bbbk, i, j \in I$, where V is a vector space with basis $x_i, i \in I$. Then

$$c(x_i \otimes x_j) = q_{ij}x_j \otimes x_i, \text{for all } i, j \in I$$

is a solution of the braid equation.

Examples of braided vector spaces come from braided categories. In this article, we are mainly concerned with examples related to the notion of Yetter–Drinfeld modules.

DEFINITION 1.2. Let H be a Hopf algebra. A (left) *Yetter–Drinfeld module* V over H is simultaneously a left H-module and a left H-comodule satisfying the compatibility condition

$$\delta(h.v) = h_{(1)}v_{(-1)}\mathcal{S}h_{(3)} \otimes h_{(2)}.v_{(0)}, \qquad v \in V, h \in H. \tag{1-4}$$

We denote by ${}^H_H\mathcal{YD}$ the category of Yetter–Drinfeld modules over H; the morphisms in this category preserve both the action and the coaction of H. The category ${}^H_H\mathcal{YD}$ is a braided monoidal category; indeed the tensor product of two Yetter–Drinfeld modules is again a Yetter–Drinfeld module, with the usual tensor product module and comodule structure. The compatibility condition (1–4) is not difficult to verify.

For any two Yetter–Drinfeld-modules M and N, the braiding $c_{M,N} : M \otimes N \to N \otimes M$ is given by

$$c_{M,N}(m \otimes n) = m_{(-1)}.n \otimes m_{(0)}, \qquad m \in M, n \in N. \qquad (1\text{–}5)$$

The subcategory of ${}^H_H\mathcal{YD}$ consisting of finite-dimensional Yetter–Drinfeld modules is rigid. Namely, if $V \in {}^H_H\mathcal{YD}$ is finite-dimensional, the dual $V^* = \operatorname{Hom}(V, \Bbbk)$ is in ${}^H_H\mathcal{YD}$ with the following action and coaction:

- $(h \cdot f)(v) = f(\mathcal{S}(h)v)$ for all $h \in H$, $f \in V^*$, $v \in V$.
- If $f \in V^*$, then $\delta(f) = f_{(-1)} \otimes f_{(0)}$ is determined by the equation

$$f_{(-1)}f_{(0)}(v) = \mathcal{S}^{-1}(v_{-1})f(v_0), \qquad v \in V.$$

Then the usual evaluation and coevaluation maps are morphisms in ${}^H_H\mathcal{YD}$.

Let V, W be two finite-dimensional Yetter–Drinfeld modules over H. We shall consider the isomorphism $\Phi : W^* \otimes V^* \to (V \otimes W)^*$ given by

$$\Phi(\varphi \otimes \psi)(v \otimes w) = \psi(v)\varphi(w), \qquad \varphi \in W^*, \psi \in V^*, v \in V, w \in W. \qquad (1\text{–}6)$$

REMARK 1.3. We see that a Yetter–Drinfeld module is a braided vector space. Conversely, a braided vector space (V, c) can be realized as a Yetter–Drinfeld module over some Hopf algebra H if and only if c is *rigid* [Tk1]. If this is the case, it can be realized in many different ways.

We recall that a Hopf bimodule over a Hopf algebra H is simultaneously a bimodule and a bicomodule satisfying all possible compatibility conditions. The category ${}^H_H\mathcal{M}^H_H$ of all Hopf bimodules over H is a braided category. The category ${}^H_H\mathcal{YD}$ is equivalent, as a braided category, to the category of Hopf bimodules. This was essentially first observed in [Wo] and then independently in [AnDe, Appendix], [Sbg], [Ro1].

If H is a finite-dimensional Hopf algebra, then the category ${}^H_H\mathcal{YD}$ is equivalent to the category of modules over the double of H [Mj1]. The braiding in ${}^H_H\mathcal{YD}$ corresponds to the braiding given by the "canonical" R-matrix of the double. In particular, if H is a semisimple Hopf algebra then ${}^H_H\mathcal{YD}$ is a semisimple category. Indeed, it is known that the double of a semisimple Hopf algebra is again semisimple.

The case of Yetter–Drinfeld modules over group algebras is especially important for the applications to pointed Hopf algebras. If $H = \Bbbk\Gamma$, where Γ is a group, then an H-comodule V is just a Γ-graded vector space: $V = \bigoplus_{g \in \Gamma} V_g$, where $V_g = \{v \in V \mid \delta(v) = g \otimes v\}$. We will write ${}^{\Gamma}_{\Gamma}\mathcal{YD}$ for the category of Yetter–Drinfeld modules over $\Bbbk\Gamma$, and say that $V \in {}^{\Gamma}_{\Gamma}\mathcal{YD}$ is a Yetter–Drinfeld module over Γ (when the field is fixed).

REMARK 1.4. Let Γ be a group, V a left $\Bbbk\Gamma$-module, and a left $\Bbbk\Gamma$-comodule with grading $V = \bigoplus_{g \in \Gamma} V_g$. We define a linear isomorphism $c : V \otimes V \to V \otimes V$ by

$$c(x \otimes y) = gy \otimes x, \quad \text{for all } x \in V_g,\ g \in \Gamma,\ y \in V. \tag{1-7}$$

Then

(a) $V \in {}^{\Gamma}_{\Gamma}\mathcal{YD}$ if and only if $gV_h \subset V_{ghg^{-1}}$ for all $g, h \in \Gamma$.

(b) If $V \in {}^{\Gamma}_{\Gamma}\mathcal{YD}$, then (V, c) is a braided vector space.

(c) Conversely, if V is a faithful Γ-module (that is, if for all $g \in \Gamma, gv = v$ for all $v \in V$, implies $g = 1$), and if (V, c) is a braided vector space, then $V \in {}^{\Gamma}_{\Gamma}\mathcal{YD}$.

PROOF. (a) is clear from the definition.

By applying both sides of the braid equation to elements of the form $x \otimes y \otimes z, x \in V_g, y \in V_h, z \in V$, it is easy to see that (V, c) is a braided vector space if and only if

$$c(gy \otimes gz) = ghz \otimes gy, \quad \text{for all } g, h \in \Gamma,\ y \in V_h,\ z \in V. \tag{1-8}$$

Let us write $gy = \sum_{a \in \Gamma} x_a$, where $x_a \in V_a$ for all $a \in \Gamma$. Then $c(gy \otimes gz) = \sum_{a \in \Gamma} agz \otimes x_a$. Hence (1–8) means that $agz = ghz$, for all $z \in V$ and $a \in \Gamma$ such that the homogeneous component x_a is not zero. This proves (b) and (c). □

REMARK 1.5. If Γ is abelian, a Yetter–Drinfeld module over $H = \Bbbk\Gamma$ is nothing but a Γ-graded Γ-module.

Assume that Γ is abelian and furthermore that the action of Γ is diagonalizable (this is always the case if Γ is finite). That is, $V = \bigoplus_{\chi \in \widehat{\Gamma}} V^\chi$, where $V^\chi = \{v \in V \mid gv = \chi(g)v \text{ for all } g \in \Gamma\}$. Then

$$V = \bigoplus_{g \in \Gamma, \chi \in \widehat{\Gamma}} V_g^\chi, \tag{1-9}$$

where $V_g^\chi = V^\chi \cap V_g$. Conversely, any vector space with a decomposition (1–9) is a Yetter–Drinfeld module over Γ. The braiding is given by

$$c(x \otimes y) = \chi(g)y \otimes x, \quad \text{for all } x \in V_g,\ g \in \Gamma,\ y \in V^\chi,\ \chi \in \widehat{\Gamma}.$$

It is useful to characterize abstractly those braided vector spaces which come from Yetter–Drinfeld modules over groups or abelian groups. The first part of the following definition is due to M. Takeuchi.

DEFINITION 1.6. Let (V, c) be a finite-dimensional braided vector space.

- (V, c) is of *group type* if there exist a basis $x_1, \ldots, x_\theta$ of V and elements $g_i(x_j) \in V$ for all i, j such that

$$c(x_i \otimes x_j) = g_i(x_j) \otimes x_i, \qquad 1 \le i, j \le \theta; \qquad (1\text{--}10)$$

necessarily $g_i \in \mathrm{GL}(V)$.
- (V, c) is of *finite group type* (resp. of *abelian group type*) if it is of group type and the subgroup of $\mathrm{GL}(V)$ generated by $g_1, \ldots, g_\theta$ is finite (resp. abelian).
- (V, c) is of *diagonal type* if V has a basis $x_1, \ldots, x_\theta$ such that

$$c(x_i \otimes x_j) = q_{ij} x_j \otimes x_i, \qquad 1 \le i, j \le \theta, \qquad (1\text{--}11)$$

for some q_{ij} in $\Bbbk$. The matrix (q_{ij}) is called the *matrix* of the braiding.
- If (V, c) is of diagonal type, then we say that it is *indecomposable* if for all $i \ne j$, there exists a sequence $i = i_1, i_2, \ldots, i_t = j$ of elements of $\{1, \ldots, \theta\}$ such that $q_{i_s, i_{s+1}} q_{i_{s+1}, i_s} \ne 1, 1 \le s \le t-1$. Otherwise, we say that the matrix is decomposable. We can also refer then to the components of the matrix.

If $V \in {}^{\Gamma}_{\Gamma}\mathcal{YD}$ is finite-dimensional with braiding c, then (V, c) is of group type by (1–5). Conversely, assume that (V, c) is a finite-dimensional braided vector space of group type. Let Γ be the subgroup of $\mathrm{GL}(V)$ generated by $g_1, \ldots, g_\theta$. Define a coaction by $\delta(x_i) = g_i \otimes x_i$ for all i. Then V is a Yetter–Drinfeld module over Γ with braiding c by Remark 1.4 (c).

A braided vector space of diagonal type is clearly of abelian group type; it is of finite group type if the q_{ij}'s are roots of one.

1.3. Braided Hopf algebras.

The notion of "braided Hopf algebra" is one of the basic features of braided categories. We will deal only with braided Hopf algebras in categories of Yetter–Drinfeld modules, mainly over a group algebra.

Let H be a Hopf algebra. First, the tensor product in ${}^{H}_{H}\mathcal{YD}$ allows us to define algebras and coalgebras in ${}^{H}_{H}\mathcal{YD}$. Namely, an algebra in the category ${}^{H}_{H}\mathcal{YD}$ is an associative algebra (R, m), where $m : R \otimes R \to R$ is the product, with unit $u : \Bbbk \to R$, such that R is a Yetter–Drinfeld module over H and both m and u are morphisms in ${}^{H}_{H}\mathcal{YD}$.

Similarly, a coalgebra in the category ${}^{H}_{H}\mathcal{YD}$ is a coassociative coalgebra (R, Δ), where $\Delta : R \to R \otimes R$ is the coproduct, with counit $\varepsilon : R \to \Bbbk$, such that R is a Yetter–Drinfeld module over H and both Δ and ε are morphisms in ${}^{H}_{H}\mathcal{YD}$.

Let now R, S be two algebras in ${}^{H}_{H}\mathcal{YD}$. Then the braiding $c : S \otimes R \to R \otimes S$ allows us to provide the Yetter–Drinfeld module $R \otimes S$ with a "twisted" algebra structure in ${}^{H}_{H}\mathcal{YD}$. Namely, the product in $R \otimes S$ is $m_{R \otimes S}$, defined as $(m_R \otimes m_S)(\mathrm{id} \otimes c \otimes \mathrm{id})$:

$$
\begin{array}{ccc}
R \otimes S \otimes R \otimes S & \longrightarrow & R \otimes S \\
{\scriptstyle \mathrm{id}\, \otimes c \otimes \mathrm{id}} \downarrow & & \uparrow {\scriptstyle =} \\
R \otimes R \otimes S \otimes S & \xrightarrow{m_R \otimes m_S} & R \otimes S.
\end{array}
$$

We shall denote this algebra by $R\underline{\otimes}S$. The difference with the usual tensor product algebra is the presence of the braiding c instead of the usual transposition τ.

DEFINITION 1.7. A *braided bialgebra in* ${}^H_H\mathcal{YD}$ is a collection $(R, m, u, \Delta, \varepsilon)$, where

- (R, m, u) is an algebra in ${}^H_H\mathcal{YD}$.
- (R, Δ, ε) is a coalgebra in ${}^H_H\mathcal{YD}$.
- $\Delta : R \to R\underline{\otimes}R$ and $\varepsilon : R \to \Bbbk$ are morphisms of algebras.

We say that it is a *braided Hopf algebra in* ${}^H_H\mathcal{YD}$ if in addition:

- The identity is convolution invertible in $\text{End}\,(R)$; its inverse is the antipode of R.

A *graded* braided Hopf algebra in ${}^H_H\mathcal{YD}$ is a braided Hopf algebra R in ${}^H_H\mathcal{YD}$ provided with a grading $R = \bigoplus_{n\geq0} R(n)$ of Yetter–Drinfeld modules, such that R is a graded algebra and a graded coalgebra.

REMARK 1.8. There is a non-categorical version of braided Hopf algebras, see [Tk1]. Any braided Hopf algebra in ${}^H_H\mathcal{YD}$ gives rise to a braided Hopf algebra in the sense of [Tk1] by forgetting the action and coaction, and preserving the multiplication, comultiplication and braiding. For the converse see [Tk1, Th. 5.7]. Analogously, one can define graded braided Hopf algebras in the spirit of [Tk1].

Let R be a finite-dimensional Hopf algebra in ${}^H_H\mathcal{YD}$. The dual $S = R^*$ is a braided Hopf algebra in ${}^H_H\mathcal{YD}$ with multiplication $\Delta_R^*\Phi$ and comultiplication $\Phi^{-1}m_R^*$, cf. (1–6); this is R^{*bop} in the notation of [AG, Section 2].

In the same way, if $R = \bigoplus_{n\geq0} R(n)$ is a graded braided Hopf algebra in ${}^H_H\mathcal{YD}$ with finite-dimensional homogeneous components, then the graded dual $S = R^* = \bigoplus_{n\geq0} R(n)^*$ is a graded braided Hopf algebra in ${}^H_H\mathcal{YD}$.

1.4. Examples. The quantum binomial formula.

We shall provide many examples of braided Hopf algebras in Chapter 2. Here we discuss a very simple class of braided Hopf algebras.

We first recall the well-known quantum binomial formula. Let U and V be elements of an associative algebra over $\Bbbk[q]$, q an indeterminate, such that $VU = qUV$. Then

$$(U + V)^n = \sum_{1\leq i\leq n} \binom{n}{i}_q U^i V^{n-i}, \qquad \text{if } n \geq 1. \qquad (1\text{–}12)$$

Here

$$\binom{n}{i}_q = \frac{(n)_q!}{(i)_q!(n-i)_q!}, \quad \text{where } (n)_q! = \prod_{1\leq i\leq n}(i)_q, \quad \text{and } (i)_q = \sum_{0\leq j\leq i-1} q^j.$$

By specialization, (1–12) holds for $q \in \Bbbk$. In particular, if U and V are elements of an associative algebra over $\Bbbk$, and q is a primitive n-th root of 1, such that $VU = qUV$ then

$$(U + V)^n = U^n + V^n. \tag{1-13}$$

EXAMPLE 1.9. Let $(q_{ij})_{1 \leq i,j \leq \theta}$ be a matrix such that

$$q_{ij}q_{ji} = 1, \quad 1 \leq i,j \leq \theta, \ i \neq j. \tag{1-14}$$

Let N_i be the order of q_{ii}, when this is finite.

Let R be the algebra presented by generators $x_1, \ldots, x_\theta$ with relations

$$x_i^{N_i} = 0, \quad \text{if ord}\, q_{ii} < \infty. \tag{1-15}$$
$$x_i x_j = q_{ij} x_j x_i, \quad 1 \leq i < j \leq \theta. \tag{1-16}$$

Given a group Γ and elements $g_1, \ldots, g_\theta$ in the center of Γ, and characters $\chi_1, \ldots, \chi_\theta$ of Γ, there exists a unique structure of Yetter–Drinfeld module over Γ on R, such that

$$x_i \in R_{g_i}^{\chi_i}, \quad 1 \leq i \leq \theta.$$

Note that the braiding is determined by

$$c(x_i \otimes x_j) = q_{ij}\, x_j \otimes x_i, \text{ where } q_{ij} = \chi_j(g_i), \quad 1 \leq i,j \leq \theta.$$

Furthermore, R is a braided Hopf algebra with the comultiplication given by $\Delta(x_i) = x_i \otimes 1 + 1 \otimes x_i$. To check that the comultiplication preserves (1–15) one uses (1–13); the verification for (1–16) is easy. We know [AS1] that dim R is infinite unless all the orders of q_{ii}'s are finite; in this last case, dim $R = \prod_{1 \leq i \leq \theta} N_i$. We also have $P(R) = \bigoplus_{1 < i \leq \theta} \Bbbk x_i$.

1.5. Biproducts, or bosonizations. Let A, H be Hopf algebras and $\pi : A \to H$ and $\iota : H \to A$ Hopf algebra homomorphisms. Assume that $\pi\iota = \text{id}_H$, so that π is surjective, and ι is injective. By analogy with elementary group theory, one would like to reconstruct A from H and the kernel of π as a semidirect product. However, the natural candidate for the kernel of π is the algebra of coinvariants

$$R := A^{\text{co}\, \pi} = \{a \in A : (\text{id} \otimes \pi)\Delta(a) = a \otimes 1\}$$

which is *not*, in general, a Hopf algebra. Instead, R is a braided Hopf algebra in $^H_H \mathcal{YD}$ with the following structure:

- The action $\cdot$ of H on R is the restriction of the adjoint action (composed with ι).
- The coaction is $(\pi \otimes \text{id})\Delta$.
- R is a subalgebra of A.
- The comultiplication is $\Delta_R(r) = r_{(1)}\iota\pi S(r_{(2)}) \otimes r_{(3)}$, for all $r \in R$.

Given a braided Hopf algebra R in $^H_H\mathcal{YD}$, one can consider the *bosonization* or *biproduct* of R by H [Ra], [Mj2]. This is a usual Hopf algebra $R\#H$, with underlying vector space $R \otimes H$, whose multiplication and comultiplication are given by

$$(r\#h)(s\#f) = r(h_{(1)} \cdot s)\#h_{(2)}f,$$
$$\Delta(r\#h) = r^{(1)}\#(r^{(2)})_{(-1)}h_{(1)} \otimes (r^{(2)})_{(0)}\#h_{(2)}. \quad (1\text{-}17)$$

The maps $\pi : R\#H \to H$ and $\iota : H \to R\#H$, $\pi(r\#h) = \varepsilon(r)h$, $\iota(h) = 1\#h$, are Hopf algebra homomorphisms; we have $R = \{a \in R\#H : (\mathrm{id} \otimes\pi)\Delta(a) = a \otimes 1\}$.

Conversely, if A and H are Hopf algebras as above and $R = A^{\mathrm{co}\,\pi}$, then $A \simeq R\#H$.

Let $\vartheta : A \to R$ be the map given by $\vartheta(a) = a_{(1)}\iota\pi S(a_{(2)})$. Then

$$\vartheta(ab) = a_{(1)}\vartheta(b)\iota\pi S(a_{(2)}), \quad (1\text{-}18)$$

for all $a, b \in A$, and $\vartheta(\iota(h)) = \varepsilon(h)$ for all $h \in H$; therefore, for all $a \in A, h \in H$, we have $\vartheta(a\iota(h)) = \vartheta(a)\varepsilon(h)$ and

$$\vartheta(\iota(h)a) = h \cdot \vartheta(a). \quad (1\text{-}19)$$

Notice also that ϑ induces a coalgebra isomorphism $A/A\iota(H)^+ \simeq R$. In fact, the isomorphism $A \to R\#H$ can be expressed explicitly as

$$a \mapsto \vartheta(a_{(1)})\#\pi(a_{(2)}), \qquad a \in A.$$

If A is a Hopf algebra, the adjoint representation ad of A on itself is given by

$$\mathrm{ad}\,x(y) = x_{(1)}yS(x_{(2)}).$$

If R is a braided Hopf algebra in $^H_H\mathcal{YD}$, then there is also a braided adjoint representation ad_c of R on itself defined by

$$\mathrm{ad}_c x(y) = \mu(\mu \otimes S_R)(\mathrm{id} \otimes c)(\Delta_R \otimes \mathrm{id})(x \otimes y),$$

where μ is the multiplication and $c \in \mathrm{End}\,(R \otimes R)$ is the braiding. Note that if $x \in \mathcal{P}(R)$ then the braided adjoint representation of x is just

$$\mathrm{ad}_c x(y) = \mu(\mathrm{id} -c)(x \otimes y) =: [x, y]_c. \quad (1\text{-}20)$$

For any $x, y \in R$, we call $[x, y]_c := \mu(\mathrm{id} -c)(x \otimes y)$ a *braided commutator*.

When $A = R\#H$, then for all $b, d \in R$,

$$\mathrm{ad}_{(b\#1)}(d\#1) = (\mathrm{ad}_c b(d))\#1. \quad (1\text{-}21)$$

1.6. Some properties of braided Hopf algebras. In this Section, we first collect several useful facts about braided Hopf algebras in the category of Yetter–Drinfeld modules over an abelian group Γ. We begin with some identities on braided commutators.

In the following two Lemmas, R denotes a braided Hopf algebra in $\,^{\Gamma}_{\Gamma}\mathcal{YD}$. Let $a_1, a_2, \cdots \in R$ be elements such that $a_i \in R_{g_i}^{\chi_i}$, for some $\chi_i \in \widehat{\Gamma}$, $g_i \in \Gamma$.

LEMMA 1.10. (a)

$$[[a_1, a_2]_c, a_3]_c + \chi_2(g_1)a_2[a_1, a_3]_c = [a_1, [a_2, a_3]_c]_c + \chi_3(g_2)[a_1, a_3]_c a_2. \quad (1\text{--}22)$$

(b) If $[a_1, a_2]_c = 0$ and $[a_1, a_3]_c = 0$ then $[a_1, [a_2, a_3]_c]_c = 0$.
(c) If $[a_1, a_3]_c = 0$ and $[a_2, a_3]_c = 0$ then $[[a_1, a_2]_c, a_3]_c = 0$.
(d) Assume that $\chi_1(g_2)\chi_2(g_1)\chi_2(g_2) = 1$. Then

$$[[a_1, a_2]_c, a_2]_c = \chi_2(g_1)\chi_1(g_2)^{-1}[a_2, [a_2, a_1]_c]_c \qquad (1\text{--}23)$$

PROOF. Left to the reader.   □

The following technical Lemma will be used at a crucial point in Section 6.1.

LEMMA 1.11. Assume that $\chi_2(g_2) \neq -1$ and

$$\chi_1(g_2)\chi_2(g_1)\chi_2(g_2) = 1, \qquad\qquad (1\text{--}24)$$
$$\chi_2(g_3)\chi_3(g_2)\chi_2(g_2) = 1. \qquad\qquad (1\text{--}25)$$

If

$$[a_2, [a_2, a_1]_c]_c = 0, \qquad\qquad (1\text{--}26)$$
$$[a_2, [a_2, a_3]_c]_c = 0, \qquad\qquad (1\text{--}27)$$
$$[a_1, a_3]_c = 0, \qquad\qquad (1\text{--}28)$$

then

$$[[[a_1, a_2]_c, a_3]_c, a_2]_c = 0. \qquad\qquad (1\text{--}29)$$

PROOF. We compute:

$$[[[a_1, a_2]_c, a_3]_c, a_2]_c = a_1 a_2 a_3 a_2 - \chi_2(g_1)\, a_2 a_1 a_3 a_2 - \chi_3(g_1)\chi_3(g_2)\, a_3 a_1 a_2^2$$
$$+ \chi_3(g_1)\chi_3(g_2)\chi_2(g_1)\, a_3 a_2 a_1 a_2$$
$$- \chi_2(g_1)\chi_2(g_2)\chi_2(g_3)\, a_2 a_1 a_2 a_3 + \chi_2(g_1)^2\chi_2(g_2)\chi_2(g_3)\, a_2^2 a_1 a_3$$
$$+ \chi_2(g_1)\chi_2(g_2)\chi_2(g_3)\chi_3(g_1)\chi_3(g_2)\, a_2 a_3 a_1 a_2$$
$$- \chi_2(g_1)^2\chi_2(g_2)\chi_2(g_3)\chi_3(g_1)\chi_3(g_2)\, a_2 a_3 a_2 a_1.$$

We index consecutively the terms in the right-hand side by roman numbers: (I), ..., (VIII). Then (II) + (VII) = 0, by (1–25) and (1–28). Now,

$$(I) = \frac{1}{\chi_3(g_2)(1 + \chi_2(g_2))} a_1 a_2^2 a_3 + \frac{\chi_2(g_2)\chi_3(g_2)}{1 + \chi_2(g_2)} a_1 a_3 a_2^2$$

$$= \frac{1}{\chi_3(g_2)(1 + \chi_2(g_2))} a_1 a_2^2 a_3 + \frac{\chi_2(g_2)\chi_3(g_2)\chi_3(g_1)}{1 + \chi_2(g_2)} a_3 a_1 a_2^2$$

$$= (Ia) + (Ib),$$

by (1–27) and (1–28). By the same equations, we also have for (VIII) the value

$$-\frac{\chi_2(g_1)^2\chi_2(g_2)\chi_2(g_3)\chi_3(g_1)}{1 + \chi_2(g_2)} a_2^2 a_3 a_1 - \frac{\chi_2(g_1)^2\chi_2(g_2)^2\chi_2(g_3)\chi_3(g_1)\chi_3(g_2)^2}{1 + \chi_2(g_2)} a_3 a_2^2 a_1$$

$$= -\frac{\chi_2(g_1)^2\chi_2(g_2)\chi_2(g_3)}{1 + \chi_2(g_2)} a_2^2 a_1 a_3 - \frac{\chi_2(g_1)^2\chi_2(g_2)^2\chi_2(g_3)\chi_3(g_1)\chi_3(g_2)^2}{1 + \chi_2(g_2)} a_3 a_2^2 a_1$$

$$= (VIIIa) + (VIIIb).$$

We next use (1–26) to show that

$$(Ia) + (V) + (VI) + (VIIIa) = 0,$$

$$(Ib) + (III) + (IV) + (VIIIb) = 0.$$

In the course of the proof of these equalities, we need (1–24) and (1–25). This finishes the proof of (1–29). □

Let H be a Hopf algebra. Then the existence of an integral for finite-dimensional braided Hopf algebras implies

LEMMA 1.12. *Let $R = \bigoplus_{n=0}^{N} R(n)$ be a finite-dimensional graded braided Hopf algebra in $\frac{H}{H}\mathcal{YD}$ with $R(N) \neq 0$. There exists $\lambda \in R(N)$ which is a left integral on R and such that*

$$R(i) \otimes R(N - i) \to \Bbbk, \quad x \otimes y \mapsto \lambda(xy),$$

is a non-degenerate pairing, for all $0 \leq i \leq N$. In particular,

$$\dim R(i) = \dim R(N - i).$$

PROOF. This is essentially due to Nichols [N, 1.5]. In this formulation, one needs the existence of non-zero integrals on R; this follows from [FMS]. See [AG, Prop. 3.2.2] for details. □

1.7. The infinitesimal braiding of Hopf algebras whose coradical is a Hopf subalgebra. For the convenience of the reader, we first recall in this Section some basic definitions from coalgebra theory.

DEFINITION 1.13. Let C be a coalgebra.

- $G(C) := \{x \in C \setminus \{0\} \mid \Delta(x) = x \otimes x\}$ is the set of all group-like elements of C.
- If $g, h \in G(C)$, then $x \in C$ is (g, h)-*skew primitive* if $\Delta(x) = x \otimes h + g \otimes x$. The space of all (g, h)-skew primitive elements of C is denoted by $\mathcal{P}(C)_{g,h}$. If C is a bialgebra or a braided bialgebra, and $g = h = 1$, then $P(C) = \mathcal{P}(C)_{1,1}$ is the space of *primitive* elements.
- The *coradical* of C is $C_0 := \sum D$, where D runs through all the simple subcoalgebras of C; it is the largest cosemisimple subcoalgebra of C. In particular, $\Bbbk G(C) \subseteq C_0$.
- C is *pointed* if $\Bbbk G(C) = C_0$.
- The *coradical filtration* of C is the ascending filtration $C_0 \subseteq C_1 \subseteq \cdots \subseteq C_j \subseteq C_{j+1} \subseteq \ldots$, defined by $C_{j+1} := \{x \in C \mid \Delta(x) \in C_j \otimes C + C \otimes C_0\}$. This is a coalgebra filtration: $\Delta C_j \subseteq \sum_{0 \le i \le j} C_i \otimes C_{j-i}$; and it is exhaustive: $C = \bigcup_{n \ge 0} C_n$.
- A *graded coalgebra* is a coalgebra G provided with a grading $G = \bigoplus_{n \ge 0} G(n)$ such that $\Delta G(j) \subseteq \sum_{0 \le i \le j} G(i) \otimes G(j - i)$ for all $j \ge 0$.
- A *coradically graded* coalgebra [CM] is a graded coalgebra $G = \bigoplus_{n \ge 0} G(n)$ such that its coradical filtration coincides with the standard ascending filtration arising from the grading: $G_n = \bigoplus_{m \le n} G(m)$. A *strictly graded* coalgebra [Sw] is a coradically graded coalgebra G such that $G(0)$ is one-dimensional.
- The graded coalgebra associated to the coalgebra filtration of C is $\operatorname{gr} C = \bigoplus_{n \ge 0} \operatorname{gr} C(n)$, where $\operatorname{gr} C(n) := C_n / C_{n-1}$, $n > 0$, $\operatorname{gr} C(0) := C_0$. It is a coradically graded coalgebra.

We shall need a basic technical fact on pointed coalgebras.

LEMMA 1.14. [M1, 5.3.3]. *A morphism of pointed coalgebras which is injective in the first term of the coalgebra filtration, is injective.*

Let now A be a Hopf algebra. We shall assume in what follows that the coradical A_0 is not only a subcoalgebra but a Hopf subalgebra of A; this is the case if A is pointed.

To study the structure of A, we consider its coradical filtration; because of our assumption on A, it is also an algebra filtration [M1]. Therefore, the associated graded coalgebra $\operatorname{gr} A$ is a graded Hopf algebra. Furthermore, $H := A_0 \simeq \operatorname{gr} A(0)$ is a Hopf subalgebra of $\operatorname{gr} A$; and the projection $\pi : \operatorname{gr} A \to \operatorname{gr} A(0)$ with kernel $\bigoplus_{n > 0} \operatorname{gr} A(n)$, is a Hopf algebra map and a retraction of the inclusion. We can then apply the general remarks of Section 1.5. Let R be the algebra of coinvariants of π; R is a braided Hopf algebra in ${}^H_H\mathcal{YD}$ and $\operatorname{gr} A$ can be reconstructed from R and H as a bosonization $\operatorname{gr} A \simeq R \# H$.

The braided Hopf algebra R is graded, since it inherits the gradation from $\operatorname{gr} A$: $R = \bigoplus_{n \ge 0} R(n)$, where $R(n) = \operatorname{gr} A(n) \cap R$. Furthermore, R is strictly graded; this means that

(a) $R(0) = \Bbbk 1$ (hence the coradical is trivial, *cf.* [Sw, Chapter 11]);

(b) $R(1) = P(R)$ (the space of primitive elements of R).

It is in general not true that a braided Hopf algebra R satisfying **(a)** and **(b)**, also satisfies

(c) R is generated as an algebra over $\Bbbk$ by $R(1)$.

A braided graded Hopf algebra satisfying **(a)**, **(b)** and **(c)** is called a Nichols algebra. In the next chapter we will discuss this notion in detail. Notice that the subalgebra R' of R generated by $R(1)$, a Hopf subalgebra of R, is indeed a Nichols algebra.

DEFINITION 1.15. The braiding

$$c : V \otimes V \to V \otimes V$$

of $V := R(1) = P(R)$ is called the *infinitesimal braiding* of A. The graded braided Hopf algebra R is called the *diagram* of A. The dimension of $V = P(R)$ is called the *rank* of A.

2. Nichols Algebras

Let H be a Hopf algebra. In this chapter, we discuss a functor $\mathfrak{B}$ from the category $^H_H \mathcal{YD}$ to the category of braided Hopf algebras in $^H_H \mathcal{YD}$; given a Yetter–Drinfeld module V, the braided Hopf algebra $\mathfrak{B}(V)$ is called the *Nichols algebra* of V.

The structure of a Nichols algebra appeared first in the paper "Bialgebras of type one" [N] of Nichols and was rediscovered later by several authors. In our language, a bialgebra of type one is just a bosonization $\mathfrak{B}(V)\#H$. Hence Nichols algebras are the H-coinvariant elements of bialgebras of type one, also called quantum symmetric algebras in [Ro2]. Several years after [N], Woronowicz defined Nichols algebras in his approach to "quantum differential calculus" [Wo]; again, they appeared as the invariant part of his "algebra of quantum differential forms". Lusztig's algebras $\mathfrak{f}$ [L3], defined by the non-degeneracy of a certain invariant bilinear form, are Nichols algebras. In fact Nichols algebras can always be defined by the non-degeneracy of an invariant bilinear form [AG]. The algebras $\mathfrak{B}(V)$ are called bitensor algebras in [Sbg]. See also [Kh; Gr; FlG].

In a sense, Nichols algebras are similar to symmetric algebras; indeed, both notions coincide in the trivial braided category of vector spaces, or more generally in any symmetric category (*e.g.* in the category of super vector spaces). But when the braiding is not a symmetry, a Nichols algebra could have a much richer structure. We hope that this will be clarified in the examples. On the other hand, Nichols algebras are also similar to universal enveloping algebras. However, in spite of the efforts of several authors, it is not clear to us how to achieve a compact, functorial definition of a "braided Lie algebra" from a Nichols algebra.

We believe that Nichols algebras are very interesting objects of an essentially new nature.

2.1. Definition of Nichols algebras. We now present one of the main notions of this survey.

DEFINITION 2.1. Let V be a Yetter–Drinfeld module over H. A braided graded Hopf algebra $R = \bigoplus_{n \geq 0} R(n)$ in $^H_H \mathcal{YD}$ is called a *Nichols algebra* of V if $\Bbbk \simeq R(0)$ and $V \simeq R(1)$ in $^H_H \mathcal{YD}$, and

$$P(R) = R(1), \tag{2-1}$$

R is generated as an algebra by $R(1)$. $\tag{2-2}$

The dimension of V will be called the *rank* of R.

We need some preliminaries to show the existence and uniqueness of the Nichols algebra of V in $^H_H \mathcal{YD}$.

Let V be a Yetter–Drinfeld module over H. Then the tensor algebra $T(V) = \bigoplus_{n \geq 0} T(V)(n)$ of the vector space V admits a natural structure of a Yetter–Drinfeld module, since $^H_H \mathcal{YD}$ is a braided category. It is then an algebra in $^H_H \mathcal{YD}$. There exists a unique algebra map $\Delta : T(V) \to T(V) \underline{\otimes} T(V)$ such that $\Delta(v) = v \otimes 1 + 1 \otimes v$, for all $v \in V$. For example, if $x, y \in V$, then

$$\Delta(xy) = 1 \otimes xy + x \otimes y + x_{(-1)} \cdot y \otimes x_{(0)} + yx \otimes 1.$$

With this structure, $T(V)$ is a graded braided Hopf algebra in $^H_H \mathcal{YD}$ with counit $\varepsilon : T(V) \to \Bbbk$, $\varepsilon(v) = 0$, if $v \in V$. To show the existence of the antipode, one notes that the coradical of the coalgebra $T(V)$ is $\Bbbk$, and uses a result of Takeuchi [M1, 5.2.10]. Hence all the braided bialgebra quotients of $T(V)$ in $^H_H \mathcal{YD}$ are braided Hopf algebras in $^H_H \mathcal{YD}$.

Let us consider the class $\mathfrak{S}$ of all $I \subset T(V)$ such that

- I is a homogeneous ideal generated by homogeneous elements of degree ≥ 2,
- I is also a coideal, *i.e.* $\Delta(I) \subset I \otimes T(V) + T(V) \otimes I$.

Note that we do *not* require that the ideals I are Yetter–Drinfeld submodules of $T(V)$. Let then $\widetilde{\mathfrak{S}}$ be the subset of $\mathfrak{S}$ consisting of all $I \in \mathfrak{S}$ which are Yetter–Drinfeld submodules of $T(V)$. The ideals

$$I(V) = \sum_{I \in \mathfrak{S}} I, \qquad \widetilde{I}(V) = \sum_{J \in \widetilde{\mathfrak{S}}} J$$

are the largest elements in $\mathfrak{S}$, respectively $\widetilde{\mathfrak{S}}$.

If $I \in \mathfrak{S}$ then $R := T(V)/I = \bigoplus_{n \geq 0} R(n)$ is a graded algebra and a graded coalgebra with

$$R(0) = \Bbbk, \qquad V \simeq R(1) \subset P(R).$$

If actually $I \in \widetilde{\mathfrak{S}}$, then R is a graded braided Hopf algebra in $^H_H \mathcal{YD}$.

We can show now existence and uniqueness of Nichols algebras.

PROPOSITION 2.2. *Let* $\mathfrak{B}(V) := T(V)/\widetilde{I}(V)$. *Then:*

(i) $V = P(\mathfrak{B}(V))$, *hence* $\mathfrak{B}(V)$ *is a Nichols algebra of* V.

(ii) $I(V) = \widetilde{I}(V)$.

(iii) *Let* $R = \bigoplus_{n\geq 0} R(n)$ *be a graded Hopf algebra in* ${}^{H}_{H}\mathcal{YD}$ *such that* $R(0) = \Bbbk 1$ *and* R *is generated as an algebra by* $V := R(1)$. *Then there exists a surjective map of graded Hopf algebras* $R \to \mathfrak{B}(V)$, *which is an isomorphism of Yetter–Drinfeld modules in degree 1.*

(iv) *Let* $R = \bigoplus_{n\geq 0} R(n)$ *be a Nichols algebra of* V. *Then* $R \simeq \mathfrak{B}(V)$ *as braided Hopf algebras in* ${}^{H}_{H}\mathcal{YD}$.

(v) *Let* $R = \bigoplus_{n\geq 0} R(n)$ *be a graded braided Hopf algebra in* ${}^{H}_{H}\mathcal{YD}$ *with* $R(0) = \Bbbk 1$ *and* $R(1) = P(R) = V$. *Then* $\mathfrak{B}(V)$ *is isomorphic to the subalgebra* $\Bbbk\langle V\rangle$ *of* R *generated by* V.

PROOF. (i) We have to show the equality $V = P(\mathfrak{B}(V))$. Let us consider the inverse image X in $T(V)$ of all homogeneous primitive elements of $\mathfrak{B}(V)$ in degree $n \geq 2$. Then X is a graded Yetter–Drinfeld submodule of $T(V)$, and for all $x \in X$, $\Delta(x) \in x \otimes 1 + 1 \otimes x + T(V) \otimes \widetilde{I}(V) + \widetilde{I}(V) \otimes T(V)$. Hence the ideal generated by $\widetilde{I}(V)$ and X is in $\widetilde{\mathfrak{S}}$, and $X \subset \widetilde{I}(V)$ by the maximality of $\widetilde{I}(V)$. Hence the image of X in $\mathfrak{B}(V)$ is zero. This proves our claim since the primitive elements form a graded submodule.

(ii) We have to show that the surjective map $\mathfrak{B}(V) \to T(V)/I(V)$ is bijective. This follows from (i) and Lemma 1.14.

(iii) The kernel I of the canonical projection $T(V) \to R$ belongs to $\widetilde{\mathfrak{S}}$; hence $I \subseteq \widetilde{I}(V)$.

(iv) follows again from Lemma 1.14, as in (ii).

(v) follows from (iv).   □

If U is a braided subspace of $V \in {}^{H}_{H}\mathcal{YD}$, that is a subspace such that $c(U \otimes U) \subset U \otimes U$, where c is the braiding of V, we can define $\mathfrak{B}(U) := T(U)/I(U)$ with the obvious meaning of $I(U)$. Then the description in Proposition 2.2 also applies to $\mathfrak{B}(U)$.

COROLLARY 2.3. *The assignment* $V \mapsto \mathfrak{B}(V)$ *is a functor from* ${}^{H}_{H}\mathcal{YD}$ *to the category of braided Hopf algebras in* ${}^{H}_{H}\mathcal{YD}$.

If U *is a Yetter–Drinfeld submodule of* V, *or more generally if* U *is a braided subspace of* V, *then the canonical map* $\mathfrak{B}(U) \to \mathfrak{B}(V)$ *is injective.*

PROOF. If $\phi : U \to V$ is a morphism in ${}^{H}_{H}\mathcal{YD}$, then $T(\phi) : T(U) \to T(V)$ is a morphism of braided Hopf algebras. Since $T(\phi)(I(U))$ is a coideal and a Yetter–Drinfeld submodule of $T(V)$, the ideal generated by $T(\phi)(\widetilde{I}(U))$ is contained in $\widetilde{I}(V)$. Hence by Proposition 2.2, $\mathfrak{B}$ is a functor.

The second part of the claim follows from Proposition 2.2(v).   □

The duality between conditions (2–1) and (2–2) in the definition of Nichols algebra, emphasized by Proposition 2.2(iii), (v), is explicitly stated in the following

LEMMA 2.4. *Let* $R = \bigoplus_{n \geq 0} R(n)$ *be a graded braided Hopf algebra in* $^H_H\mathcal{YD}$; *suppose the homogeneous components are finite-dimensional and* $R(0) = \Bbbk 1$. *Let* $S = \bigoplus_{n \geq 0} R(n)^*$ *be the graded dual of* R. *Then* $R(1) = P(R)$ *if and only if* S *is generated as an algebra by* $S(1)$.

PROOF. See for instance [AS2, Lemma 5.5]. □

EXAMPLE 2.5. Let F be a field of positive characteristic p. Let S be the (usual) Hopf algebra $F[x]/\langle x^{p^2} \rangle$ with $x \in P(S)$. Then $x^p \in P(S)$. Hence S satisfies (2–2) but not (2–1).

EXAMPLE 2.6. Let $S = \Bbbk[X] = \bigoplus_{n \geq 0} S(n)$ be a polynomial algebra in one variable. We consider S as a braided Hopf algebra in $^H_H\mathcal{YD}$, where $H = \Bbbk\Gamma$, Γ an infinite cyclic group with generator g, with action, coaction and comultiplication given by

$$\delta(X^n) = g^n \otimes X^n, \quad g \cdot X = qX, \quad \Delta(X) = X \otimes 1 + 1 \otimes X.$$

Here $q \in \Bbbk$ is a root of 1 of order N. That is, S is a so-called quantum line. Then S satisfies (2–2) but not (2–1) since X^N is also primitive. Hence the graded dual $R = S^d = \bigoplus_{n \geq 0} S(n)^*$ is a braided Hopf algebra satisfying (2–1) but not (2–2).

However, in characteristic 0 we do not know any finite-dimensional example of a braided Hopf algebra satisfying (2–1) but not (2–2).

CONJECTURE 2.7. [AS2, Conjecture 1.4] *Any finite-dimensional braided Hopf algebra in* $^H_H\mathcal{YD}$ *satisfying* (2–1) *also satisfies* (2–2). (*Recall that the base field* $\Bbbk$ *has characteristic zero.*)

The compact description of $\mathfrak{B}(V)$ in Lemma 2.2 shows that it depends only on the algebra and coalgebra structure of $T(V)$. Since the comultiplication of the tensor algebra was defined using the "twisted" multiplication of $T(V)\underline{\otimes}T(V)$, we see that $\mathfrak{B}(V)$ *depends as an algebra and coalgebra only on the braiding of* V. The explicit formula for the comultiplication of $T(V)$ leads to the following alternative description of $\mathfrak{B}(V)$.

2.2. Skew-derivations and bilinear forms. We want to describe two important techniques to prove identities in Nichols algebras even without knowing the defining relations explicitly.

The first technique was introduced by Nichols [N, 3.3] to deal with $\mathfrak{B}(V)$ over group algebras $\Bbbk\Gamma$ using skew-derivations. Let $V \in {}^\Gamma_\Gamma\mathcal{YD}$ be of finite dimension θ. We choose a basis $x_i \in V_{g_i}$ with $g_i \in \Gamma$, $1 \leq i \leq \theta$, of Γ-homogeneous elements. Let $I \in \mathfrak{S}$ and $R = T(V)/I$ (see Section 2.1). Then R is a graded Hopf algebra in ${}^\Gamma_\Gamma\mathcal{YD}$ with $R(0) = \Bbbk 1$ and $R(1) = V$. For all $1 \leq i \leq \theta$ let $\sigma_i : R \to R$ be the algebra automorphism given by the action of g_i.

Recall that if $\sigma : R \to R$ is an algebra automorphism, an (id, σ)-derivation $D : R \to R$ is a $\Bbbk$-linear map such that

$$D(xy) = xD(y) + D(x)\sigma(y), \qquad \text{for all } x, y \in R.$$

PROPOSITION 2.8. (1) For all $1 \le i \le \theta$, there exists a uniquely determined (id, σ_i)-derivation $D_i : R \to R$ with $D_i(x_j) = \delta_{i,j}$ (Kronecker δ) for all j.
(2) $I = I(V)$, that is $R = \mathfrak{B}(V)$, if and only if $\bigcap_{i=1}^{\theta} \mathrm{Ker}(D_i) = \Bbbk 1$.

PROOF. See for example [MiS, 2.4].   □

Let us illustrate this Proposition in a very simple case.

EXAMPLE 2.9. Let V be as above and assume that $g_i \cdot x_i = q_i x_i$, for some $q_i \in \Bbbk^\times$, $1 \le i \le \theta$. Then for any $n \in \mathbb{N}$,
 (a) $D_i(x_i^n) = (n)_{q_i} x_i^{n-1}$.
 (b) $x_i^n \ne 0$ if and only if $(n)_{q_i}! \ne 0$.

PROOF. (a) follows by induction on n since D_i is a skew-derivation; (b) follows from (a) and Proposition 2.8, since D_j vanishes on any power of x_i, for $j \ne i$. □

The second technique was used by Lusztig [L3] to prove very deep results about quantum enveloping algebras using a canonical bilinear form.

Let (V, c) be a braided vector space of diagonal type as in (1–11) and assume that $q_{ij} = q_{ji}$ for all i, j. Let Γ be the free abelian group of rank θ with basis $g_1, \ldots, g_\theta$. We define characters $\chi_1, \ldots, \chi_\theta$ of Γ by

$$\chi_i(g_j) = q_{ji}, \qquad 1 \le i, j \le \theta.$$

We consider V as a Yetter–Drinfeld module over $\Bbbk\Gamma$ by defining $x_i \in V_{g_i}^{\chi_i}$ for all i.

PROPOSITION 2.10. Let $B_1, \ldots, B_\theta$ be non-zero elements in $\Bbbk$. There is a unique bilinear form $(\,|\,) : T(V) \times T(V) \to \Bbbk$ such that $(1|1) = 1$ and

$$(x_j|x_j) = \delta_{ij} B_i, \qquad \text{for all } i, j; \tag{2-3}$$

$$(x|yy') = (x_{(1)}|y)(x_{(2)}|y'), \qquad \text{for all } x, y, y' \in T(V); \tag{2-4}$$

$$(xx'|y) = (x|y_{(1)})(x'|y_{(2)}), \qquad \text{for all } x, x', y \in T(V). \tag{2-5}$$

This form is symmetric and also satisfies

$$(x|y) = 0, \qquad \text{for all } x \in T(V)_g, y \in T(V)_h, g \ne h \in \Gamma. \tag{2-6}$$

The homogeneous components of $T(V)$ with respect to its usual $\mathbb{N}$-grading are also orthogonal with respect to $(\,|\,)$.

The quotient $T(V)/I(V)$, where $I(V) = \{x \in T(V) : (x|y) = 0 \forall y \in T(V)\}$ is the radical of the form, is canonically isomorphic to the Nichols algebra of V. Thus, $(\,|\,)$ induces a non-degenerate bilinear form on $\mathfrak{B}(V)$, denoted by the same name.

PROOF. The existence and uniqueness of the form, and the claims about symmetry and orthogonality, are proved exactly as in [L3, 1.2.3]. It follows from the properties of the form that $I(V)$ is a Hopf ideal. We now check that $T(V)/I(V)$ is the Nichols algebra of V; it is enough to verify that the primitive elements of $T(V)/I(V)$ are in V. Let x be a primitive element in $T(V)/I(V)$, homogeneous of degree $n \geq 2$. Then $(x|yy') = 0$ for all y, y' homogeneous of degrees $m, m' \geq 1$ with $m + m' = n$; thus $x = 0$. $\qquad\square$

A generalization of the preceding result, valid for any finite-dimensional Yetter–Drinfeld module over any group, can be found in [AG, 3.2.17].

The Proposition shows that Lusztig's algebra $\mathbf{f}$ [L3, Chapter 1] is the Nichols algebra of V over the field of rational functions $\mathbb{Q}(v)$, with $q_{ij} = v^{i \cdot j}$ if $I = \{1, \dots, \theta\}$ and $(I, \cdot)$ a Cartan datum. In particular, we can take a generalized symmetrizable Cartan matrix (a_{ij}), $0 < d_i \in \mathbb{N}$ for all i with $d_i a_{ij} = d_j a_{ji}$ for all i, j and define $i \cdot j := d_i a_{ij}$.

2.3. The braid group. Let us recall that the braid group $\mathbb{B}_n$ is presented by generators $\sigma_1, \dots, \sigma_{n-1}$ with relations

$$\sigma_i \sigma_{i+1} \sigma_i = \sigma_{i+1} \sigma_i \sigma_{i+1}, \quad 1 \leq i \leq n - 2,$$
$$\sigma_i \sigma_j = \sigma_j \sigma_i,, \quad 1 \leq i, j \leq n - 2, \ |i - j| > 1.$$

Here are some basic well-known facts about the braid group.

There is a natural projection $\pi : \mathbb{B}_n \to \mathbb{S}_n$ sending σ_i to the transposition $\tau_i := (i, i+1)$ for all i. The projection π admits a set-theoretical section $s : \mathbb{S}_n \to \mathbb{B}_n$ determined by

$$s(\tau_i) = \sigma_i, \quad 1 \leq i \leq n - 1,$$
$$s(\tau\omega) = s(\tau)s(\omega), \quad \text{if } l(\tau\omega) = l(\tau) + l(\omega).$$

Here l denotes the length of an element of $\mathbb{S}_n$ with respect to the set of generators $\tau_1, \dots, \tau_{n-1}$. The map s is called the Matsumoto section. In other words, if $\omega = \tau_{i_1} \dots \tau_{i_M}$ is a reduced expression of $\omega \in \mathbb{S}_n$, then $s(\omega) = \sigma_{i_1} \dots \sigma_{i_M}$.

Let $q \in \Bbbk$, $q \neq 0$. The quotient of the group algebra $\Bbbk(\mathbb{B}_n)$ by the two-sided ideal generated by the relations

$$(\sigma_i - q)(\sigma_i + 1), \quad 1 \leq i \leq n - 1,$$

is the so-called *Hecke algebra* of type A_n, denoted by $\mathcal{H}_q(n)$.

Using the section s, the following distinguished elements of the group algebra $\Bbbk\mathbb{B}_n$ are defined:

$$\mathfrak{S}_n := \sum_{\sigma \in \mathbb{S}_n} s(\sigma), \qquad \mathfrak{S}_{i,j} := \sum_{\sigma \in X_{i,j}} s(\sigma);$$

here $X_{i,j} \subset \mathbb{S}_n$ is the set of all (i,j)-shuffles. The element $\mathfrak{S}_n$ is called the quantum symmetrizer.

Given a braided vector space (V, c), there are representations of the braid groups $\rho_n : \mathbb{B}_n \to \operatorname{Aut}(V^{\otimes n})$ for any $n \geq 0$, given by

$$\rho_n(\sigma_i) = \operatorname{id} \otimes \cdots \otimes \operatorname{id} \otimes c \otimes \operatorname{id} \otimes \cdots \otimes \operatorname{id},$$

where c acts in the tensor product of the i and $i + 1$ copies of V. By abuse of notation, we shall denote by $\mathfrak{S}_n$, $\mathfrak{S}_{i,j}$ also the corresponding endomorphisms $\rho(\mathfrak{S}_n)$, $\rho(\mathfrak{S}_{i,j})$ of $V^{\otimes n} = T^n(V)$.

If $C = \bigoplus_{n \geq 0} C(n)$ is a graded coalgebra with comultiplication Δ, we denote by $\Delta_{i,j} : C(i + j) \to C(i) \otimes C(j)$, $i, j \geq 0$, the (i, j)-graded component of the map Δ.

PROPOSITION 2.11. *Let $V \in {}_H^H\mathcal{YD}$. Then*

$$\Delta_{i,j} = \mathfrak{S}_{i,j}, \tag{2-7}$$
$$\mathfrak{B}(V) = \bigoplus_{n \geq 0} T^n(V)/\ker(\mathfrak{S}_n). \tag{2-8}$$

PROOF. See for instance [Sbg].   □

This description of the relation of $\mathfrak{B}(V)$ does not mean that the relations are known. In general it is very hard to compute the kernels of the maps $\mathfrak{S}_n$ in concrete terms. For any braided vector space (V, c), we may define $\mathfrak{B}(V)$ by (2-8).

Using the action of the braid group, $\mathfrak{B}(V)$ can also be described as a subalgebra of the quantum shuffle algebra [N; Ro1; Ro2; Sbg].

2.4. Invariance under twisting. Twisting is a method to construct new Hopf algebras by "deforming" the comultiplication; originally due to Drinfeld [Dr2], it was adapted to Hopf algebras in [Re].

Let A be a Hopf algebra and $F \in A \otimes A$ be an invertible element. Let $\Delta_F := F\Delta F^{-1} : A \to A \otimes A$; it is again an algebra map. If

$$(1 \otimes F)(\operatorname{id} \otimes \Delta)(F) = (F \otimes 1)(\Delta \otimes \operatorname{id})(F), \tag{2-9}$$
$$(\operatorname{id} \otimes \varepsilon)(F) = (\varepsilon \otimes \operatorname{id})(F) = 1, \tag{2-10}$$

then A_F (the same algebra, but with comultiplication Δ_F) is again a Hopf algebra. We shall say that A_F is obtained from A via twisting by F; F is a cocycle in a suitable sense.

There is a dual version of the twisting operation, which amounts to a twist of the multiplication [DT]. Let A be a Hopf algebra and let $\sigma : A \times A \to \Bbbk$ be an invertible 2-cocycle[1], that is

$$\sigma(x_{(1)}, y_{(1)})\sigma(x_{(2)}y_{(2)}, z) = \sigma(y_{(1)}, z_{(1)})\sigma(x, y_{(2)}z_{(2)}),$$
$$\sigma(x, 1) = \sigma(1, x) = \varepsilon(x),$$

for all $x, y, z \in A$. Then A_σ – the same A but with the multiplication $\cdot_\sigma$ below – is again a Hopf algebra, where

$$x \cdot_\sigma y = \sigma(x_{(1)}, y_{(1)})x_{(2)}y_{(2)}\sigma^{-1}(x_{(3)}, y_{(3)}).$$

For details, see for instance [KS, 10.2.3 and 10.2.4].

Assume now that H is a Hopf algebra, R is a braided Hopf algebra in ${}^H_H\mathcal{YD}$, and $A = R\#H$. Let $\pi : A \to H$ and $\iota : H \to A$ be the canonical projection and injection. Let $\sigma : H \times H \to \Bbbk$ be an invertible 2-cocycle, and define $\sigma_\pi : A \times A \to \Bbbk$ by

$$\sigma_\pi := \sigma(\pi \otimes \pi);$$

σ_π is an invertible 2-cocycle, with inverse $(\sigma^{-1})_\pi$. The maps $\pi : A_{\sigma_\pi} \to H_\sigma$, $\iota : H_\sigma \to A_{\sigma_\pi}$ are still Hopf algebra maps. Because the comultiplication is not changed, the space of coinvariants of π is R; this is a subalgebra of A_{σ_π} that we denote R_σ; the multiplication in R_σ is given by

$$x._\sigma y = \sigma(x_{(-1)}, y_{(-1)})x_{(0)}y_{(0)}, \qquad x, y \in R = R_\sigma. \tag{2-11}$$

Equation (2–11) follows easily using (1–17). Clearly, R_σ is a Yetter–Drinfeld Hopf algebra in ${}^{H_\sigma}_{H_\sigma}\mathcal{YD}$. The coaction of H_σ on R_σ is the same as the coaction of H on R, since the comultiplication was not altered. The explicit formula for the action of H_σ on R_σ can be written down; we shall do this only in the setting we are interested in.

Let $H = \Bbbk\Gamma$ be a group algebra; an invertible 2-cocycle $\sigma : H \times H \to \Bbbk$ is uniquely determined by its restriction $\sigma : \Gamma \times \Gamma \to \Bbbk^\times$, a group 2-cocycle with respect to the trivial action.

LEMMA 2.12. *Let Γ be an abelian group and let R be a braided Hopf algebra in ${}^\Gamma_\Gamma\mathcal{YD}$. Let $\sigma : \Gamma \times \Gamma \to \Bbbk^\times$ be a 2-cocycle. Let S be the subalgebra of R generated by $P(R)$. In the case $y \in S_h^\eta$, for some $h \in \Gamma$ and $\eta \in \widehat{\Gamma}$, the action of $H = H_\sigma$ on R_σ is*

$$g \to_\sigma y = \sigma(g, h)\sigma^{-1}(h, g)\eta(g)y, \qquad g \in \Gamma. \tag{2-12}$$

Hence, the braiding c_σ in R_σ is given in this case by

$$c_\sigma(x \otimes y) = \sigma(g, h)\sigma^{-1}(h, g)\eta(g)\, y \otimes x, \qquad x \in R_g, g \in \Gamma. \tag{2-13}$$

Therefore, for such x and y, we have

$$[x, y]_{c_\sigma} = \sigma(g, h)[x, y]_c. \tag{2-14}$$

---

[1] Here "invertible" means that the associated linear map $\sigma : A \otimes A \to \Bbbk$ is invertible with respect to the convolution product.

PROOF. To prove (2–12), it is enough to assume $y \in P(R)_h^\eta$.

Let $A = R\#H$; in A_{σ_π}, we have

$$g \cdot_\sigma y = \sigma(g, \pi(y)) \, g \, \sigma^{-1}(g, 1) + \sigma(g, h) \, gy \, \sigma^{-1}(g, 1) + \sigma(g, h) \, gh \, \sigma^{-1}(g, \pi(y))$$
$$= \sigma(g, h) gy;$$
$$y \cdot_\sigma g = \sigma(\pi(y), g) \, g \, \sigma^{-1}(1, g) + \sigma(h, g) \, yg \, \sigma^{-1}(1, g) + \sigma(h, g) \, hg \, \sigma^{-1}(\pi(y), g)$$
$$= \sigma(h, g) yg;$$

hence

$$g \cdot_\sigma y = \sigma(g, h) gy = \sigma(g, h) \eta(g) yg = \sigma(g, h) \sigma^{-1}(h, g) \eta(g) y \cdot_\sigma g,$$

which is equivalent to (2–12). Now (2–13) follows at once, and (2–14) follows from (2–11) and (2–13):

$$[x, y]_{c_\sigma} = x \cdot_\sigma y - \cdot_\sigma c_\sigma(x \otimes y)$$
$$= \sigma(g, h) xy - \sigma(g, h) \sigma^{-1}(h, g) \eta(g) \, \sigma(h, g) yx = \sigma(g, h)[x, y]_c. \quad \square$$

The proof of the following lemma is clear, since the comultiplication of a Hopf algebra is not changed by twisting.

LEMMA 2.13. *Let H be a Hopf algebra and let R be a braided Hopf algebra in $_H^H\mathcal{YD}$. Let $\sigma : H \times H \to \Bbbk$ be an invertible 2-cocycle. If $R = \bigoplus_{n \geq 0} R(n)$ is a braided graded Hopf algebra in $_H^H\mathcal{YD}$, then R_σ is a braided graded Hopf algebra in $_{H_\sigma}^{H_\sigma}\mathcal{YD}$ with $R(n) = R_\sigma(n)$ as vector spaces for all $n \geq 0$. Also R is a Nichols algebra if and only if R_σ is a Nichols algebra in $_{H_\sigma}^{H_\sigma}\mathcal{YD}$.*

3. Types of Nichols Algebras

We now discuss several examples of Nichols algebras. We are interested in explicit presentations, *e.g.* by generators and relations, of $\mathfrak{B}(V)$, for braided vector spaces in suitable classes, for instance, those of group type. We would also like to determine when $\mathfrak{B}(V)$ has finite dimension, or polynomial growth.

3.1. Symmetries and braidings of Hecke type. We begin with the simplest class of braided vector spaces.

EXAMPLE 3.1. Let $\tau : V \otimes V \to V \otimes V$ be the usual transposition; the braided vector space (V, τ) can be realized as a Yetter–Drinfeld module over any Hopf algebra H, with trivial action and coaction. Then $\mathfrak{B}(V) = \mathrm{Sym}\,(V)$, the symmetric algebra of V.

The braided vector space $(V, -\tau)$, which can be realized *e.g.* in $_\mathbb{Z}^\mathbb{Z}\mathcal{YD}$, has $\mathfrak{B}(V) = \Lambda(V)$, the exterior algebra of V.

EXAMPLE 3.2. Let $V = \bigoplus_{i \in \mathbb{Z}/2} V(i)$ be a super vector space and let $c : V \otimes V \to V \otimes V$ be the supersymmetry:

$$c(v \otimes w) = (-1)^{i \cdot j} w \otimes v \qquad v \in V(i), \, w \in V(j).$$

Clearly, V can be realized as a Yetter–Drinfeld module over $\mathbb{Z}/2$. Then $\mathfrak{B}(V) \simeq$ $\mathrm{Sym}\,(V(0)) \otimes \Lambda(V(1))$, the super-symmetric algebra of V.

The simple form of $\mathfrak{B}(V)$ in these examples can be explained in the following context.

DEFINITION 3.3. We say that a braided vector space (V, c) is of *Hecke-type* with label $q \in \Bbbk$, $q \neq 0$, if

$$(c - q)(c + 1) = 0.$$

In this case, the representation of the braid group $\rho_n : \mathbb{B}_n \to \mathrm{Aut}\,(V^{\otimes n})$ factorizes through the Hecke algebra $\mathcal{H}_q(n)$, for all $n \geq 0$; cf. Section 2.3.

If $q = 1$, one says that c is a *symmetry*. Then ρ_n factorizes through the symmetric group $\mathbb{S}_n$, for all $n \geq 0$. The categorical version of symmetries is that of symmetric categories, see Section 1.1.

PROPOSITION 3.4. *Let (V, c) be a braided vector space of Hecke-type with label q, which is either 1 or not a root of 1. Then $\mathfrak{B}(V)$ is a quadratic algebra; that is, the ideal $I(V)$ is generated by $I(V)(2) = \mathrm{Ker}\,\mathfrak{S}_2$.*

Moreover, $\mathfrak{B}(V)$ is a Koszul algebra and its Koszul dual is the Nichols algebra $\mathfrak{B}(V^*)$ corresponding to the braided vector space $(V^*, q^{-1}c^t)$.

A nice exposition on Koszul algebras is [BGS, Chapter 2].

PROOF. The argument for the first claim is taken from [AA, Prop. 3.3.1]. The image of the quantum symmetrizer $\mathfrak{S}_n$ in the Hecke algebra $\mathcal{H}_q(n)$ is $[n]_q! M_\varepsilon$, where M_ε satisfies the following properties:

$$M_\varepsilon^2 = M_\varepsilon, \quad M_\varepsilon c_i = c_i M_\varepsilon = q M_\varepsilon, \quad 1 \leq i \leq n - 1.$$

See for instance [HKW]. Now, we have to show that $\mathrm{Ker}\,\mathfrak{S}_n = T^n(V) \cap I$, where I is the ideal generated by $\mathrm{Ker}\,\mathfrak{S}_2 = \mathrm{Ker}\,(c + 1) = \mathrm{Im}\,(c - q)$; but clearly $T^n(V) \cap I = \sum_i I^{n,i}$, where

$$I^{n,i} = T^{i-1}(V) \otimes \mathrm{Im}\,(c - q) \otimes T^{n-i-1}(V) = \mathrm{Im}\,(c_i - q).$$

It follows that $T^n(V) \cap I \subseteq \mathrm{Ker}\,\mathfrak{S}_n$, a fact that we already know from the general theory. But moreover, $T^n(V) \cap I$ is a $\mathcal{H}_q(n)$-submodule of $T^n(V)$ since

$$c_j(c_i - q) = (c_j - q)(c_i - q) + q(c_i - q).$$

This computation also shows that the action of $\mathcal{H}_q(n)$ on the quotient module $T^n(V)/T^n(V) \cap I$ is via the character that sends σ_i to q; hence M_ε acts on $T^n(V)/T^n(V) \cap I$ by an automorphism, and *a fortiori* $T^n(V) \cap I \supseteq \mathrm{Ker}\,\mathfrak{S}_n$. Having shown the first claim, the second claim is essentially a result from [Gu; Wa]; see also the exposition in [AA, Sections 3.3 and 3.4]. $\square$

EXAMPLE 3.5. Let $q \in \Bbbk^\times$, q is not a root of 1. The braided vector space $(V, q\tau)$ can be realized in $\frac{\mathbb{Z}}{\mathbb{Z}}\mathcal{YD}$. It can be shown that $\mathfrak{B}(V) = T(V)$, the tensor algebra of V, for all q in an open set.

It would be interesting to know whether other conditions on the minimal polynomial of a braiding have consequences on the structure of the corresponding Nichols algebra. The first candidate should be a braiding of BMW-type.

3.2. Braidings of diagonal type. In this section, (V, c) denotes a finite-dimensional braided vector space of diagonal type; that is, V has a basis $x_1, \ldots, x_\theta$ such that (1–11) holds for some non-zero q_{ij} in $\Bbbk$. Our first goal is to determine polynomial relations on the generators $x_1, \ldots, x_\theta$ that should hold in $\mathfrak{B}(V)$. We look at polynomial expressions in these generators which are homogeneous of degree ≥ 2, and give rise to primitive elements in any braided Hopf algebra containing V inside its primitive elements. For related material, see [Kh].

LEMMA 3.6. *Let R be a braided Hopf algebra in $^H_H\mathcal{YD}$, for some Hopf algebra H, such that $V \subseteq P(R)$ as braided vector spaces.*

(a) *If q_{ii} is a root of 1 of order $N > 1$ for some $i \in \{1, \ldots, \theta\}$, then $x_i^N \in P(R)$.*
(b) *Let $1 \leq i, j \leq \theta, i \neq j$, such that $q_{ij}q_{ji} = q_{ii}^r$, where $0 \leq -r < \operatorname{ord} q_{ii}$ (which could be infinite). Then $(\operatorname{ad}_c x_i)^{1-r}(x_j)$ is primitive in R.*

PROOF. (a) and (b) are consequences of the quantum binomial formula, see *e.g.* [AS2, Appendix] for (b).   □

We apply these first remarks to $\mathfrak{B}(V)$ and see how conditions on the Nichols algebra induce conditions on the braiding.

LEMMA 3.7. *Let $R = \mathfrak{B}(V)$.*

(a) *If q_{ii} is a root of 1 of order $N > 1$ then $x_i^N = 0$. In particular, if $\mathfrak{B}(V)$ is an integral domain, then $q_{hh} = 1$ or it is not a root of 1, for all h.*
(b) *If $i \neq j$, then $(\operatorname{ad}_c x_i)^r(x_j) = 0$ if and only if $(r)!_{q_{ii}} \prod_{0 \leq k \leq r-1} \left(1 - q_{ii}^k q_{ij}q_{ji}\right)$ vanishes.*
(c) *If $i \neq j$ and $q_{ij}q_{ji} = q_{ii}^r$, where $0 \leq -r < \operatorname{ord} q_{ii}$ (which could be infinite), then $(\operatorname{ad}_c x_i)^{1-r}(x_j) = 0$.*
(d) *If $\mathfrak{B}(V)$ has finite Gelfand–Kirillov dimension, then for all $i \neq j$, there exists $r_{ij} > 0$ such that $(\operatorname{ad}_c x_i)^{r_{ij}}(x_j) = 0$.*

PROOF. Parts (a) and (c) follow from Lemma 3.6; part (a) is also a special case of Example 2.9; and part (c) also follows from (b). Part (b) is stated in [Ro2, Lemma 14]. It can be shown using the skew-derivations D_j of Section 2.2. Indeed, we first claim that $D_j\left((\operatorname{ad}_c x_i)^r(x_j)\right) = \prod_{0 \leq k \leq r-1}\left(1 - q_{ii}^k q_{ij}q_{ji}\right) x_i^r$. We set $z_r = (\operatorname{ad}_c x_i)^r(x_j)$ and compute

$$D_j\left(\operatorname{ad}_c x_i(z_r)\right) = D_j(x_i z_r - (g_i \cdot z_r)x_i)$$
$$= D_j(x_i z_r - q_{ii}^r q_{ij} z_r x_i)$$
$$= x_i D_j(z_r) - q_{ii}^r q_{ij}q_{ji} D_j(z_r)x_i$$

and the claim follows by induction. Thus, by Example 2.9, $D_j \left((\text{ad }_c x_i)^r (x_j)\right) = 0$ if and only if $(r)!_{q_{ii}} \prod_{0 \le k \le r-1} \left(1 - q_{ii}^k q_{ij} q_{ji}\right) = 0$. We next claim that

$$D_i \left((\text{ad }_c x_i)^r (x_j)\right) = 0.$$

We compute

$$D_i \left(\text{ad }_c x_i(z_r)\right) = D_i(x_i z_r - (g_i \cdot z_r) x_i)$$
$$= x_i D_i(z_r) + g_i \cdot z_r - g_i \cdot z_r - D_i(g_i \cdot z_r) g_i \cdot x_i$$

and the claim follows by induction. Finally, it is clear that $D_l \left((\text{ad }_c x_i)^r (x_j)\right) = 0$, for all $l \ne i, j$. Part (b) follows then from Proposition 2.8.

Part(d) is an important result of Rosso [Ro2, Lemma 20]. $\qquad\square$

We now discuss how the twisting operation, *cf.* Section 2.4, affects Nichols algebras of diagonal type.

DEFINITION 3.8. We shall say that two braided vector spaces (V, c) and (W, d) of diagonal type, with matrices (q_{ij}) and $(\hat{q}_{ij})$, are *twist-equivalent* if $\dim V = \dim W$ and, for all i, j, $q_{ii} = \hat{q}_{ii}$ and

$$q_{ij} q_{ji} = \hat{q}_{ij} \hat{q}_{ji}. \tag{3-1}$$

PROPOSITION 3.9. *Let (V, c) and (W, d) be two twist-equivalent braided vector spaces of diagonal type, with matrices (q_{ij}) and $(\hat{q}_{ij})$, say with respect to basis $x_1, \ldots, x_\theta$ and $\hat{x}_1, \ldots, \hat{x}_\theta$. Then there exists a linear isomorphism $\psi : \mathfrak{B}(V) \to \mathfrak{B}(W)$ such that*

$$\psi(x_i) = \hat{x}_i, \qquad 1 \le i \le \theta. \tag{3-2}$$

PROOF. Let Γ be the free abelian group of rank θ, with basis $g_1, \ldots, g_\theta$. We define characters $\chi_1, \ldots, \chi_\theta$, $\hat{\chi}_1, \ldots, \hat{\chi}_\theta$ of Γ by

$$\chi_i(g_j) = q_{ji}, \qquad \hat{\chi}_i(g_j) = \hat{q}_{ji}, \qquad 1 \le i, j \le \theta.$$

We consider V, W as Yetter–Drinfeld modules over Γ by declaring $x_i \in V_{g_i}^{\chi_i}$, $\hat{x}_i \in V_{g_i}^{\hat{\chi}_i}$. Hence, $\mathfrak{B}(V), \mathfrak{B}(W)$ are braided Hopf algebras in $_\Gamma^\Gamma \mathcal{YD}$.

Let $\sigma : \Gamma \times \Gamma \to \Bbbk^\times$ be the unique bilinear form such that

$$\sigma(g_i, g_j) = \begin{cases} \hat{q}_{ij} q_{ij}^{-1}, & i \le j, \\ 1, & i > j; \end{cases} \tag{3-3}$$

it is a group cocycle. We claim that $\varphi : W \to \mathfrak{B}(V)_\sigma(1)$, $\varphi(\hat{x}_i) = x_i$, $1 \le i \le \theta$, is an isomorphism in $_\Gamma^\Gamma \mathcal{YD}$. It clearly preserves the coaction; for the action, we assume $i \le j$ and compute

$$g_j \cdot_\sigma x_i = \sigma(g_j, g_i) \sigma^{-1}(g_i, g_j) \chi_i(g_j) x_i = (\hat{q}_{ij})^{-1} q_{ij} q_{ji} x_i = \hat{q}_{ji} x_i,$$
$$g_i \cdot_\sigma x_j = \sigma(g_i, g_j) \sigma^{-1}(g_j, g_i) \chi_j(g_i) x_j = \hat{q}_{ij} q_{ij}^{-1} q_{ij} x_j = \hat{q}_{ij} x_j,$$

where we have used (2–12) and the hypothesis (3–1). This proves the claim. By Proposition 2.2, φ extends to an isomorphism $\varphi : \mathfrak{B}(W) \to \mathfrak{B}(V)_\sigma$; $\psi = \varphi^{-1}$ is the map we are looking for. □

REMARKS 3.10. (i) The map ψ defined in the proof is much more than just linear; by (2–11) and (2–14), we have for all $g, h \in \Gamma$,

$$\psi(xy) = \sigma^{-1}(g,h)\psi(x)\psi(y), \qquad x \in \mathfrak{B}(V)_g, \quad y \in \mathfrak{B}(V)_h; \qquad (3\text{--}4)$$

$$\psi([x,y]_c) = \sigma^{-1}(g,h)[\psi(x),\psi(y)]_d, \qquad x \in \mathfrak{B}(V)_g^\chi, \quad y \in \mathfrak{B}(V)_h^\eta. \qquad (3\text{--}5)$$

(ii) A braided vector space (V,c) of diagonal type, with matrix (q_{ij}), is twist-equivalent to (W,d), with a *symmetric* matrix $(\hat{q}_{ij})$.

Twisting is a very important tool. For many problems, twisting allows to reduce to the case when the diagonal braiding is symmetric; then the theory of quantum groups can be applied.

3.3. Braidings of diagonal type but not Cartan. In the next chapter, we shall concentrate on braidings of Cartan type. There are a few examples of Nichols algebras $\mathfrak{B}(V)$ of finite group type and rank 2, which are *not* of Cartan type, but where we know that the dimension is finite. We now list the examples we know, following [N; Gñ3]. The braided vector space is necessarily of diagonal type; we shall give the matrix Q of the braiding, the constraints on their entries and the dimension d of $\mathfrak{B}(V)$. Below, ω, resp. ζ, denotes an arbitrary primitive third root of 1, resp. different from 1.

$$\begin{pmatrix} q_{11} & q_{12} \\ q_{21} & -1 \end{pmatrix}; \quad q_{11}^{-1} = q_{12}q_{21} \neq 1; \qquad d = 4\,\mathrm{ord}(q_{12}q_{21}). \qquad (3\text{--}6)$$

$$\begin{pmatrix} q_{11} & q_{12} \\ q_{21} & \omega \end{pmatrix}; \quad q_{11}^{-1} = q_{12}q_{21} \neq \pm 1, \omega^{-1}; \quad d = 9\,\mathrm{ord}(q_{11})\,\mathrm{ord}(q_{12}q_{21}\omega). \qquad (3\text{--}7)$$

$$\begin{pmatrix} -1 & q_{12} \\ q_{21} & \omega \end{pmatrix}; \quad q_{12}q_{21} = -1; \qquad d = 108. \qquad (3\text{--}8)$$

$$\begin{pmatrix} -1 & q_{12} \\ q_{21} & \omega \end{pmatrix}; \quad q_{12}q_{21} = \omega; \qquad d = 72. \qquad (3\text{--}9)$$

$$\begin{pmatrix} -1 & q_{12} \\ q_{21} & \omega \end{pmatrix}; \quad q_{12}q_{21} = -\omega; \qquad d = 36. \qquad (3\text{--}10)$$

$$\begin{pmatrix} -1 & q_{12} \\ q_{21} & \zeta \end{pmatrix}; \quad q_{12}q_{21} = \zeta^{-2}; \qquad d = 4\,\mathrm{ord}(\zeta)\,\mathrm{ord}(-\zeta^{-1}). \qquad (3\text{--}11)$$

3.4. Braidings of finite non-abelian group type. We begin with a class of examples studied in [MiS].

Let Γ be a group and $T \subset \Gamma$ a subset such that for all $g \in \Gamma, t \in T, gtg^{-1} \in T$. Thus T is a union of conjugacy classes of Γ. Let $\phi : \Gamma \times T \to \Bbbk \setminus \{0\}$ be a function such that for all $g, h \in \Gamma$ and $t \in T$,

$$\phi(1, t) = 1, \tag{3-12}$$

$$\phi(gh, t) = \phi(g, hth^{-1})\phi(h, t). \tag{3-13}$$

We can then define a Yetter–Drinfeld module $V = V(\Gamma, T, \phi)$ over Γ with $\Bbbk$-basis $x_t, t \in T$, and action and coaction of Γ given by

$$gx_t = \phi(g, t)x_{gtg^{-1}}, \tag{3-14}$$

$$\delta(x_t) = t \otimes x_t \tag{3-15}$$

for all $g \in \Gamma, t \in T$.

Conversely, if the function ϕ defines a Yetter–Drinfeld module on the vector space V by (3-14), (3-15), then ϕ satisfies (3-12), (3-13).

Note that the braiding c of $V(\Gamma, T, \phi)$ is determined by

$$c(x_s \otimes x_t) = \phi(s, t)x_{sts^{-1}} \otimes x_t \text{ for all } s, t \in T,$$

hence by the values of ϕ on $T \times T$.

The main examples come from the theory of Coxeter groups ([BL, Chapitre IV]). Let S be a subset of a group W of elements of order 2. For all $s, s' \in S$ let $m(s, s')$ be the order of ss'. (W, S) is called a *Coxeter system* and W a *Coxeter group* if W is generated by S with defining relations $(ss')^{m(s,s')} = 1$ for all $s, s' \in S$ such that $m(s, s')$ is finite.

Let (W, S) be a Coxeter system. For any $g \in W$ there is a sequence $(s_1, \ldots, s_q)$ of elements in S with $g = s_1 \cdots \cdots s_q$. If q is minimal among all such representations, then $q = l(g)$ is called the *length* of g, and $(s_1, \ldots, s_q)$ is a *reduced representation* of g.

DEFINITION 3.11. Let (W, S) be a Coxeter system, and $T = \{gsg^{-1} \mid g \in W, s \in S\}$. Define $\phi : W \times T \to \Bbbk \setminus \{0\}$ by

$$\phi(g, t) = (-1)^{l(g)} \text{ for all } g \in W, t \in T. \tag{3-16}$$

This ϕ satisfies (3-12) and (3-13). Thus we have associated to each Coxeter group the Yetter–Drinfeld module $V(W, T, \phi) \in {}^W_W \mathcal{YD}$.

The functions ϕ satisfying (3-12), (3-13) can be constructed up to a diagonal change of the basis from characters of the centralizers of elements in the conjugacy classes. This is a special case of the description of the simple modules in ${}^\Gamma_\Gamma \mathcal{YD}$ (see [W] and also [L4]); the equivalent classification of the simple Hopf bimodules over Γ was obtained in [DPR] (over $\Bbbk$) and then in [Ci] (over any field).

Let t be an element in Γ. We denote by $\mathcal{O}_t$ and Γ^t the conjugacy class and the centralizer of t in Γ. Let U be any left $\Bbbk\Gamma^t$-module. It is easy to see that the induced representation $V = \Bbbk\Gamma \otimes_{\Bbbk\Gamma^t} U$ is a Yetter–Drinfeld module over Γ with the induced action of Γ and the coaction

$$\delta : V \to \Bbbk\Gamma \otimes V, \quad \delta(g \otimes u) = gtg^{-1} \otimes g \otimes u \text{ for all } g \in \Gamma, u \in U.$$

We will denote this Yetter–Drinfeld module over Γ by $M(t, U)$.

Assume that Γ is finite. Then $V = M(t, U)$ is a simple Yetter–Drinfeld module if U is a simple representation of Γ^t, and each simple module in $^{\Gamma}_{\Gamma}\mathcal{YD}$ has this form. If we take from each conjugacy class one element t and non-isomorphic simple Γ^t-modules, any two of these simple Yetter–Drinfeld modules are non-isomorphic.

Let s_i, $1 \le i \le \theta$, be a complete system of representatives of the residue classes of Γ^t. We define $t_i = s_i t s_i^{-1}$ for all $1 \le i \le \theta$. Thus

$$\Gamma/\Gamma^t \to \mathcal{O}_t, \quad s_i \Gamma^t \mapsto t_i, 1 \le i \le \theta,$$

is bijective, and as a vector space, $V = \bigoplus_{1 \le i \le \theta} s_i \otimes U$. For all $g \in \Gamma$ and $1 \le i \le \theta$, there is a uniquely determined $1 \le j \le \theta$ with $s_j^{-1} g s_i \in \Gamma^t$, and the action of g on $s_i \otimes u, u \in U$, is given by

$$g(s_i \otimes u) = s_j \otimes (s_j^{-1} g s_i)u.$$

In particular, if U is a one-dimensional Γ^t-module with basis u and action $hu = \chi(h)u$ for all $h \in \Gamma^t$ defined by the character $\chi : \Gamma^t \to \Bbbk \setminus \{0\}$, then V has a basis $x_i = s_i \otimes u, 1 \le i \le \theta$, and the action and coaction of Γ are given by

$$gx_i = \chi(s_j^{-1} g s_i)x_j \quad \text{and} \quad \delta(x_i) = t_i \otimes x_i,$$

if $s_j^{-1} g s_i \in \Gamma^t$. Note that $g t_i g^{-1} = t_j$. Hence the module we have constructed is $V(\Gamma, T, \phi)$, where T is the conjugacy class of t, and ϕ is given by $\phi(g, t_i) = \chi(s_j^{-1} g s_i)$.

We now construct another example of a function ϕ satisfying (3–12), (3–13).

DEFINITION 3.12. Let T be the set of all transpositions in the symmetric group $\mathbb{S}_n$. Define $\phi : \mathbb{S}_n \times T \to \Bbbk \setminus \{0\}$ for all $g \in \mathbb{S}_n$, $1 \le i < j \le n$, by

$$\phi(g, (ij)) = \begin{cases} 1 & , \text{ if } g(i) < g(j), \\ -1 & , \text{ if } g(i) > g(j). \end{cases} \tag{3–17}$$

Let $t = (12)$. The centralizer of t in $\mathbb{S}_n$ is

$$\langle (34), (45), \ldots, (n-1, n) \rangle \cup \langle (34), (45), \ldots, (n-1, n) \rangle (12).$$

Let χ be the character of $(\mathbb{S}_n)^t$ with $\chi((ij)) = 1$ for all $3 \le i < j \le n$, and $\chi((12)) = -1$. Then the function ϕ defined by (3–17) is given by the character χ as described above.

Up to base change we have found all functions ϕ satisfying (3–12), (3–13) for $\Gamma = \mathbb{S}_n$, where T is the conjugacy class of all transpositions, and $\phi(t, t) = -1$ for all $t \in T$. The case $\phi(t, t) = 1$ for some $t \in T$ would lead to a Nichols algebra $\mathfrak{B}(V)$ of infinite dimension.

To determine the structure of $\mathfrak{B}(V)$ for the Yetter–Drinfeld modules defined by the functions ϕ in (3–16) and (3–17) seems to be a fundamental and very hard combinatorial problem. Only a few partial results are known [MiS], [FK], [FP].

We consider some special cases; here the method of skew-derivations is applied, see Proposition 2.8.

EXAMPLE 3.13. Let $W = \mathbb{S}_n$, $n \geq 2$, and $T = \{(ij) \mid 1 \leq i < j \leq n\}$ the set of all transpositions. Define ϕ by (3–16) and let $V = V(W, T, \phi)$. Then the following relations hold in $\mathfrak{B}(V)$ for all $1 \leq i < j \leq n$, $1 \leq k < l \leq n$:

$$x_{(ij)}^2 = 0. \qquad (3\text{–}18)$$

$$\text{If } \{i, j\} \cap \{k, l\} = \varnothing, \text{ then} \qquad x_{(ij)}x_{(kl)} + x_{(kl)}x_{(ij)} = 0. \qquad (3\text{–}19)$$

$$\text{If } i < j < k, \text{ then} \qquad x_{(ij)}x_{(jk)} + x_{(jk)}x_{(ik)} + x_{(ik)}x_{(ij)} = 0, \qquad (3\text{–}20)$$

$$x_{(jk)}x_{(ij)} + x_{(ik)}x_{(jk)} + x_{(ij)}x_{(ik)} = 0.$$

EXAMPLE 3.14. Let $W = \mathbb{S}_n$, $n \geq 2$, and $T = \{(ij) \mid 1 \leq i < j \leq n\}$ the set of all transpositions. Define ϕ by (3–17) and let $V = V(W, T, \phi)$. Then the following relations hold in $\mathfrak{B}(V)$ for all $1 \leq i < j \leq n$, $1 \leq k < l \leq n$:

$$x_{(ij)}^2 = 0. \qquad (3\text{–}21)$$

$$\text{If } \{i, j\} \cap \{k, l\} = \varnothing, \text{then} \qquad x_{(ij)}x_{(kl)} - x_{(kl)}x_{(ij)} = 0. \qquad (3\text{–}22)$$

$$\text{If } i < j < k, \text{then} \qquad x_{(ij)}x_{(jk)} - x_{(jk)}x_{(ik)} - x_{(ik)}x_{(ij)} = 0, \qquad (3\text{–}23)$$

$$x_{(jk)}x_{(ij)} - x_{(ik)}x_{(jk)} - x_{(ij)}x_{(ik)} = 0.$$

The algebras $\widetilde{\mathfrak{B}}(V)$ generated by all $x_{(ij)}$, $1 \leq i < j \leq n$, with the quadratic relations in the examples 3.13 resp. 3.14 are braided Hopf algebras in the category of Yetter–Drinfeld modules over $\mathbb{S}_n$. $\widetilde{\mathfrak{B}}(V)$ in example 3.14 is the algebra $\mathcal{E}_n$ introduced by Fomin and Kirillov in [FK] to describe the cohomology ring of the flag variety. We believe that indeed the quadratic relations in the examples 3.13 and 3.14 are defining relations for $\mathfrak{B}(V)$, that is $\widetilde{\mathfrak{B}}(V) = \mathfrak{B}(V)$ in these cases.

It was noted in [MiS] that the conjecture in [FK] about the "Poincaré-duality" of the dimensions of the homogeneous components of the algebras $\mathcal{E}_n$ (in case they are finite-dimensional) follows from the braided Hopf algebra structure as a special case of Lemma 1.12.

Another result about the algebras $\mathcal{E}_n$ by Fomin and Procesi [FP] says that $\mathcal{E}_{n+1}$ is a free module over $\mathcal{E}_n$, and $P_{\mathcal{E}_n}$ divides $P_{\mathcal{E}_{n+1}}$, where P_A denotes the Hilbert series of a graded algebra A. The proof in [FP] used the relations in Example 3.14.

This result is in fact a special case of a very general splitting theorem for braided Hopf algebras in [MiS, Theorem3.2] which is an application of the fundamental theorem for Hopf modules in the braided situation. This splitting theorem generalizes the main result of [Gñ2].

In [MiS] some partial results are obtained about the structure of the Nichols algebras of Coxeter groups. In particular

THEOREM 3.15. [MiS, Corollary 5.9]*Let* (W, S) *be a Coxeter system,* T *the set of all W-conjugates of elements in* S, ϕ *defined by* (3-16), $V = V(W, T, \phi)$ *and* $R = \mathfrak{B}(V)$. *For all* $g \in W$, *choose a reduced representation* $g = s_1 \cdots s_q$, $s_1, \cdots, s_q \in S$, *of* g, *and define*

$$x_g = x_{s_1} \cdots x_{s_q}.$$

Then the subalgebra of R generated by all $x_s, s \in S$ *has the $\Bbbk$-basis* $x_g, g \in W$. *For all* $g \in W$, *the g-homogeneous component* R_g *of R is isomorphic to* R_1.
If R is finite-dimensional, then W is finite and $dim(R) = ord(W) dim(R_1)$.

This theorem holds for more general functions ϕ, in particular for S_n and ϕ defined in (3–17).

Let (W, S) be a Coxeter system and $V = V(W, T, \phi)$ as in Theorem [MiS]. Then $\mathfrak{B}(V)$ was computed in [MiS] in the following cases:

- $W = \mathbb{S}_3$, $S = \{(12), (23)\}$: The relations of $\mathfrak{B}(V)$ are the quadratic relations in Example 3.13, and $\dim \mathfrak{B}(V) = 12$.
- $W = \mathbb{S}_4$, $S = \{(12), (23), (34)\}$: The relations of $\mathfrak{B}(V)$ are the quadratic relations in Example 3.13, and $\dim \mathfrak{B}(V) = 24 \cdot 24$.
- $W = D_4$, the dihedral group of order 8, $S = \{t, t'\}$, where t, t' are generators of D_4 of order 2 such that tt' is of order 4. There are quadratic relations and relations of order 4 defining $\mathfrak{B}(V)$, and $\dim \mathfrak{B}(V) = 64$.

In all three cases the integral, which is the longest non-zero word in the generators x_t, can be described in terms of the longest element in the Coxeter group. In all the other cases it is not known whether $\mathfrak{B}(V)$ is finite-dimensional.

In [FK] it is shown that

- $\dim(\mathcal{E}_3) = 12$.
- $\dim(\mathcal{E}_4) = 24 \cdot 24$.
- $\dim(\mathcal{E}_5)$ is finite by using a computer program.

Again, for the other cases $n > 5$ it is not known whether $\mathcal{E}_n$ is finite-dimensional.

In [Gñ3, 5.3.2] another example of a finite-dimensional Nichols algebra of a braided vector space (V, c) of finite group type is given with $\dim(V) = 4$ and $\dim(\mathfrak{B}(V)) = 72$. The defining relations of $\mathfrak{B}(V)$ are quadratic and of order 6.

By a result of Montgomery [M2], any pointed Hopf algebra B can be decomposed as a crossed product

$$B \simeq A \#_\sigma \Bbbk G, \quad \sigma \text{ a 2-cocycle}$$

of A, its link-indecomposable component containing 1 (a Hopf subalgebra) and a group algebra $\Bbbk G$. However, the structure of such link-indecomposable Hopf algebras A, in particular in the case when A is finite-dimensional and the group of its group-like elements $G(A)$ is non-abelian, is not known. To define *link-indecomposable pointed Hopf algebras*, we recall the definition of the *quiver* of A in [M2]. The vertices of the quiver of A are the elements of the group $G(A)$; for $g, h \in G(A)$, there exists an arrow from h to g if $P_{g,h}(A)$ is non-trivial, that is if $\Bbbk(g - h) \subsetneq P_{g,h}(A)$. The Hopf algebra A is called link-indecomposable, if its quiver is connected as an undirected graph.

DEFINITION 3.16. *Let Γ be a finite group and $V \in {}_{\Gamma}^{\Gamma}\mathcal{YD}$. V is called link-indecomposable if the group Γ is generated by the elements g with $V_g \neq 0$.*

By [MiS, 4.2], $V \in {}_{\Gamma}^{\Gamma}\mathcal{YD}$ is link-indecomposable if and only if the Hopf algebra $\mathfrak{B}(V) \# \Bbbk\Gamma$ is link-indecomposable.

Thus, by the examples constructed above, there are link-indecomposable, finite-dimensional pointed Hopf algebras A with $G(A)$ isomorphic to $S_n, 3 \leq n \leq 5$, or to D_4.

QUESTION 3.17. Which finite groups are isomorphic to $G(A)$ for some finite-dimensional, link-indecomposable pointed Hopf algebra A? Are there finite groups which do not occur in this form?

Finally, let us come back to the simple Yetter–Drinfeld modules $V = M(t, U) \in {}_{\Gamma}^{\Gamma}\mathcal{YD}$, where $t \in \Gamma$ and U is a simple left Γ^t-module of dimension > 1. In this case, strong restrictions are known for $\mathfrak{B}(V)$ to be finite-dimensional. By Schur's lemma, t acts as a scalar q on U.

PROPOSITION 3.18. [Gñ3, 3.1] *Assume that* $\dim \mathfrak{B}(V)$ *is finite. If* $\dim U \geq 3$, *then* $q = -1$; *and if* $\dim U = 2$, *then* $q = -1$ *or q is a root of unity of order three.*

In the proof of Proposition 3.18, a result of Lusztig on braidings of Cartan type (see [AS2, Theorem 3.1]) is used. In a similar way Graña showed

PROPOSITION 3.19. [Gñ3, 3.2] *Let Γ be a finite group of odd order, and $V \in {}_{\Gamma}^{\Gamma}\mathcal{YD}$. Assume that $\mathfrak{B}(V)$ is finite-dimensional. Then the multiplicity of any simple Yetter–Drinfeld module over Γ as a direct summand in V is at most 2.*

In particular, up to isomorphism there are only finitely many Yetter–Drinfeld modules $V \in {}_{\Gamma}^{\Gamma}\mathcal{YD}$ such that $\mathfrak{B}(V)$ is finite-dimensional.

The second statement in Proposition 3.19 was a conjecture in a preliminary version of [AS2].

3.5. Braidings of (infinite) group type. We briefly mention Nichols algebras over a free abelian group of finite rank with a braiding which is not diagonal.

EXAMPLE 3.20. Let $\Gamma = \langle g \rangle$ be a free group in one generator. Let $\mathcal{V}(t, 2)$ be the Yetter–Drinfeld module of dimension 2 such that $\mathcal{V}(t, 2) = \mathcal{V}(t, 2)_g$ and the action of g on $\mathcal{V}(t, 2)$ is given, in a basis x_1, x_2, by

$$g \cdot x_1 = tx_1, \qquad g \cdot x_2 = tx_2 + x_1.$$

Here $t \in \mathbb{k}^\times$. Then:

(a) If t is not a root of 1, then $\mathfrak{B}(\mathcal{V}(t, 2)) = T(\mathcal{V}(t, 2))$.

(b) If $t = 1$, then $\mathfrak{B}(\mathcal{V}(1, 2)) = \mathbb{k}\langle x_1, x_2 | x_1 x_2 = x_2 x_1 + x_1^2 \rangle$; this is the well-known Jordanian quantum plane.

EXAMPLE 3.21. More generally, if $t \in \mathbb{k}^\times$, let $\mathcal{V}(t, \theta)$ be the Yetter–Drinfeld module of dimension $\theta \geq 2$ such that $\mathcal{V}(t, \theta) = \mathcal{V}(t, \theta)_g$ and the action of g on $\mathcal{V}(t, \theta)$ is given, in a basis $x_1, \ldots, x_\theta$, by

$$g \cdot x_1 = tx_1, \qquad g \cdot x_j = tx_j + x_{j-1}, \quad 2 \leq j \leq \theta.$$

Note there is an inclusion of Yetter–Drinfeld modules $\mathcal{V}(t, 2) \hookrightarrow \mathcal{V}(t, \theta)$; hence, if t is not a root of 1, $\mathfrak{B}(\mathcal{V}(t, \theta))$ has exponential growth.

QUESTION 3.22. Compute $\mathfrak{B}(\mathcal{V}(1, \theta))$; does it have finite growth?

4. Nichols Algebras of Cartan Type

We now discuss fundamental examples of Nichols algebras of diagonal type that come from the theory of quantum groups.

We first need to fix some notation. Let $A = (a_{ij})_{1 \leq i,j \leq \theta}$ be a generalized symmetrizable Cartan matrix [K]; let $(d_1, \ldots, d_\theta)$ be positive integers such that $d_i a_{ij} = d_j a_{ji}$. Let $\mathfrak{g}$ be the Kac–Moody algebra corresponding to the Cartan matrix A. Let $\mathcal{X}$ be the set of connected components of the Dynkin diagram corresponding to it. For each $I \in \mathcal{X}$, we let $\mathfrak{g}_I$ be the Kac–Moody Lie algebra corresponding to the generalized Cartan matrix $(a_{ij})_{i,j \in I}$ and $\mathfrak{n}_I$ be the Lie subalgebra of $\mathfrak{g}_I$ spanned by all its positive roots. We omit the subindex I when $I = \{1, \ldots, \theta\}$. We assume that for each $I \in \mathcal{X}$, there exist c_I, d_I such that $I = \{j : c_I \leq j \leq d_I\}$; that is, after reordering the Cartan matrix is a matrix of blocks corresponding to the connected components. Let $I \in \mathcal{X}$ and $i \sim j$ in I; then $N_i = N_j$, hence $N_I := N_i$ is well defined. Let Φ_I, resp. Φ_I^+, be the root system, resp. the subset of positive roots, corresponding to the Cartan matrix $(a_{ij})_{i,j \in I}$; then $\Phi = \bigcup_{I \in \mathcal{X}} \Phi_I$, resp. $\Phi^+ = \bigcup_{I \in \mathcal{X}} \Phi_I^+$ is the root system, resp. the subset of positive roots, corresponding to the Cartan matrix $(a_{ij})_{1 \leq i,j \leq \theta}$. Let $\alpha_1, \ldots, \alpha_\theta$ be the set of simple roots.

Let $\mathcal{W}_I$ be the Weyl group corresponding to the Cartan matrix $(a_{ij})_{i,j \in I}$; we identify it with a subgroup of the Weyl group $\mathcal{W}$ corresponding to the Cartan matrix (a_{ij}).

If (a_{ij}) is of finite type, we fix a reduced decomposition of the longest element $\omega_{0,I}$ of $\mathcal{W}_I$ in terms of simple reflections. Then we obtain a reduced decomposition of the longest element $\omega_0 = s_{i_1} \ldots s_{i_P}$ of $\mathcal{W}$ from the expression of ω_0 as product of the $\omega_{0,I}$'s in some fixed order of the components, say the order arising from the order of the vertices. Therefore $\beta_j := s_{i_1} \ldots s_{i_{j-1}}(\alpha_{i_j})$ is a numeration of Φ^+.

EXAMPLE 4.1. Let $q \in \Bbbk$, $q \neq 0$, and consider the braided vector space $(\mathbb{V}, c)$, where $\mathbb{V}$ is a vector space with a basis $x_1, \ldots, x_\theta$ and the braiding c is given by

$$c(x_i \otimes x_j) = q^{d_i a_{ij}} x_j \otimes x_i, \qquad (4\text{--}1)$$

THEOREM 4.2. [L3, 33.1.5] *Let* $(\mathbb{V}, c)$ *be a braided vector space with braiding matrix* (4–1). *If q is not algebraic over* $\mathbb{Q}$, *then*

$$\mathfrak{B}(\mathbb{V}) = \Bbbk\langle x_1, \ldots, x_\theta | \mathrm{ad}\,_c(x_i)^{1-a_{ij}} x_j = 0, \quad i \neq j\rangle.$$

The theorem says that $\mathfrak{B}(\mathbb{V})$ is the well-known "positive part" $U_q^+(g)$ of the Drinfeld–Jimbo quantum enveloping algebra of g.

To state the following important theorem, we recall the definition of braided commutators (1–20). Lusztig defined root vectors $X_\alpha \in \mathfrak{B}(\mathbb{V})$, $\alpha \in \Phi^+$ [L2]. One can see from [L1; L2] that, up to a non-zero scalar, each root vector can be written as an iterated braided commutator in some sequence $X_{l_1}, \ldots, X_{l_a}$ of simple root vectors such as $[[X_{l_1}, [X_{l_2}, X_{l_3}]_c]_c, [X_{l_4}, X_{l_5}]_c]_c$. See also [Ri].

THEOREM 4.3 [L1; L2; L3; dCP; Ro1; Mu]. *Let* $(\mathbb{V}, c)$ *be a braided vector space with braiding matrix* (4–1). *Assume that q is a root of 1 of odd order N; and that 3 does not divide N if there exists $I \in \mathcal{X}$ of type G_2.*

The algebra $\mathfrak{B}(\mathbb{V})$ *is finite-dimensional if and only if* (a_{ij}) *is a finite Cartan matrix.*

If this happens, then $\mathfrak{B}(\mathbb{V})$ *can be presented by generators X_i, $1 \leq i \leq \theta$, and relations*

$$\mathrm{ad}\,_c(X_i)^{1-a_{ij}}(X_j) = 0, \qquad i \neq j, \qquad (4\text{--}2)$$

$$X_\alpha^N = 0, \qquad \alpha \in \Phi^+. \qquad (4\text{--}3)$$

Moreover, the following elements constitute a basis of $\mathfrak{B}(\mathbb{V})$:

$$X_{\beta_1}^{h_1} X_{\beta_2}^{h_2} \ldots X_{\beta_P}^{h_P}, \qquad 0 \leq h_j \leq N-1, \quad 1 \leq j \leq P.$$

The theorem says that $\mathfrak{B}(\mathbb{V})$ is the well-known "positive part" $u_q^+(g)$ of the so-called Frobenius–Lusztig kernel of g.

Motivated by the preceding theorems and results, we introduce the following notion, generalizing [AS2] (see also [FG]).

DEFINITION 4.4. Let (V, c) a braided vector space of diagonal type with basis $x_1, \ldots, x_\theta$, and matrix (q_{ij}), that is

$$c(x_i \otimes x_j) = q_{i,j} x_j \otimes x_i, \text{ for all } 1 \leq i, j \leq \theta.$$

We shall say that (V, c) is *of Cartan type* if $q_{ii} \neq 1$ for all i, and there are integers a_{ij} with $a_{ii} = 2$, $1 \leq i \leq \theta$, and $0 \leq -a_{ij} < \operatorname{ord} q_{ii}$ (which could be infinite), $1 \leq i \neq j \leq \theta$, such that

$$q_{ij} q_{ji} = q_{ii}^{a_{ij}}, \qquad 1 \leq i, j \leq \theta.$$

Since clearly $a_{ij} = 0$ implies that $a_{ji} = 0$ for all $i \neq j$, (a_{ij}) is a generalized Cartan matrix in the sense of the book [K]. We shall adapt the terminology from generalized Cartan matrices and Dynkin diagrams to braidings of Cartan type. For instance, we shall say that (V, c) is of *finite Cartan type* if it is of Cartan type and the corresponding GCM is actually of finite type, *i.e.* a Cartan matrix associated to a finite-dimensional semisimple Lie algebra. We shall say that a Yetter–Drinfeld module V is *of Cartan type* if the matrix (q_{ij}) as above is of Cartan type.

DEFINITION 4.5. Let (V, c) be a braided vector space of Cartan type with Cartan matrix (a_{ij}). We say that (V, c) is of *FL-type* (or Frobenius–Lusztig type) if there exist positive integers $d_1, \ldots, d_\theta$ such that

For all i, j, $d_i a_{ij} = d_j a_{ji}$ (thus (a_{ij}) is symmetrizable).   (4–4)

There exists a root of unity $q \in \Bbbk$ such that $q_{ij} = q^{d_i a_{ij}}$ for all i, j.   (4–5)

We call (V, c) *locally of FL-type* if any principal 2×2 submatrix of (q_{ij}) defines a braiding of FL-type.

We now fix for each $\alpha \in \Phi^+$ such a representation of X_α as an iterated braided commutator. For a general braided vector space (V, c) of finite Cartan type, we define root vectors x_α in the tensor algebra $T(V)$, $\alpha \in \Phi^+$, as the same formal iteration of braided commutators in the elements $x_1, \ldots, x_\theta$ instead of $X_1, \ldots, X_\theta$ but with respect to the braiding c given by the general matrix (q_{ij}).

THEOREM 4.6. [AS2, Th. 1.1], [AS4, Th. 4.5]. *Let (V, c) be a braided vector space of Cartan type. We also assume that q_{ij} has odd order for all i, j.*

(i) *Assume that (V, c) is locally of FL-type and that, for all i, the order of q_{ii} is relatively prime to 3 whenever $a_{ij} = -3$ for some j, and is different from 3, 5, 7, 11, 13, 17. If $\mathfrak{B}(V)$ is finite-dimensional, then (V, c) is of finite Cartan type.*

(ii) *If (V, c) is of finite Cartan type, then $\mathfrak{B}(V)$ is finite dimensional, and if moreover 3 does not divide the order of q_{ii} for all i in a connected component of the Dynkin diagram of type G_2, then*

$$\dim \mathfrak{B}(V) = \prod_{I \in \mathfrak{X}} N_I^{\dim \mathfrak{n}_I},$$

where $N_I = \operatorname{ord}(q_{ii})$ *for all* $i \in I$ *and* $I \in \mathfrak{X}$. *The Nichols algebra* $\mathfrak{B}(V)$ *is presented by generators* x_i, $1 \leq i \leq \theta$, *and relations*

$$\operatorname{ad}_c(x_i)^{1-a_{ij}}(x_j) = 0, \qquad i \neq j, \qquad (4\text{--}6)$$

$$x_\alpha^{N_I} = 0, \qquad \alpha \in \Phi_I^+, \ I \in \mathfrak{X}. \qquad (4\text{--}7)$$

Moreover, the following elements constitute a basis of $\mathfrak{B}(V)$:

$$x_{\beta_1}^{h_1} x_{\beta_2}^{h_2} \dots x_{\beta_P}^{h_P}, \qquad 0 \leq h_j \leq N_I - 1, \ \text{if} \ \beta_j \in I, \quad 1 \leq j \leq P.$$

Let $\widehat{\mathfrak{B}}(V)$ be the braided Hopf algebra in $\substack{\Gamma \\ \Gamma}\mathcal{YD}$ generated by $x_1, \dots, x_\theta$ with relations (4–6), where the x_i's are primitive. Let $\mathcal{K}(V)$ be the subalgebra of $\widehat{\mathfrak{B}}(V)$ generated by $x_\alpha^{N_I}$, $\alpha \in \Phi_I^+$, $I \in \mathfrak{X}$; it is a Yetter–Drinfeld submodule of $\widehat{\mathfrak{B}}(V)$.

THEOREM 4.7. [AS4, Th. 4.8] $\mathcal{K}(V)$ *is a braided Hopf subalgebra in* $\substack{\Gamma \\ \Gamma}\mathcal{YD}$ *of* $\widehat{\mathfrak{B}}(V)$.

5. Classification of Pointed Hopf Algebras by the Lifting Method

5.1. Lifting of Cartan type. We propose subdividing the classification problem for finite-dimensional pointed Hopf algebras into the following problems:

(a) Determine all braided vector spaces V of group type such that $\mathfrak{B}(V)$ is finite-dimensional.

(b) Given a finite group Γ, determine all realizations of braided vector spaces V as in (a) as Yetter–Drinfeld modules over Γ.

(c) The lifting problem: For V as in (b), compute all Hopf algebras A such that $\operatorname{gr} A \simeq \mathfrak{B}(V) \# H$.

(d) Investigate whether any finite-dimensional pointed Hopf algebra is generated as an algebra by its group-like and skew-primitive elements.

Problem (a) was discussed in Chapters 3 and 4. We have seen the very important class of braidings of finite Cartan type and some isolated examples where the Nichols algebra is finite-dimensional. But the general case of problem (a) seems to require completely new ideas.

Problem (b) is of a computational nature. For braidings of finite Cartan type with Cartan matrix $(a_{ij})_{1 \leq i,j \leq \theta}$ and an abelian group Γ we have to compute elements $g_1, \dots, g_\theta \in \Gamma$ and characters $\chi_1, \dots, \chi_\theta \in \widehat{\Gamma}$ such that

$$\chi_i(g_j)\chi_j(g_i) = \chi_i(g_i)^{a_{ij}}, \text{ for all } 1 \leq i, j \leq \theta. \qquad (5\text{--}1)$$

To find these elements one has to solve a system of quadratic congruences in several unknowns. In many cases they do not exist. In particular, if $\theta > 2(\operatorname{ord}\Gamma)^2$, then the braiding cannot be realized over the group Γ. We refer to [AS2, Section 8] for details.

38 NICOLÁS ANDRUSKIEWITSCH AND HANS-JÜRGEN SCHNEIDER

Problem (d) is the subject of Section 5.4.

We will now discuss the lifting problem (c).

The coradical filtration $\Bbbk\Gamma = A_0 \subset A_1 \subset \cdots$ of a pointed Hopf algebra A is stable under the adjoint action of the group. For abelian groups Γ and finite-dimensional Hopf algebras, the following stronger result holds. It is the starting point of the lifting procedure, and we will use it several times.

If M is a $\Bbbk\Gamma$-module, we denote by $M^\chi = \{m \in M \mid gm = \chi(g)m$ for all $g \in \Gamma\}$, $\chi \in \widehat{\Gamma}$, the isotypic component of type χ.

LEMMA 5.1. *Let A be a finite-dimensional Hopf algebra with abelian group* $G(A) = \Gamma$ *and diagram R. Let $V = R(1) \in {}^{\Gamma}_{\Gamma}\mathcal{YD}$ with basis $x_i \in V_{g_i}^{\chi_i}, g_i \in \Gamma, \chi_i \in \widehat{\Gamma}, 1 \leq i \leq \theta$.*

(a) The isotypic component of trivial type of A_1 is A_0. Therefore, $A_1 = A_0 \oplus (\bigoplus_{\chi \neq \varepsilon}(A_1)^\chi)$ and

$$\bigoplus_{\chi \neq \varepsilon}(A_1)^\chi \xrightarrow{\sim} A_1/A_0 \xleftarrow{\sim} V\#\Bbbk\Gamma. \tag{5-2}$$

(b) For all $g \in \Gamma$, $\chi \in \widehat{\Gamma}$ with $\chi \neq \varepsilon$,

$$\mathcal{P}_{g,1}(A)^\chi \neq 0 \iff \text{there is some } 1 \leq l \leq \theta : g = g_l, \chi = \chi_l; \tag{5-3}$$

$$\mathcal{P}_{g,1}(A)^\varepsilon = \Bbbk(1-g). \tag{5-4}$$

PROOF. (a) follows from [AS1, Lemma 3.1] and implies (b). See [AS1, Lemma 5.4]. □

We assume that A is a finite-dimensional pointed Hopf algebra with abelian group $G(A) = \Gamma$, and that

$$\operatorname{gr} A \simeq \mathfrak{B}(V)\#\Bbbk\Gamma,$$

where $V \in {}^{\Gamma}_{\Gamma}\mathcal{YD}$ is a given Yetter–Drinfeld module with basis $x_i \in V_{g_i}^{\chi_i}$, $g_1 \ldots, g_\theta \in \Gamma$, $\chi_1, \ldots, \chi_\theta \in \widehat{\Gamma}$, $1 \leq i \leq \theta$.

We first lift the basis elements x_i. Using (5-2), we choose $a_i \in \mathcal{P}(A)_{g_i,1}^{\chi_i}$ such that the canonical image of a_i in A_1/A_0 is x_i (which we identify with $x_i\#1$), $1 \leq i \leq \theta$. Since the elements x_i together with Γ generate $\mathfrak{B}(V)\#\Bbbk\Gamma$, it follows from a standard argument that $a_1, \ldots, a_\theta$ and the elements in Γ generate A as an algebra.

Our aim is to find relations between the $a_i's$ and the elements in Γ which define a quotient Hopf algebra of the correct dimension $\dim \mathfrak{B}(V) \cdot \operatorname{ord}(\Gamma)$. The idea is to "lift" the relations between the $x_i's$ and the elements in Γ in $\mathfrak{B}(V)\#\Bbbk\Gamma$.

We now assume moreover that V is of finite Cartan type with Cartan matrix (a_{ij}) with respect to the basis $x_1, \ldots, x_\theta$, that is (5-1) holds. We also assume

$$\operatorname{ord}(\chi_j(g_i)) \text{ is odd for all } i, j, \tag{5-5}$$

$$N_i = \operatorname{ord}(\chi_i(g_i)) \text{ is prime to 3 for all } i \in I, I \in \mathcal{X} \text{ of type } G_2. \tag{5-6}$$

We fix a presentation $\Gamma = \langle y_1 \rangle \oplus \cdots \oplus \langle y_\sigma \rangle$, and denote by M_l the order of y_l, $1 \leq l \leq \sigma$. Then Theorem 4.6 and formulas (1–17) imply that $\mathfrak{B}(V)\#\Bbbk\Gamma$ can be presented by generators h_l, $1 \leq l \leq \sigma$, and x_i, $1 \leq i \leq \theta$ with defining relations

$$h_l^{M_l} = 1, \quad 1 \leq l \leq \sigma; \tag{5–7}$$

$$h_l h_t = h_t h_l, \quad 1 \leq t < l \leq \sigma; \tag{5–8}$$

$$h_l x_i = \chi_i(y_l) x_i h_l, \quad 1 \leq l \leq \sigma, \quad 1 \leq i \leq \theta; \tag{5–9}$$

$$x_\alpha^{N_I} = 0, \quad \alpha \in \Phi_I^+, I \in \mathfrak{X}; \tag{5–10}$$

$$\mathrm{ad}\,(x_i)^{1-a_{ij}}(x_j) = 0, \quad i \neq j, \tag{5–11}$$

and where the Hopf algebra structure is determined by

$$\Delta(h_l) = h_l \otimes h_l, \quad 1 \leq l \leq \sigma; \tag{5–12}$$

$$\Delta(x_i) = x_i \otimes 1 + g_i \otimes x_i, \quad 1 \leq i \leq \theta. \tag{5–13}$$

Thus A is generated by the elements a_i, $1 \leq i \leq \theta$, and h_l, $1 \leq l \leq \sigma$. By our previous choice, relations (5–7), (5–8), (5–9) and (5–12), (5–13) all hold in A with the $x_i's$ replaced by the $a_i's$.

The remaining problem is to lift the quantum Serre relations (5–11) and the root vector relations (5–10). We will do this in the next two Sections.

5.2. Lifting the quantum Serre relations. We divide the problem into two cases.

- Lifting of the "quantum Serre relations" $x_i x_j - \chi_j(g_i) x_j x_i = 0$, when $i \neq j$ are in different components of the Dynkin diagram.
- Lifting of the "quantum Serre relations" $\mathrm{ad}\,_c(x_i)^{1-a_{ij}}(x_j) = 0$, when $i \neq j$ are in the same component of the Dynkin diagram.

The first case is settled in the next result from [AS4, Theorem 6.8 (a)].

LEMMA 5.2. *Assume that $1 \leq i, j \leq \theta, i < j$ and $i \not\sim j$. Then*

$$a_i a_j - \chi_j(g_i) a_j a_i = \lambda_{ij}(1 - g_i g_j), \tag{5–14}$$

where λ_{ij} is a scalar in $\Bbbk$ which can be chosen such that

λ_{ij} *is arbitrary if $g_i g_j \neq 1$ and $\chi_i \chi_j = \varepsilon$, but 0 otherwise.* $\tag{5–15}$

PROOF. It is easy to check that $a_i a_j - \chi_j(g_i) a_j a_i \in \mathcal{P}(A)_{g_i g_j, 1}^{\chi_i \chi_j}$. Suppose that $\chi_i \chi_j \neq \varepsilon$ and $a_i a_j - \chi_j(g_i) a_j a_i \neq 0$. Then by (5–3), $\chi_i \chi_j = \chi_l$ and $g_i g_j = g_l$ for some $1 \leq l \leq \theta$.

Substituting g_l and χ_l in $\chi_i(g_l)\chi_l(g_i) = \chi_i(g_i)^{a_{il}}$ and using $\chi_i(g_j)\chi_j(g_i) = 1$ (since $a_{ij} = 0$, i and j lie in different components), we get $\chi_i(g_i)^2 = \chi_i(g_i)^{a_{il}}$.

Thus we have shown that $a_{il} \equiv 2 \mod N_i$, and in the same way $a_{jl} \equiv 2 \mod N_j$. Since $i \not\sim j$, a_{il} or a_{jl} must be 0, and we obtain the contradiction $N_i = 2$ or $N_j = 2$.

Therefore $\chi_i \chi_j = \varepsilon$, and the claim follows from (5–4), or $a_i a_j - \chi_j(g_i) a_j a_i = 0$, and the claim is trivial. $\qquad\qquad\qquad\qquad\qquad\qquad\qquad\qquad\qquad\qquad\qquad$ $\square$

Lemma 5.2 motivates the following notion.

DEFINITION 5.3. [AS4, Definition 5.1] We say that two vertices i and j are *linkable* (or that i is *linkable to* j) if

$$i \not\sim j, \tag{5–16}$$

$$g_i g_j \neq 1 \text{ and} \tag{5–17}$$

$$\chi_i \chi_j = \varepsilon. \tag{5–18}$$

The following elementary properties are easily verified:

If i is linkable to j, then $\chi_i(g_j)\chi_j(g_i) = 1$, $\quad \chi_j(g_j) = \chi_i(g_i)^{-1}$. $\qquad$ (5–19)

If i and k, resp. j and l, are linkable, then $a_{ij} = a_{kl}$, $a_{ji} = a_{lk}$. $\qquad$ (5–20)

A vertex i can not be linkable to two different vertices j and h. $\qquad$ (5–21)

A *linking datum* is a collection $(\lambda_{ij})_{1 \le i < j \le \theta, i \not\sim j}$ of elements in k such that λ_{ij} is arbitrary if i and j are linkable but 0 otherwise. Given a linking datum, we say that two vertices i and j are *linked* if $\lambda_{ij} \neq 0$.

The notion of a linking datum encodes the information about lifting of relations in the first case.

DEFINITION 5.4. The collection $\mathcal{D}$ formed by a finite Cartan matrix (a_{ij}), and $g_1, \ldots, g_\theta \in \Gamma, \chi_1, \ldots, \chi_\theta \in \widehat{\Gamma}$ satisfying (5–1), (5–5) and (5–6), and a linking datum $(\lambda_{ij})_{1 \le i < j \le \theta, i \not\sim j}$ will be called a *linking datum of finite Cartan type* for Γ. We define the Yetter–Drinfeld module $V \in {}^{\Gamma}_{\Gamma}\mathcal{YD}$ of $\mathcal{D}$ as the vector space with basis $x_1, \ldots, x_\theta$ with $x_i \in V_{g_i}^{\chi_i}$ for all i.

If $\mathcal{D}$ is a linking datum of finite Cartan type for Γ, we define the Hopf algebra $\mathfrak{u}(\mathcal{D})$ by generators a_i, $1 \le i \le \theta$, and h_l, $1 \le l \le \sigma$ and the relations (5–7),(5–8),(5–9),(5–10), the quantum Serre relations (5–11) for $i \neq j$ and $i \sim j$, (5–12),(5–13) with the x_i's replaced by the a_i's, and the lifted quantum Serre relations (5–14). We formally include the case when $\theta = 0$ and $\mathfrak{u}(\mathcal{D})$ is a group algebra.

In the definition of $\mathfrak{u}(\mathcal{D})$ we could always assume that the linking datum contains only elements $\lambda_{ij} \in \{0, 1\}$ (by multiplying the generators a_i with non-zero scalars).

THEOREM 5.5. [AS4, Th. 5.17] *Let Γ be a finite abelian group and $\mathcal{D}$ a linking datum of finite Cartan type for Γ with Yetter–Drinfeld module V. Then $\mathfrak{u}(\mathcal{D})$ is a finite-dimensional pointed Hopf algebra with* $\operatorname{gr} \mathfrak{u}(\mathcal{D}) \simeq \mathfrak{B}(V) \# k\Gamma$.

The proof of the theorem is by induction on the number of irreducible components of the Dynkin diagram. In the induction step a new Hopf algebra is constructed by twisting the multiplication of the tensor product of two Hopf algebras by a 2-cocycle. The 2-cocycle is defined in terms of the linking datum.

Note that the Frobenius–Lusztig kernel $u_q(\mathfrak{g})$ of a semisimple Lie algebra $\mathfrak{g}$ is a special case of $u(\mathcal{D})$. Here the Dynkin diagram of $\mathcal{D}$ is the disjoint union of two copies of the Dynkin diagram of $\mathfrak{g}$, and corresponding points are linked pairwise. But many other linkings are possible, for example 4 copies of A_3 linked in a circle [AS4, 5.13]. See [D] for a combinatorial description of all linkings of Dynkin diagrams.

Let us now turn to the second case. Luckily it turns out that (up to some small order exceptions) in the second case the Serre relations simply hold in the lifted situation without any change.

THEOREM 5.6. [AS4, Theorem 6.8]. *Let $I \in \mathcal{X}$. Assume that $N_I \neq 3$. If I is of type B_n, C_n or F_4, resp. G_2, assume further that $N_I \neq 5$, resp. $N_I \neq 7$. Then the quantum Serre relations hold for all $i, j \in I, i \neq j, i \sim j$.*

5.3. Lifting the root vector relations. Assume first that the root α is simple and corresponds to a vertex i. It is not difficult to see, using the quantum binomial formula, that $a_i^{N_i}$ is a $(g_i^{N_i}, 1)$-skew-primitive. By Lemma 5.1, we have

$$a_i^{N_i} = \mu_i \left(1 - g_i^{N_i}\right), \qquad (5\text{–}22)$$

for some scalar μ_i; this scalar can be chosen so that

$$\mu_i \text{ is arbitrary if } g_i^{N_i} \neq 1 \text{ and } \chi_i^{N_i} = 1 \text{ but 0 otherwise.} \qquad (5\text{–}23)$$

Now, if the root α is *not* simple then $a_i^{N_i}$ is not necessarily a skew-primitive, but a skew-primitive "modulo root vectors of shorter length".

In general, we define the root vector a_α for $\alpha \in \Phi_I^+, I \in \mathcal{X}$, by replacing the x_i by a_i in the formal expression for x_α as a braided commutator in the simple root vectors. Then $a_\alpha^{N_I}, \alpha \in I$, should be an element u_α in the group algebra of the subgroup generated by the N_I-th powers of the elements in Γ.

Finally, the Hopf algebra generated by a_i, $1 \le i \le \theta$, and h_l, $1 \le l \le \sigma$ with the relations (5–7),(5–8),(5–9) (with a_i instead of x_i),

- the lifted root vector relations $a_\alpha^{N_I} = u_\alpha, \alpha \in \Phi_I^+, I \in \mathcal{X}$,
- the quantum Serre relations (5–11) for $i \neq j$ and $i \sim j$ (with a_i instead of x_i),
- the lifted quantum Serre relations (5–14),

should have the correct dimension $\dim(\mathfrak{B}(V)) \cdot \mathrm{ord}(\Gamma)$.

We carried out all the steps of this program in the following cases:

(a) All connected components of the Dynkin diagram are of type A_1 [AS1].
(b) The Dynkin diagram is of type A_2, and $N > 3$ is odd [AS3].
(c) The Dynkin diagram is arbitrary, but we assume $g_i^{N_i} = 1$ for all i [AS4].

(d) The Dynkin diagram is of type A_n, any $n \geq 2$, and $N > 3$, see Section 6 of this paper.

The cases $A_2, N = 3$ and B_2, N odd and $\neq 5$, were recently done in [BDR]. Here N denotes the common order of $\chi_i(g_i)$ for all i when the Dynkin diagram is connected.

5.4. Generation in degree one. Let us now discuss step (d) of the Lifting method.

It is not difficult to show that our conjecture 2.7 about Nichols algebras, in the setting of $H = \Bbbk\Gamma$, is equivalent to

CONJECTURE 5.7. [AS3]. *Any pointed finite-dimensional Hopf algebra over* $\Bbbk$ *is generated by group-like and skew-primitive elements.*

We have seen in Section 2.1 that the corresponding conjecture is false when the Hopf algebra is infinite-dimensional or when the Hopf algebra is finite-dimensional and the characteristic of the field is > 0. A strong indication that the conjecture is true is given by:

THEOREM 5.8. [AS4, Theorem 7.6]. *Let A be a finite-dimensional pointed Hopf algebra with coradical* $\Bbbk\Gamma$ *and diagram R, that is*

$$\operatorname{gr} A \simeq R \# \Bbbk\Gamma.$$

Assume that $R(1)$ *is a Yetter–Drinfeld module of finite Cartan type with braiding* $(q_{ij})_{1 \leq i,j \leq \theta}$. *For all i, let* $q_i = q_{ii}, N_i = \operatorname{ord}(q_i)$. *Assume that* $\operatorname{ord}(q_{ij})$ *is odd and* N_i *is not divisible by 3 and* > 7 *for all* $1 \leq i, j \leq \theta$.

(i) *For any* $1 \leq i \leq \theta$ *contained in a connected component of type* B_n, C_n *or* F_4
    resp. G_2, *assume that* N_i *is not divisible by 5 resp. by 5 or 7.*
(ii) *If i and j belong to different components, assume* $q_i q_j = 1$ *or* $\operatorname{ord}(q_i q_j) = N_i$.

Then R is generated as an algebra by $R(1)$, *that is A is generated by skew-primitive and group-like elements.*

Let us discuss the idea of the proof of Theorem 5.8. At one decisive point, we use our previous results about braidings of Cartan type of rank 2.

Let S be the graded dual of R. By the duality principle in Lemma 2.4, S is generated in degree one since $P(R) = R(1)$. Our problem is to show that R is generated in degree one, that is S is a Nichols algebra.

Since S is generated in degree one, there is a surjection of graded braided Hopf algebras $S \to \mathfrak{B}(V)$, where $V = S(1)$ has the same braiding as $R(1)$. But we know the defining relations of $\mathfrak{B}(V)$, since it is of finite Cartan type. So we have to show that these relations also hold in S.

In the case of a quantum Serre relation $\operatorname{ad}_c(x_i)^{1-a_{ij}}(x_j) = 0$, $i \neq j$, we consider the Yetter–Drinfeld submodule W of S generated by x_i and $\operatorname{ad}_c(x_i)^{1-a_{ij}}(x_j)$ and assume that $\operatorname{ad}_c(x_i)^{1-a_{ij}}(x_j) \neq 0$. The assumptions (1) and (2) of the theorem guarantee that W also is of Cartan type, but not of finite Cartan type. Thus $\operatorname{ad}_c(x_i)^{1-a_{ij}}(x_j) = 0$ in S.

Since the quantum Serre relations hold in S, the root vector relations follow automatically from the next Lemma which is a consequence of Theorem 4.7.

LEMMA 5.9. [AS4, Lemma 7.5] *Let* $S = \bigoplus_{n \geq 0} S(n)$ *be a finite-dimensional graded Hopf algebra in* $_\Gamma^\Gamma \mathcal{YD}$ *such that* $S(0) = \Bbbk 1$. *Assume that* $V = S(1)$ *is of Cartan type with basis* $(x_i)_{1 \leq i, j \leq \theta}$ *as described in the beginning of this Section. Assume the Serre relations*

$$(\operatorname{ad}_c x_i)^{1-a_{ij}} x_j = 0 \text{ for all } 1 \leq i, j \leq \theta, i \neq j \text{ and } i \sim j.$$

Then the root vector relations

$$x_\alpha^{N_I} = 0, \quad \alpha \in \Phi_I^+, \quad I \in \mathcal{X},$$

hold in S.

Another result supporting Conjecture 5.7 is:

THEOREM 5.10. [AEG, 6.1] *Any finite-dimensional cotriangular pointed Hopf algebra is generated by skew-primitive and group-like elements.*

5.5. Applications. As a special case of the theory explained above we obtain a complete answer to the classification problem in a significant case.

THEOREM 5.11. [AS4, Th. 1.1] *Let p be a prime > 17, $s \geq 1$, and $\Gamma = (\mathbb{Z}/(p))^s$. Up to isomorphism there are only finitely many finite-dimensional pointed Hopf algebras A with $G(A) \simeq \Gamma$. They all have the form*

$$A \simeq \mathfrak{u}(\mathcal{D}), \text{ where } \mathcal{D} \text{ is a linking datum of finite Cartan type for } \Gamma.$$

If we really want to write down all these Hopf algebras we still have to solve the following serious problems:

- Determine all Yetter–Drinfeld modules V over $\Gamma = (\mathbb{Z}/(p))^s$ of finite Cartan type.
- Determine all the possible linkings for the modules V over $(\mathbb{Z}/(p))^s$ in (a).

By [AS2, Proposition 8.3], $\dim V \leq 2s\frac{p-1}{p-2}$, for all the possible V in (a). This proves the finiteness statement in Theorem 5.11.

Note that we have precise information about the dimension of the Hopf algebras in 5.11:

$$\dim \mathfrak{u}(\mathcal{D}) = p^{s|\phi^+|},$$

where $|\phi^+|$ is the number of the positive roots of the root system of rank $\theta \leq 2s\frac{p-1}{p-2}$ of the Cartan matrix of $\mathcal{D}$.

For arbitrary finite abelian groups Γ, there usually are infinitely many non-isomorphic pointed Hopf algebras of the same finite dimension. The first examples were found in 1997 independently in [AS1], [BDG], [G]. Now it is very easy to construct lots of examples by lifting. Using [AS3, Lemma 1.2] it is possible to decide when two liftings are non-isomorphic.

But we have a bound on the dimension of A:

THEOREM 5.12. [AS4, Th. 7.9] *For any finite (not necessarily abelian) group* Γ *of odd order there is a natural number* $n(\Gamma)$ *such that*

$$\dim A \leq n(\Gamma)$$

for any finite-dimensional pointed Hopf algebra A with $G(A) = \mathbb{k}\Gamma$.

REMARK 5.13. As a corollary of Theorem 5.11 and its proof, we get the complete classification of all finite-dimensional pointed Hopf algebras with coradical of prime dimension p, $p \neq 2, 5, 7$. By [AS2, Theorem 1.3], the only possibilities for the Cartan matrix of $\mathcal{D}$ with Γ of odd prime order p are

(a) A_1 and $A_1 \times A_1$,
(b) A_2, if $p = 3$ or $p \equiv 1 \mod 3$,
(c) B_2, if $p \equiv 1 \mod 4$,
(d) G_2, if $p \equiv 1 \mod 3$,
(e) $A_2 \times A_1$ and $A_2 \times A_2$, if $p = 3$.

The Nichols algebras over $\mathbb{Z}/(p)$ for these Cartan matrices are listed in [AS2, Theorem 1.3]. Hence we obtain from Theorem 5.11 for $p \neq 2, 5, 7$ the bosonizations of the Nichols algebras, the liftings in case (a), that is quantum lines and quantum planes [AS1], and the liftings of type A_2 [AS3] in case (b).

This result was also obtained by Musson [Mus], using the lifting method and [AS2].

The case $p = 2$ was already done in [N]. In this case the dimension of the pointed Hopf algebras with 2-dimensional coradical is not bounded.

Let us mention briefly some classification results for Hopf algebras of special order which can be obtained by the methods we have described. Let $p > 2$ be a prime. Then all pointed Hopf algebras A of dimension p^n, $1 \leq n \leq 5$ are known. If the dimension is p or p^2, then A is a group algebra or a Taft Hopf algebra. The cases of dimension p^3 and p^4 were treated in [AS1] and [AS3], and the classification of dimension p^5 follows from [AS4] and [Gñ1]. Independently and by other methods, the case p^3 was also solved in [CD] and [SvO].

See [A] for a discussion of what is known on classification of finite-dimensional Hopf algebras.

5.6. The infinite-dimensional case. Our methods are also useful in the infinite-dimensional case. Let us introduce the analogue to FL-type for infinite-dimensional Hopf algebras.

DEFINITION 5.14. Let (V, c) be a braided vector space of Cartan type with Cartan matrix (a_{ij}). We say that (V, c) is of *DJ-type* (or Drinfeld–Jimbo type) if there exist positive integers $d_1, \ldots, d_\theta$ such that

for all i, j, $d_i a_{ij} = d_j a_{ji}$ (thus (a_{ij}) is symmetrizable); $\qquad\qquad$ (5–24)

there exists $q \in \Bbbk$, not a root of unity, such that $q_{ij} = q^{d_i a_{ij}}$ for all i, j. (5–25)

To formulate a classification result for infinite-dimensional Hopf algebras, we now assume that $\Bbbk$ is the field of complex numbers and we introduce a notion from [AS5].

DEFINITION 5.15. The collection $\mathcal{D}$ formed by a *free abelian group* Γ of finite rank, a finite Cartan matrix $(a_{ij})_{1 \le i,j \le \theta}$, $g_1, \ldots, g_\theta \in \Gamma$, $\chi_1, \ldots, \chi_\theta \in \widehat{\Gamma}$, and a linking datum $(\lambda_{ij})_{1 \le i < j \le \theta,\, i \not\sim j}$, will be called a *positive datum of finite Cartan type* if

$$\chi_i(g_j)\chi_j(g_i) = \chi_i(g_i)^{a_{ij}}, \text{ and } 1 \ne \chi_i(g_i) > 0, \text{ for all } 1 \le i, j, \le \theta.$$

Notice that the restriction of the braiding of a positive datum of finite Cartan type to each connected component is twist-equivalent to a braiding of DJ-type.

If $\mathcal{D}$ is a positive datum we define the Hopf algebra $U(\mathcal{D})$ by generators a_i, $1 \le i \le \theta$, and $h_l^{\pm}$, $1 \le l \le \sigma$ and the relations $h_m^{\pm} h_l^{\pm} = h_l^{\pm} h_l^{\pm}, h_l^{\pm} h_l^{\mp} = 1$, for all $1 \le l, m \le \sigma$, defining the free abelian group of rank σ, and (5–9), the quantum Serre relations (5–11) for $i \ne j$ and $i \sim j$, (5–12),(5–13) (with a_i instead of x_i), and the lifted quantum Serre relations (5–14). We formally include the case when $\theta = 0$ and $U(\mathcal{D})$ is the group algebra of a free abelian group of finite rank.

If (V, c) is a finite-dimensional braided vector space, we will say that the *braiding is positive* if it is diagonal with matrix (q_{ij}), and the scalars q_{ii} are positive and different from 1, for all i.

The next theorem follows from a result of Rosso [Ro2, Theorem 21] and the theory described in the previous Sections.

THEOREM 5.16. [AS5] *Let A be a pointed Hopf algebra with abelian group $\Gamma = G(A)$ and diagram R. Assume that $R(1)$ has finite dimension and positive braiding. Then the following are equivalent:*

(a) A is a domain of finite Gelfand–Kirillov dimension, and the adjoint action of $G(A)$ on A (or on A_1) is semisimple.

(b) The group Γ is free abelian of finite rank, and

$$A \simeq U(\mathcal{D}), \text{ where } \mathcal{D} \text{ is a positive datum of finite Cartan type for } \Gamma.$$

It is likely that the positivity assumption on the infinitesimal braiding in the last theorem is related to the existence of a compact involution.

6. Pointed Hopf Algebras of Type A_n

In this chapter, we develop from scratch, *i.e.* without using Lusztig's results, the classification of all finite-dimensional pointed Hopf algebras whose infinitesimal braiding is of type A_n. The main results of this chapter are new.

6.1. Nichols algebras of type A_n. Let N be an integer, $N > 2$, and let q be a root of 1 of order N. For the case $N = 2$, see [AnDa].

Let q_{ij}, $1 \leq i, j \leq n$, be roots of 1 such that

$$q_{ii} = q, \quad q_{ij}q_{ji} = \begin{cases} q^{-1}, & \text{if } |i - j| = 1, \\ 1, & \text{if } |i - j| \geq 2. \end{cases} \tag{6-1}$$

for all $1 \leq i, j \leq n$. For convenience, we denote

$$B_{p,r}^{i,j} := \prod_{i \leq l \leq j-1,\, p \leq h \leq r-1} q_{l,h},$$

for any $1 \leq i < j \leq n+1$, $1 \leq p < r \leq n+1$. Then we have the following identities, whenever $i < s < j$, $p < t < r$:

$$B_{p,r}^{i,s} B_{p,r}^{s,j} = \prod_{i \leq l \leq s-1,\, p \leq h \leq r-1} q_{l,h} \prod_{s \leq l \leq j-1,\, p \leq h \leq r-1} q_{l,h} = B_{p,r}^{i,j}; \tag{6-2}$$

$$B_{p,t}^{i,j} B_{t,r}^{i,j} = B_{p,r}^{i,j}; \tag{6-3}$$

also,

$$B_{j,j+1}^{i,j} B_{i,j}^{j,j+1} = \prod_{i \leq l \leq j-1} q_{l,j} \prod_{i \leq h \leq j-1} q_{j,h} = q^{-1}; \tag{6-4}$$

$$B_{i,j}^{i,j} = q. \tag{6-5}$$

We consider in this Section a vector space $V = V_n$ with a basis $x_1, \ldots, x_n$ and braiding determined by:

$$c(x_i \otimes x_j) = q_{ij}\, x_j \otimes x_i, \quad 1 \leq i, j \leq n;$$

that is, V is of type A_n.

REMARK 6.1. Let Γ be a group, $g_1, \ldots, g_n$ in the center of Γ, and $\chi_1, \ldots, \chi_n$ in $\widehat{\Gamma}$ such that

$$q_{ij} = \langle \chi_j, g_i \rangle, \quad 1 \leq i, j \leq n.$$

Then V can be realized as a Yetter–Drinfeld module over Γ by declaring

$$x_i \in V_{g_i}^{\chi_i}, \quad 1 \leq i \leq n. \tag{6-6}$$

For example, we could consider $\Gamma = (\mathbb{Z}/P)^n$, where P is divisible by the orders of all the q_{ij}'s; and take $g_1, \ldots, g_n$ as the canonical basis of Γ.

We shall consider a braided Hopf algebra R provided with an inclusion of braided vector spaces $V \to P(R)$. We identify the elements $x_1, \ldots, x_n$ with their images in R. Distinguished examples of such R are the tensor algebra $T(V)$ and the Nichols algebra $\mathfrak{B}(V)$. Additional hypotheses on R will be stated when needed.

We introduce the family $(e_{ij})_{1 \leq i < j \leq n+1}$ of elements of R as follows:

$$e_{i,i+1} := x_i; \tag{6-7}$$

$$e_{i,j} := [e_{i,j-1}, e_{j-1,j}]_c, \quad 1 \leq i < j \leq n+1, \ j-i \geq 2. \tag{6-8}$$

The braiding between elements of this family is given by:

$$c(e_{i,j} \otimes e_{p,r}) = B_{p,r}^{i,j} e_{p,r} \otimes e_{i,j}, \quad 1 \leq i < j \leq n+1, \ 1 \leq p < r \leq n+1. \tag{6-9}$$

In particular,

$$e_{i,j} = e_{i,j-1} e_{j-1,j} - B_{j-1,j}^{i,j-1} e_{j-1,j} e_{i,j-1}.$$

REMARK 6.2. When V is realized as a Yetter–Drinfeld module over Γ as in Remark 6.1, we have $e_{i,j} \in R_{g_{i,j}}^{\chi_{i,j}}$, where

$$\chi_{i,j} = \prod_{i \leq l \leq j-1} \chi_l, \qquad g_{i,j} = \prod_{i \leq l \leq j-1} g_l, \qquad 1 \leq i < j \leq n+1. \tag{6-10}$$

LEMMA 6.3. (a) *If R is finite-dimensional or $R \simeq \mathfrak{B}(V)$, then*

$$e_{i,i+1}^N = 0, \quad if \quad 1 \leq i \leq n. \tag{6-11}$$

(b) *Assume that $R \simeq \mathfrak{B}(V)$. Then*

$$[e_{i,i+1}, e_{p,p+1}]_c = 0, \quad that \ is \quad e_{i,i+1} e_{p,p+1} = q_{ip} e_{p,p+1} e_{i,i+1}, \tag{6-12}$$

if $1 \leq i < p \leq n$, $p - i \geq 2$.
(c) *Assume that $R \simeq \mathfrak{B}(V)$. Then*

$$[e_{i,i+1}, [e_{i,i+1}, e_{i+1,i+2}]_c]_c = 0, \quad if \quad 1 \leq i < n; \tag{6-13}$$

$$[e_{i+1,i+2}, [e_{i+1,i+2}, e_{i,i+1}]_c]_c = 0, \quad if \quad 1 \leq i < n; \tag{6-14}$$

that is

$$e_{i,i+1} e_{i,i+2} = B_{i,i+2}^{i,i+1} e_{i,i+2} e_{i,i+1}, \tag{6-15}$$

$$e_{i,i+2} e_{i+1,i+2} = B_{i+1,i+2}^{i,i+2} e_{i+1,i+2} e_{i,i+2}. \tag{6-16}$$

PROOF. (a) This follows from Lemma 3.6 (a), use $c(e_{i,i+1}^N \otimes e_{i,i+1}^N) = e_{i,i+1}^N \otimes e_{i,i+1}^N$ in the finite-dimensional case.

(b), (c) By Lemma 3.6 (b), the elements

$$[e_{i,i+1}, e_{p,p+1}]_c, \quad [e_{i,i+1}, [e_{i,i+1}, e_{i+1,i+2}]_c]_c \quad and \quad [e_{i+1,i+2}, [e_{i+1,i+2}, e_{i,i+1}]_c]_c$$

are primitive. Since they are homogeneous of degree 2, respectively of degree 3, they should be 0. To derive (6–16) from (6–14), use (1–23). $\square$

LEMMA 6.4. *Assume that (6–12) holds in R. Then*

$$[e_{i,j}, e_{p,r}]_c = 0, \qquad if \quad 1 \le i < j < p < r \le n+1. \tag{6–17}$$
$$[e_{p,r}, e_{i,j}]_c = 0, \qquad if \quad 1 \le i < j < p < r \le n+1. \tag{6–18}$$
$$[e_{i,p}, e_{p,j}]_c = e_{i,j}, \qquad if \quad 1 \le i < p < j \le n+1. \tag{6–19}$$

PROOF. (6–17). For $j = i+1$ and $r = p+1$, this is (6–12); the general case follows recursively using (1–22). (6–18) follows from (6–17), since $B_{p,r}^{i,j} B_{i,j}^{p,r} = 1$ in this case.

(6–19). By induction on $j - p$; if $p = j - 1$ then (6–19) is just (6–8). For $p < j$, we have

$$e_{i,j+1} = [e_{i,j}, e_{j,j+1}]_c = [[e_{i,p}, e_{p,j}]_c, e_{j,j+1}]_c$$
$$= [e_{i,p}, [e_{p,j}, e_{j,j+1}]_c]_c = [e_{i,p}, e_{p,j+1}]_c$$

by (1–22), since $[e_{i,p}, e_{j,j+1}]_c = 0$ by (6–17). $\square$

LEMMA 6.5. *Assume that (6–12) holds in R. Then for any $1 \le i < j \le n+1$,*

$$\Delta(e_{i,j}) = e_{i,j} \otimes 1 + 1 \otimes e_{i,j} + (1 - q^{-1}) \sum_{i<p<j} e_{i,p} \otimes e_{p,j}. \tag{6–20}$$

PROOF. We proceed by induction on $j - i$. If $j - i = 1$, the formula just says that the x_i's are primitive. For the inductive step, we compute $\Delta(e_{i,j} e_{j,j+1})$ to be

$$\left(e_{i,j} \otimes 1 + 1 \otimes e_{i,j} + (1-q^{-1}) \sum_{i<p<j} e_{i,p} \otimes e_{p,j} \right) (e_{j,j+1} \otimes 1 + 1 \otimes e_{j,j+1})$$

$$= e_{i,j} e_{j,j+1} \otimes 1 + B_{j,j+1}^{i,j} e_{j,j+1} \otimes e_{i,j} + (1-q^{-1}) \sum_{i<p<j} B_{j,j+1}^{p,j} e_{i,p} e_{j,j+1} \otimes e_{p,j}$$

$$+ e_{i,j} \otimes e_{j,j+1} + 1 \otimes e_{i,j} e_{j,j+1} + (1-q^{-1}) \sum_{i<p<j} e_{i,p} \otimes e_{p,j} e_{j,j+1};$$

and $\Delta(e_{j,j+1} e_{i,j})$ to be

$$(e_{j,j+1} \otimes 1 + 1 \otimes e_{j,j+1}) \left(e_{i,j} \otimes 1 + 1 \otimes e_{i,j} + (1-q^{-1}) \sum_{i<p<j} e_{i,p} \otimes e_{p,j} \right)$$

$$= e_{j,j+1} e_{i,j} \otimes 1 + e_{j,j+1} \otimes e_{i,j} + (1-q^{-1}) \sum_{i<p<j} e_{j,j+1} e_{i,p} \otimes e_{p,j}$$

$$+ B_{i,j}^{j,j+1} e_{i,j} \otimes e_{j,j+1} + 1 \otimes e_{j,j+1} e_{i,j} + (1-q^{-1}) \sum_{i<p<j} (B_{j,j+1}^{i,p})^{-1} e_{i,p} \otimes e_{j,j+1} e_{p,j}.$$

Hence

$$\Delta(e_{i,j+1}) = e_{i,j+1} \otimes 1 + 1 \otimes e_{i,j+1} + (1 - B_{j,j+1}^{ij} B_{ij}^{j,j+1}) e_{i,j} \otimes e_{j,j+1}$$

$$+ (1 - q^{-1}) \sum_{i<p<j} \left(B_{j,j+1}^{p,j} B_{j,j+1}^{i,p} - B_{j,j+1}^{i,j} \right) e_{j,j+1} e_{i,p} \otimes e_{p,j}$$

$$+ (1 - q^{-1}) \sum_{i<p<j} e_{i,p} \otimes \left(e_{p,j} e_{j,j+1} - B_{j,j+1}^{i,j} (B_{j,j+1}^{i,p})^{-1} e_{j,j+1} e_{p,j} \right)$$

$$= e_{i,j+1} \otimes 1 + 1 \otimes e_{i,j+1} + (1 - q^{-1}) \sum_{i<p<j+1} e_{i,p} \otimes e_{p,j+1};$$

by (6–4), (6–17) and the hypothesis. $\qquad\square$

REMARK 6.6. Let Γ be a group with $g_1,\ldots,g_n$ in the center of Γ, $\chi_1,\ldots,\chi_n$ in $\widehat{\Gamma}$, as in 6.1. Let R be a braided Hopf algebra in $_{\Gamma}^{\Gamma}\mathcal{YD}$ such that (6–12) holds in R. It follows from (6–20) and the reconstruction formulas for the bosonization (1–17) that

$$\Delta_{R\#\Bbbk\Gamma}(e_{i,j}) = e_{i,j} \otimes 1 + g_{i,j} \otimes e_{i,j} + (1 - q^{-1}) \sum_{i<p<j} e_{i,p} g_{p,j} \otimes e_{p,j}. \qquad (6\text{--}21)$$

LEMMA 6.7. *Assume that* (6–12), (6–13), (6–14) *hold in* R. *Then*

$$[e_{i,j}, e_{p,r}]_c = 0, \qquad if \quad 1 \le i < p < r < j \le n+1; \qquad (6\text{--}22)$$

$$[e_{i,j}, e_{i,r}]_c = 0, \qquad if \quad 1 \le i < j < r \le n+1; \qquad (6\text{--}23)$$

$$[e_{i,j}, e_{p,j}]_c = 0, \qquad if \quad 1 \le i < p < j \le n+1. \qquad (6\text{--}24)$$

PROOF. (a) We prove (6–22) by induction on $j - i$. If $j - i = 3$, then

$$[e_{i,i+3}, e_{i+1,i+2}]_c = [[e_{i,i+2}, e_{i+2,i+3}]_c, e_{i+1,i+2}]_c$$

$$= [[[e_{i,i+1}, e_{i+1,i+2}]_c, e_{i+2,i+3}]_c, e_{i+1,i+2}]_c = 0,$$

by Lemma 1.11. If $j - i > 3$ we argue by induction on $r - p$. If $r - p = 1$, then there exists an index h such that either $i < h < p < r = p+1 < j$ or $i < p < r = p+1 < h < j$. In the first case, by (6–19), we have

$$[e_{i,j}, e_{p,r}]_c = [[e_{i,h}, e_{h,j}]_c, e_{p,p+1}]_c = 0;$$

the last equality follows from Lemma 1.10 (c), because of (6–17) and the induction hypothesis. In the second case, we have

$$[e_{i,j}, e_{p,r}]_c = [[e_{i,h}, e_{h,j}]_c, e_{p,p+1}]_c = 0;$$

the last equality follows from Lemma 1.10 (c), because of the induction hypothesis and (6–18). Finally, if $r - p > 1$ then

$$[e_{i,j}, e_{p,r}]_c = [e_{i,j}, [e_{p,r-1}, e_{r-1,r}]_c]_c = 0$$

by Lemma 1.10 (b) and the induction hypothesis.

(b) We prove (6–23) by induction on $r - i$. If $r - i = 2$, then the claimed equality is just (6–13). If $r - i > 2$ we argue by induction on $j - i$. If $j - i = 1$ we have

$$[e_{i,i+1}, e_{i,r}]_c = [e_{i,i+1}, [e_{i,r-1}, e_{r-1,r}]_c]_c = [[e_{i,i+1}, e_{i,r-1}]_c, e_{r,r-1}]_c = 0$$

by (1–22), since $[e_{i,i+1}, e_{r-1,r}]_c = 0$ by (6–17). If $j - i > 2$, we have

$$[e_{i,j}, e_{i,r}]_c = [[e_{i,j-1}, e_{j-1,j}]_c, e_{i,r}]_c = 0$$

by Lemma 1.10 (c), because of the induction hypothesis and (6–22).

The proof of (6–24) is analogous to the proof of (6–23), using (6–14) instead of (6–13).   □

LEMMA 6.8. *Assume that* (6–12), (6–13), (6–14) *hold in* R. *Then*

$$[e_{i,j}, e_{p,r}]_c = B_{jr}^{pj}(q-1)e_{ir}e_{pj}, \qquad if \quad 1 \le i < p < j < r \le n+1. \quad (6\text{--}25)$$

PROOF. We compute:

$$
\begin{aligned}
[e_{i,j}, e_{p,r}]_c &= [[e_{i,p}, e_{p,j}]_c, e_{p,r}]_c \\
&= [e_{i,p}, [e_{p,j}, e_{p,r}]_c]_c + B_{p,r}^{p,j}[e_{i,p}, e_{p,r}]_c e_{p,j} - B_{p,j}^{i,p} e_{p,j}[e_{i,p}, e_{p,r}]_c \\
&= B_{p,r}^{p,j} e_{i,r} e_{p,j} - B_{p,j}^{i,p} e_{p,j} e_{i,r} = \left(B_{p,r}^{p,j} - B_{p,j}^{i,p}(B_{p,j}^{i,r})^{-1}\right) e_{i,r} e_{p,j} \\
&= \left(B_{p,r}^{p,j} - (B_{p,j}^{p,r})^{-1}\right) e_{i,r} e_{p,j} = B_{jr}^{pj}(q-1)e_{ir}e_{pj}.
\end{aligned}
$$

Here, the first equality is by (6–19); the second, by Lemma 1.10 (a); the third, by (6–23) and by (6–19); the fourth, by (6–22).   □

LEMMA 6.9. *Assume that* (6–12), (6–13) *and* (6–14) *hold in* R. *For any* $1 \le i < j \le n+1$ *we have*

$$\Delta(e_{i,j}^N) = e_{i,j}^N \otimes 1 + 1 \otimes e_{i,j}^N + (1-q^{-1})^N \sum_{i<p<j} (B_{i,p}^{p,j})^{N(N-1)/2} e_{i,p}^N \otimes e_{p,j}^N. \quad (6\text{--}26)$$

PROOF. By (6–20), and using several times the quantum binomial formula (1–13), we have

$$
\begin{aligned}
\Delta(e_{i,j}^N) &= \left(e_{i,j} \otimes 1 + (1-q^{-1})\sum_{i<p<j} e_{i,p} \otimes e_{p,j}\right)^N + 1 \otimes e_{i,j}^N \\
&= e_{i,j}^N \otimes 1 + (1-q^{-1})^N \left(\sum_{i<p<j} e_{i,p} \otimes e_{p,j}\right)^N + 1 \otimes e_{i,j}^N \\
&= e_{i,j}^N \otimes 1 + (1-q^{-1})^N \sum_{i<p<j} (e_{i,p} \otimes e_{p,j})^N + 1 \otimes e_{i,j}^N \\
&= e_{i,j}^N \otimes 1 + (1-q^{-1})^N \sum_{i<p<j} (B_{i,p}^{p,j})^{N(N-1)/2} e_{i,p}^N \otimes e_{p,j}^N + 1 \otimes e_{i,j}^N.
\end{aligned}
$$

Here, in the first equality we use that $(1 \otimes e_{i,j})(e_{i,j} \otimes 1) = q(e_{i,j} \otimes 1)(1 \otimes e_{i,j})$ and $(1 \otimes e_{i,j})(e_{i,p} \otimes e_{p,j}) = q(e_{i,p} \otimes e_{p,j})(1 \otimes e_{i,j})$, this last by (6–24); in the second, we use $(e_{i,p} \otimes e_{p,j})(e_{i,j} \otimes 1) = q(e_{i,j} \otimes 1)(e_{i,p} \otimes e_{p,j})$, which follows from (6–23); the third, that $(e_{i,p} \otimes e_{p,j})(e_{i,s} \otimes e_{s,j}) = q^2 (e_{i,s} \otimes e_{s,j})(e_{i,p} \otimes e_{p,j})$ for $p < s$, which follows from (6–23) and (6–24); the fourth, from $(e_{i,p} \otimes e_{p,j})^h = (B_{i,p}^{p,j})^{h(h-1)/2} e_{i,p}^h \otimes e_{p,j}^h$. $\qquad \square$

REMARK 6.10. Let Γ be a group with $g_1, \dots, g_n$ in the center of Γ, $\chi_1, \dots, \chi_n$ in $\widehat{\Gamma}$, as in 6.1. Let R be a braided Hopf algebra in $_\Gamma^\Gamma \mathcal{YD}$ such that (6–12), (6–13) and (6–14) hold in R. By (6–26) and the reconstruction formulas (1–17), we have

$$\Delta_{R\#k\Gamma}(e_{i,j}^N) = e_{i,j}^N \otimes 1 + g_{i,j}^N \otimes e_{i,j}^N + (1-q^{-1})^N \sum_{i<p<j} \left(B_{i,p}^{p,j} \right)^{N(N-1)/2} e_{i,p}^N g_{p,j}^N \otimes e_{p,j}^N.$$

$$(6\text{--}27)$$

LEMMA 6.11. *Assume that* $R = \mathfrak{B}(V)$. *Then*

$$e_{i,j}^N = 0, \qquad 1 \le i < j \le n+1. \qquad (6\text{--}28)$$

PROOF. This follows from Lemma 6.9 by induction on $j - i$, the case $j - i = 1$ being Lemma 6.3 (c). $\qquad \square$

LEMMA 6.12. *Assume that* (6–12), (6–13), (6–14), (6–28), *hold in* R. *Assume, furthermore, that* R *is generated as an algebra by the elements* $x_1, \dots, x_n$. *Then the algebra* R *is spanned as a vector space by the elements*

$$e_{1,2}^{\varepsilon_{1,2}} e_{1,3}^{\varepsilon_{1,3}} \cdots e_{1,n+1}^{\varepsilon_{1,n+1}} e_{2,3}^{\varepsilon_{2,3}} \cdots e_{2,n+1}^{\varepsilon_{2,n+1}} \cdots e_{n,n+1}^{\varepsilon_{n,n+1}}, \qquad \text{with } \varepsilon_{i,j} \in \{0,1,\dots,N-1\}.$$

$$(6\text{--}29)$$

PROOF. We order the family (e_{ij}) by

$$e_{1,2} \prec e_{1,3} \prec \dots e_{1,n+1} \prec e_{2,3} \prec \dots e_{2,n+1} \prec \dots e_{n,n+1};$$

this induces an ordering in the monomials (6–29). If M is an ordered monomial, we set $\sigma(M) := e_{r,s}$ if $e_{r,s}$ is the first element appearing in M. Let B be the subspace generated by the monomials in (6–29). We show by induction on the length that, for any ordered monomial M and for any i, $e_{i,i+1}M \in B$ and it is 0 or a combination of monomials N with $\sigma(N) \succeq \min\{e_{i,i+1}, \sigma(M)\}$, length of $N \le$ length of $M + 1$. The statement is evident if the length of M is 0; so that assume that the length is positive. Write $M = e_{p,q}M'$ where $e_{p,q} \preceq M'$. We have several cases:

If $i < p$ or $i = p$ and $i + 1 < q$, $e_{i,i+1} \prec e_{p,q}$ and we are done.

If $i = p$ and $i + 1 = q$, then the claim is clear.

If $p < q < i$ then $e_{i,i+1}e_{p,q} = B_{p,q}^{i,i+1} e_{p,q}e_{i,i+1}$ by (6–18); hence $e_{i,i+1}M = e_{i,i+1}e_{p,q}M' = B_{p,q}^{i,i+1} e_{p,q}e_{i,i+1}M'$; by the inductive hypothesis and the fact that $e_{p,q} \preceq \min\{e_{i,i+1}, \sigma(M')\}$, the claim follows.

If $p < i = q$ then $e_{i,i+1}e_{p,i} = (B_{i,i+1}^{p,i})^{-1}(e_{p,i}e_{i,i+1} - e_{p,i+1})$ by (6–19); again, the inductive hypothesis and $e_{p,q} \preceq \min\{e_{i,i+1}, \sigma(M')\}$ imply that $e_{p,i}e_{i,i+1}M'$

has the form we want. To see that $e_{p,i+1}M''$ satisfies the claim when $e_{p,i} = \sigma(M')$, we use $e_{p,i+1}e_{p,i} = (B_{p,i+1}^{p,i})_{-1}e_{p,i}e_{p,i+1}$ by (6–23).

If $p < i < q$ then $e_{i,i+1}e_{p,q} = (B_{i,i+1}^{p,q})^{-1}e_{p,q}e_{i,i+1}$ by (6–22) or (6–24); we then argue as in the two preceding cases.

Therefore, $B = R$ since it is a left ideal containing 1. $\square$

We shall say that the elements $e_{1,2}, e_{1,3}, \ldots, e_{1,n+1}e_{2,3} \cdots e_{2,n+1} \cdots e_{n,n+1}$, in this order, form a *PBW-basis* for R if the monomials (6–29) form a basis of R. Then we can prove, as in [AnDa]:

THEOREM 6.13. *The elements $e_{1,2}, e_{1,3}, \ldots, e_{1,n+1}e_{2,3} \cdots e_{2,n+1} \cdots e_{n,n+1}$, in this order, form a PBW basis for $\mathfrak{B}(V_n)$. In particular,*

$$\dim \mathfrak{B}(V_n) = N^{\frac{n(n+1)}{2}}.$$

PROOF. We proceed by induction on n. The case $n = 1$ is clear, see [AS1, Section 3] for details. We assume the statement for $n - 1$. We consider V_n as a Yetter–Drinfeld module over $\Gamma = (\mathbb{Z}/P)^n$, as explained in Remark 6.1. Let $Z_n = \mathfrak{B}(V_n)\#k\Gamma$. Let $i_n : V_{n-1} \to V_n$ be given by $x_i \mapsto x_i$ and $p_n : V_n \to V_{n-1}$ by $x_i \mapsto x_i$, $1 \le i \le n-1$ and $x_n \mapsto 0$. The splitting of Yetter–Drinfeld modules $\mathrm{id}_{V_{n-1}} = p_n i_n$ gives rise to a splitting of Hopf algebras $\mathrm{id}_{Z_{n-1}} = \pi_n \iota_n$, where $\iota_n : Z_{n-1} \to Z_n$ and $\pi_n : Z_n \to Z_{n-1}$ are respectively induced by i_n, p_n. Let

$$R_n = Z_n^{\mathrm{co}\,\pi_n} = \{z \in Z_n : (\mathrm{id} \otimes \pi_n)\Delta(z) = z \otimes 1\}.$$

Then R_n is a braided Hopf algebra in the category ${}_{Z_{n-1}}^{Z_{n-1}}\mathcal{YD}$; we shall denote by c_{R_n} the corresponding braiding of R_n. We have $Z_n \simeq R_n\#Z_{n-1}$ and in particular $\dim Z_n = \dim R_n \dim Z_{n-1}$.

For simplicity, we denote $h_i = e_{i,n+1}$, $1 \le i \le n$. We have $h_i h_j = B_{j,n+1}^{i,n+1}h_j h_i$, for $i < j$, by (6–24). We claim that $h_1, \ldots, h_n$ are linearly independent primitive elements of the braided Hopf algebra R_n.

Indeed, it follows from (6–8) that $\pi_n(h_i) = 0$; by (6–20), we conclude that $h_i \in R_n$. We prove by induction on $j = n + 1 - i$ that h_i is a primitive element of R_n, the case $j = 1$ being clear. Assume the statement for j. Now

$$x_{i-1} \rightharpoonup h_i = x_{i-1}h_i + g_{i-1}h_i S(x_{i-1}) = x_{i-1}h_i - g_{i-1}h_i g_{i-1}^{-1}x_{i-1}$$
$$= x_{i-1}h_i - B_{i,n+1}^{i-1,i}h_i x_{i-1} = [x_{i-1}, h_i]_c = h_{i-1}.$$

So

$$\Delta_{R_n}(h_{i-1}) = \Delta_{R_n}(x_{i-1} \rightharpoonup h_i) = x_{i-1} \rightharpoonup \Delta(h_i)$$
$$= g_{i-1} \rightharpoonup 1 \otimes x_{i-1} \rightharpoonup h_i + x_{i-1} \rightharpoonup h_i \otimes 1 = h_{i-1} \otimes 1 + 1 \otimes h_{i-1}.$$

We prove also by induction on $j = n + 1 - i$ that $h_i \ne 0$ using (6–20) and the induction hypothesis on Z_{n-1}. Since h_i is homogeneous of degree j (with respect to the grading of Z_n), we conclude that $h_1, \ldots, h_n$ are linearly independent.

We next claim that $c_{R_n}(h_i \otimes h_j) = B^{i,n+1}_{j,n+1} h_j \otimes h_i$, for any $i > j$.
By (6-27), the coaction of Z_{n-1} on R_n satisfies

$$\delta(h_i) = g_{i,n+1} \otimes e_{i,n+1} + (1 - q^{-1}) \sum_{i < p < n+1} e_{i,p} g_{p,n+1} \otimes e_{p,n+1}.$$

If $j < i$, we compute the action on R_n:

$$e_{i,p} \rightharpoonup h_j = e_{i,p} h_j + g_{i,p} h_j \mathcal{S}(e_{i,n+1}) + (1 - q^{-1}) \sum_{i < t < p} e_{i,t} g_{t,p} h_j \mathcal{S}(e_{t,p})$$

$$= (B^{j,n+1}_{i,p})^{-1} h_j e_{i,p} + B^{i,p}_{j,n+1} h_j g_{i,p} \mathcal{S}(e_{i,n+1})$$

$$+ h_j (1 - q^{-1}) \sum_{i < t < p} B^{t,p}_{j,n+1} (B^{j,n+1}_{i,t})^{-1} e_{i,t} g_{t,p} \mathcal{S}(e_{t,p})$$

$$= B^{i,p}_{j,n+1} h_j e_{i,p(1)} \mathcal{S}(e_{i,p(2)}) = 0,$$

by (6-24). Thus

$$c_{R_n}(h_i \otimes h_j) = g_{i,n+1} \rightharpoonup h_j \otimes h_i = B^{i,n+1}_{j,n+1} h_j \otimes h_i.$$

We next claim that the dimension of the subalgebra of R_n spanned by h_1, $\ldots, h_n$ is $\geq N^n$.
We already know that

$$\Delta(h_j^{m_j}) = \sum_{0 \leq i_j \leq m_j} \binom{m_j}{i_j}_q h_j^{i_j} \otimes h_j^{m_j - i_j}, \qquad m_j \leq N.$$

Set $\mathbf{m} = (m_1, \ldots, m_j, \ldots, m_n)$, $\mathbf{1} = (1, \ldots, 1, \ldots, 1)$, $\mathbf{N} = (N, \ldots, N)$. We consider the partial order $\mathbf{i} \leq \mathbf{m}$, if $i_j \leq m_j$, $j = 1, \ldots, n$. We set $h^{\mathbf{m}} := h_n^{m_n} \ldots h_j^{m_j} \ldots h_1^{m_1}$. From the preceding claim, we deduce that

$$\Delta(h^{\mathbf{m}}) = h^{\mathbf{m}} \otimes 1 + 1 \otimes h^{\mathbf{m}} + \sum_{0 \leq \mathbf{i} \leq \mathbf{m}, \ 0 \neq \mathbf{i} \neq \mathbf{m}} c_{\mathbf{m},\mathbf{i}} h^{\mathbf{i}} \otimes h^{\mathbf{m}-\mathbf{i}}, \quad \mathbf{m} \leq \mathbf{N} - \mathbf{1};$$

where $c_{\mathbf{m},\mathbf{i}} \neq 0$ for all $\mathbf{i}$. We then argue recursively as in the proof of [AS1, Lemma 3.3] to conclude that the elements $h^{\mathbf{m}}$, $\mathbf{m} \leq \mathbf{N} - \mathbf{1}$, are linearly independent; hence the dimension of the subalgebra of R_n spanned by $h_1, \ldots, h_n$ is $\geq N^n$, as claimed.

We can now finish the proof of the theorem. Since $\dim Z_n \leq N^{\frac{n(n+1)}{2}}$ by Lemma 6.12 and $\dim Z_{n-1} = N^{\frac{n(n-1)}{2}}$ by the induction hypothesis, we have $\dim R_n \leq N^n$. By what we have just seen, this dimension is exactly N^n. Therefore, $\dim Z_n = N^{\frac{n(n+1)}{2}}$; in presence of Lemma 6.12, this implies the theorem. $\square$

THEOREM 6.14. *The Nichols algebra $\mathfrak{B}(V)$ can be presented by generators $e_{i,i+1}$,* $1 \leq i \leq n$, *and relations* (6-12), (6-13), (6-14) *and* (6-28).

PROOF. Let $\mathfrak{B}'$ be the algebra presented by generators $e_{i,i+1}$, $1 \leq i \leq n$, and relations (6–12), (6–13), (6–14) and (6–28). We claim that $\mathfrak{B}'$ is is a braided Hopf algebra with the $e_{i,i+1}$'s primitive. Indeed, the claim follows without difficulty; use Lemma 6.9 for relations (6–28).

By Lemma 6.12, we see that the monomials (6–29) span $\mathfrak{B}'$ as a vector space, and in particular that $\dim \mathfrak{B}' \leq N^{\frac{n(n+1)}{2}}$. By Lemmas 6.3 and 6.11, there is a surjective algebra map $\psi : \mathfrak{B}' \to \mathfrak{B}(V)$. By Theorem 6.13, ψ is an isomorphism.

$\square$

6.2. Lifting of Nichols algebras of type A_n.

We fix in this Section a finite *abelian* group Γ such that our braided vector space V can be realized in $^{\Gamma}_{\Gamma}\mathcal{YD}$, as in Remark 6.1. That is, we have $g_1, \ldots, g_n$ in Γ, $\chi_1, \ldots, \chi_n$ in $\widehat{\Gamma}$, such that $q_{ij} = \langle \chi_j, g_i \rangle$ for all i,j, and V can be realized as a Yetter–Drinfeld module over Γ by (6–6).

We also fix a finite-dimensional pointed Hopf algebra A such that $G(A)$ is isomorphic to Γ, and the infinitesimal braiding of A is isomorphic to V as a Yetter–Drinfeld module over Γ. That is, $\mathrm{gr}\, A \simeq R \# \Bbbk\Gamma$, and the subalgebra R' of R generated by $R(1)$ is isomorphic to $\mathfrak{B}(V)$. We choose elements $a_i \in (A_1)^{\chi_i}_{g_i}$ such that $\pi(a_i) = x_i$, $1 \leq i \leq n$.

We shall consider, more generally, Hopf algebras H provided with

- a group isomorphism $\Gamma \to G(H)$;
- elements $a_1, \ldots, a_n$ in $\mathcal{P}(H)^{\chi_i}_{g_i,1}$.

Further hypotheses on H will be stated when needed. The examples of such H we are thinking of are the Hopf algebra A, and any bosonization $R \# \Bbbk\Gamma$, where R is any braided Hopf algebra in $^{\Gamma}_{\Gamma}\mathcal{YD}$ provided with a monomorphism of Yetter–Drinfeld modules $V \to P(R)$; so that $a_i := x_i \# 1$, $1 \leq i \leq n$. This includes notably the Hopf algebras $T(V) \# \Bbbk\Gamma$, $\widehat{\mathfrak{B}}(V) \# \Bbbk\Gamma$, $\mathfrak{B}(V) \# \Bbbk\Gamma$.

Here $\widehat{\mathfrak{B}}(V)$ is the braided Hopf algebra in $^{\Gamma}_{\Gamma}\mathcal{YD}$ generated by $x_1, \ldots, x_\theta$ with relations (6–12), (6–13) and (6–14).

We introduce inductively the following elements of H:

$$E_{i,i+1} := a_i; \tag{6–30}$$

$$E_{i,j} := \mathrm{ad}\,(E_{i,j-1})(E_{j-1,j}), \quad 1 \leq i < j \leq n+1, \; j-i \geq 2. \tag{6–31}$$

Assume that $H = R \# \Bbbk\Gamma$ as above. Then, by the relations between braided commutators and the adjoint (1–21), the relations (6–12), (6–13) and (6–14) translate respectively to

$$\mathrm{ad}\,E_{i,i+1}(E_{p,p+1}) = 0; \quad 1 \leq i < p \leq n, \quad p - i \geq 2; \tag{6–32}$$

$$(\mathrm{ad}\,E_{i,i+1})^2(E_{i+1,i+2}) = 0, \quad 1 \leq i < n; \tag{6–33}$$

$$(\mathrm{ad}\,E_{i+1,i+2})^2(E_{i,i+1}) = 0, \quad 1 \leq i < n. \tag{6–34}$$

REMARK 6.15. Relations (6–32), (6–33) and (6–34) can be considered, more generally, in any H as above. If these relations hold in H, then we have a Hopf algebra map $\pi_H : \widehat{\mathfrak{B}}(V)\#\Bbbk\Gamma \to H$. On the other hand, we know by Remark 6.10 that the comultiplication of the elements E_{ij}^N is given by (6–27). Hence, the same formula is valid in H, provided that relations (6–32), (6–33) and (6–34) hold in it. In particular, the subalgebra of H generated by the elements E_{ij}^N, $g_{i,j}^N$, $1 \leq i < j \leq n+1$, is a Hopf subalgebra of H.

LEMMA 6.16. *Relations* (6–32), (6–33) *and* (6–34) *hold in A if $N > 3$.*

PROOF. This is a particular case of Theorem 5.6; we include the proof for completeness. We know, by Lemma 2.13, that

$$\text{ad}\, E_{i,i+1}(E_{p,p+1}) \in \mathcal{P}_{g_i g_p,1}(A)^{\chi_i \chi_p}, \qquad 1 \leq i < p \leq n, \qquad p - i \geq 2,$$

$$(\text{ad}\, E_{i,i+1})^2(E_{p,p+1}) \in \mathcal{P}_{g_i^2 g_p,1}(A)^{\chi_i^2 \chi_p}, \qquad 1 \leq i,p \leq n, \qquad |p - i| = 1.$$

Assume that $\text{ad}\, E_{i,i+1}(E_{p,p+1}) \neq 0$, and $\chi_i \chi_p \neq \varepsilon$, where $1 \leq i < p \leq n$, $p - i \geq 2$. By Lemma 5.1, there exists l, $1 \leq l \leq n$, such that $g_i g_p = g_l$, $\chi_i \chi_p = \chi_l$. But then

$$q = \chi_l(g_l) = \chi_i(g_i)\chi_i(g_p)\chi_p(g_i)\chi_p(g_p) = q^2.$$

Hence $q = 1$, a contradiction.

Assume next that $\text{ad}\, E_{i,i+1}^2(E_{p,p+1}) \neq 0$, $|p - i| = 1$. and $\chi_i^2 \chi_p \neq \varepsilon$. By Lemma 5.1 , there exists l, $1 \leq l \leq n$, such that $g_i^2 g_p = g_l$, $\chi_i^2 \chi_p = \chi_l$. But then

$$q = \chi_l(g_l) = \chi_i(g_i)^4 \chi_i(g_p)^2 \chi_p(g_i)^2 \chi_p(g_p) = q^3.$$

Hence $q = \pm 1$, a contradiction (we assumed $N > 2$).

It remains to exclude the cases $\chi_i \chi_p = \varepsilon, |p - i| \geq 2$, and $\chi_i^2 \chi_p = \varepsilon, |p - i| = 1$. The first case leads to the contradiction $N = 3$. In the second case it follows from the connectivity of A_n that N would divide 2 which is also impossible. $\square$

LEMMA 6.17. *If $H = A$, then $E_{i,j}^N \in \Bbbk\Gamma^N$, for any $1 \leq i < j \leq n+1$.*

PROOF. We first show that $E_{i,j}^N \in \Bbbk\Gamma$, $1 \leq i < j \leq n+1$. (For our further purposes, this is what we really need).

Let $i < j$. We claim that there exists no l, $1 \leq l \leq n$, such that $g_{i,j}^N = g_l$, $\chi_{i,j}^N = \chi_l$. Indeed, otherwise we would have

$$q = \chi_l(g_l) = \chi_{i,j}(g_{i,j})^{N^2} = q^{N^2} = 1.$$

By Lemma 6.16 and Remark 6.15, we have

$$\Delta(E_{i,j}^N) = E_{i,j}^N \otimes 1 + g_{i,j}^N \otimes E_{i,j}^N + (1 - q^{-1})^N \sum_{i < p < j} \left(B_{i,p}^{p,j} \right)^{N(N-1)/2} E_{i,p}^N g_{p,j}^N \otimes E_{p,j}^N.$$

$$(6\text{–}35)$$

We proceed by induction on $j - i$. If $j - i = 1$, then, by Lemma 5.1 , either $E_{i,i+1}^N \in \Bbbk\Gamma$ or $E_{i,i+1}^N \in \mathcal{P}_{g_i^N,1}(A)^{\chi_i^N}$ and $\chi_i^N \neq \varepsilon$, hence $g_i^N = g_l, \chi_i^N = \chi_l$ for some l; but this last possibility contradicts the claim above. Assume then that $j - i > 1$. By the induction hypothesis, $\Delta(E_{i,j}^N) = E_{i,j}^N \otimes 1 + g_{i,j}^N \otimes E_{i,j}^N + u$, for some $u \in k\Gamma \otimes k\Gamma$. In particular, we see that $E_{i,j}^N \in (A_1)^{\chi_i^N}$. Then, by Lemma 5.1, either $\chi_i^N = \varepsilon$ and hence $E_{i,i+1}^N \in k\Gamma$, or else $\chi_i^N \neq \varepsilon$, which implies $u = 0$ and $E_{i,i+1}^N \in \mathcal{P}_{g_i^N,1}(A)^{\chi_i^N}$. Again, this last possibility contradicts the claim above.

Finally, let C be the subalgebra of A generated by the elements E_{ij}^N, $g_{i,j}^N$, $1 \leq i < j \leq n+1$, which is a Hopf subalgebra of H. Since $E_{i,j}^N \in k\Gamma \cap C$, we conclude that $E_{i,j}^N \in C_0 = k\Gamma^N$.   □

To solve the lifting problem, we see from Lemma 6.17 that we first have to answer a combinatorial question in the group algebra of an abelian group. To simplify the notation we define

$$h_{ij} = g_{i,j}^N, \qquad C_{i,p}^j = (1 - q^{-1})^N \left(B_{i,p}^{p,j}\right)^{N(N-1)/2}.$$

We are looking for families $(u_{ij})_{1 \leq i < j \leq n+1}$ of elements in $k\Gamma$ such that

$$\Delta(u_{ij}) = u_{ij} \otimes 1 + h_{i,j} \otimes u_{ij} + \sum_{i<p<j} C_{i,p}^j u_{i,p} h_{p,j} \otimes u_{p,j}, \text{ for all } 1 \leq i < j \leq n+1.$$

$$(6\text{--}36)$$

The coefficients $C_{i,p}^j$ satisfy the rule

$$C_{is}^j C_{st}^j = C_{is}^t C_{it}^j, \text{ for all } 1 \leq i < s < t < j \leq n+1. \qquad (6\text{--}37)$$

This follows from (6–2) and (6–3) since

$$B_{is}^{sj} B_{st}^{tj} = B_{is}^{st} B_{is}^{tj} B_{st}^{tj} = B_{is}^{st} B_{it}^{tj}.$$

THEOREM 6.18. *Let Γ be a finite abelian group and $h_{ij} \in \Gamma$, $1 \leq i < j \leq n+1$, a family of elements such that*

$$h_{ij} = h_{i,p} h_{p,j}, \qquad if \quad i < p < j. \qquad (6\text{--}38)$$

Let $C_{i,p}^j \in k^\times$, $1 \leq i < p < j \leq n+1$, be a family of elements satisfying (6–37). Then the solutions $(u_{ij})_{1 \leq i < j \leq n+1}$ of (6–36), $u_{ij} \in k\Gamma$ for all $i < j$, have the form $(u_{ij}(\gamma))_{1 \leq i < j \leq n+1}$ where $\gamma = (\gamma_{ij})_{1 \leq i < j \leq n+1}$ is an arbitrary family of scalars $\gamma_{ij} \in k$ such that

$$for \ all \ 1 \leq i < j \leq n+1, \ \gamma_{ij} = 0 \ if \ h_{ij} = 1, \qquad (6\text{--}39)$$

and where the elements $u_{ij}(\gamma)$ are defined by induction on $j - i$ by

$$u_{ij}(\gamma) = \gamma_{ij}(1 - h_{ij}) + \sum_{i<p<j} C_{ip}^j \gamma_{ip} u_{pj}(\gamma) \ for \ all \ 1 \leq i < j \leq n+1. \quad (6\text{--}40)$$

PROOF. We proceed by induction on k. We claim that the solutions $u_{ij} \in k\Gamma$, $1 \le i < j \le n+1, j-i \le k$, of (6–36) for all $i < j$ with $j - i \le k$ are given by arbitrary families of scalars γ_{ij}, $1 \le i < j \le n+1, j-i \le k$ such that

$$u_{ij} = \gamma_{ij}(1 - h_{ij}) + \sum_{i<p<j} C_{ip}^j \gamma_{ip} u_{pj} \text{ for all } 1 \le i < j \le n+1 \text{ with } j-i \le k.$$

Suppose $k = 1$. For any $1 \le i < n, j = i+1$, $u_{i,i+1}$ is a solution of (6–36) if and only if $u_{i,i+1}$ is $(h_{i,i+1}, 1)$-primitive in $k\Gamma$, that is $u_{i,i+1} = \gamma_{i,i+1}(1 - h_{i,i+1})$ for some $\gamma_{i,i+1} \in k$. We may assume that $\gamma_{i,i+1} = 0$, if $h_{i,i+1} = 1$.

For the induction step, let $k > 2$. We assume that $\gamma_{ab} \in k$, $1 \le a < b \le n+1, b-a \le k-1$, is a family of scalars with $\gamma_{ab} = 0$, if $h_{ab} = 1$, and that the family $u_{ab} \in k\Gamma$, $1 \le a < b \le n+1, b-a \le k-1$, defined inductively by the γ_{ab} by (6–40) is a solution of (6–36). Let $1 \le i < j \le n+1$, and $j - i = k$. We have to show that

$$\Delta(u_{ij}) = u_{ij} \otimes 1 + h_{ij} \otimes u_{ij} + \sum_{i<p<j} C_{ip}^j u_{ip} h_{pj} \otimes u_{pj} \qquad (6\text{–}41)$$

is equivalent to

$$u_{ij} = \gamma_{ij}(1 - h_{ij}) + \sum_{i<p<j} C_{ip}^j \gamma_{ip} u_{pj} \text{ for some } \gamma_{ij} \in k. \qquad (6\text{–}42)$$

We then may define $\gamma_{ij} = 0$ if $h_{ij} = 1$.

We denote

$$z_{ij} := u_{ij} - \sum_{i<p<j} C_{ip}^j \gamma_{ip} u_{pj}.$$

Then (6–42) is equivalent to

$$\Delta(z_{ij}) = z_{ij} \otimes 1 + h_{ij} \otimes z_{ij}. \qquad (6\text{–}43)$$

For all $i < p < j$ we have $\Delta(u_{pj}) = u_{pj} \otimes 1 + h_{pj} \otimes u_{pj} + \sum_{p<s<j} C_{ps}^j u_{ps} h_{sj} \otimes u_{sj}$, since $j - p < k$. Using this formula for $\Delta(u_{pj})$ we compute

$$\Delta(z_{ij}) - z_{ij} \otimes 1 - h_{ij} \otimes z_{ij}$$

$$= \Delta(u_{ij}) - \sum_{i<p<j} C_{ip}^j \gamma_{ip} \Delta(u_{pj}) - z_{ij} \otimes 1 - h_{ij} \otimes z_{ij}$$

$$= \Delta(u_{ij}) - \sum_{i<p<j} C_{ip}^j \gamma_{ip} \left(u_{pj} \otimes 1 + h_{pj} \otimes u_{pj} + \sum_{p<s<j} C_{ps}^j u_{ps} h_{sj} \otimes u_{sj} \right)$$

$$- (u_{ij} - \sum_{i<p<j} C_{ip}^j \gamma_{ip} u_{pj}) \otimes 1 - h_{ij} \otimes (u_{ij} - \sum_{i<p<j} C_{ip}^j \gamma_{ip} u_{pj})$$

$$= \Delta(u_{ij}) - u_{ij} \otimes 1 - h_{ij} \otimes u_{ij} + \sum_{i<p<j} C_{ip}^j \gamma_{ip} h_{ij} \otimes u_{pj}$$

$$- \sum_{i<p<j} C_{ip}^j \gamma_{ip} h_{pj} \otimes u_{pj} - \sum_{i<p<s<j} C_{ip}^j C_{ps}^j \gamma_{ip} u_{ps} h_{sj} \otimes u_{sj}.$$

Therefore, (6–41) and (6–42) are equivalent if and only if the identity

$$\sum_{i<p<j} C_{ip}^j \gamma_{ip} h_{ij} \otimes u_{pj} - \sum_{i<p<j} C_{ip}^j \gamma_{ip} h_{pj} \otimes u_{pj} - \sum_{i<p<s<j} C_{ip}^j C_{ps}^j \gamma_{ip} u_{ps} h_{sj} \otimes u_{sj}$$

$$(6\text{–}44)$$

$$= - \sum_{i<p<j} C_{ip}^j u_{ip} h_{pj} \otimes u_{pj}$$

holds.

To prove (6–44) we use (6–40) for all $i < p$, where $i < p < j$, that is $u_{ip} = \gamma_{ip}(1 - h_{ip}) + \sum_{i<s<p} C_{is}^p \gamma_{is} u_{sp}$. Then

$$\sum_{i<p<j} C_{ip}^j h_{pj} u_{ip} \otimes u_{pj} + \sum_{i<p<j} C_{ip}^j \gamma_{ip} h_{ij} \otimes u_{pj} - \sum_{i<p<j} C_{ip}^j \gamma_{ip} h_{pj} \otimes u_{pj}$$

$$= \sum_{i<p<j} (C_{ip}^j h_{pj} (\gamma_{ip}(1 - h_{ip}) + \sum_{i<s<p} C_{is}^p \gamma_{is} u_{sp}) + C_{ip}^j \gamma_{ip}(h_{ij} - h_{pj})) \otimes u_{pj}$$

$$= \sum_{i<p<j} C_{ip}^j h_{pj} \sum_{i<s<p} C_{is}^p \gamma_{is} u_{sp} \otimes u_{pj}, \text{ since } h_{pj}(1 - h_{ip}) = h_{pj} - h_{ij} \text{ by (6–38)},$$

$$= \sum_{i<s<p<j} C_{is}^j C_{sp}^j \gamma_{is} u_{sp} h_{pj} \otimes u_{pj}, \text{ since } C_{ip}^j C_{is}^p = C_{is}^j C_{sp}^j \text{ by (6–37)}.$$

This proves (6–44) by interchanging s and p. $\qquad\qquad\square$

REMARKS 6.19. (1) Let $\gamma = (\gamma_{ij})_{1 \leq 1qi<j \leq n+1}$ be an arbitrary family of scalars. Then it is easy to see that the family $u_{ij}(\gamma) \in \mathbb{k}\Gamma$, $1 \leq i < j \leq n + 1$, can be defined explicitly as follows:

$$u_{ij}(\gamma) = \sum_{i \leq p < j} \phi_{ip}^j(\gamma)(1 - h_{pj}) \text{ for all } i < j,$$

where

$$\phi_{ip}^j(\gamma) = \sum_{i = i_1 < \cdots < i_k = p, k \geq 1} C_{i_1, i_2}^j \cdots C_{i_{k-1}, i_k}^j \gamma_{i_1, i_2} \cdots \gamma_{i_{k-1}, i_k} \gamma_{pj} \text{ for all } i \leq p < j$$

is a polynomial of degree p in the free variables $(\gamma_{ij})_{1 \leq i < j \leq n+1}$.

(2) Let $\gamma = (\gamma_{ij})_{1 \leq i < j \leq n+1}$ and $\tilde{\gamma} = (\tilde{\gamma}_{ij})_{1 \leq i < j \leq n+1}$ be families of scalars in $\mathbb{k}$ satisfying (6–39). Assume that for all $i < j$, $u_{ij}(\gamma) = u_{ij}(\tilde{\gamma})$. Then $\gamma = \tilde{\gamma}$. This follows easily by induction on $j - i$ from (6–40).

LEMMA 6.20. Assume the situation of Theorem 6.18. Let $\gamma = (\gamma_{ij})_{1 \leq i < j \leq n+1}$ be a family of scalars in $\mathbb{k}$ satisfying (6–39) and define $u_{ij} = u_{ij}(\gamma)$ for all $1 \leq i < j \leq n + 1$ by (6–40).

(1) The following are equivalent:

(a) For all $i < j$, $u_{ij} = 0$ if $\chi_{ij}^N \neq \varepsilon$.

(b) For all $i < j$, $\gamma_{ij} = 0$ if $\chi_{ij}^N \neq \varepsilon$.

(2) *Assume that $h_{ij} = g_{ij}^N$ for all $i < j$. Then the following are equivalent:*

(a) *For all $i < j, u_{ij} = 0$ if $\chi_{ij}^N(g_l) \neq 1$ for some $1 \leq l \leq n$.*

(b) *For all $i < j, \gamma_{ij} = 0$ if $\chi_{ij}^N(g_l) \neq 1$ for some $1 \leq l \leq n$.*

(c) *The elements u_{ij}, $1 \leq i < j \leq n+1$, are central in $\widehat{\mathfrak{B}}(V)\#\mathbb{k}\Gamma$.*

PROOF. (1) follows by induction on $j - i$.

Suppose $j = i + 1$. Then $u_{i,i+1} = \gamma_{i,i+1}(1 - h_{i,i+1})$. If $h_{i,i+1} = 1$, then both $u_{i,i+1}$ and $\gamma_{i,i+1}$ are 0. If $h_{i,i+1} \neq 1$, then $u_{i,i+1} = 0$ if and only if $\gamma_{i,i+1} = 0$.

The induction step follows in the same way from (6–40), since for all $i < p < j$, if $\chi_{ij}^N \neq \varepsilon$, then $\chi_{ip}^N \neq \varepsilon$ or $\chi_{pj}^N \neq \varepsilon$, hence by induction $\gamma_{ip} = 0$ or $u_{pj} = 0$, and $u_{ij} = \gamma_{ij}(1 - h_{ij})$.

(2) Suppose that for all $i < p < j$, u_{pj} is central in $\widehat{\mathfrak{B}}(V)\#\mathbb{k}\Gamma$, and let $1 \leq l \leq n$. Then

$$h_{ij}x_l = x_l\chi_l(h_{ij})h_{ij},$$

and we obtain from (6–40)

$$u_{ij}x_l = x_l\gamma_{ij}(1 - \chi_l(h_{ij})h_{ij}) + x_l \sum_{i<p<j} C_{ip}^j\gamma_{ip}u_{pj}.$$

Hence u_{ij} is central in $\widehat{\mathfrak{B}}(V)\#\mathbb{k}\Gamma$ if and only if $\gamma_{ij} = \gamma_{ij}\chi_l(h_{ij})$ for all $1 \leq l \leq n$. Since the braiding is of type A_n and the order of $q = \chi_l(g_l)$ is N,

$$\chi_l(h_{ij}) = \chi_l(g_{ij}^N) = \chi_{ij}^{-N}(g_l),$$

and the equivalence of (b) and (c) follows by induction on $j - i$. The equivalence of (a) and (b) is shown as in (1). □

6.3. Classification of pointed Hopf algebras of type A_n.

Using the previous results we will now determine exactly all finite-dimensional pointed Hopf algebras of type A_n (up to some exceptional cases). We will find a big new class of deformations of $u_q^{\geq 0}(sl_{n+1})$.

As before, we fix a natural number n, a finite abelian group Γ, an integer $N > 2$, a root of unity q of order N, $g_1, \ldots, g_n \in \Gamma, \chi_1, \ldots, \chi_n \in \widehat{\Gamma}$ such that $q_{ij} = \chi_j(g_i)$ for all i, j satisfy (6–1), and $V \in {}_{\Gamma}^{\Gamma}\mathcal{YD}$ with basis $x_i \in V_{g_i}^{\chi_i}, 1 \leq i \leq n$.

Recall that $\widehat{\mathfrak{B}}(V)$ is the braided Hopf algebra in ${}_{\Gamma}^{\Gamma}\mathcal{YD}$ generated by $x_1, \ldots, x_n$ with the quantum Serre relations (6–12), (6–13) and (6–14).

In $\widehat{\mathfrak{B}}(V)$ we consider the iterated braided commutators $e_{i,j}$, $1 \leq i < j \leq n+1$ defined inductively by (6–8) beginning with $e_{i,i+1} = x_i$ for all i.

Let $\mathbb{A}$ be the set of all families $(a_{i,j})_{1 \leq i < j \leq n+1}$ of integers $a_{i,j} \geq 0$ for all $1 \leq i < j \leq n+1$. For any $a \in \mathbb{A}$ we define

$$e^a := (e_{1,2})^{a_{1,2}}(e_{1,3})^{a_{1,3}} \cdots (e_{n,n+1})^{a_{n,n+1}},$$

where the order in the product is the lexicographic order of the index pairs. We begin with the PBW-theorem for $\widehat{\mathfrak{B}}(V)$.

THEOREM 6.21. *The elements $e^a, a \in \mathbb{A}$, form a basis of the $\mathbb{k}$-vector space $\widehat{\mathfrak{B}}(V)$.*

PROOF. The proof is similar to the proof of Theorem 6.13. For general finite Cartan type the theorem can be derived from the PBW-basis of $U_q(\mathfrak{g})$ (see [L3]) by changing the group and twisting as described in [AS4, Section 4.2]. □

The following commutation rule for the elements e_{ij}^N is crucial.

LEMMA 6.22. *For all $1 \leq i < j \leq n+1$, $1 \leq s < t \leq n+1$,*

$$[e_{i,j}, e_{s,t}^N]_c = 0, \text{ that is } e_{i,j}e_{s,t}^N = \chi_{s,t}^N(g_{i,j})e_{s,t}^N e_{i,j}.$$

PROOF. Since $c_{i,j}$ is a linear combination of elements of the form $x_{i_1} \ldots x_{i_k}$ with $k = j - i$ and $g_{i_1} \ldots g_{i_k} = g_{ij}$, it is enough to consider the case when $j = i+1$.

To show $[e_{i,i+1}, e_{s,t}^N]_c = 0$, we will distinguish several cases.

First assume that $(i, i+1) < (s, t)$. If $i+1 < s$ resp. $i = s$ and $i+1 < t$, then $[e_{i,i+1}, e_{s,t}]_c = 0$ by (6–17) resp. (6–23), and the claim follows.

If $i+1 = s$, we denote $x = e_{i+1,t}, y = e_{i,i+1}, z = e_{it}$ and $\alpha = \chi_{i+1,t}(g_i), \beta = \chi_{i+1,t}(g_{i,t})$. Then

$$yx = \alpha xy + z, \text{ by } (6\text{–}19), \text{ and } zx = \beta xz, \text{ by } (6\text{–}24).$$

Moreover, $\alpha = \chi_{i+1,t}(g_i) \neq \beta = \chi_{i+1,t}(g_{i,t}) = \chi_{i+1,t}(g_i)\chi_{i+1,t}(g_{i+1,t})$, and $\alpha^N = \beta^N$, since $\chi_{i+1,t}(g_{i+1,t}) = q$ by (6–5). Therefore it follows from [AS4, Lemma 3.4] that $yx^N = \alpha^N x^N y$, which was to be shown.

The claim is clear if $i = s$, and $i+1 = t$, since $\chi_i^N(g_i) = 1$.

It remains to consider the case when $(i, i+1) > (s, t)$. If $s < i$ and $t = i+1$, then $e_{s,t}e_{i,i+1} = \chi_i(g_{s,t})e_{i,i+1}e_{s,t}$ by (6–24). If $s < i$ and $i+1 < t$, the same result is obtained from (6–22), and if $s = i$ and $t < i+1$, from (6–23). Hence in all cases, $e_{i,i+1}e_{s,t}^N = \chi_i^{-N}(g_{s,t})e_{s,t}^N e_{i,i+1}$. This proves the claim $[e_{i,i+1}, e_{s,t}^N]_c = 0$, since $\chi_i^{-N}(g_{s,t}) = \chi_{s,t}^N(g_i)$. □

We want to compute the dimension of certain quotient algebras of $\widehat{\mathfrak{B}}(V)\#\mathbb{k}\Gamma$. Since this part of the theory works for any finite Cartan type, we now consider more generally a left $\mathbb{k}\Gamma$-module algebra R over any abelian group Γ and assume that there are integers P and $N_i > 1$, elements $y_i \in R$, $h_i \in \Gamma$, $\eta_i \in \widehat{\Gamma}$, $1 \leq i \leq P$, such that

$$g \cdot y_i = \eta_i(g)y_i, \text{ for all } g \in \Gamma, 1 \leq i \leq P. \tag{6–45}$$

$$y_i y_j^{N_j} = \eta_j^{N_j}(h_i)y_j^{N_j} y_i \text{ for all } 1 \leq i, j \leq P. \tag{6–46}$$

The elements $y_1^{a_1} \ldots y_P^{a_P}, a_1, \ldots, a_P \geq 0$, form a $\mathbb{k} - $ basis of R. (6–47)

Let $\mathbb{L}$ be the set of all $l = (l_i)_{1 \leq i \leq P} \in \mathbb{N}^P$ such that $0 \leq l_i < N_i$ for all $1 \leq i \leq P$. For $a = (a_i)_{1 \leq i \leq P} \in \mathbb{N}^P$, we define

$$y^a = y_1^{a_1} \ldots y_P^{a_P}, \text{ and } aN = (a_i N_i)_{1 \leq i \leq P}.$$

Then by (6–46), (6–47), the elements

$$y^l y^{aN}, \qquad l \in \mathbb{L}, \ a \in \mathbb{N}^P,$$

form a $\Bbbk$-basis of R.

In the application to $\widehat{\mathfrak{B}}(V) \# \Bbbk\Gamma$, P is the number of positive roots, and the y_i play the role of the root vectors $e_{i,j}$.

To simplify the notation in the smash product algebra $R \# \Bbbk\Gamma$, we identify $r \in R$ with $r \# 1$ and $v \in \Bbbk\Gamma$ with $1 \# v$. For $1 \leq i \leq P$, let $\widetilde{\eta}_i : \Bbbk\Gamma \to \Bbbk\Gamma$ be the algebra map defined by $\widetilde{\eta}_i(g) = \eta_i(g)g$ for all $g \in \Gamma$. Then

$$vy_i = y_i \widetilde{\eta}_i(v) \text{ for all } v \in \Bbbk\Gamma.$$

We fix a family u_i, $1 \leq i \leq P$, of elements in $\Bbbk\Gamma$, and denote

$$u^a := \prod_{1 \leq i \leq P} u_i^{a_i}, \text{ if } a = (a_i)_{1 \leq i \leq P} \in \mathbb{N}^P.$$

Let M be a free right $\Bbbk\Gamma$-module with basis $m(l), l \in \mathbb{L}$. We then define a right $\Bbbk\Gamma$-linear map

$$\varphi : R \# \Bbbk\Gamma \to M \text{ by } \varphi(y^l y^{aN}) := m(l)u^a \text{ for all } l \in \mathbb{L}, a \in \mathbb{N}^P.$$

LEMMA 6.23. *Assume that*

- u_i *is central in* $R \# \Bbbk\Gamma$, *for all* $1 \leq i \leq P$, *and*
- $u_i = 0$ *if* $\eta_i^{N_i}(h_j) \neq 1$ *for some* $1 \leq j \leq P$.

Then the kernel of φ *is a right ideal of* $R \# \Bbbk\Gamma$ *containing* $y_i^{N_i} - u_i$ *for all* $1 \leq i \leq P$.

PROOF. By definition, $\varphi(y_i^{N_i}) = m(0)u_i = \varphi(u_i)$.

To show that the kernel of φ is a right ideal, let

$$z = \sum_{l \in \mathbb{L}, a \in \mathbb{N}^P} y^l y^{aN} v_{l,a}, \text{ where } v_{l,a} \in \Bbbk\Gamma, \text{for all } l \in \mathbb{L}, a \in \mathbb{N}^P,$$

be an element with $\varphi(z) = 0$. Then $\varphi(z) = \sum_{l,a} m(l)u^a v_{l,a} = 0$, hence

$$\sum_{a \in \mathbb{N}^P} u^a v_{l,a} = 0, \text{for all } l \in \mathbb{L}.$$

Fix $1 \leq i \leq P$. We have to show that $\varphi(zy_i) = 0$.

For any $l \in \mathbb{L}$, we have the basis representation

$$y^l y_i = \sum_{t \in \mathbb{L}, b \in \mathbb{N}^P} \alpha_{t,b}^l y^t y^{bN}, \text{ where } \alpha_{t,b}^l \in \Bbbk \text{ for all } t \in \mathbb{L}, b \in \mathbb{N}^P.$$

Since u^a is central in $R \# \Bbbk\Gamma$,

$$u^a = \widetilde{\eta}_i(u^a) \text{ for all } a \in \mathbb{N}^P. \tag{6–48}$$

For any $a = (a_i)_{1 \leq i \leq P} \in \mathbb{N}^P$ and any family $(g_i)_{1 \leq i \leq P}$ of elements in Γ we define $\eta^{aN}((g_i)) = \prod_i \eta_i{}^{a_i N_i}(g_i)$. Then by (6–46), for all $a, b \in \mathbb{N}^P$,

$$y^{aN} y_i = y_i y^{aN} \eta^{aN}(g^a), \text{ and } y^{bN} y^{aN} = y^{(a+b)N} \eta^{aN}(g^b), \qquad (6\text{--}49)$$

for some families of elements g^a, g^b in Γ.

By a reformulation of our assumption,

$$u^a \eta^{aN}((g_i)) = u^a \text{ for any } a \in \mathbb{N}^P \text{ and family } (g_i) \text{ in } \Gamma. \qquad (6\text{--}50)$$

Using (6–49) we now can compute

$$
\begin{aligned}
zy_i &= \sum_{l,a} y^l y^{aN} v_{l,a} y_i = \sum_{l,a} y^l y^{aN} y_i \widetilde{\eta}_i(v_{l,a}) \\
&= \sum_{l,a} y^l y_i y^{aN} \eta^{aN}(g^a) \widetilde{\eta}_i(v_{l,a}) \\
&= \sum_{l,a} \sum_{t,b} \alpha_{t,b}^l y^t y^{bN} e^{aN} \eta^{aN}(g^a) \widetilde{\eta}_i(v_{l,a}) \\
&= \sum_{l,a} \sum_{t,b} \alpha_{t,b}^l y^t y^{(a+b)N} \eta^{aN}(g^a) \eta^{aN}(g^b) \widetilde{\eta}_i(v_{l,a}).
\end{aligned}
$$

Therefore

$$
\begin{aligned}
\varphi(zy_i) &= \sum_t m(t) \sum_{l,a,b} \alpha_{t,b}^l u^{a+b} \eta^{aN}(g^a) \eta^{aN}(g^b) \widetilde{\eta}_i(v_{l,a}) \\
&= \sum_t m(t) \sum_{l,a,b} \alpha_{t,b}^l u^{a+b} \widetilde{\eta}_i(v_{l,a}), \text{ by (6–50)}, \\
&= \sum_t m(t) \sum_{b,l} \alpha_{t,b}^l u^b \sum_a u^a \widetilde{\eta}_i(v_{l,a}) \\
&= \sum_t m(t) \sum_{b,l} \alpha_{t,b}^l u^b \widetilde{\eta}_i (\sum_a u^a v_{l,a}), \text{ by (6–48)}, \\
&= 0, \text{ since } \sum_a u^a v_{l,a} = 0.
\end{aligned}
$$

$\square$

THEOREM 6.24. *Let u_i, $1 \leq i \leq P$, be a family of elements in $\mathbb{k}\Gamma$, and I the ideal in $R\#\mathbb{k}\Gamma$ generated by all $y_i^{N_i} - u_i$, $1 \leq i \leq P$. Let $A = (R\#\mathbb{k}\Gamma)/I$ be the quotient algebra. Then the following are equivalent:*
  (1) *The residue classes of $y^l g, l \in \mathbb{L}, g \in \Gamma$, form a $\mathbb{k}$-basis of A.*
  (2) *u_i is central in $R\#\mathbb{k}\Gamma$ for all $1 \leq i \leq P$, and $u_i = 0$ if $\eta_i^{N_i} \neq \varepsilon$.*

PROOF. (1) $\Rightarrow$ (2) : For all i and $g \in \Gamma$, $g y_i^{N_i} = \eta_i^{N_i}(g) y_i^{N_i} g$, hence $u_i g = g u_i \equiv \eta_i^{N_i}(g) u_i g \mod I$. Since by assumption, $\mathbb{k}\Gamma$ is a subspace of A, we conclude that $u_i = \eta_i^{N_i}(g) u_i$, and $u_i = 0$ if $\eta_i^{N_i} \neq \varepsilon$.

Similarly, for all $1 \leq i, j \leq n$, $y_i y_j^{N_j} = \eta_j^{N_j}(h_i) y_j^{N_j} y_i$ by (6–46), hence $y_i u_j \equiv \eta_j^{N_j}(h_i) u_j y_i \mod I$. Since we already know that $u_i = 0$ if $\eta_j^{N_j} \neq \varepsilon$, we see that

$y_i u_j \equiv u_j y_i \mod I$. On the other hand $u_j y_i = y_i \tilde{\eta}_i(u_j)$. Then our assumption in (1) implies that $\tilde{\eta}_i(u_j) = u_j$. In other words, u_j is central in $R \# \Bbbk \Gamma$.

(2) $\Rightarrow$ (1): Let J be the right ideal of $R \# \Bbbk \Gamma$ generated by all $y_i^{N_i} - u_i$, $1 \leq i \leq P$. For any $1 \leq i \leq P$ and $g \in \Gamma$,

$$g(y_i^{N_i} - u_i) = y_i^{N_i} g \eta_i^{N_i}(g) - g u_i = (y_i^{N_i} - u_i) \eta_i^{N_i}(g) g,$$

since $g u_i = u_i \eta_i^{N_i}(g) g$ by (2).

And for all $1 \leq i, j \leq P$,

$$y_i(y_j^{N_j} - u_j) = \eta_j^{N_j}(h_i) y_j^{N_j} y_i - y_i u_j = (y_j^{N_j} - u_j) \eta_j^{N_j}(h_i) y_i,$$

since by (2) $y_i u_j = u_j y_i = u_j \eta_j^{N_j}(h_i) y_i$.

This proves $J = I$.

It is clear that the images of all $y^l g, l \in \mathbb{L}, g \in \Gamma$, generate the vector space A. To show linear independence, suppose

$$\sum_{l \in \mathbb{L}, g \in \Gamma} \alpha_{l,g} y^l g \in I, \quad \text{with } \alpha_{l,g} \in \Bbbk \text{ for all } l \in \mathbb{L}, g \in \Gamma.$$

Since $I = J$, we obtain from Lemma 6.23 that $\varphi(I) = 0$. Therefore,

$$0 = \varphi\Big(\sum_{l \in \mathbb{L}, g \in \Gamma} \alpha_{l,g} y^l g \Big) = \sum_{l \in \mathbb{L}, g \in \Gamma} \alpha_{l,g} m(l) g,$$

hence $\alpha_{l,g} = 0$ for all l, g. $\qquad \square$

We come back to A_n. Our main result in this chapter is

THEOREM 6.25. (i) *Let* $\gamma = (\gamma_{i,j})_{1 \leq i < j \leq n+1}$ *be any family of scalars in* $\Bbbk$ *such that for any* $i < j$, $\gamma_{i,j} = 0$ *if* $g_{i,j}^N = 1$ *or* $\chi_{i,j}^N \neq \varepsilon$. *Define* $u_{i,j} = u_{i,j}(\gamma) \in \Bbbk \Gamma$, $1 \leq i < j \leq n+1$, *by* (6-40). *Then*

$$A_\gamma := (\mathfrak{B}(V) \# \Bbbk \Gamma) / (e_{i,j}^N - u_{i,j} \mid 1 \leq i < j \leq n+1)$$

is a pointed Hopf algebra of dimension $N^{n(n+1)/2} \operatorname{ord}(\Gamma)$ *satisfying* $\operatorname{gr} A_\gamma \simeq \mathfrak{B}(V) \# \Bbbk \Gamma$.

(ii) *Conversely, let* A *be a finite-dimensional pointed Hopf algebra such that either*

(a) $\operatorname{gr} A \simeq \mathfrak{B}(V) \# \Bbbk \Gamma$, *and* $N > 3$, *or*

(b) *the infinitesimal braiding of* A *is of type* A_n *with* $N > 7$ *and not divisible by* 3.

Then A *is isomorphic to a Hopf algebra* A_γ *in* (i).

PROOF. (i) By Lemma 6.20, the elements $u_{i,j}$ are central in $\widehat{U} := \widehat{\mathfrak{B}}(V) \# \Bbbk \Gamma$, and $u_{i,j} = 0$ if $g_{i,j}^N = 1$ or $\chi_{i,j}^N \neq \varepsilon$. Hence the residue classes of the elements $e^l g, l \in \mathbb{A}, 0 \leq l_{i,j} < N$ for all $1 \leq i < j \leq n+1, g \in \Gamma$, form a basis of A_γ

by Theorem 6.24. By Theorem 6.18, the $u_{i,j}$ satisfy (6–36). The ideal I of $\widehat{U}$ generated by all $e_{i,j}^N - u_{i,j}$ is a biideal, since

$$\Delta(e_{i,j}^N - u_{i,j}) = (e_{i,j}^N - u_{i,j}) \otimes 1 + g_{i,j}^N \otimes (e_{i,j}^N - u_{i,j})$$
$$+ \sum_{i<p<j} C_{i,p}^j ((e_{i,p}^N - u_{i,p})g_{p,j}^N \otimes e_{p,j}^N + u_{i,p}g_{p,j}^N \otimes (e_{p,j}^N - u_{p,j}))$$
$$\in I \otimes \widehat{U} + \widehat{U} \otimes I,$$

by (6–35) and (6–36).

Since A_γ is generated by group-like and skew-primitive elements, and the group-like elements form a group, A_γ is a Hopf algebra.

For all $1 \leq i \leq n$, let $a_i \in \mathrm{gr}\,(A_\gamma)(1)$ be the residue class of $x_i \in (A_\gamma)_1$. Define root vectors $a_{i,j} \in \mathrm{gr}\,(A_\gamma)$, $1 \leq i < j \leq N+1$ inductively as in (6–30) and (6–31). Then $a_{i,j}^N = 0$ in $\mathrm{gr}\,(A_\gamma)$ since $e_{i,j}^N = 0$ in A_γ. Therefore, by Theorem 6.14, there is a surjective Hopf algebra map

$$\mathfrak{B}(V)\#\Bbbk\Gamma \to \mathrm{gr}\,(A_\gamma) \text{ mapping } x_i\#g \text{ onto } a_ig, \quad 1 \leq i \leq n, g \in \Gamma.$$

This map is an isomorphism, since $\dim(\mathrm{gr}\,(A_\gamma)) = \dim(A_\gamma) = N^{n(n+1)/2}\,|\Gamma| = \dim(\mathfrak{B}(V)\#\Bbbk\Gamma)$ by Theorem 6.13.

(ii). As in Section 6.2, we choose elements $a_i \in (A_1)_{g_i}^{\chi_i}$ such that $\pi(a_i) = x_i$, $1 \leq i \leq n$. By assumption resp. by Lemma 6.16, there is a Hopf algebra map

$$\phi : \widehat{\mathfrak{B}}(V)\#\Bbbk\Gamma \to A, \phi(x_i\#g) = a_ig, \quad 1 \leq i \leq n, g \in \Gamma.$$

By Theorem 5.8, A is generated in degree one, hence ϕ is surjective. We define the root vector $E_{i,j} \in A$, $1 \leq i < j \leq n+1$, by (6–30), (6–31). By Lemma 6.17, $E_{i,j}^N =: u_{i,j} \in \Bbbk\Gamma$ for all $1 \leq i < j \leq n+1$. Then for all $g \in \Gamma$ and $i < j$, $gE_{i,j}^N = \chi_{i,j}^N(g)E_{i,j}^Ng$, hence $gu_{i,j} = \chi_{i,j}^N(g)u_{i,j}g$, and $u_{i,j} = \chi_{i,j}^N(g)u_{i,j}$. By (6–35) and Theorem 6.18 we therefore know that $u_{i,j} = u_{i,j}(\gamma)$ for all $i < j$, for some family $\gamma = (\gamma_{i,j})_{1 \leq i \leq J \leq n+1}$ of scalars in $\Bbbk$ such that for all $i < j$, $\gamma_{i,j} = 0$ if $g_{i,j}^N = 1$ or $\chi_{i,j}^N = \varepsilon$. Hence ϕ indices a surjective Hopf algebra map $A_\gamma \to A$ which is an isomorphism since $\dim(A_\gamma) = N^{n(n+1)/2}\,\mathrm{ord}(\Gamma) = \dim(A)$ by (1). $\square$

REMARK 6.26. Up to isomorphism, A_γ does not change if we replace each x_i by a non-zero scalar multiple of itself. Hence in the definition of A_γ we may always assume that

$$\gamma_{i,i+1} = 0 \text{ or } 1 \text{ for all } 1 \leq i \leq n.$$

We close the paper with a very special case of Theorem 6.25. We obtain a large class of non-isomorphic Hopf algebras which have exactly the same infinitesimal braiding as $u_q^{\geq 0}(sl_n)$. Here q has order N, but the group is $\prod_{i=1}^n \mathbb{Z}/(Nm_i)$ and not $(\mathbb{Z}/(N))^n$ as for $u_q^{\geq 0}(sl_{n+1})$.

EXAMPLE 6.27. Let N be > 2, q a root of unity of order N, and $m_1, \ldots, m_n$ integers > 1 such that $m_i \neq m_j$ for all $i \neq j$. Let Γ be the commutative group generated by $g_1, \ldots, g_n$ with relations $g_i^{Nm_i} = 1$, $1 \leq i \leq n$. Define $\chi_1, \ldots, \chi_n \in \widehat{\Gamma}$ by

$$\chi_j(g_i) = q^{a_{ij}}, \text{ where } a_{ii} = 2 \text{ for all } i, a_{ij} = -1 \text{ if } |i-j| = 1, a_{ij} = 0 \text{ if } |i-j| \geq 2.$$

Then $\chi_{i,j}^N = \varepsilon$ and $g_{i,j}^N \neq 1$ for all $i < j$. Thus for any family $\gamma = (\gamma_{i,j})_{1 \leq i < j \leq n+1}$ of scalars in $\Bbbk$, A_γ in Theorem 6.25 has infinitesimal braiding of type A_n.

Moreover, if $\gamma, \widetilde{\gamma}$ are arbitrary such families with $\gamma_{i,i+1} = 1 = \widetilde{\gamma}_{i,i+1}$ for all $1 \leq i \leq n$, then

$$A_\gamma \not\cong A_{\widetilde{\gamma}}, \text{ if } \gamma \neq \widetilde{\gamma}.$$

PROOF. We let $\widetilde{x}_i$ and $\widetilde{e}_{i,j}$ denote the elements of $A_{\widetilde{\gamma}}$ corresponding to x_i and $e_{i,j}$ in A_γ as above, for all i and $i < j$. Suppose $\phi : A_\gamma \to A_{\widetilde{\gamma}}$ is a Hopf algebra isomorphism. By Lemma [AS3, Lemma 1.2] there exist non-zero scalars $\alpha_1, \ldots, \alpha_n \in \Bbbk$ and a permutation $\sigma \in \mathbb{S}_n$ such that $\phi(g_i) = g_{\sigma(i)}$ and $\phi(x_i) = \alpha_i \widetilde{x}_{\sigma(i)}$ for all i. Since $\mathrm{ord}(g_i) = m_i N \neq m_j N = \mathrm{ord}(g_j)$ for all $i \neq j$, σ must be the identity, and ϕ induces the identity on Γ by restriction. In particular, $1 - g_i^N = \phi(x_i^N) = \alpha_i^N \widetilde{x}_i^N = \alpha_i^N(1 - g_i^N)$, and $\alpha_i^N = 1$ for all i. Therefore we obtain for all $i < j$,

$$u_{i,j}(\gamma) = \phi(e_{i,j}^N) = \alpha_i^N \alpha_{i+1}^N \cdots \alpha_{j-1}^N \widetilde{e}_{i,j}^N = u_{i,j}(\widetilde{\gamma}),$$

and by Remark 6.19(2), $\gamma = \widetilde{\gamma}$. $\qquad\square$

References

[A] N. Andruskiewitsch, *About finite-dimensional Hopf algebras*, Contemp. Math., to appear.

[AA] A. Abella and N. Andruskiewitsch, *Compact quantum groups arising from the FRT construction*, Bol. Acad. Ciencias (Córdoba) **63** (1999), 15–44.

[AEG] N. Andruskiewitsch, P. Etingof and S. Gelaki, *Triangular Hopf Algebras With The Chevalley Property*, Michigan Math. J. **49** (2001), 277–298.

[AG] N. Andruskiewitsch and M. Graña, *Braided Hopf algebras over non-abelian groups*, Bol. Acad. Ciencias (Córdoba) **63** (1999), 45–78.

[AnDa] N. Andruskiewitsch and S. Dăscălescu, *On quantum groups at −1*, Algebr. Represent. Theory, to appear.

[AnDe] N. Andruskiewitsch and J. Devoto, *Extensions of Hopf algebras*, Algebra i Analiz **7** (1995), 17–52.

[AS1] N. Andruskiewitsch and H.-J. Schneider, *Lifting of Quantum Linear Spaces and Pointed Hopf Algebras of order p^3*, J. Algebra **209** (1998), 658–691.

[AS2] ———, *Finite quantum groups and Cartan matrices*, Adv. Math. **154** (2000), 1–45.

[AS3] ———, *Lifting of Nichols algebras of type A_2 and Pointed Hopf Algebras of order p^4*, in "Hopf algebras and quantum groups", eds. S. Caeneppel and F. van Oystaeyen, M. Dekker, 1–16.

[AS4] _____, *Finite quantum groups over abelian groups of prime exponent*, Ann. Sci. Ec. Norm. Super., to appear.

[AS5] _____, *A chacterization of quantum groups*, in preparation.

[B] N. Bourbaki, *Commutative algebra. Chapters 1–7*, Springer-Verlag (1989).

[BDG] M. Beattie, S. Dăscălescu, and L. Grünenfelder, *On the number of types of finite-dimensional Hopf algebras*, Inventiones Math. **136** (1999), 1–7.

[BDR] M. Beattie, S. Dăscălescu, and S. Raianu, *Lifting of Nichols algebras of type B_2*, preprint (2001).

[BGS] A. Beilinson, V. Ginsburg and W. Sörgel, *Koszul duality patterns in representation theory*, J. of Amer. Math. Soc. **9** (1996), 473–526.

[BL] N. Bourbaki, *Groupes et algèbres de Lie. Chap. IV, V, VI*, Hermann, Paris, 1968.

[CD] S. Caenepeel and S. Dăscălescu, *Pointed Hopf algebras of dimension p^3*, J. Algebra **209** (1998), 622–634.

[Ci] C. Cibils, *Tensor products of Hopf bimodules over a group algebra*, Proc. A. M. S. **125** (1997), 1315–1321.

[CM] W. Chin and I. Musson, *The coradical filtration for quantized universal enveloping algebras*, J. London Math. Soc. **53** (1996), 50–67.

[D] D. Didt, *Linkable Dynkin diagrams*, preprint (2001).

[dCP] C. de Concini and C. Procesi, *Quantum Groups*, in "D-modules, Representation theory and Quantum Groups", 31–140, Lecture Notes in Math. **1565** (1993), Springer.

[DPR] R. Dijkgraaf, V. Pasquier and P. Roche, *Quasi Hopf algebras, group cohomology and orbifold models*, Nuclear Phys. B Proc. Suppl. **18B** (1991), 60–72.

[Dr1] V. Drinfeld, *Quantum groups*, Proceedings of the ICM Berkeley 1986, Amer .Math. Soc.

[Dr2] _____, *Quasi-Hopf algebras*, Leningrad Math. J. **1** (1990), 1419–1457.

[DT] Y. Doi and M. Takeuchi, *Multiplication alteration by two-cocycles. The quantum version*, Commun. Algebra **22**, No.14, (1994), 5715–5732.

[FG] C. Fronsdal and A. Galindo, *The Ideals of Free Differential Algebras*, J. Algebra, **222** (1999), 708–746.

[FlG] D. Flores de Chela and J. Green, *Quantum symmetric algebras*, Algebr. Represent. Theory **4** (2001), 55–76.

[FMS] D. Fischman, S. Montgomery, H.-J. Schneider, *Frobenius extensions of subalgebras of Hopf algebras*, Trans. Amer. Math. Soc. **349** (1997), 4857–4895.

[FK] S. Fomin and K. N. Kirillov, *Quadratic algebras, Dunkl elements, and Schubert calculus*, Progr. Math. **172**, Birkhauser, (1999), 146–182.

[FP] S. Fomin and C. Procesi, *Fibered quadratic Hopf algebras related to Schubert calculus*, J. Alg. **230** (2000), 174–183.

[G] S. Gelaki, *On pointed Hopf algebras and Kaplansky's tenth conjecture*, J. Algebra **209** (1998), 635–657.

[Gñ1] M. Graña, *On Pointed Hopf algebras of dimension p^5*, Glasgow Math. J. **42** (2000), 405–419.

[Gñ2] _____, *A freeness theorem for Nichols algebras*, J. Algebra **231** (2000), 235–257.

[Gñ3] _____, *On Nichols algebras of low dimension*, Contemp. Math. **267** (2000), 111–134.

[Gr] J. Green, *Quantum groups, Hall algebras and quantized shuffles*, in "Finite reductive groups" (Luminy, 1994), Progr. Math. **141**, Birkhäuser, (1997), 273–290.

[Gu] D. Gurevich, *Algebraic aspects of the quantum Yang–Baxter equation*, Leningrad J. M. **2** (1991), 801–828.

[HKW] P. de la Harpe, M. Kervaire and C. Weber, *On the Jones polynomial*, L'Ens. Math. **32** (1987), 271–335.

[JS] Joyal, A. and Street, R., *Braided Tensor Categories*, Adv. Math. **102** (1993), 20–78.

[K] V. Kac, *Infinite-dimensional Lie algebras*, Cambridge Univ. Press, Third edition, 1995.

[Ka] C. Kassel, *Quantum groups*, Springer-Verlag (1995).

[Kh] V. Kharchenko, *An Existence Condition for Multilinear Quantum Operations*, J. Alg. **217** (1999), 188–228.

[KS] A. Klimyk and K. Schmüdgen, *Quantum groups and their representations*, Texts and Monographs in Physics, Springer-Verlag, Berlin, (1997).

[L1] G. Lusztig, *Finite dimensional Hopf algebras arising from quantized universal enveloping algebras*, J. Amer. Math. Soc. **3** 257–296.

[L2] _____, *Quantum groups at roots of 1*, Geom. Dedicata **35** (1990), 89–114.

[L3] _____, *Introduction to quantum groups*, Birkhäuser, 1993.

[L4] _____, *Exotic Fourier transform*, Duke Math. J. **73** (1994), 227–241.

[M1] S. Montgomery, *Hopf algebras and their actions on rings*, CBMS Lecture Notes 82, Amer. Math. Soc., 1993.

[M2] _____, *Indecomposable coalgebras, simple comodules, and pointed Hopf algebras*, Proc. AMS, **123** (1995), 2343–2351.

[MiS] A. Milinski and H-J. Schneider, *Pointed Indecomposable Hopf Algebras over Coxeter Groups*, Contemp. Math. **267** (2000), 215–236.

[Mj1] S. Majid, *Foundations of Quantum Group Theory*. Cambridge Univ. Press, 1995.

[Mj2] _____, *Crossed products by braided groups and bosonization*, J.Algebra **163** (1994), 165–190.

[Mu] E. Müller, *Some topics on Frobenius-Lusztig kernels, I*, J. Algebra **206** (1998), 624–658.

[Mu2] E. Müller, *Some topics on Frobenius-Lusztig kernels, II*, J. Algebra **206** (1998), 659–681.

[Mus] I. Musson, *Finite Quantum Groups and Pointed Hopf Algebras*, preprint (1999).

[N] W.D. Nichols, *Bialgebras of type one*, Commun. Alg. **6** (1978), 1521–1552.

[Ra] D. Radford, *Hopf algebras with projection*, J. Algebra **92** (1985), 322–347.

[Re] N. Reshetikhin, *Multiparameter quantum groups and twisted quasitriangular Hopf algebras*, Lett. Math. Phys. **20**, (1990), 331–335.

68   NICOLÁS ANDRUSKIEWITSCH AND HANS-JÜRGEN SCHNEIDER

[Ri]   C. Ringel, *PBW-bases of quantum groups*, J. Reine Angew. Math. **470** (1996), 51–88.

[Ro1] M. Rosso, *Groupes quantiques et algebres de battage quantiques*, C.R.A.S. (Paris) **320** (1995), 145–148.

[Ro2] _____, *Quantum groups and quantum shuffles*, Inventiones Math. **133** (1998), 399–416.

[Sbg] P. Schauenburg, *A characterization of the Borel-like subalgebras of quantum enveloping algebras*, Comm. in Algebra **24** (1996), 2811–2823.

[SvO] D. Stefan and F. van Oystaeyen, *Hochschild cohomology and coradical filtration of pointed Hopf algebras*, J. Algebra **210** (1998), 535–556.

[Sw]   M.E. Sweedler, *Hopf algebras*, Benjamin, New York, 1969.

[Tk1] M. Takeuchi, *Survey of braided Hopf algebras*, Contemp. Math. **267** (2000), 301–324.

[W]   S. Witherspoon, *The representation ring of the quantum double of a finite group*, J. Alg. **179** (1996), 305–329.

[Wa]   M. Wambst, *Complexes de Koszul quantiques*, Ann. Inst. Fourier (Grenoble) **43** (1993), 1089–1156.

[Wo]   S. L. Woronowicz, *Differential calculus on compact matrix pseudogroups (quantum groups)*, Comm. Math. Phys. **122** (1989), 125–170.

NICOLÁS ANDRUSKIEWITSCH
FACULTAD DE MATEMÁTICA, ASTRONOMÍA Y FÍSICA
UNIVERSIDAD NACIONAL DE CÓRDOBA
(5000) CIUDAD UNIVERSITARIA
CÓRDOBA
ARGENTINA
andrus@mate.uncor.edu

HANS-JÜRGEN SCHNEIDER
MATHEMATISCHES INSTITUT
UNIVERSITÄT MÜNCHEN
THERESIENSTRASSE 39
D-80333 MÜNCHEN
GERMANY
hanssch@rz.mathematik.uni-muenchen.de

New Directions in Hopf Algebras
MSRI Publications
Volume **43**, 2002

On the Classification of Finite-Dimensional Triangular Hopf Algebras

SHLOMO GELAKI

ABSTRACT. A fundamental problem in the theory of Hopf algebras is the classification and construction of finite-dimensional quasitriangular Hopf algebras over $\mathbb{C}$. Quasitriangular Hopf algebras constitute a very important class of Hopf algebras, introduced by Drinfeld. They are the Hopf algebras whose representations form a braided tensor category. However, this intriguing problem is extremely hard and is still widely open. Triangular Hopf algebras are the quasitriangular Hopf algebras whose representations form a symmetric tensor category. In that sense they are the closest to group algebras. The structure of triangular Hopf algebras is far from trivial, and yet is more tractable than that of general Hopf algebras, due to their proximity to groups. This makes triangular Hopf algebras an excellent testing ground for general Hopf algebraic ideas, methods and conjectures. A general classification of triangular Hopf algebras is not known yet. However, the problem was solved in the semisimple case, in the minimal triangular pointed case, and more generally for triangular Hopf algebras with the Chevalley property. In this paper we report on all of this, and explain in full details the mathematics and ideas involved in this theory. The classification in the semisimple case relies on Deligne's theorem on Tannakian categories and on Movshev's theory in an essential way. We explain Movshev's theory in details, and refer to [G5] for a detailed discussion of the first aspect. We also discuss the existence of grouplike elements in quasitriangular semisimple Hopf algebras, and the representation theory of cotriangular semisimple Hopf algebras. We conclude the paper with a list of open problems; in particular with the question whether any finite-dimensional triangular Hopf algebra over $\mathbb{C}$ has the Chevalley property.

1. Introduction

A fundamental problem in the theory of Hopf algebras is the classification and construction of finite-dimensional quasitriangular Hopf algebras (A, R) over an algebraically closed field k. Quasitriangular Hopf algebras constitute a very important class of Hopf algebras, which were introduced by Drinfeld [Dr1] in order to supply solutions to the quantum Yang-Baxter equation that arises in mathematical physics. Quasitriangular Hopf algebras are the Hopf algebras whose finite-dimensional representations form a braided rigid tensor category, which naturally relates them to low dimensional topology. Furthermore, Drinfeld showed that *any* finite-dimensional Hopf algebra can be embedded in a finite-dimensional quasitriangular Hopf algebra, known now as its Drinfeld double or quantum double. However, this intriguing problem turns out to be extremely hard and it is still widely open. One can hope that resolving this problem first in the *triangular* case would contribute to the understanding of the general problem.

Triangular Hopf algebras are the Hopf algebras whose representations form a symmetric tensor category. In that sense, they are the class of Hopf algebras closest to group algebras. The structure of triangular Hopf algebras is far from trivial, and yet is more tractable than that of general Hopf algebras, due to their proximity to groups and Lie algebras. This makes triangular Hopf algebras an excellent testing ground for general Hopf algebraic ideas, methods and conjectures. A general classification of triangular Hopf algebras is not known yet. However, there are two classes that are relatively well understood. One of them is semisimple triangular Hopf algebras over k (and cosemisimple if the characteristic of k is positive) for which a complete classification is given in [EG1, EG4]. The key theorem about such Hopf algebras states that each of them is obtained by twisting a group algebra of a finite group [EG1, Theorem 2.1] (see also [G5]).

Another important class of Hopf algebras is that of *pointed* ones. These are Hopf algebras whose all simple comodules are 1-dimensional. Theorem 5.1 in [G4] (together with [AEG, Theorem 6.1]) gives a classification of *minimal* triangular pointed Hopf algebras.

Recall that a finite-dimensional algebra is called *basic* if all of its simple modules are 1-dimensional (i.e., if its dual is a pointed coalgebra). The same Theorem 5.1 of [G4] gives a classification of minimal triangular basic Hopf algebras, since the dual of a minimal triangular Hopf algebra is again minimal triangular.

Basic and semisimple Hopf algebras share a common property. Namely, the Jacobson radical Rad(A) of such a Hopf algebra A is a Hopf ideal, and hence the quotient $A/\text{Rad}(A)$ (the semisimple part) is itself a Hopf algebra. The representation-theoretic formulation of this property is: The tensor product of two simple A-modules is semisimple. A remarkable classical theorem of Chevalley [C, p. 88] states that, in characteristic 0, this property holds for the group algebra of any (not necessarily finite) group. So we called this property of A **the Chevalley property** [AEG].

In [AEG] it was proved that any finite-dimensional triangular Hopf algebra with the Chevalley property is obtained by twisting a finite-dimensional triangular Hopf algebra with R-matrix of rank ≤ 2, and that any finite-dimensional triangular Hopf algebra with R-matrix of rank ≤ 2 is a suitable modification of a finite-dimensional cocommutative Hopf superalgebra (i.e., the group algebra of a finite supergroup). On the other hand, by a theorem of Kostant [Ko], a finite supergroup is a semidirect product of a finite group with an odd vector space on which this group acts. Moreover, the converse result that any such Hopf algebra does have the Chevalley property is also proved in [AEG]. As a corollary, we proved that any finite-dimensional triangular Hopf algebra whose coradical is a Hopf subalgebra (e.g., pointed) is obtained by twisting a triangular Hopf algebra with R-matrix of rank ≤ 2.

The purpose of this paper is to present all that is currently known to us about the classification and construction of finite-dimensional triangular Hopf algebras, and to explain the mathematics and ideas involved in this theory.

The paper is organized as follows. In Section 2 we review some necessary material from the theory of Hopf algebras. In particular the important notion of a twist for Hopf algebras, which was introduced by Drinfeld [Dr1].

In Section 3 we explain in details the theory of Movshev on twisting in group algebras of finite groups [Mov]. The results of [EG4, EG5] (described in Sections 4 and 5 below) rely, among other things, on this theory in an essential way.

In Section 4 we concentrate on the theory of triangular semisimple and co-semisimple Hopf algebras. We first describe the classification and construction of triangular semisimple and cosemisimple Hopf algebras over *any* algebraically closed field k, and then describe some of the consequences of the classification theorem, in particular the one concerning the existence of grouplike elements in triangular semisimple and cosemisimple Hopf algebras over k [EG4]. The classification uses, among other things, Deligne's theorem on Tannakian categories [Del] in an essential way. We refer the reader to [G5] for a detailed discussion of this aspect. The proof of the existence of grouplike elements relies on a theorem from [HI] on central type groups being solvable, which is proved using the classification of finite simple groups. The classification in positive characteristic relies also on the lifting functor from [EG5].

In Section 5 we concentrate on the dual objects of Section 4; namely, on semisimple and cosemisimple *cotriangular* Hopf algebras over k, studied in [EG3]. We describe the representation theory of such Hopf algebras, and in particular obtain that Kaplansky's 6th conjecture [Kap] holds for them (i.e., they are of Frobenius type).

In Section 6 we concentrate on the pointed case, studied in [G4] and [AEG, Theorem 6.1]. The main result in this case is the classification of minimal triangular pointed Hopf algebras.

In Section 7 we generalize and concentrate on the classification of finite-dimensional triangular Hopf algebras with the Chevalley property, given in [AEG]. We note that similarly to the case of semisimple Hopf algebras, the proof of the main result of [AEG] is based on Deligne's theorem [Del]. In fact, we used Theorem 2.1 of [EG1] to prove the main result of this paper.

In Section 8 we conclude the paper with a list of relevant questions raised in [AEG] and [G4].

Throughout the paper the ground field k is assumed to be algebraically closed. The symbol $\mathbb{C}$ will always denote the field of complex numbers. For a Hopf (super)algebra A, $\mathbf{G}(A)$ will denote its group of grouplike elements.

Acknowledgment. The work described in Sections 3–5 is joint with Pavel Etingof, whom I am grateful to for his help in reading the manuscript. The work described in Subsection 4.5 is joint also with Robert Guralnick and Jan Saxl. The work described in Section 7 is joint with Nicholas Andruskiewitsch and Pavel Etingof.

2. Preliminaries

In this Section we recall the necessary background needed for this paper. We refer the reader to the books [ES, Kass, Mon, Sw] for the general theory of Hopf algebras and quantum groups.

2.1. Quasitriangular Hopf algebras. We recall Drinfeld's notion of a (quasi)triangular Hopf algebra [Dr1]. Let $(A, m, 1, \Delta, \varepsilon, S)$ be a finite-dimensional Hopf algebra over k, and let $R = \sum_i a_i \otimes b_i \in A \otimes A$ be an invertible element. Define a linear map $f_R : A^* \to A$ by $f_R(p) = \sum_i \langle p, a_i \rangle b_i$ for $p \in A^*$. The tuple $(A, m, 1, \Delta, \varepsilon, S, R)$ is said to be a *quasitriangular* Hopf algebra if the following axioms hold:

$$(\Delta \otimes \mathrm{Id})(R) = R_{13}R_{23}, \quad (\mathrm{Id} \otimes \Delta)(R) = R_{13}R_{12} \qquad (2\text{--}1)$$

where Id is the identity map of A, and

$$\Delta^{\mathrm{cop}}(a)R = R\Delta(a) \text{ for any } a \in A; \qquad (2\text{--}2)$$

or, equivalently, if $f_R : A^* \to A^{\mathrm{cop}}$ is a Hopf algebra map and (2–2) is satisfied. The element R is called an R-matrix. Observe that using Sweedler's notation for the comultiplication [Sw], (2–2) is equivalent to

$$\sum \langle p_{(1)}, a_{(2)} \rangle a_{(1)} f_R(p_{(2)}) = \sum \langle p_{(2)}, a_{(1)} \rangle f_R(p_{(1)}) a_{(2)} \qquad (2\text{--}3)$$

for any $p \in A^*$ and $a \in A$.

A quasitriangular Hopf algebra (A, R) is called *triangular* if $R^{-1} = R_{21}$; or equivalently, if $f_R * f_{R_{21}} = \varepsilon$ in the convolution algebra $\mathrm{Hom}_k(A^*, A)$, i.e. (using Sweedler's notation again)

$$\sum f_R(p_{(1)}) f_{R_{21}}(p_{(2)}) = \langle p, 1 \rangle 1 \text{ for any } p \in A^*. \qquad (2\text{--}4)$$

Let

$$u := \sum_i S(b_i) a_i \qquad (2\text{--}5)$$

be the *Drinfeld element* of (A, R). Drinfeld showed [Dr2] that u is invertible and that

$$S^2(a) = uau^{-1} \text{ for any } a \in A. \qquad (2\text{--}6)$$

He also showed that (A, R) is triangular if and only if u is a grouplike element [Dr2].

Suppose also that $(A, m, 1, \Delta, \varepsilon, S, R)$ is *semisimple and cosemisimple* over k.

LEMMA 2.1.1. *The Drinfeld element u is central, and*

$$u = S(u). \tag{2-7}$$

PROOF. By [LR1] in characteristic 0, and by [EG5, Theorem 3.1] in positive characteristic, $S^2 = I$. Hence by (2-6), u is central. Now, we have $(S \otimes S)(R) = R$ [Dr2], so $S(u) = \sum_i S(a_i)S^2(b_i) = \sum_i a_i S(b_i)$. This shows that $\mathrm{tr}(u) = \mathrm{tr}(S(u))$ in every irreducible representation of A. But u and $S(u)$ are central, so they act as scalars in this representation, which proves (2-7). □

LEMMA 2.1.2. *In particular,*

$$u^2 = 1. \tag{2-8}$$

PROOF. Since $S(u) = u^{-1}$, the result follows from (2-7). □

Let us demonstrate that it is always possible to replace R with a new R-matrix $\tilde{R}$ so that the new Drinfeld element $\tilde{u}$ equals 1. Indeed, if k does not have characteristic 2, set

$$R_u := \tfrac{1}{2}(1 \otimes 1 + 1 \otimes u + u \otimes 1 - u \otimes u). \tag{2-9}$$

If k is of characteristic 2 (in which case $u = 1$ by semisimplicity), set $R_u := 1$. Set $\tilde{R} := R R_u$.

LEMMA 2.1.3. $(A, \tilde{R})$ *is a triangular semisimple and cosemisimple Hopf algebra with Drinfeld element* 1.

PROOF. Straightforward. □

This observation allows to reduce questions about triangular semisimple and cosemisimple Hopf algebras over k to the case when the Drinfeld element is 1.

Let (A, R) be *any* triangular Hopf algebra over k. Write $R = \sum_{i=1}^{n} a_i \otimes b_i$ in the shortest possible way, and let A_m be the Hopf subalgebra of A generated by the a_i's and b_i's. Following [R2], we will call A_m the *minimal part* of A. We will call $n = \dim(A_m)$ the *rank* of the R-matrix R. It is straightforward to verify that the corresponding map $f_R : A_m^{*\mathrm{cop}} \to A_m$ defined by $f_R(p) = (p \otimes I)(R)$ is a Hopf algebra isomorphism. This property of minimal triangular Hopf algebras will play a central role in our study of the pointed case (see Section 6 below). It implies in particular that $\mathbf{G}(A_m) \cong \mathbf{G}((A_m)^*)$, and hence that the group $\mathbf{G}(A_m)$ is *abelian* (see e.g., [G2]). Thus, $\mathbf{G}(A_m) \cong \mathbf{G}(A_m)^\vee$ (where $\mathbf{G}(A_m)^\vee$ denotes the character group of $\mathbf{G}(A_m)$), and we can identify the Hopf algebras $k[\mathbf{G}(A_m)^\vee]$ and $k[\mathbf{G}(A_m)]^*$. Also, if (A, R) is minimal triangular and pointed then f_R being an isomorphism implies that A^* is pointed as well.

Note that if (A, R) is (quasi)triangular and $\pi : A \to A'$ is a surjective map of Hopf algebras, then (A', R') is (quasi)triangular as well, where $R' := (\pi \otimes \pi)(R)$.

2.2. Hopf superalgebras.

2.2.1. Supervector spaces.
We start by recalling the definition of the category of supervector spaces. A Hopf algebraic way to define this category is as follows. Let us assume that $k = \mathbb{C}$.

Let u be the generator of the order-two group $\mathbb{Z}_2$, and let $R_u \in \mathbb{C}[\mathbb{Z}_2] \otimes \mathbb{C}[\mathbb{Z}_2]$ be as in (2–9). Then $(\mathbb{C}[\mathbb{Z}_2], R_u)$ is a minimal triangular Hopf algebra.

DEFINITION 2.2.1.1. The category of supervector spaces over $\mathbb{C}$ is the symmetric tensor category $\mathrm{Rep}(\mathbb{C}[\mathbb{Z}_2], R_u)$ of representations of the triangular Hopf algebra $(\mathbb{C}[\mathbb{Z}_2], R_u)$. This category will be denoted by SuperVect.

For $V \in$ SuperVect and $v \in V$, we say that v is even if $uv = v$ and odd if $uv = -v$. The set of even vectors in V is denoted by V_0 and the set of odd vectors by V_1, so $V = V_0 \oplus V_1$. We define the parity of a vector v to be $p(v) = 0$ if v is even and $p(v) = 1$ if v is odd (if v is neither odd nor even, $p(v)$ is not defined).

Thus, as an ordinary tensor category, SuperVect is equivalent to the category of representations of $\mathbb{Z}_2$, but the commutativity constraint is different from that of $\mathrm{Rep}(\mathbb{Z}_2)$ and equals $\beta := R_u P$, where P is the permutation of components. In other words, we have

$$\beta(v \otimes w) = (-1)^{p(v)p(w)} w \otimes v, \qquad (2\text{–}10)$$

where both v, w are either even or odd.

2.2.2. Hopf superalgebras.
Recall that in any symmetric (more generally, braided) tensor category, one can define an algebra, coalgebra, bialgebra, Hopf algebra, triangular Hopf algebra, etc, to be an object of this category equipped with the usual structure maps (morphisms in this category), subject to the same axioms as in the usual case. In particular, any of these algebraic structures in the category SuperVect is usually identified by the prefix "super". For example:

DEFINITION 2.2.2.1. A Hopf superalgebra is a Hopf algebra in SuperVect.

More specifically, a Hopf superalgebra $\mathcal{A}$ is an ordinary $\mathbb{Z}_2$-graded associative unital algebra with multiplication m, equipped with a coassociative map

$$\Delta : \mathcal{A} \to \mathcal{A} \otimes \mathcal{A}$$

(a morphism in SuperVect) which is multiplicative in the super-sense, and with a counit and antipode satisfying the standard axioms. Here multiplicativity in the super-sense means that Δ satisfies the relation

$$\Delta(ab) = \sum (-1)^{p(a_2)p(b_1)} a_1 b_1 \otimes a_2 b_2 \qquad (2\text{–}11)$$

for all $a, b \in \mathcal{A}$ (where $\Delta(a) = \sum a_1 \otimes a_2$, $\Delta(b) = \sum b_1 \otimes b_2$). This is because the tensor product of two algebras A, B in SuperVect is defined to be $A \otimes B$ as a vector space, with multiplication

$$(a \otimes b)(a' \otimes b') := (-1)^{p(a')p(b)} aa' \otimes bb'. \tag{2–12}$$

REMARK 2.2.2.2. Hopf superalgebras appear in [Ko], under the name of "graded Hopf algebras".

Similarly, a (quasi)triangular Hopf superalgebra $(\mathcal{A}, \mathcal{R})$ is a Hopf superalgebra with an R-matrix (an *even* element $\mathcal{R} \in \mathcal{A} \otimes \mathcal{A}$) satisfying the usual axioms. As in the even case, an important role is played by the Drinfeld element u of $(\mathcal{A}, \mathcal{R})$:

$$u := m \circ \beta \circ (\mathrm{Id} \otimes S)(\mathcal{R}). \tag{2–13}$$

For instance, $(\mathcal{A}, \mathcal{R})$ is triangular if and only if u is a grouplike element of $\mathcal{A}$.

As in the even case, the tensorands of the R-matrix of a (quasi)triangular Hopf superalgebra $\mathcal{A}$ generate a finite-dimensional sub Hopf superalgebra $\mathcal{A}_m$, called the *minimal part of* $\mathcal{A}$ (the proof does not differ essentially from the proof of the analogous fact for Hopf algebras). A (quasi)triangular Hopf superalgebra is said to be minimal if it coincides with its minimal part. The dimension of the minimal part is the *rank* of the R-matrix.

2.2.3. Cocommutative Hopf superalgebras.

DEFINITION 2.2.3.1. We will say that a Hopf superalgebra $\mathcal{A}$ is commutative (resp. cocommutative) if $m = m \circ \beta$ (resp. $\Delta = \beta \circ \Delta$).

EXAMPLE 2.2.3.2 [Ko]. Let G be a group, and $\mathbf{g}$ a Lie superalgebra with an action of G by automorphisms of Lie superalgebras. Let $\mathcal{A} := \mathbb{C}[G] \ltimes U(\mathbf{g})$, where $U(\mathbf{g})$ denotes the universal enveloping algebra of $\mathbf{g}$. Then $\mathcal{A}$ is a cocommutative Hopf superalgebra, with $\Delta(x) = x \otimes 1 + 1 \otimes x$, $x \in \mathbf{g}$, and $\Delta(g) = g \otimes g$, $g \in G$. In this Hopf superalgebra, we have $S(g) = g^{-1}$, $S(x) = -x$, and in particular $S^2 = \mathrm{Id}$.

The Hopf superalgebra $\mathcal{A}$ is finite-dimensional if and only if G is finite, and $\mathbf{g}$ is finite-dimensional and purely odd (and hence commutative). Then $\mathcal{A} = \mathbb{C}[G] \ltimes \Lambda V$, where $V = \mathbf{g}$ is an odd vector space with a G-action. In this case, $\mathcal{A}^*$ is a commutative Hopf superalgebra.

REMARK 2.2.3.3. We note that as in the even case, it is convenient to think about $\mathcal{A}$ and $\mathcal{A}^*$ in geometric terms. Consider, for instance, the finite-dimensional case. In this case, it is useful to think of the "affine algebraic supergroup" $\tilde{G} := G \ltimes V$. Then one can regard $\mathcal{A}$ as the group algebra $\mathbb{C}[\tilde{G}]$ of this supergroup, and $\mathcal{A}^*$ as its function algebra $F(\tilde{G})$. Having this in mind, we will call the algebra $\mathcal{A}$ **a supergroup algebra**.

It turns out that like in the even case, any cocommutative Hopf superalgebra is of the type described in Example 2.2.3.2. Namely, we have the following theorem.

THEOREM 2.2.3.4. ([**Ko**], **Theorem** 3.3) *Let $\mathcal{A}$ be a cocommutative Hopf super-algebra over $\mathbb{C}$. Then $\mathcal{A} = \mathbb{C}[\mathbf{G}(\mathcal{A})] \ltimes U(P(\mathcal{A}))$, where $U(P(\mathcal{A}))$ is the universal*

enveloping algebra of the Lie superalgebra of primitive elements of A, and $\mathbf{G}(A)$
is the group of grouplike elements of A.

In particular, in the finite-dimensional case we get:

COROLLARY 2.2.3.5. *Let A be a finite-dimensional cocommutative Hopf super-*
algebra over $\mathbb{C}$. Then $A = \mathbb{C}[\mathbf{G}(A)] \ltimes \Lambda V$, where V is the space of primitive
elements of A (regarded as an odd vector space) and $\mathbf{G}(A)$ is the finite group of
grouplikes of A. In other words, A is a supergroup algebra.

2.3. Twists. Let $(A, m, 1, \Delta, \varepsilon, S)$ be a Hopf algebra over a field k. We recall
Drinfeld's notion of a *twist* for A [Dr1].

DEFINITION 2.3.1. A quasitwist for A is an element $J \in A \otimes A$ which satisfies

$$\left.\begin{array}{c} (\Delta \otimes \mathrm{Id})(J)(J \otimes 1) = (\mathrm{Id} \otimes \Delta)(J)(1 \otimes J), \\ (\varepsilon \otimes \mathrm{Id})(J) = (\mathrm{Id} \otimes \varepsilon)(J) = 1. \end{array}\right\} \qquad (2\text{--}14)$$

An invertible quasitwist for A is called a *twist*.

Given a twist J for A, one can define a new Hopf algebra structure

$$(A^J, m, 1, \Delta^J, \varepsilon, S^J)$$

on the algebra $(A, m, 1)$ as follows. The coproduct is determined by

$$\Delta^J(a) = J^{-1}\Delta(a)J \text{ for any } a \in A, \qquad (2\text{--}15)$$

and the antipode is determined by

$$S^J(a) = Q^{-1}S(a)Q \text{ for any } a \in A, \qquad (2\text{--}16)$$

where $Q := m \circ (S \otimes \mathrm{Id})(J)$. If A is (quasi)triangular with the universal R-matrix
R, then so is A^J, with the universal R-matrix $R^J := J_{21}^{-1}RJ$.

EXAMPLE 2.3.2. Let G be a finite *abelian* group, and $G^\vee$ its character group.
Then the set of twists for $A := k[G]$ is in one to one correspondence with the
set of 2-cocycles c of $G^\vee$ with coefficients in k^*, such that $c(0,0) = 1$. Indeed,
let J be a twist for A, and define $c : G^\vee \times G^\vee \to k^*$ via $c(\chi, \psi) := (\chi \otimes \psi)(J)$.
Then it is straightforward to verify that c is a 2-cocycle of $G^\vee$ (see e.g., [Mov,
Proposition 3]), and that $c(0,0) = 1$.

Conversely, let $c : G^\vee \times G^\vee \to k^*$ be a 2-cocycle of $G^\vee$ with coefficients
in k^*, such that $c(0,0) = 1$. Note that the 2-cocycle condition implies that
$c(0,\chi) = 1 = c(\chi, 0)$ for all $\chi \in G^\vee$. For $\chi \in G^\vee$, let $E_\chi := |G|^{-1}\sum_{g \in G}\chi(g)g$
be the associated idempotent of A. Then it is straightforward to verify that
$J := \sum_{\chi, \psi \in G^\vee} c(\chi, \psi)E_\chi \otimes E_\psi$ is a twist for A (see e.g., [Mov, Proposition 3]).
Moreover it is easy to check that the above two assignments are inverse to each
other.

REMARK 2.3.3. Unlike for finite abelian groups, the study of twists for finite non-abelian groups is much more involved. This was done in [EG2, EG4, Mov] (see Section 4 below).

If J is a (quasi)twist for A and x is an invertible element of A such that $\varepsilon(x) = 1$, then

$$J^x := \Delta(x)J(x^{-1} \otimes x^{-1}) \qquad (2\text{--}17)$$

is also a (quasi)twist for A. We will call the (quasi)twists J and J^x *gauge equivalent*. Observe that if (A, R) is a (quasi)triangular Hopf algebra, then the map $(A^J, R^J) \to (A^{J^x}, R^{J^x})$ determined by $a \mapsto xax^{-1}$ is an isomorphism of (quasi)triangular Hopf algebras.

Let A be a group algebra of a finite group. We will say that a twist J for A is *minimal* if the right (and left) components of the R-matrix $R^J := J_{21}^{-1}J$ span A, i.e., if the corresponding triangular Hopf algebra $(A^J, J_{21}^{-1}J)$ is minimal.

A twist for a Hopf algebra in *any symmetric tensor category* is defined in the same way as in the usual case. For instance, if $\mathcal{A}$ is a Hopf superalgebra then a twist for $\mathcal{A}$ is an invertible *even* element $\mathcal{J} \in \mathcal{A} \otimes \mathcal{A}$ satisfying (2–14).

2.4. Projective representations and central extensions.
Here we recall some basic facts about projective representations and central extensions. They can be found in textbooks, e.g. [CR, Section 11E].

A projective representation over k of a group H is a vector space V together with a homomorphism of groups $\pi_V : H \to \mathrm{PGL}(V)$, where $\mathrm{PGL}(V) \cong \mathrm{GL}(V)/k^*$ is the projective linear group.

A linearization of a projective representation V of H is a central extension $\hat{H}$ of H by a central subgroup ζ together with a linear representation $\tilde{\pi}_V : \hat{H} \to \mathrm{GL}(V)$ which descends to π_V. If V is a finite-dimensional projective representation of H then there exists a linearization of V such that ζ is finite (in fact, one can make $\zeta = \mathbb{Z}/(\dim(V))\mathbb{Z}$).

Any projective representation V of H canonically defines a cohomology class $[V] \in H^2(H, k^*)$. The representation V can be lifted to a linear representation of H if and only if $[V] = 0$.

2.5. Pointed Hopf algebras.
The Hopf algebras which are studied in Section 6 are pointed. Recall that a Hopf algebra A is *pointed* if its simple subcoalgebras are all 1-dimensional or equivalently (when A is finite-dimensional) if the irreducible representations of A^* are all 1-dimensional (i.e., A^* is basic). For any $g, h \in \mathbf{G}(A)$, we denote the vector space of $g : h$ *skew primitives* of A by $P_{g,h}(A) := \{x \in A \mid \Delta(x) = x \otimes g + h \otimes x\}$. Thus the classical *primitive* elements of A are $P(A) := P_{1,1}(A)$. The element $g - h$ is always $g : h$ skew primitive. Let $P'_{g,h}(A)$ denote a complement of $sp_k\{g - h\}$ in $P_{g,h}(A)$. Taft-Wilson theorem

[TW] states that the first term A_1 of the coradical filtration of A is given by:

$$A_1 = k[\mathbf{G}(A)] \bigoplus \left(\bigoplus_{g,h \in \mathbf{G}(A)} P'_{g,h}(A) \right). \tag{2-18}$$

In particular, if A is *not* cosemisimple then there exists $g \in \mathbf{G}(A)$ such that $P'_{1,g}(A) \neq 0$.

If A is a Hopf algebra over the field k, which is generated as an algebra by a subset S of $\mathbf{G}(A)$ and by $g : g'$ skew primitive elements, where g, g' run over S, then A is pointed and $\mathbf{G}(A)$ is generated as a group by S (see e.g., [R4, Lemma 1]).

3. Movshev's Theory on the Algebra Associated with a Twist

In this section we describe Movshev's theory on twisting in group algebras of finite groups [Mov]. Our classification theory of triangular semisimple and cosemisimple Hopf algebras [EG4] (see Section 4 below), and our study of the representation theory of cotriangular semisimple and cosemisimple Hopf algebras [EG3] (see Section 5 below) rely, among other things, on this theory in an essential way.

Let k be an algebraically closed field whose characteristic is relatively prime to $|G|$. Let $A := k[G]$ be the group algebra of a finite group G, equipped with the usual multiplication, unit, comultiplication, counit and antipode, denoted by m, 1, Δ, ε and S respectively. Let $J \in A \otimes A$. Movshev had the following nice idea of characterizing quasitwists [Mov]. Let $(A_J, \Delta_J, \varepsilon)$ where $A_J = A$ as vector spaces, and Δ_J is the map

$$\Delta_J : A \to A \otimes A, \ a \mapsto \Delta(a)J. \tag{3-1}$$

PROPOSITION 3.1. $(A_J, \Delta_J, \varepsilon)$ *is a coalgebra if and only if J is a quasitwist for A.*

PROOF. Straightforward.   □

Regard A as the left regular representation of G. Then $(A_J, \Delta_J, \varepsilon)$ is a G-coalgebra (i.e., $\Delta_J(ga) = (g \otimes g)\Delta_J(a)$ and $\varepsilon(ga) = \varepsilon(a)$ for all $g \in G$, $a \in A$). In fact, we have the following important result.

PROPOSITION 3.2 [Mov, Proposition 5]. *Suppose that $(C, \tilde{\Delta}, \tilde{\varepsilon})$ is a G-coalgebra which is isomorphic to the regular representation of G as a G-module. Then there exists a quasitwist $J \in A \otimes A$ such that $(C, \tilde{\Delta}, \tilde{\varepsilon})$ and $(A, \Delta_J, \varepsilon)$ are isomorphic as G-coalgebras. Moreover, J is unique up to gauge equivalence.*

PROOF. We can choose an element $\lambda \in C$ such that the set $\{g \cdot \lambda \mid g \in G\}$ forms a basis of C, and $\tilde{\varepsilon}(\lambda) = 1$. Now, write $\tilde{\Delta}(\lambda) = \sum_{a,b \in G} \gamma(a,b)a \cdot \lambda \otimes b \cdot \lambda$, and set

$$J := \sum_{a,b \in G} \gamma(a,b)a \otimes b \in A \otimes A. \tag{3-2}$$

We have to show that J is a quasitwist for A. Indeed, let $f : A \to C$ be determined by $f(a) = a \cdot \lambda$. Clearly, f is an isomorphism of G-modules which satisfies $\tilde{\Delta}(f(a)) = (f \otimes f)\Delta_J(a)$, $a \in A$. Therefore $(A_J, \Delta_J, \varepsilon)$ is a coalgebra, which is equivalent to saying that J is a quasitwist by Proposition 3.1. This proves the first claim.

Suppose that $(A_{J'}, \Delta_{J'}, \varepsilon)$ and $(A_J, \Delta_J, \varepsilon)$ are isomorphic as G-coalgebras via $\phi : A \to A$. We have to show that J, J' are gauge equivalent. Indeed, ϕ is given by right multiplication by an invertible element $x \in A$, $\phi(a) = ax$. On one hand, $(\phi \otimes \phi)(\Delta_J(1)) = J(x \otimes x)$, and on the other hand, $\Delta_{J'}(\phi(1)) = \Delta(x)J'$. The equality between the two right hand sides implies the desired result. $\square$

We now focus on the dual algebra $(A_J)^*$ of the coalgebra $(A_J, \Delta_J, \varepsilon)$, and summarize Movshev's results about it [Mov]. Note that $(A_J)^*$ is a G-algebra which is isomorphic to the regular representation of G as a G-module.

PROPOSITION 3.3 [Mov, Propositions 6 and 7]. 1. *The algebra* $(A_J)^*$ *is semi-simple.*

2. *There exists a subgroup* St *of* G *(the stabilizer of a maximal two sided ideal* I *of* $(A_J)^*$*) such that* $(A_J)^*$ *is isomorphic to the algebra of functions from the set* $G/$ St *to the matrix algebra* $M_{|St|^{1/2}}(k)$.

Note that, in particular, the group St acts on the matrix algebra $(A_J)^*/I \cong M_{|St|^{1/2}}(k)$. Hence this algebra defines a projective representation $T :$ St $\to$ PGL$(|St|^{1/2}, k)$ (since Aut$(M_{|St|^{1/2}}(k)) =$ PGL$(|St|^{1/2}, k)$).

PROPOSITION 3.4 [Mov, Propositions 8 and 9]. T *is irreducible, and the associated 2-cocycle* $c :$ St $\times$ St $\to k^*$ *is nontrivial.*

Consider the twisted group algebra $k[St]^c$. This algebra has a basis $\{X_g \mid g \in St\}$ with relations $X_g X_h = c(g, h)X_{gh}$, and a natural structure as a St-algebra given by $a \cdot X_g := X_a X_g (X_a)^{-1}$ for all $a \in$ St (see also [Mov, Proposition 10]). Recall that c is called *nondegenerate* if for all $1 \neq g \in$ St, the map $C_{St}(g) \to k^*$, $m \mapsto c(m, g)/c(g, m)$ is a nontrivial homomorphism of the centralizer of g in St to k^*. In [Mov, Propositions 11,12] Movshev reproduces the following well known criterion for $k[St]^c$ to be a simple algebra (i.e., isomorphic to the matrix algebra $M_{|St|^{1/2}}(k)$).

PROPOSITION 3.5. *The twisted group algebra* $k[St]^c$ *is simple if and only if* c *is nondegenerate. Furthermore, if this is the case, then* $k[St]^c$ *is isomorphic to the regular representation of* St *as a* St-module.

Assume c is nondegenerate. By Proposition 3.2, the simple St-coalgebra $(k[St]^c)^*$ is isomorphic to the St-coalgebra $(k[St]_{\tilde{J}}, \Delta_{\tilde{J}})$ for some unique (up to gauge equivalence) quasitwist $\tilde{J} \in k[St] \otimes k[St]$.

PROPOSITION 3.6 [Mov, Propositions 13 and 14]. $\tilde{J}$ *is in fact a twist for* $k[St]$ *(i.e., it is invertible). Furthermore,* J *is the image of* $\tilde{J}$ *under the coalgebra embedding* $(k[St]^c)^* \hookrightarrow A_J$.

PROOF. We only reproduce here the proof of the invertibility of $\tilde{J}$ (in a slightly expanded form). Set $C := k[\mathrm{St}]_{\tilde{j}}$. Suppose on the contrary that $\tilde{J}$ is not invertible. Then there exists $0 \neq L \in C^* \otimes C^*$ such that $\tilde{J}L = 0$. Let $F : C \otimes C \to C \otimes C$ be defined by $F(x) = xL$. Clearly, F is a morphism of St $\times$ St-representations, and $F \circ \Delta_{\tilde{j}} = 0$. Thus the image $\mathrm{Im}(F^*)$ of the morphism of St $\times$ St-representations $F^* : C^* \otimes C^* \to C^* \otimes C^*$ is contained in the kernel of the multiplication map $m := (\Delta_{\tilde{j}})^*$. Let $U := (C^* \otimes 1)\mathrm{Im}(F^*)(1 \otimes C^*)$. Clearly, U is contained in the kernel of m too. But, for any $x \in U$ and $g \in \mathrm{St}$, $(1 \otimes X_g)x(1 \otimes X_g)^{-1} \in U$. Thus, U is a left $C^* \otimes C^*$-module under left multiplication. Similarly, it is a right module over this algebra under right multiplication. So, it is a bimodule over $C^* \otimes C^*$. Since $U \neq 0$, this implies that $U = C^* \otimes C^*$. This is a contradiction, since we get that $m = 0$. Hence $\tilde{J}$ is invertible as desired.   □

REMARK 3.7. In the paper [Mov] it is assumed that the characteristic of k is equal to 0, but all the results generalize in a straightforward way to the case when the characteristic of k is positive and relatively prime to the order of the group G.

4. The Classification of Triangular Semisimple and Cosemisimple Hopf Algebras

In this section we describe the classification of triangular semisimple and cosemisimple Hopf algebras over *any* algebraically closed field k, given in [EG4].

4.1. Construction of triangular semisimple and cosemisimple Hopf algebras from group-theoretical data.

Let H be a finite group such that $|H|$ is not divisible by the characteristic of k. Suppose that V is an irreducible projective representation of H over k satisfying $\dim(V) = |H|^{1/2}$. Let $\pi : H \to \mathrm{PGL}(V)$ be the projective action of H on V, and let $\tilde{\pi} : H \to \mathrm{SL}(V)$ be any lifting of this action ($\tilde{\pi}$ need not be a homomorphism). We have $\tilde{\pi}(x)\tilde{\pi}(y) = c(x,y)\tilde{\pi}(xy)$, where c is a 2-cocycle of H with coefficients in k^*. This cocycle is nondegenerate (see Section 3) and hence the representation of H on $\mathrm{End}_k(V)$ is isomorphic to the regular representation of H (see e.g., Proposition 3.5). By Propositions 3.2 and 3.6, this gives rise to a twist $J(V)$ for $k[H]$, whose equivalence class is canonically associated to (H, V).

Now, for any group $G \supseteq H$, whose order is relatively prime to the characteristic of k, define a triangular semisimple Hopf algebra

$$F(G, H, V) := \left(k[G]^{J(V)}, J(V)_{21}^{-1} J(V)\right). \tag{4-1}$$

We wish to show that it is also cosemisimple.

LEMMA 4.1.1. *The Drinfeld element of the triangular semisimple Hopf algebra* $(A, R) := F(G, H, V)$ *equals* 1.

PROOF. The Drinfeld element u is a grouplike element of A, and for any finite-dimensional A-module V one has $\operatorname{tr}|_V(u) = \dim_{\operatorname{Rep}_k(A)}(V) = \dim(V)$ (since $\operatorname{Rep}_k(A)$ is equivalent to $\operatorname{Rep}_k(G)$, see e.g., [G5]). In particular, we can set V to be the regular representation, and find that $\operatorname{tr}|_A(u) = \dim(A) \neq 0$ in k. But it is clear that if g is a nontrivial grouplike element in any finite-dimensional Hopf algebra A, then $\operatorname{tr}|_A(g) = 0$. Thus, $u = 1$. $\qquad\square$

REMARK 4.1.2. Lemma 4.1.1 fails for infinite-dimensional *cotriangular* Hopf algebras, which shows that this lemma can not be proved by an explicit computation. We refer the reader to [EG6] for the study of infinite-dimensional cotriangular Hopf algebras which are obtained from twisting in function algebras of affine proalgebraic groups.

COROLLARY 4.1.3. *The triangular semisimple Hopf algebra* $(A, R) := F(G, H, V)$ *is also cosemisimple.*

PROOF. Since $u = 1$, one has $S^2 = \operatorname{Id}$ and so A is cosemisimple (as $\dim A \neq 0$). $\qquad\square$

Thus we have assigned a triangular semisimple and cosemisimple Hopf algebra with Drinfeld element $u = 1$ to any triple (G, H, V) as above.

4.2. The classification in characteristic 0. In this subsection we assume that k is of characteristic 0. We first recall Theorem 2.1 from [EG1] and Theorem 3.1 from [EG4], and state them in a single theorem which is the key structure theorem for triangular semisimple Hopf algebras over k.

THEOREM 4.2.1. *Let (A, R) be a triangular semisimple Hopf algebra over an algebraically closed field k of characteristic 0, with Drinfeld element u. Set $\bar{R} := RR_u$. Then there exist a finite group G, a subgroup $H \subseteq G$ and a minimal twist $J \in k[H] \otimes k[H]$ such that $(A, \bar{R})$ and $(k[G]^J, J_{21}^{-1}J)$ are isomorphic as triangular Hopf algebras. Moreover, the data (G, H, J) is unique up to isomorphism of groups and gauge equivalence of twists. That is, if there exist a finite group G', a subgroup $H' \subseteq G'$ and a minimal twist $J' \in k[H'] \otimes k[H']$ such that $(A, \bar{R})$ and $(k[G']^{J'}, J_{21}'^{-1}J')$ are isomorphic as triangular Hopf algebras, then there exists an isomorphism of groups $\phi : G \to G'$ such that $\phi(H) = H'$ and $(\phi \otimes \phi)(J)$ and J' are gauge equivalent as twists for $k[H']$.*

The proof of this theorem relies, among other things, on the following (special case of a) deep theorem of Deligne on Tannakian categories [De1] in an essential way.

THEOREM 4.2.2. *Let k be an algebraically closed field of characteristic 0, and $(\mathcal{C}, \otimes, 1, a, l, r, c)$ a k-linear abelian symmetric rigid category with $\operatorname{End}(1) = k$, which is semisimple with finitely many irreducible objects. If categorical dimensions of objects are nonnegative integers, then there exist a finite group G and an equivalence of symmetric rigid categories $F : \mathcal{C} \to \operatorname{Rep}_k(G)$.*

We refer the reader to [G5, Theorems 5.3, 6.1, Corollary 6.3] for a complete and detailed proof of [EG1, Theorem 2.1] and [EG4, Theorem 3.1], along with a discussion of Tannakian categories.

Let (A, R) be a triangular semisimple Hopf algebra over k whose Drinfeld element u is 1, and let (G, H, J) be the associated group-theoretic data given in Theorem 4.2.1.

PROPOSITION 4.2.3. *The H-coalgebra $(k[H]_J, \Delta_J)$ (see (3–1)) is simple, and is isomorphic to the regular representation of H as an H-module.*

PROOF. By Proposition 3.6, J is the image of $\tilde{J}$ under the embedding $k[\mathrm{St}]_{\tilde{J}} \hookrightarrow k[H]_J$. Since J is minimal and $\mathrm{St} \subseteq H$, it follows that $\mathrm{St} = H$. Hence the result follows from the discussion preceding Proposition 3.6. $\square$

We are now ready to prove our first classification result.

THEOREM 4.2.4. *The assignment $F : (G, H, V) \mapsto (A, R)$ is a bijection between*

1. *isomorphism classes of triples (G, H, V) where G is a finite group, H is a subgroup of G, and V is an irreducible projective representation of H over k satisfying $\dim(V) = |H|^{1/2}$, and*
2. *isomorphism classes of triangular semisimple Hopf algebras over k with Drinfeld element $u = 1$.*

PROOF. We need to construct an assignment F' in the other direction, and check that both $F' \circ F$ and $F \circ F'$ are the identity assignments.

Let (A, R) be a triangular semisimple Hopf algebra over k whose Drinfeld element u is 1, and let (G, H, J) be the associated group-theoretic data given in Theorem 4.2.1. By Proposition 4.2.3, the H-algebra $(k[H]_J)^*$ is simple. So we see that $(k[H]_J)^*$ is isomorphic to $\mathrm{End}_k(V)$ for some vector space V, and we have a homomorphism $\pi : H \to \mathrm{PGL}(V)$. Thus V is a projective representation of H. By Proposition 3.4, this representation is irreducible, and it is obvious that $\dim(V) = |H|^{1/2}$.

It is clear that the isomorphism class of the representation V does not change if J is replaced by a twist J' which is gauge equivalent to J as twists for $k[H]$. Thus, to any isomorphism class of triangular semisimple Hopf algebras (A, R) over k, with Drinfeld element 1, we have assigned an isomorphism class of triples (G, H, V). Let us write this as

$$F'(A, R) := (G, H, V). \tag{4–2}$$

The identity $F \circ F' = id$ follows from Proposition 3.2. Indeed, start with $(A, R) \cong (k[G]^J, J_{21}^{-1} J)$, where J is a minimal twist for $k[H]$, H a subgroup of G. Then by Proposition 4.2.3, we have that $(k[H]_J, \Delta_J)$ is a simple H-coalgebra which is isomorphic to the regular representation of H. Now let V be the associated irreducible projective representation of H, and $J(V)$ the associated twist as in (4–1). Then $(k[H]_J, \Delta_J)$ and $(k[H]_{J(V)}, \Delta_{J(V)})$ are isomorphic as H-coalgebras, and the claim follows from Proposition 3.2.

The identity $F' \circ F = id$ follows from the uniqueness part of Theorem 4.2.1. $\square$

REMARK 4.2.5. Observe that it follows from Theorem 4.2.4 that the twist $J(V)$ associated to (H, V) is *minimal*.

Now let (G, H, V, u) be a quadruple, in which (G, H, V) is as above, and u is a central element of G of order ≤ 2. We extend the map F to quadruples by setting

$$F(G, H, V, u) := (A, RR_u) \text{ where } (A, R) := F(G, H, V). \qquad (4\text{–}3)$$

THEOREM 4.2.6. *The assignment F given in (4–3) is a bijection between*

1. *isomorphism classes of quadruples (G, H, V, u) where G is a finite group, H is a subgroup of G, V is an irreducible projective representation of H over k satisfying $\dim(V) = |H|^{1/2}$, and $u \in G$ is a central element of order ≤ 2, and*
2. *isomorphism classes of triangular semisimple Hopf algebras over k.*

PROOF. Define F' by $F'(A, R) := (F'(A, RR_u), u)$, where $F'(A, RR_u)$ is defined in (4–2). Using Theorem 4.2.4, it is straightforward to see that both $F' \circ F$ and $F \circ F'$ are the identity assignments. $\square$

Theorem 4.2.6 implies the following classification result for *minimal* triangular semisimple Hopf algebras over k.

PROPOSITION 4.2.7. $F(G, H, V, u)$ *is minimal if and only if G is generated by H and u.*

PROOF. As we have already pointed out in Remark 4.2.5, if $(A, R) := F(G, H, V)$ then the sub Hopf algebra $k[H]^J \subseteq A$ is minimal triangular. Therefore, if $u = 1$ then $F(G, H, V)$ is minimal if and only if $G = H$. This obviously remains true for $F(G, H, V, u)$ if $u \neq 1$ but $u \in H$. If $u \notin H$ then it is clear that the R-matrix of $F(G, H, V, u)$ generates $k[H']$, where $H' = H \cup uH$. This proves the proposition. $\square$

REMARK 4.2.8. As was pointed out already by Movshev, the theory developed in [Mov] and extended in [EG4] is an analogue, for finite groups, of the theory of quantization of skew-symmetric solutions of the classical Yang-Baxter equation, developed by Drinfeld [Dr3]. In particular, the operation F is the analogue of the operation of quantization in [Dr3].

4.3. The classification in positive characteristic. In this subsection we assume that k is of positive characteristic p, and prove an analogue of Theorem 4.2.6 by using this theorem itself and the lifting techniques from [EG5].

We first recall some notation from [EG5]. Let $\mathcal{O} := W(k)$ be the ring of Witt vectors of k (see e.g., [Se, Sections 2.5, 2.6]), and K the field of fractions of $\mathcal{O}$. Recall that $\mathcal{O}$ is a local complete discrete valuation ring, and that the characteristic of K is zero. Let $\mathfrak{m}$ be the maximal ideal in $\mathcal{O}$, which is generated by p. One has $\mathfrak{m}^n/\mathfrak{m}^{n+1} = k$ for any $n \geq 0$ (here $\mathfrak{m}^0 := \mathcal{O}$).

Let F be the assignment defined in (4–3). We now have the following classification result.

THEOREM 4.3.1. *The assignment F is a bijection between*

1. *isomorphism classes of quadruples (G, H, V, u) where G is a finite group of order prime to p, H is a subgroup of G, V is an irreducible projective representation of H over k satisfying $\dim(V) = |H|^{1/2}$, and $u \in G$ is a central element of order ≤ 2, and*
2. *isomorphism classes of triangular semisimple and cosemisimple Hopf algebras over k.*

PROOF. As in the proof of Theorem 4.2.6 we need to construct the assignment F'.

Let (A, R) be a triangular semisimple and cosemisimple Hopf algebra over k. Lift it (see [EG5]) to a triangular semisimple Hopf algebra $(\bar{A}, \bar{R})$ over K. By Theorem 4.2.6, we have that $(\bar{A} \otimes_K \bar{K}, \bar{R}) = F(G, H, V, u)$. We can now reduce V "$\mathrm{mod}\, p$" to get V_p which is an irreducible projective representation of H over the field k. This can be done since V is defined by a nondegenerate 2-cocycle c (see Section 3) with values in roots of unity of degree $|H|^{1/2}$ (as the only irreducible representation of the simple H-algebra with basis $\{X_h \mid h \in H\}$, and relations $X_g X_h = c(g, h)X_{gh}$). This cocycle can be reduced $\mathrm{mod}\, p$ and remains nondegenerate (since the groups of roots of unity of order $|H|^{1/2}$ in k and K are naturally isomorphic), so it defines an irreducible projective representation V_p. Define $F'(A, R) := (G, H, V_p, u)$. It is shown like in characteristic 0 that $F \circ F'$ and $F' \circ F$ are the identity assignments.   □

The following is the analogue of Theorem 4.2.1 in positive characteristic.

COROLLARY 4.3.2. *Let (A, R) be a triangular semisimple and cosemisimple Hopf algebra over any algebraically closed field k, with Drinfeld element u. Set $\tilde{R} := RR_u$. Then there exist a finite group G, a subgroup $H \subseteq G$ and a minimal twist $J \in k[H] \otimes k[H]$ such that $(A, \tilde{R})$ and $(k[G]^J, J_{21}^{-1}J)$ are isomorphic as triangular Hopf algebras. Moreover, the data (G, H, J) is unique up to isomorphism of groups and gauge equivalence of twists.*

PROPOSITION 4.3.3. *Proposition 4.2.7 holds in positive characteristic as well.*

PROOF. As before, if $(A, R) := F(G, H, V)$, the sub Hopf algebra $k[H]^J \subseteq A$ is minimal triangular. This follows from the facts that it is true in characteristic 0, and that the rank of a triangular structure does not change under lifting. Thus, Proposition 4.2.7 holds in characteristic p.   □

REMARK. The class of finite-dimensional triangular cosemisimple Hopf algebras over k is invariant under twisting (see Remark 3.7 in [AEGN]). Using this and

Theorem 4.2.6, we were able in Theorem 6.3 of [AEGN] to classify the isomorphism classes of finite-dimensional triangular cosemisimple Hopf algebras over k. Namely, we proved that the classification is the same as in characteristic 0 (see Theorem 4.2.6) except that the subgroup H has to be of order coprime to the characteristic of k.

4.4. The solvability of the group underlying a minimal triangular semisimple Hopf algebra.

In this subsection we consider finite groups which admit a minimal twist as studied in [EG4]. We also consider the existence of nontrivial grouplike elements in triangular semisimple and cosemisimple Hopf algebras, following [EG4].

A classical fact about complex representations of finite groups is that the dimension of any irreducible representation of a finite group K does not exceed $|K : Z(K)|^{1/2}$, where $Z(K)$ is the center of K. Groups of central type are those groups for which this inequality is in fact an equality. More precisely, a finite group K is said to be of *central type* if it has an irreducible representation V such that $\dim(V)^2 = |K : Z(K)|$ (see e.g., [HI]). We shall need the following theorem (conjectured by Iwahori and Matsumoto in 1964) whose proof uses the classification of finite simple groups.

THEOREM 4.4.1 [HI, Theorem 7.3]. *Any group of central type is solvable.*

As corollaries, we have the following results.

COROLLARY 4.4.2. *Let H be a finite group which admits a minimal twist. Then H is solvable.*

PROOF. We may assume that k has characteristic 0 (otherwise we can lift to characteristic 0). As we showed in the proof of Theorem 4.2.4, H has an irreducible projective representation V with $\dim(V) = |H|^{1/2}$. Let K be a finite central extension of H with central subgroup Z, such that V lifts to a linear representation of K. We have $\dim(V)^2 = |K : Z|$. Since $\dim(V)^2 \le |K : Z(K)|$ we get that $Z = Z(K)$ and hence that K is a group of central type. But by Theorem 4.4.1, K is solvable and hence $H \cong K/Z(K)$ is solvable as well. □

REMARK 4.4.3. Movshev conjectures in the introduction to [Mov] that any finite group with a nondegenerate 2-cocycle is solvable. As explained in the Proof of Corollary 4.4.2, this result follows from Theorem 4.4.1.

COROLLARY 4.4.4. *Let A be a triangular semisimple and cosemisimple Hopf algebra over k of dimension bigger than 1. Then A has a nontrivial grouplike element.*

PROOF. We can assume that the Drinfeld element u is equal to 1 and that A is not cocommutative. Let A_m be the minimal part of A. By Corollary 4.4.2, $A_m = k[H]^J$ for a solvable group H, $|H| > 1$. Therefore, A_m has nontrivial 1-dimensional representations. Since $A_m \cong A_m^{*op}$ as Hopf algebras, we get that A_m, and hence A, has nontrivial grouplike elements. □

4.5. Biperfect quasitriangular semisimple Hopf algebras. Corollary 4.4.4 motivates the following question. Let (A, R) be a *quasitriangular* semisimple Hopf algebra over k with characteristic 0 (e.g., the quantum double of a semisimple Hopf algebra), and let $\dim(A) > 1$. Is it true that A possesses a nontrivial grouplike element? We now follow [EGGS] and show that the answer to this question is *negative*.

Let G be a finite group. If G_1 and G_2 are subgroups of G such that $G = G_1 G_2$ and $G_1 \cap G_2 = 1$, we say that $G = G_1 G_2$ is an *exact factorization*. In this case G_1 can be identified with G/G_2, and G_2 can be identified with G/G_1 as sets, so G_1 is a G_2-set and G_2 is a G_1-set. Note that if $G = G_1 G_2$ is an exact factorization, then $G = G_2 G_1$ is also an exact factorization by taking the inverse elements.

Following Kac [KaG] and Takeuchi [T], one constructs a semisimple Hopf algebra from these data, as follows: Take the vector space $H := \mathbb{C}[G_2]^* \otimes \mathbb{C}[G_1]$. Introduce a product on H by:

$$(\varphi \otimes a)(\psi \otimes b) = \varphi(a \cdot \psi) \otimes ab \qquad (4\text{--}4)$$

for all $\varphi, \psi \in \mathbb{C}[G_2]^*$ and $a, b \in G_1$. Here $\cdot$ denotes the associated action of G_1 on the algebra $\mathbb{C}[G_2]^*$, and $\varphi(a \cdot \psi)$ is the multiplication of φ and $a \cdot \psi$ in the algebra $\mathbb{C}[G_2]^*$.

Identify the vector spaces

$$H \otimes H = (\mathbb{C}[G_2] \otimes \mathbb{C}[G_2])^* \otimes (\mathbb{C}[G_1] \otimes \mathbb{C}[G_1])$$
$$= \mathrm{Hom}_{\mathbb{C}}(\mathbb{C}[G_2] \otimes \mathbb{C}[G_2], \mathbb{C}[G_1] \otimes \mathbb{C}[G_1])$$

in the usual way, and introduce a coproduct on H by:

$$(\Delta(\varphi \otimes a))(b \otimes c) = \varphi(bc) a \otimes b^{-1} \cdot a \qquad (4\text{--}5)$$

for all $\varphi \in \mathbb{C}[G_2]^*$, $a \in G_1$ and $b, c \in G_2$. Here $\cdot$ denotes the action of G_2 on G_1.

Introduce a counit on H by:

$$\varepsilon(\varphi \otimes a) = \varphi(1_G) \qquad (4\text{--}6)$$

for all $\varphi \in \mathbb{C}[G_2]^*$ and $a \in G_1$.

Finally, identify the vector spaces $H = \mathbb{C}[G_2]^* \otimes \mathbb{C}[G_1] = \mathrm{Hom}_{\mathbb{C}}(\mathbb{C}[G_2], \mathbb{C}[G_1])$ in the usual way, and introduce an antipode on H by

$$S\left(\sum_{a \in G_1} \varphi_a \otimes a \right)(x) = \sum_{a \in G_1} \varphi_{(x^{-1} \cdot a)^{-1}} \left((a^{-1} \cdot x)^{-1} \right) a \qquad (4\text{--}7)$$

for all $\sum_{a \in G_1} \varphi_a \otimes a \in H$ and $x \in G_2$, where the first $\cdot$ denotes the action of G_2 on G_1, and the second one denotes the action of G_1 on G_2.

THEOREM 4.5.1 [KaG, T]. *The multiplication, comultiplication, counit and antipode described in* (4–4)-(4–7) *determine a semisimple Hopf algebra structure on the vector space* $H := \mathbb{C}[G_2]^* \otimes \mathbb{C}[G_1]$.

The Hopf algebra H is called the *bicrossproduct* Hopf algebra associated with G, G_1, G_2, and is denoted by $H(G, G_1, G_2)$.

THEOREM 4.5.2 [Ma2]. $H(G, G_2, G_1) \cong H(G, G_1, G_2)^*$ *as Hopf algebras.*

Let us call a Hopf algebra H *biperfect* if the groups $G(H)$, $G(H^*)$ are both trivial. We are ready now to prove:

THEOREM 4.5.3. $H(G, G_1, G_2)$ *is biperfect if and only if G_1, G_2 are self normalizing perfect subgroups of G.*

PROOF. It is well known that the category of finite-dimensional representations of $H(G, G_1, G_2)$ is equivalent to the category of G_1-equivariant vector bundles on G_2, and hence that the irreducible representations of $H(G, G_1, G_2)$ are indexed by pairs (V, x) where x is a representative of a G_1-orbit in G_2, and V is an irreducible representation of $(G_1)_x$, where $(G_1)_x$ is the isotropy subgroup of x. Moreover, the dimension of the corresponding irreducible representation is

$$\frac{\dim(V)|G_1|}{|(G_1)_x|}.$$

Thus, the 1-dimensional representations of $H(G, G_1, G_2)$ are indexed by pairs (V, x) where x is a fixed point of G_1 on $G_2 = G/G_1$ (i.e., $x \in N_G(G_1)/G_1$), and V is a 1-dimensional representation of G_1. The result follows now using Theorem 4.5.2. □

By Theorem 4.5.3, in order to construct an example of a biperfect semisimple Hopf algebra, it remains to find a finite group G which admits an exact factorization $G = G_1 G_2$, where G_1, G_2 are self normalizing perfect subgroups of G. Amazingly the Mathieu simple group $G := M_{24}$ of degree 24 provides such an example!

THEOREM 4.5.4. *The group G contains a subgroup $G_1 \cong \mathrm{PSL}(2, 23)$, and a subgroup $G_2 \cong A_7 \ltimes (\mathbb{Z}_2)^4$ where A_7 acts on $(\mathbb{Z}_2)^4$ via the embedding $A_7 \subset A_8 = \mathrm{SL}(4, 2) = \mathrm{Aut}((\mathbb{Z}_2)^4)$ (see [AT]). These subgroups are perfect, self normalizing and G admits an exact factorization $G = G_1 G_2$. In particular, $H(G, G_1, G_2)$ is biperfect.*

We suspect that not only is M_{24} the smallest example but it may be the only finite simple group with a factorization with all the needed properties.

Clearly, the Drinfeld double $D(H(G, G_1, G_2))$ is an example of a biperfect quasitriangular semisimple Hopf algebra.

4.6. Minimal triangular Hopf algebras constructed from a bijective 1-cocycle.
In this subsection we describe an explicit way of constructing minimal twists for certain solvable groups (hence of central type groups) given in [EG2]. For simplicity we let $k := \mathbb{C}$.

DEFINITION 4.6.1. Let G, A be finite groups and $\rho : G \to \mathrm{Aut}(A)$ a homomorphism. By a 1-cocycle of G with coefficients in A we mean a map $\pi : G \to A$ which satisfies the equation

$$\pi(gg') = \pi(g)(g \cdot \pi(g')), \quad g, g' \in G, \tag{4-8}$$

where $\rho(g)(x) = g \cdot x$ for $g \in G, x \in A$.

We will be interested in the case when π is a *bijection* (so in particular, $|G| = |A|$), because of the following proposition.

PROPOSITION 4.6.2. *Let G, A be finite groups, $\pi : G \to A$ a bijective 1-cocycle, and J a twist for $\mathbb{C}[A]$ which is G-invariant. Then $\bar{J} := (\pi^{-1} \otimes \pi^{-1})(J)$ is a quasitwist for $\mathbb{C}[G]$.*

PROOF. It is obvious that the second equation of (2–14) is satisfied for $\bar{J}$. So we only have to prove the first equation of (2–14) for $\bar{J}$. Write $J = \sum a_{xy} x \otimes y$. Then

$$(\pi \otimes \pi \otimes \pi)((\Delta \otimes \mathrm{Id})(\bar{J})(\bar{J} \otimes 1))$$

$$= \sum_{x,y,z,t \in A} a_{xy} a_{zt} \pi(\pi^{-1}(x)\pi^{-1}(z)) \otimes \pi(\pi^{-1}(x)\pi^{-1}(t)) \otimes \pi(\pi^{-1}(y))$$

$$= \sum_{x,y,z,t \in A} a_{xy} a_{zt} x(\pi^{-1}(x)z) \otimes x(\pi^{-1}(x)t) \otimes y.$$

Using the G-invariance of J, we can remove the $\pi^{-1}(x)$ in the last expression and get

$$(\pi \otimes \pi \otimes \pi)((\Delta \otimes \mathrm{Id})(\bar{J})(\bar{J} \otimes 1)) = (\Delta \otimes \mathrm{Id})(J)(J \otimes 1). \tag{4-9}$$

Similarly,

$$(\pi \otimes \pi \otimes \pi)((\mathrm{Id} \otimes \Delta)(\bar{J})(1 \otimes \bar{J})) = (\mathrm{Id} \otimes \Delta)(J)(1 \otimes J). \tag{4-10}$$

But J is a twist, so the right hand sides of (4–9) and (4–10) are equal. Since π is bijective, this implies equation (2–14) for $\bar{J}$. $\qquad\square$

Now, given a quadruple (G, A, ρ, π) as above such that A is *abelian*, define $\tilde{G} := G \ltimes A^\vee$, $\tilde{A} := A \times A^\vee$, $\tilde{\rho} : \tilde{G} \to \mathrm{Aut}(\tilde{A})$ by $\tilde{\rho}(g) = \rho(g) \times \rho^*(g)^{-1}$, and $\tilde{\pi} : \tilde{G} \to \tilde{A}$ by $\tilde{\pi}(a^* g) = \pi(g) a^*$ for $a^* \in A^\vee$, $g \in G$. It is straightforward to check that $\tilde{\pi}$ is a bijective 1-cocycle. We call the quadruple $(\tilde{G}, \tilde{A}, \tilde{\rho}, \tilde{\pi})$ the *double* of (G, A, ρ, π).

Consider the element $J \in \mathbb{C}[\tilde{A}] \otimes \mathbb{C}[\tilde{A}]$ given by

$$J := |A|^{-1} \sum_{x \in A, y^* \in A^\vee} e^{(x,y^*)} x \otimes y^*,$$

where $(,)$ is the duality pairing between A and $A^\vee$. It is straightforward to check that J is a twist for $\mathbb{C}[\tilde{A}]$, and that it is G-invariant. This allows to construct the corresponding element

$$\bar{J} := |A|^{-1} \sum_{x \in A, y^* \in A^\vee} e^{(x,y^*)} \pi^{-1}(x) \otimes y^* \in \mathbb{C}[\tilde{G}] \otimes \mathbb{C}[\tilde{G}]. \tag{4-11}$$

PROPOSITION 4.6.3. $\bar{J}$ is a twist for $\mathbb{C}[\tilde{G}]$, and

$$\bar{J}^{-1} = |A|^{-1} \sum_{z \in A, t^* \in A^\vee} e^{-(z,t^*)} \pi^{-1}(T(z)) \otimes t^*, \qquad (4\text{-}12)$$

where $T : A \to A$ is a bijective map (not a homomorphism, in general) defined by

$$\pi^{-1}(x^{-1})\pi^{-1}(T(x)) = 1.$$

PROOF. Denote the right hand side of (4–12) by J'. We need to check that $J' = \bar{J}^{-1}$. It is enough to check it after evaluating any $a \in A$ on the second component of both sides. We have

$$(1 \otimes a)(\bar{J}) = |A|^{-1} \sum_{x,y^*} e^{(xa,y^*)} \pi^{-1}(x) = \pi^{-1}(a^{-1})(1 \otimes a)(J')$$

$$= |A|^{-1} \sum_{x,y^*} e^{-(xa^{-1},y^*)} \pi^{-1}(T(x)) = \pi^{-1}(T(a)).$$

This concludes the proof of the proposition. □

We can now prove:

THEOREM 4.6.4. Let $\bar{J}$ be as in (4–11). Then $\bar{J}$ is a minimal twist for $\mathbb{C}[\tilde{G}]$, and it gives rise to a minimal triangular semisimple Hopf algebra $(\mathbb{C}[\tilde{G}]^J, R^J)$, with universal R-matrix

$$R^J = |A|^{-2} \sum_{\substack{x,y \in A \\ x^*, y^* \in A^\vee}} e^{(x,y^*)-(y,x^*)} x^* \pi^{-1}(x) \otimes \pi^{-1}(T(y))y^*.$$

PROOF. Minimality of $\bar{J}$ follows from the fact that $\{x^* \pi^{-1}(x) \mid x^* \in A^\vee, x \in A\}$ and $\{\pi^{-1}(T(y))y^* \mid y \in A, y^* \in A^\vee\}$ are bases of $\mathbb{C}[\tilde{G}]$, and the fact that the matrix $c_{xx^*,yy^*} = e^{(x,y^*)-(y,x^*)}$ is invertible (because it is proportional to the matrix of Fourier transform on $A \times A^\vee$). □

REMARK 4.6.5. By Theorem 4.6.4, every bijective 1-cocycle $\pi : G \to A$ gives rise to a minimal triangular structure on $\mathbb{C}[G \ltimes A^\vee]$. So it remains to construct a supply of bijective 1-cocycles. This was done in [ESS]. The theory of bijective 1-cocycles was developed in [ESS], because it was found that they correspond to set-theoretical solutions of the quantum Yang-Baxter equation. In particular, many constructions of these 1-cocycles were found. We refer the reader to [ESS] for further details.

We now give two examples of nontrivial minimal triangular semisimple Hopf algebras. The first one has the least possible dimension; namely, dimension 16, and the second one has dimension 36.

EXAMPLE 4.6.6. Let $G := \mathbb{Z}_2 \times \mathbb{Z}_2$ with generators x, y, and $A := \mathbb{Z}_4$ with generator a. Define an action of G on A by letting x act trivially, and y act as an automorphism via $y \cdot a = a^{-1}$. Eli Aljadeff pointed out to us that the group $\tilde{G} := (\mathbb{Z}_2 \times \mathbb{Z}_2) \ltimes \mathbb{Z}_4$ has a 2-cocycle c with coefficients in $\mathbb{C}^*$, such that the twisted group algebra $\mathbb{C}[\tilde{G}]^c$ is simple. This implies that $\tilde{G}$ has a minimal twist (see Subsection 4.1). We now use Theorem 4.6.4 to explicitly construct our example.

Define a bijective 1-cocycle $\pi : G \to A$ as follows: $\pi(1) = 1$, $\pi(x) = a^2$, $\pi(y) = a$ and $\pi(xy) = a^3$. Then by Theorem 4.6.4, $\mathbf{C}[\tilde{G}]^J$ is a non-commutative and non-cocommutative minimal triangular semisimple Hopf algebra of dimension 16.

We remark that it follows from the classification of semisimple Hopf algebras of dimension 16 [Kash], that the Hopf algebra $\mathbf{C}[\tilde{G}]^J$ constructed above, appeared first in [Kash]. However, our triangular structure on this Hopf algebra is new. Indeed, Kashina's triangular structure on this Hopf algebra is not minimal, since it arises from a twist of a subgroup of $\tilde{G}$ which is isomorphic to $\mathbb{Z}_2 \times \mathbb{Z}_2$.

EXAMPLE 4.6.7. Let $G := S_3$ be the permutation group of three letters, and $A := \mathbb{Z}_2 \times \mathbb{Z}_3$. Define an action of G on A by $s(a,b) = (a, (-1)^{sign(s)}b)$ for $s \in G$, $a \in \mathbb{Z}_2$ and $b \in \mathbb{Z}_3$. Define a bijective 1-cocycle $\pi = (\pi_1, \pi_2) : G \to A$ as follows: $\pi_1(s) = 0$ if s is even and $\pi_1(s) = 1$ if s is odd, and $\pi_2(id) = 0$, $\pi_2((123)) = 1$, $\pi_2((132)) = 2$, $\pi_2((12)) = 2$, $\pi_2((13)) = 0$ and $\pi_2((23)) = 1$. Then by Theorem 4.6.4, $\mathbf{C}[\tilde{G}]^J$ is a noncommutative and noncocommutative minimal triangular semisimple Hopf algebra of dimension 36.

We now wish to determine the group-theoretical data corresponding, under the bijection of the classification given in Theorem 4.2.6, to the minimal triangular semisimple Hopf algebras constructed in Theorem 4.6.4.

Let $H := G \ltimes A^\vee$. Following Theorem 4.6.4, we can associate to this data the element

$$J := |A|^{-1} \sum_{g \in G, b \in A^\vee} e^{(\pi(g),b)} b \otimes g$$

(for convenience we use the opposite element to the one we used before). We proved that this element is a minimal twist for $\mathbb{C}[H]$, so $\mathbb{C}[H]^J$ is a minimal triangular semisimple Hopf algebra with Drinfeld element $u = 1$. Now we wish to find the irreducible projective representation V of H which corresponds to $\mathbb{C}[H]^J$ under the correspondence of Theorem 4.2.6.

Let $V := \mathrm{Fun}(A, \mathbb{C})$ be the space of $\mathbb{C}$-valued functions on A. It has a basis $\{\delta_a \mid a \in A\}$ of characteristic functions of points. Define a projective action ϕ of H on V by

$$\phi(b)\delta_a = e^{-(a,b)}\delta_a, \quad \phi(g)\delta_a = \delta_{g \cdot a + \pi(g)} \text{ and } \phi(bg) = \phi(b)\phi(g) \qquad (4\text{-}13)$$

for $g \in G$ and $b \in A^\vee$. It is straightforward to verify that this is indeed a projective representation.

PROPOSITION 4.6.8. *The representation V is irreducible, and corresponds to $\mathbb{C}[H]^J$ under the bijection of the classification given in Theorem 4.2.6.*

PROOF. It is enough to show that the H-algebras $(\mathbb{C}[H]_J)^*$ and $\mathrm{End}_{\mathbb{C}}(V)$ are isomorphic.

Let us compute the multiplication in the algebra $(\mathbb{C}[H]_J)^*$. We have

$$\Delta_J(bg) = |A|^{-1} \sum_{g' \in G, b' \in A^\vee} e^{(\pi(g'), b')} b(g \cdot b') g \otimes bgg'. \qquad (4\text{--}14)$$

Let $\{Y_{bg}\}$ be the dual basis of $(\mathbb{C}[H]_J)^*$ to the basis $\{bg\}$ of $\mathbb{C}[H]_J$. Let $*$ denote the multiplication law dual to the coproduct Δ_J. Then, dualizing equation (4–14), we have

$$Y_{b_2 g_2} * Y_{b_1 g_1} = e^{(\pi(g_1) - \pi(g_2), b_2 - b_1)} Y_{b_1 g_2} \qquad (4\text{--}15)$$

for $g_1, g_2 \in G$ and $b_1, b_2 \in A^\vee$ (here for convenience we write the operations in A and $A^\vee$ additively). Define $Z_{bg} := e^{(\pi(g), b)} Y_{bg}$ for $g \in G$ and $b \in A^\vee$. In the basis $\{Z_{bg}\}$ the multiplication law in $(\mathbb{C}[H]_J)^*$ is given by

$$Z_{b_2 g_2} * Z_{b_1 g_1} = e^{(\pi(g_1), b_2)} Z_{b_1 g_2}. \qquad (4\text{--}16)$$

Now let us introduce a left action of $(\mathbb{C}[H]_J)^*$ on V. Set

$$Z_{bg} \delta_a := e^{(a, b)} \delta_{\pi(g)}. \qquad (4\text{--}17)$$

It is straightforward to check, using (4–16), that (4–17) is indeed a left action. It is also straightforward to compute that this action is H-invariant. Thus, (4–17) defines an isomorphism $(\mathbb{C}[H]_J)^* \to \mathrm{End}_k(V)$ as H-algebras, which proves the proposition. $\qquad \square$

5. The Representation Theory of Cotriangular Semisimple and Cosemisimple Hopf Algebras

If (A, R) is a minimal triangular Hopf algebra then A and $A^{*\mathrm{op}}$ are isomorphic as Hopf algebras. But any nontrivial triangular semisimple and cosemisimple Hopf algebra A, over any algebraically closed field k, which is *not* minimal, gives rise to a new Hopf algebra A^*, which is also semisimple and cosemisimple. These are very interesting semisimple and cosemisimple Hopf algebras which arise from finite groups, and they are abundant by the constructions given in [EG2, EG4] (see Section 4). Generally, the dual Hopf algebra of a triangular Hopf algebra is called *cotriangular* in the literature.

In this section we explicitly describe the representation theory of cotriangular semisimple and cosemisimple Hopf algebras $A^* = (k[G]^J)^*$ studied in [EG3], in terms of representations of some associated groups. As a corollary we prove that Kaplansky's 6th conjecture [Kap] holds for A^*; that is, that the dimension of any irreducible representation of A^* divides the dimension of A.

5.1. The algebras associated with a twist. Let $A := k[H]$ be the group algebra of a finite group H whose order is relatively prime to the characteristic of k. Let $J \in A \otimes A$ be a *minimal* twist, and $A_1 := (A_J, \Delta_J, \varepsilon)$ be as in (3–1). Similarly, we define the coalgebra $A_2 := ({}_J A, {}_J \Delta, \varepsilon)$, where ${}_J A = k[H]$ as vector spaces, and

$$ {}_J \Delta : A \to A \otimes A, \quad {}_J \Delta(a) = J^{-1} \Delta(a) \tag{5-1} $$

for all $a \in A$. Note that since J is a twist, ${}_J \Delta$ is indeed coassociative. For $h \in H$, let $\delta_h : k[H] \to k$ be the linear map determined by $\delta_h(h) = 1$ and $\delta_h(h') = 0$ for $h \neq h' \in H$. Clearly the set $\{\delta_h \mid h \in H\}$ forms a linear basis of the dual algebras $(A_1)^*$ and $(A_2)^*$.

THEOREM 5.1.1. 1. $(A_1)^*$ and $(A_2)^*$ are H-algebras via

$$ \rho_1(h)\delta_y = \delta_{hy}, \quad \rho_2(h)\delta_y = \delta_{yh^{-1}} $$

respectively.

2. $(A_1)^* \cong (A_2)^{*op}$ as H-algebras (where H acts on $(A_2)^{*op}$ as it does on $(A_2)^*$).
3. *The algebras* $(A_1)^*$ *and* $(A_2)^*$ *are simple, and are isomorphic as H-modules to the regular representation* R_H *of* H.

PROOF. The proof of part 1 is straightforward.

The proof of part 3 follows from Proposition 4.2.3 and Part 2.

Let us prove part 2. It is straightforward to verify that $(S \otimes S)(J) = (Q \otimes Q)J_{21}^{-1}\Delta(Q)^{-1}$ where Q is as in (2–16) (see e.g., (2.17) in [Ma1, Section 2.3]). Hence the map $(A_2)^* \to (A_1)^{*op}$, $\delta_x \mapsto \delta_{S(x)Q^{-1}}$ determines an H-algebra isomorphism. $\square$

Since the algebras $(A_1)^*$, $(A_2)^*$ are simple, the actions of H on $(A_1)^*$, $(A_2)^*$ give rise to projective representations $H \to \mathrm{PGL}(|H|^{1/2}, k)$. We will denote these projective representations by V_1, V_2 (they can be thought of as the simple modules over $(A_1)^*$, $(A_2)^*$, with the induced projective action of H). Note that Part 2 of Theorem 5.1.1 implies that V_1, V_2 are dual to each other, hence that $[V_1] = -[V_2]$.

5.2. The main result. Let (A, R) be a triangular semisimple and cosemisimple Hopf algebra over k, with Drinfeld element $u = 1$, and let H, G and J be as before. Consider the dual Hopf algebra A^*. It has a basis of δ-functions δ_g. The first simple but important fact about the structure of A^* as an algebra is the following:

PROPOSITION 5.2.1. *Let Z be a double coset of H in G, and $(A^*)_Z := \oplus_{g \in Z} k\delta_g \subset A^*$. Then $(A^*)_Z$ is a subalgebra of A^*, and $A^* = \oplus_Z (A^*)_Z$ as algebras.*

PROOF. Straightforward. $\square$

Thus, to study the representation theory of A^*, it is sufficient to describe the representations of $(A^*)_Z$ for any Z.

Let Z be a double coset of H in G, and let $g \in Z$. Let $K_g := H \cap gHg^{-1}$, and define the embeddings $\theta_1, \theta_2 : K_g \to H$ given by $\theta_1(a) = g^{-1}ag$, $\theta_2(a) = a$. Denote by W_i the pullback of the projective H-representation V_i to K_g by means of θ_i, $i = 1, 2$.

Our main result is the following theorem, which is proved in the next subsection.

THEOREM 5.2.2. *Let W_1, W_2 be as above, and let $(\hat{K}_g, \tilde{\pi}_W)$ be any linearization of the projective representation $W := W_1 \otimes W_2$ of K_g. Let ζ be the kernel of the projection $\hat{K}_g \to K_g$, and $\chi : \zeta \to k^*$ be the character by which ζ acts in W. Then there exists a 1–1 correspondence between*

1. *isomorphism classes of irreducible representations of $(A^*)_Z$ and*
2. *isomorphism classes of irreducible representations of $\hat{K}_g$ with ζ acting by χ.*

Moreover, if a representation Y of $(A^*)_Z$ corresponds to a representation X of $\hat{K}_g$, then

$$\dim(Y) = \frac{|H|}{|K_g|} \dim(X).$$

As a corollary we get Kaplansky's 6th conjecture [Kap] for cotriangular semisimple and cosemisimple Hopf algebras.

COROLLARY 5.2.3. *The dimension of any irreducible representation of a cotriangular semisimple and cosemisimple Hopf algebra over k divides the dimension of the Hopf algebra.*

PROOF. Since $\dim(X)$ divides $|K_g|$ (see e.g., [CR, Proposition 11.44]), we have

$$\frac{|G|}{\frac{|H|}{|K_g|} \dim(X)} = \frac{|G|}{|H|} \frac{|K_g|}{\dim(X)}$$

and the result follows. □

In some cases the classification of representations of $(A^*)_Z$ is even simpler. Namely, let $\bar{g} \in \operatorname{Aut}(K_g)$ be given by $\bar{g}(a) = g^{-1}ag$. Then we have:

COROLLARY 5.2.4. *If the cohomology class $[W_1]$ is $\bar{g}$-invariant then irreducible representations of $(A^*)_Z$ correspond in a 1–1 manner to irreducible representations of K_g, and if Y corresponds to X, then*

$$\dim(Y) = \frac{|H|}{|K_g|} \dim(X).$$

PROOF. plus 3mu For any $\alpha \in \operatorname{Aut}K_g$ and $f \in \operatorname{Hom}((K_g)^n, k^*)$, let $\alpha \circ f \in \operatorname{Hom}((K_g)^n, k^*)$ be given by $(\alpha \circ f)(h_1, \ldots, h_n) = f(\alpha(h_1), \ldots, \alpha(h_n))$ (which determines the action of α on $H^i(K_g, k^*)$). Then it follows from the identity $[V_1] = -[V_2]$ that $[W_1] = -\bar{g} \circ [W_2]$. Thus, in our situation $[W] = 0$, hence W comes from a linear representation of K_g. Thus, we can set $\hat{K}_g = K_g$ in the theorem, and the result follows. □

EXAMPLE 5.2.5. Let $k := \mathbb{C}$. Let $p > 2$ be a prime number, and $H :=$ $(\mathbb{Z}/p\mathbb{Z})^2$ with the standard symplectic form $(\cdot,\cdot) : H \times H \to k^*$ given by $((x,y),(x',y')) = e^{2\pi i(xy'-yx')/p}$. Then the element $J := p^{-2}\sum_{a,b\in H}(a,b)a \otimes b$ is a minimal twist for $\mathbb{C}[H]$. Let $g \in \mathrm{GL}_2(\mathbb{Z}/p\mathbb{Z})$ be an automorphism of H, and G_0 be the cyclic group generated by g. Construct the group $G := G_0 \ltimes H$. It is easy to see that in this case, the double cosets are ordinary cosets $g^k H$, and $K_{g^k} = H$. Moreover, one can show either explicitly or using Proposition 3.4, that $[W_1]$ is a generator of $H^2(H,\mathbb{C}^*)$ which is isomorphic to $\mathbb{Z}/p\mathbb{Z}$. The element g^k acts on $[W_1]$ by multiplication by $\det(g^k)$. Therefore, by Corollary 5.2.4, the algebra $(A^*)_{g^k H}$ has p^2 1-dimensional representations (corresponding to linear representations of H) if $\det(g^k) = 1$.

However, if $\det(g^k) \neq 1$, then $[W]$ generates $H^2(H,\mathbb{C}^*)$. Thus, W comes from a linear representation of the Heisenberg group $\hat{H}$ (a central extension of H by $\mathbb{Z}/p\mathbb{Z}$) with some central character χ. Thus, $(A^*)_{g^k H}$ has one p-dimensional irreducible representation, corresponding to the unique irreducible representation of $\hat{H}$ with central character χ (which is W).

5.3. Proof of Theorem 5.2.2.

Let $Z \subset G$ be a double coset of H in G. For any $g \in Z$ define the linear map

$$F_g : (A^*)_Z \to (A_2)^* \otimes (A_1)^*, \quad \delta_y \mapsto \sum_{h,h'\in H: y=hgh'} \delta_h \otimes \delta_{h'}.$$

PROPOSITION 5.3.1. Let ρ_1, ρ_2 be as in Theorem 5.1.1. Then:

1. The map F_g is an injective homomorphism of algebras.
2. $F_{aga'}(\varphi) = (\rho_2(a) \otimes \rho_1(a')^{-1})F_g(\varphi)$ for any $a, a' \in H$, $\varphi \in (A^*)_Z$.

PROOF. 1. It is straightforward to verify that the map $(F_g)^* : A_2 \otimes A_1 \to A_Z$ is determined by $h \otimes h' \mapsto hgh'$, and that it is a surjective homomorphism of coalgebras. Hence the result follows.

  2. Straightforward. ☐

For any $a \in K_g$ define $\rho(a) \in \mathrm{Aut}((A_2)^* \otimes (A_1)^*)$ by $\rho(a) = \rho_2(a) \otimes \rho_1(a^g)$, where $a^g := g^{-1}ag$ and ρ_1, ρ_2 are as in Theorem 5.1.1. Then ρ is an action of K_g on $(A_2)^* \otimes (A_1)^*$.

PROPOSITION 5.3.2. Let $U_g := ((A_2)^*\otimes(A_1)^*)^{\rho(K_g)}$ be the algebra of invariants. Then $\mathrm{Im}(F_g) = U_g$, so $(A^*)_Z \cong U_g$ as algebras.

PROOF. It follows from Proposition 5.3.1 that $\mathrm{Im}(F_g) \subseteq U_g$, and $\mathrm{rk}(F_g) = \dim((A^*)_Z) = |H|^2/|K_g|$. On the other hand, by Theorem 5.1.1, $(A_1)^*, (A_2)^*$ are isomorphic to the regular representation R_H of H. Thus, $(A_1)^*$ and $(A_2)^*$ are isomorphic to $(|H|/|K_g|)R_{K_g}$ as representations of K_g, via $\rho_1(a)$ and $\rho_2(a^g)$. Thus,

$$(A_2)^* \otimes (A_1)^* \cong \frac{|H|^2}{|K_g|^2}(R_{K_g} \otimes R_{K_g}) \cong \frac{|H|^2}{|K_g|}R_{K_g}.$$

So U_g has dimension $|H|^2/|K_g|$, and the result follows. ☐

Now we are in a position to prove Theorem 5.2.2. Since $W_1 \otimes W_1^* \cong (A_1)^*$ and $W_2 \otimes W_2^* \cong (A_2)^*$, it follows from Theorem 5.1.1 that $W_1 \otimes W_2 \otimes W_1^* \otimes W_2^* \cong (|H|^2/|K_g|)R_{K_g}$ as $\hat{K}_g$ modules. Thus, if χ_W is the character of $W := W_1 \otimes W_2$ as a $\hat{K}_g$ module then

$$|\chi_W(x)|^2 = 0, \, x \notin \zeta \text{ and } |\chi_W(x)|^2 = |H|^2, \, x \in \zeta.$$

Therefore,

$$\chi_W(x) = 0, \, x \notin \zeta \text{ and } \chi_W(x) = |H| \cdot x_W, \, x \in \zeta,$$

where x_W is the root of unity by which x acts in W. Now, it is clear from the definition of U_g (see Proposition 5.3.2) that $U_g = \text{End}_{\hat{K}_g}(W)$. Thus if $W = \bigoplus_{M \in \text{Irr}(\hat{K}_g)} W(M) \otimes M$, where $W(M) := \text{Hom}_{\hat{K}_g}(M, W)$ is the multiplicity space, then

$$U_g = \bigoplus_{M : W(M) \neq 0} \text{End}_k(W(M)).$$

So $\{W(M) \mid W(M) \neq 0\}$ is the set of irreducible representations of U_g. Thus the following result implies the theorem:

LEMMA. 1. $W(M) \neq 0$ if and only if for all $x \in \zeta$, $x_{|M} = x_{|W}$.

2. If $W(M) \neq 0$ then $\dim(W(M)) = \dfrac{|H|}{|K_g|} \dim(M)$.

PROOF. The "only if" part of 1 is clear. For the "if" part compute $\dim(W(M))$ as the inner product (χ_W, χ_M). We have

$$(\chi_W, \chi_M) = \sum_{x \in \zeta} \frac{|H|}{|\hat{K}_g|} x_{|W} \cdot \dim(M) \cdot \bar{x}_{|M}.$$

If $x_{|M} = x_{|W}$ then

$$(\chi_W, \chi_M) = \sum_{x \in \zeta} \frac{|H|}{|\hat{K}_g|} \dim(M) = \frac{|H||\zeta|}{|\hat{K}_g|} \dim(M) = \frac{|H|}{|K_g|} \dim(M).$$

This proves part 2 as well.   □

This concludes the proof of the theorem.   □

6. The Pointed Case

In this section we consider finite-dimensional triangular pointed Hopf algebras over an algebraically closed field k of characteristic 0, and in particular describe the classification and explicit construction of minimal triangular pointed Hopf algebras, given in [G4]. Throughout the section, unless otherwise specified, the ground field k will be assumed to be algebraically closed with characteristic 0.

6.1. The antipode of triangular pointed Hopf algebras. In this subsection we prove that the fourth power of the antipode of any triangular pointed Hopf algebra (A, R) is the identity. Along the way we prove that the group algebra of the group of grouplike elements of A_R (which must be abelian) admits a minimal triangular structure and consequently that A has the structure of a biproduct [R1].

THEOREM 6.1.1. *Let (A, R) be a minimal triangular pointed Hopf algebra over k with Drinfeld element u, and set $K := k[\mathbf{G}(A)]$. Then there exists a projection of Hopf algebras $\pi : A \to K$, and consequently $A = B \times K$ is a biproduct where $B := \{x \in A \mid (I \otimes \pi)\Delta(x) = x \otimes 1\} \subseteq A$. Moreover, K admits a minimal triangular structure with Drinfeld element $u_K = u$.*

PROOF. Since $\mathbf{G}(A)$ is abelian, $K^* \cong K$ and $K \cong k[\mathbf{G}(A^{*\mathrm{cop}})]$ as Hopf algebras. Hence, $\dim(K^*) = \dim(k[\mathbf{G}(A^*)])$. Consider the series of Hopf algebra homomorphisms

$$K \overset{i}{\hookrightarrow} A^{\mathrm{cop}} \overset{(f_R)^{-1}}{\longrightarrow} A^* \overset{i^*}{\longrightarrow} K^*,$$

where i is the inclusion map. Since A^* is pointed it follows from the above remarks that $i^*_{|k[\mathbf{G}(A^*)]} : k[\mathbf{G}(A^*)] \to K^*$ is an isomorphism of Hopf algebras (see e.g., [Mon, 5.3.5]), and hence that $i^* \circ (f_R)^{-1} \circ i$ determines a minimal quasitriangular structure on K^*. This structure is in fact triangular since $(f_R)^{-1}$ determines a triangular structure on A^*. Clearly, $(i^* \circ (f_R)^{-1} \circ i)(u) = (u_{K^*})^{-1} = u_{K^*}$ is the Drinfeld element of K^*. Since K and K^* are isomorphic as Hopf algebras we conclude that K admits a minimal triangular structure with Drinfeld element $u_K = u$.

Finally, set $\varphi := i^* \circ (f_R)^{-1} \circ i$ and $\pi := \varphi^{-1} \circ i^* \circ (f_R)^{-1}$. Then $\pi : A \to K$ is onto, and moreover $\pi \circ i = \varphi^{-1} \circ i^* \circ (f_R)^{-1} \circ i = \varphi^{-1} \circ \varphi = id_K$. Hence π is a projection of Hopf algebras and by [R1], $A = B \times K$ is a biproduct where $B := \{x \in A \mid (I \otimes \pi)\Delta(x) = x \otimes 1\}$ as desired. This concludes the proof of the theorem. $\qquad\square$

THEOREM 6.1.2. *Let (A, R) be any triangular pointed Hopf algebra with antipode S and Drinfeld element u over any field k of characteristic 0. Then $S^4 = \mathrm{Id}$. If in addition A_m is not semisimple and A is finite-dimensional then $\dim(A)$ is divisible by 4.*

PROOF. We may assume that k is algebraically closed. By (2–6), $S^2(a) = uau^{-1}$ for all $a \in A$. Let $K := k[\mathbf{G}(A_m)]$. Since $u \in A_m$, and by Theorem 6.1.1, $u = u_K$ and $(u_K)^2 = 1$, we have that $S^4 = \mathrm{Id}$.

In order to prove the second claim, we may assume that (A, R) is minimal (since by [NZ], $\dim(A_m)$ divides $\dim(A)$). Since A is not semisimple it follows from [LR1] that $S^2 \neq \mathrm{Id}$, and hence that $u \neq 1$. In particular, $|\mathbf{G}(A)|$ is even. Now, let B be as in Theorem 6.1.1. Since $S^2(B) = B$, B has a basis $\{a_i, b_j \mid S^2(a_i) = a_i, S^2(b_j) = -b_j, 1 \leq i \leq n, 1 \leq j \leq m\}$. Hence by Theorem

6.1.1,

$$\{a_i g, b_j g \mid g \in \mathbf{G}(A), 1 \leq i \leq n, 1 \leq j \leq m\}$$

is a basis of A. Since by [R3], $\operatorname{tr}(S^2) = 0$, we have $0 = \operatorname{tr}(S^2) = |\mathbf{G}(A)|(n - m)$, which implies that $n = m$, and hence that $\dim(B)$ is even as well. $\qquad\square$

In fact, the first part of Theorem 6.1.2 can be generalized.

THEOREM 6.1.3. *Let (A, R) be a finite-dimensional quasitriangular Hopf algebra with antipode S over any field k of characteristic 0, and suppose that the Drinfeld element u of A acts as a scalar in any irreducible representation of A (e.g., when A^* is pointed). Then $u = S(u)$ and in particular $S^4 = \mathrm{Id}$.*

PROOF. We may assume that k is algebraically closed. In any irreducible representation V of A, $\operatorname{tr}_{|V}(u) = \operatorname{tr}_{|V}(S(u))$ (see Subsection 2.1). Since $S(u)$ also acts as a scalar in V (the dual of $S(u)_{|V}$ equals $u_{|V^*}$) it follows that $u = S(u)$ in any irreducible representation of A. Therefore, there exists a basis of A in which the operators of left multiplication by u and $S(u)$ are represented by upper triangular matrices with the same main diagonal. Hence the special grouplike element $uS(u)^{-1}$ is unipotent. Since it has a finite order we conclude that $uS(u)^{-1} = 1$, and hence that $S^4 = \mathrm{Id}$. $\qquad\square$

REMARK 6.1.4. If (A, R) is a minimal triangular pointed Hopf algebra then all its irreducible representations are 1-dimensional. Hence Theorem 6.1.3 is applicable, and the first part of Theorem 6.1.2 follows.

EXAMPLE 6.1.5. Let A be Sweedler's 4-dimensional Hopf algebra [Sw]. It is generated as an algebra by a grouplike element g and a $1 : g$ skew primitive element x satisfying the relations $g^2 = 1$, $x^2 = 0$ and $gx = -xg$. It is known that A admits minimal triangular structures all of which with g as the Drinfeld element [R2]. In this example, $K = k[\langle g \rangle]$ and $B = sp\{1, x\}$. Note that g is central in K but is not central in A, so $(S_{|K})^2 = \mathrm{Id}$ but $S^2 \neq \mathrm{Id}$ in A. However, $S^4 = \mathrm{Id}$.

6.2. Construction of minimal triangular pointed Hopf algebras. In this section we give a method for the construction of minimal triangular pointed Hopf algebras which are *not* necessarily semisimple.

Let G be a finite abelian group, and $F : G \times G \to k^*$ be a non-degenerate skew symmetric bilinear form on G. That is, $F(xy, z) = F(x, z)F(y, z)$, $F(x, yz) = F(x, y)F(x, z)$, $F(1, x) = F(x, 1) = 1$, $F(x, y) = F(y, x)^{-1}$ for all $x, y, z \in G$, and the map $f : G \to G^\vee$ defined by $\langle f(x), y \rangle = F(x, y)$ for all $x, y \in G$ is an isomorphism. Let $U_F : G \to \{-1, 1\}$ be defined by $U_F(g) = F(g, g)$. Then U_F is a homomorphism of groups. Denote $U_F^{-1}(-1)$ by I_F.

DEFINITION 6.2.1. Let k be an algebraically closed field of characteristic zero. A datum $\mathcal{D} = (G, F, n)$ is a triple where G is a finite abelian group, $F : G \times G \to k^*$ is a non-degenerate skew symmetric bilinear form on G, and n is a non-negative integer function $I_F \to \mathbb{Z}^+$, $g \mapsto n_g$.

REMARK 6.2.2. (1) The map $f : k[G] \to k[G^\vee]$ determined by $\langle f(g), h \rangle = F(g, h)$ for all $g, h \in G$ determines a minimal triangular structure on $k[G^\vee]$.
(2) If I_F is not empty then G has an even order.

To each datum $\mathcal{D}$ we associate a Hopf algebra $H(\mathcal{D})$ in the following way. For each $g \in I_F$, let V_g be a vector space of dimension n_g, and let $\mathcal{B} = \bigoplus_{g \in I_F} V_g$. Then $H(\mathcal{D})$ is generated as an algebra by $G \cup \mathcal{B}$ with the following additional relations (to those of the group G and the vector spaces V_g's):

$$xy = F(h, g)yx \quad \text{and} \quad xa = F(a, g)ax \qquad (6\text{–}1)$$

for all $g, h \in I_F$, $x \in V_g$, $y \in V_h$ and $a \in G$.

The coalgebra structure of $H(\mathcal{D})$ is determined by letting $a \in G$ be a grouplike element and $x \in V_g$ be a $1 : g$ skew primitive element for all $g \in I_F$. In particular, $\varepsilon(a) = 1$ and $\varepsilon(x) = 0$ for all $a \in G$ and $x \in V_g$.

In the special case where $G = \mathbb{Z}_2 = \{1, g\}$, $F(g, g) = -1$ and $n := n_g$, the associated Hopf algebra will be denoted by $H(n)$. Clearly, $H(0) = k\mathbb{Z}_2$, $H(1)$ is Sweedler's 4-dimensional Hopf algebra, and $H(2)$ is the 8-dimensional Hopf algebra studied in [G1, Section 2.2] in connection with KRH invariants of knots and 3-manifolds. We remark that the Hopf algebras $H(n)$ are studied in [PO1, PO2] where they are denoted by $E(n)$.

For a finite-dimensional vector space V we let $\bigwedge V$ denote the exterior algebra of V. Set $B := \bigotimes_{g \in I_F} \bigwedge V_g$.

PROPOSITION 6.2.3. 1. *The Hopf algebra $H(\mathcal{D})$ is pointed and $\mathbf{G}(H(\mathcal{D})) = G$.*
2. *$H(\mathcal{D}) = B \times k[G]$ is a biproduct.*
3. *$H(\mathcal{D})_1 = k[G] \bigoplus (k[G]\mathcal{B})$, and $P_{a,b}(H(\mathcal{D})) = sp\{a-b\} \bigoplus aV_{a^{-1}b}$ for all $a, b \in G$ (here we agree that $V_{a^{-1}b} = 0$ if $a^{-1}b \notin I_F$).*

PROOF. Part 1 follows since (by definition) $H(\mathcal{D})$ is generated as an algebra by grouplike elements and skew primitive elements. Now, it is straightforward to verify that the map $\pi : H(\mathcal{D}) \to k[G]$ determined by $\pi(a) = a$ and $\pi(x) = 0$ for all $a \in G$ and $x \in \mathcal{B}$, is a projection of Hopf algebras. Since $B = \{x \in H(\mathcal{D}) \mid (I \otimes \pi)\Delta(x) = x \otimes 1\}$, Part 2 follows from [R1]. Finally, by Part 2, B is a braided graded Hopf algebra in the Yetter-Drinfeld category ${}^{k[G]}_{k[G]}\mathcal{YD}$ (see e.g., [AS]) with respect to the grading where the elements of $\mathcal{B}$ are homogeneous of degree 1. Write $B = \bigoplus_{n \geq 0} B(n)$, where $B(n)$ denotes the homogeneous component of degree n. Then, $B(0) = k1 = B_1$ (since $B \cong H(\mathcal{D})/H(\mathcal{D})k[G]^+$ as coalgebras, it is connected). Furthermore, by similar arguments used in the proof of [AS, Lemma 3.4], $P(B) = B(1) = \mathcal{B}$. But then by [AS, Lemma 2.5], $H(\mathcal{D})$ is coradically graded (where the *n*th component $H(\mathcal{D})(n)$ is just $B(n) \times k[G]$) which means by definition that $H(\mathcal{D})_1 = H(\mathcal{D})(0) \bigoplus H(\mathcal{D})(1) = k[G] \bigoplus (k[G]\mathcal{B})$ as desired. The second statement of Part 3 follows now, using (1), by counting dimensions. $\square$

In the following we determine *all* the minimal triangular structures on $H(\mathcal{D})$. Let $f : k[G] \to k[G^\vee]$ be the isomorphism from Remark 6.2.2(1), and set $I'_F :=$ $\{g \in I_F \mid n_g \neq 0\}$. Let Φ be the set of all isomorphisms $\varphi : G^\vee \to G$ satisfying $\varphi^*(\alpha) = \varphi(\alpha^{-1})$ for all $\alpha \in G^\vee$, and $(\varphi \circ f)(g) = g$ for all $g \in I'_F$ (here we identify G with $G^{\vee\vee}$).

Extend any $\alpha \in G^\vee$ to an algebra homomorphism $H(\mathcal{D}) \to k$ by setting $\alpha(z) = 0$ for all $z \in \mathcal{B}$. Extend any $x \in V_g^*$ to $P_x \in H(\mathcal{D})^*$ by setting $\langle P_x, ay \rangle = 0$ for all $a \in G$ and $y \in \bigotimes_{g \in I_F} \bigwedge V_g$ of degree different from 1, and $< P_x, ay > = \delta_{g,h} \langle x, y \rangle$ for all $a \in G$ and $y \in V_h$. We shall identify the vector spaces V_g^* and $\{P_x \mid x \in V_g^*\}$ via the map $x \mapsto P_x$.

For $g \in I'_F$, let $S_g(k)$ be the set of all isomorphisms $M_g : V_g^* \to V_{g^{-1}}$. Let $S(k) \subseteq \times_{g \in I'_F} S_g(k)$ be the set of all tuples (M_g) satisfying $M_g^* = M_{g^{-1}}$ for all $g \in I'_F$.

THEOREM 6.2.4. (1) *For each $T := (\varphi, (M_g)) \in \Phi \times S(k)$, there exists a unique Hopf algebra isomorphism $f_T : H(\mathcal{D})^{*\mathrm{cop}} \to H(\mathcal{D})$ determined by $\alpha \mapsto \varphi(\alpha)$ and $P_x \mapsto M_g(x)$ for $\alpha \in G^\vee$ and $x \in V_g^*$.*
(2) *There is a one to one correspondence between $\Phi \times S(k)$ and the set of minimal triangular structures on $H(\mathcal{D})$ given by $T \mapsto f_T$.*

PROOF. We first show that f_T is a well defined isomorphism of Hopf algebras. Using Proposition 6.2.3(2), it is straightforward to verify that

$$\Delta(P_x) = \varepsilon \otimes P_x + P_x \otimes f(g^{-1}),$$
$$P_x \alpha = \langle \alpha, g \rangle \alpha P_x, \text{ and}$$
$$P_x P_y = F(h, g) P_y P_x$$

for all $\alpha \in G^\vee$, $g, h \in I_F$, $x \in V_g^*$ and $y \in V_h^*$. Let $\mathcal{B}^* := \{P_x \mid x \in V_g^*, g \in I_F\}$, and H be the sub Hopf algebra of $H(\mathcal{D})^{*\mathrm{cop}}$ generated as an algebra by $G^\vee \cup \mathcal{B}^*$. Then, using (4) and our assumptions on T, it is straightforward to verify that the map $f_T^{-1} : H(\mathcal{D}) \to H$ determined by $a \mapsto \varphi^{-1}(a)$ and $z \mapsto M_g^{-1}(z)$ for $a \in G$ and $z \in V_{g^{-1}}$, is a surjective homomorphism of Hopf algebras. Let us verify for instance that $f_T^{-1}(za) = F(a, g) f_T^{-1}(az)$. Indeed, this is equivalent to $\langle \varphi^{-1}(a), g \rangle = \langle f(a), g \rangle$ which in turn holds by our assumptions on φ. Now, using Proposition 4.3(3), it is straightforward to verify that f_T^{-1} is injective on $P_{a,b}(H(\mathcal{D}))$ for all $a, b \in G$. Since $H(\mathcal{D})$ is pointed, f_T^{-1} is also injective (see e.g., [Mon, Corollary 5.4.7]). This implies that $H = H(\mathcal{D})^{*\mathrm{cop}}$, and that $f_T : H(\mathcal{D})^{*\mathrm{cop}} \to H(\mathcal{D})$ is an isomorphism of Hopf algebras as desired. Note that in particular, $G^\vee = G(H(\mathcal{D})^*)$.

The fact that f_T satisfies (2-3) follows from a straightforward computation (using (6-1)) since it is enough to verify it for algebra generators $p \in G^\vee \cup \mathcal{B}^*$ of $H(\mathcal{D})^{*\mathrm{cop}}$, and $a \in G \cup \mathcal{B}$ of $H(\mathcal{D})$.

We have to show that f_T satisfies (2.3.1). Indeed, it is straightforward to verify that $f_T^* : H(\mathcal{D})^{*\mathrm{op}} \to H(\mathcal{D})$ is determined by $\alpha \mapsto \varphi(\alpha^{-1})$ and $P_x \mapsto g M_g(x)$

for $\alpha \in G^\vee$ and $x \in V_g^*$. Hence, $f_T^* = f_T \circ S$, where S is the antipode of $H(\mathcal{D})^*$, as desired.

We now have to show that *any* minimal triangular structure on $H(\mathcal{D})$ comes from f_T for some T. Indeed, let $\mathbf{f} : H(\mathcal{D})^{*\mathrm{cop}} \to H(\mathcal{D})$ be any Hopf isomorphism. Then $\mathbf{f}$ must map $G^\vee$ onto G, $\{f(g^{-1}) \mid g \in I_F'\}$ onto I_F', and $P_{f(g^{-1}),\varepsilon}(H(\mathcal{D})^{*\mathrm{cop}})$ bijectively onto $P_{1,\varphi(f(g^{-1}))}(H(\mathcal{D}))$. Therefore there exists an invertible operator $M_g : V_g^* \to V_{\varphi(f(g^{-1}))}$ such that $\mathbf{f}$ is determined by $\alpha \mapsto \varphi(\alpha)$ and $P_x \mapsto M_g(x)$. Suppose $\mathbf{f}$ satisfies (2–3). Then letting $p = P_x$ and $a \in G$ in (2–3) yields $af(P_x) = F(a,g)\mathbf{f}(P_x)a$ for all $a \in G$. But by (6–1), this is equivalent to $(\varphi \circ f)(g) = g$ for all $g \in I_F'$. Since by Theorem 6.1.1, $\varphi : k[G^\vee] \to k[G]$ determines a minimal triangular structure on $k[G]$ it follows that $\varphi \in \Phi$. Since $\mathbf{f} : H(\mathcal{D})^{*\mathrm{cop}} \to H(\mathcal{D})$ satisfies (2.3.1), $(M_g) \in S(k)$, and hence $\mathbf{f}$ is of the form f_T for some T as desired. $\square$

For a triangular structure on $H(\mathcal{D})$ corresponding to the map f_T, we let R_T denote the corresponding R-matrix.

REMARK 6.2.5. Note that if $n_{g^{-1}} \neq n_g$ for some $g \in I_F'$, then $S(k)$ is empty and $H(\mathcal{D})$ does *not* have a minimal triangular structure.

6.3. The classification of minimal triangular pointed Hopf algebras.

In this subsection we use Theorems 6.1.1, 6.1.2 and [AEG, Theorem 6.1] to classify minimal triangular pointed Hopf algebras. Namely, we prove:

THEOREM 6.3.1. *Let (A, R) be a minimal triangular pointed Hopf algebra over an algebraically closed field k of characteristic 0. There exist a datum $\mathcal{D}$ and $T \in \Phi \times S(k)$ such that $(A, R) \cong (H(\mathcal{D}), R_T)$ as triangular Hopf algebras.*

Before we prove Theorem 6.3.1 we need to fix some notation and prove a few lemmas.

In what follows, (A, R) will always be a minimal triangular pointed Hopf algebra over k, $G := \mathbf{G}(A)$ and $K := k[\mathbf{G}(A)]$. For any $g \in G$, $P_{1,g}(A)$ is a $\langle g \rangle$-module under conjugation, and $sp\{1 - g\}$ is a submodule of $P_{1,g}(A)$. Let $V_g \subset P_{1,g}(A)$ be its complement, and set $n_g := \dim(V_g)$.

By Theorem 6.1.1, $A = B \times K$ where $B = \{x \in A \mid (I \otimes \pi)\Delta(x) = x \otimes 1\} \subseteq A$ is a left coideal subalgebra of A (equivalently, B is an object in the Yetter-Drinfeld category $^{k[G]}_{k[G]}\mathcal{YD}$). Note that $B \cap K = k1$. Let $\rho : B \to K \otimes B$ be the associated comodule structure and write $\rho(x) = \sum x^1 \otimes x^2$, $x \in B$. By [R1], $B \cong A/AK^+$ as coalgebras, hence B is a connected pointed coalgebra. Let $P(B) := \{x \in B \mid \Delta_B(x) = x \otimes 1 + 1 \otimes x\}$ be the space of primitive elements of B.

LEMMA 6.3.2. *For any $g \in G$, $V_g = \{x \in P(B) \mid \rho(x) = g \otimes x\}$.*

PROOF. Let $x \in V_g$. Since g acts on V_g by conjugation we may assume by [G1, Lemma 0.2], that $gx = \omega xg$ for some $1 \neq \omega \in k$. Since $\pi(x)$ and $\pi(g) = g$ commute we must have that $\pi(x) = 0$. But then $(I \otimes \pi)\Delta(x) = x \otimes 1$ and hence

$x \in B$. Since $\Delta(x) = \sum x_1 \times x_2^1 \otimes x_2^2 \times 1$, applying the maps $\varepsilon \otimes I \otimes I \otimes \varepsilon$ and $I \otimes \varepsilon \otimes I \otimes \varepsilon$ to both sides of the equation $\sum x_1 \times x_2^1 \otimes x_2^2 \times 1 = x \times 1 \otimes 1 \times 1 + 1 \times g \otimes x \times 1$, yields that $x \in P(B)$ and $\rho(x) = g \otimes x$ as desired.

Suppose that $x \in P(B)$ satisfies $\rho(x) = g \otimes x$. Since $\Delta(x) = x \otimes 1 + \rho(x)$, it follows that $x \in V_g$ as desired. $\qquad\square$

LEMMA 6.3.3. *For every $x \in V_g$, $x^2 = 0$ and $gx = -xg$.*

PROOF. Suppose $V_g \neq 0$ and let $0 \neq x \in V_g$. Then $S^2(x) = g^{-1}xg$, $g^{-1}xg \neq x$ by [G1, Lemma 0.2], and $g^{-1}xg \in V_g$. Since by Theorem 6.1.2, $S^4 = \mathrm{Id}$ it follows that g^2 and x commute, and hence that $gx = -xg$ for every $x \in V_g$.

Second we wish to show that $x^2 = 0$. By Lemma 6.3.2, $x \in B$ and hence $x^2 \in B$ (B is a subalgebra of A). Since $\Delta(x^2) = x^2 \otimes 1 + g^2 \otimes x^2$, and x^2 and g^2 commute, it follows from [G1, Lemma 0.2] that $x^2 = \alpha(1 - g^2) \in K$ for some $\alpha \in k$. We thus conclude that $x^2 = 0$, as desired. $\qquad\square$

Recall that the map $f_R : A^{*\mathrm{cop}} \to A$ is an isomorphism of Hopf algebras, and let $F : G \times G \to k^*$ be the associated non-degenerate skew symmetric bilinear form on G defined by $F(g,h) := \langle f_R^{-1}(g), h \rangle$ for every $g, h \in G$.

LEMMA 6.3.4. *For any $x \in V_g$ and $y \in V_h$, $xy = F(h,g)yx$.*

PROOF. If either $V_g = 0$ or $V_h = 0$, there is nothing to prove. Suppose $V_g, V_h \neq 0$, and let $0 \neq x \in V_g$ and $0 \neq y \in V_h$. Set $P := f_R^{-1}(x)$. Then $P \in P_{f_R^{-1}(g),\varepsilon}(A^{*\mathrm{cop}})$. Substituting $p := P$ and $a := y$ in equation (2–3) yields $yx - F(g,h)xy = \langle P, y \rangle(1 - gh)$. Since $\langle P, y \rangle(1 - gh) \in B \cap K$, it is equal to 0, and hence $yx = F(g,h)xy$. $\qquad\square$

LEMMA 6.3.5. *For any $a \in G$ and $x \in V_g$, $xa = F(a,g)ax$.*

PROOF. Set $P := f_R^{-1}(x)$. Then the result follows by letting $p := P$ and $a \in G$ in (2–3), and noting that $P \in P_{f_R^{-1}(g),\varepsilon}(A^{*\mathrm{cop}})$. $\qquad\square$

We can now prove Theorem 6.3.1.

PROOF OF THEOREM 6.3.1. Let $n : I_F \to \mathbb{Z}^+$ be the nonnegative integer function defined by $n(g) = n_g$, and let $\mathcal{D} := (G, F, n)$. By [AEG, Theorem 6.1], A is generated as an algebra by $G \cup (\bigoplus_{g \in I_F} V_g)$. By Lemmas 6.3.3-6.3.5, relations (6–1) are satisfied. Therefore there exists a surjection of Hopf algebras $\varphi : H(\mathcal{D}) \to A$. Using Proposition 4.3(3), it is straightforward to verify that φ is injective on $P_{a,b}(H(\mathcal{D}))$ for all $a, b \in G$. Since $H(\mathcal{D})$ is pointed, φ is also injective (see e.g., [Mon, Corollary 5.4.7]). Hence φ is an isomorphism of Hopf algebras. The rest of the theorem follows now from Theorem 6.2.4. $\qquad\square$

REMARK 6.3.6. Theorem 6.1 in [AEG] states that a finite-dimensional cotriangular pointed Hopf algebra is generated by its grouplike and skewprimitive elements. This confirms the conjecture that this is the case for *any* finite-dimensional pointed Hopf algebra over $\mathbb{C}$ [AS2], in the cotriangular case. The proof uses a categorical point of view (or, alternatively, the Lifting method; see [AS1, AS2]).

7. Triangular Hopf Algebras with the Chevalley Property

As we said in the introduction, semisimple cosemisimple triangular Hopf algebras and minimal triangular pointed Hopf algebras share in common the Chevalley property. In this section we describe the classification of finite-dimensional triangular Hopf algebras with the Chevalley property, given in [AEG].

7.1. Triangular Hopf algebras with Drinfeld element of order ≤ 2. We start by classifying triangular Hopf algebras with R-matrix of rank ≤ 2. We show that such a Hopf algebra is a suitable modification of a cocommutative Hopf superalgebra (i.e., the group algebra of a supergroup). On the other hand, by Corollary 2.2.3.5, a finite supergroup is a semidirect product of a finite group with an odd vector space on which this group acts.

7.1.1. The correspondence between Hopf algebras and superalgebras. We start with a correspondence theorem between Hopf algebras and Hopf superalgebras.

THEOREM 7.1.1.1. *There is a one to one correspondence between*

1. *isomorphism classes of pairs (A, u) where A is an ordinary Hopf algebra, and u is a grouplike element in A such that $u^2 = 1$, and*
2. *isomorphism classes of pairs $(\mathcal{A}, g)$ where $\mathcal{A}$ is a Hopf superalgebra, and g is a grouplike element in $\mathcal{A}$ such that $g^2 = 1$ and $gxg^{-1} = (-1)^{p(x)}x$ (i.e., g acts on x by its parity),*

such that the tensor categories of representations of A and $\mathcal{A}$ are equivalent.

PROOF. Let (A, u) be an ordinary Hopf algebra with comultiplication Δ, counit ε, antipode S, and a grouplike element u such that $u^2 = 1$. Let $\mathcal{A} = A$ regarded as a superalgebra, where the $\mathbb{Z}_2$-grading is given by the adjoint action of u. For $a \in A$, let us define Δ_0, Δ_1 by writing $\Delta(a) = \Delta_0(a) + \Delta_1(a)$, where $\Delta_0(a) \in A \otimes A_0$ and $\Delta_1(a) \in A \otimes A_1$. Define a map $\tilde{\Delta} : \mathcal{A} \to \mathcal{A} \otimes \mathcal{A}$ by $\tilde{\Delta}(a) := \Delta_0(a) - (-1)^{p(a)}(u \otimes 1)\Delta_1(a)$. Define $\tilde{S}(a) := u^{p(a)}S(a)$, $a \in A$. Then it is straightforward to verify that $(\mathcal{A}, \tilde{\Delta}, \varepsilon, \tilde{S})$ is a Hopf superalgebra.

The element u remains grouplike in the new Hopf superalgebra, and acts by parity, so we can set $g := u$.

Conversely, suppose that $(\mathcal{A}, g)$ is a pair where $\mathcal{A}$ is a Hopf superalgebra with comultiplication $\tilde{\Delta}$, counit ε, antipode $\tilde{S}$, and a grouplike element g, with $g^2 = 1$, acting by parity. For $a \in \mathcal{A}$, let us define $\tilde{\Delta}_0, \tilde{\Delta}_1$ by writing $\tilde{\Delta}(a) = \tilde{\Delta}_0(a) + \tilde{\Delta}_1(a)$, where $\tilde{\Delta}_0(a) \in \mathcal{A} \otimes \mathcal{A}_0$ and $\tilde{\Delta}_1(a) \in \mathcal{A} \otimes \mathcal{A}_1$. Let $A = \mathcal{A}$ as algebras, and define a map $\Delta : A \to A \otimes A$ by $\Delta(a) := \tilde{\Delta}_0(a) - (-1)^{p(a)}(g \otimes 1)\tilde{\Delta}_1(a)$. Define $S(a) := g^{p(a)}\tilde{S}(a)$, $a \in A$. Then it is straightforward to verify that $(A, \Delta, \varepsilon, S)$ is an ordinary Hopf algebra, and we can set $u := g$.

It is obvious that the two assignments constructed above are inverse to each other. The equivalence of tensor categories is straightforward to verify. The theorem is proved. □

Theorem 7.1.1.1 implies the following. Let $\mathcal{A}$ be *any* Hopf superalgebra, and $\mathbb{C}[\mathbb{Z}_2] \ltimes \mathcal{A}$ be the semidirect product, where the generator g of $\mathbb{Z}_2$ acts on $\mathcal{A}$ by $gxg^{-1} = (-1)^{p(x)}x$. Then we can define an ordinary Hopf algebra $\overline{A}$, which is the one corresponding to $(\mathbb{C}[\mathbb{Z}_2] \ltimes \mathcal{A}, g)$ under the correspondence of Theorem 7.1.1.1.

The constructions of this subsection have the following explanation in terms of Radford's biproduct construction [R1]. Namely $\mathcal{A}$ is a Hopf algebra in the Yetter-Drinfeld category of $\mathbb{C}[\mathbb{Z}_2]$, so Radford's biproduct construction yields a Hopf algebra structure on $\mathbb{C}[\mathbb{Z}_2] \otimes \mathcal{A}$, and it is straightforward to see that this Hopf algebra is exactly $\overline{A}$. Moreover, it is clear that for any pair (A, u) as in Theorem 7.1.1.1, gu is central in $\overline{A}$ and $A = \overline{A}/(gu - 1)$.

7.1.2. Correspondence of twists. Let us say that a twist J for a Hopf algebra A with an involutive grouplike element g is *even* if it is invariant under $\mathrm{Ad}(g)$.

PROPOSITION 7.1.2.1. *Let (A, g) be a pair as in Theorem 7.1.1.1, and let A be the associated ordinary Hopf algebra. Let $\mathcal{J} \in \mathcal{A} \otimes \mathcal{A}$ be an even element. Write $\mathcal{J} = \mathcal{J}_0 + \mathcal{J}_1$, where $\mathcal{J}_0 \in \mathcal{A}_0 \otimes \mathcal{A}_0$ and $\mathcal{J}_1 \in \mathcal{A}_1 \otimes \mathcal{A}_1$. Define $J := \mathcal{J}_0 - (g \otimes 1)\mathcal{J}_1$. Then J is an even twist for A if and only if $\mathcal{J}$ is a twist for $\mathcal{A}$. Moreover, $\mathcal{A}^{\mathcal{J}}$ corresponds to A^J under the correspondence in Theorem 7.1.1.1. Thus, there is a one to one correspondence between even twists for A and twists for $\mathcal{A}$, given by $J \to \mathcal{J}$.*

PROOF. Straightforward. □

7.1.3. The Correspondence between triangular Hopf algebras and superalgebras. Let us now return to our main subject, which is triangular Hopf algebras and superalgebras. For triangular Hopf algebras whose Drinfeld element u is involutive, we will make the natural choice of the element u in Theorem 7.1.1.1, namely define it to be the Drinfeld element of A.

THEOREM 7.1.3.1. *The correspondence of Theorem 7.1.1.1 extends to a one to one correspondence between*

1. *isomorphism classes of ordinary triangular Hopf algebras A with Drinfeld element u such that $u^2 = 1$, and*
2. *isomorphism classes of pairs $(\mathcal{A}, g)$ where $\mathcal{A}$ is a triangular Hopf superalgebra with Drinfeld element 1 and g is an element of $\mathbf{G}(\mathcal{A})$ such that $g^2 = 1$ and $gxg^{-1} = (-1)^{p(x)}x$.*

PROOF. Let (A, R) be a triangular Hopf algebra with $u^2 = 1$. Since $(S \otimes S)(R) = R$ and $S^2 = \mathrm{Ad}(u)$ [Dr2], $u \otimes u$ and R commute. Hence we can write $R = R_0 + R_1$, where $R_0 \in A_0 \otimes A_0$ and $R_1 \in A_1 \otimes A_1$. Let $\mathcal{R} := (R_0 + (1 \otimes u)R_1)R_u$. Then $\mathcal{R}$

is even. Indeed, since

$$R_0 = 1/2(R + (u \otimes 1)R(u \otimes 1)) \text{ and}$$
$$R_1 = 1/2(R - (u \otimes 1)R(u \otimes 1)),$$

$u \otimes u$ and $\mathcal{R}$ commute.

It is now straightforward to show that $(\mathcal{A}, \mathcal{R})$ is triangular with Drinfeld element 1. Let us show for instance that $\mathcal{R}^{-1} = \mathcal{R}_{21}$. Let us use the notation $a * b, X^{21}$ for multiplication and opposition in the tensor square of a superalgebra, and the notation ab, X^{op} for usual algebras. Then,

$$\mathcal{R} * \mathcal{R}_{21} = (R_0 + (1 \otimes u)R_1)R_u * (R_0^{\mathrm{op}} - (u \otimes 1)R_1^{\mathrm{op}})R_u.$$

Since, $R_u R_0 = R_0 R_u$, $R_u R_1 = -(u \otimes u)R_1 R_u$, we get that the RHS equals

$$(R_0 + (1 \otimes u)R_1) * (R_0^{\mathrm{op}} + (1 \otimes u)R_1^{\mathrm{op}}) = R_0 R_0^{\mathrm{op}} + R_1 R_1^{\mathrm{op}} + (1 \otimes u)(R_1 R_0^{\mathrm{op}} + R_0 R_1^{\mathrm{op}}).$$

But, $R_0 R_0^{\mathrm{op}} + R_1 R_1^{\mathrm{op}} = 1$ and $(1 \otimes u)(R_1 R_0^{\mathrm{op}} + R_0 R_1^{\mathrm{op}}) = 0$, since $RR^{\mathrm{op}} = 1$, so we are done.

Conversely, suppose that $(\mathcal{A}, g)$ is a pair where $\mathcal{A}$ is a triangular Hopf superalgebra with R-matrix $\mathcal{R}$ and Drinfeld element 1. Let $\mathcal{R} = \mathcal{R}_0 + \mathcal{R}_1$, where $\mathcal{R}_0$ has even components, and $\mathcal{R}_1$ has odd components. Let $R := (\mathcal{R}_0 + (1 \otimes g)\mathcal{R}_1)R_g$. Then it is straightforward to show that $(\mathcal{A}, R)$ is triangular with Drinfeld element $u = g$. The theorem is proved.   □

COROLLARY 7.1.3.2. *If* $(\mathcal{A}, \mathcal{R})$ *is a triangular Hopf superalgebra with Drinfeld element 1, then the Hopf algebra* $\overline{\mathcal{A}}$ *is also triangular, with the R-matrix*

$$\overline{R} := (\mathcal{R}_0 + (1 \otimes g)\mathcal{R}_1)R_g, \qquad\qquad (7\text{--}1)$$

where g is the grouplike element adjoined to $\mathcal{A}$ to obtain $\overline{\mathcal{A}}$. Moreover, $\mathcal{A}$ is minimal if and only if so is $\overline{\mathcal{A}}$.

PROOF. Clear.   □

The following corollary, combined with Kostant's theorem, gives a classification of triangular Hopf algebras with R-matrix of rank ≤ 2 (i.e., of the form R_u as in (2–9), where u is a grouplike of order ≤ 2).

COROLLARY 7.1.3.3. *The correspondence of Theorem 7.1.3.1 restricts to a one to one correspondence between*

1. *isomorphism classes of ordinary triangular Hopf algebras with R-matrix of rank ≤ 2, and*

2. *isomorphism classes of pairs $(\mathcal{A}, g)$ where $\mathcal{A}$ is a cocommutative Hopf superalgebra and g is an element of $\mathbf{G}(\mathcal{A})$ such that $g^2 = 1$ and $gxg^{-1} = (-1)^{p(x)}x$.*

PROOF. Let (A, R) be an ordinary triangular Hopf algebra with $R-$ matrix of rank ≤ 2. In particular, the Drinfeld element u of A satisfies $u^2 = 1$, and $R = R_u$. Hence by Theorem 7.1.3.1, $(\mathcal{A}, \tilde{\Delta}, \mathcal{R})$ is a triangular Hopf superalgebra. Moreover, it is cocommutative since $\mathcal{R} = R_u R_u = 1$.

Conversely, for any $(\mathcal{A}, g)$, by Theorem 7.1.3.1, the pair (A, R_g) is an ordinary triangular Hopf algebra, and clearly the rank of R_g is ≤ 2. $\square$

In particular, Corollaries 2.2.3.5 and 7.1.3.3 imply that finite-dimensional triangular Hopf algebras with R-matrix of rank ≤ 2 correspond to supergroup algebras. In view of this, we make the following definition.

DEFINITION 7.1.3.4. A finite-dimensional triangular Hopf algebra with R-matrix of rank ≤ 2 is called a modified supergroup algebra.

7.1.4. Construction of twists for supergroup algebras.

PROPOSITION 7.1.4.1. Let $\mathcal{A} = \mathbb{C}[G] \ltimes \Lambda V$ be a supergroup algebra. Let $r \in S^2 V$. Then $\mathcal{J} := e^{r/2}$ is a twist for $\mathcal{A}$. Moreover, $((\Lambda V)^{\mathcal{J}}, \mathcal{J}_{21}^{-1} \mathcal{J})$ is minimal triangular if and only if r is nondegenerate.

PROOF. Straightforward. $\square$

EXAMPLE 7.1.4.2. Let G be the group of order 2 with generator g. Let $V := \mathbb{C}$ be the nontrivial 1-dimensional representation of G, and write $\Lambda V = sp\{1, x\}$. Then the associated ordinary triangular Hopf algebra to $(\mathcal{A}, g) := (\mathbb{C}[G] \ltimes \Lambda V, g)$ is Sweedler's 4-dimensional Hopf algebra A [Sw] (see Example 6.1.5) with the triangular structure R_g. It is known [R2] that the set of triangular structures on A is parameterized by $\mathbb{C}$; namely, R is a triangular structure on A if and only if

$$R = R_\lambda := R_g - \frac{\lambda}{2}(x \otimes x - gx \otimes x + x \otimes gx + gx \otimes gx), \quad \lambda \in \mathbb{C}.$$

Clearly, (A, R_λ) is minimal if and only if $\lambda \neq 0$.

Let $r \in S^2 V$ be defined by $r := \lambda x \otimes x$, $\lambda \in \mathbb{C}$. Set $\mathcal{J}_\lambda := e^{r/2} = 1 + \frac{1}{2}\lambda x \otimes x$; it is a twist for $\mathcal{A}$. Hence, $J_\lambda := 1 - \frac{1}{2}\lambda gx \otimes x$ is a twist for A. It is easy to check that $R_\lambda = (J_\lambda)_{21}^{-1} R_g J_\lambda$. Thus, $(A, R_\lambda) = (A, R_0)^{J_\lambda}$.

REMARK 7.1.4.3. In fact, Radford's classification of triangular structures on A can be easily deduced from Lemma 7.3.2.6 below.

7.2. The Chevalley property. Recall that in the introduction we made the following definition.

DEFINITION 7.2.1. A Hopf algebra A over $\mathbb{C}$ is said to have the Chevalley property if the tensor product of any two simple A-modules is semisimple. More generally, let us say that a tensor category has the Chevalley property if the tensor product of two simple objects is semisimple.

Let us give some equivalent formulations of the Chevalley property.

PROPOSITION 7.2.2. *Let A be a finite-dimensional Hopf algebra over $\mathbb{C}$. The following conditions are equivalent*:

1. *A has the Chevalley property.*
2. *The category of (right) A^*-comodules has the Chevalley property.*
3. *$\mathrm{Corad}(A^*)$ is a Hopf subalgebra of A^*.*
4. *$\mathrm{Rad}(A)$ is a Hopf ideal and thus $A/\mathrm{Rad}(A)$ is a Hopf algebra.*
5. *$S^2 = \mathrm{Id}$ on $A/\mathrm{Rad}(A)$, or equivalently on $\mathrm{Corad}(A^*)$.*

PROOF. $(1 \Leftrightarrow 2)$ Clear, since the categories of left A-modules and right A^*-comodules are equivalent.

$(2 \Rightarrow 3)$ Recall the definition of a matrix coefficient of a comodule V over A^*. If $\rho : V \to V \otimes A^*$ is the coaction, $v \in V$, $\alpha \in V^*$, then

$$\phi_{v,\alpha}^V := (\alpha \otimes \mathrm{Id})\rho(v) \in A^*.$$

It is well-known that:

(a) The coradical of A^* is the linear span of the matrix coefficients of all simple A^*-comodules.

(b) The product in A^* of two matrix coefficients is a matrix coefficient of the tensor product. Specifically,

$$\phi_{v,\alpha}^V \phi_{w,\beta}^W = \phi_{v\otimes w,\alpha\otimes\beta}^{V\otimes W}.$$

It follows at once from (a) and (b) that $\mathrm{Corad}(A^*)$ is a subalgebra of A^*. Since the coradical is stable under the antipode, the claim follows.

$(3 \Leftrightarrow 4)$ To say that $\mathrm{Rad}(A)$ is a Hopf ideal is equivalent to saying that $\mathrm{Corad}(A^*)$ is a Hopf algebra, since $\mathrm{Corad}(A^*) = (A/\mathrm{Rad}(A))^*$.

$(4 \Rightarrow 1)$ If V, W are simple A-modules then they factor through $A/\mathrm{Rad}(A)$. But $A/\mathrm{Rad}(A)$ is a Hopf algebra, so $V \otimes W$ also factors through $A/\mathrm{Rad}(A)$, so it is semisimple.

$(3 \Rightarrow 5)$ Clear, since a cosemisimple Hopf algebra is involutory.

$(5 \Rightarrow 3)$ Consider the subalgebra B of A^* generated by $\mathrm{Corad}(A^*)$. This is a Hopf algebra, and $S^2 = \mathrm{Id}$ on it. Thus, B is cosemisimple and hence $B = \mathrm{Corad}(A^*)$ is a Hopf subalgebra of A^*. $\square$

REMARK 7.2.3. The assumption that the base field has characteristic 0 is needed only in the proof of $(5 \Leftrightarrow 3)$

7.3. The classification of triangular Hopf algebras with the Chevalley property.

7.3.1. The main theorem. The main result of Section 7 is the following theorem.

THEOREM 7.3.1.1. *Let A be a finite-dimensional triangular Hopf algebra over $\mathbb{C}$. Then the following are equivalent*:

1. *A is a twist of a finite-dimensional triangular Hopf algebra with R-matrix of rank ≤ 2 (i.e., of a modified supergroup algebra).*

2. A has the Chevalley property.

7.3.2. Proof of the main theorem. The proof will take the remainder of this subsection. We shall need the following result, whose proof is given in [AEG].

THEOREM 7.3.2.1. *Let A be a local finite-dimensional Hopf superalgebra (not necessarily supercommutative). Then $A = \Lambda V^*$ for a finite-dimensional vector space V. In other words, A is the function algebra of an odd vector space V.*

REMARK 7.3.2.2. Note that in the commutative case Theorem 7.3.2.1 is a special case of Proposition 3.2 of [Ko].

We start by giving a super-analogue of Theorem 3.1 in [G4].

LEMMA 7.3.2.3. *Let A be a minimal triangular pointed Hopf superalgebra. Then $\mathrm{Rad}(A)$ is a Hopf ideal, and $A/\mathrm{Rad}(A)$ is minimal triangular.*

PROOF. The proof is a tautological generalization of the proof of Theorem 3.1 in [G4] to the super case.

First of all, it is clear that $\mathrm{Rad}(A)$ is a Hopf ideal, since its orthogonal complement (the coradical of A^*) is a sub Hopf superalgebra (as A^* is isomorphic to A^{cop} as a coalgebra, and hence is pointed). Thus, it remains to show that the triangular structure on A descends to a minimal triangular structure on $A/\mathrm{Rad}(A)$. For this, it suffices to prove that the composition of the Hopf superalgebra maps

$$\mathrm{Corad}(A^{*\mathrm{cop}}) \hookrightarrow A^{*\mathrm{cop}} \to A \to A/\mathrm{Rad}(A)$$

(where the middle map is given by the R-matrix) is an isomorphism. But this follows from the fact that for any surjective coalgebra map $\eta : C_1 \to C_2$, the image of the coradical of C_1 contains the coradical of C_2 (see e.g., [Mon, Corollary 5.3.5]): One needs to apply this statement to the map $A^{*\mathrm{cop}} \to A/\mathrm{Rad}(A)$. □

LEMMA 7.3.2.4. *Let A be a minimal triangular pointed Hopf superalgebra, such that the R-matrix $\mathcal{R}$ of A is unipotent (which is to say, $\mathcal{R} - 1 \otimes 1$ is 0 in $A/\mathrm{Rad}(A) \otimes A/\mathrm{Rad}(A)$). Then $A = \Lambda V$ as a Hopf superalgebra, and $\mathcal{R} = e^r$, where $r \in S^2 V$ is a nondegenerate symmetric (in the usual sense) bilinear form on V^*.*

PROOF. By Lemma 7.3.2.3, $\mathrm{Rad}(A)$ is a Hopf ideal, and $A/\mathrm{Rad}(A)$ is minimal triangular. But the R-matrix of $A/\mathrm{Rad}(A)$ must be $1 \otimes 1$, so $A/\mathrm{Rad}(A)$ is 1-dimensional. Hence A is local, so by Theorem 7.3.2.1, $A = \Lambda V$. If $\mathcal{R}$ is a triangular structure on A then it comes from an isomorphism $\Lambda V^* \to \Lambda V$ of Hopf superalgebras, which is induced by a linear isomorphism $r : V^* \to V$. So $\mathcal{R} = e^r$, where r is regarded as an element of $V \otimes V$. Since $\mathcal{R}\mathcal{R}_{21} = 1$, we have $r + r^{21} = 0$ (where $r^{21} = -r^{\mathrm{op}}$ is the opposite of r in the supersense), so $r \in S^2 V$. □

REMARK 7.3.2.5. The classification of pointed finite-dimensional Hopf algebras with coradical of dimension 2 is known [CD, N]. In [AEG] we used the Lifting method [AS1, AS2] to give an alternative proof. Below we shall need the following more precise version of this result in the triangular case.

LEMMA 7.3.2.6. *Let A be a minimal triangular pointed Hopf algebra, whose coradical is $\mathbb{C}[\mathbb{Z}_2] = sp\{1, u\}$, where u is the Drinfeld element of A. Then $A = \overline{(\Lambda V)^{\mathcal{J}}}$ with the triangular structure of Corollary 7.1.3.2, where $\mathcal{J} = e^{r/2}$, with $r \in S^2 V$ a nondegenerate element. In particular, A is a twist of a modified supergroup algebra.*

PROOF. Let $\mathcal{A}$ be the associated triangular Hopf superalgebra to A, as described in Theorem 7.1.3.1. Then the R-matrix of $\mathcal{A}$ is unipotent, because it turns into $1 \otimes 1$ after killing the radical.

Let $\mathcal{A}_m$ be the minimal part of $\mathcal{A}$. By Lemma 7.3.2.4, $\mathcal{A}_m = \Lambda V$ and $\mathcal{R} = e^r$, $r \in S^2 V$. So if $\mathcal{J} := e^{r/2}$ then $\mathcal{A}^{\mathcal{J}^{-1}}$ has R-matrix equal to $1 \otimes 1$. Thus, $\mathcal{A}^{\mathcal{J}^{-1}}$ is cocommutative, so by Corollary 2.2.3.5, it equals $\mathbb{C}[\mathbb{Z}_2] \ltimes \Lambda V$. Hence $\mathcal{A} = \mathbb{C}[\mathbb{Z}_2] \ltimes (\Lambda V)^{\mathcal{J}}$, and the result follows from Proposition 7.1.2.1. $\square$

We shall need the following lemma.

LEMMA 7.3.2.7. *Let $B \subseteq A$ be finite-dimensional associative unital algebras. Then any simple B-module is a constituent (in the Jordan-Holder series) of some simple A-module.*

PROOF. Since A, considered as a B-module, contains B as a B-module, any simple B-module is a constituent of A.

Decompose A (in the Grothendieck group of A) into simple A-modules: $A = \sum V_i$. Further decomposing as B-modules, we get $V_i = \sum W_{ij}$, and hence $A = \sum_i \sum_j W_{ij}$. Now, by Jordan-Holder theorem, since A (as a B-module) contains all simple B-modules, any simple B-module X is in $\{W_{ij}\}$. Thus, X is a constituent of some V_i, as desired. $\square$

PROPOSITION 7.3.2.8. *Any minimal triangular Hopf algebra A with the Chevalley property is a twist of a triangular Hopf algebra with R-matrix of rank ≤ 2.*

PROOF. By Proposition 7.2.2, the coradical A_0 of A is a Hopf subalgebra, since $A \cong A^{*\mathrm{cop}}$, being minimal triangular. Consider the Hopf algebra map $\varphi : A_0 \to A^{*\mathrm{cop}}/\mathrm{Rad}(A^{*\mathrm{cop}})$, given by the composition of the following maps:

$$A_0 \hookrightarrow A \cong A^{*\mathrm{cop}} \to A^{*\mathrm{cop}}/\mathrm{Rad}(A^{*\mathrm{cop}}),$$

where the second map is given by the R-matrix. We claim that φ is an isomorphism. Indeed, A_0 and $A^{*\mathrm{cop}}/\mathrm{Rad}(A^{*\mathrm{cop}})$ have the same dimension, since $\mathrm{Rad}(A^{*\mathrm{cop}}) = (A_0)^{\perp}$, and φ is injective, since A_0 is semisimple by [LR]. Let $\pi : A \to A_0$ be the associated projection.

We see, arguing exactly as in [G4, Theorem 3.1], that A_0 is also minimal triangular, say with R-matrix R_0.

Now, by [EG1, Theorem 2.1], we can find a twist J in $A_0 \otimes A_0$ such that $(A_0)^J$ is isomorphic to a group algebra and has R-matrix $(R_0)^J$ of rank ≤ 2. Notice that here we are relying on Deligne's theorem, as mentioned in the introduction.

Let us now consider J as an element of $A_0 \otimes A_0$ and the twisted Hopf algebra A^J, which is again triangular.

The projection $\pi : A^J \rightarrow (A_0)^J$ is still a Hopf algebra map, and sends R^J to $(R_0)^J$. It induces a projection $(A^J)_m \rightarrow \mathbb{C}[\mathbb{Z}_2]$, whose kernel K_m is contained in the kernel of π. Because any simple $(A^J)_m$-module is contained as a constituent in a simple A-module (see Lemma 7.3.2.7), $K_m = \mathrm{Rad}((A^J)_m)$. Hence, $(A^J)_m$ is minimal triangular and $(A^J)_m/\mathrm{Rad}((A^J)_m) = (\mathbb{C}[\mathbb{Z}_2], R_u)$. It follows, again by minimality, that $(A^J)_m$ is also pointed with coradical isomorphic to $\mathbb{C}[\mathbb{Z}_2]$. So by Lemma 7.3.2.6, $(A^J)_m$, and hence A^J, can be further twisted into a triangular Hopf algebra with R-matrix of rank ≤ 2, as desired.   □

Now we can prove the main theorem.

PROOF OF THEOREM 7.3.1.1. $(2 \Rightarrow 1)$ By Proposition 7.2.2, $A/\mathrm{Rad}(A)$ is a semisimple Hopf algebra. Let A_m be the minimal part of A, and A'_m be the image of A_m in $A/\mathrm{Rad}(A)$. Then A'_m is a semisimple Hopf algebra.

Consider the kernel K of the projection $A_m \rightarrow A'_m$. Then $K = \mathrm{Rad}(A) \cap A_m$. This means that any element $k \in K$ is zero in any simple A-module. This implies that k acts by zero in any simple A_m-module, since by Lemma 7.3.2.7, any simple A_m-module occurs as a constituent of some simple A-module. Thus, K is contained in $\mathrm{Rad}(A_m)$. On the other hand, A_m/K is semisimple, so $K = \mathrm{Rad}(A_m)$. This shows that $\mathrm{Rad}(A_m)$ is a Hopf ideal. Thus, A_m is minimal triangular satisfying the conditions of Proposition 7.3.2.8. By Proposition 7.3.2.8, A_m is a twist of a triangular Hopf algebra with R-matrix of rank ≤ 2. Hence A is a twist of a triangular Hopf algebra with R-matrix of rank ≤ 2 (by the same twist), as desired.

$(1 \Rightarrow 2)$ By assumption, $\mathrm{Rep}(A)$ is equivalent to $\mathrm{Rep}(\tilde{G})$ for some supergroup $\tilde{G}$ (as a tensor category without braiding). But we know that supergroup algebras have the Chevalley property, since, modulo their radicals, they are group algebras. This concludes the proof of the main theorem.   □

REMARK 7.3.2.9. Notice that it follows from the proof of the main theorem that any triangular Hopf algebra with the Chevalley property can be obtained by twisting of a triangular Hopf algebra with R-matrix of rank ≤ 2 by an *even* twist.

DEFINITION 7.3.2.10. If a triangular Hopf algebra A over $\mathbb{C}$ satisfies condition 1. or 2. of Theorem 7.3.1.1, we will say that H is of supergroup type.

7.3.3. Corollaries of the main theorem.

COROLLARY 7.3.3.1. *A finite-dimensional triangular Hopf algebra A is of supergroup type if and only if so is its minimal part A_m.*

PROOF. If A is of supergroup type then $\mathrm{Rad}(A)$ is a Hopf ideal, so like in the proof of Theorem 7.3.1.1 ($2 \Rightarrow 1$) we conclude that $\mathrm{Rad}(A_m)$ is a Hopf ideal, i.e., A_m is of supergroup type.

Conversely, if A_m is of supergroup type then A_m is a twist of a triangular Hopf algebra with R-matrix of rank ≤ 2. Hence A is a twist of a triangular Hopf algebra with R-matrix of rank ≤ 2 (by the same twist), so A is of supergroup type.   □

COROLLARY 7.3.3.2. *A finite-dimensional triangular Hopf algebra whose coradical is a Hopf subalgebra is of supergroup type. In particular, this is the case for finite-dimensional triangular pointed Hopf algebras.*

PROOF. This follows from Corollary 7.3.3.1.   □

COROLLARY 7.3.3.3. *Any finite-dimensional triangular basic Hopf algebra is of supergroup type.*

PROOF. A basic Hopf algebra automatically has the Chevalley property since all its irreducible modules are 1-dimensional. Hence the result follows from the main theorem.   □

REMARK 7.3.3.4. The classification in Theorem 7.3.1.1 can be made more effective and explicit. Indeed, Theorem 2.2 in [EG7] states a bijection between the set of isomorphism classes of finite-dimensional triangular Hopf algebras with the Chevalley property and the set of isomorphism classes of septuples (G, W, H, Y, B, V, u) where G is a finite group, W is a finite-dimensional representation of G, H is a subgroup of G, Y is an H−invariant subspace of W, B is an H−invariant non-degenerate element in $S^2 Y$, V is an irreducible projective representation of H of dimension $|H|^{1/2}$, and $u \in G$ is a central element of order ≤ 2 acting by -1 on W. In the semisimple case, the septuples reduce to the quadruples of Theorem 4.2.6. In the minimal pointed case, we recover Theorem 6.3.1.

7.4. Categorical dimensions in symmetric categories with finitely many irreducibles are integers. In [AEG] we classified finite-dimensional triangular Hopf algebras with the Chevalley property. We also gave one result that is valid in a greater generality for any finite-dimensional triangular Hopf algebra, and even for any symmetric rigid category with finitely many irreducible objects.

Let $\mathcal{C}$ be a $\mathbb{C}$-linear abelian symmetric rigid category with $\mathbf{1}$ as its unit object, and suppose that $\mathrm{End}(\mathbf{1}) = \mathbb{C}$. Recall that there is a natural notion of dimension in $\mathcal{C}$, generalizing the ordinary dimension of an object in Vect, and having the properties of being additive and multiplicative with respect to the tensor product. Let β denote the commutativity constraint in $\mathcal{C}$, and for an object V, let ev_V, $coev_V$ denote the associated evaluation and coevaluation morphisms.

DEFINITION 7.4.1 [DM]. The categorical dimension $\dim_c(V) \in \mathbb{C}$ of $V \in \mathcal{C}$ is the morphism

$$\dim_c(V) : \mathbf{1} \xrightarrow{ev_V} V \otimes V^* \xrightarrow{\beta_{V,V^*}} V^* \otimes V \xrightarrow{coev_V} \mathbf{1}. \tag{7-2}$$

The main result of this subsection is the following:

THEOREM 7.4.2. *In any $\mathbb{C}$-linear abelian symmetric rigid tensor category $\mathcal{C}$ with finitely many irreducible objects, the categorical dimensions of objects are integers.*

PROOF. First note that the categorical dimension of any object V of $\mathcal{C}$ is an algebraic integer. Indeed, let $V_1 \ldots, V_n$ be the irreducible objects of $\mathcal{C}$. Then $\{V_1 \ldots, V_n\}$ is a basis of the Grothendieck ring of $\mathcal{C}$. Write $V \otimes V_i = \sum_j N_{ij}(V) V_j$ in the Grothendieck ring. Then $N_{ij}(V)$ is a matrix with integer entries, and $\dim_c(V)$ is an eigenvalue of this matrix. Thus, $\dim_c(V)$ is an algebraic integer.

Now, if $\dim_c(V) = d$ then it is easy to show (see e.g. [Del]) that

$$\dim_c(S^k V) = d(d+1) \cdots (d+k-1)/k!,$$

and

$$\dim_c(\Lambda^k V) = d(d-1) \cdots (d-k+1)/k!,$$

hence they are also algebraic integers. So the theorem follows from:

LEMMA. *Suppose d is an algebraic integer such that $d(d+1) \cdots (d+k-1)/k!$ and $d(d-1) \cdots (d-k+1)/k!$ are algebraic integers for all k. Then d is an integer.*

PROOF. Let Q be the minimal monic polynomial of d over $\mathbb{Z}$. Since

$$d(d-1) \cdots (d-k+1)/k!$$

is an algebraic integer, so are the numbers $d'(d'-1) \cdots (d'-k+1)/k!$, where d' is any algebraic conjugate of d. Taking the product over all conjugates, we get that

$$N(d)N(d-1) \cdots N(d-k+1)/(k!)^n$$

is an integer, where n is the degree of Q. But $N(d-x) = (-1)^n Q(x)$. So we get that $Q(0)Q(1) \cdots Q(k-1)/(k!)^n$ is an integer. Similarly from the identity for $S^k V$, it follows that $Q(0)Q(-1) \cdots Q(1-k)/(k!)^n$ is an integer. Now, without loss of generality, we can assume that $Q(x) = x^n + ax^{n-1} + \cdots$, where $a \leq 0$ (otherwise replace $Q(x)$ by $Q(-x)$; we can do it since our condition is symmetric under this change). Then for large k, we have $Q(k-1) < k^n$, so the sequence $b_k := Q(0)Q(1) \cdots Q(k-1)/k!^n$ is decreasing in absolute value or zero starting from some place. But a sequence of integers cannot be strictly decreasing in absolute value forever. So $b_k = 0$ for some k, hence Q has an integer root. This means that d is an integer (i.e., Q is linear), since Q must be irreducible over the rationals. This concludes the proof of the lemma, and hence of the theorem. $\square$

$\square$

COROLLARY 7.4.3. *For any triangular Hopf algebra A (not necessarily finite-dimensional), the categorical dimensions of its finite-dimensional representations are integers.*

PROOF. It is enough to consider the minimal part A_m of A which is finite-dimensional, since $\dim_c(V) = \mathrm{tr}(u_{|V})$ for any module V (where u is the Drinfeld element of A), and $u \in A_m$. Hence the result follows from Theorem 7.4.2. $\square$

REMARK 7.4.4. Theorem 7.4.2 is false without the finiteness conditions. In fact, in this case any complex number can be a dimension, as is demonstrated in examples constructed by Deligne [De2, p. 324–325]. Also, it is well known that the theorem is false for ribbon, nonsymmetric categories (e.g., for fusion categories of semisimple representations of finite-dimensional quantum groups at roots of unity [L], where dimensions can be irrational algebraic integers).

REMARK 7.4.5. In any rigid braided tensor category with finitely many irreducible objects, one can define the Frobenius-Perron dimension of an object V, $\mathrm{FPdim}(V)$, to be the largest positive eigenvalue of the matrix of multiplication by V in the Grothendieck ring. This dimension is well defined by the Frobenius-Perron theorem, and has the usual additivity and multiplicativity properties. For example, for the category of representations of a quasi-Hopf algebra, it is just the usual dimension of the underlying vector space. If the answer to Question 8.7 is positive then $\mathrm{FPdim}(V)$ for symmetric categories is always an integer, which is equal to $\dim_c(V)$ modulo 2. It would be interesting to check this, at least in the case of modules over a triangular Hopf algebras, when the integrality of FPdim is automatic (so only the mod 2 congruence has to be checked).

8. Questions

We conclude the paper with some natural questions motivated by the above results [AEG, G4].

QUESTION 8.1. Let (A, R) be *any* finite-dimensional triangular Hopf algebra with Drinfeld element u. Is it true that $S^4 = \mathrm{Id}$? Does u satisfy $u^2 = 1$? Is it true that $S^4 = \mathrm{Id}$ implies $u^2 = 1$?

REMARK 8.2. A positive answer to the second question in Question 8.1 will imply that an odd-dimensional triangular Hopf algebra must be semisimple.

Note that if A is of supergroup type, then the answer to Question 8.1 is positive. Indeed, since $S^2 = \mathrm{Id}$ on the semisimple part of A, u acts by a scalar in any irreducible representation of A. In fact, since $\mathrm{tr}(u) = \mathrm{tr}(u^{-1})$, we have that $u = 1$ or $u = -1$ on any irreducible representation of A, and hence $u^2 = 1$ on any irreducible representation of A. Thus, u^2 is unipotent. But it is of finite order (as it is a grouplike element), so it is equal to 1 as desired.

QUESTION 8.3. Does any finite-dimensional triangular Hopf algebra over $\mathbb{C}$ have the Chevalley property (i.e., is of supergroup type)? Is it true under the assumption that $S^4 = \mathrm{Id}$ or at least under the assumption that $u^2 = 1$?

REMARK 8.4. Note that the answer to question 8.3 is negative in the infinite dimensional case. Namely, although the answer is positive in the cocommutative case (by [C]), it is negative already for triangular Hopf algebras with R-matrix of rank 2, which correspond to cocommutative Hopf superalgebras. Indeed, let us take the cocommutative Hopf superalgebra $\mathcal{A} := \mathrm{U}(\mathrm{gl}(n|n))$ (for the definition of the Lie superalgebra $\mathrm{gl}(n|n)$, see [KaV, p. 29]). The associated triangular Hopf algebra $\overline{\mathcal{A}}$ does not have the Chevalley property, since it is well known that Chevalley theorem fails for Lie superalgebras (e.g., $\mathrm{gl}(n|n)$); more precisely, already the product of the vector and covector representations for this Lie superalgebra is not semisimple.

REMARK 8.5. It follows from Corollary 7.3.3.1 that a positive answer to Question 8.3 in the minimal case would imply the general positive answer.

Here is a generalization of Question 8.3.

QUESTION 8.6. Does any $\mathbb{C}$-linear abelian symmetric rigid tensor category, with $\mathrm{End}(1) = \mathbb{C}$ and finitely many simple objects, have the Chevalley property?

Even a more ambitious question:

QUESTION 8.7. Is such a category equivalent to the category of representations of a finite-dimensional triangular Hopf algebra with R-matrix of rank ≤ 2? In particular, is it equivalent to the category of representations of a supergroup, as a category without braiding? Are these statements valid at least for categories with Chevalley property? For semisimple categories?

REMARK 8.8. Note that Theorem 7.4.2 can be regarded as a piece of supporting evidence for a positive answer to Question 8.7.

References

[AEG] N. Andruskiewitsch, P. Etingof and S. Gelaki, Triangular Hopf Algebras with the Chevalley property, *submitted*, math.QA/0008232.

[AEGN] E. Aljadeff, P. Etingof, S. Gelaki and D. Nikshych, On twisting of finite-dimensional Hopf algebras, *Michigan Journal of Mathematics* **49** (2001), 277–298.

[AS1] N. Andruskiewitsch and H.-J. Schneider, Lifting of quantum linear spaces and pointed Hopf algebras of order p^3, *J. Algebra* **209** (1998), 658–691.

[AS2] N. Andruskiewitsch and H.-J. Schneider, Finite quantum groups and Cartan matrices, *Adv. in Math.* **154** (2000), 1–45.

[AT] J. H. Conway, R. T. Curtis, S. P. Norton, R. A. Parker and R. A. Wilson, Atlas of finite groups, *Clarendon Press, Oxford* (1985).

[C] C. Chevalley, Theory of Lie groups, v. III, 1951 (in French).

[CD] S. Caenepeel and S. Dăscălescu, On pointed Hopf algebras of dimension 2^n, *Bull. London Math. Soc.* **31** (1999), 17–24

[CR] C. Curtis and I. Reiner, Methods of representation theory 1, John Wiley & Sons, Inc. (1981).

[De1] P. Deligne, Categories tannakiennes, In The Grothendick Festschrift, Vol. II, Prog. Math. **87** (1990), 111–195.

[De2] P. Deligne, La série exceptionnelle de groupes de Lie. (French) [The exceptional series of Lie groups], *C. R. Acad. Sci. Paris Sor. I Math.* **322** (1996), no.4, 321–326.

[DM] P. Deligne and J. Milne, Tannakian categories, Lecture Notes in Mathematics **900**, 101–228, 1982.

[Dr1] V. Drinfeld, Quantum groups, *Proceedings of the International Congress of Mathematics, Berkeley* (1987), 798–820.

[Dr2] V. Drinfeld, On almost cocommutative hopf algebras, *Leningrad Mathematics Journal* **1** (1990), 321–342.

[Dr3] V. Drinfeld, Constant quasiclassical solutions of the quantum Yang-Baxter equation, *Dokl. Acad. Nauk. SSSR* **273 No. 3** (1983), 531–535.

[EG1] P. Etingof and S. Gelaki, Some properties of finite-dimensional semisimple Hopf algebras, *Mathematical Research Letters* **5** (1998), 191–197.

[EG2] P. Etingof and S. Gelaki, A method of construction of finite-dimensional triangular semisimple Hopf algebras, *Mathematical Research Letters* **5** (1998), 551–561.

[EG3] P. Etingof and S. Gelaki, The representation theory of cotriangular semisimple Hopf algebras, *International Mathematics Research Notices* **7** (1999), 387–394.

[EG4] P. Etingof and S. Gelaki, The classification of triangular semisimple and cosemisimple Hopf algebras over an algebraically closed field, *International Mathematics Research Notices* **5** (2000), 223–234.

[EG5] P. Etingof and S. Gelaki, On finite-dimensional semisimple and cosemisimple Hopf algebras in positive characteristic, *International Mathematics Research Notices* **16** (1998), 851–864.

[EG6] P. Etingof and S. Gelaki, On cotriangular Hopf algebras, *American Journal of Mathematics* **123** (2001), 699–713.

[EG7] P. Etingof, S. Gelaki, Classification of finite-dimensional triangular Hopf algebras with the Chevalley property, *Mathematical Research Letters* **8** (2001), 331–345.

[EGGS] P. Etingof, S. Gelaki, R. Guralnick and J. Saxl, Biperfect Hopf algebras, *Journal of Algebra* **232** (2000), 331–335.

[ES] P. Etingof and O. Schiffmann, Lectures on quantum groups, Lectures in Mathematical Physics, *International Press, Boston, MA* (1998).

[ESS] P. Etingof, T. Schedler and A. Soloviev, Set-theoretic solutions to the quantum Yang–Baxter equation, *Duke Math. J.* **100** (1999), 169–209.

[G1] S. Gelaki, On pointed ribbon Hopf algebras, *Journal of Algebra* **181** (1996), 760–786.

[G2] S. Gelaki, Quantum groups of dimension pq^2, *Israel Journal of Mathematics* **102** (1997), 227–267.

[G3] S. Gelaki, Pointed Hopf algebras and Kaplansky's 10th Conjecture, *Journal of Algebra* **209** (1998), 635–657.

[G4] S. Gelaki, Some examples and properties of triangular pointed Hopf algebras, *Mathematical Research Letters* **6** (1999), 563–572; see corrected version posted at math.QA/9907106.

[G5] S. Gelaki, Semisimple triangular Hopf algebras and Tannakian categories, *submitted.*

[HI] R. B. Howlett and I. M. Isaacs, On groups of central type, *Mathematische Zeitschrift* **179** (1982), 555–569.

[KaG] G. I. Kac, Extensions of groups to ring groups, *Math. USSR sbornik* **5** No. 3 (1968).

[KaV] V. Kac, Lie superalgebras, *Advances in Math.* **26**, No.1, 1977.

[Kap] I. Kaplansky, Bialgebras, University of Chicago, 1975.

[Kash] Y. Kashina, Classification of semisimple Hopf algebras of dimension 16, *Journal of Algebra* **232** (2000), 617–663.

[Kass] C. Kassel, Quantum Groups, Springer, New York, 1995.

[Ko] B. Kostant, Graded manifolds, graded Lie theory, and prequantization, Differ. geom. Meth. math. Phys., Proc. Symp. Bonn 1975, *Lect. Notes Math.* **570** (1977), 177–306.

[L] G. Lusztig, Finite dimensional Hopf algebras arising from quantized universal enveloping algebras, *J. of the A. M. S* Vol.3, No.1 (1990), 257–296.

[LR1] R. Larson and D. Radford, Semisimple Cosemisimple Hopf algebras, *American Journal of Mathematics* **110** (1988), 187–195.

[LR2] R. G. Larson and D. E. Radford, Finite-dimensional cosemisimple Hopf algebras in characteristic 0 are semisimple, *J. Algebra* **117** (1988), 267–289.

[Ma1] S. Majid, Foundations of quantum group theory, Cambridge University Press, 1995.

[Ma2] S. Majid, Physics for algebraists: Non-commutative and non-cocommutative Hopf algebras by a bicrossproduct construction, *J. Algebra* **130** (1990), 17–64.

[Mon] S. Montgomery, Hopf algebras and their actions on rings, *CBMS Lecture Notes* **82**, AMS, 1993.

[Mov] M. Movshev, Twisting in group algebras of finite groups, *Func. Anal. Appl.* **27** (1994), 240–244.

[NZ] W. D. Nichols and M. B. Zoeller, A Hopf algebra freeness theorem, *American Journal of Mathematics* **111** (1989), 381–385.

[PO1] F. Panaite and F. V. Oystaeyen, Quasitriangular structures for some pointed Hopf algebras of dimension 2^n, *Communications in Algebra* **27** (1999), 4929–4942.

[PO2] F. Panaite and F. V. Oystaeyen, Clifford-type algebras as cleft extensions for some pointed Hopf algebras, *Communications in Algebra* **28** (2000), 585–600.

[R1] D. E. Radford, The structure of Hopf algebras with a projection, *Journal of Algebra* **2** (1985), 322–347.

[R2] D. E. Radford, Minimal quasitriangular Hopf algebras, *Journal of Algebra* **157** (1993), 285–315.

[R3] D. E. Radford, The trace function and Hopf Algebras *Journal of Algebra* **163** (1994), 583–622.

[R4] D. E. Radford, On Kauffman's knot invariants arising from finite-dimensional Hopf algebras, Advances in Hopf algebras, 158, 205–266, Lectures Notes in Pure and Applied Mathematics, Marcel Dekker, N. Y., 1994.

[Se] J-P. Serre, Local fields, Graduate Texts in Mathematics, **67**.

[Sw] M. Sweedler, Hopf algebras, Benjamin Press, 1968.

[T] M. Takeuchi, Matched pairs of groups and bismash products of Hopf algebras, *Comm. Algebra* **9**, No. 8 (1981), 841–882.

[TW] E. J. Taft and R. L. Wilson, On antipodes in pointed Hopf algebras, *Journal of Algebra* **29** (1974), 27–32.

SHLOMO GELAKI
TECHNION – ISRAEL INSTITUTE OF TECHNOLOGY
DEPARTMENT OF MATHEMATICS
HAIFA 32000, ISRAEL
   gelaki@math.technion.ac.il

New Directions in Hopf Algebras
MSRI Publications
Volume 43, 2002

Coideal Subalgebras and Quantum Symmetric Pairs

GAIL LETZTER

ABSTRACT. Coideal subalgebras of the quantized enveloping algebra are surveyed, with selected proofs included. The first half of the paper studies generators, Harish-Chandra modules, and associated quantum homogeneous spaces. The second half discusses various well known quantum coideal subalgebras and the implications of the abstract theory on these examples. The focus is on the locally finite part of the quantized enveloping algebra, analogs of enveloping algebras of nilpotent Lie subalgebras, and coideals used to form quantum symmetric pairs. The last family of examples is explored in detail. Connections are made to the construction of quantum symmetric spaces.

The introduction of quantum groups in the early 1980's has had a tremendous influence on the theory of Hopf algebras. Indeed, quantum groups provide a source of new and interesting examples. We shall discuss the reverse impact: the theory of quantum groups uses the Hopf structure extensively. This special structure is often hidden in the classical setting, while it is prominent and fundamental for quantum analogs.

Let $\mathbf{g}$ be a semisimple Lie algebra and write G for the corresponding connected, simply connected algebraic group. There are two standard types of quantum groups associated to $\mathbf{g}$ and G. The first is the quantized enveloping algebra which is a quantum analog of the enveloping algebra of $\mathbf{g}$. The second is the quantized function algebra which is a quantum analog of the algebra of regular functions on G. We will be focusing on a particular aspect of the Hopf theory of both types of quantum groups: the study of (one-sided) coideal subalgebras.

One of the reasons coideal subalgebras are so important in the study of quantum groups is that quantum groups do not have "enough" Hopf subalgebras. This shortage of Hopf subalgebras is especially notable for quantized enveloping algebras. Consider a Lie subalgebra $\mathbf{t}$ of the Lie algebra $\mathbf{g}$. The enveloping algebra $U(\mathbf{t})$ of $\mathbf{t}$ is a Hopf subalgebra of $U(\mathbf{g})$. However, upon passage to the

Supported by NSA grant MDA 904-99-1-0033.

AMS subject classification: 17B37.

quantum case, $U_q(\mathbf{t})$, even when it is defined, is often not isomorphic to a Hopf subalgebra of $U_q(\mathbf{g})$. In many cases, there are subalgebras of $U_q(\mathbf{g})$ which are not Hopf subalgebras but are still good quantum analogs of $U(\mathbf{t})$. Moreover, these subalgebras often turn out to be coideals. For example, let $\mathbf{g} = \mathbf{n}^- \oplus \mathbf{h} \oplus \mathbf{n}^+$ be the triangular decomposition of $\mathbf{g}$. There is a natural subalgebra U^+ of $U_q(\mathbf{g})$ which is an analog of the subalgebra $U(\mathbf{n}^+)$ of $U(\mathbf{g})$. This subalgebra U^+ is a coideal but is not a Hopf subalgebra of the quantizing enveloping algebra. Of more interest to us is the fixed Lie subalgebra $\mathbf{g}^\theta$ corresponding to an involution θ of $\mathbf{g}$. In the classical case, the symmetric pair $\mathbf{g}^\theta, \mathbf{g}$ is used to form symmetric spaces. However, in the quantum case, $U_q(\mathbf{g}^\theta)$ does not usually embed inside of $U_q(\mathbf{g})$. Thus it was initially unclear how to develop the theory of quantum symmetric spaces. In [K], Koornwinder constructed two-sided coideal analogs of $\mathbf{g}^\theta$ in type A_1 and used them to produce quantum symmetric spaces. More families of coideal analogs were discovered in [N], [NS],[DN], and [L1]. In [L2], a uniform approach was developed in the maximally split case using one-sided coideal subalgebras of the quantized enveloping algebra. The one-sided coideal condition turned out to be critical in characterizing these quantum analogs of $U(\mathbf{g}^\theta)$.

Quantum symmetric spaces were first defined using the quantized function algebra. (See for example [KS, 11.6.3 and 11.6.4].) Koornwinder's work [K] inspired the development of a quantum symmetric space theory using analogs of $\mathbf{g}^\theta$ contained in $U_q(\mathbf{g})$. The axiomatic theory of quantum symmetric spaces (see [Di]) proceeded more rapidly than the discovery of a general way to construct examples. Indeed, Dijkhuizen [Di, end of Section 3] outlined the desirable properties that analogs of $\mathbf{g}^\theta$ contained in $U_q(\mathbf{g})$ should have in order to form "nice" quantum symmetric spaces. As in Koornwinder's work [K], one of the key properties is the coideal condition. Another crucial property of an analog is that its finite-dimensional spherical modules be characterized in a similar way to the characterization in the classical case. This is obtained in [L3] for the coideal subalgebras of [L2]. The proof uses quantum Harish-Chandra modules associated to quantum symmetric pairs. The coideal condition plays a prominent role in defining and developing the theory of quantum Harish-Chandra modules ([L3]).

This paper is based on a talk given at the MSRI Hopf Algebra Workshop. It offers a panorama of the use of coideal subalgebras in constructing quantum symmetric pairs, in forming quantum Harish-Chandra modules, and in producing quantum symmetric spaces. In the first half of the paper, we present topics in the general theory of quantum coideal subalgebras. Section 1 sets notation and presents some basic facts about coideal subalgebras inside arbitrary Hopf algebras. In Section 2, we define Harish-Chandra modules associated to quantum "reductive" pairs. We prove a basic result: every $U_q(\mathbf{g})$ module contains a large Harish-Chandra module associated to a quantum reductive pair. In Section 3, we discuss how coideal subalgebras of the quantized enveloping algebra can be used in the dual quantum function algebra setting. Connections are made to the theory of quantum homogeneous spaces. Section 4 studies filtrations on the quantized enveloping algebra and their impact on coideal subalgebras. As a result, we obtain a nice description of the generators of a coideal subalgebra under mild restrictions.

The final three sections are devoted to specific coideal subalgebras of the quantized enveloping algebra. Section 5 discusses the locally finite part, $F(U)$, of $U_q(\mathbf{g})$. It is well known that the classical enveloping algebra $U(\mathbf{g})$ can be written as a direct sum of finite-dimensional ad $\mathbf{g}$ modules. This result plays an important role in understanding the structure of $U(\mathbf{g})$ and classifying its primitive ideals. Unfortunately, the quantized enveloping algebra contains infinite dimensional $U_q(\mathbf{g})$ modules with respect to the adjoint action. Thus it is often necessary to use the locally finite part, $F(U)$, which is the maximal subalgebra of $U_q(\mathbf{g})$ that can be written as a direct sum of finite-dimensional simple ad $U_q(\mathbf{g})$ modules. This algebra $F(U)$ is not a Hopf subalgebra of $U_q(\mathbf{g})$, but it is a coideal. The structure of this coideal subalgebra is briefly reviewed with some consideration for the implications of the results of Section 4. Certain quantum Harish-Chandra modules defined originally in [JL3] using $F(U)$ are elucidated in terms of the general approach presented in Section 2. Section 6 considers coideal subalgebra analogs of enveloping algebras of nilpotent and parabolic Lie subalgebras of $\mathbf{g}$. Much of the material in this section is based on [Ke]. The last part, Section 7, is devoted to the theory of quantum symmetric pairs. This material is largely drawn from [L2] and [L3]. However, since the papers appeared, we have found simpler approaches which are presented here with many proofs included. We show how to lift a maximally split involution θ of $\mathbf{g}$ to the quantum setting. Exploiting this lift, we define a coideal subalgebra B_θ of $U_q(\mathbf{g})$. As in [L2], B_θ is characterized as the "unique" maximal coideal subalgebra of $U_q(\mathbf{g}^\theta)$ which specializes to $U(\mathbf{g}^\theta)$ as q goes to 1. Using the results of Section 4, we give a new proof of this uniqueness theorem which does not involve the intricate specialization arguments found in [L2]. We also take the opportunity to make some corrections in the case work necessary to make the uniqueness tight. Results on the Harish-Chandra module and quantum symmetric space theory associated to these pairs are described.

Acknowledgement. Part of this paper was written while the author spent a month as a visiting professor at the University of Rheims. The author would like to thank the mathematics department there for their hospitality and J. Alev for his valuable comments. The author would also like to thank the referee for a painstakingly careful reading of the first draft and many useful suggestions. Finally, the author would like to thank Dan Farkas whose support transcends multiple revisions.

1. Background and Notation

Let H be a Hopf algebra over a field k with comultiplication Δ, antipodal map σ, and counit ε. Given any $a \in H$, write $\Delta(a) = \sum a_{(1)} \otimes a_{(2)}$ using Sweedler notation. A vector subspace I of H is called a left coideal if

$$\Delta(I) \subset H \otimes I.$$

Similarly, I is called a right coideal if $\Delta(I) \subset I \otimes H$. In particular, a left (resp. right) coideal is a left (resp. right) H comodule contained in H. If I is both a left (resp. right) coideal and a subalgebra of H, then we simply say that I is a left (resp. right) coideal subalgebra. There is also a notion of two sided coideals but those are generally not considered here. We will usually choose to discuss left coideals and left coideal subalgebras; analogous results can be proved for the right-handed versions.

We first present two general results about coideals inside of an arbitrary Hopf algebra. First, assume that H contains the group algebra kG of a group G. Choose a vector space complement Y to kG in H. Let P be the projection map of H onto kG as vector spaces using the decomposition $H = kG \oplus Y$. Assume that H is a left kG comodule where the comodule structure comes from the comultiplication and the projection P. In particular, H is the direct sum of vector subspaces H_g where

$$(P \otimes \mathrm{Id})\Delta(H_g) \subset g \otimes H_g. \tag{1.1}$$

Given any left coideal I of H, set $I_g = I \cap H_g$.

LEMMA 1.1. *A left coideal I contained in H is equal to a direct sum of the vector spaces I_g. Thus I is a left kG comodule.*

PROOF. Write $a \in I$ as $a = \sum_{g \in G} a_g$ where each $a_g \in H_g$. The lemma follows if we show that each $a_g \in I$. By (1.1),

$$\Delta(a) \in \sum_{g \in G} g \otimes a_g + Y \otimes H.$$

The coideal property now ensures that each $a_g \in I$. □

Every Hopf algebra H comes equipped with a (left) adjoint action. Using this adjoint action, H becomes an $(\mathrm{ad}\, H)$ module. In particular, given $a, b \in H$,

$$(\mathrm{ad}\, a)b = \sum a_{(1)} b \sigma(a_{(2)}). \tag{1.2}$$

In the quantum case, it is often interesting to consider ad-invariant coideals. The following result (which is basically [Jo, Lemma 1.3.5]) is particularly useful.

LEMMA 1.2. *Let I be a left coideal in H and let M be a Hopf subalgebra of H. Then $(\mathrm{ad}\, M)I$ is an ad M invariant left coideal of H.*

PROOF. First note that ([Jo, 1.1.10])

$$\Delta(\sigma(a)) = \sum \sigma(a_{(2)}) \otimes \sigma(a_{(1)}). \qquad (1.3)$$

Hence

$$\Delta((\mathrm{ad}\, a)b) = \Delta(\sum a_{(1)} b\sigma(a_{(2)})) = \sum (a_{(1)} b_{(1)} \sigma(a_{(4)})) \otimes (a_{(2)} b_{(2)} \sigma(a_{(3)})).$$

The result follows from the fact that $\Delta(a_{(2)}) = \sum a_{(2)} \otimes a_{(3)}$. $\square$

Before defining the quantized enveloping algebra, we recall some basic facts about semisimple Lie algebras. Denote the set of nonnegative integers by $\mathbb{N}$, the complex numbers by $\mathbb{C}$, and the real numbers by $\mathbb{R}$. Let $\mathbf{g}$ be a complex semisimple Lie algebra with triangular decomposition $\mathbf{n}^- \oplus \mathbf{h} \oplus \mathbf{n}^+$. Write $h_1, \ldots, h_n$ for a basis of $\mathbf{h}^*$. Let Δ denote the root system of $\mathbf{g}$ and write Δ^+ for the set of positive roots. Recall that Δ is a subset of $\mathbf{h}^*$. Furthermore, $\mathbf{n}^+$ (resp. $\mathbf{n}^-$) has a basis of root vectors $\{e_\beta | \beta \in \Delta^+\}$ (resp. $\{f_{-\beta} | \beta \in \Delta^+\}$). These root vectors are common eigenvectors, called weight vectors, for the adjoint action of $\mathbf{h}$ on $\mathbf{g}$. In particular, $(\mathrm{ad}\, h_i)e_\beta = [h_i, e_\beta] = \beta(h_i)e_\beta$ and $(\mathrm{ad}\, h_i)f_{-\beta} = [h_i, f_{-\beta}] = -\beta(h_i)f_{-\beta}$ for each $\beta \in \Delta^+$. We further assume that $\{e_\beta, f_{-\beta} | \beta \in \Delta^+\} \cup \{h_1, \ldots, h_n\}$ is a Chevalley basis for $\mathbf{g}$ ([H, Theorem 25.2]). Let $\pi = \{\alpha_1, \ldots, \alpha_n\}$ denote the set of simple roots in Δ^+ and $(\,,\,)$ denote the Cartan inner product on $\mathbf{h}^*$. Recall further that $(\,,\,)$ is positive definite on the real vector space spanned by π. The set π is a basis for $\mathbf{h}^*$. Given $\alpha_i \in \pi$, we write e_i (resp. f_i) for e_{α_i} (resp. $f_{-\alpha_i}$). The Cartan matrix associated to the root system Δ is the matrix with entries $a_{ij} = 2(\alpha_i, \alpha_j)/(\alpha_i, \alpha_i)$. (The reader is referred to [H, Chapters II and III] for additional information on semisimple Lie algebras and root systems.)

Let q be an indeterminate and set $q_i = q^{(\alpha_i, \alpha_i)/2}$. Let $[m]_q$ denote the q number $(q^m - q^{-m})/(q - q^{-1})$ and $[m]_q!$ denote the q factorial $[m]_q[m-1]_q \cdots [1]_q$. The q binomial coefficients are defined by

$$\begin{bmatrix} m \\ j \end{bmatrix}_q = \frac{[m]_q!}{[j]_q![m-j]_q!}.$$

The quantized enveloping algebra $U = U_q(\mathbf{g})$ is generated by $x_1, \ldots, x_n, t_1^{\pm 1}, \ldots, t_n^{\pm 1}, y_1, \ldots, y_n$ over $\mathbb{C}(q)$ with the relations listed below (see for example [Jo, 3.2.9] or [DK, Section 1]).

(1.4) $x_i y_j - y_j x_i = \delta_{ij}(t_i - t_i^{-1})/(q_i - q_i^{-1})$ for each $1 \le i \le n$.

(1.5) The $t_1^{\pm 1}, \ldots, t_n^{\pm 1}$ generate a free abelian group T of rank n.

(1.6) $t_i x_j = q^{(\alpha_i, \alpha_j)} x_j t_i$ and $t_i y_j = q^{-(\alpha_i, \alpha_j)} y_j t_i$ for all $1 \le i, j \le n$.

(1.7) The quantum Serre relations:

$$\sum_{m=0}^{1-a_{ij}} (-1)^m \begin{bmatrix} 1 - a_{ij} \\ m \end{bmatrix}_{q_i} x_i^{1-a_{ij}-m} x_j x_i^m = 0$$

and

$$\sum_{m=0}^{1-a_{ij}} (-1)^m \begin{bmatrix} 1 - a_{ij} \\ m \end{bmatrix}_{q_i} y_i^{1-a_{ij}-m} y_j y_i^m = 0$$

for all $1 \leq i, j \leq n$ with $i \neq j$.

The algebra U is a Hopf algebra with comultiplication Δ, antipode σ, and counit ε defined on generators as follows.

(1.8) $\Delta(t) = t \otimes t \qquad \varepsilon(t) = 1 \qquad \sigma(t) = t^{-1}$ for all t in T

(1.9) $\Delta(x_i) = x_i \otimes 1 + t_i \otimes x_i \qquad \varepsilon(x_i) = 0 \qquad \sigma(x_i) = -t_i^{-1}x_i$

(1.10) $\Delta(y_i) = y_i \otimes t_i^{-1} + 1 \otimes y_i \qquad \varepsilon(y_i) = 0 \qquad \sigma(y_i) = -y_i t_i$

for $1 \leq i \leq n$.

It is well known that the algebra U specializes to $U(\mathbf{g})$ as q goes to 1. This can be made more precise as follows. Set A equal to $\mathbb{C}[q, q^{-1}]_{(q-1)}$. Let $\hat{U}$ be the A subalgebra of U generated by $x_i, y_i, t_i^{\pm 1}$, and $(t_i - 1)/(q - 1)$ for $1 \leq i \leq n$. Then $\hat{U} \otimes_A \mathbf{C}$ is isomorphic to $U(\mathbf{g})$. (See for example [L2, beginning of Section 2]). Given a subalgebra S of U, set $\hat{S} = S \cap \hat{U}$. We say that S specializes to the subalgebra $\bar{S}$ of $U(\mathbf{g})$ if the image of $\hat{S}$ in $\hat{U} \otimes_A \mathbf{C}$ is $\bar{S}$.

Set $Q(\pi)$ equal to the integral lattice generated by π. Let $Q^+(\pi)$ (resp. $Q^-(\pi)$) be the subset of $Q(\pi)$ consisting of nonnegative (resp. nonpositive) integer linear combinations of elements in π. The standard partial ordering on the root lattice $Q(\pi)$ is defined by $\lambda \geq \mu$ provided $\lambda - \mu$ is in $Q^+(\pi)$. Let $P^+(\pi)$ denote the set of dominant integral weights associated to π. In particular, $\lambda \in \mathbf{h}^*$ is an element of $P^+(\pi)$ if and only if $2(\lambda, \alpha_i)/(\alpha_i, \alpha_i)$ is a nonnegative integer for all $1 \leq i \leq n$. There is an isomorphism τ of abelian groups from $Q(\pi)$ to T defined by $\tau(\alpha_i) = t_i$, for $1 \leq i \leq n$. Using this isomorphism, we can replace condition (1.6) with

$$\tau(\lambda)x_i\tau(\lambda)^{-1} = q^{(\lambda,\alpha_i)}x_i \text{ and } \tau(\lambda)y_i\tau(\lambda)^{-1} = q^{-(\lambda,\alpha_i)}y_i \qquad (1.11)$$

for all $\tau(\lambda) \in T$ and $1 \leq i \leq n$.

Let M be a U module. A nonzero vector $v \in U$ has weight $\gamma \in \mathbf{h}^*$ provided that $\tau(\lambda) \cdot v = q^{(\lambda,\gamma)}v$ for all $\tau(\lambda) \in T$. Given a subspace $V \subset M$, the subspace of V spanned by the γ weight vectors is called the γ weight space of V and denoted by V_γ. Now U can be given the structure of a U module using the quantum adjoint action (1.2). Let v be an element of U. We say that v has weight γ provided that it is a γ weight vector in terms of this adjoint action. In particular, v has weight γ if $\tau(\lambda)v\tau(\lambda)^{-1} = q^{(\lambda,\gamma)}v$ for all $\tau(\lambda) \in T$.

Let G^+ be the subalgebra of U generated by $x_1 t_1^{-1}, \ldots, x_n t_n^{-1}$. Similarly, let U^- be the subalgebra of U generated by $y_1, \ldots, y_n$. Let U° be the group algebra of T. It is well known that both U^- and G^+ are a direct sum of their weight spaces. The quantized enveloping algebra U admits a triangular decomposition. More precisely, there is an isomorphism of vector spaces using the multiplication

map ([R]):
$$U \cong U^- \otimes U^o \otimes G^+. \tag{1.12}$$
It follows that there is a direct sum decomposition
$$U = \bigoplus_{t \in T} U^- G^+ t. \tag{1.13}$$

Let G_+^+ (resp. U_+^-) denote the augmentation ideal of G^+ (resp. U^-) and set Y equal to the vector space $(U_+^- G^+ U^o + U^- G_+^+ U^o)$. The direct sum decomposition (1.13) implies that
$$U = U^o \oplus Y. \tag{1.14}$$

Using the definition of the comultiplication of U, it is straightforward to check that for any $b \in U^- G^+ t$
$$\Delta(b) \in t \otimes b + Y \otimes U.$$
Thus the projection of $\Delta(U)$ onto $U^o \otimes U$ makes U into a left U^o comodule with $U_t = U^- G^+ t$. Hence we have the following version of Lemma 1.1 for quantized enveloping algebras.

LEMMA 1.3. *Let I be left coideal of U. Then*
$$I = \bigoplus_{t \in T} (I \cap U^- G^+ t).$$

2. Harish-Chandra Modules

Consider a Lie subalgebra $\mathbf{k}$ of the semisimple Lie algebra $\mathbf{g}$. A Harish-Chandra module associated to the pair $\mathbf{g}, \mathbf{k}$ is a $\mathbf{g}$ module which can be written as a direct sum of finite-dimensional simple $\mathbf{k}$ modules. Harish-Chandra modules are an important tool in classical representation theory. This is especially true when $\mathbf{g}, \mathbf{k}$ is a symmetric pair. Harish-Chandra modules associated to symmetric pairs provide an algebraic approach to the representation theory of real reductive Lie groups.

There is a nice introduction to the theory of Harish-Chandra modules presented in [D, Chapter 9] from an algebraic point of view. The first section of [D, Chapter 9] only assumes that $\mathbf{k}$ is reductive in $\mathbf{g}$. A basic result which is used repeatedly in this chapter of [D] is the following.

THEOREM 2.1 [D, Proposition 1.7.9]. *The direct sum of all the finite-dimensional simple $\mathbf{k}$ modules inside a $\mathbf{g}$ module is a Harish-Chandra module for the pair $\mathbf{g}, \mathbf{k}$.*

This theorem allows one to find large Harish-Chandra modules inside of infinite-dimensional $\mathbf{g}$ modules. Its proof uses the fact that $\mathbf{k}$ is reductive in $\mathbf{g}$ and that $U(\mathbf{g})$ is a locally finite ad $U(\mathbf{g})$ module.

In the quantum setting, $U_q(\mathbf{k})$ is not always a subalgebra of $U_q(\mathbf{g})$ when $\mathbf{k}$ is a Lie subalgebra of $\mathbf{g}$. However, one often finds a quantum analog of $U(\mathbf{k})$ which is a one-sided coideal subalgebra of $U_q(\mathbf{g})$. Thus any good theory of quantum

Harish-Chandra modules must work for pairs $U_q(\mathbf{g})$, I where I is a one-sided coideal subalgebra of $U_q(\mathbf{g})$. In order to begin such a theory, it is necessary to have an analog of Theorem 2.1. This presents two difficulties. The first is that U, in contrast to the classical situation, is not a locally finite ad U module. (We will return to this obstruction in Section 4.) The second is: what does it mean for a coideal subalgebra to be reductive in U?

In this section, we present a quantum version of Theorem 2.1 using the locally finite part $F(U)$ of U and a certain condition on coideal subalgebras which substitutes for reductivity. The material of this section is based on [L1, Section 4] and [L3, Section 3]. This result sets the stage for the development of a quantum Harish-Chandra module theory. Indeed, the author has checked that many of the results of [D, Section 9.1] and their proofs carry over to coideal subalgebras satisfying this quantum version of Theorem 2.1. Some properties of quantum principal series modules analogous to those in [D, Section 9.3] are proved in [L3, Section 6]. Spherical modules (see [D, 9.5.4]) have been classified in the quantum case (see Section 7, Theorem 7.7 and [L3, Section 4]). This is discussed further in Section 7.

Recall the definition of the adjoint action (1.2) and define

$$F(U) = \{v \in U \,|\, \dim(\mathrm{ad}\,U)v < \infty\}. \tag{2.1}$$

By [JL1, Corollary 2.3, Theorem 5.12, and Theorem 6.4], $F(U)$ is an algebra, it can be written as a direct sum of finite-dimensional simple U modules, and it is "large" in U. It is also true that $F(U)$ is a left coideal of U, a subject we will return to in Section 5.

Fix a left coideal subalgebra I of U. Note that

$$F(U)I = \{\textstyle\sum f_i r_i \,|\, f_i \in F(U), r_i \in I\}$$

is also a subalgebra of U. This follows from the fact that $rf = \sum r_{(1)}\varepsilon(r_{(2)})f = \sum((\mathrm{ad}\,r_{(1)})f)r_{(2)}$ for any $r \in I$ and $f \in F(U)$. Since I is a left coideal, each $r_{(2)} \in I$. Furthermore the ad-invariance of $F(U)$ implies that $(\mathrm{ad}\,r_{(1)})f$ is in $F(U)$. We use $F(U)I$ to define Harish-Chandra modules.

DEFINITION 2.2. A *Harish-Chandra module* for the pair U, I is an $F(U)I$ module which is a direct sum of finite-dimensional simple I modules.

Let us take a closer look at the condition that $\mathbf{k}$ is reductive in $\mathbf{g}$. Reductivity means that $(\mathrm{ad}\,\mathbf{k})$ acts semisimply on $\mathbf{g}$. This assumption is enough to prove that $\mathbf{k}$ is itself reductive and that the center of $\mathbf{k}$ can be extended to a Cartan subalgebra of $\mathbf{g}$. It is unclear what the corresponding assumption in the quantum case, namely that $(\mathrm{ad}\,I)$ acts semisimply on $F(U)$, implies. It seems unlikely that this assumption alone will yield an analog of Theorem 2.1.

Of course, there would be no problem if I acted semisimply on all finite-dimensional I modules. When I turns out to be a Hopf subalgebra of U isomorphic to a quantized enveloping algebra of a semisimple Lie subalgebra of $\mathbf{g}$, this is certainly true. However, complete reducibility does not hold in general for the large class of coideal subalgebras considered in Section 7. Thus we need a replacement for the notion of reductive in $\mathbf{g}$. This substitute is invariance under the action of a certain conjugate linear antiautomorphism of U.

Let κ denote the conjugate linear form of the quantum Chevalley antiautomorphism. In particular, let $U_{\mathbb{R}(q)}$ denote the $\mathbb{R}(q)$ subalgebra of U generated by $x_i, y_i, t_i^{\pm 1}$, for $1 \leq i \leq n$. The antiautomorphism κ of $U_{\mathbb{R}(q)}$ is defined by $\kappa(x_i) = y_i t_i$, $\kappa(y_i) = t_i^{-1} x_i$ and $\kappa(t) = t$ for all $t \in T$. We then extend κ to U using conjugation. More precisely, given $a \in \mathbb{C}$, write $\bar{a}$ for the complex conjugate of a. Set $\bar{q} = q$. Then $\kappa(au) = \bar{a}\kappa(u)$ for all $u \in U_{\mathbb{R}(q)}$.

It is straightforward to check using (1.8), (1.9), and (1.10) that

$$\Delta(\kappa(b)) = (\kappa \otimes \kappa)\Delta(b) \qquad (2.2)$$

for all $b \in U$. Moreover κ gives U the structure of a Hopf $*$ algebra where $* = \kappa$ ([CP, Section 4.1F]).

For the remainder of this section, we assume that I is a left coideal subalgebra such that $\kappa(I) = I$. Thus one can think of I as a $*$ subalgebra of U.

The field $\mathbb{R}(q)$ can be made into a real ordered field ([J, Section 11.1]) where the positive elements are defined as follows. Write a polynomial $f(q)$ in the form $(q-1)^s(f_m(q-1)^m + \cdots + f_1(q-1) + f_0)$ where each $f_i \in \mathbb{R}$ and $m, s \in \mathbb{N}$. Then $f(q)$ is positive if and only if $f_0 > 0$. An element $h \in \mathbb{R}(q)$ is positive if and only if h can be written as a quotient of positive polynomials. This induces a total order on $\mathbb{R}(q)$.

We next specify a class of "nice" finite-dimensional I modules.

DEFINITION 2.3. An I module W is called **unitary** if it admits a sesquilinear form S_W (i.e. linear in the first variable and conjugate linear in the second variable) such that

(i) $S_W(av, w) = S_W(v, \kappa(a)w)$ for all $a \in I$ and v, w in W
(ii) $S_W(v, v)$ is a positive element of $\mathbb{R}(q)$ for each nonzero vector $v \in W$
(iii) $S_W(v, w) = \overline{S_W(w, v)}$ for all v and w in W.

Let W be a finite-dimensional unitary I module. Choose a nonzero vector $v \in W$ such that $S_W(v, v) = 1$. Now suppose that $w \in W$ such that $S_W(w, v) = 0$. By Definition 2.3(iii), it follows that $S_W(v, w)$ also equals zero. Hence one can show using induction that W has an orthonormal basis with respect to S_W.

The following result and its corollary show that I has an extensive family of unitary modules, namely the finite-dimensional simple I submodules of any finite-dimensional simple U module.

THEOREM 2.4. *Every finite-dimensional unitary I module can be written as a direct sum of finite-dimensional simple unitary I modules.*

PROOF. Let W be a finite-dimensional unitary I module with sesquilinear form $S = S_W$ as in Definition 2.3. Let V be a finite-dimensional simple I submodule inside of W. By Definition 2.3, the restriction of S to W again satisfies conditions (i), (ii), and (iii). Furthermore, Definition 2.3(i) implies that the orthogonal complement $W^\perp$ of W with respect to S is an I module. Hence $V \cong W \oplus W^\perp$, a direct sum of unitary I modules with smaller dimension. The proof follows by induction on $\dim V$.   □

COROLLARY 2.5. *Every finite-dimensional simple U module V is a Harish-Chandra module for the pair U, I.*

PROOF. Let V be a finite-dimensional simple U module. It is well known that finite-dimensional U modules are a direct sum of their weight spaces. Moreover, the weight space of maximal weight is one dimensional. Let v be a basis vector for this highest weight space and note that v generates V as a U module. The vector v is called a highest weight generating vector of V. Recall that U_+^- denotes the augmentation ideal of U^-. Note that $U_+^- v$ is the subspace of V spanned by those weight vectors whose weights are strictly less than that of v. Furthermore, V is the direct sum of $\mathbb{C}(q)v$ and $U_+^- v$.

Let φ be the projection of U onto U^o using the direct sum decomposition $U = U^o \oplus Y$ (1.14). Define a sesquilinear form S on V by $S(v, v) = 1$ and $S(av, bv) = S(v, \varphi(\kappa(a)b)v)$ for all $a, b \in U$. Since $\varphi(b) = 0$ for all b in U_+^-, it follows that $S(v, U_+^- v) = 0$.

Note that S satisfies Definition 2.3(i). As in say ([L1, Lemma 4.2]), S specializes to the classical positive definite Shapovalov form of [Ka, 11.5 and Theorem 11.7]. Thus ([L1, Lemma 4.2]) $S(w, w) \neq 0$ for any nonzero vector $w \in V$. It is straightforward to check that S restricts to a $\mathbb{R}(q)$ bilinear form on $U_{\mathbb{R}(q)}v$. Moreover, S restricted to $\hat{U}_{\mathbb{R}(q)}$ takes values in $\mathbb{R}[q, q^{-1}]_{(q-1)}$. Let w be an element in $\hat{U}_{\mathbb{R}(q)}v$. We can write $S(w, w) = f(q)$ with $f(q)$ in $\mathbb{R}[q, q^{-1}]_{(q-1)}$. Since the specialization of S is positive definite, we must have that $f(1) \geq 0$. It follows that $f(q) \geq 0$. This fact and the nondegeneracy property implies that S satisfies Definition 2.3(ii).

Recall the direct sum decomposition (1.14) of U. Note that $\kappa(Y) = Y$ and $\kappa(a) = a$ for all $a \in U^o \cap U_{\mathbb{R}(q)}$. Therefore $\varphi(\kappa(b)) = \varphi(b)$ for all $b \in U_{\mathbb{R}(q)}$. It follows that S is symmetric when restricted to $U_{\mathbb{R}(q)}v$. In particular, S satisfies Definition 2.3(iii). Thus V is a unitary I module. The result now follows from Theorem 2.4.   □

Note that we cannot expect $\Delta(I)$ to be a subset of $I \otimes I$. Hence the tensor product of two I modules does not necessarily admit an action of I via the comultiplication of U. However, since I is a left coideal, the tensor product $V \otimes W$ of a U module V with an I module W is an I module. In particular, $a(v \otimes w) = \sum a_{(1)}v \otimes a_{(2)}w$ for all $v \otimes w \in V \otimes W$ and $a \in I$. The next lemma shows that the notion of unitary behaves well with respect to the tensor product of a U module with an I module.

LEMMA 2.6. *Let V be a finite-dimensional U module and let W be a finite-dimensional unitary I module. Then $V \otimes W$ is a finite-dimensional unitary I module.*

PROOF. Let S_V (resp. S_W) denote the sesquilinear form on V (resp. W) satisfying the conditions of Definition 2.3. Define a sesquilinear form $S = S_{V \otimes W}$ on $V \otimes W$ by setting $S(a \otimes b, a' \otimes b') = S_V(a, a')S_W(b, b')$. It is easy to check Definition 2.3(iii) holds for S. Let $\{v_i\}$ be an orthonormal basis for V with respect to S_V and let $\{w_i\}$ be an orthonormal basis for W with respect to S_W. Then $S(\sum b_{ij}v_i \otimes w_j, \sum b_{ij}v_i \otimes w_j) = \sum b_{ij}\bar{b}_{ij}$. Thus Definition 2.3(ii) holds for S. Condition (2.2) on κ ensures that S satisfies Definition 2.3(i). In particular, for $c \in I$, we have $S(c(a \otimes b), a' \otimes b') = S(\sum c_{(1)}a \otimes c_{(2)}b, a' \otimes b') = S(a \otimes b, \sum \kappa(c_{(1)})a' \otimes \kappa(c_{(2)})b') = S(a \otimes b, \kappa(c)(a' \otimes b'))$. $\square$

We now obtain a quantum analog of Theorem 2.1.

THEOREM 2.7. *The sum of all the finite-dimensional simple unitary I modules inside of the $F(U)I$ module M is a Harish-Chandra module for the pair U, I.*

PROOF. Assume that W is a finite-dimensional simple unitary I module contained in M. It suffices to show that the $F(U)I$ module generated by W is a direct sum of finite-dimensional simple unitary modules. Note that $F(U)IW = F(U)W = IF(U)W$ is an I module. The vector space $F(U) \otimes W$ is also an I module where the action is given by

$$a \cdot (f \otimes w) = \sum (\mathrm{ad}\, a_{(1)})f \otimes a_{(2)}w$$

for all $f \in F(U)$, $w \in W$, and $a \in I$. Furthermore, $F(U)W$ is a homomorphic image of the I module $F(U) \otimes W$. Recall that $F(U)$ is a direct sum of finite-dimensional simple $(\mathrm{ad}\, U)$ modules. By Corollary 2.5, each finite-dimensional simple $(\mathrm{ad}\, U)$ module is a unitary I module. Thus by Lemma 2.6, $F(U) \otimes W$, and hence $F(U)W$, splits into a direct sum of finite-dimensional simple unitary I modules. $\square$

Let $\mathcal{H}_\mathbb{R}$ be the set of all Hopf algebra automorphisms of U which restrict to a Hopf algebra automorphism of $U_{\mathbb{R}(q)}$. Let $\Upsilon \in \mathcal{H}_\mathbb{R}$. Suppose that I is a left coideal subalgebra such that $\Upsilon^{-1}\kappa\Upsilon(I) = I$. Then the results of this section hold for I where we define unitary I modules using $\Upsilon^{-1}\kappa\Upsilon$ instead of κ.

3. The Dual Picture

In this section, we consider the connection between coideal subalgebras of U and coideal subalgebras inside the Hopf dual of U. The results presented in this section are well known and are related to the theory of quantum homogeneous spaces. A good reference for most of the material presented here and for other basic results about quantum homogeneous spaces is [KS, Chapter 11] (see also [Jo, 1.4.15]).

Let $R_q[G]$ denote the quantized function algebra of the connected, simply connected algebraic Lie group G with Lie algebra $\mathbf{g}$. (See [Jo, Section 9.1] for a precise definition.) Note that up to a finite group, $R_q[G]$ is the Hopf dual of U. Furthermore, $R_q[G]$ satisfies a Peter–Weyl theorem ([Jo, 9.1.1 and 1.4.13]). That is, there is an isomorphism of U bimodules

$$R_q[G] \cong \bigoplus_{\lambda \in P^+(\pi)} L(\lambda) \otimes L(\lambda)^*. \tag{3.1}$$

Here $L(\lambda)$ is the (left) finite-dimensional simple U module with highest weight λ contained in the set $P^+(\pi)$ of dominant integral weights. Moreover, $L(\lambda)^*$ is thought of as a right U module. Thus, the right U module action on $R_q[G]$ comes from the action of U on $L(\lambda)^*$, while the left action comes from the action of U on $L(\lambda)$.

Given a left coideal I of U and a (left) U module M, a (left) invariant is an $m \in M$ such that $am = \varepsilon(a)m$ for all $a \in I$. Write M_I^I for the collection of all left invariants in M. Equivalently, M_I^I is equal to the elements of M annihilated (on the left) by the augmentation ideal of I. Consider the special case where I is the quantum analog of the enveloping algebra of a Lie subalgebra of $\mathbf{g}$ corresponding to a subgroup H of G. Then $R_q[G]_I^I$ is often written as $R_q[G/H]$. In particular, $R_q[G/H]$ is thought of as the quantized function algebra on the quotient space G/H.

THEOREM 3.1. *For any left coideal I of U, $R_q[G]_I^I$ is a left coideal subalgebra of $R_q[G]$.*

PROOF. Let ϕ, ϕ' be elements of $R_q[G]_I^I$ and r an element of I. We first show that $\phi\phi'$ is also in $R_q[G]_I^I$. To see this, consider

$$r \cdot (\phi\phi') = \sum (r_{(1)} \cdot \phi)(r_{(2)} \cdot \phi')$$
$$= \sum (r_{(1)} \cdot \phi)\varepsilon(r_{(2)})\phi' = (r \cdot \phi)\phi' = \varepsilon(r)\phi\phi'.$$

We now check the coideal condition. One can show using the precise definition of the action of U on $R_q[G]$ and the coalgebra structure of $R_q[G]$ that

$$\Delta(r \cdot \phi) = (1 \otimes r)\Delta(\phi) = \sum \phi_{(1)} \otimes r \cdot \phi_{(2)}. \tag{3.2}$$

On the other hand,

$$\Delta(r \cdot \phi) = \Delta(\varepsilon(r)\phi) = \sum \phi_{(1)} \otimes \varepsilon(r)\phi_{(2)}. \tag{3.3}$$

Since we can choose the $\phi_{(1)}$ to be linearly independent, (3.2) and (3.3) force $r \cdot \phi_{(2)} = \varepsilon(r)\phi_{(2)}$. Thus each $\phi_{(2)} \in R_q[G]_I^I$. $\quad\square$

In [KS, Chapter 11.6], a quantum homogeneous space associated to $R_q[G]$ is defined up to isomorphism as a one-sided coideal subalgebra of $R_q[G]$. (Note that quantum homogeneous spaces are actually quantum analogs of the algebra of regular functions on classical homogeneous spaces.) Thus the theorem above shows there is a left quantum homogeneous space, $R_q[G]_l^I$, associated to each left coideal subalgebra I. Using the Peter–Weyl decomposition (3.1), we obtain the following nice description of $R_q[G]_l^I$.

$$R_q[G]_l^I \cong \bigoplus_{\lambda \in P^+(\pi)} L(\lambda)_l^I \otimes L(\lambda)^*. \tag{3.4}$$

Now $R_q[G]_l^I$ is the set of left I invariants of $R_q[G]$. We may similarly define the set of right I invariants $R_q[G]_r^I$. Using the fact that the right action satisfies $\Delta(\phi \cdot r) = \Delta(\phi) \cdot (r \otimes 1)$, the same argument as in the proof of Theorem 3.1 shows that $R_q[G]_r^I$ is a right coideal subalgebra of $R_q[G]$. One may also study the set of bi-invariants $R_q[G]_{bi}^I = R_q[G]_l^I \cap R_q[G]_r^I$. As above, $R_q[G]_{bi}^I$ is a subalgebra of $R_q[G]^I$. However, it is not a coideal.

4. Generators and Filtrations

Consider the Hopf algebra $U(L)$, the universal enveloping algebra of a complex Lie algebra L. Since $U(L)$ is cocommutative, the one-sided coideal subalgebras of $U(L)$ are exactly the subbialgebras of $U(L)$. It is very easy to understand the coideal subalgebras of $U(L)$. Indeed, the next observation is well known. It follows from say [Mo, Theorem 5.6.5] and the fact that every subcoalgebra of $U(L)$ is connected ([Mo, Definition 5.1.5 and Lemma 5.1.9]). (Theorem 5.6.5 of [Mo] is stated for Hopf algebras, however, the proof also works for bialgebras.)

THEOREM 4.1. *The set of (left) coideal subalgebras of $U(L)$ is the set of enveloping algebras $U(L')$ of Lie subalgebras L' of L.*

An immediate consequence of the above result is that any coideal subalgebra of $U(L)$ is generated by elements of the underlying Lie algebra L. We would like to obtain a similar result for coideal subalgebras of the quantized enveloping algebra. However, passing to the quantum case, the situation becomes more complicated. Indeed the coalgebra structure is not cocommutative for quantized enveloping algebras. So the set of coideal subalgebras of the quantized enveloping algebra is much larger than the set of subbialgebras. By analyzing and deepening Lemma 1.3 and studying the comultiplication of U, we are able to obtain detailed information about coideal subalgebras and their generators.

The next result is known as well. It describes the coideal subalgebras of a group algebra.

LEMMA 4.2. *Let I be a (left) coideal subalgebra of the group algebra of the group G. Then $I \cap G$ is a semigroup and $I \cap kG$ is spanned by $I \cap G$ as a vector space.*

We introduce two subalgebras of U which are similar to U^- and G^+. Let U^+ be the subalgebra of U generated by $x_1, \ldots, x_n$ and G^- be the subalgebra of U generated by $y_1 t_1, \ldots, y_n t_n$. Once again, we have that U^+ and G^- are a direct sum of their weight spaces. We may replace U^- by G^- and G^+ by U^+ to obtain the following version of the triangular decomposition.

$$U \cong G^- \otimes U^o \otimes U^+. \tag{4.1}$$

In this section, we show how to break up a coideal subalgebra into three parts corresponding to coideal subalgebras of G^-, U^o, and U^+ respectively. First, however, we obtain basic properties of coideal subalgebras of G^- and U^+.

Using the formulas for comultiplicaton (1.8), (1.9), and (1.10), it is straightforward to check that G^- and U^+ are left coideal subalgebras of U. Consider now an arbitrary coideal subalgebra J of U which is either a subset of G^- or U^+. Note that if J is also an $\operatorname{ad} T$ module, then J can be written as a direct sum of its weight spaces. We obtain a nice result on the generators of J analogous to Theorem 4.1 when J is an $\operatorname{ad} T$ module.

LEMMA 4.3. *Let J be an $\operatorname{ad} T$ submodule and a coideal subalgebra of G^- (resp. U^+). Then there exists a subset Δ' of Δ^+ and weight vectors $\tilde{f}_{-\gamma}$ of weight $-\gamma$, $\gamma \in \Delta'$ (resp. $\tilde{e}_\gamma$ of weight $\gamma \in \Delta'$) which generate J as an algebra. Moreover, each $\tilde{f}_{-\gamma}$ (resp. $\tilde{e}_\gamma$) specializes to the root vector $f_{-\gamma}$ (resp. e_γ) as q goes to 1.*

PROOF. Note that the weight spaces of G^- are finite-dimensional. Hence J has finite-dimensional weight spaces. Let $\bar{J}$ denote the specialization of J as q goes to 1. Consider a weight space J_μ of J. We have that $\hat{J}_\mu = \hat{U} \cap J_\mu = \hat{G}^- \cap J_\mu$. Also, G^- is a free A module and A is a principal ideal domain with unique maximal ideal generated by $(q-1)$. Hence one can find a basis for J_μ which is a subset of $\hat{J}_\mu$ and remains linearly independent modulo $(q-1)\hat{U}$. In particular, the specialization of this basis as q goes to 1 is a basis for $\bar{J}_\mu$. Hence the weight spaces of $\bar{J}$ have the same dimension as the weight spaces of J.

Note that the comultiplication of U specializes to the comultiplication of $U(\mathbf{g})$. Hence $\bar{J}$ is a coideal subalgebra of $U(\mathbf{n}^-)$. By Theorem 4.1, $\bar{J}$ is an enveloping algebra of a Lie subalgebra, say $\mathbf{a}$, of $\mathbf{n}^-$. Now $\bar{J}$ is a direct sum of its weight spaces. Hence there exists a subset Δ' of Δ^+ such that the set $\{f_{-\gamma} | \gamma \in \Delta'\}$ is a basis of $\mathbf{a}$. Thus for each $\gamma \in \Delta'$, we can find a vector $\tilde{f}_{-\gamma}$ of weight $-\gamma$ in J such the image of $\tilde{f}_{-\gamma}$ under specialization is $f_{-\gamma}$. Write $\Delta' = \{\gamma_1, \dots, \gamma_m\}$. A standard argument shows that the set

$$\mathcal{B}_\eta = \{f_{-\gamma_1}^{i_1} \cdots f_{-\gamma_m}^{i_m} | i_j \in \mathbb{N} \text{ for } 1 \leq j \leq m \text{ and } i_1\gamma_1 + \dots + i_m\gamma_m = \eta\}$$

is a basis for the $-\eta$ weight space of $U(\mathbf{a})$. Furthermore $\mathcal{B} = \cup_\eta \mathcal{B}_\eta$ is a basis for $U(\mathbf{a})$. Since the dimensions of the $-\eta$ weight spaces of $U(\mathbf{a})$ and J agree, the corresponding set $\tilde{\mathcal{B}}_\eta$ with $\tilde{f}$ playing the role of f is a basis of J_η. Thus the set $\tilde{\mathcal{B}} = \cup_\eta \tilde{\mathcal{B}}_\eta$ is a basis for J. It follows that the $\tilde{f}_{-\gamma}$, $\gamma \in \Delta'$ generate J as an algebra.

The same analysis applies to coideal subalgebras of U^+. $\square$

The direct sum decomposition (1.13) can be made finer using weight spaces. It is well known that the set of weights of G^+ and U^+ equals $Q^+(\pi)$ and the set of weights of G^- and U^- equals $Q^-(\pi)$. Thus there are direct sum decompositions

$$U = \oplus_{\lambda,\mu} U^-_{-\lambda} G^+_\mu U^\circ \text{ and } U = \oplus_{\lambda,\mu,t} U^-_{-\lambda} G^+_\mu t \qquad (4.2)$$

where λ and μ run over elements of $Q^+(\pi)$ and t runs over elements in T. Let $\pi_{\lambda,\mu}$ be the projection of U onto the subspace $U^-_{-\lambda} G^+_\mu U^\circ$. Write $[\lambda,\mu]$ for a typical element in $Q(\pi) \times Q(\pi)$ (so as to avoid confusion with the Cartan inner product).

Consider elements $c \in U^-_{-\lambda}$ and $d \in G^+_\mu$. It follows from the definition of the comultiplication map on the generators of U ((1.8), (1.9), and (1.10)) that

$$(\pi_{\lambda,\mu} \otimes \text{Id})\Delta(cd) = cd \otimes \tau(-\lambda - \mu), \qquad (4.3)$$

$$(\pi_{\lambda,0} \otimes \text{Id})\Delta(cd) = \sum c \otimes \tau(-\lambda)d, \qquad (4.4)$$

$$(\pi_{0,\mu} \otimes \text{Id})\Delta(cd) = \sum d \otimes c\tau(-\mu). \qquad (4.5)$$

In the next theorem, we consider coideal subalgebras of U which behave rather nicely in terms of the second decomposition in (4.2).

THEOREM 4.4. *Let I be a left coideal subalgebra of U such that*

$$I = \sum_{\lambda,\mu,t} (I \cap U^-_{-\lambda} G^+_\mu t) \qquad (4.6)$$

and $I \cap T$ is a group. Then $I \cap G^-$, $I \cap U^\circ$, and $I \cap U^+$ are ad T submodules and left coideal subalgebras of I. Moreover, the multiplication map induces an isomorphism

$$I \cong (I \cap G^-) \otimes (I \cap U^+) \otimes (I \cap U^\circ)$$

of vector spaces.

PROOF. Since I, G^-, U°, and U^+ are all left coideal subalgebras, so are $I \cap G^-$, $I \cap U^\circ$, and $I \cap U^+$. Note that every element in $U^-_{-\lambda} G^+_\mu t$ is a weight vector of weight $-\lambda + \mu$. Thus I is a direct sum of its weight spaces and I is an ad T module. It follows that $I \cap G^-$, $I \cap U^\circ$, and $I \cap U^+$ are all ad T modules.

The triangular decomposition of U (4.1) ensures that the multiplication map induces an injection

$$(I \cap G^-) \otimes (I \cap U^+) \otimes (I \cap U^\circ) \to I$$

of vector spaces. We obtain an isomorphism by showing that each element of I is contained in $(I \cap G^-)(I \cap U^+)(I \cap U^\circ)$.

Recall the direct sum decomposition of I given in Lemma 1.3. Let b be an element of $I \cap U^-_{-\lambda} G^+_\mu t$ where $t \in T$. There exists $c_i \in U^-_{-\lambda}$ and $d_i \in G^+_\mu$ so that $b = \sum_i c_i d_i t$. We may further assume that $\{c_i\}$ and $\{d_i\}$ are each linearly independent sets. By (1.8) and (4.3), $(\pi_{\lambda,\mu} \otimes \text{Id})\Delta(b) = b \otimes \tau(-\lambda - \mu)t$. Hence

$\tau(-\lambda - \mu)t$ is an element of $I \cap T$. Since $I \cap T$ is a group, $\tau(\lambda + \mu)t^{-1}$ is also contained in $I \cap T$.

Equation (4.4) implies that

$$(\pi_{\lambda,0} \otimes \mathrm{Id})\Delta(b) = \sum_i c_i t \otimes \tau(-\lambda)d_i t.$$

Hence each $\tau(-\lambda)d_i t \in I$. Recall that d_i is a weight vector of weight μ in G^+. Thus, multiplying $\tau(-\lambda)d_i t$ by $\tau(\lambda + \mu)t^{-1}$ yields that $d_i \tau(\mu)$ is an element of $U_\mu^+ \cap I$. Similarly, (4.5) ensures that

$$(\pi_{0,\mu} \otimes \mathrm{Id})\Delta(b) = \sum_i d_i t \otimes c_i \tau(-\mu)t.$$

So $c_i \tau(-\mu)t \in I$ and hence $c_i \tau(-\mu)t\tau(\lambda+\mu)t^{-1} = c_i \tau(\lambda)$ is an element of $I \cap G^-$. Therefore,

$$b = \sum_i c_i d_i t = \sum_i q^{(-\lambda,\mu)}(c_i \tau(\lambda))(d_i \tau(\mu))\tau(-\lambda - \mu)t$$

$$\in (I \cap G^-)(I \cap U^+)(I \cap U^\circ). \qquad \square$$

Let I be a left coideal subalgebra of U such that $I \cap T$ is a group and I satisfies (4.6). Then Theorem 4.4 combined with Lemmas 4.2 and 4.3 imply that I is generated by $I \cap T$ and quantum analogs of root vectors in G^- and U^+. This description of the generators can be thought of as an analog of Theorem 4.1 for these special coideal subalgebras. Below, we generalize these results to other coideal subalgebras by introducing filtrations and associated gradings of U.

Filtration I. Define the filtration $\mathcal{F}$ on U using the degree function:

$$\deg x_i t_i^{-1} = \deg y_i = 1 \qquad \deg t_i = -1$$

for all $1 \le i \le n$. Write $\mathrm{gr}_{\mathcal{F}} U$ for the associated graded algebra of this filtration. This filtration is invariant under the adjoint action and used to understand the locally finite part of U (see [JL2, Section 2.2]). (It should be noted that the quantized enveloping algebra is defined in a different though equivalent manner in [JL2]. So the x_i (resp. t_i) in this paper corresponds to $x_i t_i$ (resp. t_i^2) in [JL2]. Furthermore the degree of an element as defined in [JL2] is twice the degree of the corresponding element given here.)

Given $\gamma = \sum_{\alpha_i \in \pi} m_i \alpha_i$, set $\mathrm{ht}(\gamma) = \sum_i m_i$. Note that any nonzero element of $U_{-\lambda}^- G_\mu^+$ has degree $\mathrm{ht}(\lambda + \mu)$. Let $x \in U$ and set $\mathrm{supp}(x) = \{[\lambda, \mu] | \pi_{\lambda,\mu}(x) \ne 0\}$. Further, for x an element of $U^- G^+ t$ for some $t \in T$, set

$$\mathrm{max_{ht}}(x) = \{[\lambda, \mu] | [\lambda, \mu] \in \mathrm{supp}(x) \text{ and } \mathrm{ht}(\lambda + \mu) = \deg(x) - \deg(t)\}.$$

The next lemma connects the filtration $\mathcal{F}$ with the height function.

LEMMA 4.5. *Let I be a left coideal of U and let b be an element of $I \cap U^- G^+ t$ for some $t \in T$. Then*

$$b = \sum_{\{[\lambda,\mu]|\ [\lambda,\mu]\in\text{max}_{\text{ht}}(b)\}} \pi_{\lambda,\mu}(b) + \text{ lower degree terms.}$$

PROOF. The lemma follows from the fact that $\deg \pi_{\lambda,\mu}(b) = \deg b$ if and only if $[\lambda, \mu] \in \text{max}_{\text{ht}}(b)$. □

By induction on $\text{ht}(\lambda + \mu)$ and the definition of the comultiplication (1.8), (1.9), and (1.10), we have the following:

$$\Delta(U_{-\lambda}^- G_\mu^+) \subset \sum_{\gamma+\beta=\lambda,\alpha+\xi=\mu} U_{-\gamma}^- G_\alpha^+ \otimes U_{-\beta}^- G_\xi^+ \tau(-\gamma - \alpha). \qquad (4.7)$$

Consider a subset S of $Q^+(\pi) \times Q^+(\pi)$. Set $|S|$ equal to the number of elements in S. We call S *transversal* if whenever both $[\lambda, \mu]$ and $[\lambda', \mu']$ are in S and $[\lambda, \mu] \neq [\lambda', \mu']$ then $\lambda \neq \lambda'$ and $\mu \neq \mu'$. Now assume that $b \in U^- G^+ t$ and that $\text{max}_{\text{ht}}(b)$ is transversal. Given $[\lambda, \mu] \in \text{max}_{\text{ht}}(b)$, find $c_i \in U_{-\lambda}^-$ and $d_i \in G_\mu^+$ such that $\pi_{\lambda,\mu}(b) = \sum_i c_i d_i t$. As in the proof of Theorem 4.4, we may further assume that $\{c_i\}$ and $\{d_i\}$ are each linearly independent sets. It follows from (4.7), (4.3), (4.4), and (4.5) that

$$(\pi_{\lambda,\mu} \otimes \text{Id})\Delta(b) = \sum_i c_i d_i t \otimes \tau(-\lambda - \mu)t, \qquad (4.8)$$

$$(\pi_{\lambda,0} \otimes \text{Id})\Delta(b) = \sum_i c_i t \otimes (\tau(-\lambda)d_i t + \text{terms of lower degree}) \qquad (4.9)$$

$$(\pi_{0,\mu} \otimes \text{Id})\Delta(b) = \sum d_i t \otimes (c_i \tau(-\mu)t + \text{terms of lower degree}) \qquad (4.10)$$

A consequence of the next lemma is that any left coideal subalgebra which is also an $\text{ad}\,T$ module has a basis $\mathcal{B}$ such that $\text{max}_{\text{ht}}(b)$ is transversal for each $b \in \mathcal{B}$. This in turn is used to generalize Theorem 4.4.

LEMMA 4.6. *Let $b \in U$ be a weight vector. Then $\text{max}_{\text{ht}}(b)$ is transveral.*

PROOF. Fix η and let b be an element in U of weight η. Note that $\pi_{\lambda,\mu}(b) \neq 0$ implies that $-\lambda+\mu = \eta$. Now assume that both $[\lambda, \mu]$ and $[\lambda', \mu']$ are in $\text{supp}(b)$. Hence $-\lambda + \mu = -\lambda' + \mu'$. Thus $\lambda = \lambda'$ if and only if $\mu = \mu'$. In particular, $\text{supp}(b)$ is transversal. The lemma now follows from the fact that $\text{max}_{\text{ht}}(b)$ is a subset of $\text{supp}(b)$. □

Given a left coideal subalgebra I of U, set I_η^- equal to the subset of G^- such that $I \cap G^- \tau(\eta) = I_\eta^- \tau(\eta)$. Similarly, set I_η^+ equal to the subset of U^+ such that $I \cap U^+ \tau(\eta) = I_\eta^+ \tau(\eta)$. The following result can be thought of as an analog of Theorem 4.4 for coideal subalgebras which admit an $\text{ad}\,T$ module structure.

THEOREM 4.7. *Let I be a left coideal subalgebra and $\text{ad}\,T$ submodule of U. Then*

$$\text{gr}_{\mathcal{F}} I \subset \sum_{\{\eta|\tau(\eta)\in I\cap T\}} \text{gr}_{\mathcal{F}} I_\eta^- I_\eta^+ \tau(\eta).$$

PROOF. Let b be a weight vector of I which is also contained in $I \cap U^- G^+ \tau(\beta)$ for some $\tau(\beta) \in T$. By Lemma 4.6, $\max_{\text{ht}}(b)$ is transversal. We prove the theorem when $\max_{\text{ht}}(b)$ contains exactly one element $[\lambda, \mu]$. The same argument works in general. We argue as in the proof of Theorem 4.4. Find $c_i \in U_{-\lambda}^-$ and $d_i \in G_\mu^+$ so that $\pi_{\lambda,\mu}(b) = \sum_i c_i d_i \tau(\beta)$. We may further assume that $\{c_i\}$ and $\{d_i\}$ are each linearly independent sets. By our assumption on $\max_{\text{ht}}(b)$ and Lemma 4.5,

$$b = \sum_i c_i d_i \tau(\beta) + \text{ lower degree terms.}$$

Set $\eta = -\lambda - \mu + \beta$. By (4.8), $\tau(\eta)$ is in I. Now (4.9) implies that there exist elements $\tau(-\lambda) D_i \tau(\beta) \in I$ such that $D_i = d_i+$ lower degree terms and

$$(\pi_{\lambda,0} \otimes \text{Id})\Delta(b) = \sum_i c_i \otimes \tau(-\lambda) D_i \tau(\beta).$$

Note that (4.7) ensures that $D_i - d_i$ is an element of $U^- G^+ \tau(-\lambda + \beta)$. Also, d_i is in $G_\mu^+ \tau(-\lambda + \beta)$. Thus d_i has degree $\text{ht}(\mu + \lambda - \beta)$. By Lemma 4.5, $[\xi, \gamma] \in \text{supp}(D_i - d_i)$ implies that $\text{ht}(\xi + \gamma) < \text{ht}(\mu)$. Since ξ is in $Q^+(\pi)$, we also have $\text{ht}(-\xi + \gamma) < \text{ht}(\mu)$ and thus $-\xi + \gamma$ is not equal to μ. Therefore, for each $[\xi, \gamma] \in \text{supp}(D_i - d_i)$, $\pi_{\xi,\gamma}(D_i - d_i)$ has weight $-\xi + \gamma$ which is different from the weight μ of d_i. Since I is an $\text{ad}\, T$ module, it follows that the μ weight term of $\tau(-\lambda) D_i \tau(\beta)$, namely $\tau(-\lambda) d_i \tau(\beta)$, is contained in I. Hence $d_i \tau(\mu) \tau(\eta) \in I \cap U^+ \tau(\eta)$ and $d_i \tau(\mu) \in I_\eta^+$. A similar argument shows that $c_i \tau(\lambda) \in I_\eta^-$. Therefore

$$\text{gr}_{\mathcal{F}} b = \text{gr}_{\mathcal{F}} \sum_i c_i d_i \tau(\eta) = \text{gr}_{\mathcal{F}} \sum_i q^{-(\lambda,\mu)} (c_i \tau(\lambda))(d_i \tau(\mu)) \tau(\eta)$$

is an element of $\text{gr}_{\mathcal{F}} I_\eta^- I_\eta^+ \tau(\eta)$.   □

Filtration II. Order the set $\mathbb{N} \times \mathbb{N}$ lexicographically from left to right. Define a filtration on U by

$$\mathcal{G}_{m,n}(U) = \{u \in U | \ (\text{ht}(\lambda), \text{ht}(\mu)) \leq (m, n) \text{ for all } [\lambda, \mu] \in \text{supp}(u)\}.$$

The associated graded algebra for this filtration is defined by setting

$$\text{gr}_{\mathcal{G}}^{m,n}(U) = \mathcal{G}_{m,n}(U) \Big/ \sum_{(m',n')<(m,n)} \mathcal{G}_{m,n}(U)$$

and

$$\text{gr}_{\mathcal{G}}(U) = \bigoplus_{m,n} \text{gr}_{\mathcal{G}}^{m,n}(U).$$

Given a subset S of $Q^+(\pi) \times Q^+(\pi)$, set $\|S\|_1 = \max_{[\lambda,\mu] \in S}\{\text{ht}(\lambda)\}$. We can define a bidegree: for x in U, we say that $\text{bideg}(x) = (m, n)$ if (m, n) is the smallest element of $\mathbb{N} \times \mathbb{N}$ such that $x \in \mathcal{G}_{m,n}(U)$. Set $\max(x) = \{[\lambda, \mu] | [\lambda, \mu] \in$

$\mathrm{supp}(x)$ and $\mathrm{bideg}(x) = (\mathrm{ht}(\lambda), \mathrm{ht}(\mu))\}$. Now consider an element $b \in U^- G^+ t$ for some $t \in T$. The inclusion (4.7) implies the following variation of (4.3):

$$(\pi_{\lambda,\mu} \otimes \mathrm{Id})(b) = \pi_{\lambda,\mu}(b) \otimes t\tau(-\lambda - \mu) \text{ for all } [\lambda, \mu] \in \max(b). \qquad (4.11)$$

LEMMA 4.8. *Let I be a left coideal subalgebra such that $I \cap T$ is a group. Then I has a basis $\mathcal{B}$ such that for each $b \in \mathcal{B}$, $\max(b)$ is transversal.*

PROOF. Recall that I is a direct sum of the subspaces $I \cap U^- G^+ t$. Let $C = \{x \in I \mid \max(x) \text{ is transversal}\}$. It is enough to show that for each $t \in T$, every element of $I \cap U^- G^+ t$ is contained in the span of C. Consider $b \in I \cap U^- G^+ t$. We prove this result under the additional assumption that $\max(b)$ consists of exactly two elements $[\lambda, \mu]$ and $[\lambda, \mu']$. A similar argument works in the general case using induction on $|\max(b)|$ and $\|\max(b)\|_1$. Note that

$$b = \pi_{\lambda,\mu}(b) + \pi_{\lambda,\mu'}(b) + \text{ terms of lower bidegree.}$$

By (4.11), $t\tau(-\lambda - \mu)$ and $t\tau(-\lambda - \mu')$ are both elements of the group $I \cap T$. Hence $\tau(\mu - \mu')$ is contained in $I \cap T$. Consider the element

$$b' = \tau(\mu - \mu')b\tau(\mu - \mu')^{-1} = q^{(-\lambda+\mu,\mu-\mu')}\pi_{\lambda,\mu}(b) + q^{(-\lambda+\mu',\mu-\mu')}\pi_{\lambda,\mu'}(b)$$

$$+ \text{ terms of lower bidegree.}$$

Now $(\mu - \mu', \mu - \mu')$ is positive since $(\,,\,)$ is positive definite on $Q(\pi)$. Hence $q^{(-\lambda+\mu,\mu-\mu')} \neq q^{(-\lambda+\mu',\mu-\mu')}$. Taking linear combinations of b and b', it follows that there exists b_1 and b_2 in $U^- G^+ t \cap I$ such that $\{[\lambda, \mu]\} = \max(b_1)$ and $\{[\lambda, \mu']\} = \max(b_2)$. In particular, both $\max(b_1)$ and $\max(b_2)$ are transversal and b is a linear combination of b_1 and b_2. $\qquad \square$

Now assume that $b \in U^- G^+ t$ and that $\max(b)$ is transversal. We have versions of (4.4) and (4.5) similar to (4.9) and (4.10) in the discussion of the first filtration. Given $[\lambda, \mu] \in \max(b)$, find $c_i \in U^-_{-\lambda}$ and $d_i \in G^+_\mu$ so that $\pi_{\lambda,\mu}(b) = \sum_i c_i d_i t$ and that $\{c_i\}$ and $\{d_i\}$ are each linearly independent sets. It follows from (4.7), (4.4), and (4.5) that

$$(\pi_{\lambda,0} \otimes \mathrm{Id})\Delta(b) = \sum_i c_i t \otimes (\tau(-\lambda)d_i t + \text{terms of lower bidegree}) \quad (4.12)$$

$$(\pi_{0,\mu} \otimes \mathrm{Id})\Delta(b) = \sum d_i t \otimes (c_i \tau(-\mu)t + \text{terms of lower bidegree}). \quad (4.13)$$

The filtration $\mathcal{G}$ restricts to filtrations on the subalgebras G^+, U^- and U°. Indeed, $U^\circ = \mathcal{G}_{0,0}(U)$ and so $\mathrm{gr}_\mathcal{G} U^\circ \cong U^\circ$ as algebras. Upon restriction to G^+, the filtration $\mathcal{G}$ becomes filtration by the degree function associated to the first filtration $\mathcal{F}$. The subalgebra of G^+ satisfies exactly the same relations as U^+. In particular, the only relations satisfied by the elements of G^+ are the quantum Serre relations (1.7) (see the discussion in Section 7 concerning (7.18)) which are homogeneous with respect to degree. Hence we have an algebra isomorphism $\mathrm{gr}_\mathcal{G} G^+ \cong G^+$. A similar argument shows that $\mathrm{gr}_\mathcal{G} U^- \cong U^-$. Since the elements in T have bidegree $(0,0)$, we further have that $\mathrm{gr}_\mathcal{G} G^- \cong G^-$ and $\mathrm{gr}_\mathcal{G} U^+ \cong U^+$. For the rest of the paper, we will often identify $\mathrm{gr}_\mathcal{G} G^-$ with G^-, $\mathrm{gr}_\mathcal{G} U^+$ with U^+, and $\mathrm{gr}_\mathcal{G} U^\circ$ with U°.

Now the images of x_i and y_j commute with each other inside the associated graded algebra of U with respect to $\mathcal{G}$. (See relation (1.4) of U.) It follows that the image of $U^- U^+$ in the associated graded algebra is isomorphic to the tensor product $U^- \otimes U^+$ as an algebra. If we replace U^- by G^-, the images of the elements x_i and $y_j t_j$ do not commute. However, they do commute up to a power of q. Thus the image of $G^- U^+$ in the associated graded algebra can be thought of as a q form of the tensor product which we write as $G^- \otimes_q U^+$.

The group algebra U° acts on weight vectors by $\tau(\lambda) \cdot a_\mu = \tau(\lambda) a_\mu \tau(\lambda)^{-1} = q^{(\lambda,\mu)} a_\mu$ for $a_\mu \in U_\mu$. Thus we obtain the following algebra isomorphism using a smash product construction:

$$\mathrm{gr}_\mathcal{G}(U) \cong U^\circ \#(G^- \otimes_q U^+). \tag{4.14}$$

(Compare this with a similar result for a different filtration in [Jo, 7.4.7].)

THEOREM 4.9. *Let I be a left coideal subalgebra such that $I \cap T$ is a group. Then*

$$\mathrm{gr}_\mathcal{G}(I) \cong (I \cap U^\circ) \#((\mathrm{gr}_\mathcal{G}(I) \cap G^-) \otimes_q ((\mathrm{gr}_\mathcal{G}(I) \cap U^+))).$$

PROOF. The proof is a graded version of the proof of Theorem 4.4. Using Lemma 1.3 and Lemma 4.8, we can find a basis $\mathcal{B}$ of I such that $\mathcal{B} = \bigcup_t (\mathcal{B} \cap U^- G^+ t)$ and $\max(b)$ is transversal for each $b \in \mathcal{B}$. By (4.14),

$$(I \cap U^\circ) \#((\mathrm{gr}_\mathcal{G}(I) \cap G)^- \otimes_q ((\mathrm{gr}_\mathcal{G}(I) \cap U^+))$$

is isomorphic to a subalgebra of $\mathrm{gr}_\mathcal{G} I$. To show this subalgebra is all of $\mathrm{gr}_\mathcal{G} I$ it is sufficient to show that each element of $\mathcal{B}$ is contained in $((\mathrm{gr}_\mathcal{G}(I) \cap G^-)((\mathrm{gr}_\mathcal{G}(I) \cap U^+))\mathrm{gr}_\mathcal{G}(I \cap U^\circ)$.

Let t be an element of T and let b be an element of $\mathcal{B} \cap U^- G^+ t$. Choose $[\lambda, \mu] \in \max(b)$. There exists $c_i \in U_{-\lambda}^-$ and $d_i \in G_\mu^+$ so that $\pi_{\lambda,\mu}(b) = \sum_i c_i d_i t$ and the $\{c_i\}$ and $\{d_i\}$ are each linearly independent sets. Using (4.11), (4.12), and (4.13) and arguing as in the proofs of Theorem 4.4 and Theorem 4.7, I contains $\tau(-\lambda - \mu)t$ and $\tau(\lambda + \mu)t^{-1}$ and elements $\tilde{d}_i$ and $\tilde{c}_i$ such that

$$\tilde{d}_i = d_i \tau(\mu) + \text{ terms of lower bidegree}$$

and

$$\tilde{c}_i = c_i \tau(\lambda) + \text{ terms of lower bidegree}.$$

Note that $\mathrm{gr}_\mathcal{G} \tilde{d}_i \in \mathrm{gr}_\mathcal{G}(I) \cap U^+$ and $\mathrm{gr}_\mathcal{G} \tilde{c}_i \in \mathrm{gr}_\mathcal{G}(I) \cap G^-$. Set

$$b' = b - \sum_i q^{-(\lambda,\mu)} \tilde{c}_i \tilde{d}_i \tau(-\mu - \lambda)t.$$

Note that b' is in I. By construction, $\pi_{\lambda,\mu}(b') = 0$. Thus either $\max(b') = \max(b) - \{[\lambda,\mu]\}$ or the bidegree of b' is strictly smaller than the bidegree of b. The theorem now follows by induction on $|\max(b)|$ and the bidegree of b. $\square$

Consider a left coideal subalgebra I such that $I \cap T$ is a group. Given x in U, set $\mathrm{tip}(x) = \sum_{[\lambda,\mu] \in \max(x)} \pi_{\lambda,\mu}(x)$. The element $\mathrm{tip}(x)$ can be thought of as the highest bidegree term of x. Note that $\mathrm{gr}_{\mathrm{g}}(I) \cap \mathrm{gr}_{\mathrm{g}}(G^-)$ identifies with $\mathrm{tip}(I) \cap G^-$ under the isomorphism $G^- \cong \mathrm{gr}_{\mathrm{g}}(G^-)$. Thus $\mathrm{tip}(I) \cap G^-$ is a subalgebra of G^-. Consider the elements $\tilde{c}_i$ in the proof of Theorem 4.9. Note that each $\mathrm{tip}(\tilde{c}_i)$ is a weight vector. In particular, it follows implicitly from the proof of Theorem 4.9 that $\mathrm{tip}(I) \cap G^-$ is spanned by weight vectors and hence is an $\mathrm{ad}\,T$ module. One can further show using (4.7) that $\mathrm{tip}(I) \cap G^-$ is a left coideal of G^-. Thus $\mathrm{tip}(I) \cap G^-$ is a left coideal subalgebra and $\mathrm{ad}\,T$ submodule of G^-. Similarly, $\mathrm{tip}(I) \cap U^+$ is a left coideal subalgebra and $\mathrm{ad}\,T$ submodule of U^+. Thus combining Theorem 4.9 with Lemmas 4.2 and 4.3 yields the following.

COROLLARY 4.10.   *Let I be a left coideal subalgebra of U such that $I \cap T$ is a subgroup of T. Then there exists subsets Δ' and Δ'' of Δ^+ such that I is generated by elements $c_{-\gamma}, \gamma \in \Delta'$; $d_\beta, \beta \in \Delta''$; and $I \cap T$. Moreover each $\mathrm{tip}(c_{-\gamma})$ (resp. $\mathrm{tip}(d_\beta)$) is a weight vector of weight $-\gamma$ (resp. β) which specializes to the root vector $f_{-\gamma}$ (resp. e_β) of $U(\mathbf{g})$.*

5. The Locally Finite Part of U

One of the most important coideal subalgebras contained in the quantized enveloping algebra is the locally finite part, $F(U)$, defined by (2.1). This subalgebra is studied extensively in [JL1] and [JL2] (see also [Jo]). Here we present some of the known results about this algebra by directly showing that $F(U)$ is a coideal subalgebra of U. We will see some of the implications of Section 4 on the structure of $F(U)$.

Recall that $F(U)$ is defined using the quantum adjoint action in Section 2. It is helpful to see how the generators of U act via the adjoint action. In particular

$$(\mathrm{ad}\ y_i)b = y_i b t_i - b y_i t_i \quad (\mathrm{ad}\ x_i)b = x_i b - t_i b t_i^{-1} x_i \quad (\mathrm{ad}\ t_i)b = t_i b t_i^{-1}$$

for all $1 \le i \le n$ and $b \in U$.

THEOREM 5.1.   *$F(U)$ is a left coideal subalgebra of U.*

PROOF.   Let $b \in F(U)$. A straightforward computation shows

$$\Delta((\mathrm{ad}\,x_i)b) = \sum x_i b_{(1)} \otimes b_{(2)} - \sum t_i b_{(1)} t_i^{-1} x_i \otimes t_i b_{(2)} t_i^{-1}$$

$$+ \sum t_i b_{(1)} \otimes (\mathrm{ad}\,x_i)b_{(2)} \quad (5.1)$$

for each $1 \le i \le n$.

We may write $\Delta(b)$ as a sum $\sum_{j=1}^{s} c_j \otimes b_j$ where the b_j, $1 \le j \le s$, are linearly independent weight vectors in U of weight λ_j respectively. Extend the standard partial ordering on the integral lattice $Q(\pi)$ to a total archimedean ordering. (This can be done by embedding $Q^+(\pi)$ in the nonnegative real numbers.) We may further suppose that $\lambda_1 \geq \lambda_2 \geq \cdots \geq \lambda_s$. For sake of simplicity, assume that these inequalities are all strict. (A similar argument works in general.)

Let U_i be the subalgebra of U generated by $x_i, y_i, t_i^{\pm 1}$. Note that U_i is isomorphic to $U_q(\mathrm{sl}\,2)$. We show below that each b_j generates a finite-dimensional $\mathrm{ad}\,U_i$ module for $1 \leq i \leq n$. By [JL1, Theorem 5.9], this forces each b_j to be an element of $F(U)$ (see also the proof of [JL1, Proposition 6.5]).

Suppose that $(\mathrm{ad}\,x_i)^m b = 0$. Using (5.1) and induction, we obtain

$$\Delta((\mathrm{ad}\,x_i)^m b) \in t_i^m c_1 \otimes (\mathrm{ad}\,x_i)^m b_1 + \sum_{\beta < \lambda_1 + m\alpha_i} U \otimes U_\beta.$$

Hence $(\mathrm{ad}\,x_i)^m b_1 = 0$. Choose r such that $(m-1)\alpha_i + \lambda_1 < (m+r)\alpha_i + \lambda_2$. We further have that

$$\Delta((\mathrm{ad}\,x_i)^{m+r} b) \in t_i^{m+r} c_2 \otimes (\mathrm{ad}\,x_i)^{m+r} b_2 + \sum_{\beta < \lambda_2 + (m+r)\alpha_i} U \otimes U_\beta.$$

Thus $(\mathrm{ad}\,x_i)^m b = 0$ also implies that $(\mathrm{ad}\,x_i)^{m+r} b_2 = 0$. By induction, it follows that there exists $M > 0$ such that $(\mathrm{ad}\,x_i)^M b_j = 0$ for all $1 \leq j \leq s$ and $1 \leq i \leq n$. One obtains a similar property for the action of each $\mathrm{ad}\,y_i$ on b. In particular, for each $1 \leq i \leq n$ and each $1 \leq j \leq s$, both $\mathrm{ad}\,y_i$ and $\mathrm{ad}\,x_i$ act nilpotently on b_j. Since b_j is a weight vector, it further follows that $\mathrm{ad}\,t_i$ acts semisimply on b_j. Thus b_j generates a finite-dimensional $\mathrm{ad}\,U_i$ module for all $1 \leq i \leq n$ and all $1 \leq j \leq s$. Therefore each $b_j \in F(U)$. $\square$

Set $T_F = T \cap F(U)$. It follows from Lemma 4.2 that the algebra generated by T_F is equal to the intersection of U° with $F(U)$. By [JL1, 6.2], $\tau(\lambda) \in F(U)$ if and only if $(\mathrm{ad}\,x_i)$ and $(\mathrm{ad}\,y_i)$ act nilpotently on $\tau(\lambda)$ for $1 \leq i \leq n$. Furthermore, ([JL1, the proof of Lemma 6.1]) s is the least positive integer such that $(\mathrm{ad}\,x_i)^s \tau(\lambda) = 0$ and $(\mathrm{ad}\,y_i)^s \tau(\lambda) = 0$ if and only if

$$\frac{(\lambda, \alpha_i)}{(\alpha_i, \alpha_i)} = -s + 1.$$

Thus, $\tau(\lambda) \in F(U)$ if and only if $(\lambda, \alpha_i)/(\alpha_i, \alpha_i)$ is a nonpositive integer for all $1 \leq i \leq n$. In particular, $-\lambda/2$ is a dominant integral weight. So $\tau(\lambda)$ is in $F(U)$ if and only if λ is in $R(\pi) := Q(\pi) \cap -2P^+(\pi)$. (This is [JL1, Lemma 6.1. Note that the notation in [JL1] is different than in this paper. In particular, t_i here corresponds to t_i^2 in [JL1]. Thus divisibility by 4 in [JL1, Lemma 6.1] corresponds to divisibility by 2 in this paper.) For example, when $\mathbf{g}$ is $\mathrm{sl}\,2$, then T_F is just the set

$$\{t^{-m} | m \in \mathbf{N}\}.$$

Note that this set is a semigroup but is not a group. This is true in general for T_F.

Recall $\mathcal{F}$, the first filtration discussed in Section 4. For each $\xi \in R(\pi)$, set K_ξ^- equal to the subspace of G^- such that $F(U) \cap G^- \tau(\xi) = K_\xi^- \tau(\xi)$. Similarly, set K_ξ^+ equal to the subspace of U^+ such that $F(U) \cap U^+ \tau(\xi) = K_\xi^+ \tau(\xi)$. It is shown in [JL2, Section 4.9, 4.10] that

$$\mathrm{gr}_{\mathcal{F}}(F(U)) = \oplus_{\xi \in R(\pi)} \mathrm{gr}_{\mathcal{F}}(K_\xi^- K_\xi^+ \tau(\xi)). \tag{5.2}$$

Note that the inclusion of the left hand side of (5.2) inside the right hand side is just Theorem 4.7 applied to $F(U)$. In particular, Theorem 4.7 gives a new proof of this inclusion. Moreover, Theorem 4.7 can be thought of as a generalization of this part of (5.2) to other left coideal subalgebras which admit an $\mathrm{ad}\,T$ module structure.

The analysis in [JL2, Section 4, see Section 4.10], shows that

$$\mathrm{gr}_{\mathcal{F}}(K_\xi^- K_\xi^+ \tau(\xi)) = (\mathrm{ad}\,U)\mathrm{gr}_{\mathcal{F}}\tau(\xi).$$

Moreover,

$$(\mathrm{ad}\,U)\mathrm{gr}_{\mathcal{F}}\tau(\xi) \cong (\mathrm{ad}\,U)\tau(\xi)$$

as $\mathrm{ad}\,U$ modules for each $\tau(\xi) \in T_F$. Thus one has the direct sum decomposition in the nongraded case [JL2, Corollary 4.11]):

$$F(U) = \oplus_{t \in T_F}(\mathrm{ad}\,U)t. \tag{5.3}$$

Now (5.3) implies that $F(U)$ is the $\mathrm{ad}\,U$ module generated by the algebra $F(U) \cap U^\circ$. The fact that $F(U)$ is a left coideal was originally proved using this fact and a weakened version of Lemma 1.2 ([JL3, Lemma 5.3]).

Note that $(\mathrm{ad}\,U)t$ is an *ad-invariant left coideal* of U for each $t \in T$. On the other hand, (4.11) guarantees that any left coideal of U contains an element of T. Thus by (5.3) the minimal ad-invariant left coideals of U contained in $F(U)$ are exactly the vector subspaces $(\mathrm{ad}\,U)t$ where $t \in T_F$. This argument and result is due to [HS, Theorem 3.9] where it is actually proved in the dual setting. The description of the minimal ad-invariant left coideals is, in turn, a crucial step in the classification of bicovariant differential calculi on the quantized function algebra $R_q[G]$.

The algebra $F(U)$ can be localized by the normal elements T_F to obtain the larger coideal subalgebra $\mathbf{F} = F(U)T_F^{-1}$. Now $\mathbf{F} \cap T$ is just the subgroup generated by T_F. It is straightforward to show that $\mathbf{F}$ is generated by $x_i, y_i t_i$, and $\mathbf{F} \cap T$ for $1 \le i \le n$. In particular, given i, there exists some $t \in T_F$ such that $(\mathrm{ad}\,x_i)t$ is a nonzero multiple of $x_i t$. Hence $x_i t \in F(U)$ and $x_i \in \mathbf{F}$. A similar argument shows that $y_i t_i \in \mathbf{F}$ for all $1 \le i \le n$. Thus $\mathbf{F}$ contains $\mathbf{F} \cap T$, x_i, and $y_i t_i$, for $1 \le i \le n$. In the notation of Corollary 4.10, we get that $\Delta' = \Delta'' = \Delta^+$, and, moreover, $\mathbf{F} \cap T$, x_i, and $y_i t_i$, $1 \le i \le n$, generate $\mathbf{F}$. It further follows that G^- and U^+ are subalgebras of $\mathbf{F}$ and that $G^- U^+ t$ is a subset of $\mathbf{F}$ for each $t \in \mathbf{F} \cap T$. Recall the notation of Theorem 4.7. Note that $\mathbf{F}_\eta^- = G^- \subset \mathbf{F}$ and

$\mathbf{F}_\eta^+ = U^+ \subset \mathbf{F}$ for all $\eta \in Q(\pi)$. Hence Theorem 4.7 implies the following direct sum decomposition of $\mathbf{F}$:

$$\mathbf{F} = \oplus_{t \in \mathbf{F} \cap T} G^- U^+ t.$$

Since $\mathbf{F} \cap T$ is a subgroup of finite index in T, we see that $\mathbf{F}$, and hence $F(U)$, is "large" in U (For a stronger version of this, see [JL1, Theorem 6.4]).

A particular type of quantum Harish-Chandra module, defined differently (and earlier) than those of Section 2, was introduced in [JL3] in order to classify the primitive ideals of U. These modules were originally specified as a subcategory of the $F(U)$ bimodules with a "compatible" $\operatorname{ad} U$ action (see [JL3, 5.4] or [Jo, 8.2.3 and 8.4.1]). In [JL3, 5.4], an $F(U)$ bimodule M has a compatible $\operatorname{ad} U$ module structure provided that

$$\sum ((\operatorname{ad} a)(b \cdot m \cdot c) = \sum (\operatorname{ad} a_{(1)}) b \cdot (\operatorname{ad} a_{(2)}) m \cdot (\operatorname{ad} a_{(3)}) c \qquad (5.5)$$

and

$$(\operatorname{ad} t) m \cdot t = t \cdot m \qquad (5.6)$$

for all $a \in U$, b and c in $F(U)$, $m \in M$, and $t \in F(U) \cap T$. A different definition of compatible is given in [Jo, 8.2.3]. In particular, the $\operatorname{ad} U$ action must satisfy the following condition in [Jo, 8.2.3]:

$$\sum ((\operatorname{ad} a_{(1)}) m) \cdot a_{(2)} = a \cdot m \qquad (5.7)$$

for all $a \in F(U)$ and $m \in M$. Note that (5.6) follows from (5.7) by setting $a = t$.

The purpose of introducing the compatibility conditions (5.5) and (5.6) was to study the specific Harish-Chandra module category $\mathcal{H}_\chi$ associated to a dominant regular weight Λ defined in [JL3, Section 5.7]. By [JL3, 5.12] and [Jo, 8.4.11], this category is the same as the one described in [Jo, 8.4.1] using condition (5.7). Hence this category consists of modules with an $F(U)$ bimodule structure and $(\operatorname{ad} U)$ module action which satisfy both (5.5) and (5.7). In this paper, we say that $F(U)$ has a compatible $\operatorname{ad} U$ module action if both (5.5) and (5.7) hold. We show here that $F(U)$ bimodules with a compatible $\operatorname{ad} U$ module action fit exactly into the framework of Section 2.

Let U^{op} denote the Hopf algebra with underlying vector space U, the opposite multiplication, the same comultiplication and counit as U, and with antipode σ^{-1} ([Jo, 1.1.12]). Note that the algebra $U \otimes U^{\mathrm{op}}$ can be made into a Hopf algebra with comultiplication $\Delta(a \otimes b) = (\operatorname{Id} \otimes \mathbf{tw} \otimes \operatorname{Id})(\Delta \otimes \Delta)(a \otimes b)$ where $\mathbf{tw}$ denotes the twist map sending $a \otimes b$ to $b \otimes a$. The other Hopf operations can be defined similarly. Observe that $U \otimes U^{\mathrm{op}}$ is isomorphic to $U_q(\mathbf{g} \oplus \mathbf{g}^*)$ as a Hopf algebra. There is an algebra embedding ψ of U into $U \otimes U^{\mathrm{op}}$ which sends an element u to $\sum u_{(1)} \otimes \sigma(u_{(2)})$. The image of U in $U \otimes U^{\mathrm{op}}$ under ψ is not a Hopf subalgebra of $U \otimes U^{\mathrm{op}}$. However, by the next lemma it is a coideal subalgebra.

LEMMA 5.2. *The algebra $\psi(U)$ is a left coideal of $U \otimes U^{\mathrm{op}}$.*

PROOF. By (1.3),

$$\Delta(\sum u_{(1)} \otimes \sigma(u_{(2)})) = \sum (u_{(1)} \otimes \sigma(u_{(4)})) \otimes (u_{(2)} \otimes \sigma(u_{(3)})).$$

Thus $\psi(U)$ is a left coideal since $\Delta(u_{(2)}) = \sum u_{(2)} \otimes u_{(3)}$. □

Let $F(U \otimes U^{\mathrm{op}})$ denote the locally finite part of $U \otimes U^{\mathrm{op}}$. We show that $F(U)$ modules with compatible $\mathrm{ad}\, U$ module action are $F(U \otimes U^{\mathrm{op}})\psi(U)$ modules. The next lemma relates $F(U \otimes U^{\mathrm{op}})$ to the locally finite part $F(U)$ of U.

LEMMA 5.3. $F(U \otimes U^{\mathrm{op}}) = F(U) \otimes F(U)^{\mathrm{op}}$

PROOF. Let $\mathrm{ad}^{\mathrm{op}}$ denote the (left) adjoint action of U^{op}. With sufficient care to indentification of elements in U and U^{op}, one checks using (1.3) that $(\mathrm{ad}^{\mathrm{op}}\sigma(a))b = (\mathrm{ad}\,a)b$. Thus $F(U^{\mathrm{op}}) = F(U)^{\mathrm{op}}$ as algebras. □

Recall that since $\psi(U)$ is a left coideal and $F(U \otimes U^{\mathrm{op}})$ is an $\mathrm{ad}\, U \otimes U^{\mathrm{op}}$ module, we have that $F(U \otimes U^{\mathrm{op}})\psi(U) = \psi(U)F(U \otimes U^{\mathrm{op}})$. The next lemma shows that $F(U)$ is a free as a left $\psi(U)$ module.

LEMMA 5.4. *The multiplication map induces an isomorphism ϕ of vector spaces*

$$\psi(U) \otimes (1 \otimes F(U)^{\mathrm{op}}) \rightarrow \psi(U)F(U \otimes U^{\mathrm{op}}).$$

PROOF. Let $a \in F(U)$. Note that

$$a \otimes 1 = \sum a_{(1)}\varepsilon(a_{(2)}) \otimes 1 = \sum a_{(1)} \otimes \varepsilon(a_{(2)})$$
$$= \sum a_{(1)} \otimes a_{(3)}\sigma(a_{(2)}) = \sum \psi(a_{(1)})(1 \otimes a_{(2)}) \qquad (5.8)$$

for all $a \in U$. It follows that

$$\psi(U)F(U \otimes U^{\mathrm{op}}) = \psi(U)(1 \otimes F(U)^{\mathrm{op}}).$$

This proves that ϕ is surjective.

Suppose that $\sum_i \psi(c_i)(1 \otimes b_i) = 0$ where the set $\{b_i\}$ is a linearly independent subset of $F(U)^{\mathrm{op}}$. We argue that each $\psi(c_i) = 0$. This in turn implies that ϕ is injective.

There is a version of Lemma 1.3 for right coideal subalgebras. In particular, $U = \oplus_t G^- U^+ t$ and each $G^- U^+ t$ is a right coideal of U. We may write $c_i = \sum_t c_{it}$ where each $c_{it} \in G^- U^+ t$. It follows that $\psi(c_{it})(1 \otimes b_i) \in G^- U^+ t \otimes U$ for each i and $t \in T$. Thus $\sum_i \psi(c_{it})(1 \otimes b_i) = 0$. This allows us to reduce to the case where there exists $t \in T$ such that c_i is in $G^- U^+ t$ for all i.

Recall the notation of Section 4, Filtration II. Let (M, N) be the maximum value of the set of bidegrees of the c_i. Reordering if necessary, we may assume that c_1 has bidegree (M, N). Choose $[\lambda, \mu] \in \max(c_1)$. By say (4.11), we have

$$(\pi_{\lambda,\mu} \otimes \mathrm{Id}) \sum_i \psi(c_i)(1 \otimes b_i) = \sum_i \pi_{\lambda,\mu}(c_i) \otimes \sigma(t)b_i.$$

Note that $\sigma(t) = t^{-1}$. Since the set $\{b_i\}$ is linearly independent, the set $\{\sigma(t)b_i\}$ is also linearly independent. Hence $\pi_{\lambda,\mu}(c_i) = 0$ for each i. The choice of $[\lambda, \mu]$ now forces $c_i = 0$ for each i.   □

The next result shows that $F(U)$ modules with compatible $\operatorname{ad} U$ modules are just $F(U \otimes U^{\mathrm{op}})\psi(U)$ modules.

THEOREM 5.5. *The set of $F(U)$ bimodules with compatible $\operatorname{ad} U$ module action can be identified with the set of $F(U \otimes U^{\mathrm{op}})\psi(U)$ modules.*

PROOF. Let M be a $F(U \otimes U^{\mathrm{op}})\psi(U)$ module. Note that M is a $F(U)$ bimodule in a natural way. In particular, set $a \cdot m \cdot b = (a \otimes b)m$ for all $a, b \in F(U)$ and $m \in M$. We define an action of $\operatorname{ad} U$ on M by setting $(\operatorname{ad} c)m = \psi(c)m$ for all $c \in U$. By (5.8), it follows that this $\operatorname{ad} U$ action satisfies (5.7). A straightforward computation shows that this action satisfies (5.5) as well.

Now let M be an $F(U)$ bimodule with compatible $\operatorname{ad} U$ module action. Make M into a $F(U \otimes U^{\mathrm{op}})\psi(U)$ module by setting

$$(a \otimes b)m = a \cdot m \cdot b \quad \text{and} \quad (\psi(c))m = (\operatorname{ad} c)m \tag{5.9}$$

for all $a \otimes b \in F(U \otimes U^{\mathrm{op}})$, $c \in U$, and $m \in M$.

One checks that $\psi(c)(1 \otimes b) = \sum (1 \otimes (\operatorname{ad} c_{(2)})b)\psi(c_{(1)})$. By (5.5), $(\operatorname{ad} c)(m \cdot b) = (\operatorname{ad} c_{(1)} \cdot m) \cdot ((\operatorname{ad} c_{(2)})b)$. Hence the action of $\psi(c)$ on $(1 \otimes b)m$ described in (5.9) agrees with the action of $(1 \otimes (\operatorname{ad} c_{(2)})b)$ on $\psi(c_{(1)})m$. Therefore, to show that the action in (5.9) is well defined it is sufficient to show that the action of an element $x \in F(U \otimes U^{\mathrm{op}})\psi(U)$ on M is independent of the way x is written as a sum of terms of the form bu where $b \in F(U \otimes U^{\mathrm{op}})$ and $u \in \psi(U)$.

The compatibility condition (5.7) ensures that

$$(\operatorname{ad} c)(a \cdot m \cdot b) = \sum (\operatorname{ad} ca_{(1)})(m \cdot ba_{(2)})$$

for all $a \otimes b \in F(U \otimes U^{\mathrm{op}})$, $c \in U$, and $m \in M$. Thus using (5.9) formally, we see that

$$\psi(c)((a \otimes b)m) = \psi(c)((a \cdot m \cdot b)) = \sum \psi(ca_{(1)})(m \cdot ba_{(2)})$$
$$= \sum \psi(ca_{(1)})((1 \otimes ba_{(2)})m).$$

In particular the action of $\psi(c)(a \otimes b)$ agrees with the action of $\sum \psi(ca_{(1)})(1 \otimes ba_{(2)})$ on M. By Lemma 5.4, every element in $F(U \otimes U^{\mathrm{op}})\psi(U)$ can be expressed uniquely in the form $\sum_i \psi(a_i)(1 \otimes b_i)$ where $\{b_i\}$ is a basis of $F(U)^{\mathrm{op}}$. The theorem now follows.   □

One can apply the results of Section 2 to the study of Harish-Chandra modules for the pair $U \otimes U^{\mathrm{op}}$, $\psi(U)$. Identify the algebra $U \otimes U^{\mathrm{op}}$ with $U_q(\mathbf{g} \oplus \mathbf{g}^*)$. Let $\tilde{\kappa}$ denote the conjugate linear Chevalley antiautomorphism of Section 2 associated here to the quantized enveloping algebra $U_q(\mathbf{g} \oplus \mathbf{g}^*)$. One can find a Hopf algebra automorphism $\Upsilon \in \mathcal{H}_{\mathbb{R}}$ such that $\Upsilon(\psi(U))$ is invariant under $\tilde{\kappa}$. Thus the results in Section 2 apply here. However, the main results of Section 2, such as Theorem 2.7, can be proved easily in this case since $\psi(U)$ is isomorphic as an algebra to $U_q(\mathbf{g})$. Thus it acts completely reducibly on all finite-dimensional $\psi(U)$ modules. Furthermore, one checks that all finite-dimensional $\psi(U)$ modules are unitary using the fact that this is true for $U_q(\mathbf{g})$.

For an example of a Harish-Chandra module associated to the pair $U \otimes U^{\mathrm{op}}$, $\psi(U)$, consider two left U modules M and N. Define the U bimodule $\mathrm{Hom}(M, N)$ by $(a \cdot f \cdot b)(m) = af(bm)$. As explained in [JL3, 5.4] and [Jo, 8.2.3], $\mathrm{Hom}(M, N)$ has a compatible $(\mathrm{ad}\, U)$ module structure in the sense of (5.5) and (5.7) given by $(\mathrm{ad}\, a)f = \sum a_{(1)} \cdot f \cdot \sigma(a_{(2)})$. Thus from the above Theorem 5.5, we see that $\mathrm{Hom}(M, N)$ is a $(F(U) \otimes F(U)^{\mathrm{op}})\psi(U)$ module. By Theorem 2.7, the sum of all finite-dimensional $\mathrm{ad}\, U$ modules $F(M, N)$ inside of $\mathrm{Hom}(M, N)$ is a Harish-Chandra module for the pair $U \otimes U^{\mathrm{op}}$, $\psi(U)$.

In [JL3, Theorem 5.13] (see also [Jo, Chapter 8]), the theory of Harish-Chandra modules associated to the pair $U \otimes U^{\mathrm{op}}$, $\psi(U)$ is used to prove an equivalence of categories between certain Harish-Chandra modules and various category $\mathcal{O}$ modules. This is critical in obtaining the quantum version of Duflo's theorem: every primitive ideal of U is the annihilator of a highest weight simple module ([JL3, Corollary 6.4] or [Jo, 8.4.17]).

6. Nilpotent and Parabolic Coideal Subalgebras

Yet another left coideal subalgebra of U is G^-, an obvious quantum analog of $U(\mathbf{n}^-)$. In this section, we consider coideal subalgebras of G^- which correspond to classical enveloping algebras of Lie subalgebras of $\mathbf{n}^-$ and related Lie subalgebras of $\mathbf{g}$. Most of the results presented here are from [Ke].

Let π' be a subset of the simple roots π of $\mathbf{g}$. There are a number of Lie subalgebras of $\mathbf{g}$ which can be associated to π'. The most obvious is the semisimple Lie subalgebra $\mathbf{m}$ of $\mathbf{g}$ generated by the e_i, f_i, h_i, for those i with $\alpha_i \in \pi'$. Since the simple roots π' associated to the root system of $\mathbf{m}$ are contained in the simple roots π of $\mathbf{g}$, the entire picture can be lifted to the quantum setting. In particular, $U_q(\mathbf{g})$ contains a Hopf subalgebra $\mathcal{M}$ isomorphic to $U_q(\mathbf{m})$ and generated by the x_i, y_i, t_i, t_i^{-1} for the same i. Set $\mathcal{M}^- = \mathcal{M} \cap G^-$ and $\mathcal{M}^+ = \mathcal{M} \cap U^+$.

Let Δ' denote the set of positive roots associated to the simple roots π'. The vector space $\mathbf{n}_{\pi'}^-$ spanned by the root vectors $f_{-\gamma}$, γ in $\Delta^+ - \Delta'$, is a second Lie subalgebra of $\mathbf{n}^-$. Let $\mathbf{m}^-$ denote the Lie subalgebra of $\mathbf{m}$ generated by the f_i for $\alpha_i \in \pi'$. Then

$$\mathbf{n}^- = \mathbf{n}_{\pi'}^- \oplus \mathbf{m}^-.$$

Thus the multiplication map defines a vector space isomorphism:

$$U(\mathbf{n}^-) \cong U(\mathbf{n}_{\pi'}^-) \otimes U(\mathbf{m}^-). \tag{6.1}$$

We shall see that the algebra $U(\mathbf{n}_{\pi'}^-)$ can be lifted to the quantum setting using a coideal subalgebra.

Let $G^-_{\pi-\pi'}$ be the subalgebra of G^- generated by the $y_i t_i$ such that α_i is in $\pi - \pi'$. Note that $G^-_{\pi-\pi'}$ is a left coideal subalgebra of G^-. Now $G^-_{\pi-\pi'}$ is generated by weight vectors and in particular, $(\operatorname{ad} T)G^-_{\pi-\pi'} = G^-_{\pi-\pi'}$. Also, $(\operatorname{ad} x_i)y_j t_j = 0$ for all $i \neq j$. Thus $(\operatorname{ad}(\mathcal{M}^+ T))G^-_{\pi-\pi'} \subset G^-_{\pi-\pi'}$. Recall that $\mathcal{M}$ is equal to the quantized enveloping $U_q(\mathbf{m})$. Hence the triangular decomposition (1.12) implies that $\mathcal{M}T = \mathcal{M}^- \mathcal{M}^+ T$. Hence $(\operatorname{ad} \mathcal{M}^-)G^-_{\pi-\pi'}$ equals $(\operatorname{ad} \mathcal{M})G^-_{\pi-\pi'}$. By Lemma 1.2, $(\operatorname{ad} \mathcal{M}^-)G^-_{\pi-\pi'}$ is a left coideal. Let $N^-_{\pi'}$ be the subalgebra of G^- generated by $(\operatorname{ad} \mathcal{M}^-)G^-_{\pi-\pi'}$. It is a left coideal subalgebra since $(\operatorname{ad} \mathcal{M}^-)G^-_{\pi-\pi'}$ is a left coideal.

By [Ke], one has a quantum analog of (6.1). Namely there is an isomorphism of vector spaces

$$G^- \cong N^-_{\pi'} \otimes \mathcal{M}^-. \qquad (6.2)$$

Kébé actually proves a stronger result with this as a consequence, namely, G^- is isomorphic to the smash product of $N^-_{\pi'}$ and $\mathcal{M}^-$.

By construction, $N^-_{\pi'}$ is generated by weight vectors and hence is a direct sum of its weight spaces. By Lemma 4.3, we can find a subset Δ_1 of Δ^+ such that $N^-_{\pi'}$ is generated by weight vectors $\tilde{f}_{-\gamma}$ of weight $\gamma \in \Delta_1$ which specialize to root vectors in $U(\mathbf{n}^+)$. By [L3, proof of Proposition 2.2], Δ_1 consists of those positive roots which are not linear combinations of roots in π'. In particular, $N^-_{\pi'}$ specializes to $U(\mathbf{n}^-_{\pi'})$ as q goes to 1 ([L3, proof of Proposition 2.2]). Thus the left coideal subalgebra $N^-_{\pi'}$ is a natural choice of quantum analog of $U(\mathbf{n}^-_{\pi'})$ inside of $U(\mathbf{g})$.

It is instructive to look at the generators of $N^-_{\pi'}$. Let I be a tuple $(i_1, \ldots, i_r)$ of (arbitrary) length r and suppose that α_{i_s} is in $\pi - \pi'$ for $1 \leq s \leq r$. By the argument in [L3, Proposition 2.2], the algebra $N^-_{\pi'}$ is generated by elements of the form

$$Y_{I,j} = (\operatorname{ad} y_{i_1} \cdots y_{i_r}) y_j t_j$$

where $\alpha_j \notin \pi'$.

Now each $Y_{I,j}$ is an element of the subcoideal $(\operatorname{ad} \mathcal{M}^-)y_j t_j$ of $N^-_{\pi'}$ as well as an element of G^-. Hence

$$(\operatorname{Id} \otimes \pi_{0,0})\Delta(Y_{I,j}) = Y_{I,j} \otimes 1.$$

Thus

$$\Delta(Y_{I,j}) = Y_{I,j} \otimes 1 + \sum Y_i \otimes Y_i'$$

where Y_i is in U and Y_i' is in $(\operatorname{ad} \mathcal{M}^-)y_j t_j$. We can actually say more about the Y_i. First recall that $Y_{I,j}$ is in G^-. Set $\lambda = \alpha_{i_1} + \cdots + \alpha_{i_r} + \alpha_j$ and note that the weight of $Y_{I,j}$ is $-\lambda$. Set $\mu = 0$. We may apply (4.7) to $Y_{I,j}\tau(-\lambda)$ using this λ and μ. By (4.7) and weight space considerations, each Y_i is in $\mathcal{M} \cap G^- U^\circ$. Furthermore, (4.7) implies that each $Y_i \in U^- \tau(\lambda)$. Since $\tau(\lambda) \in \mathcal{M}t_j$, it follows that each Y_i is an element of $(\mathcal{M} \cap U^- U^o) t_j$. In particular, we get that (see [AJS, Proposition C.5])

$$\Delta(Y_{I,j}) \in Y_{I,j} \otimes 1 + (\mathcal{M} \cap U^- U^o) t_j \otimes (\operatorname{ad} \mathcal{M}^-)(y_j t_j). \qquad (6.3)$$

The elements $Y_{I,j}$ also satisfy a uniqueness property. In particular, by [L2, Proposition 4.1], if Y is an element of G^- of weight $-\lambda$ such that

$$\Delta(Y) \in Y \otimes 1 + (\mathcal{M} \cap U^- U^o) t_j \otimes (\operatorname{ad} \mathcal{M}^-)(y_j t_j)$$

then Y is a nonzero scalar multiple of $Y_{I,j}$. This uniqueness property will be used in the uniqueness result Theorem 7.5 concerning quantum symmetric pairs.

Let $\mathbf{n}_{\pi'}^+$ be the Lie subalgebra of $\mathbf{n}^+$ spanned by the root vectors e_γ, where γ runs over $\Delta^+ - \Delta'$. One can similarly define left coideal subalgebras $N_{\pi'}^+$ of U^+ which are analogs of $U(\mathbf{n}_{\pi'}^+)$. These can be constructed directly using the same methods described above for $N_{\pi'}^-$.

Of course, one could take the perspective of right coideal subalgebras instead of left coideal subalgebras. This will be useful in the next section. For example, right coideal analogs of $U(\mathbf{n}_{\pi'}^+)$ are subalgebras of G^+ defined using the right adjoint action,

$$(\operatorname{ad}_r a) b = \sum \sigma(a_{(1)}) b a_{(2)} \qquad (6.4)$$

for all a and b in U. Let $G_{\pi - \pi'}^+$ be the subalgebra of G^+ generated by the $x_i t_i^{-1}$ for all i such that $\alpha_i \in \pi - \pi'$. Then the subalgebra $N_{\pi',r}^+$ generated by $(\operatorname{ad}_r \mathcal{M}^+) G_{\pi - \pi'}^+$ is a right coideal subalgebra of G^+ and an analog of $U(\mathbf{n}_{\pi'}^+)$. The algebra $N_{\pi',r}^+$ is generated by elements of the form

$$X_{I,j} = (\operatorname{ad}_r x_{i_1} \cdots x_{i_r}) x_j t_j^{-1}$$

where each $\alpha_{i_s} \in \pi'$ and $\alpha_j \notin \pi'$. Moreover the comultiplication of these elements is similar to that of the $Y_{I,j}$, e.g.,

$$\Delta(X_{I,j}) \in 1 \otimes X_{I,j} + (\operatorname{ad}_r \mathcal{M}^+)(x_j t_j^{-1}) \otimes (\mathcal{M} \cap G^+ U^o) t_j^{-1}. \qquad (6.5)$$

Using $N_{\pi'}^-$, $N_{\pi'}^+$, and $\mathcal{M}^-$, one can construct what are called generalized Verma modules. Let $\mathcal{P}$ be the subalgebra of U generated by $\mathcal{M}$, U^o, and $N_{\pi'}^+$. Note that $\mathcal{P}$ is a left coideal subalgebra since it is generated by left coideal subalgebras. It is an analog of the enveloping algebra of the parabolic Lie subalgebra $(\mathbf{m} + \mathbf{h}) \oplus \mathbf{n}_{\pi'}^+$. Using (6.2), one obtains an isomorphism of vector spaces via the multiplication map

$$U \cong N_{\pi'}^- \otimes \mathcal{P}. \qquad (6.6)$$

Let W be a finite-dimensional simple $\mathcal{M}$ module. Extend the action of $\mathcal{M}$ on W to U^o by insisting that the highest weight generating vector of W is a weight vector of say weight Λ with respect to the action of T. Extend further the action on W to $N_{\pi'}^+$ by insisting that the augmentation ideal of $N_{\pi'}^+$ acts as zero on all vectors in W. These extensions make W into a $\mathcal{P}$ module. The generalized Verma module $M_{\pi'}(\Lambda)$ is defined to be $U \otimes_{\mathcal{P}} W$. In particular, elements of U

act by left multiplication and $pu \otimes w = \sum (\operatorname{ad} p_{(1)}) u \otimes p_{(2)} w$ for all $p \in \mathcal{P}$, $u \in U$, and $w \in W$. As a left $N_{\pi'}^{-}$ module, $U \otimes_{\mathcal{P}} W \cong N_{\pi'}^{-} \otimes W$. Furthermore, the action of $\mathcal{M}$ on $\mathbb{N}_{\pi'}^{-}$ is both locally finite and semisimple. Hence the generalized Verma module $M_{\pi'}(\Lambda)$ is a Harish-Chandra module for the pair U, $\mathcal{M}$.

Using the coideal subalgebras discussed in this section, one can form quantized homogenous spaces as in Section 3. For example, the homogeneous space associated to G^{-}, $R_q[G/N] = R_q[G]_l^{G^{-}}$ is studied in [Jo, Chapter 9] where it is used to obtain the complete description of the prime and primitive spectra of the quantized function algebra $R_q[G]$.

7. Quantum Symmetric Pairs

We turn now to the theory of quantum symmetric pairs. First, we present the construction and characterization of the coideal subalgebras used to form such pairs. The results are drawn from [L2] and [L3], but the methods in this paper are often simpler. The involutions used to construct these algebras are given in a concrete fashion here. The relations for the coideal subalgebras as algebras are also presented more explicitly. Moreover, using the results of Section 4, we give a new, less intricate, proof of the uniqueness characterization for the subalgebras used to form quantum symmetric pairs (see Theorem 7.5 below.) The Harish-Chandra module and symmetric space theory associated to these pairs is also described with the aid of Sections 2 and 3.

A symmetric pair is defined for each Lie algebra involution (equivalently, a Lie algebra automorphism of order 2) of $\mathbf{g}$. More precisely, let θ be a Lie algebra involution of $\mathbf{g}$. Write $\mathbf{g}^{\theta}$ for the Lie subalgebra of $\mathbf{g}$ consisting of elements fixed by θ. The pair $\mathbf{g}, \mathbf{g}^{\theta}$ is a classical symmetric pair. A classification of involutions and classical symmetric pairs up to isomorphism can be found in [He1, Chapter 10, Sections 2, 5, and 6] and [OV, Section 4.1.4].

Let $\mathbf{p} = \{v \in \mathbf{g} | \theta(v) = -v\}$. A commutative Lie subalgebra of $\mathbf{g}$ which is reductive in $\mathbf{g}$ and is equal to its centralizer in $\mathbf{p}$ is called a Cartan subspace of $\mathbf{p}$ (see [D, 1.13.5].) A Cartan subalgebra $\mathbf{h}'$ of $\mathbf{g}$ is called maximally split ([V, Section 0.4.1]) with respect to θ provided that $\mathbf{h}' \cap \mathbf{p}$ is a Cartan subspace of $\mathbf{p}$. By [D, 1.13.6, 1.13.7], $\mathbf{p}$ contains Cartan subspaces and moreover each Cartan subspace can be extended to a Cartan subalgebra of $\mathbf{g}$.

Recall that we have already specified a Cartan subalgebra $\mathbf{h}$ of $\mathbf{g}$. Let θ be an involution of $\mathbf{g}$ such that $\mathbf{h}$ is maximally split with respect to θ. Let $\mathcal{L}$ be the set of Lie algebra automorphisms ψ of $\mathbf{g}$ such that $\psi(\mathbf{p} \cap \mathbf{h})$ is a subset of $\mathbf{h}$. If $\psi \in \mathcal{L}$ then $\mathbf{h}$ is also maximally split with respect to the involution $\psi \theta \psi^{-1}$. By [D, 1.13.7 and 1.13.8], one can replace θ by $\psi \theta \psi^{-1}$ for some $\psi \in \mathcal{L}$ so that θ also satisfies the following conditions:

$$\theta(\mathbf{h}) = \mathbf{h}; \tag{7.1}$$

$$\text{if } \theta(h_i) = h_i \text{ then } \theta(e_i) = e_i \text{ and } \theta(f_i) = f_i; \tag{7.2}$$

if $\theta(h_i) \neq h_i$ then $\theta(e_i)$ is a nonzero root vector in $\mathbf{n}^-$
and $\theta(f_i)$ is a nonzero root vector in $\mathbf{n}^+$. (7.3)

By [D, 1.13.8], θ also induces an automorphism Θ of the root system Δ.

if $\theta(h_i) \neq h_i$ then $\theta(e_i)$ is a nonzero root vector in $\mathbf{n}^-$

Now consider an arbitrary involution θ' of $\mathbf{g}$. One can find a Lie algebra automorphism Υ of $\mathbf{g}$ so that $\mathbf{h}$ is maximally split with respect to the involution $\Upsilon\theta'\Upsilon^{-1}$. In the quantum case, we do not have as much flexibility in "moving" involutions around using an automorphism of U. In particular, there is only one choice of quantum Cartan subalgebra, since the only invertible elements of U are the nonzero scalars and the elements of T. Hence any automorphism of U restricts to an automorphism of T. Thus the relationship between an involution of $\mathbf{g}$ and the particular Cartan subalgebra $\mathbf{h}$ is important in lifting the involution to the quantum case. In this section, we call an involution θ of $\mathbf{g}$ a maximally split involution if $\mathbf{h}$ is maximally split with respect to θ and θ satisfies (7.1), (7.2), and (7.3). (Similar terminology was introduced in [Di, Section 5].) We discuss lifts of maximally split involutions and the associated quantum symmetric pairs. There are also a few scattered results on quantum symmetric pairs when the involution is not maximally split. The reader is referred to [G] and [BF] for more information.

For the remainder of this section, let θ be a maximally split involution with respect to the fixed Cartan subalgebra $\mathbf{h}$. Consider the Cartan subspace $\mathbf{a} = \mathbf{p} \cap \mathbf{h}$ of $\mathbf{p}$. Since $\mathbf{a}$ is subset of $\mathbf{h}$, the action of $\mathrm{ad}\,\mathbf{a}$ on $\mathbf{g}$ is semisimple. Given $\lambda \in \mathbf{a}^*$, set

$$\mathbf{g}_\lambda = \{x \in \mathbf{g}|(\mathrm{ad}\,a)x = \lambda(a)x \text{ for all } a \in \mathbf{a}\}.$$

Let

$$\Sigma = \{\lambda \in \mathbf{a}^*| \ \mathbf{g}_\lambda \neq 0\}.$$

We can write $\mathbf{g} = \oplus_{\lambda \in \Sigma} \mathbf{g}_\lambda$. Furthermore, by [OV, Theorem 3.4.2], Σ is an abstract root system called the restricted root system associated to θ (or more precisely, to $\mathbf{g}, \mathbf{a}$.) A classification of restricted root systems associated to involutions can be found in [Kn, Chapter VI, Section 11] (see also [He1, Chapter X, Section F under Exercises and Further Results]). Note that an abstract root system is slightly more general than an ordinary root system (often called a reduced root system) described in [H, Chapter III]. Good references for abstract root systems are [Kn, Chapter II, Section 5] and [OV, Chapter 3, Section 1.1]. The abstract root systems have been classified as the set of reduced root systems and one additional nonreduced family referred to as type BC ([Kn, Chapter II, Section 8]).

Before discussing the quantum case, we further describe the action of θ on the generators of $\mathbf{g}$. Set $\Delta_\Theta = \{\alpha \in \Delta | \Theta(\alpha) = \alpha\}$ where Θ is the associated root system automorphism. This is the root system for the semisimple Lie subalgebra $\mathbf{m}$ of $\mathbf{g}$ generated by the e_i, f_i, h_i with $\theta(h_i) = h_i$. Write $\mathbf{m} = \mathbf{m}^- \oplus \mathbf{m}^o \oplus \mathbf{m}^+$ for the obvious triangular decomposition of $\mathbf{m}$. Set $\pi_\Theta = \Delta_\Theta \cap \pi$. Note that π_Θ

is a set of positive simple roots for the root system Δ_Θ. Write $Q(\pi_\Theta)$ for the lattice of integral linear combinations of the simple roots in π_Θ. Let $Q^+(\pi_\Theta)$ be the set of nonnegative integral linear combinations of the elements in π_Θ.

Note that $\pi_\Theta = \Theta(\pi) \cap \pi$. Also, $\Theta(-\alpha_i) \in \Delta^+$ for all $\alpha_i \notin \pi_\Theta$ by (7.3). It follows that

$$\Theta(-\alpha_i) \in \sum_{\alpha_j \notin \pi_\Theta} \mathbf{N}\alpha_j + Q^+(\pi_\Theta) \tag{7.4}$$

for each $\alpha_i \notin \pi_\Theta$. Since Θ is a root system automorphism, every element of Δ can be written as an integral linear combination of roots in $\{\Theta(\alpha_i) | \alpha_i \in \pi\}$ where either all the coefficients are positive or all the coefficients are negative. Hence each $\alpha_i \notin \pi_\Theta$ can be written as a linear combination of elements in $\{\Theta(\alpha_i) | \alpha_i \notin \pi_\Theta\} \cup \pi_\Theta$ with just negative integers as coefficients. Observation (7.4) thus implies that there exists a permutation p on the set $\{i \, | \alpha_i \in \pi - \pi_\Theta\}$ such that for each $\alpha_i \in \pi - \pi_\Theta$,

$$\Theta(-\alpha_i) - \alpha_{p(i)} \in Q^+(\pi_\Theta). \tag{7.5}$$

Choose a maximal subset π^* of $\pi - \pi_\Theta$ such that if $j = p(j)$ then $\alpha_j \in \pi^*$ and if $j \neq p(j)$, then exactly one of the pair $\alpha_j, \alpha_{p(j)}$ is in π^*. Consider i such that $\alpha_i \in \pi^*$. The root vector $e_{p(i)}$ associated to the simple root $\alpha_{p(i)}$ satisfies $(\mathrm{ad}\, f_j)e_{p(i)} = [f_j, e_{p(i)}] = 0$ for all $\alpha_j \in \pi_\Theta$. Thus $e_{p(i)}$ is a lowest weight vector for the action of $\mathrm{ad}\,\mathbf{m}^-$. Let V be the corresponding simple $\mathrm{ad}\,\mathbf{m}$ module generated by $e_{p(i)}$. By (7.3), $\theta(f_i)$ is a root vector in $\mathbf{n}^+$. Furthermore (7.5) implies that the weight of this root vector is $\alpha_{p(i)}$ plus some element in $Q^+(\pi_\Theta)$. Thus $\theta(f_i)$ can be written as a bracket $[a_1[a_2, \ldots, [a_{s-1}, a_s]\ldots]$ where exactly one of the a_j equals $e_{p(i)}$ and the others are elements of $\mathbf{m}^+$. Using the Jacobi identity, we see that $\theta(f_i)$ is an element of $(\mathrm{ad}\,\mathbf{m}^+)e_{p(i)}$. In particular $\theta(f_i)$ is an element of V. Furthermore, since elements of $\mathbf{m}^+$ commute with f_i and thus with $\theta(f_i)$, we see that $\theta(f_i)$ must be a highest weight vector of V. Thus we can find a sequence of elements $\alpha_{i_1}, \ldots, \alpha_{i_r}$ in π_Θ and a sequence of positive integers $m_1, \ldots, m_r$ such that (up to a slight adjustment of θ)

$$\theta(f_i) = (\mathrm{ad}\, e_{i_1}^{(m_1)} \cdots e_{i_r}^{(m_r)})e_{p(i)}. \tag{7.6}$$

Here $e_j^{(m)} = e_j^m/m!$. We may further assume that both the sequence of roots and the sequence of integers are chosen so that each $(\mathrm{ad}\, e_{i_s}^{(m_s)} \cdots e_{i_r}^{(m_r)})e_{p(i)}$, $1 \leq s \leq r$, is an extreme vector of V. (In particular, $(\mathrm{ad}\, e_{i_s}^{(m_s)} \cdots e_{i_r}^{(m_r)})e_{p(i)}$ is a highest weight vector for the action of $\mathrm{ad}\, e_{i_s}$ and $(\mathrm{ad}\, e_{i_{s-1}}^{(m_{s-1})} \cdots e_{i_r}^{(m_r)})e_{p(i)}$ is a lowest weight vector for the action of $\mathrm{ad}\, f_{i_s}$.) Suppose that the sequence $\alpha_{j_1}, \ldots, \alpha_{j_s}$ of elements in π_Θ and the positive integers $n_1, \ldots, n_s$ also satisfy this condition on extreme vectors and that $\sum_k m_k \alpha_{i_k} = \sum_k n_k \alpha_{j_k}$. By [Ve], $(\mathrm{ad}\, e_{i_1}^{(m_1)} \cdots e_{i_r}^{(m_r)})e_{p(i)} = (\mathrm{ad}\, e_{j_1}^{(n_1)} \cdots e_{j_s}^{(n_s)})e_{p(i)}$. Thus (7.6) is independent of the choice of such sequences.

Using lowest weight vectors instead of highest weight vectors, we obtain

$$\theta(e_{p(i)}) = (\operatorname{ad} f_{i_r}^{(m_r)} \cdots f_{i_1}^{(m_1)}) f_i \qquad (7.7)$$

up to a nonzero scalar. A straightforward **sl** 2 computation shows that

$$(\operatorname{ad} e_{i_1}^{(m_1)} \cdots e_{i_r}^{(m_r)})[(\operatorname{ad} f_{i_r}^{(m_r)} \cdots f_{i_1}^{(m_1)}) f_i] = f_i$$

and

$$(\operatorname{ad} f_{i_r}^{(m_r)} \cdots f_{i_1}^{(m_1)})[(\operatorname{ad} e_{i_1}^{(m_1)} \cdots e_{i_r}^{(m_r)}) e_{p(i)}] = e_{p(i)}. \qquad (7.8)$$

Since θ^2 is the identity, the scalar in (7.7) must be 1.

Set $m(i) = m_1 + \ldots + m_r$. Now $[\theta(e_i), \theta(f_i)] = \theta(h_i)$ is an element of **h** by (7.1). Furthermore, by (7.2) and (7.3), $\theta(h_i)$ must be the coroot $h_{\Theta(\alpha_i)}$ associated to the root $\Theta(\alpha_i)$. The description of the Chevalley basis for **g** given in [H, Proposition 25.2 and Theorem 25.2] ensures that both $\theta(e_i)$ and $\theta(f_i)$ are Chevalley basis vectors up to a sign. Furthermore, by [H, Proposition 25.2(b)] and (7.6),we must have

$$\theta(e_i) = (-1)^{m(i)} (\operatorname{ad} f_{i_1}^{(m_1)} \cdots f_{i_r}^{(m_r)}) f_{p(i)}.$$

Similarly, by [H, Proposition 25.2(b)] and (7.7)

$$\theta(f_{p(i)}) = (-1)^{m(i)} (\operatorname{ad} e_{i_r}^{(m_r)} \cdots e_{i_1}^{(m_1)}) e_i.$$

Note that when $p(i) = i$, we have

$$(\operatorname{ad} e_{i_r}^{(m_r)} \cdots e_{i_1}^{(m_1)}) e_i = (\operatorname{ad} e_{i_1}^{(m_1)} \cdots e_{i_r}^{(m_r)}) e_i.$$

Hence $m(i)$ is even in this case.

The above analysis allows us to better describe the root space automorphism Θ. Let W' denote the Weyl group associated to the root system Δ_Θ of **m** considered as a subgroup of the Weyl group of Δ. Let w_o denote the longest element of W'. Note that w_o is a product of reflections in W' but can also be considered as an element of W. Let d be the diagram automorphism on π_Θ such that $d = -w_o$ when restricted to π_Θ. Note that d induces a permutation on the set $\{i | \alpha_i \in \pi_\Theta\}$ which we also denote by d. In particular, given $\alpha_i \in \pi_\Theta$, $d(\alpha_i) = \alpha_{d(i)}$. Extend d to a function on π, and thus to Δ, by setting $d(\alpha_i) = \alpha_{p(i)}$ for $\alpha_i \notin \pi_\Theta$. It follows that $\Theta = -w_o d$. Note that this forces d to be a diagram automorphism of the larger root system Δ.

Before lifting θ to the quantum case, we recall and introduce more notation. The right adjoint action is defined by (6.4). This action on the generators of U is given by:

$$(\operatorname{ad}_r y_i) b = b y_i - y_i t_i b t_i^{-1} \quad (\operatorname{ad}_r x_i) b = t_i^{-1} b x_i - t_i^{-1} x_i b \quad (\operatorname{ad}_r t_i) b = t_i^{-1} b t_i$$

for $1 \le i \le n$. Recall the definitions of $[m]_q$ and q_i used to define the quantized enveloping algebra ((1.4)-(1.7)). The divided powers of x_i and y_i are defined by $x_i^{(m)} = x_i^m / [m]_{q_i}!$ and $y_i^{(m)} = y_i^m / [m]_{q_i}!$. (Note that these are quantum analogs of the divided power $e_i^{(m)}$.) Let $\mathcal{M}$ denote the subalgebra of U generated by the corresponding elements x_i, y_i, t_i, t_i^{-1} where $\theta(h_i) = h_i$. Note that $\mathcal{M}$ is just a copy of the quantized enveloping algebra $U_q(\mathbf{m})$ so this notation is consistent with that of Section 6. Let ι be the $\mathbb{C}$ algebra automorphism of U fixing $x_i t_i^{-1}$ and $t_i y_i$ for $1 \leq i \leq n$, sending t to t^{-1} for all $t \in T$ and q to q^{-1}. Recall the sequences $\alpha_{i_1}, \ldots, \alpha_{i_r}$ and $m_1, \ldots, m_r$ used in (7.6) and (7.7). (As in the classical case, using [Lu, Proposition 39.3.7], the description of $\tilde{\theta}(y_i)$ in (7.12) below is independent of the choice of such sequences.)

In the next theorem, we lift θ to a $\mathbb{C}$ algebra automorphism of U. This is in the spirit of [L2, Theorem 3.1]. The main difference here is that we do not insist that $\tilde{\theta}$ is a $\mathbb{C}$ algebra involution on all of U.

THEOREM 7.1. *There exists a $\mathbb{C}$ algebra automorphism $\tilde{\theta}$ on U such that:*

$$\tilde{\theta}(x_i) = x_i \text{ and } \tilde{\theta}(y_i) = y_i \text{ for all } \alpha_i \in \pi_\Theta. \tag{7.9}$$

$$\tilde{\theta}(\tau(\lambda)) = \tau(\Theta(-\lambda)) \text{ for all } \tau(\lambda) \in T. \tag{7.10}$$

$$\tilde{\theta}(q) = q^{-1}. \tag{7.11}$$

For $\alpha_i \in \pi^*$, we have $\tilde{\theta}(y_i) = [(\mathrm{ad}_r x_{i_1}^{(m_1)} \cdots x_{i_r}^{(m_r)}) t_{p(i)}^{-1} x_{p(i)}]$ and

$$\tilde{\theta}(y_{p(i)}) = (-1)^{m(i)} [(\mathrm{ad}_r x_{i_r}^{(m_r)} \cdots x_{i_1}^{(m_1)}) t_i^{-1} x_i]. \tag{7.12}$$

Furthermore, $\tilde{\theta}^2$ is the identity when restricted to $\mathcal{M}$ and to T. Finally, $\tilde{\theta}$ specializes to θ as q goes to 1.

PROOF. To show that $\tilde{\theta}$ extends to a $\mathbb{C}$ algebra automorphism of U, we relate it to Lusztig's automorphisms. Let T_{w_o} be Lusztig's automorphism associated to w_o, the longest element of W'. We follow the notation of [DK, Section 1.6]. Fix $\alpha_i \in \pi_\Theta$. Recall that $-w_o(\alpha_i) = \alpha_{d(i)}$. By [DK, Section 1.6 and Proposition 1.6], T_{w_o} sends y_i to a nonzero scalar multiple of $x_{d(i)} t_{d(i)}^{-1}$, sends x_i to a nonzero scalar multiple of $y_{d(i)} t_{d(i)}$, and sends t_i to $t_{d(i)}^{-1}$. Furthermore, one checks using [DK, Remark 1.6] that for each $\alpha_i \notin \pi_\Theta$ the composition

$$(\iota \circ T_{w_o})(t_{p(i)}^{-1} x_{p(i)}) = u_i [(\mathrm{ad}_r x_{i_1}^{(m_1)} \cdots x_{i_r}^{(m_r)}) t_{p(i)}^{-1} x_{p(i)}]$$

for some nonzero scalar u_i.

Define a function $\tilde{\theta}$ on the generators of U using (7.9), (7.10), (7.11), (7.12), and setting

$$\tilde{\theta}(x_i) = u_i^{-1} (\iota \circ T_{w_o})(y_{p(i)} t_{p(i)}) \text{ and } \tilde{\theta}(x_{p(i)}) = (-1)^{m(i)} u_{p(i)}^{-1} (\iota \circ T_{w_o})(y_i t_i)$$

for each $\alpha_i \in \pi^*$. It is clear from (7.9) and (7.10) that $\tilde{\theta}$ extends to a $\mathbf{C}$ algebra automorphism on both $\mathcal{M}$ and T. Now $\tilde{\theta}^2$ is clearly the identity on $\mathcal{M}$. Since Θ is an involution on the root system of $\mathbf{g}$, condition (7.10) ensures that $\tilde{\theta}$ also restricts to an involution on the group T.

We check that $\tilde{\theta}$ extends to a $\mathbb{C}$ algebra automorphism of U. In particular, $\tilde{\theta}(y_i)\tilde{\theta}(x_i) - \tilde{\theta}(x_i)\tilde{\theta}(y_i) = (\iota \circ T_{w_o})(y_{p(i)} x_{p(i)} - x_{p(i)} y_{p(i)}) = \tilde{\theta}(y_i x_i - x_i y_i)$ for

$\alpha_i \notin \pi_\Theta$. Furthermore, for $\alpha_i \in \pi_\Theta$, $(\iota \circ T_{w_o})(t_{d(i)}x_{d(i)}) = y_i = \tilde{\theta}(y_i)$ up to some nonzero scalar. Hence the $\tilde{\theta}(y_i)$, $1 \leq i \leq n$ satisfy the quantum Serre relations (1.7). Similarly, $(\iota \circ T_{w_o})(y_i t_i) = x_{d(i)} = \tilde{\theta}(x_{d(i)})$ up to a nonzero scalar when $\alpha_i \in \pi_\Theta$. It follows that the $\tilde{\theta}(x_i)$ for $1 \leq i \leq n$ satisfy the quantum Serre relations (1.7). Moreover, $\tilde{\theta}$ preserves the relations between the x_i and the y_j for $1 \leq i, j \leq n$. Thus $\tilde{\theta}$ extends to a $\mathbb{C}$ algebra automorphism $\tilde{\theta}$ of U.

Now consider an element b in $\mathcal{M} \cup T \cup \{y_i | 1 \leq i \leq n\}$ and write $\bar{b}$ for its specialization as q goes to 1. Note that the specialization of $\tilde{\theta}(b)$ is just $\theta(\bar{b})$. This is enough to force $\tilde{\theta}$ to specialize to θ. $\square$

We are now ready to introduce the quantum analog of $U(\mathbf{g}^\theta)$. Set

$$T_\Theta = \{\tau(\lambda) | \Theta(\lambda) = \lambda\},$$

a subgroup of T. Let $B = B_{\tilde{\theta}}$ be the subalgebra of U generated by $\mathcal{M}$, T_Θ, and the elements

$$B_i = y_i t_i + \tilde{\theta}(y_i) t_i$$

for $\alpha_i \notin \pi_\Theta$. The next result shows that B is a coideal subalgebra of U. This fact combined with the results of Sections 1 and 4 is used below to describe the relations satisfied by these generators. As a consequence, we show below that B specializes to $U(\mathbf{g}^\theta)$ as q goes to 1.

THEOREM 7.2. *B is a left coideal subalgebra of U.*

PROOF. We need to check that

$$\Delta(b) \in U \otimes B \tag{7.13}$$

for all $b \in B$. Since Δ is an algebra homomorphism from U to $U \otimes U$, it is sufficient to check (7.13) for a set of generators of B. Now B is generated by the elements B_i, for $\alpha_i \notin \pi_\Theta$, and two Hopf algebras: $\mathcal{M}$ and the group algebra generated by T_Θ. In particular, each $b \in \mathcal{M}$ and each $b \in T_\Theta$ satisfies (7.13). Hence it is sufficient to check (7.13) holds for the remaining generators, namely when $b = B_i$ for $\alpha_i \notin \pi_\Theta$.

Set $\mathcal{M}^+ = U^+ \cap \mathcal{M}$. Note that $t_i t_{p(i)}^{-1}$ is in T_Θ for all i with $\alpha_i \notin \pi_\Theta$. Thus by (6.5) and the definition of $\tilde{\theta}$, the element $\tilde{\theta}(y_i) t_i$ satisfies the following nice property with respect to the comultiplication of U:

$$\begin{aligned}
\Delta(\tilde{\theta}(y_i)t_i) &\in t_i \otimes \tilde{\theta}(y_i)t_i + U \otimes (\mathcal{M} \cap G^+ U^\circ)t_{p(i)}^{-1} t_i \\
&\subset t_i \otimes \tilde{\theta}(y_i)t_i + U \otimes \mathcal{M}^+ T_\Theta.
\end{aligned} \tag{7.14}$$

This combined with the formula for $\Delta(y_i t_i)$ (see (1.8) and (1.10)) yields

$$\Delta(B_i) \in t_i \otimes B_i + U \otimes \mathcal{M}^+ T_\Theta \subset U \otimes B \tag{7.15}$$

and the theorem follows. $\square$

We turn now to understanding the relations satisfied by the generators of B. The elements B_i have already been defined when $\alpha_i \notin \pi_\Theta$. Set $B_i = y_i t_i$ for $\alpha_i \in \pi_\Theta$. Given a tuple $I = (i_1, \ldots, i_r)$, set $|I| = r$, $\mathrm{wt}(I) = \alpha_{i_1} + \ldots + \alpha_{i_r}$, $B_I = B_{i_1} \cdots B_{i_r}$, and $Y_I = y_{i_1} t_{i_1} \cdots y_{i_r} t_{i_r}$.

Recall ((1.4) and (1.11)) that $x_i y_j t_j = q^{(-\alpha_i, \alpha_j)} y_j t_j x_i$ whenever $j \neq i$. By Theorem 7.1, $\tilde{\theta}(x_i) = x_i$ whenever $\alpha_i \in \pi_\Theta$. Furthermore, $\tilde{\theta}(y_j)$ and y_j have the same weight with respect to the adjoint action of T_Θ. Hence

$$x_i B_j = q^{(-\alpha_i, \alpha_j)} B_j x_i \text{ and } \tau(\lambda) B_j = q^{(\lambda, -\alpha_j)} B_j \tau(\lambda) \qquad (7.16)$$

for all $\alpha_i \in \pi_\Theta$ with $\alpha_j \notin \pi_\Theta$, and $\tau(\lambda) \in T_\Theta$. It follows that

$$B = \sum_I B_I \mathcal{M}^+ T_\Theta. \qquad (7.17)$$

Let $\mathcal{J}$ be a set such that $\{Y_J | J \in \mathcal{J}\}$ is a basis for G^-. Note that $B_J = Y_J +$ (terms of higher weight) for each tuple J. The triangular decomposition (4.1) of U implies that the subspaces $\{Y_J \mathcal{M}^+ T_\Theta | J \in \mathcal{J}\}$, and hence the subspaces $\{B_J \mathcal{M}^+ T_\Theta | J \in \mathcal{J}\}$, are linearly independent.

Let F_{ij} be the function in two variables X_1 and X_2 defined by

$$F_{ij}(X_1, X_2) = \sum_{m=0}^{1-a_{ij}} (-1)^m \begin{bmatrix} 1 - a_{ij} \\ m \end{bmatrix}_{q_i} X_1^{1-a_{ij}-m} X_2 X_1^m.$$

The quantum Serre relations (1.7) are the set of equations $F_{ij}(y_i, y_j) = 0$ for $i \neq j$. A straightforward computation shows that if $(\lambda_i, \alpha_j) = (\lambda_j, \alpha_i)$ then $F_{ij}(y_i \tau(\lambda_i), y_j \tau(\lambda_j)) = 0$. Hence

$$F_{ij}(y_i t_i, y_j t_j) = 0. \qquad (7.18)$$

It follows that the generators $y_i t_i$ of G^- satisfy the same relations as the generators of U^-. Furthermore, since $(\Theta(-\alpha_i), \alpha_j) = (\Theta(-\alpha_j), \alpha_i)$, we have

$$F_{ij}(\tilde{\theta}(y_i) t_i, \tilde{\theta}(y_j) t_j) = 0. \qquad (7.19)$$

We show below that the B_i for $1 \leq i \leq n$ satisfy relations which come from the quantum Serre relations on G^-. First, we consider the evaluation of the function F_{ij} at B_i, B_j in a few special cases.

If both α_i and α_j are in π_Θ, then $F_{ij}(B_i, B_j) = F_{ij}(y_i t_i, y_j t_j)$. Similarly, if $\alpha_i \in \pi_\Theta$ and $\alpha_j \notin \pi_\Theta$, then $F_{ij}(B_i, B_j) = F_{ij}(y_i t_i, y_j t_j) + F_{ij}(\tilde{\theta}(y_i) t_i, \tilde{\theta}(y_j) t_j)$. Hence (7.18) and (7.19) imply that

$$\text{if } \alpha_i \in \pi_\Theta \text{ then } F_{ij}(B_i, B_j) = 0. \qquad (7.20)$$

Now suppose that i and j are chosen such that $\pi_{0,0}(Y_{ij})$ is nonzero. It follows that Y_{ij} must have a zero weight summand. Checking the possibilities for the quantum Serre relations, we must have $a_{ij} = 0$ and $\Theta(\alpha_i) = -\alpha_j$. In particular, $B_i = y_i t_i + q_i^{-2} x_j t_j^{-1} t_i$ and $B_j = y_j t_j + q_i^{-2} x_i t_i^{-1} t_j$. A straightforward computation shows that

$$a_{ij} = 0 \text{ and } \Theta(\alpha_i) = -\alpha_j \text{ imply } F_{ij}(B_i, B_j) = B_i B_j - B_j B_i = \frac{t_i^{-1} t_j - t_j^{-1} t_i}{q_i - q_i^{-1}}.$$

(7.21)

Given $\lambda \in Q(\pi)$, let P_λ be the projection of B onto $U^- G^+ \tau(\lambda)$ with respect to the direct sum decomposition of Lemma 1.3 applied to the coideal B. The next lemma provides more detailed information about $F_{ij}(B_i, B_j)$.

LEMMA 7.3. *Let $Y_{ij} = F_{ij}(B_i, B_j)$ for $i \neq j$ and $\lambda_{ij} = (1 - a_{ij})\alpha_i + \alpha_j$. If $(\pi_{\beta,\gamma} \circ P_{\lambda_{ij}})(Y_{ij}) \neq 0$ then $[\beta, \gamma] \neq 0$, $\tau(\lambda_{ij} - \beta) \notin T_\Theta$, and $\tau(\lambda_{ij} - \gamma) \notin T_\Theta$.*

PROOF. Set $P_{ij} = P_{\lambda_{ij}}$. Suppose that $(\pi_{0,0} \circ P_{ij})(Y_{ij}) \neq 0$. It follows that $\pi_{0,0}(Y_{ij}) \neq 0$. Hence $a_{ij} = 0$ and $\Theta(\alpha_i) = -\alpha_j$. Now $\lambda_{ij} = \alpha_i + \alpha_j$ in this case. By (7.21) $P_{ij}(Y_{ij}) = P_{ij}(t_i^{-1} t_j - t_j^{-1} t_i) = 0$. Therefore, $\pi_{0,0}(Y_{ij}) = 0$ for all choices of i and j.

By (7.20), we may assume that α_i is not in π_Θ. Assume that β and γ are chosen so that $\pi_{\beta,\gamma}(Y_{ij}) \neq 0$. Note that Y_{ij} can be written as a sum of monomials in $2 - a_{ij}$ terms where $1 - a_{ij}$ of those terms are from the set $\{y_i t_i, \tilde{\theta}(y_i) t_i\}$ and the other term is from the set $\{y_j t_j, \tilde{\theta}(y_j) t_j\}$. It follows that $\gamma = s_1 \alpha_{p(i)} + s_2 \alpha_{p(j)} + \eta$ for some $\eta \in Q^+(\pi_\Theta)$ and nonnegative integers s_1 and s_2 such that $s_1 \leq 1 - a_{ij}$ and $s_2 \leq 1$. Set $\gamma' = s_1 \alpha_i + s_2 \alpha_j$ and note that $\tau(\gamma - \gamma')$ is in T_Θ. The above description of the monomials which add to Y_{ij} further implies that $\gamma' + \beta \leq \lambda_{ij}$. Moreover, by (7.18), $0 \leq \beta < \lambda_{ij}$ and by (7.19), $0 \leq \gamma' < \lambda_{ij}$. Now β and γ' are both linear combinations of α_i and α_j. Thus the lemma follows if neither α_i nor α_j are elements of $Q^+(\pi_\Theta)$. In the case when $\alpha_j \in \pi_\Theta$, (7.19) further implies that that $0 \leq \gamma' < \lambda_{ij} - \alpha_j$ and $0 \leq \beta < \lambda_{ij} - \alpha_i$. The lemma thus follows in this case as well. $\square$

The next result gives a description of the generators and relations of B.

THEOREM 7.4. *Let $\tilde{B}$ be the algebra freely generated over $\mathcal{M}^+ T_\Theta$ by the elements $\tilde{B}_i$, $1 \leq i \leq n$. Then there exist elements $c_j^{ij} \in \mathcal{M}^+ T_\Theta$ such that $B \cong \tilde{B}/L$ where L is the ideal generated by the following elements:*

(i) $\tau(\lambda)\tilde{B}_i \tau(-\lambda) - q^{-(\lambda, \alpha_i)} \tilde{B}_i$ *for all $\tau(\lambda) \in T_\Theta$ and $\alpha_i \notin \pi_\Theta$.*

(ii) $t_j^{-1} x_j \tilde{B}_i - \tilde{B}_i t_j^{-1} x_j - \delta_{ij}(t_j - t_j^{-1})/(q_j - q_j^{-1})$ *for all $\alpha_j \in \pi_\Theta$ and $1 \leq i \leq n$.*

(iii)

$$\sum_{m=0}^{1-a_{ij}} (-1)^m \begin{bmatrix} 1 - a_{ij} \\ m \end{bmatrix}_{q_i} \tilde{B}_i^{1-a_{ij}-m} \tilde{B}_j \tilde{B}_i^m - \sum_{\{J \in \mathcal{J} \,|\, \mathrm{wt}(J) < (1-a_{ij})\alpha_i + \alpha_j\}} \tilde{B}_J c_J^{ij}$$

for each $i \neq j$, $1 \leq i, j \leq n$.

PROOF. Relations (i) and (ii) follow from (7.16) and (1.4). We now show that the B_i, $1 \leq i \leq n$, satisfy the relations described in (iii). Fix a quantum Serre relation $Y = F_{ij}(B_i, B_j)$ for given α_i, α_j with $i \neq j$. Set $\lambda = (1 - a_{ij})\alpha_i + \alpha_j$ and $Z = P_\lambda(Y)$. By (4.7), $((P_\lambda \circ \pi_{0,0}) \otimes \mathrm{Id})\Delta(Y) = (\pi_{0,0} \otimes \mathrm{Id})\Delta(Z)$. Moreover, (4.7) ensures that

$$(\pi_{0,0} \otimes \mathrm{Id})\Delta(Z) = \tau(\lambda) \otimes Z.$$

By (7.15) and (7.17), we have

$$\Delta(Y) \in \tau(\lambda) \otimes Y + \sum_{\{J| \ \mathrm{wt}(J) < \lambda\}} U \otimes B_J \mathcal{M}^+ T_\Theta. \tag{7.22}$$

Now if J has weight less than λ, one checks from (1.7) that there is no quantum Serre relation of weight greater than or equal to $-\lambda$. Hence if $\mathrm{wt}(J) < \lambda$ then J is an element of the set $\mathcal{J}$. Now, (7.22) implies that

$$((P_\lambda \circ \pi_{0,0}) \otimes \mathrm{Id})\Delta(Y) \in \tau(\lambda) \otimes (Y + \sum_{\{J \in \mathcal{J}| \ \mathrm{wt}(J) < \lambda\}} B_J \mathcal{M}^+ T_\Theta)$$

Thus we can find $X \in \sum_{\{J \in \mathcal{J}|\mathrm{wt}(J) < \lambda\}} B_J \mathcal{M}^+ T_\Theta$ such that $Y + X = Z$. We obtain a relation of the form described in (iii) by proving $Z = 0$.

Recall the notation of Section 4, Filtration II. Assume that Z is nonzero and hence $\max(Z)$ is nonempty. Choose $[\beta, \gamma] \in \max(Z)$. It follows that $\pi_{\beta, \gamma}(Z) \neq 0$. By Lemma 7.3, $[\beta, \gamma] \neq [0, 0]$, and neither $\tau(\lambda - \beta)$ nor $\tau(\lambda - \gamma)$ is an element of T_Θ. Write $(\pi_{\beta, 0} \otimes \mathrm{Id})(\Delta(Z)) = \sum v_i \otimes u_i$ where the $v_i \in U_{-\beta}^- \tau(\lambda)$ and the $u_i \in G^+ T$. We may assume that the v_i are linearly independent elements of $U_{-\beta}^- \tau(\lambda)$. Note that at least one of the u_i has a (nonzero) summand of weight γ in $G_\gamma^+ \tau(\lambda - \beta)$. By (7.19), the maximality of $[\beta, \gamma]$, and the fact that $\beta \neq 0$, each u_i is in $G^+ T \cap \sum_{\{J \in \mathcal{J}|\mathrm{wt}(J) < \lambda\}} B_J \mathcal{M}^+ T_\Theta$. This intersection is just $\mathcal{M}^+ T_\Theta$. Hence $\tau(\lambda - \beta) \in T_\Theta$, a contradiction. This forces $\beta = 0$. It follows that $(\pi_{0,\gamma} \otimes \mathrm{Id})(\Delta(Z)) \in U \otimes \tau(\lambda - \gamma)$. Again $\tau(\lambda - \gamma)$ must be in T_Θ. This contradiction forces $\max(Z)$ to be empty. In particular, $Z = 0$.

We have shown that B is isomorphic to a homomorphic image of $\tilde{B}/L$. A consequence of relations (i), (ii), and (iii) is that $\mathcal{M}^+ T_\Theta \tilde{B}_I \subset \sum_{J \in \mathcal{J}} \tilde{B}_J \mathcal{M}^+ T_\Theta + L$ for each tuple I. Thus

$$\tilde{B}/L = \bigoplus_{J \in \mathcal{J}} (\tilde{B}_J \mathcal{M}^+ T_\Theta + L).$$

Since the elements B_i in B satisfy the relations (i), (ii), (iii), we also have the following direct sum decomposition:

$$B = \bigoplus_{J \in \mathcal{J}} (B_J \mathcal{M}^+ T_\Theta).$$

Therefore $B \cong \tilde{B}/L$. $\square$ $\square$

Note that (7.20) and (7.21) both provide examples of the relations described in Theorem 7.4 (iii). We illustrate how to compute the c_j^{ij} in a more complicated example. Consider the case where $\Theta(\alpha_i) = -\alpha_i$, $\Theta(\alpha_j) = -\alpha_j$ and $a_{ij} = -1$. So $B_i = y_i t_i + q_i^{-2} x_i$ and $B_j = y_j t_j + q_j^{-2} x_j$ and $Y = B_i^2 B_j - (q_i + q_i^{-1}) B_i B_j B_i + B_j B_i^2$. Thus by (1.9) and (1.10),

$$\Delta(B_r) = B_r \otimes 1 + t_r \otimes B_r$$

for $r = i, j$. It follows that

$$\Delta(Y) = t_i^2 t_j \otimes Y + (B_i^2 t_j - (q_i + q_i^{-1}) B_i t_j B_i + t_j B_i^2) \otimes B_j + W \otimes B$$

for some W which satisfies $\pi_{0,0}(W) = 0$. A straightforward computation using the relations of U shows that $P_\lambda \circ \pi_{0,0}((B_i^2 t_j - (q_i + q_i^{-1}) B_i t_j B_i + t_j B_i^2) = -q_i^{-1} t_i^2 t_j$. Thus

$$0 = (P_\lambda \circ \pi_{0,0}) \otimes \mathrm{Id}(\Delta(Y) = t_i^2 t_j \otimes Y - q_i^{-1} t_i^2 t_j \otimes B_j.$$

It follows that

$$B_i^2 B_j - (q_i + q_i^{-1}) B_i B_j B_i + B_j B_i^2 = -q_i^2 B_j.$$

(This relation is also computed in [L1, Lemma 2.2 (2.2)]. The generators for U and B are somewhat different in [L1]. In particular, when $\alpha_i = -\Theta(\alpha_i)$, B_i in [L1] is equal to $y_i t_i + x_i$ in the notation of this paper. Thus using a Hopf algebra automorphism of U, the B_i in [L1] corresponds to q_i^{-1} times the B_i defined in this paper. This explains the difference in coefficient of B_j found in the two papers.) Note that a similar argument shows that $c_j^{ij} = 0$ whenever $-\alpha_i \neq \alpha_{p(i)}$ and $-\alpha_i \neq \alpha_{p(j)}$. The c_j^{ij} are computed in [L1, Lemma 2.2] for the cases when $a_{ij} \geq -2$ and $\Theta(-\alpha_i) = \alpha_{p(i)}$.

Note that the generators of B specialize to the generators of $U(\mathbf{g}^\theta)$ as q goes to 1. Thus the specialization of $B_J \mathcal{M}^+ T_\Theta$ is contained in $U(\mathbf{g}^\theta)$. Moreover the set of spaces $\{B_J \mathcal{M}^+ T_\Theta, J \in \mathcal{J}\}$ remain linearly independent after specialization. As q goes to 1, since these spaces span B, we conclude that B specializes to $U(\mathbf{g}^\theta)$.

The algebra B also satisfies a maximality condition. Indeed, suppose that C is a subalgebra of U containing B and that C also specializes to $U(\mathbf{g}^\theta)$. Then by [L2, Theorem 4.9], $C = B$. The proof in [L2] uses a quantum version of the Iwasawa decomposition. The result also follows directly from Theorem 7.4. The idea is as follows. Recall the notation of Section 6. Set $N_\Theta^+ = N_{\pi_\Theta}^+$. By (6.6) (interchanging the roles of $N_{\pi'}^+$ with $N_{\pi'}^-$), we have

$$U = \sum_{J \in \mathcal{J}} Y_J \mathcal{M}^+ T N_\Theta^+.$$

By induction on $|J|$ (as in [L2, Lemma 4.3]), one can show that U is spanned by the spaces B; Bt, $t \notin T_\Theta$; and $BT(N_\Theta^+)_+$ where $(N_\Theta^+)_+$ is the augmentation ideal of N_{Θ}^{+}. Let X be in C. Subtracting an element of B if necessary, we may assume that X is a linear combination of elements in $B(t-1)/(q-1)$ for $t \notin T_{\Theta}$, and $BT(N_{\Theta}^{+})_{+}$. Assume that X is nonzero. Rescale X by a power of $(q-1)$ so that it is an element of $\hat{C} - (q-1)\hat{C}$. It follows that X does not specialize to an element of $U(\mathbf{g}^{\theta})$. This contradiction forces $X = 0$ and thus $B = C$.

We have shown that the algebra B satisfies the following properties.

B is a left coideal in U. $\hfill (7.23)$

B specializes to $U(\mathbf{g}^{\theta})$. $\hfill (7.24)$

If $B \subset C$ and C is a subalgebra of U specializing to $U(\mathbf{g}^{\theta})$ then $B = C$. $\hfill (7.25)$

We now turn to characterizing all subalgebras of U which satisfy (7.23), (7.24), and (7.25). First, we present two variations which satisfy these conditions as well.

Variation 1. For sake of simplicity, we assume first that $\mathbf{g}$ is simple. Recall the permutation p used in (7.5). Suppose that there exists an $r \in \{1, 2, \ldots, n\}$ such that $\alpha_r \notin \pi_{\Theta}$ and $p(r) \neq r$. Assume further that $(\alpha_r, \Theta(\alpha_r)) \neq 0$. Recall the Cartan subspace $\mathbf{a} = \{x \in \mathbf{h} | \theta(x) = -x\}$ and the restricted root system Σ associated to θ introduced at the beginning of this section. Let β be the restricted root corresponding to e_r. Note that β is just the restriction of $\alpha_r \in \mathbf{h}^*$ to $\mathbf{a}^*$. Furthermore, $(\operatorname{ad} a)[e_r, \Theta(f_r)] = 2\beta(a)[e_r, \Theta(f_r)]$ for all $a \in \mathbf{a}$. In particular, the restricted root system Σ contains both β and 2β. Thus Σ is nonreduced and hence must be of type BC. One can choose the positive roots of Σ so that each α_j restricted to $\mathbf{a}^*$ is either zero or a simple positive root in Σ. Furthermore, α_r and α_j restrict to the same root if and only if $j = r$ or $j = p(r)$. It follows from [Kn, Chapter II, Section 8] that there is exactly one positive simple root in Σ such that twice this root is also in Σ. Hence r and $p(r)$ are the only values of j such that $(\alpha_j, \Theta(\alpha_j)) \neq 0$.

Let c be an element in $A = \mathbb{C}[q, q^{-1}]_{(q-1)}$ that specializes to 1 as q goes to 1. Define the $\mathbb{C}$ algebra automorphism $\tilde{\theta}_c$ of U by requiring that

$$\tilde{\theta}_c(y_r) = c^{-1}\tilde{\theta}(y_r), \quad \tilde{\theta}_c(x_r) = c\tilde{\theta}(x_r),$$

and that $\tilde{\theta}_c$ agrees with $\tilde{\theta}$ on all other generators of U. Note that $\tilde{\theta}_c$ is also a $\mathbb{C}$ algebra automorphism of U which specializes to θ and restricts to $\tilde{\theta}$ on MT_{Θ}. Define $B_{\tilde{\theta}_c}$ in the same way as $B_{\tilde{\theta}}$ using $\tilde{\theta}_c$ instead of $\tilde{\theta}$. Thus $B_{\tilde{\theta}_c}$ is generated by $\mathcal{M}$, T_{Θ}, and elements $B_i^c = y_i t_i + \tilde{\theta}_c(y_i) t_i$ for $\alpha_i \notin \pi_{\Theta}$. Moreover, $B_i^c = B_i$ for $i \neq r$. Since $\tilde{\theta}_c(y_i)$ is a scalar multiple of $\tilde{\theta}(y_i)$ for all i, the proof of Theorem 7.2 also works for $B_{\tilde{\theta}_c}$. Hence $B_{\tilde{\theta}_c}$ is a left coideal subalgebra of U. Consider a quantum Serre relation $F_{ij}(y_i, y_j)$ where either i or j equals r. Note that if $\Theta(\alpha_i) = -\alpha_j$ then, by the assumptions on r, $\{i, j\} = \{r, p(r)\}$ and $(\alpha_i, \alpha_j) \neq 0$. Thus as in the proof of Lemma 7.3, $(\pi_{0,0} \circ P_{ij})(F_{ij}(B_i^c, B_j^c)) \neq 0$ whenever $i \neq j$. Hence the arguments for $B_{\tilde{\theta}}$ used to prove Lemma 7.3 and Theorem 7.4 work for $B_{\tilde{\theta}_c}$ as well. In particular, $B_{\tilde{\theta}_c}$ satisfies conditions (7.23), (7.24), and (7.25).

Note that $B_{\tilde{\theta}_c}$ is not isomorphic to $B_{\tilde{\theta}}$ via a Hopf algebra automorphism of U for $c \neq 1$. It appears unlikely in general that two such algebras are isomorphic using just an algebra isomorphism. It should be noted that the existence of this one parameter family of analogs is implicit in the proof of [L2, Theorem 5.8]. However, it was mistakenly concluded in the paragraph directly preceding [L2, Theorem 5.8] that all the analogs of Variation 1 were isomorphic to $B_{\tilde{\theta}}$ via a Hopf algebra automorphism.

In the general semisimple case, the one parameter c is replaced by a multipa-rameter $\mathbf{c}$. In particular, each parameter corresponds to a pair of roots $\alpha_{i_j}, \alpha_{p(i_j)}$ such that $(\alpha_{i_j}, \Theta(\alpha_{i_j})) \neq 0$. The automorphism $\tilde{\theta}_{\mathbf{c}}$ is defined in a similar fash-ion to $\tilde{\theta}_c$. Let $[\Theta]$ be the set of automorphisms of the form $\tilde{\theta}_{\mathbf{c}}$. Following the convention in [L3], we refer to $B_{\tilde{\theta}'}$, $\tilde{\theta}' \in [\Theta]$ as a standard analog of $U(\mathbf{g}^\theta)$.

Variation 2. Let S_1 be the subset of $\pi - \pi_\Theta$ consisting of those roots α_i such that $\Theta(\alpha_i) = -\alpha_i$. Let S be the subset of S_1 such that if $\alpha_i \in S$ and $\alpha_j \in S_1$ then $2(\alpha_i, \alpha_j)/(\alpha_j, \alpha_j)$ is even. Let $\mathbf{S}$ be the set of n tuples $\mathbf{s} = (s_1, \ldots, s_n)$ such that each s_i is in $A = \mathbf{C}[q, q^{-1}]_{(q-1)}$ and $s_i \neq 0$ implies $\alpha_i \in S$. Given $\hat{\theta} \in [\Theta]$, let $B_{\hat{\theta}, \mathbf{s}}$ be the subalgebra of U generated by $T_\Theta, \mathcal{M}$, the B_i for $\alpha_i \in \pi - S$, and the $B_{i,\mathbf{s}}$ defined by

$$B_{i,\mathbf{s}} = y_i t_i + q^{-(\alpha_i, \alpha_i)} x_i + s_i t_i$$

for $\alpha_i \in S$. In particular, when the entries of $\mathbf{s}$ are all zero, $B_{i,\mathbf{s}}$ is just equal to B_i.

Note that

$$\Delta(B_{i,\mathbf{s}}) = t_i \otimes B_i + (y_i t_i + q^{-(\alpha_i, \alpha_i)} x_i) \otimes 1.$$

Thus by the same arguments as in Theorem 7.2, $B_{\hat{\theta}, \mathbf{s}}$ is a left coideal subalgebra of U. Note that if $\tau(\lambda) \in T_\Theta$ and $\alpha_i \in S$ then $(\lambda, \alpha_i) = 0$. It follows that $aB_{i,\mathbf{s}} = B_{i,\mathbf{s}}a$ for all $\alpha_i \in S$ and $a \in \mathcal{M}T_\Theta$. Recall the notation of Lemma 7.3. To show that $B_{\hat{\theta}, \mathbf{s}}$ satisfies the conditions (7.23), (7.24), and (7.25), it suffices to check for all i, j, $i \neq j$, that $(\pi_{0,0} \circ P_{ij})(F_{ij}(B_{i,\mathbf{s}}, B_{j,\mathbf{s}})) = 0$. A lengthy but routine computation shows that this holds exactly when the n tuple $\mathbf{s}$ is in $\mathbf{S}$.

Following the convention in [L3], the $B_{\hat{\theta}, \mathbf{s}}$, $\hat{\theta} \in [\Theta]$ are called nonstandard analogs of $U(\mathbf{g}^\theta)$. A nonstandard analog $B_{\hat{\theta}, \mathbf{s}}$ is not isomorphic to a standard analog using a Hopf algebra automorphism of U. However, ([L2, Lemma 5.7]) $B_{\hat{\theta}, \mathbf{s}}$ is isomorphic as an algebra to $B_{\hat{\theta}}$.

Nonstandard analogs were first observed (to the suprise of the author) in [L2, Section 5]. In [L3, Section 2], nonstandard analogs were claimed to exist when $\mathbf{S}$ is defined using the larger set S_1 instead of S. (See in particular the definition of $\mathbf{S}$ given following [L3, (2.11) and Theorem 2.1].) Our analysis in Variation 2 corrects this point.

We are now ready to show that the only possible subalgebras of U which satisfy (7.23), (7.24), and (7.25) are our standard and nonstandard analogs associated to an automorphism in $[\Theta]$. In particular, we give a new proof of [L2, Theorem 5.8] using the approach and results of Section 4. Note that when the restricted roots Σ associated to the involution θ do not contain a component of type BC, then all the analogs described below are isomorphic to each other as algebras. This is precisely what Theorem 5.8 in [L2] states. On the other hand, by the discussion of Variation 1, if Σ contains m components of type BC, then there is an m parameter family of analogs up to algebra isomorphism.

THEOREM 7.5. *A subalgebra B of U satisfies* (7.23), (7.24), *and* (7.25) *if and only if B is isomorphic as an algebra to $B_{\hat\theta}$ for some $\hat\theta \in [\Theta]$. In particular, B is isomorphic to a standard or nonstandard analog of $U(\mathbf{g}^\theta)$ corresponding to an element $\hat\theta$ in $[\Theta]$ and an element $\mathbf{s}$ in $\mathcal{S}$ via a Hopf algebra automorphism of U.*

PROOF. We use the notation of the second filtration introduced in Section 4. Let B be a subalgebra of U which satisfies (7.23), (7.24), and (7.25). The proof of this theorem has three steps:

(i) $B \cap T = T_\Theta$
(ii) $B \cap U^+ = \mathcal{M}^+$
(iii) $\mathrm{gr}_\mathfrak{g} B \cap G^- = G^-$.

More precisely, we first prove that $B \cap T$ is a subgroup of T_Θ and $B \cap U^\circ U^+$ is a coideal subalgebra of $\mathcal{M}^+ T_\Theta$. We then use the second filtration introduced in Section 4 to analyze $\mathrm{gr}_\mathfrak{g} B \cap G^-$ and thus prove (iii). This information is then used to show that $B \cap U^+ U^\circ$ specializes to $U(\mathbf{g}^\theta) \cap U(\mathbf{n}^+ + \mathbf{h})$. Next we obtain (i) and (ii). The last part of the proof takes a closer look at the generators of B whose tip is in G^- and show they are of the desired form. The details follow.

Consider the set $B \cap T$. By (7.24), $B \cap T$ is a subset of T_Θ. Hence $B \cap T = B \cap T_\Theta$. Note that any element of B can be written as a direct sum of weight vectors with respect to $B \cap T$. Hence by (7.25), we may assume that $B \cap T_\Theta$ is a group. Condition (7.23) and Lemma 4.2 ensure that $B \cap U^\circ$ is the group algebra generated by $B \cap T_\Theta$. Since T_Θ is free abelian of finite rank, $B \cap T_\Theta$ is free abelian of rank at most the rank of T_Θ.

Consider the coideal subalgebra $B \cap U^+ U^\circ$ of B. We show that $B \cap U^+ U^\circ$ is a subalgebra of $\mathcal{M}^+ T_\Theta$. By Lemma 1.3, $B \cap U^\circ U^+$ is a direct sum of the vector spaces $B \cap G^+ \tau(\mu)$, where $\tau(\mu) \in T$. Suppose that $c \in B \cap G^+ \tau(\mu)$. Choose γ maximal with respect to the standard partial ordering on $Q^+(\pi)$ so that $\pi_{0,\gamma}(c) \neq 0$ and $\gamma \in Q^+(\pi_\Theta)$. Then by (4.7),

$$\pi_{0,\gamma}(c) \in G_\gamma^+ \tau(\mu) \otimes Y$$

where $Y \in \tau(\mu - \gamma) + \sum_{\gamma' > \gamma} G_{\gamma'-\gamma}^+ \tau(\mu - \gamma)$. Since B is a coideal, Y is an element of B. Rescaling if necessary, we may assume that Y is in $\hat{B} - (q-1)\hat{B}$. Hence Y specializes to a nonzero element in $U(\mathbf{g}^\theta)$. The choice of γ implies that $\gamma' - \gamma \notin Q^+(\pi_\Theta)$ for all γ' which appear in the definition of Y. Hence,

$Y \in \tau(\mu - \gamma) + (q - 1)\sum_{\gamma' > \gamma} \hat{G}^+_{\gamma' - \gamma}\tau(\mu - \gamma)$. But then $(q - 1)^{-1}(Y - 1)$ is also in $\hat{B}$ and thus specializes to an element of $U(\mathbf{g}^\theta)$. This forces $\tau(\mu - \gamma)$, and thus $\tau(\mu)$, to be in T_Θ. Now consider λ maximal such that $\pi_{0,\lambda}(c) \neq 0$. Then by (4.7), $\tau(\mu - \lambda) \in T_\Theta$. Hence $\lambda \in Q^+(\pi_\Theta)$. Note that if $\lambda' \in Q^+(\pi)$ and $\lambda' < \lambda$ then λ' is also in $Q^+(\pi_\Theta)$. It follows that c is a sum of weight vectors with weights in $Q^+(\pi_\Theta)$. In particular, $c \in \mathcal{M}^+ T_\Theta$ and $B \cap U^o U^+$ is a subalgebra of $\mathcal{M}^+ T_\Theta$.

We next analyze the part of B whose top degree terms are in G^-. To do this, we introduce the left B module B/N where N is the left ideal $B(B \cap (U^+ U^o)_+)$ of B. (Here $(U^+ U^o)_+$ is equal to the augmentation ideal of $U^+ U^o$.) The filtration $\mathcal{G}$ on B induces a filtration which we also denote by $\mathcal{G}$ on B/N which makes $\mathrm{gr}_\mathcal{G} B/N$ into a $\mathrm{gr}_\mathcal{G} B$ module. By Theorem 4.9, the only important contributions to this graded module occur in bidegree $(m, 0)$ for $m \geq 0$. In particular, $\mathrm{gr}_\mathcal{G} B/N$ is spanned by elements $b + N$ where $b \in B$ and $\mathrm{tip}(b) \in G^-$. Note that the subspace of G^- of elements of bidegree less than or equal to $(m, 0)$ is finite dimensional. Thus the filtration on B/N is a finite discrete filtration. Moreover, $\mathrm{gr}_\mathcal{G} B$ is finitely generated by the image of the generators of B described in Corollary 4.10. Hence we have equality of Gelfand Kirillov dimension: $\mathrm{GKdim}\ \mathrm{gr}_\mathcal{G}(B/N) = \mathrm{GKdim}\ B/N$ ([KL, Prop. 6.6]). Now $\mathrm{gr}_\mathcal{G} B/N$ identifies with $\mathrm{gr}_\mathcal{G}(B) \cap G^-$ as a left $\mathrm{gr}_\mathcal{G}(B) \cap G^-$ module. It is straightforward to check that the GK dimension of $\mathrm{gr}_\mathcal{G} B/N$ as a $\mathrm{gr}_\mathcal{G} B$ module is equal to the GK dimension of $\mathrm{gr}_\mathcal{G} B/N$ as a $\mathrm{gr}_\mathcal{G}(B) \cap G^-$ module. Hence, the form of the generators of B given in Corollary 4.10 implies that $\mathrm{GKdim}\ \mathrm{gr}_\mathcal{G} B/N \leq \dim \mathbf{n}^-$.

Let $\mathbf{r}$ be a Lie subalgebra of $\mathbf{g}^\theta$. A standard argument similar to the argument in the previous paragraph yields that the $U(\mathbf{g}^\theta)$ module $U(\mathbf{g}^\theta)/(U(\mathbf{g}^\theta)\mathbf{r})$ has GK dimension equal to $\dim \mathbf{g}^\theta - \dim \mathbf{r}$. (This follows for example from [D, Proposition 2.2.7].) Consider the $\hat{B}$ module $\hat{B}/\hat{N}$. Write $\bar{N}$ for the specialization of N at $q = 1$. By Theorem 4.1, $B \cap (U^+ U^o)$ specializes to the enveloping algebra of a Lie subalgebra, say $\mathbf{s}$, of $\mathbf{g}^\theta$. Note that $\bar{N} = U(\mathbf{g}^\theta)\mathbf{s}$. Now $B \cap U^+ U^o \subset \mathcal{M}^+ T_\Theta$. Hence $\mathbf{s}$ is a Lie subalgebra of $\mathbf{m}^+ + (\mathbf{g}^\theta \cap \mathbf{h})$. The map which sends each $b + \hat{N}$ in $\hat{B}/\hat{N}$ to $\bar{b} + \bar{N}$ in $U(\mathbf{g}^\theta)/\bar{N}$ allows us to specialize the left $\hat{B}$ module $\hat{B}/\hat{N}$ to the $U(\mathbf{g}^\theta)$ module $U(\mathbf{g}^\theta)/\bar{N}$ at $q = 1$. We can choose generating sets for B and B/N which specialize to generating sets of $U(\mathbf{g}^\theta)$ and $U(\mathbf{g}^\theta)/\bar{N}$ respectively. Hence $\mathrm{GKdim}\ B/N \geq \mathrm{GKdim}\ U(\mathbf{g}^\theta)/\bar{N}$. Note that

$$\mathrm{GKdim}\ U(\mathbf{g}^\theta)/\bar{N} = \dim \mathbf{g}^\theta - \dim \mathbf{s} \geq \dim \mathbf{g}^\theta - \dim(\mathbf{m}^+ + (\mathbf{g}^\theta \cap \mathbf{h})) = \dim \mathbf{n}^-.$$

By the previous paragraph, this inequality is an equality. Hence

$$\mathrm{GKdim}\ U(\mathbf{g}^\theta)/\bar{N} = \mathrm{GKdim}\ \mathbf{n}^- = \mathrm{GKdim}\ G^-.$$

Moreover $\dim \mathbf{s} = \dim(\mathbf{m}^+ + (\mathbf{g}^\theta \cap \mathbf{h}))$. Since $\mathbf{s}$ is a subalgebra of $\mathbf{m}^+ + (\mathbf{g}^\theta \cap \mathbf{h})$, it follows that $\mathbf{s} = \mathbf{m}^+ + (\mathbf{g}^\theta \cap \mathbf{h})$. Thus $B \cap U^+ U^o$ specializes to $U(\mathbf{m}^+ + (\mathbf{g}^\theta \cap \mathbf{h}))$.

Recall the set Δ' of Corollary 4.10 used to define the generators of B whose top degree term is in G^-. The description of the generators of B in Corollary 4.10 implies that $\mathrm{GKdim}\ B/N$ is equal to the number of elements in Δ'. Since the number of elements in Δ^+ is just the dimension of $\mathbf{n}^-$, it follows that $\Delta' = \Delta^+$. Hence by Corollary 4.10, B contains elements $y_i t_i + b_i$, $1 \leq i \leq n$, where b_i is in $U^+ U^o$. It follows that $\text{tip}(B) \cap G^- = G^-$. This proves (iii).

Let N' be the left ideal of $B \cap U^+ U^o$ generated by the augmentation ideal of $B \cap U^o$. We can analyze the left $B \cap U^+ U^o$ module $(B \cap U^+ U^o)/N'$ in a similar fashion to the analysis of B/N. It follows that $B \cap U^+ U^o = B \cap \mathcal{M}^+ T_\Theta$ contains elements $x_i + c_i \in \hat{B}$ for each $\alpha_i \in \pi_\Theta$. Furthermore, $c_i \in U^o$ and $B \cap U^o$ specializes to $U(\mathbf{g}^\theta \cap \mathbf{h})$. Now $B \cap U^o$ is just the group algebra generated by $B \cap T_\Theta$. Therefore, rank $B \cap T_\Theta = $ rank T_Θ. Hence we can find generators of T_Θ such that a power of each generator lies in B. This in turn implies that B can be written as a direct sum of T_Θ weight spaces. By the maximality condition (7.25) of B, we obtain $B \cap T_\Theta = T_\Theta$. This completes the proof of step (i).

Since $T_\Theta \subset B$, any element in $U^+ U^o \cap B = \mathcal{M}^+ T_\Theta \cap B$ is a sum of T_Θ weight vectors contained in B. Thus $x_i + c_i \in B$ implies $x_i \in B$. In particular B contains x_i for all $\alpha_i \in \pi_\Theta$. Hence $B \cap U^+ U^o = \mathcal{M}^+ T_\Theta$ and (ii) follows.

Fix i and consider again the element $y_i t_i + b_i$ in B where $b_i \in U^+ U^o$. Replacing b_i by another element in $U^+ U^o$ if necessary, we may assume that $y_i t_i + b_i$ is a weight vector for the action of T_Θ. By Lemma 1.3, we may further assume that $b_i \in G^+ t_i$. First consider the case when $\alpha_i \in \pi_\Theta$. Choose β maximal with respect to the standard partial ordering on $Q(\pi)$ such that $\pi_{0,\beta}(b_i) \neq 0$. By (4.7), $(\pi_{0,\beta} \otimes \text{Id})\Delta(y_i t_i + b_i)$ is a nonzero element of $G_\beta^+ t_i \otimes \tau(-\beta) t_i$. Hence $\tau(-\beta) t_i \in T_\Theta$ and $\beta \in Q^+(\pi_\Theta)$. If $0 < \gamma < \beta$, then γ is also in $Q^+(\pi_\Theta)$. Thus $\text{supp}(b_i)$ is a subset of $\{0\} \times Q^+(\pi_\Theta)$. This forces b_i to be an element of $\mathcal{M}^+ T_\Theta$ and so $y_i t_i \in B$.

Now assume that $\alpha_i \notin \pi_\Theta$. Choose β such that $[0, \beta] \in \max(b)$. Then by (4.11), $(\pi_{0,\beta} \otimes \text{Id})\Delta(y_i t_i + b_i)$ is a nonzero element of $G_\beta^+ t_i \otimes \tau(-\beta) t_i$. In particular, $\tau(-\beta) t_i = \tau(-\beta + \alpha_i)$ is in T_Θ. Since $\beta \in Q^+(\pi)$, it follows that $\beta \in \alpha_i + Q^+(\pi_\Theta)$ or $\beta \in \alpha_{p(i)} + Q^+(\pi_\Theta)$. However, β must also be of the same T_Θ weight as $-\alpha_i$. The only possibility is $\beta = \Theta(-\alpha_i)$. By the uniqueness property of the $Y_{I,j}$ and $X_{I,j}$ discussed in Section 6 (see (6.3) and the following discussion), the β weight term is a scalar multiple of $\tilde{\theta}(y_i) t_i$. Indeed this is necessary in order for $\Delta(y_i t_i + b_i)$ to be an element of $U \otimes B$. Therefore $b_i = c\tilde{\theta}(y_i) t_i + d t_i$ for some scalar c and element $d \in G^+$ of bidegree less than $\text{bideg}(\tilde{\theta}(y_i) t_i)$. By (7.23),

$$\Delta(y_i t_i + b_i) \in t_i \otimes (y_i t_i + b_i) + U \otimes B.$$

By (1.8), (1.10), and (7.14), it follows that

$$\Delta(d t_i) \in t_i \otimes d t_i + U \otimes \mathcal{M}^+ T_\Theta.$$

Since $\alpha_i \notin \pi_\Theta$, this forces $d t_i$ to be a scalar multiple of t_i. Hence, up to a Hopf algebra automorphism of U, the only possibility for B is one of the standard or nonstandard analogs of $U(\mathbf{g}^\theta)$. $\qquad\square$

Let us return for now to our first analog $B_{\tilde{\theta}}$. Recall the definition of the anti-automorphism κ. One checks that $\kappa((\mathrm{ad}_r x_j)b) = -((\mathrm{ad}_r y_j)\kappa(b))$ for any $b \in U$ and $1 \le j \le n$. Recall that $m(i) = m_1 + \cdots + m_r$. Hence

$$\kappa[(\mathrm{ad}_r x_{i_1}^{(m_1)} \cdots x_{i_r}^{(m_r)})t_{p(i)}^{-1}x_{p(i)}] = (-1)^{m(i)}(\mathrm{ad}_r y_{i_1}^{(m_1)} \cdots y_{i_r}^{(m_r)})y_{p(i)}.$$

A straightforward $U_q(\mathbf{sl}\,2)$ computation as in the classical case (see (7.8)) yields

$$(\mathrm{ad}_r y_{i_1}^{(m_1)} \cdots y_{i_r}^{(m_r)})(\mathrm{ad}_r x_{i_r}^{(m_r)} \cdots x_{i_1}^{(m_1)})t_i^{-1}x_i = t_i^{-1}x_i.$$

Set $y_j \cdot bt_{p(i)} = bt_{p(i)}q^{(\alpha_{p(i)}, \alpha_j)}y_j - y_j t_j bt_{p(i)}t_j^{-1}$ for any $b \in U$ and $1 \le i, j \le n$. Note that $y_j \cdot bt_{p(i)} = ((\mathrm{ad}_r y_j)b)t_{p(i)}$. Recall the definition of π^* immediately following (7.5). We have

$$(-1)^{m(i)}[y_{i_1}^{(m_1)} \cdots y_{i_r}^{(m_r)} \cdot B_{p(i)}]t_{p(i)}^{-1}t_i$$
$$= ([(-1)^{m(i)}(\mathrm{ad}_r y_{i_1}^{(m_1)} \cdots y_{i_r}^{(m_r)})y_{p(i)}]t_i + t_i^{-1}x_i t_i$$
$$= \kappa(\tilde{\theta}(y_i)t_i) + q^{-(\alpha_i, \alpha_i)}\kappa(y_i t_i)$$

is an element of $B_{\tilde{\theta}}$ for each $\alpha_i \in \pi^*$. A similar argument shows that

$$\kappa(\tilde{\theta}(y_{p(i)})t_{p(i)}) + q^{-(\alpha_i, \alpha_i)}\kappa(y_{p(i)}t_{p(i)})$$

is also in $B_{\tilde{\theta}}$ for each $\alpha_i \in \pi^*$. Thus one can find a Hopf algebra automorphism Υ in $\mathcal{H}_\mathbb{R}$ such that Υ restricts to the identity on $\mathcal{M}$ and T_Θ and $\Upsilon(B)$ contains $\kappa(\Upsilon(B_i))$ for each $\alpha_i \notin \pi_\Theta$. Furthermore, one can show that $\Upsilon(B)$ is generated by $\mathcal{M}$, T_Θ, and the $\kappa(\Upsilon(B_i))$, $\alpha_i \notin \pi_\Theta$. Now $\kappa(\Upsilon(\mathcal{M})) = \mathcal{M}$ and $\kappa(\Upsilon(T_\Theta)) = T_\Theta$. It follows that $\kappa(\Upsilon(B)) = \Upsilon(B)$. Hence the results of Section 2 hold for B.

The same argument works for analogs of Variations 1 and 2 provided that all entries of the tuples involved are from $\mathbb{R}(q)$. In particular, let $[\Theta]_r$ be the set $\{\theta_\mathbf{b}|$ all entries of $\mathbf{b}$ are in $\mathbb{R}(q)\}$. We refer to analogs of the form $B_{\theta_\mathbf{b}, \mathbf{s}}$ for $\theta_\mathbf{b} \in [\Theta]_r$ and all entries of $\mathbf{s}$ are in $\mathbb{R}(q)$ as real analogs of $U(\mathbf{g}^\theta)$. Given $\theta_\mathbf{b} \in [\Theta]_r$, one can find $\Upsilon \in \mathcal{H}_\mathbb{R}$ such that $\Upsilon^{-1}\kappa\Upsilon(B_{\theta_\mathbf{b}}) = B_{\theta_\mathbf{b}}$. Furthermore, for any $\mathbf{s}$ such that all of its entries are in $\mathbb{R}(q)$, we also have that $\Upsilon^{-1}\kappa\Upsilon(B_{\theta_\mathbf{b}, \mathbf{s}}) = B_{\theta_\mathbf{b}, \mathbf{s}}$. Hence, we may apply results of Section 2 to all real analogs of $U(\mathbf{g}^\Theta)$.

Consider a real analog B of $U(\mathbf{g}^\theta)$. Given a U module M, set $X(M)$ equal to the sum of all the finite-dimensional unitary B submodules of M. The next result on basic Harish-Chandra modules associated to the pair U, B follows from Section 2.

THEOREM 7.6. *Let B be a real analog of $U(\mathbf{g}^\theta)$ and let M be a U module. Then any finite-dimensional U module is a B unitary module and a Harish-Chandra module for the pair U, B. Furthermore both $F(U)$ and $X(M)$ are Harish-Chandra modules for the pair U, B.*

We continue the assumption that B is a real analog of $U(\mathbf{g}^\theta)$. Using the approach of Section 3, we can define the quantum homogeneous space associated to B. The left invariants $R_q[G]_l^B$ are often referred to as $R_q[G/K]$ (or $\mathcal{A}_q[G/K]$) in the literature (see for example [NS,(2.5)]). Here K can be thought of merely as a symbol or as the complexification of the compact Lie group in G with Lie algebra $\mathbf{g}^\theta$. Thus the homogeneous space G/K is a symmetric space. The notation $R_q[G/K]$ suggests that the right B invariants of $R_q[G]$ is the quantum analog of the ring of regular functions on G/K. In [L3], it is shown that B is a "good" analog of $U(\mathbf{g}^\theta)$ for constructing quantum symmetric spaces in the sense of [Di, end of Section 3]. In particular, $R_q[G/K]$ has the same left U module structure as its classical counterpart (see Theorem 7.8 below). We summarize this and related results here. A good survey on how to construct quantum symmetric spaces which includes a description of the classical situation is [Di]. For further information about classical symmetric spaces, the reader is referred to [He1] and [He2].

A finite-dimensional U module V is called a spherical module for B if the space of invariants V^B has dimension 1. Recall the notion of Cartan subspace and restricted root system introduced at the beginning of this section. Let $\mathbf{a}$ be the Cartan subspace $\{x \in \mathbf{h} | \theta(x) = -x\}$ and let Σ be the associated restricted root system. Let P_Θ^+ be the subset of $P^+(\pi)$ containing those λ such that

(i) $(\lambda, \beta) = 0$ for all $\beta \in Q(\pi)$ such that $\Theta(\beta) = \beta$;
(ii) the restriction $\tilde\lambda$ of λ to $\mathbf{a}^*$ satisfies $(\tilde\lambda, \beta)/(\beta, \beta)$ is an integer for every restricted root β.

The set P_Θ^+ is exactly the set of dominant integral weights such that the corresponding finite-dimensional simple $\mathbf{g}$ module is spherical ([Kn, Theorem 8.49]). By [L3, Theorem 4.2 and Theorem 4.3] we have the same classification in the quantum case.

THEOREM 7.7. *Let $L(\lambda)$ be a finite-dimensional U module with highest weight λ up to some possible roots of unity. Then*

$$\dim L(\lambda)^B \leq 1. \tag{i}$$

Moreover,

$$\dim L(\lambda)^B = 1 \text{ if and only if } \lambda \in P_\Theta^+. \tag{ii}$$

SKETCH OF PROOF. (See [L3] for full details.) Let v_λ denote the highest weight generating vector of $L(\lambda)$. Recall that for each $y \in G^-$ there exists a $b \in B$ such that $b = y+$ higher weight terms. Now $L(\lambda)$ is spanned by weight vectors of the form yv_λ where $y \in G^-$. Hence $\dim L(\lambda)/B^+v_\lambda \leq 1$ where B^+ is the augmentation ideal of B. Statement (i) follows from the fact that $B^+v_\lambda \cap L(\lambda)^B$ is empty. A careful analysis using the form of the generators of B further shows that $v_\lambda \in B^+v_\lambda$ if and only if $\lambda \notin P_\Theta^+$. This in turn implies (ii). The argument turns out to be much more delicate when B is a nonstandard analog. $\square$

Theorem 7.7, the Peter–Weyl decomposition of $R_q[G]$, (3.1), and (3.4) imply the following characterization of $R_q[G]_l^B$ as a right U module.

THEOREM 7.8. *There is an isomorphism of right U modules*

$$R_q[G]_l^B \cong \bigoplus_{\lambda \in P_\Theta^+} L(\lambda)^*$$

There is an analogous statement for the right B invariants of $R_q[G]$. One can also describe the B bi-invariants in a nice way. Identifying $L(\lambda)$ with a subspace of $R_q[G]_l^B$, set $\mathcal{H}(\lambda) = R_q[G]_{bi}^B \cap L(\lambda)$. Note that $\mathcal{H}(\lambda)$ is a trivial left and right B module. Moreover, by Theorem 7.7, $\mathcal{H}(\lambda)$ is one-dimensional if $\lambda \in P_\Theta^+$ and zero otherwise. The following direct sum decomposition into trivial one-dimensional B bimodules is thus an immediate consequence of Theorem 7.8.

$$R_q[G]_{bi}^B \cong \bigoplus_{\lambda \in P_\Theta^+} \mathcal{H}(\lambda).$$

Let $\mathcal{A}$ be the subgroup of T consisting of those elements $\tau(\lambda)$ such that $\Theta(\lambda) = -\lambda$. Thus $\mathcal{A}$ can be thought of as a quantum version of $\mathbf{a}$. Let W_Θ denote the Weyl group associated to the restricted root system Σ. Since $\Sigma \subset \mathbf{a}^*$, $\mathbf{a}$ and hence $\mathcal{A}$ inherit an action of W_Θ. The author has recently shown that, $R_q[G]_{bi}^B$ is commutative and moreover is isomorphic to $\mathbb{C}(q)[\mathcal{A}]^{W_\Theta}$. Thus, the $\mathcal{H}(\lambda)$ are natural choices of quantum zonal spherical functions (see [Di, the discussion concerning (3.4)]). In special cases, these quantum zonal spherical functions have been determined to be Macdonald polynomials or other q hypergeometric series (see for example [K], [N], [DN], [NS]). Preliminary work by the author suggests that this should be true in general.

It should be noted that these papers use analogs of $U(\mathbf{g}^\theta)$ whose definition differs from the definition of the $B_{\tilde{\theta}}$ and its variations found in this paper. In [NM], one-sided coideal subalgebras are used. By [L2, Section 6], using Theorem 7.5, these are shown to be examples of the analogs presented here. In other papers, two-sided coideals analogs of $\mathbf{g}^\theta$ are used. The specialization of these two-sided coideals generate a much larger subalgebra than $U(\mathbf{g}^\theta)$. The important object in these papers, used to define quantum symmetric spaces, is the left ideal generated by these two-sided coideals analogs of $\mathbf{g}^\theta$. It seems likely that these left ideals can be shown to be generated by the augmentation ideal of one of the analogs presented here. This is certainly true for the left coideals studied in [K] and also for those in [N] (combine [N, Section 2.4] with [L2, Section 6]).

References

[AJS] H. H. Andersen, J. C. Jantzen, and W. Soergel, Representations of quantum groups at a p-th root of unity and of semisimple groups in characteristic p: Independence of p, *Astérisque* **220**, Soc. Math. France, Paris (1994).

[BF] W. Baldoni and P. M. Frajria, The quantum analog of a symmetric pair: a construction in type $(C_n, A_1 \times C_{n-1})$, *Trans. Amer. Math. Soc.* **8** (1997), 3235-3276.

[CP] V. Chari and A. Pressley, *A Guide to Quantum Groups*, Cambridge University Press, Cambridge, (1995).

[DK] C. DeConcini and V. G. Kac, Representations of quantum groups at roots of 1, In: *Operator Algebras, Unitary Representations, Enveloping Algebras, and Invariant Theory*, Progress in Math. **92**, Birkhäuser, Boston (1990), 471-506.

[Di] M. S. Dijkhuizen, Some remarks on the construction of quantum symmetric spaces, In: *Representations of Lie Groups, Lie Algebras and Their Quantum Analogues*, Acta Appl. Math. **44** (1996), no. 1-2, 59-80.

[DN] M. S. Dijkhuizen and M. Noumi, A family of quantum projective spaces and related q-hypergeometric orthogonal polynomials, *Trans. Amer. Math. Soc.* **350** (1998), no. 8, 3269-3296.

[D] J. Dixmier, *Algèbres Enveloppantes*, Cahiers Scientifiques, XXXVII, Gauthier-Villars, Paris (1974).

[G] V. Guizzi, A classification of unitary highest weight modules of the quantum analogue of the symmetric pair (A_n, A_{n-1}), *J. Algebra* **192** (1997), 102-129.

[H] J. E. Humphreys, *Introduction to Lie Algebras and Representation Theory*, Springer-Verlag, New York (1972).

[HS] I. Heckenberger and K. Schmudgen, Classification of bicovariant differential calculi on the quantum groups $SL_q(n+1)$ and $Sp_q(2n)$, *J. Reine Angew. Math.* **502** (1998), 141-162.

[He1] S. Helgason, *Differential Geometry, Lie Groups, and Symmetric Spaces*, Pure and Applied Mathematics **80**, Academic Press, New York (1978).

[He2] S. Helgason, *Groups and Geometric Analysis, Integral Geometry, Invariant Differential Operators, and Spherical Functions*, Pure and Applied Mathematics **113**, Academic Press, Inc., Orlando (1984).

[J] N. Jacobson, *Basic Algebra II*, W. H. Freeman and Company, San Francisco (1980).

[JL1] A. Joseph and G. Letzter, Local finiteness of the adjoint action for quantized enveloping algebras, *J. of Algebra* **153** (1992), 289 -318.

[JL2] A. Joseph and G. Letzter, Separation of variables for quantized enveloping algebras, *American Journal of Math.* **116** (1994), 127-177.

[JL3] A. Joseph and G. Letzter, Verma module annihilators for quantized enveloping algebras, *Ann. Sci. Ecole Norm. Sup.*(4) **28** (1995), no. 4, 493-526.

[Jo] A. Joseph, *Quantum Groups and Their Primitive Ideals*, Springer-Verlag, New York (1995).

[Ka] V. G. Kac, *Infinite-Dimensional Lie Algebras*, Third ed., Cambridge University Press, Cambridge (1990).

[Ke] M. S. Kébé, Ο-algèbres quantiques, *C. R. Acad. Sci. Paris, Ser. I Math.* **322** (1996), no. 1, 1-4.

[KS] A. Klimyk and K. Schmüdgen, *Quantum Groups and Their Representations*, Texts and Monographs in Physics, Springer-Verlag, Berlin (1997).

[Kn] A. W. Knapp, *Lie Groups Beyond an Introduction*, Progress in Math. **140**, Birkhäuser, Boston (1996).

[K]   T. Koornwinder, Askey-Wilson polynomials as zonal spherical functions on the SU(2) quantum group. *SIAM J. Math. Anal.* **24** (1993), no. 3, 795–813.

[KL]   G. R. Krause and T. H. Lenagan, *Growth of Algebras and Gelfand-Kirillov Dimension*, Research Notes in Mathematics **116**, Pitman, London (1985).

[L1]   G. Letzter, Subalgebras which appear in quantum Iwasawa decompositions, *Canadian Journal of Math.* **49** (1997), no. 6, 1206-1223.

[L2]   G. Letzter, Symmetric pairs for quantized enveloping algebras, *J. Algebra* **220** (1999), no. 2, 729-767.

[L3]   G. Letzter, Harish-Chandra modules for quantum symmetric pairs, *Representation Theory, An Electronic Journal of the AMS* **4** (1999) 64-96.

[Lu]   G. Lusztig, *Introduction to Quantum Groups,* Progress in Math. **110**, Birkhäuser, Boston (1994).

[M]   S. Montgomery, *Hopf Algebras and Their Actions on Rings,* CBMS Regional Conference Series in Mathematics **82**, American Mathematical Society, Providence (1993).

[N]   M. Noumi, Macdonald's symmetric polynomials as zonal spherical functions on some quantum homogeneous spaces, *Advances in Mathematics* **123** (1996), no. 1, 16-77.

[NS]   M. Noumi and T. Sugitani, Quantum symmetric spaces and related q-orthogonal polynomials, In: *Group Theoretical Methods in Physics (ICGTMP)* (Toyonaka, Japan, 1994), World Sci. Publishing, River Edge, N. J. (1995) 28-40.

[OV]   A. L. Onishchik and E. B. Vinberg, *Lie Groups and Lie Algebras III: Structure of Lie Groups and Lie Algebras,* Springer-Verlag, Berlin (1994).

[R]   M. Rosso, Groupes Quantiques, Représentations Linéaires et Applications, Thèse, Paris 7 (1990).

[Ve]   D. N. Verma, Structure of certain induced representations of complex semisimple Lie algebras, *Bulletin of the American Mathematical Society* **74** (1968), 160-166.

[V]   D. Vogan, *Representations of Real Reductive Lie Groups,* Progress in Math. **15**, Birkhäuser, Boston (1981).

GAIL LETZTER
MATHEMATICS DEPARTMENT
VIRGINIA POLYTECHNIC INSTITUTE STATE UNIVERSITY
BLACKSBURG, VA 24061
letzter@math.vt.edu

New Directions in Hopf Algebras
MSRI Publications
Volume **43**, 2002

Hopf Algebra Extensions and Cohomology

AKIRA MASUOKA

ABSTRACT. This is an expository paper on 'abelian' extensions of (quasi-) Hopf algebras, which can be managed by the abelian cohomology, with emphasis on the author's recent results which are motivated by an exact sequence due to George Kac. The cohomology plays here an important role in constructing and classifying those extensions, and even their cocycle deformations. We see also a strong connection of Hopf algebra extensions arising from a (matched) pair of Lie algebras with Lie bialgebra extensions.

Introduction

Let us first recall the theory of group extensions with abelian kernel [Mac, Chap. IV, Sections 3,4]. Each extension $M \to \Sigma \to \Pi$ of a group Π by an abelian group M gives rise to a Π-module structure on M. Those extensions which give rise to a fixed Π-module structure form an abelian group, $\mathrm{Opext}(\Pi, M)$, which is isomorphic to the cohomology $H^2(\Pi, M)$. The results were generalized by Singer [S] (1972) and Hofstetter [H] (1994) for those Hopf algebra extensions

$$K \to A \to H$$

which are abelian in the sense that H is cocommutative, K is commutative and A is cleft as an H-comodule algebra: each such extension gives rise to some structure, called a Singer pair structure (see Definition 2.2), on (H, K), and those extensions which give rise to a fixed Singer pair structure form an abelian group, $\mathrm{Opext}(H, K)$, which is isomorphic to some cohomology group. But, Kac [K] (1968) had already obtained these results in the case when $H = kF$ (group algebra), $K = k^G \ (= (kG)^*)$ with F, G finite groups, and further proved an interesting, exact cohomology sequence involving $\mathrm{Opext}(kF, k^G)$, which we

The author gratefully acknowledges that many of his own results presented here were obtained while he was staying at the University of Munich as a Humboldt Research Fellow (November 1996 – September 1997 and June 1998) and visiting the National University of Córdoba, Argentina (October 1997). He thanks the referee for kind suggestions that improved the exposition.

call the Kac exact sequence. Unfortunately, Kac' work on extensions had long
been overlooked by most of Hopf algebraists (perhaps, including Singer and
Hofstetter), and especially his exact sequence had been peculiar to the restricted
case as above. But, in these several years his work has been applied especially for
the classification problem of semisimple Hopf algebras; see [M1], [N], [Ka], and
also [IK]. In addition, the author [M] recently established the formulation of the
Kac exact sequence in the Lie algebra case (see below), and then Schauenburg
[Sb2] proved the sequence for general H and finite-dimensional K (see Remark
1.11 (3)).

This is an expository paper on abelian extensions of (quasi-)Hopf algebras
with emphasis on the author's recent results which are motivated by the Kac exact sequence. We will see the cohomology plays an important role in constructing
and classifying those extensions, and even their cocycle deformations.

The paper consists of two parts. Part I (Sections 1–4) begins with an elementary exposition of the Kac theory, which is followed by the generalized results
from [S], [H]. Then Section 3 is devoted to the study of cocycle deformations of
the middle term A by 2-cocycles for H (or by their liftings to A), with sample
computations. In Section 4, we study quasi-Hopf algebra extensions of a similar
form as above, but A is replaced by a quasi-Hopf algebra with Drinfeld associator in $K^{\otimes 3}$; parallel results to the Hopf algebra case are proved, including
the modified Kac exact sequence. In Part II (Sections 5–8), we let $\mathfrak{f}$, $\mathfrak{g}$ denote
finite-dimensional Lie algebras in characteristic zero, and see a strong connection
between Hopf algebra extensions of the form $(U\mathfrak{g})^{\circ} \to A \to U\mathfrak{f}$ and Lie bialgebra
extensions of the form $\mathfrak{g}^{*} \to \mathfrak{l} \to \mathfrak{f}$; the result proves a Lie algebra version of the
Kac exact sequence. Similar results are proved when $\mathfrak{g}$ is nilpotent and the Hopf
dual $(U\mathfrak{g})^{\circ}$ of the universal envelope $U\mathfrak{g}$ is replaced by its irreducible component
$(U\mathfrak{g})'$ containing 1. In the final section these parallel results are unified and
generalized by introducing some topology onto $U\mathfrak{g}$.

This paper may be regarded as an enlarged version of the Córdoba lecture
notes [M4], though some computational results are omitted here. Instead, proofs
of the results in Sections 7 and 8 are included. Sections 3 and 4 are also added;
they include unpublished results which are first presented with proofs. Following M. Takeuchi's suggestion, we emphasize categorical treatment of extensions,
which seems new.

We work over a fixed ground field k. Tensor products $\otimes$ and exterior products
$\wedge$ are taken over k, unless otherwise stated. For vector spaces V, W, we let
$\mathrm{Hom}(V,W)$ denote the vector space of all linear maps $V \to W$, and write $V^{*} =
\mathrm{Hom}(V,k)$, the dual vector space. By a module (resp., comodule), we mean a
left module (resp., right comodule) unless otherwise stated. The coproduct, the
counit and the antipode of a Hopf algebra are denoted, as usual, by Δ, ε and S,
respectively. We use the Sweedler notation such as $\Delta(a) = \sum a_1 \otimes a_2$.

PART I: KAC THEORY, ITS GENERALIZATION AND VARIATION

In this part, F and G denote groups, which are supposed to be finite unless otherwise stated.

1. The Kac Theory

The theory will be reviewed from modern Hopf-algebraic view-point, but as elementarily as possible.

DEFINITION 1.1 [T1, Def. 2.1].   A *matched pair (of groups)* is a pair (F, G) together with group actions $G \xleftarrow{\lhd} G \times F \xrightarrow{\rhd} F$ on the sets such that

$$x \rhd ab = (x \rhd a)((x \lhd a) \rhd b),$$
$$xy \lhd a = (x \lhd (y \rhd a))(y \lhd a)$$

for $a, b \in F$, $x, y \in G$, or equivalently such that the cartesian product $F \times G$ forms a group under the product

$$(a, x)(b, y) = (a(x \rhd b), (x \lhd b)y).$$

We denote this group by $F \bowtie G$, following [Mj]. (It was originally denoted by $F \, \mathbb{X} \, G$; see [T1, p. 842].)

The group $F \bowtie G$ includes subgroups $F = F \times 1$ and $G = 1 \times G$ so that the product map $F \times G \to F \bowtie G$ is a bijection. Conversely, if a group Σ includes F, G as subgroups so that the product $F \times G \to \Sigma$ is a bijection, then the actions $G \xleftarrow{\lhd} G \times F \xrightarrow{\rhd} F$ determined by

$$ax = (a \rhd x)(a \lhd x) \ (a \in F, \ x \in G)$$

make (F, G) matched so that $F \bowtie G \cong \Sigma$.

Let kG denote the group Hopf algebra in which each element in G is grouplike. Let k^G denote the dual Hopf algebra $(kG)^*$ of kG; it is spanned by the orthogonal idempotents e_x defined by $\langle e_x, y \rangle = \delta_{x,y}$, where $x, y \in G$. We may suppose that k^G is the algebra consisting of all maps $G \to k$ which has the pointwise product, and so that the abelian group $(k^G)^\times$ of units in k^G consists of all maps $G \to k^\times = k \setminus 0$.

By a G-module, we mean a module over the integral group ring $\mathbb{Z}G$, as usual.

Let $\lhd: G \times F \to G$ be an action on the set G, which corresponds to an action $\rightharpoonup: F \times k^G \to k^G$ of algebra automorphisms so that

$$a \rightharpoonup e_x = e_{x \lhd a^{-1}} \ (a \in F, \ x \in G).$$

Let $\sigma : F \times F \to (k^G)^\times$ be a 'normalized' 2-cocycle of the group F with coefficients in $(k^G)^\times$, which is an F-module under the action induced by $\rightharpoonup$. We identify σ naturally with the map $(x, a, b) \mapsto \sigma(a, b)(x)$, $G \times F \times F \to k^\times$, and denote the last value by $\sigma(x; a, b)$. Then the 2-cocycle condition for σ is given by

$$\sigma(x \triangleleft a; b, c)\sigma(x; a, bc) = \sigma(x; a, b)\sigma(x; ab, c),$$

while the normalization condition means here

$$\sigma(1; a, b) = \sigma(x; 1, b) = \sigma(x; a, 1) = 1,$$

where $a, b, c \in F$ and $x \in G$. The familiar construction of crossed product makes the tensor product $k^G \otimes kF = \bigoplus_{a \in F} k^G a$ into an algebra with unit $1 \otimes 1$, whose product is given by

$$(e_x a)(e_y b) = e_x(a \rightharpoonup e_y)\sigma(a, b)ab = \delta_{x \triangleleft a, y}\sigma(x; a, b)e_x ab,$$

where $a, b \in F$ and $x, y \in G$.

Suppose also that we are given an action $\triangleright: G \times F \to F$ and a normalized 2-cocycle $\tau: G \times G \times F \to k^\times$ of G with coefficients in the right G-module $(k^F)^\times$. They make $kG \otimes k^F$ into an algebra of right crossed product, and so by duality make $k^G \otimes kF$ into a coalgebra. One sees that the coalgebra structure is given by

$$\Delta(e_x a) = \sum_{y \in G} \tau(xy^{-1}, y; a)e_{xy^{-1}}(y \triangleright a) \otimes e_y a,$$
$$\varepsilon(e_x a) = \delta_{1,x}.$$

Let $k^G \#_{\sigma, \tau} kF$ denote the tensor product $k^G \otimes kF$ with the described algebra and coalgebra structures.

LEMMA 1.2. $k^G \#_{\sigma, \tau} kF$ is a bialgebra if and only if $(F, G, \triangleleft, \triangleright)$ is a matched pair and

$\sigma(xy; a, b)\tau(x, y; ab)$

$\quad = \sigma(x; y \triangleright a, (y \triangleleft a) \triangleright b)\sigma(y; a, b)\tau(x, y; a)\tau(x \triangleleft (y \triangleright a), y \triangleleft a; b)$

for all $a, b \in F$, $x, y \in G$. In this case, $k^G \#_{\sigma, \tau} kF$ is necessarily a Hopf algebra.

The proof is straightforward; see the proof of [M4, Prop. 4.7]. If the conditions given above are satisfied, the maps $\iota: k^G \to k^G \#_{\sigma, \tau} kF$, $\iota(e_x) = e_x 1$ and $\pi: k^G \#_{\sigma, \tau} kF \to kF$, $\pi(e_x a) = \delta_{1,x} a$ are obviously Hopf algebra maps. Further, we will see

$$(k^G \#_{\sigma, \tau} kF) = k^G \xrightarrow{\iota} k^G \#_{\sigma, \tau} kF \xrightarrow{\pi} kF \qquad (1.3)$$

is a Hopf algebra extension.

Let $(A) = K \xrightarrow{\iota} A \xrightarrow{\pi} H$ be a sequence of finite-dimensional Hopf algebras. Let K^+ denote the kernel $\text{Ker}(\varepsilon: K \to k)$ of the counit. Regarding A as a right (or left) H-comodule along π, let A^{coH} (or ^{coH}A) denote the subalgebra of H-coinvariants. Thus, A^{coH} consists of $a \in A$ such that $\sum a_1 \otimes \pi(a_2) = a \otimes \pi(1)$.

DEFINITION 1.4. Suppose ι is an injection and π is a surjection, so that we may regard ι as an inclusion and π as a quotient. The sequence (A) is called an *extension of H by K* if it satisfies the following equivalent conditions: (a) $A/K^+A = H$; (b) $A/AK^+ = H$; (c) $K = A^{coH}$; (d) $K = {}^{coH}A$. For two extensions (A), (A') of H by K, an *equivalence* $(A) \to (A')$ is a Hopf algebra map $f : A \to A'$ which induces the identity maps on H and K. If such exists, we say that (A) and (A') are *equivalent*.

One sees easily from [Sw2, Lemmas 16.0.2–3] that Conditions (a)–(d) are equivalent. We see easily that the sequence (1.3) is an extension of kF by k^G.

If (A) is an extension of H by K, then $A \supset K$ is a right H-Galois extension in the sense of [Mo, Def. 8.1.1]. Since A is right K-free by the Nichols-Zöller theorem, it follows that the map f giving an equivalence is necessarily an isomorphism, which justifies the term.

By [Sd, Thm. 2.4] or [MD, Thm. 3.5], a finite-dimensional extension (A) is necessarily cleft (see Definition 2.5 below) in the sense there is a K-linear and H-colinear isomorphism $A \cong K \otimes H$ which preserves unit and counit. This is easily proved in the special case when $H = kF$, $K = k^G$, since then A is a strictly graded F-algebra with neutral component k^G so that it is necessarily a crossed product. This proves also the following.

PROPOSITION 1.5. *Any extension (A) of kF by k^G is equivalent to an extension of the form* (1.3).

For another choice $(A) \sim (k^G \#_{\sigma',\tau'} kF)$ of equivalence, the same matched pair $(F, G, \lhd, \rhd)$ forms the Hopf algebras $k^G \#_{\sigma,\tau} kF$ and $k^G \#_{\sigma',\tau'} kF$, since the neutral components in A and in A^* are commutative.

DEFINITION 1.6. We say that (A) *is associated with* the matched pair $(F, G, \lhd, \rhd)$ thus uniquely determined by (A).

Two equivalent extensions of kF by k^G are associated with the same matched pair. In what follows, we fix a matched pair $(F, G, \lhd, \rhd)$. We denote by

$$\text{Opext}(kF, k^G)$$

the set of all equivalence classes of extensions associated with it. (The notation stems from $\text{Opext}(\Pi, A, \varphi)$ [Mac, Chap. IV, Sect. 3] for the group extensions of a group Π by an abelian kernel A with fixed operators $\varphi : \Pi \to \text{Aut}\, A$.)

We will give a cohomological description of $\text{Opext}(kF, k^G)$. Let $\Sigma = F \bowtie G$ denote the group constructed by the fixed matched pair.

Let $0 \leftarrow \mathbb{Z} \leftarrow B.$ be the normalized bar resolution of the trivial F-module $\mathbb{Z}$. Thus,

$$B. = 0 \longleftarrow B_0 \xleftarrow{d_1} B_1 \xleftarrow{d_2} B_2 \xleftarrow{d_3} \cdots$$

consists of the free F-modules B_p with basis $[a_1|\cdots|a_p]$, where $1 \neq a_i \in F$, and the differentials d_p are defined by

$$d_p[a_1|\cdots|a_p] = a_1[a_2|\cdots|a_p] + \sum_{i=1}^{p-1}(-1)^i[a_1|\cdots|a_ia_{i+1}|\cdots|a_p] + (-1)^p[a_1|\cdots|a_{p-1}].$$

The augmentation $\varepsilon : B_0 = \mathbb{Z}F \to \mathbb{Z}$ is given by $\varepsilon(a) = 1$ for $a \in F$. Define an action of G on the canonical $\mathbb{Z}$-free basis of B_p by

$$x(a[a_1|\cdots|a_p]) = x \triangleright a[(x \triangleleft a) \triangleright a_1|(x \triangleleft aa_1) \triangleright a_2|\cdots|(x \triangleleft aa_1 \cdots a_{p-1}) \triangleright a_p],$$

where $x \in G$ and $1 \neq a_i$, $a \in F$. Then one sees that this together with the original F-action makes B_p into a Σ-module, and that d_p, ε are Σ-linear, where $\mathbb{Z}$ is a trivial Σ-module. So, $0 \leftarrow \mathbb{Z} \leftarrow B.$ turns to be a complex of Σ-modules.

    The symmetric argument using a mirror makes the normalized bar resolution $0 \leftarrow \mathbb{Z} \xleftarrow{\varepsilon'} B'.$ of the trivial right G-module $\mathbb{Z}$ into a complex of right Σ-modules. Regard it as a complex of left Σ-modules by twisting the action through the inverse of Σ, and tensor it with $B.$ over $\mathbb{Z}$. Then we obtain the double complex

$$
B' \otimes_{\mathbb{Z}} B. =
\begin{array}{ccccc}
\vdots & & \vdots & & \\
\downarrow & & \downarrow & & \\
B'_1 \otimes_{\mathbb{Z}} B_0 & \xleftarrow{1 \otimes d_1} & B'_1 \otimes_{\mathbb{Z}} B_1 & \longleftarrow & \cdots \\
\downarrow{\scriptstyle d'_1 \otimes 1} & & \downarrow{\scriptstyle -d'_1 \otimes 1} & & \\
B'_0 \otimes_{\mathbb{Z}} B_0 & \xleftarrow{1 \otimes d_1} & B'_0 \otimes_{\mathbb{Z}} B_1 & \longleftarrow & \cdots
\end{array}
$$

of Σ-modules, where Σ acts diagonally on each term. Here and in what follows, when we construct a double complex, we resort such a trick (sign trick) that changes the sign of the differentials in odd columns (see above) unless otherwise stated. One sees that each Σ-module $B'_q \otimes_{\mathbb{Z}} B_p$ is free with basis $[x_q|\cdots|x_1] \otimes [a_1|\cdots|a_p]$, where $1 \neq a_i \in F$, $1 \neq x_i \in G$. This implies the following.

LEMMA 1.7. *The total complex of $B'. \otimes_{\mathbb{Z}} B.$ gives a Σ-free resolution of $\mathbb{Z}$ via the augmentation $\varepsilon' \otimes \varepsilon : B'_0 \otimes_{\mathbb{Z}} B_0 \to \mathbb{Z}$.*

Regard $k^\times$ as a trivial Σ-module, and form the double complex

$$D^{..} = \mathrm{Hom}_\Sigma(B'. \otimes_{\mathbb{Z}} B., k^\times).$$

Since $B'_q \otimes_{\mathbb{Z}} B_p$ has the Σ-free basis noted above, $\mathrm{Hom}_\Sigma(B'_q \otimes_{\mathbb{Z}} B_p, k^\times)$ is identified with the abelian group $\mathrm{Map}_+(G^q \times F^p, k^\times)$ of all maps $G^q \times F^p \to k^\times$ satisfying the normalization condition, where X^r denotes the cartesian product of r copies of $X = F, G$. Thus, $D^{..}$ looks as follows.

$$D^{..} = \quad \vdots \qquad\qquad \vdots$$

$$\text{Map}_+(G, k^\times) \longrightarrow \text{Map}_+(G \times F, k^\times) \longrightarrow \cdots$$

$$k^\times \longrightarrow \text{Map}_+(F, k^\times) \longrightarrow \cdots$$

Note that the edges in $D^{..}$ consist of the standard complexes for computing the group cohomologies $H^{.}(F, k^\times)$, $H^{.}(G, k^\times)$. Remove these edges from $D^{..}$ to obtain the following double complex.

$$A^{..} = \quad \vdots \qquad\qquad \vdots$$

$$\text{Map}_+(G^2 \times F, k^\times) \longrightarrow \text{Map}_+(G^2 \times F^2, k^\times) \longrightarrow \cdots$$

$$\text{Map}_+(G \times F, k^\times) \longrightarrow \text{Map}_+(G \times F^2, k^\times) \longrightarrow \cdots$$

For example, the horizontal and vertical differentials ∂, ∂' going into $\text{Map}_+(G^2 \times F^2, k^\times)$ are given by

$$\partial\tau(x, y; a, b) = \tau(x \triangleleft (y \triangleright a), y \triangleleft a; b)\tau(x, y; ab)^{-1}\tau(x, y; a)$$
$$\partial'\sigma(x, y; a, b) = \sigma(y; a, b)\sigma(xy; a, b)^{-1}\sigma(x; y \triangleright a, (y \triangleleft a) \triangleright b),$$

where $a, b \in F$, $x, y \in G$, $\sigma \in \text{Map}_+(G \times F^2, k^\times)$ and $\tau \in \text{Map}_+(G^2 \times F, k^\times)$. Let $\text{Tot}\, A^{..}$ denote the total complex of $A^{..}$.

PROPOSITION 1.8 (cf. [K, Thm. 5]). *For a total 1-cocycle (σ, τ) in $A^{..}$, we have an extension $(k^G \#_{\sigma,\tau} kF)$ associated with the fixed matched pair $(F, G, \triangleleft, \triangleright)$. The assignment $(\sigma, \tau) \mapsto (k^G \#_{\sigma,\tau} kF)$ induces a bijection*

$$H^1(\text{Tot}\, A^{..}) \cong \text{Opext}(kF, k^G).$$

PROOF. By Lemma 1.2 and Proposition 1.5, the assignment gives a surjection $Z^1(\text{Tot}\, A^{..}) \to \text{Opext}(kF, k^G)$ from the group of total 1-cocycles.

Let (σ, τ), (σ', τ') be total 1-cocycles. If $\nu : G \times F \to k^\times$ is a 0-cochain such that $\sigma' = \sigma\partial\nu$, $\tau' = \tau\partial'\nu$, then $e_x a \mapsto \nu(x; a)e_x a$ gives an equivalence $(k^G \#_{\sigma',\tau'} kF) \xrightarrow{\cong} (k^G \#_{\sigma,\tau} kF)$. Conversely, one sees that any equivalence is given in this way by some ν, in which case we have $\sigma' = \sigma\partial\nu$, $\tau' = \tau\partial'\nu$ by simple computation. This proves the injectivity of the induced map. $\square$

The Baer product of group crossed products gives rise to a product on the set $\mathrm{Opext}(kF, k^G)$, which forms thereby an abelian semigroup; see Section 2 for more general treatment. One sees that the bijection in the last proposition preserves product, whence $\mathrm{Opext}(kF, k^G)$ is a group. The unit is represented by the extension $(k^G \,\#_{1,1}\, kF)$ which is given by the constant cocycles $(\sigma, \tau) = (1, 1)$ with value 1. We write simply $(k^G \,\#\, kF)$ for this extension, and let $\mathrm{Aut}(k^G \,\#\, kF)$ denote the group of its auto-equivalences.

PROPOSITION 1.9. *For a total 0-cocycle ν in $A^{..}$, $e_x a \mapsto \nu(x; a) e_x a$ gives an auto-equivalence of $(k^G \,\#\, kF)$. This gives an isomorphism*

$$H^0(\mathrm{Tot}\, A^{..}) \cong \mathrm{Aut}(k^G \,\#\, kF).$$

PROOF. This follows by the argument given in the second paragraph of the last proof if we suppose $\sigma = \sigma' = 1$, $\tau = \tau' = 1$. □

THEOREM 1.10 (cf. [K, (3.14)]). *We have an exact sequence*

$$0 \to H^1(F \bowtie G, k^\times) \to H^1(F, k^\times) \oplus H^1(G, k^\times) \to \mathrm{Aut}(k^G \,\#\, kF)$$
$$\to H^2(F \bowtie G, k^\times) \to H^2(F, k^\times) \oplus H^2(G, k^\times) \to \mathrm{Opext}(kF, k^G)$$
$$\to H^3(F \bowtie G, k^\times) \to H^3(F, k^\times) \oplus H^3(G, k^\times),$$

where $H^{.}$ denotes the group cohomology with coefficients in the trivial module $k^\times$.

PROOF. Since $A^{..}$ is regarded (by dimension shift) as a double subcomplex of $D^{..}$ such that the cokernel is the edges $E^{..}$ in $D^{..}$, we have a short exact sequence $0 \to A^{..} \to D^{..} \to E^{..} \to 0$ of double complexes, which induces a long exact sequence of total cohomologies. It gives the desired sequence by Propositions 1.8 and 1.9, since we have also

$$H^n(\mathrm{Tot}\, D^{..}) = H^n(F \bowtie G, k^\times), \qquad (n > 0),$$
$$H^n(\mathrm{Tot}\, E^{..}) = H^n(F, k^\times) \oplus H^n(G, k^\times) \quad (n > 0). \qquad \square$$

REMARK 1.11. (1) Using the Kac exact sequence just given, it can be proved that the abelian group $\mathrm{Opext}(kF, k^G)$ is torsion, but not necessarily finite; it is finite if k is algebraically closed. On the other hand, $\mathrm{Aut}(k^G \,\#\, kF)$ is always finite. See [M4, Props. 7.7–8].

(2) Suppose $k = \mathbb{C}$. Kac [K] actually worked on Hopf $*$-algebra (today called Kac algebra) extensions of $\mathbb{C}F$ of $\mathbb{C}^G$, where $*$-structures are given to $\mathbb{C}F$ by $a^* = a^{-1}$ ($a \in F$), and to $\mathbb{C}^G$ by $(e_x)^* = e_x$ ($x \in G$). See [IK] for new achievement. Let $\mathrm{Opext}^*(\mathbb{C}F, \mathbb{C}^G)$, $\mathrm{Aut}^*(\mathbb{C}^G \,\#\, \mathbb{C}F)$ denote the groups of all $*$-equivalence classes of Hopf $*$-algebra extensions, and of all $*$-auto-equivalences, respectively. These are described cohomologically by the double complex which modifies $A^{..}$ with $\mathbb{C}^\times$ replaced by $\mathbb{T} = \{z \in \mathbb{C} \mid |z| = 1\}$. In the same way as above, we have an exact sequence consisting of these groups and group cohomologies with coefficients in

T. By comparing it with the original sequence, it follows that the natural group maps

$$\mathrm{Opext}^*(\mathbb{C}F, \mathbb{C}^G) \to \mathrm{Opext}(\mathbb{C}F, \mathbb{C}^G),$$

$$\mathrm{Aut}^*(\mathbb{C}^G \# \mathbb{C}F) \hookrightarrow \mathrm{Aut}(\mathbb{C}^G \# \mathbb{C}F)$$

are both isomorphisms, since the universal coefficient theorem implies that, for Γ a finite group, $H^n(\Gamma, \mathbb{T}) \cong H^n(\Gamma, \mathbb{C}^\times)$ for $n > 0$. See [M3, Remark 2.4].

(3) Recently, Schauenburg [Sb2] proved the Kac exact sequence for cleft Hopf algebra extensions $K \to A \to H$ (see Definition 2.5 below) at least when H is cocommutative and K is commutative and finite-dimensional; the group cohomologies were then replaced by the Sweedler cohomologies [Sw1] of the cocommutative Hopf algebras H, K^* and $H \bowtie K^*$ with coefficients in k. He actually deduced the sequence from nice, general results on monoidal equivalences of (generalized) Hopf bimodule categories; see also [Sb4], a very readable article. But, we do not discuss these interesting results any more in this paper.

2. Generalities on Hopf Algebra Extensions

Let H, J be cocommutative Hopf algebras.

DEFINITION 2.1 (cf. [Kas, Def. IX 2.2]). A *matched pair of* (*cocommutative*) *Hopf algebras* is a pair (H, J) together with actions $J \overset{\triangleleft}{\longleftarrow} J \otimes H \overset{\triangleright}{\longrightarrow} H$ such that $(H, \triangleright)$ is a left J-module coalgebra, $(J, \triangleleft)$ is a right H-module coalgebra, and

$$x \triangleright ab = \sum (x_1 \triangleright a_1)((x_2 \triangleleft a_2) \triangleright b),$$

$$xy \triangleleft a = \sum (x \triangleleft (y_1 \triangleright a_1))(y_2 \triangleleft a_2)$$

for $a, b \in H$, $x, y \in J$. The conditions are equivalent to that the tensor product coalgebra $H \otimes J$ is a bialgebra, which is necessarily a Hopf algebra, with unit $1 \otimes 1$ under the product

$$(a \otimes x)(b \otimes y) = \sum a(x_1 \triangleright b_1) \otimes (x_2 \triangleleft b_2)y.$$

The Hopf algebra is denoted by $H \bowtie J$, whose antipode is given by

$$S(a \otimes x) = \sum (S(x_2) \triangleright S(a_2)) \otimes (S(x_1) \triangleleft S(a_1)).$$

If $H = kF$ and $J = kG$ are group Hopf algebras, then the structures of matched pair of groups on (F, G) are obviously in 1-1 correspondence with the structures of matched pair of Hopf algebras on (kF, kG).

In what follows, we suppose H is a cocommutative Hopf algebra and K is a commutative Hopf algebra.

DEFINITION 2.2. A *Singer pair of Hopf algebras* is a pair (H, K) together with an action and a coaction,

$$\rightharpoonup: H \otimes K \to K \quad \text{and} \quad \rho: H \to H \otimes K, \quad \rho(a) = \sum a_H \otimes a_K,$$

such that $(K, \rightharpoonup)$ is an H-module algebra, (H, ρ) is a K-comodule coalgebra, and

$$\rho(ab) = \sum \rho(a_1)(b_H \otimes (a_2 \rightharpoonup b_K)),$$
$$\Delta(a \rightharpoonup t) = \sum ((a_1)_H \rightharpoonup t_1) \otimes (a_1)_K (a_2 \rightharpoonup t_2)$$

for $a, b \in H$, $t \in K$.

The notion was introduced by Singer [S] under the name 'abelian matched pair'. We propose the term given above to avoid confusion with the notion defined by Definition 2.2.

DEFINITION 2.3 [S, Def. 3.3]. Given a Singer pair $(H, K, \rightharpoonup, \rho)$, we define a category $C = C(H, K, \rightharpoonup, \rho)$ as follows. An object in C is an H-module M equipped with a K-comodule structure $\lambda : M \to M \otimes K$, $\lambda(m) = \sum m_0 \otimes m_1$ such that

$$\lambda(am) = \sum (a_1)_H m_0 \otimes (a_1)_K (a_2 \rightharpoonup m_1)$$

for $a \in H$, $m \in M$. A morphism in C is an H-linear and K-colinear map. In fact, C forms a k-abelian category.

REMARK 2.4. Suppose J is a finite-dimensional cocommutative Hopf algebra. There is a 1-1 correspondence between the matched pair structures $(\lhd, \rhd)$ on (H, J) and the Singer pair structures $(\rightharpoonup, \rho)$ on (H, J^*); it is given by the two familiar correspondences between module actions $\lhd : J \otimes H \to J$ and $\rightharpoonup : H \otimes J^* \to J^*$, and between module actions $\rhd : J \otimes H \to H$ and comodule coactions $\rho : H \to H \otimes J^*$. Similarly we see that the module category $H \bowtie J$-Mod arising from a matched pair $(H, J, \lhd, \rhd)$ is isomorphic to the category C arising from the corresponding Singer pair $(H, J^*, \rightharpoonup, \rho)$.

DEFINITION 2.5. A sequence $(A) = K \xrightarrow{\iota} A \xrightarrow{\pi} H$ of Hopf algebras is called a *cleft extension of H by K* if the following equivalent conditions (see [MD, Prop. 3.2]) are satisfied, where A is regarded as a K-module along ι, and as an H-comodule along π.

(a) There is a left K-linear and right H-colinear isomorphism $\xi : A \xrightarrow{\cong} K \otimes H$;

(b) There is such an isomorphism ξ as in (a) which also preserves unit and counit;

(c) A is right H-cleft [Mo, Def. 7.2.1] in the sense that there is a (convolution-) invertible right H-colinear map $H \to A$ which preserves unit and counit, and ι induces an isomorphism $K \xrightarrow{\cong} A^{coH}$;

(d) A is left K-cocleft in the sense that there is an invertible left K-linear map $A \to K$ which preserves unit and counit, and π induces an isomorphism $A/K^+A \xrightarrow{\cong} H$.

An *equivalence* between cleft extensions of H by K is defined in the same way as in Definition 1.4.

REMARK 2.6. (1) It follows by [Sb3] that, if (A) is a cleft extension of H by K, the antipode of A is necessarily bijective since those of H and K are. Hence,

Conditions (a)–(d) are equivalent to those conditions obtained by exchanging 'left' and 'right', since the antipode gives a right K-linear and left H-colinear isomorphism $A \xrightarrow{\cong} A^{op,cop}$.

(2) Suppose a sequence $(A) = K \to A \to H$ is given. If (A) is a cleft extension, then A is injective as a right H-comodule and $K \cong A^{coH}$. One sees that the converse holds true if H is irreducible, or in particular if H is the universal envelope $U\mathfrak{f}$ of a Lie algebra $\mathfrak{f}$, as will be the case in Part II.

Let $\rightharpoonup: H \otimes K \to K$ be an action which makes K into an H-module algebra. Let $\sigma : H \otimes H \to K$ be an invertible linear map which satisfies the 2-cocycle condition that

$$\sum [a_1 \rightharpoonup \sigma(b_1, c_1)]\sigma(a_2, b_2 c_2) = \sum \sigma(a_1, b_1)\sigma(a_2 b_2, c) \qquad (2.7)$$

and the normalization condition that

$$\sigma(1, a) = \varepsilon(a)1 = \sigma(a, 1), \ \ \varepsilon \circ \sigma(a, b) = \varepsilon(a)\varepsilon(b), \qquad (2.8)$$

where $a, b, c \in H$. They make the tensor product $K \otimes H$ into an algebra of crossed product [Mo, Def. 7.1.1]; it has unit $1 \otimes 1$ and its product is given by

$$(s \otimes a)(t \otimes b) = \sum s(a_1 \rightharpoonup t)\sigma(a_2, b_1) \otimes a_3 b_2, \qquad (2.9)$$

where $a, b \in H$ and $s, t \in K$.

Dually, let $\rho : H \to H \otimes K$ be a coaction which makes H into a K-comodule coalgebra, and let $\tau : H \to K \otimes K$ be an invertible linear map satisfying the dual 2-cocycle condition and the normalization condition. They make $K \otimes H$ into a coalgebra of crossed coproduct. We denote by $K \#_{\sigma,\tau} H$ the tensor product $K \otimes H$ with the described algebra and coalgebra structures.

LEMMA 2.10. $K \#_{\sigma,\tau} H$ is a bialgebra if and only if $(H, K, \rightharpoonup, \rho)$ is a Singer pair and (σ, τ) is a total 1-cocycle in the double complex $A_0^{\cdot}$ defined below. In this case, $K \#_{\sigma,\tau} H$ is necessarily a Hopf algebra with the antipode S given by

$$\begin{aligned} S(t \,\#\, a) \\ = \sum (\sigma^{-1}(S(a_{1H1}), a_{1H2}) \,\#\, S(a_{1H3}))(S(ta_{1K})(1, S) \circ \tau^{-1}(a_2) \,\#\, 1), \end{aligned} \qquad (2.11)$$

where $(1, S)(s \otimes t) = sS(t)$.

This follows by [H, Props. 3.3, 3.8 and 3.13]. If the conditions given above are satisfied, we obviously have a cleft extension

$$(K \#_{\sigma,\tau} H) = K \xrightarrow{\iota} K \#_{\sigma,\tau} H \xrightarrow{\pi} H \qquad (2.12)$$

of H by K, where $\iota(t) = t \,\#\, 1$, $\pi(t \,\#\, a) = \varepsilon(t)a$. Conversely, it follows from [Mo, Prop. 7.2.3] and the dual result (see also [H, Prop. 3.6]) that any cleft extension (A) of H by K is equivalent to some extension of the form (2.12), since we have such an isomorphism $\xi : A \xrightarrow{\cong} K \otimes H$ as in Condition (b) in Definition 2.5.

Here the Singer pair $(H, K, \rightharpoonup, \rho)$ which together with σ, τ forms $K \#_{\sigma,\tau} H$ is uniquely determined by (A), being independent of choice of ξ.

DEFINITION 2.13. In this case, we say that (A) *is associated with* the Singer pair $(H, K, \rightharpoonup, \rho)$. We denote by

$$\mathcal{O}\!peat\,(H, K) = \mathcal{O}\!peat\,(H, K, \rightharpoonup, \rho)$$

the category of cleft extensions associated with a fixed Singer pair $(H, K, \rightharpoonup, \rho)$, whose morphisms are equivalences between extensions so that this is a groupoid.

REMARK 2.14. If $H = kF$, $K = k^G$, then the Singer pair structures on (kF, k^G) are in 1-1 correspondence with the matched pair structures on (F, G). A cleft extension associated with a Singer pair $(kF, k^G, \rightharpoonup, \rho)$ is precisely an extension associated with the corresponding matched pair $(F, G, \lhd, \rhd)$ as defined by Definition 1.6. For the 1-1 correspondence above, F may be infinite. We see also that all results, except Remark 1.11, in the preceding section hold true even if F is infinite. In particular, Conditions (a)–(d) in Definition 1.4 are still equivalent for a sequence $(A) = k^G \to A \to kF$ with F infinite and, if they are satisfied, (A) is necessarily a cleft extension.

In what follows we fix a Singer pair $(H, K, \rightharpoonup, \rho)$. Recall H is cocommutative and K is commutative by assumption.

Let (A_1), (A_2) be in $\mathcal{O}\!peat\,(H, K)$. Form the tensor product $A_1 \otimes_K A_2$ of the left K-modules, on which two right H-comodule structures arise from the factors A_1 and A_2. Take the cotensor product of these comodule structures. Then we obtain the bi-tensor product $A_1 \otimes_K^H A_2$ as defined in [H, Sect. 4], which forms a Hopf algebra with the structure induced from the Hopf algebra $A_1 \otimes A_2$ of tensor product. Further, it forms naturally an extension $(A_1 \otimes_K^H A_2)$ in $\mathcal{O}\!peat\,(H, K)$. We write

$$(A_1) * (A_2) = (A_1 \otimes_K^H A_2).$$

One sees that $(K \#_{\sigma_1,\tau_1} H) * (K \#_{\sigma_2,\tau_2} H) = (K \#_{\sigma,\tau} H)$, where $\sigma = \sigma_1\sigma_2$, $\tau = \tau_1\tau_2$, convolution products.

If $\sigma : H \otimes H \to K$ and $\tau : H \to K \otimes K$ are trivial so that $\sigma(a, b) = \varepsilon(a)\varepsilon(b)1$, $\tau(a) = \varepsilon(a)1 \otimes 1$, then $K \#_{\sigma,\tau} H$ is the Hopf algebra of bi-smash product [T1, p.849], for which we write simply $K \# H$. This forms an extension $(K \# H)$ in $\mathcal{O}\!peat\,(H, K)$.

PROPOSITION 2.15. $\mathcal{O}\!peat\,(H, K)$ *forms a symmetric monoidal groupoid with tensor product* $*$ *and unit object* $(K \# H)$.

This is essentially proved in [H, Sect. 5]. The associativity constraint and the symmetry are induced from the obvious isomorphisms $(A_1 \otimes A_2) \otimes A_3 \xrightarrow{\cong} A_1 \otimes (A_2 \otimes A_3)$ and $A_1 \otimes A_2 \xrightarrow{\cong} A_2 \otimes A_1$, respectively.

We denote by

$$\mathrm{Opext}(H, K) = \mathrm{Opext}(H, K, \rightharpoonup, \rho)$$

the set of all isomorphism (or equivalence) classes in $\mathcal{O}\!pext\,(H,K)$, which is a monoid under the product arising from $*$. Each object (A) in $\mathcal{O}\!pext\,(H,K)$ has inverse (A^{-1}) in the sense $(A) * (A^{-1})$ is isomorphic to the unit object $(K \# H)$, since we have $(A^{-1}) = (K \#_{\sigma^{-1},\tau^{-1}} H)$ if $(A) \sim (K \#_{\sigma,\tau} H)$. Hence, $\mathrm{Opext}(H,K)$ is in fact an abelian group.

We denote by $\mathrm{Aut}(K \# H)$ the group of auto-equivalences of $(K \# H)$. The group of auto-equivalences of any (A) in $\mathcal{O}\!pext\,(H,K)$ is canonically isomorphic to $\mathrm{Aut}(K \# H)$, since $(A)*$ gives a category equivalence.

In general, if $\mathcal{M}$ is a symmetric monoidal category with a small skeleton, the groups $K_0(\mathcal{M})$ and $K_1(\mathcal{M})$ of $\mathcal{M}$ are defined; see [B, Chap. VII, Sect. 1]. Suppose each object in $\mathcal{M}$ has inverse in the sense as above. This is equivalent to saying that all isomorphism classes of the objects in $\mathcal{M}$ form an abelian group under the product arising from the tensor product. Then, $K_0(\mathcal{M})$ is canonically isomorphic to this abelian group, while $K_1(\mathcal{M})$ is isomorphic to the automorphism group of the unit object (or any object). Therefore those groups of $\mathcal{O}\!pext\,(H,K)$ are given by

$$K_0 = \mathrm{Opext}(H,K), \quad K_1 = \mathrm{Aut}(K \# H).$$

We follow Singer [S] to give cohomological description of these groups by technique of simplicial homology. Recall first the category $\mathcal{C} = \mathcal{C}(H,K,\rightharpoonup,\rho)$ is defined by Definition 2.3. We will denote by $V^{\otimes n} = V \otimes \cdots \otimes V$ the n-fold tensor product of a vector space V.

Let Comod-K denote the category of K-comodules. We define a functor

$$\mathbb{F} : \text{Comod-}K \to \mathcal{C}, \quad \mathbb{F}(P) = H \otimes P \tag{2.16}$$

by endowing the H-module $H \otimes P$ with the K-comodule structure $a \otimes p \mapsto \sum (a_1)_H \otimes p_0 \otimes (a_1)_K (a_2 \rightharpoonup p_1)$, $H \otimes P \to H \otimes P \otimes K$. Since one sees that this is left adjoint to the forgetful functor $\mathbb{U} : \mathcal{C} \to \text{Comod-}K$, it follows by [W, 8.6.2, p.280] that the functor $\mathbb{F} \circ \mathbb{U} : \mathcal{C} \to \mathcal{C}$, which we denote simply by $\mathbb{F}$, forms a cotriple $(\mathbb{F}, \varepsilon, \delta)$ on $\mathcal{C}$, where $\varepsilon : \mathbb{F} \to \mathrm{id}$, $\delta : \mathbb{F} \to \mathbb{F}^2$ are the natural transformations defined by

$$\varepsilon_M : H \otimes M \to M, \qquad \varepsilon_M(a \otimes m) = am,$$
$$\delta_M : H \otimes M \to H \otimes H \otimes M, \quad \delta_M(a \otimes m) = a \otimes 1 \otimes m$$

for $M \in \mathcal{C}$. Regard k as an object in $\mathcal{C}$ with the trivial structure. Then we have a simplicial object $\Phi_{\cdot}(k) = \{\mathbb{F}^{p+1}(k)\}_{p \geq 0}$ in $\mathbb{C}$; accompanied with the face and degeneracy operators determined by ε and δ, it looks like

$$\Phi_{\cdot}(k) = H \;\rightleftarrows\; H^{\otimes 2} \;\rightleftarrows\; H^{\otimes 3} \;\rightleftarrows\; \cdots \tag{2.17}$$

Dually we define a functor

$$\mathbb{G} : H\text{-Mod} \to \mathcal{C}, \quad \mathbb{G}(Q) = Q \otimes K \qquad (2.18)$$

by endowing the K-comodule $Q \otimes K$ with an H-module structure via $a(q \otimes t) = \sum (a_1)_H q \otimes (a_1)_K (a_2 \rightharpoonup t)$, where $a \in H$, $q \otimes t \in Q \otimes K$. Since this is right adjoint to the forgetful functor $\mathbb{U} : \mathcal{C} \to H\text{-Mod}$, we have a triple $(\mathbb{G} = \mathbb{G} \circ \mathbb{U}, \eta, \mu)$ on $\mathcal{C}$, where $\eta : \text{id} \to \mathbb{G}$, $\mu : \mathbb{G}^2 \to \mathbb{G}$ are the natural transformations defined by

$$\eta_M : M \to M \otimes K, \qquad \eta_M(m) = \sum m_0 \otimes m_1,$$
$$\mu_M : M \otimes K \otimes K \to M \otimes K, \quad \mu_M(m \otimes s \otimes t) = m \otimes \varepsilon(s)t.$$

We have also a cosimplicial object $\Psi^{\cdot}(k) = \{\mathbb{G}^{q+1}(k)\}_{q \geq 0}$ in $\mathcal{C}$, which looks like

$$\Psi^{\cdot}(k) = K \rightrightarrows K^{\otimes 2} \mathrel{\substack{\longrightarrow\\\longrightarrow\\\longrightarrow}} K^{\otimes 3} \cdots \qquad (2.19)$$

If $M, N \in \mathcal{C}$, then $M \otimes N$ is an object in $\mathcal{C}$ with the diagonal H-action and K-coaction. Thus, $\mathcal{C} = (\mathcal{C}, \otimes, k)$ forms a symmetric monoidal category with the obvious symmetry. Let $\mathcal{C}_c$ denote the category of cocommutative coalgebras in $\mathcal{C}$. Since H is in $\mathcal{C}_c$, it follows that for $C \in \mathcal{C}_c$, the coalgebra $\mathbb{F}(C) = H \otimes C$ of tensor product is in $\mathcal{C}_c$. Therefore, $(\mathbb{F}, \varepsilon, \delta)$ is regarded as a cotriple on $\mathcal{C}_c$ so that $\Phi_{\cdot}(k)$ is a simplicial object in $\mathcal{C}_c$. Similarly, $(\mathbb{G}, \eta, \mu)$ is regarded as a triple on the category $\mathcal{C}_a$ of commutative algebras in $\mathcal{C}$ so that $\Psi^{\cdot}(k)$ is a cosimplicial object in $\mathcal{C}_a$.

Let Reg_H^K (resp., Reg) denote the abelian group of (convolution-)invertible, H-linear and K-colinear (resp., k-linear) maps. If $C \in \mathcal{C}_c$, $A \in \mathcal{C}_a$, then an isomorphism

$$\text{Reg}_H^K(\mathbb{F}(C), \mathbb{G}(A)) \cong \text{Reg}(C, A) \qquad (2.20)$$

is given by $f \mapsto (c \mapsto (1 \otimes \varepsilon) \circ f(1 \otimes c))$.

Form the double cosimplicial object $\text{Reg}_H^K(\Phi_{\cdot}(k), \Psi^{\cdot}(k))$ in the category of abelian groups; by (2.20), it looks like

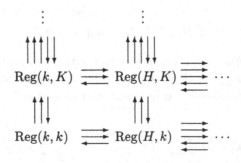

Further, form the associated normalized double complex, whose term

$$\mathrm{Reg}_+(H^{\otimes p}, K^{\otimes q})$$

in the (p, q)-th position consists of the invertible linear maps $H^{\otimes p} \to K^{\otimes q}$ satisfying the normalization condition such as given in (2.8). Remove the edges from the double complex just formed to obtain

$$
\begin{array}{ccc}
\vdots & & \vdots \\
\uparrow & & \uparrow \\
A_0^{\cdot\cdot} = \quad \mathrm{Reg}_+(H, K^{\otimes 2}) & \longrightarrow & \mathrm{Reg}_+(H^{\otimes 2}, K^{\otimes 2}) \longrightarrow \cdots \\
\uparrow & & \uparrow \\
\mathrm{Reg}_+(H, K) & \longrightarrow & \mathrm{Reg}_+(H^{\otimes 2}, K) \longrightarrow \cdots
\end{array}
$$

One sees that, if $H = kF$, $K = k^G$ in particular, $A_0^{\cdot\cdot}$ is identified with $A^{\cdot\cdot}$ which was defined in the preceding section.

PROPOSITION 2.20 [H, Props. 3.15, 6.5].
(1) *The assignment* $(\sigma, \tau) \mapsto (K \#_{\sigma,\tau} H)$, *where* (σ, τ) *is a total 1-cocycle in* $A_0^{\cdot\cdot}$, *induces an isomorphism*

$$H^1(\mathrm{Tot}\, A_0^{\cdot\cdot}) \cong \mathrm{Opext}(H, K).$$

(2) *For a total 0-cocycle* $\nu : H \to K$ *in* $A_0^{\cdot\cdot}$, $t \# a \mapsto \sum t\nu(a_1) \# a_2$ *gives an auto-equivalence of* $(K \# H)$. *This gives an isomorphism*

$$H^0(\mathrm{Tot}\, A_0^{\cdot\cdot}) \cong \mathrm{Aut}(K \# H).$$

3. Cocycle Deformations Arising in Extensions

Let A be a Hopf algebra. A (*normalized*) 2-*cocycle for* A is an invertible linear map $\sigma : A \otimes A \to k$ which satisfies the conditions given by (2.7) and (2.8) if we suppose therein $H = A$ and $K = k$, the trivial A-module algebra. The *cocycle deformation* A^σ by such σ is the coalgebra A endowed with the twisted product $\cdot$ defined by

$$a \cdot b = \sum \sigma(a_1, b_1) a_2 b_2 \sigma^{-1}(a_3, b_3),$$

where $a, b \in A$; this is in fact a Hopf algebra with the same unit 1 and the twisted antipode S^σ given by

$$S^\sigma(a) = \sum \sigma(a_1, S(a_2)) S(a_3) \sigma^{-1}(S(a_4), a_5),$$

where $a \in A$; see [D2, Thm. 1.6]. If $B = A^\sigma$, then σ^{-1} is regarded as a 2-cocycle for B, and we have $B^{\sigma^{-1}} = A$. This allows us to say that A and B are cocycle deformations of each other. If this is the case, the right (or equivalently left) comodule categories Comod-A and Comod-B are k-linearly monoidally equivalent. The converse holds true, if A or B, then necessarily both are finite-dimensional or pointed. See [Sb1, Sect. 5]. Here recall that the comodules over a bialgebra form a monoidal category in the obvious manner.

Let $(H, K, \rightharpoonup, \rho)$ be a Singer pair of Hopf algebras. Let $(A) = K \xrightarrow{\iota} A \xrightarrow{\pi} H$ be a cleft extension associated with the pair. A 2-cocycle $\theta : H \otimes H \to k$ for H is regarded as a 2-cocycle for A, composed with $\pi \otimes \pi$. We see that the cocycle deformation A^θ by such θ forms a cleft extension $(A^\theta) = K \xrightarrow{\iota} A^\theta \xrightarrow{\pi} H^\theta = H$ associated with the same Singer pair. Here note $H^\theta = H$, since H is cocommutative. (The deformation, as above, by lifted 2-cocycles dualizes the construction in [L], [EV].)

The 2nd Sweedler cohomology $H^2(H, k)$ [Sw1, Sect. 2] with coefficients in the trivial H-module algebra k is the 2nd cohomology of the bottom complex which was removed when we constructed $A_0^{\cdot\cdot}$; see Section 2. The removed vertical differential $\partial' : \mathrm{Reg}_+(H^{\otimes 2}, k) \to \mathrm{Reg}_+(H^{\otimes 2}, K)$ is given by

$$\partial'\theta(a, b) = \sum \theta(a_{1H}, b_{1H}) a_{1K} (a_2 \rightharpoonup b_{1K}) \theta^{-1}(a_3, b_2),$$

where $\theta \in \mathrm{Reg}_+(H^{\otimes 2}, k)$. We see that, if θ is a 2-cocycle for H, then $(\partial'\theta, \varepsilon)$, where ε is the identity in $\mathrm{Reg}_+(H, K^{\otimes 2})$, is a total 1-cocycle in $A_0^{\cdot\cdot}$, and that $\theta \mapsto (\partial'\theta, \varepsilon)$ induces a group map $H^2(H, k) \to H^1(\mathrm{Tot}\, A_0^{\cdot\cdot}) \cong \mathrm{Opext}(H, K)$, which we denote by

$$\delta : H^2(H, k) \to \mathrm{Opext}(H, K).$$

PROPOSITION 3.1. *Let* (A), (A') *be in* $\mathcal{O}pext(H, K)$. *There is a 2-cocycle* θ *for* H *such that* (A^θ) *is equivalent to* (A') *if and only if* (A) *and* (A') *are equal in the cokernel* $\mathrm{Opext}(H, K)/\mathrm{Im}\,\delta$ *of* δ.

PROOF. This follows since we see that, if (A) is given by a total 1-cocycle (σ, τ) in $A_0^{\cdot\cdot}$, then (A^θ) is given by $(\sigma\partial'\theta, \tau)$. $\square$

We will give some results of sample computations in the special case when $H = kF$, $K = k^G$. The map δ is then identified with

$$\delta : H^2(F, k^\times) \to \mathrm{Opext}(kF, k^G)$$

which arises from $\partial' : \mathrm{Map}_+(F^2, k^\times) \to \mathrm{Map}_+(G \times F^2, k^\times)$ given by

$$\partial'\theta(x; a, b) = \theta(x \triangleright a, (x \triangleleft a) \triangleright b)\theta(a, b)^{-1},$$

where $a, b \in F$, $x \in G$ and $\theta \in \mathrm{Map}_+(F^2, k^\times)$. Note that this δ is involved in the Kac exact sequence.

If $\mathrm{triv} : G \times F \to G$ denotes the trivial action, any action $\triangleright : G \times F \to F$ by group automorphisms forms a matched pair $(F, G, \mathrm{triv}, \triangleright)$ of groups, so that $F \bowtie G = F \rtimes G$, the semi-direct product given by $\triangleright$.

Fix an integer $n > 1$. Suppose $F = \mathbb{Z}_n \oplus \mathbb{Z}_n$, where $\mathbb{Z}_n$ denotes the additive group of integers modulo n. Fix a matched pair $(F, \mathbb{Z}_2, \mathrm{triv}, \triangleright)$ of groups, where

$\triangleright : \mathbb{Z}_2 \times F \to F$ is defined by

$$0 \triangleright (i,j) = (i,j), \quad 1 \triangleright (i,j) = (j,i).$$

Let $\mu_n(k)$ denote the group of all n-th roots of 1 in k. Let $\zeta \in \mu_n(k)$. Then,

$$\theta_\zeta : F \times F \to k^\times, \quad \theta_\zeta((i,j),(k,l)) = \zeta^{il}$$

is a group 2-cocycle. We define a Hopf algebra A_ζ including $k^{\mathbb{Z}_2}$ as a central Hopf subalgebra as follows; A_ζ is generated by elements a_+, a_- over $k^{\mathbb{Z}_2}$, and is defined by the relations

$$a_\pm^n = 1, \quad a_- a_+ = (e_0 + \zeta e_1)a_+ a_-$$

together with the structures

$$\Delta(a_\pm) = a_\pm \otimes e_0 a_\pm + a_\mp \otimes e_1 a_\pm, \quad \varepsilon(a_\pm) = 1, \quad S(a_\pm) = e_0 a_\pm^{-1} + e_1 a_\mp^{-1}.$$

Thus, A_ζ is not cocommutative, and is commutative only if $\zeta = 1$. If $\pi : A_\zeta \to kF$ denotes the Hopf algebra map determined by $\pi(e_0) = 1$, $\pi(e_1) = 0$, $\pi(a_+) = (1,0)$ and $\pi(a_-) = (0,1)$, then we see that $(A_\zeta) = k^{\mathbb{Z}_2} \hookrightarrow A_\zeta \xrightarrow{\pi} kF$ is an extension associated with the fixed matched pair.

PROPOSITION 3.2. *Suppose* $(k^\times)^n = k^\times$. *Then,* $\zeta \mapsto \theta_\zeta$ *and* $\zeta \mapsto (A_\zeta)$ *induce isomorphisms*

$$\mu_n(k) \cong H^2(F, k^\times), \quad \mu_n(k) \cong \mathrm{Opext}(kF, k^{\mathbb{Z}_2}),$$

respectively. The map $\delta : H^2(F, k^\times) \to \mathrm{Opext}(kF, k^{\mathbb{Z}_2})$ *is induced by* $\theta_\zeta \mapsto (A_{\zeta^2})$. *Therefore, if n is odd, then δ is an isomorphism so that every A_ζ is a cocycle deformation of the commutative Hopf algebra* A_1.

PROOF. It is easy to see the first isomorphism. The group $\mathrm{Opext}(k\mathbb{Z}_2, k^F)$ associated with the matched pair $(\mathbb{Z}_2, F, \triangleright, \mathrm{triv})$ is computed by [M2, Thm. 2.1 and corrigendum], whose proof gives the second isomorphism since $(A) \mapsto (A^*)$ induces an isomorphism $\mathrm{Opext}(kF, k^{\mathbb{Z}_2}) \cong \mathrm{Opext}(k\mathbb{Z}_2, k^F)$; see [M4, Exercise 5.5]. Since we compute

$$\partial' \theta_\zeta(1;(0,1),(1,0))/\partial' \theta_\zeta(1;(1,0),(0,1)) = \zeta^2,$$

it follows that $\delta\theta_\zeta$ is equivalent to (A_{ζ^2}). $\square$

Next, we suppose $F = \mathbb{Z} \oplus \mathbb{Z}$. By a slight modification we define the matched pair $(F, \mathbb{Z}_2, \mathrm{triv}, \triangleright)$, the group 2-cocycle θ_ζ for F, and the extension (A_ζ) associated with the matched pair, where $\zeta \in k^\times$. The modification will be obvious except that we replace the relation $a_\pm^n = 1$ for A_ζ by the condition that $a_\pm$ are invertible. We see that $\zeta \mapsto \theta_\zeta$ induces an isomorphism $k^\times \cong H^2(F, k^\times)$. A slight modification of the last proof proves the following.

PROPOSITION 3.3. *Suppose* $(k^\times)^2 = k^\times$. *Then,* $\zeta \mapsto (A_\zeta)$ *induces an isomorphism* $k^\times \cong \text{Opext}(kF, k^{\mathbb{Z}_2})$. *The map* $\delta : H^2(F, k^\times) \to \text{Opext}(kF, k^{\mathbb{Z}_2})$ *is a surjection, induced by* $\theta_\zeta \mapsto (A_{\zeta^2})$. *Therefore every* A_ζ *is a cocycle deformation of the commutative Hopf algebra* A_1.

REMARK 3.4. Suppose $k = \mathbb{C}$, and $\zeta \in \mathbb{C}$ with $|\zeta| = 1$. Then, A_ζ is a Hopf *-algebra with the *-structure $e_i^* = e_i$ ($i = 0, 1$), $a_\pm^* = a_\pm^{-1}$. By (the proof of) the last proposition, A_ζ is isomorphic to the cocycle deformation $(A_1)^\sigma$ by $\sigma = \theta_{\sqrt{\zeta}}$ via $e_i \mapsto e_i$, $a_\pm \mapsto a_\pm$, which preserves *-structure, too. Suppose $\zeta = q^2$ with $q \in \mathbb{C}$. Then, A_{q^2} is isomorphic to the coordinate Hopf *-algebra $A(DT_q^2)$ of the quantum double torus DT_q^2 due to Hajac and Masuda [HM]: in fact, $e_0 \mapsto D^{-1}ad$, $a_+ \mapsto a + c$, $a_- \mapsto b + d$ give an isomorphism $A_{q^2} \cong A(DT_q^2)$. This concludes that their results [HM, Sect. 3] on unitary representations of DT_q^2 coincide completely with the results for the 'classical' double torus DT^2 with coordinate Hopf *-algebra A_1.

4. Quasi-Hopf Algebras Obtained by Extension

A *quasi-bialgebra* is a quadruple $(A, \Delta, \varepsilon, \Phi)$ which consists of an algebra A, algebra maps $\Delta : A \to A \otimes A$, $\varepsilon : A \to k$, and an invertible element $\Phi \in A \otimes A \otimes A$, called the *Drinfeld associator*, such that Conditions (2.1)–(2.4) in [BP, Def. 2.1] are fulfilled, or equivalently such that the module category A-Mod forms a monoidal category, where the tensor product is the usual one $V \otimes W$ of vector spaces on which A acts through Δ, the unit object is k on which A acts through ε, the left and right unit constraints are the obvious isomorphisms $k \otimes V = V = V \otimes k$, and the associativity constraint is given by $u \otimes v \otimes w \mapsto \Phi(u \otimes v \otimes w)$, $(U \otimes V) \otimes W \xrightarrow{\cong} U \otimes (V \otimes W)$. Among the conditions just refered to, (2.2) is the usual counit property, and (2.1) is

$$(1 \otimes \Delta) \circ \Delta(a) = \Phi(\Delta \otimes 1) \circ \Delta(a)\Phi^{-1} \ (a \in A).$$

Hence a commutative quasi-bialgebra is necessarily an ordinary bialgebra.

The notion was first introduced by Drinfeld, which we define here in a stricter sense than original, assuming that the unit constraints are given by the obvious isomorphisms; cf [Kas, Prop. XV 1.2].

Let $A = (A, \Delta, \varepsilon, \Phi)$ be a quasi-bialgebra. A *gauge transformation* on A is an invertible element $\varphi \in A \otimes A$ satisfying the normalization condition

$$(1 \otimes \varepsilon)(\varphi) = 1 \otimes 1 = (\varepsilon \otimes 1)(\varphi).$$

By such φ, one constructs a new quasi-bialgebra $A_\varphi = (A, \Delta_\varphi, \varepsilon, \Phi_\varphi)$, where Δ_φ and Φ_φ are defined by

$$\Delta_\varphi(a) = \varphi\Delta(a)\varphi^{-1} \ (a \in A),$$
$$\Phi_\varphi = (1 \otimes \varphi)(1 \otimes \Delta)(\varphi)\Phi(\Delta \otimes 1)(\varphi^{-1})(\varphi^{-1} \otimes 1),$$

respectively, so that A-Mod and A_φ-Mod are monoidally equivalent in a natural manner; see [Kas, Prop. XV 3.2, Thm. XV 3.5]. This generalizes the dual notion of cocycle deformation (see Section 3) of an ordinary bialgebra.

A *map* $f : A \to A'$ *of quasi-bialgebras* is a linear map which preserves the structures, and so in particular $(f \otimes f \otimes f)(\Phi) = \Phi'$.

Let $(H, K, \to, \rho)$ be a Singer pair of Hopf algebras. Let

$$\sigma : H \otimes H \to K, \ \tau : H \to K \otimes K, \ \Phi \in K \otimes K \otimes K$$

be two invertible linear maps and an invertible element which all satisfy the normalization condition. On the tensor product $K \otimes H$, $\to$ and σ define such a product with unit $1 \otimes 1$ as defined by the formula (2.9); ρ and τ define a coproduct with counit $\varepsilon \otimes \varepsilon$ dually by the crossed coproduct construction. Let $K \#_{\sigma,\tau} H$ denote $K \otimes H$ with these structures. Identify Φ with its natural image $(\iota \otimes \iota \otimes \iota)(\Phi)$ in $(K \#_{\sigma,\tau} H)^{\otimes 3}$, where $\iota(t) = t \# 1$ for $t \in K$, and denote the image by Φ, too. We see directly the following.

LEMMA 4.1 (cf. [BP, Remark 3.2]). $K \#_{\sigma,\tau} H$ *is a quasi-bialgebra with Drinfeld associator* Φ, *if and only if* (σ, τ, Φ) *is a total 2-cocycle in the double complex* $A_1^{\cdots}$ *given below. In this case,* $K \#_{\sigma,\tau} H$ *is necessarily a quasi-Hopf algebra in the sense defined by* [Kas, Def. XV 5.1], *whose requirement is fulfilled by the map* S *defined by the same formula as* (2.11), *and by* $\alpha = \sum S(\Phi_1)\Phi_2 S(\Phi_3) \# 1$, $\beta = 1 \# 1$, *if we write* $\Phi = \sum \Phi_1 \otimes \Phi_2 \otimes \Phi_3$.

The double complex

$$
A_1^{\cdots} =
\begin{array}{ccccc}
\vdots & & & & \\
\uparrow & & & & \\
\mathrm{Reg}_+(k, K^{\otimes 3}) \longrightarrow & & \raisebox{0.5ex}{\vdots} \cdots & & \\
\uparrow & & \uparrow & & \\
\mathrm{Reg}_+(k, K^{\otimes 2}) \longrightarrow & \mathrm{Reg}_+(H, K^{\otimes 2}) \longrightarrow & \raisebox{0.5ex}{\vdots} \cdots & & \\
\uparrow & \uparrow & & \uparrow & \\
\mathrm{Reg}_+(k, K) \longrightarrow & \mathrm{Reg}_+(H, K) \longrightarrow & \mathrm{Reg}_+(H^{\otimes 2}, K) \longrightarrow & \cdots &
\end{array}
$$

enlarges $A_0^{\cdots}$ by joining the most left vertical complex which was removed when we constructed $A_0^{\cdots}$. Note that Φ is regarded as an element in $\mathrm{Reg}_+(k, K^{\otimes 3})$. The joined complex is the standard complex for computing the Doi cohomology $H^{\cdot}(k, K)$ [D1, Sect. 2.6], where k is the trivial K-comodule coalgebra. (Note $H^{\cdot}(k, K)$ is denoted by Coalg-$H^{\cdot}(k, K)$ in [D1].)

Since K is commutative, a Drinfeld associator of K is none other than an element in $\mathrm{Reg}_+(k, K^{\otimes 3})$ which vanishes through the vertical differential. Therefore, if (σ, τ, Φ) is a total 2-cocycle in $A_1^{..}$, then $\iota : (K, \Phi) \to (K \#_{\sigma,\tau} H, \Phi)$, $\iota(t) = t \# 1$ is a quasi-bialgebra map. We regard H as a quasi-bialgebra with trivial Drinfeld associator, so that $\pi : K \#_{\sigma,\tau} H \to H$, $\pi(t \# a) = \varepsilon(t)a$ is a quasi-bialgebra map. We let

$$(K \#_{\sigma,\tau} H, \Phi) = K \xrightarrow{\iota} K \#_{\sigma,\tau} H \xrightarrow{\pi} H$$

denote the sequence of quasi-bialgebras thus obtained.

DEFINITION 4.2. A sequence $(A, \Phi) = K \to A \to H$ of quasi-bialgebras together with a Drinfeld associator Φ of K is called a *cleft extension of H by K*, if it is equivalent to some $(K \#_{\sigma,\tau} H, \Phi)$ in the sense that there is a quasi-bialgebra isomorphism $A \xrightarrow{\cong} K \#_{\sigma,\tau} H$ which induces the identity maps on H and K. We say that (A, Φ) *is associated with* the Singer pair $(H, K, \rightharpoonup, \rho)$ which forms $K \#_{\sigma,\tau} H$. (This is well-defined; see Proposition 4.5 below.)

If (A, Φ) is a cleft extension of H by K, then A is an (ordinary) H-comodule algebra as well as a K-module, and satisfies Conditions (a), (b) in Definition 2.5. Suppose $\Phi = 1$, the unit of $K^{\otimes 3}$. Then, $(A, 1)$ is a cleft quasi-bialgebra extension in the sense above if and only if it is a cleft Hopf algebra extension in the sense of Definition 2.5.

Let $(A, \Phi) = K \xrightarrow{\iota} A \xrightarrow{\pi} H$ be a cleft extension, and let φ be a gauge transformation on K. Then, $(\iota \otimes \iota)(\varphi)$ is a gauge transformation on A, which we denote by φ, too. We see easily the following.

LEMMA 4.3. $(A_\varphi, \Phi_\varphi) = K \xrightarrow{\iota} A_\varphi \xrightarrow{\pi} H$ *is a cleft extension associated with the same Singer pair as* (A, Φ). *(Note* $\Phi_\varphi = \Phi \partial' \varphi$.*)*

DEFINITION 4.4. A *quasi-equivalence* between cleft extensions is a pair $(f, \varphi) : (A, \Phi) \to (A', \Phi')$, where $f : A \to A'$ is a linear map and φ is a gauge transformation on K, such that f gives an equivalence $(A_\varphi, \Phi_\varphi) \xrightarrow{\cong} (A', \Phi')$. Its *composite* with another quasi-equivalence $(f', \varphi') : (A', \Phi') \to (A'', \Phi'')$ is the quasi-equivalence $(A, \Phi) \to (A'', \Phi'')$ defined by

$$(f', \varphi') \circ (f, \varphi) = (f' \circ f, \varphi' \varphi).$$

The quasi-equivalence defines a equivalence relation among all cleft extensions of H by K. It is easy to see the following.

PROPOSITION 4.5. *Two cleft extensions of H by K are associated with the same Singer pair if they are quasi-equivalent to each other.*

Let γ be an invertible element in K such that $\varepsilon(\gamma) = 1$. Its image under the vertical differential $\partial' : \mathrm{Reg}_+(k, K) \to \mathrm{Reg}_+(k, K^{\otimes 2})$ in $A_1^{..}$ is given by

$$\partial' \gamma = (\gamma \otimes \gamma) \Delta(\gamma^{-1}),$$

which is hence a gauge transformation on K.

LEMMA 4.6. *For any cleft extension* (A, Φ) *of* H *by* K, $(\mathrm{inn}\, \gamma, \partial'\gamma) : (A, \Phi) \xrightarrow{\cong}$ (A, Φ) *is a quasi-auto-equivalence, where* $\mathrm{inn}\, \gamma(a) = \gamma^{-1} a \gamma$ *for* $a \in A$.

PROOF. This follows easily if one notices that $\Phi_{\partial'\gamma} = \Phi\, \partial'^2 \gamma = \Phi$. □

DEFINITION 4.7 (cf. [Sb2, Def. 6.2.5]). Two quasi-equivalences $(f, \varphi), (f', \varphi')$: $(A, \Phi) \to (A', \Phi')$ *are said to be* cohomologous *if there is* $\gamma \in \mathrm{Reg}_+(k, K)$ *such that* $(f', \varphi') = (\mathrm{inn}\, \gamma, \partial'\gamma) \circ (f, \varphi)$.

This defines an equivalence relation among quasi-equivalences, which is compatible with the composition since we have $(f, \varphi) \circ (\mathrm{inn}\, \gamma, \partial'\gamma) = (\mathrm{inn}\, \gamma, \partial'\gamma) \circ (f, \varphi)$.

DEFINITION 4.8. We denote by

$$\mathcal{O}\!pext'(H, K) = \mathcal{O}\!pext'(H, K, \rightharpoonup, \rho)$$

the groupoid of cleft extensions of quasi-bialgebras which are associated with a fixed Singer pair $(H, K, \rightharpoonup, \rho)$; the morphisms are cohomology classes of quasi-equivalences.

For (A_i, Φ_i) in $\mathcal{O}\!pext'(H, K)$, where $i = 1, 2$, we define by using the bi-tensor product $\otimes_K^H$

$$(A_1, \Phi_1) * (A_2, \Phi_2) = (A_1 \otimes_K^H A_2, \Phi_1 \Phi_2),$$

which is naturally an object in $\mathcal{O}\!pext'(H, K)$. As a variation of Proposition 2.15 we have the following.

PROPOSITION 4.9. $\mathcal{O}\!pext'(H, K)$ *forms a symmetric monoidal groupoid with tensor product* $*$ *and unit object* $(K \# H, 1)$, *in which each object has inverse.*

Fix a Singer pair $(H, K, \rightharpoonup, \rho)$. We denote by

$$\mathrm{Opext}'(H, K), \quad \mathrm{Aut}'(K \# H, 1)$$

the group of all quasi-equivalence classes in $\mathcal{O}\!pext'(H, K)$ and the cohomology group of all quasi-auto-equivalences of $(K \# H, 1)$, respectively. They give respectively the K_0 and K_1 groups of $\mathcal{O}\!pext'(H, K)$.

PROPOSITION 4.10.
(1) *The assignment*

$$(\sigma, \tau, \Phi) \mapsto (K \underset{\sigma, \tau}{\#} H, \Phi),$$

where (σ, τ, Φ) *is a total* 2-*cocycle in* $A_1^{\cdot\cdot}$, *induces an isomorphism*

$$H^2(\mathrm{Tot}\, A_1^{\cdot\cdot}) \cong \mathrm{Opext}'(H, K).$$

(2) *To a total* 1-*cocycle* (ν, φ) *in* $A_1^{\cdot\cdot}$, *there is assigned a quasi-auto-equivalence* (f_ν, φ) *of* $(K \# H, 1)$, *where* $f_\nu(t \# a) = \sum t\nu(a_1) \# a_2$. *This assignment induces an isomorphism*

$$H^1(\mathrm{Tot}\, A_1^{\cdot\cdot}) \cong \mathrm{Aut}'(K \# H, 1).$$

The proof is straightforward.

Note that $A_0^{\cdot}$ is regarded as a double subcomplex of $A_1^{\cdot}$ such that the cokernel is the standard complex for computing the Doi cohomology $H^{\cdot}(k, K)$. The short exact sequence of complexes thus obtained gives rise to the following.

THEOREM 4.11. *We have an exact sequence*

$$1 \to G(K)^H \quad \to H^1(k, K) \quad \to \operatorname{Aut}(K \# H) \to \operatorname{Aut}'(K \# H, 1)$$
$$\to H^2(k, K) \to \operatorname{Opext}(H, K) \to \operatorname{Opext}'(H, K) \to H^3(k, K),$$

where $G(K)^H$ denotes the group of H-invariant grouplikes in K.

Let us call the last group map

$$\beta : \operatorname{Opext}'(H, K) \to H^3(k, K),$$

which is induced by $(K \#_{\sigma, \tau} H, \Phi) \mapsto \Phi$. We will see that β is a split surjection in some special case, though it is not even a surjection in general. Suppose H is finite-dimensional, and $K = H^*$. Let $(H, H, \lhd, \operatorname{triv})$ be the matched pair of Hopf algebras defined by the adjoint action $\lhd: H \otimes H \to H$, $x \lhd a = \sum S(a_1) x a_2$ and the trivial action $\operatorname{triv} : H \otimes H \to H$, $x \rhd a = \varepsilon(x) a$. By Remark 2.4, it gives rise to a Singer pair $(H, H^*, \frown, \operatorname{triv})$. We have an identification $\operatorname{Reg}_+(k, (H^*)^{\otimes 3}) = \operatorname{Reg}_+(H^{\otimes 3}, k)$. For $\Phi \in \operatorname{Reg}_+(H^{\otimes 3}, k)$, define $\sigma_\Phi, \tau_\Phi \in \operatorname{Reg}_+(H^{\otimes 3}, k)$ by

$$\sigma_\Phi(x; a, b) = \sum \Phi^{-1}(x, a, b) \Phi(a, x \lhd a, b) \Phi^{-1}(a, b, x \lhd ab)$$
$$\tau_\Phi(x, y; a) = \sum \Phi^{-1}(x, y, a) \Phi(x, a, y \lhd a) \Phi^{-1}(a, x \lhd a, y \lhd a),$$

where $a, b, x, y \in H$. Here we wrote as $\Delta(a) = \sum a \otimes a$, omitting the subscripts of numbers; it would be allowed since H is cocommutative. Further, identify σ_Φ, τ_Φ with

$$H \otimes H \to H^*, \qquad\qquad a \otimes b \mapsto (x \mapsto \sigma_\Phi(x; a, b)),$$
$$H \to H^* \otimes H^* = (H \otimes H)^*, a \mapsto (x \otimes y \mapsto \tau_\Phi(x, y; a)),$$

respectively. Then we see $\sigma_\Phi \in \operatorname{Reg}_+(H^{\otimes 2}, H^*)$, $\tau_\Phi \in \operatorname{Reg}_+(H, (H^*)^{\otimes 2})$.

PROPOSITION 4.12. *If Φ is a Drinfeld associator on H^*, then $(\sigma_\Phi, \tau_\Phi, \Phi)$ is a total 2-cocycle in the double complex $A_1^{\cdot}$ defined by the Singer pair $(H, H^*, \frown, \operatorname{triv})$. The assignment $\Phi \mapsto (H^* \#_{\sigma_\Phi, \tau_\Phi} H, \Phi)$ induces a group map $\bar\beta : H^3(k, H^*) \to \operatorname{Opext}'(H, H^*)$ such that $\beta \circ \bar\beta = 1$.*

This is a reformulation of part of [BP, Theorem 3.1]; it proves further that the quasi-Hopf algebra $(H^* \#_{\sigma_\Phi, \tau_\Phi} H, \Phi)$ is quasi-triangular, generalizing [DPR] in which $H = kG$, a finite group Hopf algebra.

Suppose a matched pair $(F, G, \lhd, \rhd)$ of groups, where G is finite, is given. By the same way of proving Theorem 1.10, we have the following variation of the Kac exact sequence.

THEOREM 4.13. *We have an exact sequence*

$$1 \to X(G)^F \qquad\qquad \to H^1(F \bowtie G, k^\times) \to H^1(F, k^\times)$$
$$\to \mathrm{Aut}'(k^G \# kF, 1) \to H^2(F \bowtie G, k^\times) \to H^2(F, k^\times)$$
$$\to \mathrm{Opext}'(kF, k^G) \quad \to H^3(F \bowtie G, k^\times) \to H^3(F, k^\times),$$

where $X(G)^F$ *denotes the group of the group maps* $f : G \to k^\times$ *such that* $f(x \triangleleft a) = f(x)$ *for all* $x \in G$, $a \in F$.

PART II: HOPF ALGEBRA EXTENSIONS ARISING FROM LIE ALGEBRAS

In this part, $\mathfrak{f}$ and $\mathfrak{g}$ denote finite-dimensional Lie algebras. The characteristic ch k of k will be supposed to be zero in Sections 6–8.

5. Lie Bialgebra Extensions

We show some results for Lie (bi)algebras that are parallel to those for groups given in Section 1.

DEFINITION 5.1 [Mj, Def. 8.3.1]. A *matched pair of Lie algebras* is a pair $(\mathfrak{f}, \mathfrak{g})$ together with Lie module actions $\mathfrak{g} \xleftarrow{\triangleleft} \mathfrak{g} \otimes \mathfrak{f} \xrightarrow{\triangleright} \mathfrak{f}$ such that

$$x \triangleright [a, b] = [x \triangleright a, b] + [a, x \triangleright b] + (x \triangleleft a) \triangleright b - (x \triangleleft b) \triangleright a,$$
$$[x, y] \triangleleft a = [x, y \triangleleft a] + [x \triangleleft a, y] + x \triangleleft (y \triangleright a) - y \triangleleft (x \triangleright a)$$

for $a, b \in \mathfrak{f}$, $x, y \in \mathfrak{g}$, or equivalently such that the direct sum $\mathfrak{f} \oplus \mathfrak{g}$ of vector spaces forms a Lie algebra under the bracket

$$[a \oplus x, b \oplus y] = ([a, b] + x \triangleright b - y \triangleright a) \oplus ([x, y] + x \triangleleft b - y \triangleleft a).$$

This Lie bialgebra is denoted by $\mathfrak{f} \bowtie \mathfrak{g}$.

The universal envelope $U\mathfrak{f}$ of $\mathfrak{f}$ forms a cocommutative Hopf algebra in which each element in $\mathfrak{f}$ is primitive.

PROPOSITION 5.2 [M, Prop. 2.4]. *Actions* $\mathfrak{g} \xleftarrow{\triangleleft} \mathfrak{g} \otimes \mathfrak{f} \xrightarrow{\triangleright} \mathfrak{f}$ *which make* $(\mathfrak{f}, \mathfrak{g})$ *into a matched pair of Lie algebras are extended uniquely to actions* $U\mathfrak{g} \xleftarrow{\triangleleft} U\mathfrak{g} \otimes U\mathfrak{f} \xrightarrow{\triangleright} U\mathfrak{f}$ *which make* $(U\mathfrak{f}, U\mathfrak{g})$ *into a matched pair of Hopf algebras. The resulting Hopf algebra* $U\mathfrak{g} \bowtie U\mathfrak{f}$ *is naturally isomorphic to* $U(\mathfrak{g} \bowtie \mathfrak{f})$. *If* ch $k = 0$, *any matched pair structure on* $(U\mathfrak{f}, U\mathfrak{g})$ *is obtained in this way.*

The last assertion follows, since in characteristic zero, the primitives in $U\mathfrak{g} \bowtie U\mathfrak{f}$ are exactly $(\mathfrak{g} \otimes k) \oplus (k \otimes \mathfrak{f}) = \mathfrak{g} \oplus \mathfrak{f}$, which forms hence a Lie algebra, and so $(\mathfrak{f}, \mathfrak{g})$ is matched in such a way that the Lie algebra equals $\mathfrak{g} \bowtie \mathfrak{f}$.

Let $\rightharpoonup: \mathfrak{f} \otimes \mathfrak{g}^* \to \mathfrak{g}^*$ be an action by which $\mathfrak{g}^*$ is an $\mathfrak{f}$-Lie module. Then its transpose $\lhd: \mathfrak{g} \otimes \mathfrak{f} \to \mathfrak{g}$ makes $\mathfrak{g}$ into a (right) $\mathfrak{f}$-Lie module. Let $\rho : \mathfrak{f} \to \mathfrak{f} \otimes \mathfrak{g}^*$, $\rho(a) = \sum a_{[0]} \otimes a_{[1]}$ be a coaction by which $\mathfrak{f}$ is a $\mathfrak{g}^*$-Lie comodule. This is equivalent to that the action $\rhd: \mathfrak{g} \otimes \mathfrak{f} \to \mathfrak{f}$ defined by $x \rhd a = \sum a_{[0]} \langle x, a_{[1]} \rangle$ ($x \in \mathfrak{g}$, $a \in \mathfrak{f}$) makes $\mathfrak{f}$ into a $\mathfrak{g}$-Lie module.

DEFINITION 5.3. $(\mathfrak{f}, \mathfrak{g}^*, \rightharpoonup, \rho)$ is called a *Singer pair of Lie bialgebras* (though called a matched pair in [Mj, p.383]), if $(\mathfrak{f}, \mathfrak{g}, \lhd, \rhd)$ is a matched pair of Lie algebras.

A finite-dimensional vector space $\mathfrak{l}$ is called a *Lie coalgebra* with co-bracket δ : $\mathfrak{l} \to \mathfrak{l} \otimes \mathfrak{l}$, if the dual space $\mathfrak{l}^*$ is a Lie algebra with bracket $\delta^* : \mathfrak{l}^* \otimes \mathfrak{l}^* = (\mathfrak{l} \otimes \mathfrak{l})^* \to \mathfrak{l}^*$. $\mathfrak{l}$ is called a *Lie bialgebra* [Dr], if it is a Lie algebra and Lie coalgebra such that

$$\delta[a, b] = a\delta(b) + \delta(a)b \ (a, b \in \mathfrak{l}),$$

where $a(x \otimes y) = [a, x] \otimes y + x \otimes [a, y]$, $(x \otimes y)b = [x, b] \otimes y + x \otimes [y, b]$.

We regard $\mathfrak{f}$ as a Lie bialgebra with zero co-bracket. Naturally, $\mathfrak{g}^*$ is a Lie coalgebra, which we regard as a Lie bialgebra with zero bracket. By a *(Lie bialgebra) extension of $\mathfrak{f}$ by $\mathfrak{g}^*$*, we mean a sequence $(\mathfrak{l}) = \mathfrak{g}^* \to \mathfrak{l} \to \mathfrak{f}$ of Lie bialgebras and Lie bialgebra maps which is a short exact sequence of vector spaces. An *equivalence* between two such extensions is defined in the obvious way.

Given an $\mathfrak{f}$-Lie module action $\rightharpoonup: \mathfrak{f} \otimes \mathfrak{g}^* \to \mathfrak{g}^*$ together with a (Lie) 2-cocycle $\sigma : \mathfrak{f} \wedge \mathfrak{f} \to \mathfrak{g}^*$ with coefficients in the $\mathfrak{f}$-Lie module $(\mathfrak{g}^*, \rightharpoonup)$, a Lie algebra $\mathfrak{g}^* \rtimes_\sigma \mathfrak{f}$ of crossed sum is constructed on the vector space $\mathfrak{g}^* \oplus \mathfrak{f}$ by the bracket

$$[s \oplus a, t \oplus b] = (a \rightharpoonup t - b \rightharpoonup s + \sigma(a, b)) \oplus [a, b].$$

Given also a right $\mathfrak{g}$-Lie module action $\leftharpoonup: \mathfrak{f}^* \otimes \mathfrak{g} \to \mathfrak{f}^*$ together with a 2-cocycle $\tau : \mathfrak{g} \wedge \mathfrak{g} \to \mathfrak{f}^*$ with coefficients in $(\mathfrak{f}^*, \leftharpoonup)$, a Lie algebra $\mathfrak{g} \ltimes_\tau \mathfrak{f}^*$ is constructed similarly, whose dual Lie coalgebra is denoted by $\mathfrak{g}^* \blacktriangleright\!\!\lessdot_\tau \mathfrak{f}$. Denote by $\mathfrak{g}^* \blacktriangleright\!\!\blacktriangleleft_{\sigma,\tau} \mathfrak{f}$ the Lie algebra and Lie coalgebra thus obtained.

LEMMA 5.4 [M, Prop. 1.8]. $\mathfrak{g}^* \blacktriangleright\!\!\blacktriangleleft_{\sigma,\tau} \mathfrak{f}$ *is a Lie bialgebra if and only if* $\rightharpoonup$ *and the dual coaction* $\rho = (\leftharpoonup)^* : \mathfrak{f} \to \mathfrak{f} \otimes \mathfrak{g}^*$ *of* $\leftharpoonup$ *make* $(\mathfrak{f}, \mathfrak{g}^*)$ *into a Singer pair and* (σ, τ) *is a total 1-cocycle in the double complex* $C_0^{\cdot\cdot}$ *defined below.*

If these conditions are satisfied, the Lie bialgebra forms a Lie bialgebra extension

$$(\mathfrak{g}^* \blacktriangleright\!\!\blacktriangleleft_{\sigma,\tau} \mathfrak{f}) = \mathfrak{g}^* \to \mathfrak{g}^* \blacktriangleright\!\!\blacktriangleleft_{\sigma,\tau} \mathfrak{f} \to \mathfrak{f},$$

in which the maps are the natural inclusion and the projection.

Any extension $(\mathfrak{l})$ of $\mathfrak{f}$ by $\mathfrak{g}^*$ is equivalent to some $(\mathfrak{g}^* \blacktriangleright\!\!\blacktriangleleft_{\sigma,\tau} \mathfrak{f})$, since an identification $\mathfrak{l} = \mathfrak{g}^* \oplus \mathfrak{f}$ of vector spaces gives rise to a 'bicrossed sum' structure. Here the Singer pair $(\mathfrak{f}, \mathfrak{g}^*, \rightharpoonup, \rho)$ which forms $\mathfrak{g}^* \blacktriangleright\!\!\blacktriangleleft_{\sigma,\tau} \mathfrak{f}$ is uniquely determined by $(\mathfrak{l})$, being independent of the way of identification $\mathfrak{l} = \mathfrak{g}^* \oplus \mathfrak{f}$.

DEFINITION 5.5. In this case, we say that $(\mathfrak{l})$ *is associated with* the Singer pair $(\mathfrak{f}, \mathfrak{g}^*, \rightharpoonup, \rho)$. We denote by

$$\mathscr{O}\!pext\,(\mathfrak{f}, \mathfrak{g}^*) = \mathscr{O}\!pext\,(\mathfrak{f}, \mathfrak{g}^*, \rightharpoonup, \rho)$$

the groupoid of Lie bialgebra extensions associated with a fixed Singer pair $(\mathfrak{f}, \mathfrak{g}^*, \rightharpoonup, \rho)$, whose morphisms are equivalences.

In what follows we fix a Singer pair $(\mathfrak{f}, \mathfrak{g}^*, \rightharpoonup, \rho)$ of Lie bialgebras.

Let $(\mathfrak{l}_1)$, $(\mathfrak{l}_2)$ be in $\mathscr{O}\!pext\,(\mathfrak{f}, \mathfrak{g}^*)$. From the direct sum $(\mathfrak{l}_1 \oplus \mathfrak{l}_2)$, form first the pullback $(\mathfrak{l}')$ along the diagonal map $a \mapsto a \oplus a$, $\mathfrak{f} \to \mathfrak{f} \oplus \mathfrak{f}$, and then the pushout $(\mathfrak{l})$ along the addition $s \oplus t \mapsto s + t$, $\mathfrak{g}^* \oplus \mathfrak{g}^* \to \mathfrak{g}^*$, as follows.

$$
\begin{array}{ccccc}
(\mathfrak{l}_1 \oplus \mathfrak{l}_2) = \mathfrak{g}^* \oplus \mathfrak{g}^* & \rightarrow & \mathfrak{l}_1 \oplus \mathfrak{l}_2 & \rightarrow & \mathfrak{f} \oplus \mathfrak{f} \\
\| & & \uparrow & \text{p.b.} & \uparrow \\
(\mathfrak{l}') = \mathfrak{g}^* \oplus \mathfrak{g}^* & \rightarrow & \mathfrak{l}' & \rightarrow & \mathfrak{f} \\
\downarrow & \text{p.o.} & \downarrow & & \| \\
(\mathfrak{l}) = \mathfrak{g}^* & \rightarrow & \mathfrak{l} & \rightarrow & \mathfrak{f}
\end{array}
$$

One sees that $(\mathfrak{l})$ is in $\mathscr{O}\!pext\,(\mathfrak{f}, \mathfrak{g}^*)$, which we denote by $(\mathfrak{l}_1) * (\mathfrak{l}_2)$. If we form first pushout and then pullback, obtained is an extension of the same kind, which is equivalent to $(\mathfrak{l})$ through the isomorphism induced from the identity map on $\mathfrak{l}_1 \oplus \mathfrak{l}_2$. We see that $(\mathfrak{g}^* \bowtie_{\sigma_1, \tau_1} \mathfrak{f}) * (\mathfrak{g}^* \bowtie_{\sigma_2, \tau_2} \mathfrak{f}) = (\mathfrak{g}^* \bowtie_{\sigma, \tau} \mathfrak{f})$, where $\sigma = \sigma_1 + \sigma_2$, $\tau = \tau_1 + \tau_2$.

If σ and τ are both zero maps, we write simply $\mathfrak{g}^* \bowtie \mathfrak{f}$ for $\mathfrak{g}^* \bowtie_{0,0} \mathfrak{f}$.

PROPOSITION 5.6. $\mathscr{O}\!pext\,(\mathfrak{f}, \mathfrak{g}^*)$ *forms a symmetric monoidal groupoid with tensor product* $*$ *and unit object* $(\mathfrak{g}^* \bowtie \mathfrak{f})$, *in which each object has inverse.*

The associativity constraint and the symmetry are induced from the obvious isomorphisms $(\mathfrak{l}_1 \oplus \mathfrak{l}_2) \oplus \mathfrak{l}_3 \xrightarrow{\cong} \mathfrak{l}_1 \oplus (\mathfrak{l}_2 \oplus \mathfrak{l}_3)$, $\mathfrak{l}_1 \oplus \mathfrak{l}_2 \xrightarrow{\cong} \mathfrak{l}_2 \oplus \mathfrak{l}_1$, respectively. If we write $\mathfrak{l}_0 = \mathfrak{g}^* \bowtie \mathfrak{f}$, the projection $\mathfrak{l}_0 \oplus \mathfrak{l}_1 \to \mathfrak{l}_1$ induces the (left) unit constraint. If $(\mathfrak{l})$ is equivalent to $(\mathfrak{g}^* \bowtie_{\sigma, \tau} \mathfrak{f})$, it has inverse $(\mathfrak{g}^* \bowtie_{-\sigma, -\tau} \mathfrak{f})$.

We denote by

$$\mathrm{Opext}(\mathfrak{f}, \mathfrak{g}^*) = \mathrm{Opext}(\mathfrak{f}, \mathfrak{g}^*, \rightharpoonup, \rho)$$

all isomorphism (or equivalence) classes in $\mathscr{O}\!pext\,(\mathfrak{f}, \mathfrak{g}^*)$, which form naturally a group. We denote by $\mathrm{Aut}(\mathfrak{g}^* \bowtie \mathfrak{f})$ the group of auto-equivalences of $(\mathfrak{g}^* \bowtie \mathfrak{f})$. The K_0 and K_1 groups of $\mathscr{O}\!pext\,(\mathfrak{f}, \mathfrak{g}^*)$ are given by

$$K_0 = \mathrm{Opext}(\mathfrak{f}, \mathfrak{g}^*), \quad K_1 = \mathrm{Aut}(\mathfrak{g}^* \bowtie \mathfrak{f}).$$

For cohomological description of the groups, note first that by Definition 5.3, a matched pair $(\mathfrak{f}, \mathfrak{g}, \triangleleft, \triangleright)$ is obtained from the fixed Singer pair. Write $H = U\mathfrak{f}$,

$J = U\mathfrak{g}$. By Proposition 5.2, we have a matched pair $(H, J, \lhd, \rhd)$ of Hopf algebras so that $H \bowtie J = U(\mathfrak{f} \bowtie \mathfrak{g})$.

Let

$$V.(\mathfrak{f}) = 0 \leftarrow H \leftarrow H \otimes \mathfrak{f} \leftarrow H \otimes \wedge^2\mathfrak{f} \leftarrow \cdots$$

be the Chevalley-Eilenberg complex; the differentials are given by

$$\partial(u\langle a_1, \ldots, a_p\rangle) = \sum_{i=1}^{p}(-1)^{i+1}ua_i\langle a_1, \ldots, \hat{a}_i, \cdots, a_p\rangle$$

$$+ \sum_{i<j}(-1)^{i+j}u\langle[a_i, a_j], a_1, \ldots, \hat{a}_i, \ldots, \hat{a}_j, \ldots, a_p\rangle$$

for $u \in H$, $\langle a_1, \ldots, a_p\rangle := a_1 \wedge \cdots \wedge a_p \in \wedge^p\mathfrak{f}$, where $\hat{a}_i$ denotes the omitted term. This gives an H-free resolution $0 \leftarrow k \leftarrow V.(\mathfrak{f})$ of the trivial H-module k, whose augmentation $H \to k$ is the counit ε of H. Regard each $P = \wedge^p\mathfrak{f}$ as a J-module with the diagonal action. As a general fact, $H \otimes P$ is an $H \bowtie J$-module, where H acts on the factor H and J acts by

$$x(a \otimes p) = \sum(x_1 \rhd a_1) \otimes (x_2 \lhd a_2)p \quad (x \in J, \ a \otimes p \in H \otimes P).$$

It follows by [M, Lemma 2.6] that ∂ and ε are $H \bowtie J$-linear, where k is the trivial $H \bowtie J$-module.

Similarly the right version

$$V.'(\mathfrak{g}) = 0 \leftarrow J \leftarrow \mathfrak{g} \otimes J \leftarrow \wedge^2\mathfrak{g} \otimes J \leftarrow \cdots$$

of the Chevalley-Eilenberg complex gives a right $H \bowtie J$-resolution of k. Regard $0 \leftarrow k \leftarrow V.'(\mathfrak{g})$ as a left $H \bowtie J$-resolution by twisting the action through the antipode, and form the double complex $V.'(\mathfrak{g}) \otimes V.(\mathfrak{f})$ with a sign trick applied as before. Each term in the double complex is of the form $(Q \otimes J) \otimes (H \otimes P)$ with $P = \wedge^p\mathfrak{f}$, $Q = \wedge^q\mathfrak{g}$; this is an $H \bowtie J$-free module whose free basis is given by any basis of the vector space $(Q \otimes k) \otimes (k \otimes P)$, so that we have

$$\mathrm{Hom}_{H \bowtie J}((Q \otimes J) \otimes (H \otimes P), M) = \mathrm{Hom}(Q \otimes P, M)$$

for an $H \bowtie J$-module M. Thus the total complex of $V.'(\mathfrak{g}) \otimes V.(\mathfrak{f})$ gives a non-standard $H \bowtie J$-free resolution of k.

Form the double complex $\mathrm{Hom}_{H \bowtie J}(V.'(\mathfrak{g}) \otimes V.(\mathfrak{f}), k)$, and then remove from it the edges, which consist of the standard complexes for computing the Lie algebra cohomologies $H^{\cdot}(\mathfrak{f}, k)$, $H^{\cdot}(\mathfrak{g}, k)$. We obtain the desired complex:

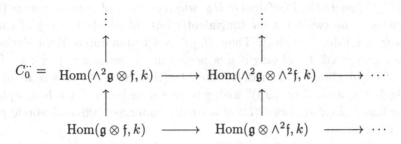

$$C_0^{\cdot\cdot} = \quad \mathrm{Hom}(\wedge^2\mathfrak{g} \otimes \mathfrak{f}, k) \longrightarrow \mathrm{Hom}(\wedge^2\mathfrak{g} \otimes \wedge^2\mathfrak{f}, k) \longrightarrow \cdots$$

$$\mathrm{Hom}(\mathfrak{g} \otimes \mathfrak{f}, k) \longrightarrow \mathrm{Hom}(\mathfrak{g} \otimes \wedge^2\mathfrak{f}, k) \longrightarrow \cdots$$

PROPOSITION 5.7 [M, Prop. 2.9].
(1) *Let* (σ, τ) *be a total 1-cocycle in* $C_0^{\cdot\cdot}$, *and identify* $\sigma : \mathfrak{g} \otimes \wedge^2\mathfrak{f} \to k$, $\tau :$ $\wedge^2\mathfrak{g} \otimes \mathfrak{f} \to k$ *naturally with linear maps* $\wedge^2\mathfrak{f} \to \mathfrak{g}^*$, $\wedge^2\mathfrak{g} \to \mathfrak{f}^*$, *respectively,* *which are indeed Lie 2-cocycles. Then,* $(\mathfrak{g}^* \bowtie_{\sigma, \tau} \mathfrak{f})$ *is in* $\mathcal{O}pext(\mathfrak{f}, \mathfrak{g}^*)$. *The* *assignment* $(\sigma, \tau) \mapsto (\mathfrak{g}^* \bowtie_{\sigma, \tau} \mathfrak{f})$ *induces an isomorphism*

$$H^1(\mathrm{Tot}\, C_0^{\cdot\cdot}) \cong \mathrm{Opext}(\mathfrak{f}, \mathfrak{g}^*).$$

(2) *Let* $\nu : \mathfrak{g} \otimes \mathfrak{f} \to k$ *be a total 0-cocycle in* $C_0^{\cdot\cdot}$, *and identify it naturally with* *a linear map* $\mathfrak{f} \to \mathfrak{g}^*$. *Then an auto-equivalence of* $(\mathfrak{g}^* \bowtie \mathfrak{f})$ *is given by* $s \oplus a \mapsto (s + \nu(a)) \oplus a$. *The assignment gives an isomorphism*

$$H^0(\mathrm{Tot}\, C_0^{\cdot\cdot}) \cong \mathrm{Aut}(\mathfrak{g}^* \bowtie \mathfrak{f}).$$

From the construction of $C_0^{\cdot\cdot}$, we see that the next result follows in the same way as in the group case.

THEOREM 5.8 [M, Thm. 2.10]. *We have an exact sequence*

$$0 \to H^1(\mathfrak{f} \bowtie \mathfrak{g}, k) \to H^1(\mathfrak{f}, k) \oplus H^1(\mathfrak{g}, k) \to \mathrm{Aut}(\mathfrak{g}^* \bowtie \mathfrak{f})$$
$$\to H^2(\mathfrak{f} \bowtie \mathfrak{g}, k) \to H^2(\mathfrak{f}, k) \oplus H^2(\mathfrak{g}, k) \to \mathrm{Opext}(\mathfrak{f}, \mathfrak{g}^*)$$
$$\to H^3(\mathfrak{f} \bowtie \mathfrak{g}, k) \to H^3(\mathfrak{f}, k) \oplus H^3(\mathfrak{g}, k),$$

where $H^{\cdot}$ *denotes the Lie algebra cohomology with coefficients in the trivial Lie* *module* k.

This exact sequence and the Hochschild-Serre spectral sequence imply the following.

COROLLARY 5.9 [M, Cor. 2.11]. *Suppose* $\mathrm{ch}\, k = 0$. *If either* (a) $\mathfrak{f}$ *is semisimple* *and* $\rightharpoonup$ *is zero or* (b) $\mathfrak{g}$ *is semisimple and* ρ *is zero, then the groups* $\mathrm{Opext}(\mathfrak{f}, \mathfrak{g}^*)$ *and* $\mathrm{Aut}(\mathfrak{g}^* \bowtie \mathfrak{f})$ *are both trivial.*

Throughout in the following sections we suppose $\mathrm{ch}\, k = 0$.

6. Hopf Algebra Extensions of $U\mathfrak{f}$ by the Full Dual $(U\mathfrak{g})^\circ$

Let $(U\mathfrak{g})^\circ$ denote the Hopf dual of $U\mathfrak{g}$, which consists of the elements in $(U\mathfrak{g})^*$ annihilating some two-sided (or equivalently one-sided) ideal in $U\mathfrak{g}$ of cofinite dimension; see [Mo, Sect. 9.1]. Thus, $(U\mathfrak{g})^\circ$ is a commutative Hopf algebra; it is finitely generated if and only if $\mathfrak{g}$ is *perfect* in the sense $\mathfrak{g} = [\mathfrak{g}, \mathfrak{g}]$; see [Ho2, p.261]. If k is algebraically closed, it follows by [Ho1, Sect. 3] that a commutative Hopf algebra is of the form $(U\mathfrak{g})^\circ$ with $\mathfrak{g}$ perfect if and only if it is isomorphic to the coordinate Hopf algebra $O(G)$ of a simply connected affine algebraic group G with $G = [G, G]$.

Given a Singer pair $(\mathfrak{f}, \mathfrak{g}^*, \rightharpoonup, \rho)$ of Lie bialgebras, we have by Definition 5.3 a matched pair $(\mathfrak{f}, \mathfrak{g}, \lhd, \rhd)$ of Lie algebras and so by Proposition 5.2 such a pair $(U\mathfrak{f}, U\mathfrak{g}, \lhd, \rhd)$ of Hopf algebras. Since the action $\rhd : U\mathfrak{g} \otimes U\mathfrak{f} \to U\mathfrak{f}$ makes $U\mathfrak{f}$ into a locally finite $U\mathfrak{g}$-module, it gives rise to a comodule coaction $\rho' : U\mathfrak{f} \to U\mathfrak{f} \otimes (U\mathfrak{g})^\circ$. By [M, Lemma 4.1], the transpose $\rightharpoonup' : U\mathfrak{f} \otimes (U\mathfrak{g})^* \to (U\mathfrak{g})^*$ of the other action $\lhd : U\mathfrak{g} \otimes U\mathfrak{f} \to U\mathfrak{g}$ stabilizes $(U\mathfrak{g})^\circ$, and the induced action $\rightharpoonup' : U\mathfrak{f} \otimes (U\mathfrak{g})^\circ \to (U\mathfrak{g})^\circ$ together with ρ' makes $(U\mathfrak{f}, (U\mathfrak{g})^\circ)$ into a Singer pair of Hopf algebras.

PROPOSITION 6.1 [M, Prop. 4.3]. *The assignment* $(\rightharpoonup, \rho) \mapsto (\rightharpoonup', \rho')$ *thus obtained gives an injection from the set of structures of Singer pair of Lie bialgebras on* $(\mathfrak{f}, \mathfrak{g}^*)$ *into the set of structures of such pair of Hopf algebras on* $(U\mathfrak{f}, (U\mathfrak{g})^\circ)$. *This is a bijection if* $\mathfrak{g}$ *is perfect.*

Write $K = (U\mathfrak{g})^\circ$, and let $\varpi : K^+ \to \mathfrak{g}^*$ denote the 'restriction' map induced from the inclusion $\mathfrak{g} \hookrightarrow U\mathfrak{g}$, which is a surjection by the Ado theorem. It follows from the proof of [M, Prop. 4.3] that a structure $(\rightharpoonup', \rho')$ arises from some $(\rightharpoonup, \rho)$ if and only if $\rightharpoonup'$ stabilizes the kernel $\operatorname{Ker} \varpi$ of ϖ, in which case the induced action $\mathfrak{f} \otimes \mathfrak{g}^* \to \mathfrak{g}^*$ is the desired $\rightharpoonup$. (Note that $\rightharpoonup'$ necessarily stabilizes K^+, since $\varepsilon(a \rightharpoonup' t) = \varepsilon(a)\varepsilon(t)$ by [T1, Lemma 1.2].) This is always the case if $\mathfrak{g}$ is perfect, since then (and only then) $\operatorname{Ker} \varpi = (K^+)^2$.

Let $(\mathfrak{f}, \mathfrak{g}^*, \rightharpoonup, \rho)$ be a Singer pair of Lie bialgebras. It gives rise to a Singer pair $(U\mathfrak{f}, (U\mathfrak{g})^\circ, \rightharpoonup', \rho')$ of Hopf algebras, as was seen above. Suppose that we are given a cleft extension (see Remark 2.6 (2))

$$(A) = (U\mathfrak{g})^\circ \xrightarrow{\iota} A \xrightarrow{\pi} U\mathfrak{f}$$

associated with the last pair. Regard A as a Lie algebra with the bracket $[a, b] = ab - ba$, as usual, and as a Lie coalgebra with the co-bracket $\delta(a) = \sum a_1 \otimes a_2 - \sum a_1 \otimes a_1$ (note that A is not necessarily a Lie bialgebra). Let L_A^+ denote the subspace of A consisting of the elements a such that $\varepsilon(a) = 0$ and $(1 \otimes \pi) \circ \Delta(a) = a \otimes 1 + 1 \otimes c$ for some $c \in \mathfrak{f}$. Then, L_A^+ is seen to be a Lie subalgebra and Lie subcoalgebra of A, and we have an extension $(L_A^+) = K^+ \xrightarrow{\iota} L_A^+ \xrightarrow{\pi} \mathfrak{f}$ of Lie algebras and at the same time of Lie coalgebras, where $K = (U\mathfrak{g})^\circ$. Since the restriction $\varpi : K^+ \to \mathfrak{g}^*$ is a surjection of Lie algebras and Lie coalgebras, we have an extension

$$(\mathfrak{l}_A) = \mathfrak{g}^* \to \mathfrak{l}_A \to \mathfrak{f},$$

where $\mathfrak{l}_A = L_A^+/\iota(\mathrm{Ker}\,\varpi)$. It is proved that $(\mathfrak{l}_A)$ is in $\mathscr{O}pext\,(\mathfrak{f},\mathfrak{g}^*,\to,\rho)$; see the paragraph preceding [M, Example 4.19].

THEOREM 6.2. *The assignment* $(A) \mapsto (\mathfrak{l}_A)$ *gives a symmetric monoidal equivalence*

$$\mathscr{O}pext\,(U\mathfrak{f},(U\mathfrak{g})^\circ,\rightharpoonup',\rho') \xrightarrow{\approx} \mathscr{O}pext\,(\mathfrak{f},\mathfrak{g}^*,\to,\rho).$$

Let (A_1), (A_2) be in $\mathscr{O}pext\,(U\mathfrak{f},(U\mathfrak{g})^\circ,\rightharpoonup',\rho')$, and form $A = A_1 \otimes_H^K A_2$. From the extension

$$K^+ \oplus K^+ \to L_{A_1}^+ \oplus L_{A_2}^+ \to \mathfrak{f} \oplus \mathfrak{f},$$

construct an extension $(L_{A_1}^+) * (L_{A_2}^+)$ of $\mathfrak{f}$ by K^+, as before, by forming first a pullback along the diagonal map $\mathfrak{f} \to \mathfrak{f} \oplus \mathfrak{f}$ and then a pushout along the addition $K^+ \oplus K^+ \to K^+$. One sees that $a \oplus b \mapsto a \otimes 1 + 1 \otimes b$, $L_{A_1}^+ \oplus L_{A_2}^+ \to A_1 \otimes A_2$ induces an equivalence $(L_{A_1}^+) * (L_{A_2}^+) \xrightarrow{\cong} (L_A)$, and it in turn induces $(\mathfrak{l}_{A_1}) * (\mathfrak{l}_{A_2}) \xrightarrow{\cong} (\mathfrak{l}_A)$; this gives the monoidal structure of the functor $(A) \mapsto (\mathfrak{l}_A)$.

We see easily that $(A) \mapsto (\mathfrak{l}_A)$ gives a symmetric monoidal functor, and so that it induces group maps

$$\kappa_0 : \mathrm{Opext}(U\mathfrak{f},(U\mathfrak{g})^\circ) \to \mathrm{Opext}(\mathfrak{f},\mathfrak{g}^*),$$
$$\kappa_1 : \mathrm{Aut}((U\mathfrak{g})^\circ \,\#\, U\mathfrak{f}) \to \mathrm{Aut}(\mathfrak{g}^* \bowtie \mathfrak{f})$$

between the K_0, K_1 groups, which will be explicitly described by Proposition 6.5. We remark that κ_0, κ_1 just given coincide respectively with κ_1, κ_0 in [M, Thm. 4.11]; thus the notations are reverse. Since κ_0 and κ_1 are isomorphisms by [M, Thm. 4.11], we see from the following standard fact that the functor is equivalence, which proves Theorem 6.2: a symmetric monoidal functor between symmetric monoidal groupoids in which each object has inverse is an equivalence, if and only if the induced maps between the mutual K_0, K_1 groups are both isomorphisms. □

Since κ_0 and κ_1 are isomorphisms, the sequence given in Theorem 5.8 remains exact if the groups $\mathrm{Opext}(\mathfrak{f},\mathfrak{g}^*)$, $\mathrm{Aut}(\mathfrak{g}^* \bowtie \mathfrak{f})$ are replaced by $\mathrm{Opext}(U\mathfrak{f},(U\mathfrak{g})^\circ)$, $\mathrm{Aut}((U\mathfrak{g})^\circ \,\#\, U\mathfrak{f})$, respectively. The exact sequence thus obtained cannot be covered by the generalized Kac exact sequence [Sb2] (see Remark 1.11 (3)), for which the kernel of extensions is supposed to be finite-dimensional, while we have the infinite-dimensional kernel $(U\mathfrak{g})^\circ$ unless $\mathfrak{g} = 0$.

Combined with Corollary 5.9, the isomorphisms prove also the following.

COROLLARY 6.3 [M, Cor. 4.13]. *The groups* $\mathrm{Opext}(U\mathfrak{f},(U\mathfrak{g})^\circ)$, $\mathrm{Aut}(U\mathfrak{f},(U\mathfrak{g})^\circ)$ *associated with a Singer pair* $(U\mathfrak{f},(U\mathfrak{g})^\circ,\to,\rho)$ *are both trivial, if either* (a) $\mathfrak{f}$ *is semisimple and* $\to$ *is trivial or* (b) $\mathfrak{g}$ *is semisimple and* ρ *is trivial.*

Combined with the last statement of Proposition 6.1, the theorem gives also the following corollary.

COROLLARY 6.4 (cf. [M, Thm. 4.14]). *If $\mathfrak{g}$ is perfect, we have a natural equivalence from the groupoid $\mathscr{E}\!xt\,(U\mathfrak{f},(U\mathfrak{g})^\circ)$ of all cleft extensions of $U\mathfrak{f}$ by $(U\mathfrak{g})^\circ$ to the groupoid $\mathscr{E}\!xt\,(\mathfrak{f},\mathfrak{g}^*)$ of all Lie bialgebra extensions of $\mathfrak{f}$ by $\mathfrak{g}^*$.*

As in Theorem 6.2, let $(U\mathfrak{f},(U\mathfrak{g})^\circ,\rightharpoonup',\rho')$ be a Singer pair which arises from such a pair $(\mathfrak{f},\mathfrak{g}^*,\rightharpoonup,\rho)$, and consider the double complexes $A_0^{\cdot\cdot}$, $C_0^{\cdot\cdot}$ defined by these pairs. Through the isomorphisms given by Propositions 2.21 and 5.7, let us regard κ_0, κ_1 as isomorphisms

$$\kappa_{1-n} : H^n(\mathrm{Tot}\,A_0^{\cdot\cdot}) \overset{\cong}{\longrightarrow} H^n(\mathrm{Tot}\,C_0^{\cdot\cdot})\ (n=0,1).$$

PROPOSITION 6.5.

(1) *Let (σ,τ) be a total 1-cocycle in $A_0^{\cdot\cdot}$ consisting of $\sigma : U\mathfrak{f} \otimes U\mathfrak{f} \to (U\mathfrak{g})^\circ$ and $\tau : U\mathfrak{f} \to (U\mathfrak{g})^\circ \otimes (U\mathfrak{g})^\circ$. Then the linear maps $\bar\sigma : \mathfrak{f} \wedge \mathfrak{f} \to \mathfrak{g}^*$, $\bar\tau : \mathfrak{f} \to (\mathfrak{g} \wedge \mathfrak{g})^*$ determined by*

$$\langle \bar\sigma(a \wedge b), x \rangle = \langle \sigma(a,b) - \sigma(b,a), x \rangle,$$
$$\langle \bar\tau(a), x \wedge y \rangle = \langle \tau(a), x \otimes y - y \otimes x \rangle,$$

where $a,b \in \mathfrak{f}$, $x,y \in \mathfrak{g}$, form a total 1-cocycle $(\bar\sigma,\bar\tau)$ in $C_0^{\cdot\cdot}$ (under the natural identification as in Proposition 5.7). The isomorphism κ_0 is induced by $(\sigma,\tau) \mapsto (\bar\sigma,\bar\tau)$.

(2) *Let $\nu : U\mathfrak{f} \to (U\mathfrak{g})^\circ$ be a total 0-cocycle in $A_0^{\cdot\cdot}$. Then the linear map $\bar\nu : \mathfrak{f} \to \mathfrak{g}^*$ determined by*

$$\langle \bar\nu(a), x \rangle = \langle \nu(a), x \rangle,$$

where $a \in \mathfrak{f}$, $x \in \mathfrak{g}$, is a total 0-cocycle in $C_0^{\cdot\cdot}$, and κ_1 is given by $\nu \mapsto \bar\nu$.

The result of Part 1 is given by the second paragraph following [M, Remark 4.15]. Part 2 is seen more easily.

7. Hopf Algebra Extensions of $U\mathfrak{f}$ by the Irreducible $(U\mathfrak{g})'$ with $\mathfrak{g}$ Nilpotent

Let $(U\mathfrak{g})'$ denote the largest irreducible subcoalgebra, necessarily a Hopf subalgebra, of $(U\mathfrak{g})^\circ$ containing 1, which consists of the elements in $(U\mathfrak{g})^*$ annihilating some power of $(U\mathfrak{g})^+$; see [Mo, Def. 9.2.2]. The next proposition follows from [Ho2, Thm. XVI 4.2].

PROPOSITION 7.1. *For a commutative Hopf algebra K, the following conditions are equivalent:*

(a) *K is of the form $(U\mathfrak{g})'$, where $\mathfrak{g}$ is nilpotent;*
(b) *K is the coordinate Hopf algebra $O(G)$ of a unipotent affine algebraic group G;*
(c) *K is finitely generated as an algebra and irreducible as a coalgebra, and the intersection of all ideals in K of codimension 1 is zero (the last condition can be removed if k is algebraically closed).*

In what follows in this section, we suppose that $\mathfrak{g}$ is nilpotent. We will show parallel results of those in the preceding section, replacing $(U\mathfrak{g})^{\circ}$ by $(U\mathfrak{g})'$.

PROPOSITION 7.2. *There is a natural 1-1 correspondence between* (a) *the set of structures* $(\rightharpoonup, \rho)$ *of Singer pair of Lie bialgebras on* $(\mathfrak{f}, \mathfrak{g}^*)$ *such that the action* $\rhd: \mathfrak{g} \otimes \mathfrak{f} \to \mathfrak{f}$ *corresponding to* ρ *is nilpotent* (*in the sense* $\mathfrak{g}^l \rhd \mathfrak{f} = 0$ *for some integer* $l > 0$) *and* (b) *the set of structures* $(\rightharpoonup', \rho')$ *of Singer pair of Hopf algebras on* $(U\mathfrak{f}, (U\mathfrak{g})')$.

PROOF. Let $(\rightharpoonup, \rho)$ be in (a), and let $(\lhd, \rhd)$ be the corresponding matched pair structure on $(\mathfrak{f}, \mathfrak{g})$, which gives rise to a matched pair structure on $(U\mathfrak{f}, U\mathfrak{g})$, by Proposition 5.2. Since $\rhd: \mathfrak{g} \otimes \mathfrak{f} \to \mathfrak{f}$ is nilpotent, $\rhd: U\mathfrak{g} \otimes U\mathfrak{f} \to U\mathfrak{f}$ is locally nilpotent. Since locally nilpotent module-actions $U\mathfrak{g} \otimes U\mathfrak{f} \to U\mathfrak{f}$ are in natural 1-1 correspondence with comodule coactions $U\mathfrak{f} \to U\mathfrak{f} \otimes (U\mathfrak{g})'$, a comodule coaction ρ' arises from the last $\rhd$. Suppose $\mathfrak{g}^l \rhd \mathfrak{f} = 0$, and write $I = (U\mathfrak{g})^+$. It follows by induction on n $(\geq l)$ that $I^{n-1} \lhd \mathfrak{f} \subset I^{n-l}$, and so that the transpose $\rightharpoonup': U\mathfrak{f} \otimes (U\mathfrak{g})^* \to (U\mathfrak{g})^*$ of $\lhd$ satisfies $\mathfrak{f} \rightharpoonup' (U\mathfrak{g}/I^{n-l})^* \subset (U\mathfrak{g}/I^{n-1})^*$; this implies that $\rightharpoonup'$ stabilizes $(U\mathfrak{g})'$. One sees as in the proof of [M, Lemma 4.1] that the pair $(\rightharpoonup', \rho')$, where $\rightharpoonup'$ is the restricted action on $(U\mathfrak{g})'$, is in (b).

Let $(\rightharpoonup', \rho')$ be in (b). Reversing the procedure, we obtain a coaction $\rho: \mathfrak{f} \to \mathfrak{f} \otimes \mathfrak{g}^*$ from ρ'. Write $K = (U\mathfrak{g})'$. Since it follows from [Ho2, Thm. XVI 4.2] that the restriction map $K^+ \to \mathfrak{g}^*$ induces an isomorphism $K^+/(K^+)^2 \cong \mathfrak{g}^*$, one sees as in the proof of [M, Prop. 4.3] that $\rightharpoonup'$ induces an action $\rightharpoonup: \mathfrak{f} \otimes \mathfrak{g}^* \to \mathfrak{g}^*$, and that $(\rightharpoonup, \rho)$ is in (a).

We see easily that the correspondences thus defined are inverses of each other.
□

Choose such $(\rightharpoonup, \rho)$, $(\rightharpoonup', \rho')$ respectively from (a), (b) that correspond to each other. Let (A) be in $\mathcal{O}pext(U\mathfrak{f}, (U\mathfrak{g})', \rightharpoonup', \rho')$. Since the restriction $K^+ \to \mathfrak{g}^*$ is a surjection as was seen above, we can construct $(\mathfrak{l}_A)$ in $\mathcal{O}pext(\mathfrak{f}, \mathfrak{g}^*, \rightharpoonup, \rho)$ in the same way as in the preceding section.

THEOREM 7.3. *Suppose* $\mathfrak{g}$ *is nilpotent. The assignment* $(A) \mapsto (\mathfrak{l}_A)$ *gives a symmetric monoidal equivalence*

$$\mathcal{O}pext(U\mathfrak{f}, (U\mathfrak{g})', \rightharpoonup', \rho') \xrightarrow{\approx} \mathcal{O}pext(\mathfrak{f}, \mathfrak{g}^*, \rightharpoonup, \rho).$$

A proof will be given in the next section in a more general context.

The induced isomorphisms between the K_0, K_1 groups together with Corollary 5.9 prove the following.

COROLLARY 7.4. *The groups* Opext$(U\mathfrak{f}, (U\mathfrak{g})')$, Aut$((U\mathfrak{g})' \# U\mathfrak{f})$ *associated with a Singer pair* $(U\mathfrak{f}, (U\mathfrak{g})', \rightharpoonup, \rho)$ *are both trivial if* $\mathfrak{f}$ *is semisimple,* $\mathfrak{g}$ *is nilpotent and* $\rightharpoonup$ *is trivial.*

EXAMPLE 7.5. Let $\mathfrak{f} = ka$, $\mathfrak{g} = kx$ be 1-dimensional (abelian) Lie algebras. Then

$$U\mathfrak{f} = k[a], \quad U\mathfrak{g} = k[x], \quad (U\mathfrak{g})' = k[t],$$

polynomial Hopf algebras with a, x, t primitive. Here t is the element in $(U\mathfrak{g})^*$ determined by $\langle t, x^n \rangle = \delta_{1,n}$ $(n = 0, 1, \ldots)$; see [Mo, Example 9.1.7]. For arbitrary ξ, η in k, the actions $\lhd$, $\rhd$ determined by

$$x \lhd a = \xi x, \quad x \rhd a = \eta a$$

make $(\mathfrak{f}, \mathfrak{g})$ matched. The action $\rhd$ is nilpotent if and only if $\eta = 0$. The action $\rightharpoonup : k[a] \otimes k[t] \rightarrow k[t]$ arising from $\lhd$ is determined by

$$a \rightharpoonup t^n = n\xi t^n \ (n = 0, 1, \ldots).$$

Hence this $\rightharpoonup$ together with the trivial coaction $k[a] \rightarrow k[a] \otimes k[t]$ gives all possible Singer pair structures on $(k[a], k[t])$. Moreover, for each such pair the group $\mathrm{Opext}(k[a], k[t])$ is trivial since obviously $H^1(\mathrm{Tot}\, C_0^{\cdot\cdot}) = 0$. Thus the equivalence classes of all cleft extensions of $k[a]$ by $k[t]$ are in 1-1 correspondence with the elements ξ in k.

EXAMPLE 7.6. Let $\mathfrak{sl}_2$ denote the special linear algebra of 2×2 matrices, with standard (Chevalley) basis x, y, h. The decomposition

$$\mathfrak{sl}_2 = \mathfrak{f} \oplus \mathfrak{g}$$

into the Lie subalgebras $\mathfrak{f} := kh + ky$ and $\mathfrak{g} := kx$ gives rise to a matched pair structure on $(\mathfrak{f}, \mathfrak{g})$, in which the action $\rhd$ is nilpotent. We have $(U\mathfrak{g})' = k[t]$ as above, so $\langle t^m, x^n/n! \rangle = \delta_{m,n}$. The matched pair structure corresponds to the Singer pair structure on $(U\mathfrak{f}, (U\mathfrak{g})')$ given by

$$\frac{y^l}{l!} h^m \rightharpoonup t^n = \begin{cases} \binom{n-1}{l} (-1)^{l+m} 2^m (n-l)^m t^{n-l} & (l < n) \\ 0 & (l \geq n), \end{cases}$$

$$\rho\left(\frac{y^l}{l!} h^m\right) = \sum_{i=0}^{l} \frac{y^{l-i}}{(l-i)!} \binom{h-l+i}{i} h^m \otimes t^i,$$

where l, m and n are non-negative integers, since we see from [Hum, Lemma 26.2] that in $U(\mathfrak{sl}_2)$,

$$\frac{x^n}{n!} \frac{y^l}{l!} h^m = \sum_{i=0}^{\min(l,n)} \frac{y^{l-i}}{(l-i)!} \binom{h-l-n+2i}{i} (h - 2(n-i))^m \frac{x^{n-i}}{(n-i)!}.$$

By using the Kac exact sequence we compute

$$\mathrm{Opext}(U\mathfrak{f}, (U\mathfrak{g})') \cong H^3(\mathfrak{sl}_2, k) = k,$$

since it is easy to see $H^n(\mathfrak{f}, k) = H^n(\mathfrak{g}, k) = 0$ for $n = 2, 3$.

REMARK 7.7. In their new paper [VV], Vaes and Vainerman study the same subject as this Part II in the framework of operator algebras, and especially give many examples. The computation of an Opext group given in [VV, Remark 5.7] (without detailed proof) seems essentially the same as our result in the preceding example.

8. Generalization by Introducing Topology

To unify and generalize the results in the preceding two sections, we let $\mathcal{I}$ be a set of (two-sided) ideals in $U\mathfrak{g}$ of cofinite dimension which satisfies the following conditions.

(i) For any $I_1, I_2 \in \mathcal{I}$, there exists $I \in \mathcal{I}$ such that $I \subset I_1 \cap I_2$;

(ii) For any $I \in \mathcal{I}$, there exists $I' \in \mathcal{I}$ such that $\Delta(I') \subset U\mathfrak{g} \otimes I + I \otimes U\mathfrak{g}$, $S(I') \subset I$;

(iii) There exists $I \in \mathcal{I}$ such that $\varepsilon(I) = 0$.

DEFINITION 8.1. Let $(U\mathfrak{g})^\circ_{\mathcal{I}}$ denote the subset of $(U\mathfrak{g})^*$ consisting of the elements f such that $f(I) = 0$ for some $I \in \mathcal{I}$. We see easily that this is a Hopf subalgebra of $(U\mathfrak{g})^\circ$.

EXAMPLE 8.2. (1) Suppose that $\mathcal{I}$ consists of only one ideal $(U\mathfrak{g})^+$. Then, $(U\mathfrak{g})^\circ_{\mathcal{I}} = k$.

(2) Suppose that $\mathcal{I}$ consists of all ideals of cofinite dimension. Then, $(U\mathfrak{g})^\circ_{\mathcal{I}} = (U\mathfrak{g})^\circ$.

(3) Suppose that $\mathcal{I}$ consists of the powers $I, I^2, \cdots$ of $I = (U\mathfrak{g})^+$. Then, $(U\mathfrak{g})^\circ_{\mathcal{I}} = (U\mathfrak{g})'$.

By a *topological vector space* [T2, p.507], we mean a vector space V with a topology such that for each $w \in V$, the translation $v \mapsto v + w$, $V \to V$ is continuous, and V has a basis of neighborhoods of 0 consisting of vector subspaces, which we call a *topological basis*.

Every vector space is a topological vector space with the discrete topology. We regard k as a discrete topological vector space.

The direct sum $\bigoplus_\lambda V_\lambda$ of topological vector spaces V_λ is a topological vector space with the direct sum topology [T2, 1.2]; it has a topological basis consisting of all $\bigoplus_\lambda W_\lambda$, where W_λ is in a fixed topological basis of V_λ. For a vector space X, we denote by $V \otimes (X)$ or $(X) \otimes V$ the topological vector space $V \otimes X$ or $X \otimes V$ which is given the direct sum topology, identified with the direct sum of $\dim X$ copies of V.

By Condition (i), $U\mathfrak{g}$ is a topological vector space with topological basis $\mathcal{I}$. Conditions (ii) and (iii) are equivalent to that the structure maps Δ, S and ε are continuous, where $U\mathfrak{g} \otimes U\mathfrak{g}$ is given the (tensor product) topology with topological basis consisting of all $U\mathfrak{g} \otimes I + I \otimes U\mathfrak{g}$ with $I \in \mathcal{I}$. We see that $(U\mathfrak{g})^\circ_{\mathcal{I}}$ equals the vector space $\mathrm{Hom}_c(U\mathfrak{g}, k)$ of continuous linear maps $U\mathfrak{g} \to k$.

We regard $U\mathfrak{f}$ as a discrete vector space, and $U\mathfrak{g} \otimes (U\mathfrak{f})$ as a topological vector space so as defined above.

THEOREM 8.3. *Fix a Singer pair* $(\mathfrak{f}, \mathfrak{g}^*, \rightharpoonup, \rho)$ *of Lie bialgebras, which by definition corresponds to a matched pair* $(\mathfrak{f}, \mathfrak{g}, \lhd, \rhd)$ *of Lie algebras. Suppose that the unique extensions*

$$U\mathfrak{g} \xleftarrow{\lhd} U\mathfrak{g} \otimes (U\mathfrak{f}) \xrightarrow{\rhd} U\mathfrak{f}$$

of the actions $\mathfrak{g} \xleftarrow{\lhd} \mathfrak{g} \otimes \mathfrak{f} \xrightarrow{\rhd} \mathfrak{f}$ *which make* $(U\mathfrak{f}, U\mathfrak{g})$ *matched* (*see Proposition 5.2*) *are both continuous.*

(1) *The extended actions give rise to an action* $\rightharpoonup': U\mathfrak{f} \otimes (U\mathfrak{g})^\circ_\mathcal{I} \to (U\mathfrak{g})^\circ_\mathcal{I}$ *and a coaction* $\rho' : U\mathfrak{f} \to U\mathfrak{f} \otimes (U\mathfrak{g})^\circ_\mathcal{I}$ *such that* $(U\mathfrak{f}, (U\mathfrak{g})_\mathcal{I})^\circ, \rightharpoonup', \rho')$ *is a Singer pair.*
(2) *There exist natural group maps*

$$\kappa_0 : \mathrm{Opext}(U\mathfrak{f}, (U\mathfrak{g})^\circ_\mathcal{I}) \to \mathrm{Opext}(\mathfrak{f}, \mathfrak{g}^*),$$

$$\kappa_1 : \mathrm{Aut}((U\mathfrak{g})^\circ_\mathcal{I} \# U\mathfrak{f}) \to \mathrm{Aut}(\mathfrak{g}^* \bowtie \mathfrak{f})$$

between the K_0, K_1 *groups of* $\mathcal{O}\!pext(U\mathfrak{f}, (U\mathfrak{g})^\circ_\mathcal{I}, \rightharpoonup', \rho')$ *and of* $\mathcal{O}\!pext(\mathfrak{f}, \mathfrak{g}^*, \rightharpoonup, \rho)$.
(3) *Suppose* $H^1(\mathfrak{g}, (U\mathfrak{g})^\circ_\mathcal{I}) = 0$, *where* $(U\mathfrak{g})^\circ_\mathcal{I}$ *is a left* (*or equivalently right*) $\mathfrak{g}$-*Lie module with the transposed action of the right* (*or left*) *multiplication on* $U\mathfrak{g}$. *Then,* κ_1 *is an isomorphism.*
(4) *If in addition* $H^2(\mathfrak{g}, (U\mathfrak{g})^\circ_\mathcal{I}) = 0$, *then* κ_0 *is also an isomorphism.*
(5) *If in addition* $I \cap \mathfrak{g} = 0$ *for some* $I \in \mathcal{I}$, *there exists a symmetric monoidal equivalence*

$$\mathcal{O}\!pext\,(U\mathfrak{f}, (U\mathfrak{g})^\circ_\mathcal{I}, \rightharpoonup', \rho') \xrightarrow{\approx} \mathcal{O}\!pext\,(\mathfrak{f}, \mathfrak{g}^*, \rightharpoonup, \rho)$$

which induces the isomorphisms κ_0, κ_1.

REMARK 8.4. We see that the following conditions including the assumption in Part 5 above are equivalent to each other.
(a) $I \cap \mathfrak{g} = 0$ for some $I \in \mathcal{I}$;
(b) $\bigcap_{I \in \mathcal{I}} I = 0$, or the topological space $U\mathfrak{g}$ is Hausdorff;
(c) The restriction map $(U\mathfrak{g})^\circ_\mathcal{I} \to \mathfrak{g}^*$ is a surjection;
(d) The canonical algebra map $U\mathfrak{g} \to [(U\mathfrak{g})^\circ_\mathcal{I}]^*$ is an injection.

To prove Part 1 of the theorem, we generalize the situation as follows. Let H, J be cocommutative Hopf algebras. Suppose we are given a set $\mathcal{I}$ of ideals in J of cofinite dimension which satisfies the same conditions as (i)–(iii) given above for $U\mathfrak{g}$; J is thus a topological vector space with topological basis $\mathcal{I}$. Let $K = J^\circ_\mathcal{I}$ denote the commutative Hopf algebra consisting of all continuous linear maps $J \to k$. We regard H as a discrete vector space. Suppose we are given also continuous actions

$$J \xleftarrow{\lhd} J \otimes (H) \xrightarrow{\rhd} H$$

which make (H, J) matched.

Note that $H \otimes K = \mathrm{Hom}_c(J, H)$, the subspace of $\mathrm{Hom}(J, H)$ consisting of all continuous linear maps $J \to H$. Since $\triangleright$ is continuous, the image of the linear map $\rho : H \to \mathrm{Hom}(J, H)$ defined by

$$\rho(a)(x) = x \triangleright a \ (a \in H, \ x \in J)$$

is included in $H \otimes K$, so that we have a coaction $\rho : H \to H \otimes K$. Since $\triangleleft$ is continuous, $\triangleleft \, a : J \to J$ is continuous for each $a \in H$. By applying $\mathrm{Hom}_c(\ , k)$, we obtain a map $a \rightharpoonup : K \to K$, and hence also an action $\rightharpoonup : H \otimes K \to K$.

LEMMA 8.5. $(H, K, \rightharpoonup, \rho)$ *forms a Singer pair.*

This follows easily as in the proof of [M, Lemma 4.1]. If J is Hausdorff, the correspondence $(\triangleleft, \triangleright) \mapsto (\rightharpoonup, \rho)$ between the sets of structures, given as above, is injective since then $\triangleleft$ is recovered from the transpose of $\rightharpoonup$ through the canonical injection $J \to K^*$; see the proof of [M, Cor. 4.2].

Part 1 of Theorem 8.3 follows from the last lemma. $\square$

Let $(H, J, \triangleleft, \triangleright)$, $(H, K, \rightharpoonup, \rho)$ be as above. From the matched pair, the Hopf algebra $H \bowtie J$ and its module category $H \bowtie J\text{-Mod}$ are constructed. From the Singer pair, the category $\mathcal{C} = \mathcal{C}(H, K, \rightharpoonup, \rho)$ is defined by Definition 2.3.

Generalizing the observation given above that the continuous action $\triangleright$ gives rise to ρ, we see that for a discrete vector space M, there is a natural 1-1 correspondence between the continuous module actions $J \otimes (M) \to M$ and the comodule coactions $M \to M \otimes K$. This proves the following.

LEMMA 8.6. $\mathcal{C}$ *is regarded as a full subcategory of* $H \bowtie J\text{-Mod}$ *which consists of the* $H \bowtie J$-*modules* M *such that the restricted action* $J \otimes (M) \to M$ *by* J *is continuous.*

Recall from (2.17), (2.19) the (co)simplicial objects $\Phi.(k)$, $\Psi^{\cdot}(k)$ in $\mathcal{C}$. The normalized (co)chain complexes associated with these objects coincide with the standard (co)free resolutions of k, if we forget K-coactions or H-actions. By removing the 0th terms H, K from them, we obtain (co)chain complexes

$$X.(H) = 0 \leftarrow H \otimes H_+ \leftarrow H \otimes H_+^{\otimes 2} \leftarrow \cdots,$$
$$Y^{\cdot}(K) = 0 \to K^+ \otimes K \to K^{+\otimes 2} \otimes K \to \cdots$$

in $\mathcal{C}$, where $H_+ = H/k1$.

Let Hom_H^K denote the vector space of H-linear and K-colinear maps. Form the double complex $B_0^{\cdot\cdot} = \mathrm{Hom}_H^K(X.(H), Y^{\cdot}(K))$. Here and in what follows when we form a double complex, we resort such a sign trick that changes the sign of differentials in even columns beginning with the 0th column. Note that each term in $X.(H)$ (resp., in $Y^{\cdot}(K)$) is of the form $\mathbb{F}(P)$ as given in (2.16) (resp., $\mathbb{G}(Q)$ as given in (2.18)). Since we have a natural isomorphism

$$\mathrm{Hom}_H^K(\mathbb{F}(P), \mathbb{G}(Q)) \cong \mathrm{Hom}(P, Q) \qquad (8.7)$$

given in the same way of (2.20), $B_0^{\cdot\cdot}$ turns to be as follows.

$$
B_0^{\cdot\cdot} = \quad
\begin{array}{ccccc}
\vdots & & \vdots & & \\
\uparrow & & \uparrow & & \\
\operatorname{Hom}(H_+, K^{+\otimes 2}) & \longrightarrow & \operatorname{Hom}(H_+^{\otimes 2}, K^{+\otimes 2}) & \longrightarrow & \cdots \\
\uparrow & & \uparrow & & \\
\operatorname{Hom}(H_+, K^+) & \longrightarrow & \operatorname{Hom}(H_+^{\otimes 2}, K^+) & \longrightarrow & \cdots
\end{array}
$$

PROPOSITION 8.8 [M, Prop. 3.14]. *Suppose $H = U\mathfrak{f}$ ($\mathfrak{f}$ can be of infinite dimension). An isomorphism $A_0^{\cdot\cdot} \cong B_0^{\cdot\cdot}$ between the double complexes of abelian groups is given by*

$$
\log : \operatorname{Reg}_+(H^{\otimes p}, K^{\otimes q}) \to \operatorname{Hom}(H_+^{\otimes p}, K^{+\otimes q}), \quad \log f = \sum_{n=1}^{\infty} \frac{(-1)^{n-1}}{n}(f - \varepsilon)^n,
$$

where ε is the identity in $\operatorname{Reg}_+(H^{\otimes p}, K^{\otimes q})$ and $(f - \varepsilon)^n = (f - \varepsilon)\cdots(f - \varepsilon)$ denotes the convolution product.

To prove Part 2 of Theorem 8.3, suppose in particular $H = U\mathfrak{f}$, $J = U\mathfrak{g}$, and so $K = (U\mathfrak{g})_{\mathcal{I}}^{\circ}$. We may suppose that the Singer pair $(H, K, \rightharpoonup, \rho)$ given by Lemma 8.5 arises from the Singer pair $(\mathfrak{f}, \mathfrak{g}^*, \rightharpoonup, \rho)$ of Lie bialgebras fixed in the theorem. Here we continue to denote the structure of the pair (H, K) by $\rightharpoonup$, ρ, instead of $\rightharpoonup'$, ρ'. The Singer pairs define the double complexes $A_0^{\cdot\cdot}$, $C_0^{\cdot\cdot}$. We will define natural group maps

$$
\kappa_{1-n} : H^n(\operatorname{Tot} A_0^{\cdot\cdot}) \to H^n(\operatorname{Tot} C_0^{\cdot\cdot}) \ (n = 0, 1),
$$

which will prove Part 2 by Propositions 2.21 and 5.7.

We obtain from the Chevalley-Eilenberg complex $V_\cdot(\mathfrak{f})$, by removing its term H, a chain complex

$$
X_\cdot(\mathfrak{f}) = 0 \leftarrow H \otimes \mathfrak{f} \leftarrow H \otimes \wedge^2 \mathfrak{f} \leftarrow \cdots
$$

in $H \bowtie J$-Mod, and so in $\mathcal{C}$ by Lemma 8.6. The well-known embedding $\varphi_\cdot : X_\cdot(\mathfrak{f}) \to X_\cdot(H)$ given by

$$
\varphi_{p-1} : X_{p-1}(\mathfrak{f}) = H \otimes \wedge^p \mathfrak{f} \to H \otimes H_+^{\otimes p} = X_{p-1}(H),
$$

$$
\varphi_{p-1}(u\langle a_1, \ldots, a_p \rangle) = \sum_{\sigma \in \mathfrak{S}_p} (\operatorname{sgn} \sigma) u \otimes \overline{a}_{\sigma(1)} \otimes \cdots \otimes \overline{a}_{\sigma(p)}
$$

is a map of complexes in $\mathcal{C}$; see [M, Lemma 4.8]. The symmetric argument using a mirror gives chain complexes

$$
X'_\cdot(J) = 0 \leftarrow J_+ \otimes J \leftarrow J_+^{\otimes 2} \otimes J \leftarrow \cdots,
$$

$$
X'_\cdot(\mathfrak{g}) = 0 \leftarrow \mathfrak{g} \otimes J \leftarrow \wedge^2 \mathfrak{g} \otimes J \leftarrow \cdots
$$

of right $H \bowtie J$-modules and an embedding $\varphi'_\cdot : X'_\cdot(\mathfrak{g}) \to X'_\cdot(J)$. Through the injection $(K^+)^{\otimes q} \otimes K \hookrightarrow (J_+)^{*\otimes q} \otimes J^* \subset (J_+^{\otimes q} \otimes J)^*$ induced from the inclusion $K \hookrightarrow J^*$, we can regard $Y^\cdot(K)$ as a subcomplex of the dual complex $X'_\cdot(J)^*$ in $H \bowtie J$-Mod.

In general, if Q is a right H-module of finite dimension, the right J-module $Q \otimes J$ is a right $H \bowtie J$-module, where H acts by

$$(q \otimes x)a = \sum q(x_1 \rhd a_1) \otimes (x_2 \lhd a_2) \ (a \in H, \ q \otimes x \in Q \otimes J),$$

so that $(Q \otimes J)^* = Q^* \otimes J^*$ is a left $H \bowtie J$-module. The object $\mathbb{G}(Q^*) = Q^* \otimes K$ in $\mathcal{C}$ given by (2.18) is the largest $H \bowtie J$-submodule of $Q^* \otimes J^*$ which is an object in $\mathcal{C}$. Therefore we have a cochain complex in $\mathcal{C}$

$$Y^\cdot(\mathfrak{g}^*) = 0 \to \mathfrak{g}^* \otimes K \to (\wedge^2 \mathfrak{g})^* \otimes K \to \cdots$$

which is a subcomplex of the dual complex $X'_\cdot(\mathfrak{g})^*$ in $H \bowtie J$-Mod.

LEMMA 8.9. $Y^\cdot(\mathfrak{g}^*)$ *is naturally isomorphic to the complex obtained by removing the term K from the standard complex*

$$0 \to K \to \mathrm{Hom}(\mathfrak{g}, K) \to \mathrm{Hom}(\wedge^2 \mathfrak{g}, K) \to \cdots$$

for computing the cohomology $H^\cdot(\mathfrak{g}, K)$ with coefficients in the right $\mathfrak{g}$-Lie module K.

PROOF.   The canonical isomorphism $\mathrm{Hom}_J(Q \otimes J, J^*) \cong Q^* \otimes J^*$ induces $\mathrm{Hom}_J(Q \otimes J, K) \cong Q^* \otimes K$, which gives rise to an isomorphism between the complexes. $\square$

Form the double complex $\mathrm{Hom}_H^K(X_\cdot(\mathfrak{f}), Y^\cdot(\mathfrak{g}^*))$, in which each term is of the form $\mathrm{Hom}(\wedge^p \mathfrak{f}, (\wedge^q \mathfrak{g})^*)$ by (8.7).

LEMMA 8.10. *The double complex just formed is naturally identified with $C_0^{\cdot\cdot}$.*

PROOF. Write $P = \wedge^p \mathfrak{f}, Q = \wedge^q \mathfrak{g}$. The natural maps

$$\mathrm{Hom}(H \otimes P, Q^* \otimes K) \hookrightarrow \mathrm{Hom}(H \otimes P, (Q \otimes J)^*) \cong \mathrm{Hom}((Q \otimes J) \otimes (H \otimes P), k)$$

are $H \bowtie J$-linear, where $H \bowtie J$ acts on the Hom spaces by conjugation. By taking $H \bowtie J$-invariants we obtain $\mathrm{Hom}(P, Q^*) \cong \mathrm{Hom}(Q \otimes P, k)$, which gives a natural identification between the double complexes. $\square$

From the dual of the embedding $\varphi'_\cdot : X'_\cdot(\mathfrak{g}) \to X'_\cdot(J)$, a map $\psi^\cdot : Y^\cdot(K) \to Y^\cdot(\mathfrak{g}^*)$ of complexes in $\mathcal{C}$ is induced. Define a map of double complexes by

$$\alpha_0^{\cdot\cdot} = \mathrm{Hom}_H^K(\varphi_\cdot, \psi^\cdot) : B_0^{\cdot\cdot} \to C_0^{\cdot\cdot}.$$

Compose this with the isomorphism $A_0^{\cdot\cdot} \xrightarrow{\cong} B_0^{\cdot\cdot}$ given by Proposition 8.8 to obtain a map $A_0^{\cdot\cdot} \to C_0^{\cdot\cdot}$ of double complexes. As desired maps, we define $\kappa_{1-n} : H^n(\mathrm{Tot}\, A_0^{\cdot\cdot}) \to H^n(\mathrm{Tot}\, C_0^{\cdot\cdot})$ $(n = 0, 1)$ to be the induced maps between the cohomology groups. This proves Part 2. $\square$

We see that the maps just defined are so as described by Proposition 6.5 if $(U\mathfrak{g})^\circ$ is replaced by $(U\mathfrak{g})^\circ_{\mathcal{I}}$.

To prove Parts 3 and 4 of Theorem 8.3, it suffices to show that the maps $H^n(\operatorname{Tot} B_0^{..}) \to H^n(\operatorname{Tot} C_0^{..})$ $(n = 0, 1)$ induced by $\alpha_0^{..}$ are isomorphisms.

Form the double complexes

$$B^{..} = \operatorname{Hom}(X.(H), Y^.(K)), \quad C^{..} = \operatorname{Hom}(X.(\mathfrak{f}), Y^.(\mathfrak{g}^*))$$

in $H \bowtie J$-Mod, which look as follows.

$$
\begin{array}{c}
\vdots \\
\uparrow \\
B^{..} = \quad \operatorname{Hom}(H \otimes H_+, K^{+\otimes 2} \otimes K) \longrightarrow \qquad \vdots \cdots \\
\uparrow \qquad\qquad\qquad\qquad\qquad \uparrow \\
\operatorname{Hom}(H \otimes H_+, K^+ \otimes K) \longrightarrow \operatorname{Hom}(H \otimes H_+^{\otimes 2}, K^+ \otimes K) \longrightarrow \cdots
\end{array}
$$

$$
\begin{array}{c}
\vdots \\
\uparrow \\
C^{..} = \quad \operatorname{Hom}(H \otimes \mathfrak{f}, (\wedge^2 \mathfrak{g})^* \otimes K) \longrightarrow \qquad \vdots \cdots \\
\uparrow \qquad\qquad\qquad\qquad\qquad \uparrow \\
\operatorname{Hom}(H \otimes \mathfrak{f}, \mathfrak{g}^* \otimes K) \longrightarrow \operatorname{Hom}(H \otimes \wedge^2 \mathfrak{f}, \mathfrak{g}^* \otimes K) \longrightarrow \cdots
\end{array}
$$

Their subcomplexes of $H \bowtie J$-invariants are precisely $B_0^{..}$ and $C_0^{..}$, respectively. The map

$$\alpha^{..} = \operatorname{Hom}(\varphi., \psi^.) : B^{..} \to C^{..}$$

is restricted to $\alpha_0^{..}$. Denote by

$$(B^., d^.) = \operatorname{Tot} B^{..}, \quad (C^., \partial^.) = \operatorname{Tot} C^{..}, \quad \alpha^. : B^. \to C^.$$

the two total complexes and the total map of $\alpha^{..}$. Then we have the following commutative diagram in $H \bowtie J$-Mod:

$$
\begin{array}{ccccccccccc}
0 & \longrightarrow & B^0 & \xrightarrow{d^0} & B^1 & \xrightarrow{d^1} & \operatorname{Im} d^1 & \longrightarrow & 0 & \longrightarrow & 0 & \longrightarrow & \cdots \\
& & \downarrow{\alpha^0} & & \downarrow{\alpha^1} & & \downarrow{\alpha^2} & & \downarrow & & \downarrow & & \\
0 & \longrightarrow & C^0 & \xrightarrow{\partial^0} & C^1 & \xrightarrow{\partial^1} & \operatorname{Im} \partial^1 & \longrightarrow & 0 & \longrightarrow & 0 & \longrightarrow & \cdots
\end{array}
\qquad (8.11)
$$

We suppose $H^1(\mathfrak{g}, K) = 0 = H^2(\mathfrak{g}, K)$, and claim the following.

CLAIM 8.12. $\alpha^0, \alpha^1, \alpha^2, 0, \ldots$ *give a homotopy equivalence between the complexes in* $H \bowtie J$-Mod.

If this is proved, we see by taking $H \bowtie J$-invariants that κ_0 and κ_1 are isomorphisms, which proves Part 4. Under the assumption $H^1(\mathfrak{g}, K) = 0$, Part 3 will be proved by similar argument with the diagram (8.11) replaced by the reduced one involving $d^0 : B^0 \to \operatorname{Im} d^0$, $\partial^0 : C^0 \to \operatorname{Im} \partial^0$ in its rows.

Since we have resolutions $0 \leftarrow H^+ \leftarrow X.(H)$ and $0 \to K_+ \to Y^\cdot(K)$ in $\mathcal{C}$, the first row in (8.11) gives a resolution of $\operatorname{Hom}(H^+, K_+)$ in $H \bowtie J$-Mod. Here the augmentations are given by the product $H \otimes H_+ \to H^+$, the coproduct $K_+ \to K^+ \otimes K$ and the map $e : \operatorname{Hom}(H^+, K_+) \to B^0$ induced by them. By Lemma 8.9 together with the assumption of cohomologies vanishing, one forms an exact sequence in $\mathcal{C}$ by splicing the first three terms $\mathfrak{g}^* \otimes K \to (\wedge^2 \mathfrak{g})^* \otimes K \to (\wedge^3 \mathfrak{g})^* \otimes K$ in $Y^\cdot(\mathfrak{g}^*)$ with the injection $0 \to K_+ \to \mathfrak{g}^* \otimes K$ induced by the coproduct of K. Since also the product $H \otimes \mathfrak{f} \to H^+$ makes $0 \leftarrow H^+ \leftarrow X.(\mathfrak{f})$ into a resolution in $\mathcal{C}$, the second row in (8.11) gives again a resolution of $\operatorname{Hom}(H^+, K_+)$ in $H \bowtie J$-Mod, whose augmentation $\eta : \operatorname{Hom}(H^+, K_+) \to C^0$ are induced from the last injection and the product. Clearly we have $\eta = \alpha^0 \circ e$.

To prove Claim 8.12, return to the general situation given after Remark 8.4, in which we are given a matched pair $(H, J, \lhd, \rhd)$ of cocommutative Hopf algebras with continuous actions.

DEFINITION 8.13 (cf. [M, Def. 6.8]). We define a category $\mathcal{D}$ as follows. An object in $\mathcal{D}$ is an $H \bowtie J$-module M, and so in particular an H- and J-module, such that

(a) M is a topological vector space with topological basis consisting of J-sub-modules,
(b) The action $J \otimes (M) \to M$ is continuous and
(c) The action $(H) \otimes M \to M$ is continuous.

A morphism in $\mathcal{D}$ is a continuous $H \bowtie J$-linear map.

One sees that $\mathcal{D}$ is a k-additive category. Let M be an object in $\mathcal{D}$, and suppose $N \subset M$ is an $H \bowtie J$-submodule. Then, N and M/N are objects in $\mathcal{D}$ respectively with the sub- and the quotient topologies (cf. [M, Prop. 6.9]), so that any morphism in $\mathcal{D}$ has kernel and cokernel. However, $\mathcal{D}$ is not abelian in general, since a monomorphism (an epimorphism) is not necessarily a (co)kernel.

Recall that the Singer pair $(H, K = J_{\mathcal{I}}^\circ, \rightharpoonup, \rho)$ arising from $(H, J, \lhd, \rhd)$ defines the category $\mathcal{C}$; see Definition 2.3.

LEMMA 8.14. $\mathcal{C}$ *is regarded as a full subcategory of* $\mathcal{D}$ *which consists of the discrete objects.*

This is shown similarly by the idea to prove Lemma 8.6.

For discrete vector spaces V and W, we regard $\mathrm{Hom}(V, W)$ as a topological vector space with topological basis consisting of $\mathrm{Hom}(V/V_\lambda, W)$, where V_λ ranges over all finite-dimensional subspaces of V. If M and N are in $\mathcal{C}$, we see that $\mathrm{Hom}(M, N)$ is an object in $\mathcal{D}$ with the conjugate action by $H \bowtie J$; cf. [M, Lemma 6.10 1)]. Thus we have a k-linear functor

$$\mathrm{Hom}(\ , \ ) : \mathcal{C}^{op} \times \mathcal{C} \to \mathcal{D},$$

so that (8.11) is a commutative diagram in $\mathcal{D}$. In particular for the objects $\mathbb{F}(P)$, $\mathbb{G}(Q)$ in $\mathcal{C}$ given by (2.16), (2.18), $L := \mathrm{Hom}(\mathbb{F}(P), \mathbb{G}(Q))$ is an object in $\mathcal{D}$. This behaves like as an injective object as seen below.

PROPOSITION 8.15. *Suppose that a morphism* $f : M \to N$ *in* $\mathcal{D}$ *is strict in the sense that the* $H \bowtie J$-*linear isomorphism* $M/\mathrm{Ker}\, f \xrightarrow{\cong} \mathrm{Im}\, f$ *induced from* f *is a homeomorphism. Then for any* $g : M \to L$ *in* $\mathcal{D}$ *with* $g(\mathrm{Ker}\, f) = 0$, *there exists* $h : N \to L$ *in* $\mathcal{D}$ *such that* $g = h \circ f$.

This is essentially the same as [M, Cor. 6.14].

Let us return to the diagram (8.11). Since one sees as in the proof of [M, Lemma 6.15] that the differentials d^0, ∂^0 and the augmentations e, η are strict, the familiar argument for uniqueness of injective resolution proves that $\alpha^0, \alpha^1, \alpha^2, 0, \ldots$ give a homotopy equivalence between the complexes in $\mathcal{D}$, so in $H \bowtie J$-Mod. This proves Claim 8.12, and so Part 4 of Theorem 8.3. $\square$

To prove Part 5 of the theorem, suppose that $I \cap \mathfrak{g} = 0$ for some $I \in \mathcal{I}$, or equivalently that the restriction $(U\mathfrak{g})^\circ_{\mathcal{I}} \to \mathfrak{g}^*$ is a surjection; see Remark 8.4. Then we construct as in Section 6 the symmetric monoidal functor $(A) \mapsto (\mathfrak{l}_A)$, $\mathcal{O}pext\,(U\mathfrak{f}, (U\mathfrak{g})^\circ_{\mathcal{I}}) \to \mathcal{O}pext\,(\mathfrak{f}, \mathfrak{g}^*)$. Since we see that the induced maps between the K_0, K_1 groups coincide with κ_0, κ_1, it follows that the functor is equivalent if (and only if) these maps are isomorphisms; see the proof of Theorem 6.2. This proves Part 5, and completes the proof of Theorem 8.3. $\square$

Let us see that the theorem just proved implies Theorems 6.2 and 7.3. For this, it suffices to prove that the cohomologies H^1 and H^2 vanish when $\mathcal{I}$ is as in (2) or (3) in Example 8.2 (and $\mathfrak{g}$ is nilpotent in the latter case).

To see H^1 vanishes, note that in either case, $\mathcal{I}$ satisfies the following condition which is stronger than (i).

(i') For any $I_1, I_2 \in \mathcal{I}$, there exists $I \in \mathcal{I}$ such that $I \subset I_1 I_2$.

Then the desired result follows from the next proposition.

PROPOSITION 8.16. *If* $\mathcal{I}$ *satisfies Condition* (i'), *then we have* $H^1(\mathfrak{g}, (U\mathfrak{g})^\circ_{\mathcal{I}}) = 0$.

PROOF. Write $K = (U\mathfrak{g})^\circ_{\mathcal{I}}$. Recalling $H^1(\mathfrak{g}, K) = \mathrm{Ext}^1_{U\mathfrak{g}}(k, K)$ by definition, we will prove that any short exact sequence $0 \to K \to M \to k \to 0$ of $U\mathfrak{g}$-modules splits. Note that K is injective as a K-comodule. Then we have only to prove that the discrete $U\mathfrak{g}$-module M is *continuous* in the sense that the action

$U\mathfrak{g} \otimes (M) \to M$ is continuous, since then the short exact sequence is that of K-comodules and hence splits. Given $U\mathfrak{g}$-modules $N \subset M$, we will prove that, if N and M/N are continuous, then M is, too. We may suppose M is finitely generated. Then so is N as well as M/N, since $U\mathfrak{g}$ is noetherian. Since M/N is continuous, $I_2 M \subset N$ for some $I_2 \in \mathcal{I}$. Since N is continuous, $I_1 N = 0$ for some $I_1 \in \mathcal{I}$. Take $I \in \mathcal{I}$ such that $I \subset I_1 I_2$. Then, $IM \subset I_1 I_2 M \subset I_1 N = 0$, so that M is continuous. $\qquad\qquad\qquad\qquad\qquad\qquad\qquad\qquad\qquad\qquad\qquad\qquad\quad\square$

The H^2 vanishes since we have the following.

THEOREM 8.17. (1) (Schneider [M, Thm. 5.2]) $H^2(\mathfrak{g}, (U\mathfrak{g})^\circ) = 0$.

(2) (Koszul [Kos, Thm. 6]) *If $\mathfrak{g}$ is nilpotent, $H^n(\mathfrak{g}, (U\mathfrak{g})') = 0$ for $n > 0$.*

We remark that $H^3(\mathfrak{g}, (U\mathfrak{g})^\circ) \neq 0$ if $\mathfrak{g}$ is semisimple; see [M, Remark 5.9].

NOTE 8.18. Suppose (H, K) is a Singer pair (Definition 2.2) with K finite-dimensional. Recall from Section 2 the double cosimplicial abelian group

$$\mathrm{Reg}_H^K(\Phi_\cdot(k), \Psi^\cdot(k)),$$

and let $D^{\cdot\cdot}$ denote the associated, normalized double complex. After this paper was accepted, I found explicit homotopy equivalences between the total complex $\mathrm{Tot}\, D^{\cdot\cdot}$ and the standard complex for computing the Sweedler cohomology $H^n(H \bowtie K^*, k)$. Obtained as a biproduct is a direct, homological proof of the generalized Kac exact sequence due to Schauenburg; see Remark 1.11 (3). The results are contained in my preprint "Cohomology and coquasi-bialgebra extensions associated to a matched pair of bialgebras".

References

[B] H. Bass, *"Algebraic K-theory"*, Benjamin, New York, 1968.

[BP] D. Bulacu and F. Panaite, *A generalization of the quasi-Hopf algebra $D^\omega(G)$*, Comm. Algebra 26(1998), 4125–4141.

[DPR] R. Dijkgraaf, V. Pasquier and P. Roche, *Quasi-Hopf algebras, group cohomology and orbifold models*, Nuclear Phys. B Proc. Suppl. 18B(1990), 60–72.

[D1] Y. Doi, *Cohomologies over commutative Hopf algebras*, J. Math. Soc. Japan 25(1973), 680–706.

[D2] Y. Doi, *Braided bialgebras and quadratic bialgebras*, Comm. Algebra 21(1993), 1731–1749.

[Dr] V.G. Drinfeld, *Quantum groups*, Proc. ICM-86, Berkeley, 1987, 798–820.

[EV] M. Enock and L. Vainerman, *Deformation of a Kac algebra by an abelian subgroup*, Commun. Math. Phys. 178(1996), 571–596.

[HM] P.M. Hajac and T. Masuda, *Quantum double-torus*, C.R. Acad. Sci. Paris 327(1998), 553–558.

[Ho1] G. Hochschild, *Algebraic groups and Hopf algebras*, Illinois J. Math. 14(1970), 52–65.

[Ho2] G. Hochschild, *"Basic Theory of Algebraic Groups and Lie Algebras"*, Graduate Texts in Math. 75, Springer-Verlag, New York, 1981.

[H] I. Hofstetter, *Extensions of Hopf algebras and their cohomological description*, J. Algebra 164(1994), 264–298.

[Hum] J. Humphreys, *"Introduction to Lie algebras and Representation Theory"*, Graduate Texts in Math. 9, Springer-Verlag, New York, 1980.

[IK] M. Izumi and H. Kosaki, *Kac algebras arising from composition of subfactors: general theory and classification*, Memoirs Amer. Math. Soc., to appear.

[K] G.I. Kac, *Extensions of groups to ring groups*, Math. USSR Sbornik 5(1968), 451–474.

[Ka] Y. Kashina, *On the classification of semisimple Hopf algebras of dimension 16*, J. Algebra 232(2000), 617–663.

[Kas] C. Kassel, *"Quantum Groups"*, Graduate Texts in Math. 155, Springer-Verlag, New-York, 1995.

[Kos] J. Koszul, *Sur les modules de représentations des algèbras de Lie résolubles*, Amer. J. Math. 76(1954), 535–554.

[L] M. Landstad, *Quantization arising from abelian subgroups*, Int. J. Math. 5(1994), 897–936.

[Mac] S. MacLane, *"Homology"*, Grundlehren der math. Wiss. 114, Springer-Verlag, Berlin, 1975.

[Mj] S. Majid, *"Foundations of Quantum Group Theory"*, Cambridge Univ. Press, Cambridge, 1995.

[M1] A. Masuoka, *Self-dual Hopf algebras of dimension p^3 obtained by extension*, J. Algebra 178(1995), 791–806.

[M2] A. Masuoka, *Calculations of some groups of Hopf algebra extensions*, J. Algebra 191(1997), 568–588; Corrigendum, J. Algebra 197(1997), 656.

[M3] A. Masuoka, *Faithfully flat forms and cohomology of Hopf algebra extensions*, Comm. Algebra 25(1997), 1169–1197.

[M4] A. Masuoka, *"Extensions of Hopf algebras" (lecture notes taken by Matías Graña)*, Notas Mat. No. 41/99, FaMAF Uni. Nacional de Córdoba, 1999.

[M] A. Masuoka, *Extensions of Hopf algebras and Lie bialgebras*, Trans. Amer. Math. Soc. 352(2000), 3837–3879.

[MD] A. Masuoka and Y. Doi, *Generalization of cleft comodule algebras*, Comm. Algebra 20(1992), 3703–3721.

[Mo] S. Montgomery, *"Hopf Algebras and Their Actions on Rings"*, CBMS Reginal Conference Series in Math. Vol. 82, Amer. Math. Soc., Providence, 1993.

[N] S. Natale, *On semisimple Hopf algebras of dimension pq^2*, J. Algebra 221(1999), 242–278.

[Sb1] P. Schauenburg, *Hopf bigalois extensions*, Comm. Algebra 24(1996), 3797–3825.

[Sb2] P. Schauenburg, *Hopf bimodules, coquasibialgebras, and an exact sequence of Kac*, Adv. in Math., to appear.

[Sb3] P. Schauenburg, *The structure of Hopf algebras with a weak projection*, Algebras and Representation Theory 3(1999), 187–211.

[Sb4] P. Scahuenburg, *Hopf algebra extensions and monoidal categories*, in this volume.

[Sd] H.-J. Schneider, *A normal basis and transitivity of crossed products for Hopf algebras*, J. Algebra 152(1992), 289–312.

[S] W.M. Singer, *Extension theory for connected Hopf algebras*, J. Algebra 21(1972), 1–16.

[Sw1] M.E. Sweedler, *Cohomology of algebras over Hopf algebras*, Trans. Amer. Math. Soc. 127(1968), 205–239.

[Sw2] M.E. Sweedler, *"Hopf Algebras"*, Benjamin, New York, 1969.

[T1] M. Takeuchi, *Matched pairs of groups and bismash products of Hopf algebras*, Comm. Algebra 9(1981), 841–882.

[T2] M. Takeuchi, *Topological coalgebras*, J. Algebra 87(1985), 505–539.

[VV] S. Vaes and L. Vainerman, *Extensions of locally compact quantum groups and the bicrossed product construction*, preprint 2001-2, Max-Planck-Institut für Mathematik, Bonn, 2001; math.QA/0101133.

[W] C.A. Weibel, *"An Introduction to Homological Algebra"*, Cambridge Univ. Press, Cambridge, 1994.

AKIRA MASUOKA
INSTITUTE OF MATHEMATICS
UNIVERSITY OF TSUKUBA
TSUKUBA, IBARAKI 305-8571
JAPAN
akira@math.tsukuba.ac.jp

New Directions in Hopf Algebras
MSRI Publications
Volume **43**, 2002

Finite Quantum Groupoids
and Their Applications

DMITRI NIKSHYCH AND LEONID VAINERMAN

ABSTRACT. We give a survey of the theory of finite quantum groupoids (weak Hopf algebras), including foundations of the theory and applications to finite depth subfactors, dynamical deformations of quantum groups, and invariants of knots and 3-manifolds.

CONTENTS

Nikshych thanks P. Etingof for numerous stimulating discussions on quantum groupoids and to MIT for their hospitality during his visit. Vainerman is grateful to M. Enock, V. Turaev and J.-M. Vallin for many valuable discussions and to l'Université Louis Pasteur (Strasbourg) for their kind hospitality during his work on this survey.

1. Introduction

By *quantum groupoids* we understand weak Hopf algebras introduced in [BNSz], [BSz1], [N]. These objects generalize Hopf algebras (in fact, many Hopf-algebraic concepts can be extended to the quantum groupoid case) and usual finite groupoids (see [NV1] for discussion). Every quantum groupoid has two canonical subalgebras that play the same role as the space of units in a groupoid and projections on these subalgebras generalizing the source and target maps in a groupoid. We use the term "quantum groupoid" instead of "weak Hopf algebra" to stress this similarity that leads to many interesting constructions and examples.

Our initial motivation for studying quantum groupoids in [NV1], [NV2],[N1] was their connection with depth 2 von Neumann subfactors, first mentioned in [O2], which was also one of the main topics of [BNSz], [BSz1], [BSz2], [NSzW]. It was shown in [NV2] that quantum groupoids naturally arise as non-commutative symmetries of subfactors, namely if $N \subset M \subset M_1 \subset M_2 \subset \cdots$ is the Jones tower constructed from a finite index depth 2 inclusion $N \subset M$ of II_1 factors, then $H = M' \cap M_2$ has a canonical structure of a quantum groupoid acting outerly on M_1 such that $M = M_1^H$ and $M_2 = M_1 \rtimes H$; moreover $\widehat{H} = N' \cap M_1$ is a quantum groupoid dual to H.

In [NV3] we extended this result to show that quantum groupoids give a description of arbitrary finite index and finite depth II_1 subfactors via a Galois correspondence and explained how to express subfactor invariants such as bimodule categories and principal graphs in quantum groupoid terms. Thus, in this respect quantum groupoids play the same role as Ocneanu's paragroups [O1], [D].

Quantum groupoids also appear naturally in the theory of dynamical deformations of quantum groups [EV], [Xu]. It was shown in [EN] that for every simple Lie algebra $\mathfrak{g}$ the corresponding Drinfeld–Jimbo quantum group $U_q(\mathfrak{g})$ for q a primitive root of unity has a family of dynamical twists that give rise to self-dual finite quantum groupoids. To every such a twist one can associate a solution of the quantum dynamical Yang–Baxter equation [ES].

As in the case of Hopf algebras, the representation category $\mathrm{Rep}(H)$ of a quantum groupoid H is a monoidal category with duality. Existence of additional (quasitriangular, ribbon, or modular) structures on H makes $\mathrm{Rep}(H)$, respectively, braided, ribbon, or modular [NTV]. It is well known [RT], [T], that such categories give rise to invariants of knots and 3-manifolds.

The survey is organized as follows.

In Section 2 we give basic definitions and examples of finite quantum groupoids, discuss their fundamental properties (following [BNSz] and [NV1]) and relations with other versions of quantum groupoids.

The exposition of the theory of integrals in Section 3 follows [BNSz] and is similar to the Hopf algebra case [M]. The main results here are extensions of the fundamental theorem for Hopf modules and the Maschke theorem. The latter gives an equivalence between existence of normalized integrals and semi-simplicity of a quantum groupoid.

In Section 4 we define the notions of action and smash product and extend the well known Blattner–Montgomery duality theorem for Hopf algebra actions to quantum groupoids (see [N2]).

In Section 5 we describe monoidal structure and duality on the representation category of a quantum groupoid ([BSz2] and [NTV]). Then we distinguish special cases of quantum groupoids leading to braided, ribbon, and modular categories of representations.

We also develop the Drinfeld double construction and emphasize the role of factorizable quantum groupoids in producing modular categories (see [NTV]). Section 6 contains an extension of the Drinfeld twisting procedure to quantum groupoids and construction of a quantum groupoid via a *dynamical twisting* of a Hopf algebra. Important concrete examples of such twistings are *dynamical quantum groups* at roots of 1 [EN].

Semisimple and C^*-quantum groupoids are analyzed in Section 7. Following [BNSz], our discussion is concentrated around the existence of the Haar integral in connection with a special implementation of the square of the antipode.

Sections 8 and 9 are devoted to the description of finite index an finite depth subfactors via crossed products of II_1 factors with C^*-quantum groupoids. These results were obtained in [NV2], [NV3] and partially in [N1], [NSzW].

2. Definitions and Examples

In this section we give definitions and discuss basic properties of finite quantum groupoids. Most of the material presented here can be found in [BNSz] and [NV1].

Throughout this paper we use Sweedler's notation for comultiplication, writing $\Delta(b) = b_{(1)} \otimes b_{(2)}$. Let k be a field.

2.1. Definition of the quantum groupoid. By *quantum groupoids* or *weak Hopf algebras* we understand the objects introduced in [BNSz], [BSz1] as a generalization of ordinary Hopf algebras.

DEFINITION 2.1.1. A *(finite) quantum groupoid* over k is a finite dimensional k-vector space H with the structures of an associative algebra $(H, m, 1)$ with multiplication $m : H \otimes_k H \to H$ and unit $1 \in H$ and a coassociative coalgebra (H, Δ, ε) with comultiplication $\Delta : H \to H \otimes_k H$ and counit $\varepsilon : H \to k$ such that:

(i) The comultiplication Δ is a (not necessarily unit-preserving) homomorphism of algebras such that

$$(\Delta \otimes \mathrm{id})\Delta(1) = (\Delta(1) \otimes 1)(1 \otimes \Delta(1)) = (1 \otimes \Delta(1))(\Delta(1) \otimes 1), \qquad (2\text{--}1)$$

(ii) The counit is a k-linear map satisfying the identity:

$$\varepsilon(fgh) = \varepsilon(fg_{(1)})\varepsilon(g_{(2)}h) = \varepsilon(fg_{(2)})\varepsilon(g_{(1)}h), \qquad (2\text{-}2)$$

for all $f, g, h \in H$.

(iii) There is a linear map $S : H \to H$, called an *antipode*, such that, for all $h \in H$,

$$m(\mathrm{id} \otimes S)\Delta(h) = (\varepsilon \otimes \mathrm{id})(\Delta(1)(h \otimes 1)), \qquad (2\text{-}3)$$

$$m(S \otimes \mathrm{id})\Delta(h) = (\mathrm{id} \otimes \varepsilon)((1 \otimes h)\Delta(1)), \qquad (2\text{-}4)$$

$$m(m \otimes \mathrm{id})(S \otimes \mathrm{id} \otimes S)(\Delta \otimes \mathrm{id})\Delta(h) = S(h). \qquad (2\text{-}5)$$

A quantum groupoid is a Hopf algebra if and only if the comultiplication is unit-preserving if and only if the counit is a homomorphism of algebras.

A *morphism* between quantum groupoids H_1 and H_2 is a map $\alpha : H_1 \to H_2$ which is both algebra and coalgebra homomorphism preserving unit and counit and which intertwines the antipodes of H_1 and H_2, i.e., $\alpha \circ S_1 = S_2 \circ \alpha$. The image of a morphism is clearly a quantum groupoid. The tensor product of two quantum groupoids is defined in an obvious way.

2.2. Counital maps and subalgebras. The linear maps defined in (2–3) and (2–4) are called *target* and *source counital maps* (see examples below for explanation of the terminology) and denoted ε_t and ε_s respectively:

$$\varepsilon_t(h) = (\varepsilon \otimes \mathrm{id})(\Delta(1)(h \otimes 1)), \qquad \varepsilon_s(h) = (\mathrm{id} \otimes \varepsilon)((1 \otimes h)\Delta(1)). \qquad (2\text{-}6)$$

In the next proposition we collect several useful properties of the counital maps.

PROPOSITION 2.2.1. *For all $h, g \in H$,*

(i) *counital maps are idempotents in* $\mathrm{End}_k(H)$:

$$\varepsilon_t(\varepsilon_t(h)) = \varepsilon_t(h), \qquad \varepsilon_s(\varepsilon_s(h)) = \varepsilon_s(h);$$

(ii) *the relations between ε_t, ε_s, and comultiplication are*

$$(\mathrm{id} \otimes \varepsilon_t)\Delta(h) = 1_{(1)}h \otimes 1_{(2)}, \qquad (\varepsilon_s \otimes \mathrm{id})\Delta(h) = 1_{(1)} \otimes h1_{(2)};$$

(iii) *the images of counital maps are characterized by*

$$h = \varepsilon_t(h) \text{ if and only if } \Delta(h) = 1_{(1)}h \otimes 1_{(2)},$$

$$h = \varepsilon_s(h) \text{ if and only if } \Delta(h) = 1_{(1)} \otimes h1_{(2)};$$

(iv) $\varepsilon_t(H)$ *and* $\varepsilon_s(H)$ *commute;*

(v) *one also has identities dual to* (ii):

$$h\varepsilon_t(g) = \varepsilon(h_{(1)}g)h_{(2)}, \qquad \varepsilon_s(g)h = h_{(1)}\varepsilon(gh_{(2)}).$$

PROOF. We prove the identities containing the target counital map, the proofs of their source counterparts are similar. Using the axioms (2–1) and (2–2) we compute

$$\varepsilon_t(\varepsilon_t(h)) = \varepsilon(1_{(1)}h)\varepsilon(1'_{(1)}1_{(2)})1'_{(2)} = \varepsilon(1_{(1)}h)\varepsilon(1_{(2)})1_{(3)} = \varepsilon_t(h),$$

where $1'$ stands for the second copy of the unit, proving (i). For (ii) we have

$$h_{(1)} \otimes \varepsilon_t(h_{(2)}) = h_{(1)}\varepsilon(1_{(1)}h_{(2)}) \otimes 1_{(2)}$$
$$= 1_{(1)}h_{(1)}\varepsilon(1_{(2)}h_{(2)}) \otimes 1_{(3)} = 1_{(1)}h \otimes 1_{(2)}.$$

To prove (iii) we observe that

$$\Delta(\varepsilon_t(h)) = \varepsilon(1_{(1)}h)1_{(2)} \otimes 1_{(3)} = \varepsilon(1_{(1)}h)1'_{(1)}1_{(2)} \otimes 1'_{(2)} = 1'_{(1)}\varepsilon_t(h) \otimes 1'_{(2)},$$

on the other hand, applying $(\varepsilon \otimes \mathrm{id})$ to both sides of $\Delta(h) = 1_{(1)}h \otimes 1_{(2)}$, we get $h = \varepsilon_t(h)$. (iv) is immediate in view of the identity $1_{(1)} \otimes 1'_{(1)}1_{(2)} \otimes 1'_{(2)} = 1_{(1)} \otimes 1_{(2)}1'_{(1)} \otimes 1'_{(2)}$. Finally, we show (v):

$$\varepsilon(h_{(1)}g)h_{(2)} = \varepsilon(h_{(1)}g_{(1)})h_{(2)}\varepsilon_t(g_{(2)}) = \varepsilon(h_{(1)}g_{(1)})h_{(2)}g_{(2)}S(g_{(3)})$$
$$= hg_{(1)}S(g_{(2)}) = h\varepsilon_t(g),$$

where the antipode axiom (2–3) is used. □

The images of the counital maps

$$H_t = \varepsilon_t(H) = \{h \in H \mid \Delta(h) = 1_{(1)}h \otimes 1_{(2)}\}, \qquad (2\text{--}7)$$
$$H_s = \varepsilon_s(H) = \{h \in H \mid \Delta(h) = 1_{(1)} \otimes h1_{(2)}\} \qquad (2\text{--}8)$$

play the role of bases of H. The next proposition summarizes their properties.

PROPOSITION 2.2.2. H_t (resp. H_s) is a left (resp. right) coideal subalgebra of H. These subalgebras commute with each other; moreover

$$H_t = \{(\phi \otimes \mathrm{id})\Delta(1) \mid \phi \in \widehat{H}\}, \qquad H_s = \{(\mathrm{id} \otimes \phi)\Delta(1) \mid \phi \in \widehat{H}\},$$

i.e., H_t (resp. H_s) is generated by the right (resp. left) tensorands of $\Delta(1)$.

PROOF. H_t and H_s are coideals by Proposition 2.2.1(iii), they commute by 2.2.1(iv). We have $H_t = \varepsilon(1_{(1)}H)1_{(2)} \subset \{(\phi \otimes \mathrm{id})\Delta(1) \mid \phi \in \widehat{H}\}$, conversely $\phi(1_{(1)})1_{(2)} = \phi(1_{(1)})\varepsilon_t(1_{(2)}) \subset H_t$, therefore $H_t = \{(\phi \otimes \mathrm{id})\Delta(1) \mid \phi \in \widehat{H}\}$. To see that it is an algebra we note that $1 = \varepsilon_t(1) \in H_t$ and for all $h, g \in H$ compute, using Proposition 2.2.1(ii) and (v):

$$\varepsilon_t(h)\varepsilon_t(g) = \varepsilon(\varepsilon_t(h)_{(1)}g)\varepsilon_t(h)_{(2)}$$
$$= \varepsilon(1_{(1)}\varepsilon_t(h)g)1_{(2)} = \varepsilon_t(\varepsilon_t(h)g) \in H_t.$$

The statements about H_s are proven similarly. □

DEFINITION 2.2.3. We call H_t (resp. H_s) a target (resp. source) counital subalgebra of H.

2.3. Properties of the antipode. The properties of the antipode of a quantum groupoid are similar to those of a finite-dimensional Hopf algebra.

PROPOSITION 2.3.1 ([BNSz], 2.10). *The antipode S is unique and bijective. Also, it is both algebra and coalgebra anti-homomorphism.*

PROOF. Let $f * g = m(f \otimes g)\Delta$ be the convolution of $f, g \in \mathrm{End}_k(H)$. Then $S * \mathrm{id} = \varepsilon_s$, $\mathrm{id} * S = \varepsilon_t$, and $S * \mathrm{id} * S = S$. If S' is another antipode of H then

$$S' = S' * \mathrm{id} * S' = S' * \mathrm{id} * S = S * \mathrm{id} * S = S.$$

To check that S is an algebra anti-homomorphism, we compute

$$S(1) = S(1_{(1)})1_{(2)}S(1_{(3)}) = S(1_{(1)})\varepsilon_t(1_{(2)}) = \varepsilon_s(1) = 1,$$
$$S(hg) = S(h_{(1)}g_{(1)})\varepsilon_t(h_{(2)}g_{(2)}) = S(h_{(1)}g_{(1)})h_{(2)}\varepsilon_t(g_{(2)})S(h_{(3)})$$
$$= \varepsilon_s(h_{(1)}g_{(1)})S(g_{(2)})S(h_{(2)}) = S(g_{(1)})\varepsilon_s(h_{(1)})\varepsilon_t(g_{(2)})S(h_{(2)}) = S(g)S(h),$$

for all $h, g \in H$, where we used Proposition 2.2.1(iv) and easy identities $\varepsilon_t(hg) = \varepsilon_t(h\varepsilon_t(g))$ and $\varepsilon_s(hg) = \varepsilon_t(\varepsilon_s(h)g)$. Dualizing the above arguments we show that S is also a coalgebra anti-homomorphism:

$$\varepsilon(S(h)) = \varepsilon(S(h_{(1)})\varepsilon_t(h_{(2)})) = \varepsilon(S(h_{(1)})h_{(2)}) = \varepsilon(\varepsilon_t(h)) = \varepsilon(h),$$
$$\Delta(S(h)) = \Delta(S(h_{(1)})\varepsilon_t(h_{(2)}))$$
$$= \Delta(S(h_{(1)}))(\varepsilon_t(h_{(2)}) \otimes 1)$$
$$= \Delta(S(h_{(1)}))(h_{(2)}S(h_{(4)}) \otimes \varepsilon_t(h_{(3)}))$$
$$= \Delta(\varepsilon_s(h_{(1)}))(S(h_{(3)}) \otimes S(h_{(2)}))$$
$$= S(h_{(3)}) \otimes \varepsilon_s(h_{(1)})S(h_{(2)}) = S(h_{(2)}) \otimes S(h_{(1)}).$$

The proof of the bijectivity of S can be found in ([BNSz], 2.10).   □

Next, we investigate the relations between the antipode and counital maps.

PROPOSITION 2.3.2. *We have $S \circ \varepsilon_s = \varepsilon_t \circ S$ and $\varepsilon_s \circ S = S \circ \varepsilon_t$. The restriction of S defines an algebra anti-isomorphism between counital subalgebras H_t and H_s.*

PROOF. Using results of Proposition 2.3.1 we compute

$$S(\varepsilon_s(h)) = S(1_{(1)})\varepsilon(h1_{(2)}) = \varepsilon(1_{(1)}S(h))1_{(2)} = \varepsilon_t(S(h)),$$

for all $h \in H$. The second identity is proven similarly. Clearly, S maps H_t to H_s and vice versa. Since S is bijective, and $\dim H_t = \dim H_s$ by Proposition 2.2.2, therefore $S|_{H_t}$ and $S|_{H_s}$ are anti-isomorphisms.   □

PROPOSITION 2.3.3 ([NV1], 2.1.12). *Any nonzero morphism $\alpha : H \to K$ of quantum groupoids preserves counital subalgebras, i.e. $H_t \cong K_t$ and $H_s \cong K_s$. Thus, quantum groupoids with a given target (source) counital subalgebra form a full subcategory of the category of all finite quantum groupoids.*

PROOF. It is clear that $\alpha|_{H_t} : H_t \to K_t$ is a homomorphism. If we write

$$\Delta(1_H) = \sum_{i=1}^{n} w_i \otimes z_i$$

with $\{w_i\}_{i=1}^n$ and $\{z_i\}_{i=1}^n$ linearly independent, then

$$\Delta(1_K) = \sum_{i=1}^{n} \alpha(w_i) \otimes \alpha(z_i).$$

By Proposition 2.2.2, $K_t = \text{span}\{\alpha(z_i)\}$, i.e., $\alpha|_{H_t}$ is surjective. Since

$$z_j = \varepsilon_t(z_j) = \sum_{i=1}^{n} \varepsilon(w_i z_j) z_i,$$

then $\varepsilon(w_i z_j) = \delta_{ij}$, therefore,

$$\dim H_t = n = \sum_{i=1}^{n} \varepsilon_H(w_i z_i) = \sum_{i=1}^{n} \varepsilon_H(w_i S(z_i))$$
$$= \varepsilon_H(\varepsilon_t(1_H)) = \varepsilon_H(1_H) = \varepsilon_K(1_K) = \dim K_t,$$

so $\alpha|_{H_t}$ is bijective. The proof for source subalgebras is similar. $\square$

Let us recall that a k-algebra A is said to be *separable* [P] if the multiplication epimorphism $m : A \otimes_k A \to A$ has a right inverse as an $A - A$ bimodule homomorphism. This is equivalent to the existence of a *separability element* $e \in A \otimes_k A$ such that $m(e) = 1$ and $(a \otimes 1)e = e(1 \otimes a)$ for all $a \in A$.

PROPOSITION 2.3.4. *The counital subalgebras H_t and H_s are separable, with separability elements $e_t - (S \otimes \text{id})\Delta(1)$ and $e_s = (\text{id} \otimes S)\Delta(1)$, respectively.*

PROOF. For all $z \in H_t$ we compute, using Propositions 2.2.1 and 2.3.2:

$$1_{(1)} S^{-1}(z) \otimes 1_{(2)} = S^{-1}(z)_{(1)} \otimes \varepsilon_t(S^{-1}(z)_{(2)})$$
$$= 1_{(1)} \otimes \varepsilon_t(1_{(2)} S^{-1}(z)) = 1_{(1)} \otimes \varepsilon_t(1_{(2)} z) = 1_{(1)} \otimes 1_{(2)} z,$$

applying $(S \otimes \text{id})$ to this identity we get $e_t z = z e_t$. Clearly, $m(e_t) = 1$, whence e_t is a separability element. The second statement follows similarly. $\square$

2.4. The dual quantum groupoid.

The set of axioms of Definition 2.1.1 is self-dual. This allows to define a natural quantum groupoid structure on the dual vector space $\widehat{H} = \text{Hom}_k(H, k)$ by "reversing the arrows":

$$\langle h, \phi\psi \rangle = \langle \Delta(h), \phi \otimes \psi \rangle, \tag{2-9}$$
$$\langle h \otimes g, \widehat{\Delta}(\phi) \rangle = \langle hg, \phi \rangle, \tag{2-10}$$
$$\langle h, \widehat{S}(\phi) \rangle = \langle S(h), \phi \rangle, \tag{2-11}$$

for all $\phi, \psi \in \widehat{H}$, $h, g \in H$. The unit $\widehat{1}$ of $\widehat{H}$ is ε and counit $\widehat{\varepsilon}$ is $\phi \mapsto \langle \phi, 1 \rangle$.

In what follows we will use the Sweedler arrows, writing

$$h \rightharpoonup \phi = \phi_{(1)} \langle h, \phi_{(2)} \rangle, \qquad \phi \leftharpoonup h = \langle h, \phi_{(1)} \rangle \phi_{(2)}, \qquad (2\text{--}12)$$

for all $h \in H, \phi \in \widehat{H}$.

The counital subalgebras of $\widehat{H}$ are canonically anti-isomorphic to those of H. More precisely, the map $H_t \ni z \mapsto (z \rightharpoonup \varepsilon) \in \widehat{H}_s$ is an algebra isomorphism with the inverse given by $\chi \mapsto (1 \leftharpoonup \chi)$. Similarly, the map $H_s \ni z \mapsto (\varepsilon \leftharpoonup z) \in \widehat{H}_t$ is an algebra isomorphism ([BNSz], 2.6).

REMARK 2.4.1. The opposite algebra H^{op} is also a quantum groupoid with the same coalgebra structure and the antipode S^{-1}. Indeed,

$$S^{-1}(h_{(2)})h_{(1)} = S^{-1}(\varepsilon_s(h)) = S^{-1}(1_{(1)})\varepsilon(h1_{(2)})$$
$$= S^{-1}(1_{(1)})\varepsilon(hS^{-1}(1_{(2)})) = \varepsilon(h1_{(1)})1_{(2)},$$
$$h_{(2)}S^{-1}(h_{(1)}) = S^{-1}(\varepsilon_t(h)) = \varepsilon(1_{(1)}h)S^{-1}(1_{(2)})$$
$$= \varepsilon(S^{-1}(1_{(1)})h)S^{-1}(1_{(2)}) = 1_{(1)}\varepsilon(1_{(2)}h),$$
$$S^{-1}(h_{(3)})h_{(2)}S^{-1}(h_{(1)}) = S^{-1}(h_{(1)}S(h_{(2)})h_{(3)}) = S^{-1}(h).$$

Similarly, the co-opposite coalgebra H^{cop} (with the same algebra structure as H and the antipode S^{-1}) and $(H^{\mathrm{op/cop}}, S)$ are quantum groupoids.

2.5. Examples: groupoid algebras and their duals.
As group algebras and their duals are the easiest examples of Hopf algebras, groupoid algebras and their duals provide examples of quantum groupoids ([NV1], 2.1.4).

Let G be a finite *groupoid* (a category with finitely many morphisms, such that each morphism is invertible), then the groupoid algebra kG (generated by morphisms $g \in G$ with the product of two morphisms being equal to their composition if the latter is defined and 0 otherwise) is a quantum groupoid via:

$$\Delta(g) = g \otimes g, \quad \varepsilon(g) = 1, \quad S(g) = g^{-1}, \quad g \in G. \qquad (2\text{--}13)$$

The counital subalgebras of kG are equal to each other and coincide with the abelian algebra spanned by the identity morphisms:

$$(kG)_t = (kG)_s = \mathrm{span}\{gg^{-1} \mid g \in G\}.$$

The target and source counital maps are given by the operations of taking the target (resp. source) object of a morphism:

$$\varepsilon_t(g) = gg^{-1} = \mathrm{id}_{target(g)} \quad \text{and} \quad \varepsilon_s(g) = g^{-1}g = \mathrm{id}_{source(g)}.$$

The dual quantum groupoid $\widehat{kG}$ is isomorphic to the algebra of functions on G, i.e., it is generated by idempotents $p_g, g \in G$ such that $p_g p_h = \delta_{g,h} p_g$, with the following structure operations

$$\Delta(p_g) = \sum_{uv=g} p_u \otimes p_v, \quad \varepsilon(p_g) = \delta_{g,gg^{-1}}, \quad S(p_g) = p_{g^{-1}}. \qquad (2\text{--}14)$$

The target (resp. source) counital subalgebra is precisely the algebra of functions constant on each set of morphisms of G having the same target (resp. source) object. The target and source maps are

$$\varepsilon_t(p_g) = \sum_{vv^{-1}=g} p_v \quad \text{and} \quad \varepsilon_s(p_g) = \sum_{v^{-1}v=g} p_v.$$

2.6. Examples: quantum transformation groupoids. It is known that any group action on a set (i.e., on a commutative algebra of functions) gives rise to a groupoid [R]. Extending this construction, we associate a quantum groupoid with any action of a Hopf algebra on a separable algebra ("finite quantum space").

Namely, let H be a Hopf algebra and B be a separable (and, therefore, finite dimensional and semisimple [P]) right H-module algebra with the action $b \otimes h \mapsto b \cdot h$, where $b \in B, h \in H$. Then B^{op}, the algebra opposite to B, becomes a left H-module algebra via $h \otimes a \mapsto h \cdot a = a \cdot S_H(h)$. One can form a *double crossed product algebra* $B^{\mathrm{op}} \rtimes H \ltimes B$ on the vector space $B^{\mathrm{op}} \otimes H \otimes B$ with the multiplication

$$(a \otimes h \otimes b)(a' \otimes h' \otimes b') = (h_{(1)} \cdot a')a \otimes h_{(2)}h'_{(1)} \otimes (b \cdot h'_{(2)})b',$$

for all $a, a' \in B^{\mathrm{op}}$, $b, b' \in B$, and $h, h' \in H$.

Assume that k is an algebraically closed field of characteristic zero and let e be the symmetric separability element of B (note that e is an idempotent when considered in $B \otimes B^{\mathrm{op}}$). Let $\omega \in B^*$ be uniquely determined by $(\omega \otimes \mathrm{id})e = (\mathrm{id} \otimes \omega)e = 1$.

One can check that ω is the trace of the left regular representation of B and verifies the following identities:

$$\omega((h \cdot a)b) = \omega(a(b \cdot h)), \qquad e^{(1)} \otimes (h \cdot e^{(2)}) = (e^{(1)} \cdot h) \otimes e^{(2)},$$

where $a \in B^{\mathrm{op}}, b \in B$, and $e = e^{(1)} \otimes e^{(2)}$.

The structure of a quantum groupoid on $B^{\mathrm{op}} \rtimes H \ltimes B$ is given by

$$\Delta(a \otimes h \otimes b) = (a \otimes h_{(1)} \otimes e^{(1)}) \otimes ((h_{(2)} \cdot e^{(2)}) \otimes h_{(3)} \otimes b), \qquad (2\text{--}15)$$

$$\varepsilon(a \otimes h \otimes b) = \omega(a(h \cdot b)) = \omega(a(b \cdot S(h))), \qquad (2\text{--}16)$$

$$S(a \otimes h \otimes b) = b \otimes S(h) \otimes a. \qquad (2\text{--}17)$$

2.7. Examples: Temperley–Lieb algebras. It was shown in [NV3] (see Section 9 for details) that any inclusion of type II_1 factors with finite index and depth ([GHJ], 4.1) gives rise to a quantum groupoid describing the symmetry of this inclusion. In the case of *Temperley–Lieb algebras* ([GHJ], 2.1) we have this way the following example.

Let $k = \mathbb{C}$ be the field of complex numbers, $\lambda^{-1} = 4\cos^2 \frac{\pi}{n+3}$ $(n \geq 2)$, and $e_1, e_2, \ldots$ be a sequence of idempotents satisfying, for all i and j, the braid-like relations

$$e_i e_{i\pm 1} e_i = \lambda e_i, \qquad e_i e_j = e_j e_i \text{ if } |i - j| \geq 2.$$

Let $A_{k,l}$ be the algebra generated by $1, e_k, e_{k+1}, \ldots e_l$ $(k \leq l)$, σ be the algebra anti-isomorphism of $H = A_{1,2n-1}$ determined by $\sigma(e_i) = e_{2n-i}$ and $P_k \in A_{2n-k,2n-1} \otimes A_{1,k}$ be the image of the symmetric separability idempotent of $A_{1,k}$ under $(\sigma \otimes \text{id})$.

Finally, we denote by τ the non-degenerate Markov trace ([GHJ], 2.1) on H and by w the index of the restriction of τ on $A_{n+1,2n-1}$ [W], i.e., the unique central element in $A_{n+1,2n-1}$ such that $\tau(w \cdot)$ is equal to the trace of the left regular representation of $A_{n+1,2n-1}$.

Then the following operations give a quantum groupoid structure on H:

$$\Delta(yz) = (z \otimes y)P_{n-1}, \qquad y \in A_{n+1,2n-1}, \quad z \in A_{1,n-1},$$
$$\Delta(e_n) = (1 \otimes w)P_n(1 \otimes w^{-1}),$$
$$S(h) = w^{-1}\sigma(h)w,$$
$$\varepsilon(h) = \lambda^{-n}\tau(hfw), \quad h \in A,$$

where in the last line

$$f = \lambda^{n(n-1)/2}(e_n e_{n-1} \cdots e_1)(e_{n+1} e_n \cdots e_2) \cdots (e_{2n-1} e_{2n-2} \cdots e_n)$$

is the Jones projection corresponding to the n-step basic construction.

The source and target counital subalgebras of $H = A_{1,2n-1}$ are

$$H_s = A_{n+1,2n-1} \quad \text{and} \quad H_t = A_{1,n-1},$$

respectively. The example corresponding to $n = 2$ is a C^*-quantum groupoid of dimension 13 with the antipode having an infinite order (it was studied in detail in [NV2], 7.3).

See Sections 8 and 9 of this survey and the Appendix of [NV3] for the explanation of how quantum groupoids can be constructed from subfactors.

2.8. Other versions of a quantum groupoid. Here we briefly discuss several notions of quantum groupoids that appeared in the literature and relations between them. All these objects generalize both usual groupoid algebras and their duals and Hopf algebras. We apologize for possible non-intentional omissions in the list below.

Face algebras of Hayashi [H1] were defined as Hopf-like objects containing an abelian subalgebra generated by "bases". Non-trivial examples of such objects and applications to monoidal categories and II_1 subfactors were considered in [H2] and [H3]. It was shown in ([N], 5.2) that face algebras are precisely quantum groupoids whose counital subalgebras are abelian.

Generalized Kac algebras of Yamanouchi [Y] were used to characterize C^*-algebras arising from finite groupoids; they are exactly C^*-quantum groupoids with $S^2 = \text{id}$, see ([NV1], 2.5) and ([N], 8.7).

The idea of a quantum transformation groupoid (see 2.6) has been explored in a different form in [Mal1] (resp., [V]), where it was shown that an action of a Hopf algebra on a commutative (resp., noncommutative) algebra gives rise to a specific quantum groupoid structure on the tensor product of these algebras. This construction served as a strong motivation for [Mal2], where a quite general approach to quantum groupoids has been developed.

The definition of *quantum groupoids* in [Lu] and [Xu] is more general than the one we use. In their approach the existence of the bases is a part of the axioms, and they do not have to be finite-dimensional. These objects give rise to Lie bialgebroids as classical limits. It was shown in [EN] that every quantum groupoid in our sense is a quantum groupoid in the sense of [Lu] and [Xu], but not vice versa.

A suitable functional analytic framework for studying quantum groupoids (not necessarily finite) in the spirit of the multiplicative unitaries of S. Baaj and G. Skandalis [BS] was developed in [EVal], [E2] in connection with depth 2 inclusions of von Neumann algebras. A closely related notion of a *Hopf bimodule* was introduced in [V1] and then studied extensively in [EVal]. In the finite-dimensional case the equivalence of these notions to that of a C^*-quantum groupoid was shown in [V2] and [BSz3] and, respectively, in [V2], [NV1].

3. Integrals and Semisimplicity

3.1. Integrals in quantum groupoids

DEFINITION 3.1.1 ([BNSz], 3.1). A left (right) *integral* in H is an element $l \in H$ ($r \in H$) such that

$$hl = \varepsilon_t(h)l, \qquad (rh = r\varepsilon_s(h)) \qquad \text{for all } h \in H. \tag{3-1}$$

These notions clearly generalize the corresponding notions for Hopf algebras ([M], 2.1.1). We denote $\int_H^l$ (respectively, $\int_H^r$) the space of left (right) integrals in H and by $\int_H = \int_H^l \cap \int_H^r$ the space of two-sided integrals.

An integral in H (left or right) is called *non-degenerate* if it defines a non-degenerate functional on $\widehat{H}$. A left integral l is called *normalized* if $\varepsilon_t(l) = 1$. Similarly, $r \in \int_H^r$ is normalized if $\varepsilon_s(r) = 1$.

A dual notion to that of left (right) integral is the left (right) invariant measure. Namely, a functional $\phi \in \widehat{H}$ is said to be a left (right) *invariant measure* on H if

$$(\text{id} \otimes \phi)\Delta = (\varepsilon_t \otimes \phi)\Delta, \qquad (\text{resp.}, (\phi \otimes \text{id})\Delta = (\phi \otimes \varepsilon_s)\Delta). \tag{3-2}$$

A left (right) invariant measure is said to be normalized if $(\text{id} \otimes \phi)\Delta(1) = 1$ (resp., $(\phi \otimes \text{id})\Delta(1) = 1$).

EXAMPLE 3.1.2. (i) Let G^0 be the set of units of a finite groupoid G, then the elements $l_e = \sum_{gg^{-1}=e} g \, (e \in G^0)$ span $\int_{kG}^l$ and elements $r_e = \sum_{g^{-1}g=e} g \, (e \in G^0)$ span $\int_{kG}^r$.

(ii) If $H = (kG)^*$ then $\int_H^l = \int_H^r = \text{span}\{p_e, e \in G^0\}$.

The next proposition gives a description of the set of left integrals.

PROPOSITION 3.1.3 ([BNSz], 3.2). *The following conditions for $l \in H$ are equivalent:*

(i) $l \in \int_H^l$,
(ii) $(1 \otimes h)\Delta(l) = (S(h) \otimes 1)\Delta(l)$ *for all* $h \in H$,
(iii) $(\mathrm{id} \otimes l)\Delta(\widehat{H}) = \widehat{H}_t$,
(iv) $(\mathrm{Ker}\ \varepsilon_t)l = 0$,
(v) $S(l) \in \int_H^r$.

PROOF. The proof is a straightforward application of Definitions 2.1.1, 3.1.1, Propositions 2.2.1 and 2.3.2, and is left as an exercise for the reader. □

3.2. Hopf modules. Since a quantum groupoid H is both algebra and coalgebra, one can consider modules and comodules over H. As in the theory of Hopf algebras, an H-Hopf module is an H-module which is also an H-comodule such that these two structures are compatible (the action "commutes" with coaction):

DEFINITION 3.2.1. A right H-Hopf module is such a k-vector space M that

(i) M is a right H-module via $m \otimes h \mapsto m \cdot h$,
(ii) M is a right H-comodule via $m \mapsto \rho(m) = m^{(0)} \otimes m^{(1)}$,
(iii) $(m \cdot h)^{(0)} \otimes (m \cdot h)^{(1)} = m^{(0)} \cdot h_{(1)} \otimes m^{(1)}h_{(2)}$

for all $m \in M$ and $h \in H$.

REMARK 3.2.2. Condition (iii) above means that ρ is a right H-module map, where $(M \otimes H)\Delta(1)$ is a right H-module via $(m \otimes g)\Delta(1) \cdot h = (m \cdot h_{(1)}) \otimes gh_{(2)}$, for all $m \in M$ and $g, h \in H$.

EXAMPLE 3.2.3. H itself is an H-Hopf module via $\rho = \Delta$.

EXAMPLE 3.2.4. The dual vector space $\widehat{H}$ becomes a right H-Hopf module:

$$\phi \cdot h = S(h) \rightharpoonup \phi \qquad \phi^{(0)}\langle \psi, \phi^{(1)} \rangle = \psi\phi,$$

for all $\phi, \psi \in \widehat{H}$, $h \in H$. Indeed, we need to check that

$$(\phi \cdot h)^{(0)} \otimes (\phi \cdot h)^{(1)} = \phi^{(0)} \cdot h_{(1)} \otimes \phi^{(1)}h_{(2)}, \qquad \phi \in \widehat{H}, h \in H.$$

Evaluating both sides against $\psi \in \widehat{H}$ in the first factor we have:

$$
\begin{aligned}
(\phi^{(0)} \cdot h_{(1)})\langle \psi, \phi^{(1)}h_{(2)} \rangle &= (\psi_{(1)}\phi) \cdot h_{(1)}\langle \psi_{(2)}, h_{(2)} \rangle \\
&= \psi_{(1)}\phi_{(1)}\langle S(\psi_{(2)}\phi_{(2)}), h_{(1)} \rangle\langle \psi_{(3)}, h_{(2)} \rangle \\
&= \psi_{(1)}\phi_{(1)}\langle S^{-1}\varepsilon_s(\psi_{(2)})\phi_{(2)}, S(h) \rangle \\
&= \psi\phi_{(1)}\langle \phi_{(2)}, S(h) \rangle \\
&= \psi(S(h) \rightharpoonup \phi) = \psi(\phi \cdot h) \\
&= (\phi \cdot h)^{(0)}\langle \psi, (\phi \cdot h)^{(1)} \rangle.
\end{aligned}
$$

where we used Proposition 2.3.4 and that $S^{-1}\varepsilon_s(\psi) \in \widehat{H}_t$ for all $\psi \in \widehat{H}$.

The fundamental theorem for Hopf modules over Hopf algebras ([M], 1.9.4) generalizes to quantum groupoids as follows:

THEOREM 3.2.5 ([BNSz], 3.9). *Let M be a right H-Hopf module and*

$$N = \text{Coinv} M = \{m \in M \mid m^{(0)} \otimes m^{(1)} = m1_{(1)} \otimes 1_{(2)}\} \qquad (3\text{-}3)$$

be the set of its coinvariants. The H_t-module tensor product $N \otimes_{H_t} H$ (where N is a right H_t-submodule) is a right H-Hopf module via

$$(n \otimes h) \cdot g = n \otimes hg, \qquad (n \otimes h)^{(0)} \otimes (n \otimes h)^{(1)} = (n \otimes h_{(1)}) \otimes h_{(2)}, \qquad (3\text{-}4)$$

for all $h, g \in H$, $n \in N$ and the map

$$\alpha : N \otimes_{H_t} H \to M : n \otimes h \mapsto n \cdot h \qquad (3\text{-}5)$$

is an isomorphism of right H-Hopf modules.

PROOF. Proposition 2.2.1(iii) implies that the H-Hopf module structure on $N \otimes_{H_t} H$ is well defined, and it is easy to check that α is a well defined homomorphism of H-Hopf modules. We will show that

$$\beta : M \to N \otimes_{H_t} H : m \mapsto (m^{(0)} \cdot S(m^{(1)})) \otimes m^{(2)} \qquad (3\text{-}6)$$

is an inverse of α. First, we observe that $m^{(0)} \cdot S(m^{(1)}) \in N$ for all $m \in M$, since

$$\rho(m^{(0)} \cdot S(m^{(1)})) = (m^{(0)} \cdot S(m^{(3)})) \otimes m^{(1)} S(m^{(2)})$$
$$= (m^{(0)} \cdot S(1_{(2)} m^{(1)})) \otimes S(1_{(1)})$$
$$= (m^{(0)} \cdot S(m^{(1)}) 1_{(1)}) \otimes 1_{(2)},$$

so β maps to $N \otimes_{H_t} H$. Next, we check that it is both module and comodule map:

$$\beta(m \cdot h) = m^{(0)} \cdot h_{(1)} S(m^{(1)} h_{(2)}) \otimes m^{(2)} h_{(3)}$$
$$= m^{(0)} \cdot \varepsilon_t(h_{(1)}) S(m^{(1)}) \otimes m^{(2)} h_{(2)}$$
$$= m^{(0)} \cdot S(m^{(1)} 1_{(1)}) \otimes m^{(2)} 1_{(2)} h = \beta(m) \cdot h,$$
$$\beta(m^{(0)}) \otimes m^{(1)} = m^{(0)} \cdot S(m^{(1)}) \otimes m^{(2)} \otimes m^{(3)}$$
$$= \beta(m)^{(0)} \otimes \beta(m)^{(1)},$$

Finally, we verify that $\alpha \circ \beta = \text{id}$ and $\beta \circ \alpha = \text{id}$:

$$\alpha \circ \beta(m) = m^{(0)} \cdot S(m^{(1)}) m^{(2)} = m^{(0)} \cdot \varepsilon_s(m^{(1)})$$
$$= m^{(0)} \cdot 1_{(1)} \varepsilon(m^{(1)} 1_{(2)}) = m^{(0)} \varepsilon(m^{(1)}) = m,$$
$$\beta \circ \alpha(n \otimes h) = \beta(n \cdot h) = \beta(n) \cdot h$$
$$= n \cdot 1_{(1)} S(1_{(2)}) \otimes 1_{(3)} h = n \otimes h,$$

which completes the proof. $\qquad \square$

REMARK 3.2.6. $\int_{\widehat{H}}^{l} = \text{Coinv}\,\widehat{H}$ where $\widehat{H}$ is an H-Hopf module as in Example 3.2.4. Indeed, the condition

$$\lambda^{(1)} \otimes \lambda^{(2)} = \lambda \cdot 1_{(1)} \otimes 1_{(2)}$$

is equivalent to $\phi\lambda = (S(1_{(1)}) \rightharpoonup \lambda)\langle \phi, 1_{(2)} \rangle = \varepsilon_t(\phi)\lambda.$

COROLLARY 3.2.7. $\widehat{H} \cong \int_{\widehat{H}}^{l} \otimes_{H_t} H$ as right H-Hopf modules. In particular, $\int_{\widehat{H}}^{l}$ is a non-zero subspace of $\widehat{H}$.

PROOF. Take $M = \widehat{H}$ in Theorem 3.2.5 and use Remark 3.2.6.   □

3.3. Maschke's Theorem. The existence of left integrals in quantum groupoids leads to the following generalization of Maschke's Theorem, well-known for Hopf algebras ([M], 2.2.1).

THEOREM 3.3.1 ([BNSz], 3.13). *Let H be a finite quantum groupoid, then the following conditions are equivalent:*

(i) *H is semisimple,*
(ii) *There exists a normalized left integral l in H,*
(iii) *H is separable.*

PROOF. (i) $\Rightarrow$ (ii): Suppose that H is semisimple, then since $\text{Ker}\,\varepsilon_t$ is a left ideal in H we have $\text{Ker}\,\varepsilon_t = Hp$ for some idempotent p. Therefore, $\text{Ker}\,\varepsilon_t(1 - p) = 0$ and $l = 1 - p$ is a left integral by Lemma 3.1.3(iv). It is normalized since $\varepsilon_t(l) = 1 - \varepsilon_t(p) = 1$. (ii) $\Rightarrow$ (iii): if l is normalized then $l_{(1)} \otimes S(l_{(2)})$ is a separability element of H by Lemma 3.1.3(ii). (iii) $\Rightarrow$ (i): this is a standard result [P].   □

COROLLARY 3.3.2. *Let G be a finite groupoid and for every $e \in G^0$, where G^0 denotes the unit space of G, let $|e| = \#\{g \in G \mid gg^{-1} = e\}$. Then kG is semisimple if and only if $|e| \neq 0$ in k for all $e \in G^0$.*

PROOF. In the notation of Example 3.1.2(i) the element $l = \sum_{e \in G^0} \frac{1}{|e|} l_e$ is a normalized integral in kG.   □

Given a left integral l, one can show ([BNSz], 3.18) that if there exists $\lambda \in \widehat{H}$ such that $\lambda \rightharpoonup l = 1$, then it is unique, it is a left integral in $\widehat{H}$ and $l \rightharpoonup \lambda = \widehat{1}$. Such a pair (l, λ) is called a *dual pair* of left integrals. One defines dual pairs of right integrals in a similar way.

4. Actions and Smash Products

4.1. Module and comodule algebras

DEFINITION 4.1.1. An algebra A is a (left) *H-module algebra* if A is a left H-module via $h \otimes a \to h \cdot a$ and

(1) $h \cdot ab = (h_{(1)} \cdot a)(h_{(2)} \cdot b),$

(2) $h \cdot 1 = \varepsilon_t(h) \cdot 1$.

If A is an H-module algebra we will also say that H acts on A.

DEFINITION 4.1.2. An algebra A is a (right) H-*comodule algebra* if A is a right H-comodule via $\rho : a \rightarrow a^{(0)} \otimes a^{(1)}$ and

(1) $\rho(ab) = a^{(0)}b^{(0)} \otimes a^{(1)}b^{(1)}$,
(2) $\rho(1) = (\mathrm{id} \otimes \varepsilon_t)\rho(1)$.

It follows immediately that A is a left H-module algebra if and only if A is a right $\widehat{H}$-comodule algebra.

EXAMPLE 4.1.3. (i) The target counital subalgebra H_t is a trivial H-module algebra via $h \cdot z = \varepsilon_t(hz)$, $h \in H$, $z \in H_t$.
   (ii) H is an $\widehat{H}$-module algebra via the dual action $\phi \rightharpoonup h = h_{(1)} \langle \phi, h_{(2)} \rangle$, $\phi \in \widehat{H}$, $h \in H$.
   (iii) Let $A = C_H(H_s) = \{ a \in H \mid ay = ya \ \forall y \in H_s \}$ be the centralizer of H_s in H, then A is an H-module algebra via the adjoint action $h \cdot a = h_{(1)}aS(h_{(2)})$.

4.2. Smash products. Let A be an H-module algebra, then a *smash product* algebra $A \# H$ is defined on a k-vector space $A \otimes_{H_t} H$, where H is a left H_t-module via multiplication and A is a right H_t-module via

$$a \cdot z = S^{-1}(z) \cdot a = a(z \cdot 1), \qquad a \in A, z \in H_t,$$

as follows. Let $a \# h$ be the class of $a \otimes h$ in $A \otimes_{H_t} H$, then the multiplication in $A \# H$ is given by the familiar formula

$$(a \# h)(b \# g) = a(h_{(1)} \# b) \# h_{(2)}g, \qquad a, b, \in A, h, g \in H,$$

and the unit of $A \# H$ is $1 \# 1$.

EXAMPLE 4.2.1. H is isomorphic to the trivial smash product algebra $H_t \# H$.

4.3. Duality for actions. An analogue of the Blattner–Montgomery duality theorem for actions of quantum groupoids was proven in [N2]. Let H be a finite quantum groupoid and A be a left H-module algebra. Then the smash product $A \# H$ is a left $\widehat{H}$-module algebra via

$$\phi \cdot (a \# h) = a \# (\phi \rightharpoonup h), \qquad \phi \in \widehat{H}, h \in H, a \in A. \qquad (4\text{--}1)$$

In the case when H is a finite dimensional Hopf algebra, there is an isomorphism $(A \# H) \# \widehat{H} \cong M_n(A)$, where $n = \dim H$ and $M_n(A)$ is an algebra of n-by-n matrices over A [BM]. This result extends to quantum groupoid action in the form $(A \# H) \# \widehat{H} \cong \mathrm{End}(A \# H)_A$, where $A \# H$ is a right A-module via multiplication (note that $A \# H$ is not necessarily a free A-module, so that we have $\mathrm{End}(A \# H)_A \ncong M_n(A)$ in general).

THEOREM 4.3.1 ([N2], 3.1, 3.2). *The map* $\alpha : (A \# H) \# \widehat{H} \to \mathrm{End}(A \# H)_A$ *defined by*

$$\alpha((x \# h) \# \phi)(y \# g) = (x \# h)(y \# (\phi \rightharpoonup g)) = x(h_{(1)} \cdot y) \# h_{(2)}(\phi \rightharpoonup g)$$

for all $x, y \in A$, $h, g \in H$, $\phi \in \widehat{H}$ *is an isomorphism of algebras.*

PROOF. A straightforward (but rather lengthly) computation shows that α is a well defined homomorphism, cf. ([N2], 3.1). Let $\{f_i\}$ be a basis of H and $\{\xi^i\}$ be the dual basis of $\widehat{H}$, i.e., such that $\langle f_i, \xi^j \rangle = \delta_{ij}$ for all i, j, then the element $\sum_i f_i \otimes \xi^i \in H \otimes_k \widehat{H}$ does not depend on the choice of $\{f_i\}$ and the following map

$$\beta : \mathrm{End}(A \# H)_A \to (A \# H) \# \widehat{H}$$
$$: T \mapsto \sum_i T(1 \# f_{i(2)})(1 \# S^{-1}(f_{i(1)})) \# \xi^i.$$

is the inverse of α. □

COROLLARY 4.3.2. $H \# \widehat{H} \cong \mathrm{End}(H)_{H_t}$, *therefore,* $H \# \widehat{H}$ *is a semisimple algebra.*

PROOF. Theorem 4.3.1 for $A = H_t$ shows that H is a projective and finitely generated H_t-module such that $\mathrm{End}(H)_{H_t} \cong H \# \widehat{H}$. Therefore, H_t and $H \# \widehat{H}$ are Morita equivalent. Since H_t is semisimple (as a separable algebra), $H \# \widehat{H}$ is semisimple. □

5. Representation Category of a Quantum Groupoid

Representation categories of quantum groupoids were studied in [BSz2] and [NTV].

5.1. Definition of Rep(H). For a quantum groupoid H let Rep(H) be the category of representations of H, whose objects are H-modules of finite rank and whose morphisms are H-linear homomorphisms. We show that, as in the case of Hopf algebras, Rep(H) has a natural structure of a monoidal category with duality.

For objects V, W of Rep(H) set

$$V \otimes W = \{x \in V \otimes_k W \mid x = \Delta(1) \cdot x\}, \tag{5--1}$$

with the obvious action of H via the comultiplication Δ (here $\otimes_k$ denotes the usual tensor product of vector spaces).

Since $\Delta(1)$ is an idempotent, $V \otimes W = \Delta(1) \cdot (V \otimes_k W)$. The tensor product of morphisms is the restriction of usual tensor product of homomorphisms. The standard associativity isomorphisms $\Phi_{U,V,W} : (U \otimes V) \otimes W \to U \otimes (V \otimes W)$ are functorial and satisfy the pentagon condition, since Δ is coassociative. We will suppress these isomorphisms and write simply $U \otimes V \otimes W$.

The target counital subalgebra $H_t \subset H$ has an H-module structure given by $h \cdot z = \varepsilon_t(hz)$, where $h \in H$, $z \in H_t$.

LEMMA 5.1.1. H_t is the unit object of $\mathrm{Rep}(H)$.

PROOF. Define a left unit homomorphism $l_V : H_t \otimes V \to V$ by

$$l_V(1_{(1)} \cdot z \otimes 1_{(2)} \cdot v) = z \cdot v, \qquad z \in H_t,\, v \in V.$$

It is an invertible H-linear map with the inverse $l_V^{-1}(v) = S(1_{(1)}) \otimes 1_{(2)} \cdot v$. Moreover, the collection $\{l_V\}_V$ gives a natural equivalence between the functor $H_t \otimes (\ )$ and the identity functor. Similarly, the right unit homomorphism $r_V : V \otimes H_t \to V$ defined by

$$r_V(1_{(1)} \cdot v \otimes 1_{(2)} \cdot z) = S(z) \cdot v, \qquad z \in H_t,\, v \in V,$$

has the inverse $r_V^{-1}(v) = 1_{(1)} \cdot v \otimes 1_{(2)}$ and satisfies the necessary properties. Finally, one can check the triangle axiom, i.e., that

$$(\mathrm{id}_V \otimes l_W) = (r_V \otimes \mathrm{id}_W)$$

for all objects V, W of $\mathrm{Rep}(H)$ and $v \in V$, $w \in W$ ([NTV], 4.1). □

Using the antipode S of H, we can provide $\mathrm{Rep}(H)$ with a duality. For any object V of $\mathrm{Rep}(H)$ define the action of H on $V^* = \mathrm{Hom}_k(V, k)$ by $(h \cdot \phi)(v) = \phi(S(h) \cdot v)$, where $h \in H, v \in V, \phi \in V^*$. For any morphism $f : V \to W$ let $f^* : W^* \to V^*$ be the morphism dual to f (see [T], I.1.8).

For any V in $\mathrm{Rep}(H)$ define the duality homomorphisms

$$d_V : V^* \otimes V \to H_t, \qquad b_V : H_t \to V \otimes V^*$$

as follows. For $\sum_j \phi^j \otimes v_j \in V^* \otimes V$ set

$$d_V\left(\sum_j \phi^j \otimes v_j\right) = \sum_j \phi^j(1_{(1)} \cdot v_j)1_{(2)}. \tag{5-2}$$

Let $\{g_i\}_i$ and $\{\gamma^i\}_i$ be bases of V and V^* respectively, dual to each other. The element $\sum_i g_i \otimes \gamma^i$ does not depend on choice of these bases; moreover, for all $v \in V, \phi \in V^*$ one has $\phi = \sum_i \phi(g_i)\gamma^i$ and $v = \sum_i g_i\gamma^i(v)$. Set

$$b_V(z) = z \cdot \sum_i g_i \otimes \gamma^i. \tag{5-3}$$

PROPOSITION 5.1.2. The category $\mathrm{Rep}(H)$ is a monoidal category with duality.

PROOF. We know already that $\mathrm{Rep}(H)$ is monoidal. One can check ([NTV], 4.2) that d_V and b_V are H-linear. To show that they satisfy the identities

$$(\mathrm{id}_V \otimes d_V)(b_V \otimes \mathrm{id}_V) = \mathrm{id}_V, \qquad (d_V \otimes \mathrm{id}_{V^*})(\mathrm{id}_{V^*} \otimes b_V) = \mathrm{id}_{V^*},$$

take $\sum_j \phi^j \otimes v_j \in V^* \otimes V, z \in H_t$. Using the isomorphisms l_V and r_V identifying $H_t \otimes V$, $V \otimes H_t$ and V, for all $v \in V$ and $\phi \in V^*$ we have:

$$
\begin{aligned}
(\mathrm{id}_V \otimes d_V)(b_V \otimes \mathrm{id}_V)(v) &= (\mathrm{id}_V \otimes d_V)(b_V(1_{(1)} \cdot 1) \otimes 1_{(2)} \cdot v) \\
&= (\mathrm{id}_V \otimes d_V)(b_V(1_{(2)}) \otimes S^{-1}(1_{(1)}) \cdot v) \\
&= \sum_i (\mathrm{id}_V \otimes d_V)(1_{(2)} \cdot g_i \otimes 1_{(3)} \cdot \gamma^i \otimes S^{-1}(1_{(1)}) \cdot v) \\
&= \sum_i 1_{(2)} \cdot g_i \otimes (1_{(3)} \cdot \gamma^i)(1'_{(1)} S^{-1}(1_{(1)}) \cdot v) 1'_{(2)} \\
&= 1_{(2)} S(1_{(3)}) 1'_{(1)} S^{-1}(1_{(1)}) \cdot v \otimes 1'_{(2)} = v,
\end{aligned}
$$
$$
\begin{aligned}
(d_V \otimes \mathrm{id}_{V^*})(\mathrm{id}_{V^*} \otimes b_V)(\phi) &= (d_V \otimes \mathrm{id}_{V^*})(1_{(1)} \cdot \phi \otimes b_V(1_{(2)})) \\
&= \sum_i (d_V \otimes \mathrm{id}_{V^*})(1_{(1)} \cdot \phi \otimes 1_{(2)} \cdot g_i \otimes 1_{(3)} \cdot \gamma^i) \\
&= \sum_i (1_{(1)} \cdot \phi)(1'_{(1)} 1_{(2)} \cdot g_i) 1'_{(2)} \otimes 1_{(3)} \cdot \gamma^i \\
&= 1'_{(2)} \otimes 1_{(3)} 1_{(1)} S(1'_{(1)} 1_{(2)}) \cdot \phi = \phi,
\end{aligned}
$$

which completes the proof. $\qquad\qquad\qquad\qquad\qquad\qquad\qquad\qquad\qquad\qquad\qquad$ □

5.2. Quasitriangular quantum groupoids.

DEFINITION 5.2.1. A quasitriangular quantum groupoid is a pair $(H, \mathcal{R})$ where H is a quantum groupoid and $\mathcal{R} \in \Delta^{\mathrm{op}}(1)(H \otimes_k H)\Delta(1)$ satisfying the following conditions:

$$\Delta^{\mathrm{op}}(h)\mathcal{R} = \mathcal{R}\Delta(h), \qquad\qquad\qquad (5\text{–}4)$$

for all $h \in H$, where Δ^{op} denotes the comultiplication opposite to Δ,

$$(\mathrm{id} \otimes \Delta)\mathcal{R} = \mathcal{R}_{13}\mathcal{R}_{12}, \qquad\qquad\qquad (5\text{–}5)$$
$$(\Delta \otimes \mathrm{id})\mathcal{R} = \mathcal{R}_{13}\mathcal{R}_{23}, \qquad\qquad\qquad (5\text{–}6)$$

where $\mathcal{R}_{12} = \mathcal{R} \otimes 1$, $\mathcal{R}_{23} = 1 \otimes \mathcal{R}$, etc. as usual, and such that there exists $\bar{\mathcal{R}} \in \Delta(1)(H \otimes_k H)\Delta^{\mathrm{op}}(1)$ with

$$\mathcal{R}\bar{\mathcal{R}} = \Delta^{\mathrm{op}}(1), \qquad \bar{\mathcal{R}}\mathcal{R} = \Delta(1). \qquad\qquad (5\text{–}7)$$

Note that $\bar{\mathcal{R}}$ is uniquely determined by $\mathcal{R}$: if $\bar{\mathcal{R}}$ and $\bar{\mathcal{R}}'$ are two elements of $\Delta(1)(H \otimes_k H)\Delta^{\mathrm{op}}(1)$ satisfying the previous equation, then

$$\bar{\mathcal{R}} = \bar{\mathcal{R}}\Delta^{\mathrm{op}}(1) = \bar{\mathcal{R}}\mathcal{R}\bar{\mathcal{R}}' = \Delta(1)\bar{\mathcal{R}}' = \bar{\mathcal{R}}'.$$

For any two objects V and W of $\mathrm{Rep}(H)$ define $c_{V,W} : V \otimes W \to W \otimes V$ as the action of R_{21}:

$$c_{V,W}(x) = R^{(2)} \cdot x^{(2)} \otimes R^{(1)} \cdot x^{(1)}, \qquad\qquad (5\text{–}8)$$

where $x = x^{(1)} \otimes x^{(2)} \in V \otimes W$ and $\mathcal{R} = \mathcal{R}^{(1)} \otimes \mathcal{R}^{(2)}$.

PROPOSITION 5.2.2. *The family of homomorphisms* $\{c_{V,W}\}_{V,W}$ *defines a braiding in* Rep(H). *Conversely, if* Rep(H) *is braided, then there exists* $\mathcal{R}$, *satisfying the properties of Definition* 5.2.1 *and inducing the given braiding.*

PROOF. Note that $c_{V,W}$ is well-defined, since $\mathcal{R}_{21} = \Delta(1)\mathcal{R}_{21}$. To prove the H-linearity of $c_{V,W}$ we observe that

$$c_{V,W}(h \cdot x) = \mathcal{R}^{(2)} h_{(2)} \cdot x^{(2)} \otimes \mathcal{R}^{(1)} h_{(1)} \cdot x^{(1)}$$
$$= h_{(1)}\mathcal{R}^{(2)} \cdot x^{(2)} \otimes h_{(2)}\mathcal{R}^{(1)} \cdot x^{(1)} = h \cdot (c_{V,W}(x)).$$

The inverse of $c_{V,W}$ is given by

$$c_{V,W}^{-1}(y) = \bar{\mathcal{R}}^{(1)} \cdot y^{(2)} \otimes \bar{\mathcal{R}}^{(2)} \cdot y^{(1)}, \qquad \text{where } y = y^{(1)} \otimes y^{(2)} \in W \otimes V,$$

therefore $c_{V,W}$ is is an isomorphism. Finally, we check the braiding identities. Let $x = x^{(1)} \otimes x^{(2)} \otimes x^{(3)} \in U \otimes V \otimes W$, then

$$(\text{id}_V \otimes c_{U,W})(c_{U,V} \otimes \text{id}_W)(x) =$$
$$= (\text{id}_V \otimes c_{U,W})(\mathcal{R}^{(2)} \cdot x^{(2)} \otimes \mathcal{R}^{(1)} \cdot x^{(1)} \otimes x^{(3)})$$
$$= \mathcal{R}^{(2)} \cdot x^{(2)} \otimes \mathcal{R}'^{(2)} \cdot x^{(3)} \otimes \mathcal{R}'^{(1)}\mathcal{R}^{(1)} \cdot x^{(1)}$$
$$= \mathcal{R}^{(2)}_{(1)} \cdot x^{(2)} \otimes \mathcal{R}^{(2)}_{(2)} \cdot x^{(3)} \otimes \mathcal{R}^{(1)} \cdot x^{(1)} = c_{U,V \otimes W}(x).$$

Similarly, we have $(c_{U,W} \otimes \text{id}_V)(\text{id}_U \otimes c_{V,W}) = c_{U \otimes V,W}$.

The third equality of this computation shows that the relations of Definition 5.2.1 are equivalent to the braiding identities. □

LEMMA 5.2.3. *Let* $(H, \mathcal{R})$ *be a quasitriangular quantum groupoid. Then* $\mathcal{R}$ *satisfies the quantum Yang–Baxter equation:*

$$\mathcal{R}_{12}\mathcal{R}_{13}\mathcal{R}_{23} = \mathcal{R}_{23}\mathcal{R}_{13}\mathcal{R}_{12}.$$

PROOF. It follows from the first two relations of Definition 5.2.1 that

$$\mathcal{R}_{12}\mathcal{R}_{13}\mathcal{R}_{23} = (\text{id} \otimes \Delta^{\text{op}})(\mathcal{R})\mathcal{R}_{23} = \mathcal{R}_{23}(\text{id} \otimes \Delta)(\mathcal{R}) = \mathcal{R}_{23}\mathcal{R}_{13}\mathcal{R}_{12}. \qquad □$$

REMARK 5.2.4. Let us define two linear maps $\mathcal{R}_1, \mathcal{R}_2 : \widehat{H} \to H$ by

$$\mathcal{R}_1(\phi) = (\text{id} \otimes \phi)(\mathcal{R}), \quad \mathcal{R}_2(\phi) = (\phi \otimes \text{id})(\mathcal{R}), \quad \text{for all } \phi \in \widehat{H}.$$

Then condition (5–5) of Definition 5.2.1 is equivalent to $\mathcal{R}_1$ being a coalgebra homomorphism and algebra anti-homomorphism and condition (5–6) is equivalent to $\mathcal{R}_2$ being an algebra homomorphism and coalgebra anti-homomorphism. In other words, $\mathcal{R}_1 : \widehat{H} \to H^{\text{op}}$ and $\mathcal{R}_2 : \widehat{H} \to H^{\text{cop}}$ are homomorphisms of quantum groupoids.

PROPOSITION 5.2.5. *For any quasitriangular quantum groupoid* $(H, \mathcal{R})$, *we have:*

$$(\varepsilon_s \otimes \text{id})(\mathcal{R}) = \Delta(1), \qquad (\text{id} \otimes \varepsilon_s)(\mathcal{R}) = (S \otimes \text{id})\Delta^{\text{op}}(1),$$
$$(\varepsilon_t \otimes \text{id})(\mathcal{R}) = \Delta^{\text{op}}(1), \qquad (\text{id} \otimes \varepsilon_t)(\mathcal{R}) = (S \otimes \text{id})\Delta(1),$$
$$(S \otimes \text{id})(\mathcal{R}) = (\text{id} \otimes S^{-1})(\mathcal{R}) = \bar{\mathcal{R}}, \qquad (S \otimes S)(\mathcal{R}) = \mathcal{R}.$$

PROOF. The proof is essentially the same as ([Ma], 2.1.5). □

PROPOSITION 5.2.6. *Let $(H, \mathcal{R})$ be a quasitriangular quantum groupoid. Then*

$$S^2(h) = uhu^{-1}$$

for all $h \in H$, where $u = S(\mathcal{R}^{(2)})\mathcal{R}^{(1)}$ is an invertible element of H such that

$$u^{-1} = \mathcal{R}^{(2)}S^2(\mathcal{R}^{(1)}), \quad \Delta(u) = \bar{\mathcal{R}}\bar{\mathcal{R}}_{21}(u \otimes u).$$

Likewise, $v = S(u) = \mathcal{R}^{(1)}S(\mathcal{R}^{(2)})$ obeys $S^{-2}(h) = vhv^{-1}$ and

$$v^{-1} = S^2(\mathcal{R}^{(1)})\mathcal{R}^{(2)}, \quad \Delta(v) = \bar{\mathcal{R}}\bar{\mathcal{R}}_{21}(v \otimes v).$$

PROOF. Note that $S(\mathcal{R}^{(2)})y\mathcal{R}^{(1)} = S(y)u$ for all $y \in H_s$. Hence, we have

$$\begin{aligned}
S(h_{(2)})uh_{(1)} &= S(h_{(2)})S(\mathcal{R}^{(2)})\mathcal{R}^{(1)}h_{(1)} = S(\mathcal{R}^{(2)}h_{(2)})\mathcal{R}^{(1)}h_{(1)} \\
&= S(h_{(1)}\mathcal{R}^{(2)})h_{(2)}\mathcal{R}^{(1)} = S(\mathcal{R}^{(2)})\varepsilon_s(h)\mathcal{R}^{(1)} \\
&= S(\varepsilon_s(h))u,
\end{aligned}$$

for all $h \in H$. Therefore,

$$\begin{aligned}
uh &= S(1_{(2)})u1_{(2)}h = S(\varepsilon_t(h_{(2)})uh_{(1)} \\
&= S(h_{(2)}S(h_{(3)}))uh_{(1)} = S^2(h_{(3)})S(h_{(2)})uh_{(1)} \\
&= S^2(h_{(2)})S(\varepsilon_s(h_{(1)}))u = S(\varepsilon_s(h_{(1)})S(h_{(2)}))u = S^2(h)u.
\end{aligned}$$

The remaining part of the proof follows the lines of ([Ma], 2.1.8). The results for v can be obtained by applying the results for u to the quasitriangular quantum groupoid $(H^{\mathrm{op/cop}}, \mathcal{R})$. □

DEFINITION 5.2.7. The element u defined in Proposition 5.2.6 is called *the Drinfeld element* of H.

COROLLARY 5.2.8. *The element $uv = vu$ is central and obeys*

$$\Delta(uv) = (\bar{\mathcal{R}}\bar{\mathcal{R}}_{21})^2(uv \otimes uv).$$

The element $uv^{-1} = vu^{-1}$ is group-like and implements S^4 by conjugation.

PROPOSITION 5.2.9. *Given a quasitriangular quantum groupoid $(H, \mathcal{R})$, consider a linear map $F : \hat{H} \to H$ given by*

$$F : \phi \mapsto (\phi \otimes \mathrm{id})(\mathcal{R}_{21}\mathcal{R}), \qquad \phi \in \hat{H}. \tag{5--9}$$

Then the range of F belongs to $C_H(H_s)$, the centralizer of H_s.

PROOF. Take $y \in H_s$. Then we have:

$$\phi(\mathcal{R}^{(2)}\mathcal{R}'^{(1)})\mathcal{R}^{(1)}\mathcal{R}'^{(2)}y = \phi(\mathcal{R}^{(2)}y\mathcal{R}'^{(1)})\mathcal{R}^{(1)}\mathcal{R}'^{(2)} = \phi(\mathcal{R}^{(2)}\mathcal{R}'^{(1)})y\mathcal{R}^{(1)}\mathcal{R}'^{(2)},$$

therefore $F(\phi) \in C_H(H_s)$, as required. □

DEFINITION 5.2.10 (CF. [Ma], 2.1.12). A quasitriangular quantum groupoid is *factorizable* if the above map $F : \widehat{H} \to C_H(H_s)$ is surjective.

The factorizability of H means that $\mathcal{R}$ is as non-trivial as possible, in contrast to *triangular* quantum groupoids, for which $\mathcal{R}_{21} = \bar{\mathcal{R}}$ and the range of F is equal to H_t.

5.3. The Drinfeld double.

To define the *Drinfeld double* $D(H)$ of a quantum groupoid H, consider on the vector space $\widehat{H}^{\mathrm{op}} \otimes_k H$ an associative multiplication

$$(\phi \otimes h)(\psi \otimes g) = \psi_{(2)}\phi \otimes h_{(2)}g \langle S(h_{(1)}), \psi_{(1)} \rangle \langle h_{(3)}, \psi_{(3)} \rangle, \qquad (5\text{–}10)$$

where $\phi, \psi \in \widehat{H}^{\mathrm{op}}$ and $h, g \in H$. Then one can verify that the linear span J of the elements

$$\phi \otimes zh - (\varepsilon \leftharpoonup z)\phi \otimes h, \quad z \in H_t, \qquad (5\text{–}11)$$

$$\phi \otimes yh - (y \rightharpoonup \varepsilon)\phi \otimes h, \quad y \in H_s, \qquad (5\text{–}12)$$

is a two-sided ideal in $\widehat{H}^{\mathrm{op}} \otimes_k H$. Let $D(H)$ be the factor-algebra $(\widehat{H}^{\mathrm{op}} \otimes_k H)/J$ and let $[\phi \otimes h]$ denote the class of $\phi \otimes h$ in $D(H)$.

PROPOSITION 5.3.1. *$D(H)$ is a quantum groupoid with the unit $[\varepsilon \otimes 1]$, and comultiplication, counit, and antipode given by*

$$\Delta([\phi \otimes h]) = [\phi_{(1)} \otimes h_{(1)}] \otimes [\phi_{(2)} \otimes h_{(2)}], \qquad (5\text{–}13)$$

$$\varepsilon([\phi \otimes h]) = \langle \varepsilon_t(a), \phi \rangle, \qquad (5\text{–}14)$$

$$S([\phi \otimes a]) = [S^{-1}(\phi_{(2)}) \otimes S(h_{(2)})]\langle h_{(1)}, \phi_{(1)} \rangle \langle S(h_{(3)}), \phi_{(3)} \rangle. \qquad (5\text{–}15)$$

PROOF. The proof is a straightforward verification that all the structure maps are well-defined and satisfy the axioms of a quantum groupoid, which is carried out in full detail in ([NTV], 6.1). □

PROPOSITION 5.3.2. *The Drinfeld double $D(H)$ has a canonical quasitriangular structure given by*

$$\mathcal{R} = \sum_i [\xi^i \otimes 1] \otimes [\varepsilon \otimes f_i], \qquad \bar{\mathcal{R}} = \sum_j [S^{-1}(\xi_j) \otimes 1] \otimes [\varepsilon \otimes f_j] \qquad (5\text{–}16)$$

where $\{f_i\}$ and $\{\xi^i\}$ are bases in H and $\widehat{H}$ such that $\langle f_i, \xi^j \rangle = \delta_{ij}$.

PROOF. The identities $(\mathrm{id} \otimes \Delta)\mathcal{R} = \mathcal{R}_{13}\mathcal{R}_{12}$ and $(\Delta \otimes \mathrm{id})\mathcal{R} = \mathcal{R}_{13}\mathcal{R}_{23}$ can be written as (identifying $[(\widehat{H})^{\mathrm{op}} \otimes 1]$ with $(\widehat{H})^{\mathrm{op}}$ and $[\varepsilon \otimes H]$ with H):

$$\sum_i \xi^i_{(1)} \otimes \xi^i_{(1)} \otimes f_i = \sum_{ij} \xi^i \otimes \xi^j \otimes f_i f_j,$$

$$\sum_i \xi^i \otimes f_{i(1)} \otimes f_{i(2)} = \sum_{ij} \xi^j \xi^i \otimes f_j \otimes f_i.$$

The above equalities can be verified, e.g., by evaluating both sides against an element $a \in H$ in the third factor (resp., against $\phi \in (\widehat{H})^{\mathrm{op}}$ in the first factor), see ([Ma], 7.1.1). It is also straightforward to check that $\mathcal{R}$ is an intertwiner between Δ and Δ^{op} and check that $\bar{\mathcal{R}}\mathcal{R} = \Delta(1)$ and $\mathcal{R}\bar{\mathcal{R}} = \Delta^{\mathrm{op}}(1)$, see ([NTV], 6.2). $\qquad\qquad\qquad\qquad\qquad\qquad\qquad\qquad\qquad\qquad\qquad\qquad\qquad\qquad\qquad$ $\square$

REMARK 5.3.3. The dual quantum groupoid $\widehat{D(H)}$ consists of all $\sum_k h_k \otimes \phi_k$ in $H \otimes_k \widehat{H}^{\mathrm{op}}$ such that

$$\sum_k (h_k \otimes \phi_k)|_J = 0.$$

The structure operations of $\widehat{D(H)}$ are obtained by dualizing those of $D(H)$:

$$\left(\sum_k h_k \otimes \phi_k\right)\left(\sum_l g_l \otimes \psi_l\right) = \sum_{kl} h_k g_l \otimes \phi_k \psi_l,$$

$$1_{\widehat{D(H)}} = 1_{(2)} \otimes (\varepsilon \leftharpoonup 1_{(1)}),$$

$$\Delta\left(\sum_k h_k \otimes \phi_k\right) = \sum_{ijk} (h_{k(2)} \otimes \xi^i \phi_{k(1)} \xi^j) \otimes (S(f_i) h_{k(1)} f_j \otimes \phi_{k(2)}),$$

$$\varepsilon\left(\sum_k h_k \otimes \phi_k\right) = \sum_k \varepsilon(h_k) \widehat{\varepsilon}(\phi_k),$$

$$S\left(\sum_k h_k \otimes \phi_k\right) = \sum_{ijk} f_i S^{-1}(h_k) S(f_j) \otimes \xi^i S(\phi_k) \xi^j,$$

for all $\sum_k h_k \otimes \phi_k, \sum_l g_l \otimes \psi_l \in \widehat{D(H)}$, where $\{f_i\}$ and $\{\xi^j\}$ are dual bases.

COROLLARY 5.3.4 ([NTV], 6.4). *The Drinfeld double $D(H)$ is factorizable in the sense of Definition 5.2.10.*

PROOF. One can use the explicit form of the R-matrix (5–16) of $D(H)$ and the description of the dual $\widehat{D(H)}$ to check that in this case the map F from (5–9) is surjective. $\qquad\qquad\qquad\qquad\qquad\qquad\qquad\qquad\qquad\qquad\qquad\qquad\qquad\qquad\qquad$ $\square$

5.4. Ribbon quantum groupoids.

DEFINITION 5.4.1. A ribbon quantum groupoid is a quasitriangular quantum groupoid with an invertible central element $\nu \in H$ such that

$$\Delta(\nu) = \mathcal{R}_{21}\mathcal{R}(\nu \otimes \nu) \quad \text{and} \quad S(\nu) = \nu. \tag{5–17}$$

The element ν is called a *ribbon element* of H.

For an object V of $\mathrm{Rep}(H)$ we define the twist $\theta_V : V \to V$ to be the multiplication by ν:

$$\theta_V(v) = \nu \cdot v, \quad v \in V. \tag{5–18}$$

PROPOSITION 5.4.2. *Let $(H, \mathcal{R}, \nu)$ be a ribbon quantum groupoid. The family of homomorphisms $\{\theta_V\}_V$ defines a twist in the braided monoidal category $\mathrm{Rep}(H)$ compatible with duality. Conversely, if $\theta_V(v) = \nu \cdot v$ is a twist in $\mathrm{Rep}(H)$, then ν is a ribbon element of H.*

PROOF. Since ν is an invertible central element of H, the homomorphism θ_V is an H-linear isomorphism. The twist identity follows from the properties of ν:

$$c_{W,V} c_{V,W} (\theta_V \otimes \theta_W)(x) = \mathcal{R}_{21}\mathcal{R}(\nu \cdot x^{(1)} \otimes \nu \cdot x^{(2)}) = \Delta(\nu) \cdot x = \theta_{V \otimes W}(x),$$

for all $x = x^{(1)} \otimes x^{(2)} \in V \otimes W$. Clearly, the identity $\mathcal{R}_{21}\mathcal{R}(\nu \otimes \nu) = \Delta(\nu)$ is equivalent to the twist property. It remains to prove that

$$(\theta_V \otimes \mathrm{id}_{V^*}) b_V(z) = (\mathrm{id}_V \otimes \theta_{V^*}) b_V(z),$$

for all $z \in H_t$, i.e., that

$$\sum_i \nu z_{(1)} \cdot \gamma^i \otimes z_{(2)} \cdot g_i = \sum_i z_{(1)} \cdot \gamma^i \otimes \nu z_{(2)} \cdot g_i,$$

where $\sum_i \gamma^i \otimes g_i$ is the canonical element in $V^* \otimes V$. Evaluating the first factors of the above equality on an arbitrary $v \in V$, we get the equivalent condition:

$$\sum_i (\nu z_{(1)} \cdot \gamma^i)(v) z_{(2)} \cdot g_i = \sum_i (z_{(1)} \cdot \gamma^i)(v) \nu z_{(2)} \cdot g_i,$$

which reduces to $z_{(2)} S(\nu z_{(1)}) \cdot v = S(z_{(1)}) \nu z_{(2)} \cdot v$. The latter easily follows from the centrality of $\nu = S(\nu)$ and properties of H_t. $\qquad\square$

PROPOSITION 5.4.3. *The category* Rep(H) *is a ribbon category if and only if H is a ribbon quantum groupoid.*

PROOF. Follows from Propositions 5.1.2, 5.2.2, and 5.4.2. $\qquad\square$

For any endomorphism f of the object V, we define, following [T], I.1.5, its *quantum trace*

$$\mathrm{tr}_q(f) = d_V c_{V,V^*} (\theta_V f \otimes \mathrm{id}_{V^*}) b_V \qquad (5\text{--}19)$$

with values in End(H_t) and the *quantum dimension* of V by $\dim_q(V) = \mathrm{tr}_q(\mathrm{id}_V)$.

COROLLARY 5.4.4 ([NTV], 7.4). *Let $(H, \mathcal{R}, \nu)$ be a ribbon quantum groupoid, f be an endomorphism of an object V in* Rep(H). *Then*

$$\mathrm{tr}_q(f)(z) = \mathrm{Tr}\,(S(1_{(1)}) u \nu f) z 1_{(2)}, \qquad \dim_q(V)(z) = \mathrm{Tr}\,(S(1_{(1)}) u \nu) z 1_{(2)},$$
$$(5\text{--}20)$$

where Tr *is the usual trace of an endomorphism, and u is the Drinfeld element.*

COROLLARY 5.4.5. *Let k be an algebraically closed field of characteristic zero. If the trivial H-module H_t is irreducible (which happens exactly when $H_t \cap Z(H) = k$, i.e., when H is connected ([N1], 3.11, [BNSz], 2.4)), then* $\mathrm{tr}_q(f)$ *and* $\dim_q(V)$ *are scalars:*

$$\mathrm{tr}_q(f) = (\dim H_t)^{-1} \mathrm{Tr}\,(u \nu f), \qquad \dim_q(V) = (\dim H_t)^{-1} \mathrm{Tr}\,(u \nu). \qquad (5\text{--}21)$$

PROOF. Any endomorphism of an irreducible module is the multiplication by a scalar, therefore, we must have $\mathrm{Tr}\,(S(1_{(1)}) u \nu f) 1_{(2)} = \mathrm{tr}_q(f) 1$. Applying the counit to both sides and using that $\varepsilon(1) = \dim H_t$, we get the result. $\qquad\square$

5.5. Towards modular categories. In [RT] a general method of constructing invariants of 3-manifolds from modular Hopf algebras was introduced. Later it became clear that the technique of Hopf algebras can be replaced by a more general technique of modular categories (see [T]). In addition to quantum groups, such categories also arise from skein categories of tangles and, as it was observed by A. Ocneanu, from certain bimodule categories of type II_1 subfactors.

The representation categories of *quantum groupoids* give quite general construction of modular categories. Recall some definitions needed here. Let $\mathcal{V}$ be a ribbon *Ab*-category over k, i.e., such that all $\mathrm{Hom}(V, W)$ are k-vector spaces (for all objects $V, W \in \mathcal{V}$) and both operations $\circ$ and $\otimes$ are k-bilinear.

An object $V \in \mathcal{V}$ is said to be *simple* if any endomorphism of V is multiplication by an element of k. We say that a family $\{V_i\}_{i \in I}$ of objects of $\mathcal{V}$ dominates an object V of $\mathcal{V}$ if there exists a finite set $\{V_{i(r)}\}_r$ of objects of this family (possibly, with repetitions) and a family of morphisms $f_r : V_{i(r)} \to V, g_r : V \to V_{i(r)}$ such that $\mathrm{id}_V = \sum_r f_r g_r$.

A modular category ([T], II.1.4) is a pair consisting of a ribbon *Ab*-category $\mathcal{V}$ and a finite family $\{V_i\}_{i \in I}$ of simple objects of $\mathcal{V}$ satisfying four axioms:

(i)   There exists $0 \in I$ such that V_0 is the unit object.
(ii) For any $i \in I$, there exists $i^* \in I$ such that V_{i^*} is isomorphic to V_i^*.
(iii) All objects of $\mathcal{V}$ are dominated by the family $\{V\}_{i \in I}$.
(iv) The square matrix $S = \{S_{ij}\}_{i,j \in I} = \{\mathrm{tr}_q(c_{V_i, V_j} \circ c_{V_j, V_i})\}_{i,j \in I}$ is invertible over k (here tr_q is the quantum trace in a ribbon category defined by (5–19)).

If a quantum groupoid H is connected and semisimple over an algebraically closed field, modularity of $\mathrm{Rep}(H)$ is equivalent to $\mathrm{Rep}(H)$ being ribbon and such that the matrix $S = \{S_{ij}\}_{i,j \in I} = \{\mathrm{tr}_q(c_{V_i, V_j} \circ c_{V_j, V_i})\}_{i,j \in I}$, where I is the set of all (equivalent classes of) irreducible representations, is invertible. The following proposition extends a result known for Hopf algebras ([EG], 1.1).

PROPOSITION 5.5.1. *If H is a connected, ribbon, factorizable quantum groupoid over an algebraically closed field k of characteristic zero possessing a normalized two-sided integral, then $\mathrm{Rep}(H)$ is a modular category.*

PROOF. Here we only need to prove the invertibility of the matrix formed by

$$
\begin{aligned}
S_{ij} &= \mathrm{tr}_q(c_{V_i, V_j} \circ c_{V_j, V_i}) \\
&= (\dim H_t)^{-1} \mathrm{Tr}\left((u\nu) \circ c_{V_i, V_j} \circ c_{V_j, V_i}\right) \\
&= (\dim H_t)^{-1} (\chi_j \otimes \chi_i)((u\nu \otimes u\nu)\mathcal{R}_{21}\mathcal{R}),
\end{aligned}
$$

where V_i are as above, $I = \{1, \dots n\}$, $\{\chi_j\}$ is a basis in the space $C(H)$ of characters of H (we used above the formula (5–21) for the quantum trace).

It was shown in ([NTV], 5.12, 8.1) that the map $F : \phi \mapsto (\phi \otimes \mathrm{id})(\mathcal{R}_{21}\mathcal{R})$ is a linear isomorphism between $C(H) \hookleftarrow u\nu$ and the center $Z(H)$ (here $(\chi \hookleftarrow u\nu)(a) = \chi(u\nu a) \; \forall a \in H, \; \chi \in C(H)$). So, there exists an invertible matrix

$T = (T_{ij})$ representing the map F in the bases $\{\chi_j\}$ of $C(H)$ and $\{e_i\}$ of $Z(H)$, i.e., such that $F(\chi_j \leftharpoondown u\nu) = \sum_i T_{ij}e_i$. Then

$$S_{ij} = (\dim H_t)^{-1}\chi_i(u\nu F(\chi_j \leftharpoondown u\nu)) = (\dim H_t)^{-1}\sum_k T_{kj}\chi_i(u\nu e_k)$$

$$= (\dim H_t)^{-1}(\dim V_i)\chi_i(u\nu)T_{ij}.$$

Therefore, $S = DT$, where $D = \operatorname{diag}\{(\dim H_t)^{-1}(\dim V_i)\chi_i(u\nu)\}$. Theorem 7.2.2 below shows that the existence of a normalized two-sided integral in H is equivalent to H being semisimple and possessing an invertible element g such that $S^2(x) = gxg^{-1}$ for all $x \in H$ and $\chi(g^{-1}) \neq O$ for all irreducible characters χ of H. Then $u^{-1}g$ is an invertible central element of H and $\chi_i(u^{-1}) \neq 0$ for all χ_i. By Corollary 5.2.8, $uS(u) = c$ is invertible central, therefore $\chi_i(u) = \chi_i(c)\chi_i(S(u^{-1}))) \neq 0$. Hence, $\chi_i(u\nu) \neq 0$ for all i and D is invertible. $\qquad\square$

6. Twisting and Dynamical Quantum Groups

In this section we present a generalization of the Drinfeld twisting construction to quantum groupoids developed in [NV1], [Xu], and [EN]. We show that *dynamical* twists of Hopf algebras give rise to quantum groupoids. An important concrete example is given by dynamical quantum groups at roots of 1 [EN], which are dynamical deformations of the Drinfeld–Jimbo–Lusztig quantum groups $U_q(\mathfrak{g})$. The resulting quantum groupoids turn out to be selfdual, which is a fundamentally new property, not satisfied by $U_q(\mathfrak{g})$.

Most of the material of this section is taken from [EN].

6.1. Twisting of quantum groupoids. We describe the procedure of constructing new quantum groupoids by twisting a comultiplication. Twisting of Hopf algebroids without an antipode was developed in [Xu] and a special case of twisting of *-quantum groupoids was considered in [NV1].

DEFINITION 6.1.1. A *twist* for a quantum groupoid H is a pair $(\Theta, \bar{\Theta})$, with

$$\Theta \in \Delta(1)(H \otimes_k H), \qquad \bar{\Theta} \in (H \otimes_k H)\Delta(1), \qquad \text{and} \qquad \Theta\bar{\Theta} = \Delta(1) \qquad (6\text{-}1)$$

satisfying the following axioms:

$$(\varepsilon \otimes \operatorname{id})\Theta = (\operatorname{id} \otimes \varepsilon)\Theta = (\varepsilon \otimes \operatorname{id})\bar{\Theta} = (\operatorname{id} \otimes \varepsilon)\bar{\Theta} = 1, \qquad (6\text{-}2)$$

$$(\Delta \otimes \operatorname{id})(\Theta)(\Theta \otimes 1) = (\operatorname{id} \otimes \Delta)(\Theta)(1 \otimes \Theta), \qquad (6\text{-}3)$$

$$(\bar{\Theta} \otimes 1)(\Delta \otimes \operatorname{id})(\bar{\Theta}) = (1 \otimes \bar{\Theta})(\operatorname{id} \otimes \Delta)(\bar{\Theta}), \qquad (6\text{-}4)$$

$$(\Delta \otimes \operatorname{id})(\bar{\Theta})(\operatorname{id} \otimes \Delta)(\Theta) = (\Theta \otimes 1)(1 \otimes \bar{\Theta}), \qquad (6\text{-}5)$$

$$(\operatorname{id} \otimes \Delta)(\bar{\Theta})(\Delta \otimes \operatorname{id})(\Theta) = (1 \otimes \Theta)(\bar{\Theta} \otimes 1). \qquad (6\text{-}6)$$

Note that for ordinary Hopf algebras the above notion coincides with the usual notion of twist and each of the four conditions (6–3) – (6–6) implies the other three. But since Θ and $\bar{\Theta}$ are, in general, not invertible we need to impose all of them.

The next Proposition extends Drinfeld's twisting construction to the case of quantum groupoids.

PROPOSITION 6.1.2. *Let $(\Theta, \bar{\Theta})$ be a twist for a quantum groupoid H. Then there is a quantum groupoid H_Θ having the same algebra structure and counit as H with a comultiplication and antipode given by*

$$\Delta_\Theta(h) = \bar{\Theta}\Delta(h)\Theta, \qquad S_\Theta(h) = v^{-1}S(h)v, \qquad (6\text{--}7)$$

for all $h \in H_\Theta$, where $v = m(S \otimes \mathrm{id})\Theta$ is invertible in H_Θ.

PROOF. The proof is a straightforward verification, see [EN] for details. □

REMARK 6.1.3. (a) One can check that the counital maps of the twisted quantum groupoid H_Θ are given by

$$(\varepsilon_t)_\Theta(h) = \varepsilon(\Theta^{(1)}h)\Theta^{(2)}, \qquad (\varepsilon_s)_\Theta(h) = \bar{\Theta}^{(1)}\varepsilon(h\bar{\Theta}^{(2)}).$$

(b) It is possible to generalize the above twisting construction by weakening the counit condition (6–2) and requiring only the existence of $u, w \in H$ such that

$$\varepsilon(\Theta^{(1)}w)\Theta^{(2)} = \Theta^{(1)}\varepsilon(\Theta^{(2)}w) = 1, \qquad \varepsilon(u\bar{\Theta}^{(1)})\bar{\Theta}^{(2)} = \bar{\Theta}^{(1)}\varepsilon(u\bar{\Theta}^{(2)}) = 1. \quad (6\text{--}8)$$

Then the counit of the twisted quantum groupoid H_Θ is

$$\varepsilon_\Theta(h) = \varepsilon(uhw), \qquad h \in H. \qquad (6\text{--}9)$$

(c) If $(\Theta, \bar{\Theta})$ is a twist for H and $x \in H$ is an invertible element such that $\varepsilon_t(x) = \varepsilon_s(x) = 1$ then $(\Theta^x, \bar{\Theta}^x)$, where

$$\Theta^x = \Delta(x)^{-1}\Theta(x \otimes x) \quad \text{and} \quad \bar{\Theta}^x = (x^{-1} \otimes x^{-1})\bar{\Theta}\Delta(x),$$

is also a twist for H. The twists $(\Theta, \bar{\Theta})$ and $(\Theta^x, \bar{\Theta}^x)$ are called *gauge equivalent* and x is called a *gauge transformation*. Given such an x, the map $h \mapsto x^{-1}hx$ is an isomorphism between quantum groupoids H_Θ and H_{Θ^x}.

(d) A twisting of a quasitriangular quantum groupoid is again quasitriangular. Namely, if $(\Theta, \bar{\Theta})$ is a twist and $(\mathcal{R}, \bar{\mathcal{R}})$ is a quasitriangular structure for H then the quasitriangular structure for H_Θ is given by $(\bar{\Theta}_{21}\mathcal{R}\Theta, \bar{\Theta}\bar{\mathcal{R}}\Theta_{21})$. The proof of this fact is exactly the same as for Hopf algebras.

EXAMPLE 6.1.4. We will show that for every (non-commutative) separable algebra B over an algebraically closed field k of characteristic zero there is a family of quantum groupoid structures H_q on the vector space $B^{\mathrm{op}} \otimes B$ considered in ([BSz2], 5.2) that can be understood in terms of twisting. Let $H = H_{1_B}$ be a quantum groupoid with the following structure operations:

$$\Delta(b \otimes c) = (b \otimes e^{(1)}) \otimes (e^{(2)} \otimes c), \qquad \varepsilon(b \otimes c) = \omega(bc), \qquad S(b \otimes c) = c \otimes b,$$

for all $b, c \in B$, where $e = e^{(1)} \otimes e^{(2)}$ is the symmetric separability idempotent of B and ω is defined by the condition $(\omega \otimes \mathrm{id})e = 1$. Note that $S^2 = \mathrm{id}$ and H, as an algebra, is generated by its counital subalgebras, i.e., it is dual to an *elementary* quantum groupoid of ([NV1], 3.2).

Observe that the pair

$$\Theta = (1 \otimes e^{(1)} q^{-1}) \otimes (e^{(2)} \otimes 1) \qquad \bar{\Theta} = (1 \otimes e^{(1)}) \otimes (e^{(2)} \otimes 1) = \Delta(1),$$

where q is an invertible element of B with $e^{(1)} q^{-1} e^{(2)} = 1$, satisfies (6–3) – (6–6) and that conditions (6–8) hold for $u = 1$ and $w = q$. According to Proposition 6.1.2 and Remark 6.1.3(b), these data define a twisting of H such that the comultiplication and antipode of the twisted quantum groupoid are given by

$$\Delta(b \otimes c) = (b \otimes e^{(1)} q^{-1}) \otimes (e^{(2)} \otimes c), \quad \varepsilon(b \otimes c) = \omega(qcb), \quad S(b \otimes c) = qcq^{-1} \otimes b,$$

for all $b, c \in B$. The target and source counital subalgebras of H_q are $B^{\mathrm{op}} \otimes 1$ and $1 \otimes B$. The square of the antipode is implemented by $g_q = q \otimes q$. Since $\mathrm{Ad}\, g_q$ is an invariant of H_q, this example shows that there can be uncountably many non-isomorphic semisimple quantum groupoids with the same underlying algebra (for noncommutative B).

6.2. Dynamical twists of Hopf algebras.

We describe a method of constructing twists of quantum groupoids, which is a finite-dimensional modification of the construction proposed in [Xu] and is dual to that of [EV], cf. [EN].

Dynamical twists first appeared in the work of Babelon [Ba], see also [BBB].

Let U be a Hopf algebra over an algebraically closed field k" and let $A = \mathrm{Map}(\mathbb{T}, k)$ be a commutative and cocommutative Hopf algebra of functions on a finite Abelian group $\mathbb{T}$ which is a Hopf subalgebra of U. Let P_μ, $\mu \in \mathbb{T}$ be the minimal idempotents in A.

DEFINITION 6.2.1. We say that an element x in $U^{\otimes n}$, $n \geq 1$, has *zero weight* if x commutes with $\Delta^n(a)$ for all $a \in A$, where $\Delta^n : A \to A^{\otimes n}$ is the iterated comultiplication.

DEFINITION 6.2.2. An invertible, zero-weight $U^{\otimes 2}$-valued function $J(\lambda)$ on $\mathbb{T}$ is called a *dynamical twist* for U if it satisfies the following functional equations:

$$(\Delta \otimes \mathrm{id}) J(\lambda)(J(\lambda + h^{(3)}) \otimes 1) = (\mathrm{id} \otimes \Delta) J(\lambda)(1 \otimes J(\lambda)), \qquad (6\text{–}10)$$

$$(\varepsilon \otimes \mathrm{id}) J(\lambda) = (\mathrm{id} \otimes \varepsilon) J(\lambda) = 1. \qquad (6\text{–}11)$$

Here and in what follows the notation $\lambda + h^{(i)}$ means that the argument λ is shifted by the weight of the i-th component, e.g., $J(\lambda + h^{(3)}) = \sum_\mu J(\lambda + \mu) \otimes P_\mu \in U^{\otimes 2} \otimes_k A$.

Note that for every fixed $\lambda \in \mathbb{T}$ the element $J(\lambda) \in U \otimes U$ does not have to be a twist for U in the sense of Drinfeld. It turns out that an appropriate object for which J defines a twisting is a certain quantum groupoid that we describe next.

Observe that the simple algebra $\mathrm{End}_k(A)$ has a natural structure of a cocommutative quantum groupoid given as follows (cf. Section 2.5):

Let $\{E_{\lambda\mu}\}_{\lambda,\mu\in\mathbb{T}}$ be a basis of $\mathrm{End}_k(A)$ such that

$$(E_{\lambda\mu}f)(\nu) = \delta_{\mu\nu}f(\lambda), \qquad f \in A, \ \lambda,\mu,\nu \in \mathbb{T}, \tag{6-12}$$

then the comultiplication, counit, and antipode of $\mathrm{End}_k(A)$ are given by

$$\Delta(E_{\lambda\mu}) = E_{\lambda\mu} \otimes E_{\lambda\mu}, \quad \varepsilon(E_{\lambda\mu}) = 1, \quad S(E_{\lambda\mu}) = E_{\mu\lambda}. \tag{6-13}$$

Define the tensor product quantum groupoid $H = \mathrm{End}_k(A) \otimes_k U$ and observe that the elements

$$\Theta = \sum_{\lambda\mu} E_{\lambda\lambda+\mu} \otimes E_{\lambda\lambda}P_\mu \quad \text{and} \quad \bar{\Theta} = \sum_{\lambda\mu} E_{\lambda+\mu\lambda} \otimes E_{\lambda\lambda}P_\mu \tag{6-14}$$

define a twist for H. Thus, according to Proposition 6.1.2, $H_\Theta = (\mathrm{End}_k(A) \otimes_k U)_\Theta$ becomes a quantum groupoid . It is non-commutative, non-cocommutative, and not a Hopf algebra if $|\mathbb{T}| > 1$.

It was shown in [EN], following [Xu], that H_Θ can be further twisted by means of a dynamical twist $J(\lambda)$ on U. Namely, if $J(\lambda)$ is a dynamical twist on U embedded in $H \otimes_k H$ as

$$J(\lambda) = \sum_\lambda E_{\lambda\lambda}J^{(1)}(\lambda) \otimes E_{\lambda\lambda}J^{(2)}(\lambda), \tag{6-15}$$

then the pair $(F(\lambda), \bar{F}(\lambda))$, where

$$F(\lambda) = J(\lambda)\Theta \quad \text{and} \quad \bar{F}(\lambda) = \bar{\Theta}J^{-1}(\lambda)$$

defines a twist for $H = \mathrm{End}_k(A) \otimes_k U$. Thus, every dynamical twist $J(\lambda)$ for a Hopf algebra U gives rise to a quantum groupoid $H_J = H_{J(\lambda)\Theta}$.

REMARK 6.2.3. According to Proposition 6.1.2, the antipode S_J of H_J is given by $S_J(h) = v^{-1}S(h)v$ for all $h \in H_J$, where S is the antipode of H and

$$v = \sum_{\lambda\mu} E_{\lambda+\mu\lambda}(S(J^{(1)})J^{(2)})(\lambda)P_\mu.$$

REMARK 6.2.4. There is a procedure dual to the one described above. It was shown in [EV] that given a dynamical twist $J(\lambda)$ it is possible to deform the multiplication on the vector space $D = \mathrm{Map}(\mathbb{T} \times \mathbb{T}, k) \otimes_k U^*$ and obtain a *dynamical quantum group* D_J. The relation between H_J and D_J in the case when $\dim U < \infty$ was established in [EN], where it was proven that D_J is isomorphic to H_J^{*op}.

6.3. Dynamical twists for $U_q(\mathfrak{g})$ at roots of 1. Suppose that $\mathfrak{g}$ is a simple Lie algebra of type A, D or E and q is a primitive ℓth root of unity in k, where $\ell \geq 3$ is odd and coprime with the determinant of the Cartan matrix $(a_{ij})_{ij=1,\ldots,m}$ of $\mathfrak{g}$.

Let $U = U_q(\mathfrak{g})$ be the corresponding quantum group which is a finite dimensional Hopf algebra with generators E_i, F_i, K_i, where $i = 1, \ldots, m$ and relations as in [L].

Let $\mathbb{T} \cong (\mathbb{Z}/\ell\mathbb{Z})^m$ be the abelian group generated by K_i, $i = 1, \ldots, m$. For any m-tuple of integers $\lambda = (\lambda_1, \ldots, \lambda_m)$ we will write $K_\lambda = K_1^{\lambda_1} \ldots K_m^{\lambda_m} \in \mathbb{T}$.

Let $\mathcal{R}$ be the universal R-matrix of $U_q(\mathfrak{g})$ and Ω be the "Cartan part" of $\mathcal{R}$ ([T]).

For arbitrary non-zero constants $\Lambda_1, \ldots, \Lambda_m$ define a Hopf algebra automorphism Λ of U by setting

$$\Lambda(E_i) = \Lambda_i E_i, \quad \Lambda(F_i) = \Lambda_i^{-1} F_i, \quad \text{and} \quad \Lambda(K_i) = K_i \quad \text{for all } i = 1, \ldots, m.$$

We will say that $\Lambda = (\Lambda_1, \ldots, \Lambda_m)$ is *generic* if the spectrum of Λ does not contain ℓth roots of unity.

Note that the algebra U is $\mathbb{Z}$-graded with

$$\deg(E_i) = 1, \quad \deg(F_i) = -1, \quad \deg(K_i) = 0, \qquad i = 1, \ldots, m, \qquad (6\text{--}16)$$

and $\deg(XY) = \deg(X) + \deg(Y)$ for all X and Y. Of course, there are only finitely many non-zero components of U since it is finite dimensional.

Let U_+ be the subalgebra of U generated by the elements E_i, K_i, $i = 1, \ldots, m$, U_- be the subalgebra generated by F_i, K_i, $i = 1, \ldots, m$, and $I_\pm$ be the kernels of the projections from $U_\pm$ to the elements of zero degree.

The next Proposition was proven in [ABRR], [ES], [ESS] for generic values of q and in [EN] for q a root of unity:

PROPOSITION 6.3.1. *For every generic* Λ *there exists a unique element* $J(\lambda) \in 1 + I_+ \otimes I_-$ *that satisfies the following ABRR relation* [ABRR], [ES], [ESS]:

$$(\operatorname{Ad} K_\lambda \circ \Lambda \otimes \operatorname{id})(\mathcal{R}J(\lambda)\Omega^{-1}) = J(\lambda), \quad \lambda \in \mathbb{T}. \qquad (6\text{--}17)$$

The element

$$\mathcal{J}(\lambda) = J(2\lambda + h^{(1)} + h^{(2)}) \qquad (6\text{--}18)$$

is a dynamical twist for $U_q(\mathfrak{g})$ *in the sense of Definition* 6.2.2.

PROOF. The existence and uniqueness of the solution of (6–17) and the fact that $\mathcal{J}(\lambda)$ satisfies (6–10) are established by induction on the degree of the first component of $J(\lambda)$, see ([EN], 5.1). $\qquad\square$

REMARK 6.3.2. (a) The reason for introducing the "shift" automorphism Λ is to avoid singularities in equation [ABRR]. Thus, we have a family of dynamical deformations of $U_q(\mathfrak{g})$ (and, therefore, a family of quantum groupoids depending on m parameters).

(b) One can generalize the above construction and associate a dynamical twist

$\mathfrak{J}_T(\lambda)$ with any *generalized Belavin–Drinfeld triple* which consists of subsets Γ_1, Γ_2 of the set $\Gamma = (\alpha_1, \ldots, \alpha_m)$ of simple roots of $\mathfrak{g}$ together with an inner product preserving bijection $T : \Gamma_1 \to \Gamma_2$, see [EN], [ESS].

EXAMPLE 6.3.3. Let us give an explicit expression for the twists $J(\lambda)$ and $\mathfrak{J}(\lambda)$ in the case $\mathfrak{g} = sl(2)$. $U_q(\mathfrak{g})$ is then generated by E, F, K with the standard relations. The element analogous to $J(\lambda)$ for generic q was computed already in [Ba] (see also [BBB]). If we switch to our conventions, this element will take the form

$$J(\lambda) = \sum_{n=0}^{\infty} q^{-n(n+1)/2} \frac{(1-q^2)^n}{[n]_q!} (E^n \otimes F^n) \prod_{\nu=1}^{n} \frac{\Lambda q^{2\lambda}}{1 - \Lambda q^{2\lambda+2\nu}(K \otimes K^{-1})}.$$

It is obvious that the formula for q being a primitive ℓ-th root of unity is simply obtained by truncating this formula:

$$J(\lambda) = \sum_{n=0}^{\ell-1} q^{-n(n+1)/2} \frac{(1-q^2)^n}{[n]_q!} (E^n \otimes F^n) \prod_{\nu=1}^{n} \frac{\Lambda q^{2\lambda}}{1 - \Lambda q^{2\lambda+2\nu}(K \otimes K^{-1})}.$$

Therefore,

$$\mathfrak{J}(\lambda) = \sum_{n=0}^{\ell-1} q^{-n(n+1)/2} \frac{(1-q^2)^n}{[n]_q!} (E^n \otimes F^n) \prod_{\nu=1}^{n} \frac{\Lambda q^{4\lambda} K \otimes K}{1 - \Lambda q^{4\lambda+2\nu}(K^2 \otimes 1)}.$$

An important new property of the resulting quantum groupoids $H_{\mathfrak{J}} = U_q(\mathfrak{g})_{\mathfrak{J}}$, compared with quantum groups $U_q(\mathfrak{g})$ is their selfduality. A twisting of a quasitriangular quantum groupoid is again quasitriangular (Remark 6.1.3(d)), so the twisted R-matrix

$$\mathcal{R}(\lambda) = \bar{\Theta}_{21} \mathcal{R}\Theta = \sum_{\lambda\mu\nu} E_{\lambda\lambda+\nu} P_\mu \mathcal{R}^{\mathfrak{J}(1)}(\lambda) \otimes E_{\lambda+\mu\lambda} \mathcal{R}^{\mathfrak{J}(2)}(\lambda) P_\nu,$$

where $\mathcal{R}^J(\lambda) = \mathfrak{J}_{21}^{-1}(\lambda) R \mathfrak{J}(\lambda)$, establishes a homomorphism between $D_{\mathfrak{J}} = H_{\mathfrak{J}}^{*op}$ and $H_{\mathfrak{J}}$. One can show that the image of $\mathcal{R}(\lambda)$ contains all the generators of $U_q(\mathfrak{g})$, i.e., that the above homomorphism is in fact an isomorphism; see ([EN], 5.3).

REMARK 6.3.4. The same selfduality result holds for any generalized Belavin–Drinfeld triple for which T is an automorphism of the Dynkin diagram of $\mathfrak{g}$ ([EN], 5.4).

7. Semisimple and C^*-Quantum Groupoids

7.1. Definitions.

DEFINITION 7.1.1. A quantum groupoid is said to be *semisimple* (resp., *∗- or C^*-quantum groupoid*) if its algebra H is semisimple (resp., a ∗-algebra over a field k with involution, or finite-dimensional C^*-algebra over the field $\mathbb{C}$ such that Δ is a ∗-homomorphism).

Groupoid algebras and their duals give examples of commutative and cocommutative semisimple quantum groupoids (Corollary 3.3.2), which are C^*-quantum groupoids if the ground field is $\mathbb{C}$ (in which case $g^* = g^{-1}$ for all $g \in G$).

We will describe the class of quantum groupoids possessing *Haar integrals*, i.e., normalized two-sided integrals (note that if such an integral exists then it is unique and is an S-invariant idempotent).

DEFINITION 7.1.2 [W, 1.2.1]. Given an inclusion of unital k-algebras $N \subset M$, a *conditional expectation* $E : M \to N$ is an $N - N$ bimodule map E such that $E(1) = 1$. A *quasi-basis* for E is an element $\sum_i x_i \otimes y_i \in M \otimes_k M$ such that

$$\sum_i E(mx_i)y_i = m = \sum_i x_i E(y_i m), \qquad \text{for all } m \in M.$$

One can check that $\text{Index } E = \sum_i x_i y_i \in Z(M)$ does not depend on a choice of a quasi-basis; this element is called the *index* of E.

One can apply this definition to a non-degenerate functional f on H. If (l, λ) is a dual pair of left integrals, then $l_{(2)} \otimes S^{-1}(l_{(1)})$ is a quasi-basis for λ and

$$\text{Index } \lambda = S^{-1} \cdot \varepsilon_t(l) \in H_s \cap Z(H).$$

So a non-degenerate left integral is normalized if and only if its dual has index 1.

7.2. Existence of the Haar integral. Given a dual bases $\{f_i\}$ and $\{\xi^i\}$ of H and $\widehat{H}$, respectively, let us consider the following canonical element

$$\chi = \sum_i (\xi^i \leftharpoonup S^{-2}(f_i)).$$

One can show that χ is a left integral in $\widehat{H}$ such that $\chi(xy) = \chi(yS^2(x))$ for all $x, y \in H$ and $l \rightharpoonup \chi = \widehat{S}^2(\widehat{1} \leftharpoonup l)$ for any left integral l in H (see [BNSz], 3.25).

LEMMA 7.2.1 ([BNSz], 3.26). (i) *The Haar integral $h \in H$ exists if and only if the above mentioned χ is non-degenerate, in which case (h, χ) is a dual pair of left integrals.*
(ii) *A left integral l is a Haar integral if and only if $\varepsilon_s(l) = 1$.*

PROOF. (ii) If $\varepsilon_s(l) = 1$, then by $l \rightharpoonup \chi = \widehat{S}^2(\widehat{1} \leftharpoonup l)$ one has $l \rightharpoonup \chi = \widehat{1}$. Therefore, (h, χ) is a dual pair of non-degenerate left integrals. The property $\chi(xy) = \chi(yS^2(x))$ is equivalent to that the quasi-basis of χ satisfies

$$l_{(2)} \otimes S^{-1}(l_{(1)}) = S(l_{(1)}) \otimes l_{(2)},$$

from where $\Delta(l) = \Delta(S(l))$ and $l = S(l)$. Furthermore, $\varepsilon_t(l) = \varepsilon_t(S(l)) = S \cdot \varepsilon_s(l) = 1$. Thus, l is a Haar integral. The inverse statement is obvious.
(i) The "only if" part follows from the proof of (ii). If χ is non-degenerate and h is its dual left integral, then, as above, $S(h) = h$; so h is a two-sided integral and since $l \rightharpoonup \chi = \widehat{S}^2(\widehat{1} \leftharpoonup l)$, it is normalized. $\qquad \square$

THEOREM 7.2.2 ([BNSz], 3.27). *Let H be a finite quantum groupoid over an algebraically closed field k. Then the following conditions are equivalent:*

(i) *There exists a Haar integral,*

(ii) *H is semisimple and there exists an invertible element $g \in H$ such that $gxg^{-1} = S^2(x)$ for all $x \in H$ and $Tr(\pi_\alpha(g^{-1})) \neq 0$ for all irreducible representations π_α of H (here Tr is the usual trace on a matrix algebra).*

PROOF. The assumption on k is used only to ensure that $H = \oplus_\alpha M_{n_\alpha}(k)$, once knowing that it is semisimple, so that there is a k-basis of matrix units $\{e_{ij}^\alpha\}$ in H.

(ii) $\Longrightarrow$ (i): Recall that any trace τ on H is completely determined by its trace vector $\tau_\alpha = \tau(1_\alpha)$. Now let $\tau : H \to k$ be the trace with trace vector $\tau_\alpha = \mathrm{tr}\pi_\alpha(g^{-1})$. Then τ is non-degenerate and has a quasi-basis

$$\sum_i x_i \otimes y_i = \sum_\alpha \frac{1}{\tau_\alpha} \sum_{i,j=1}^{n_\alpha} e_{ij}^\alpha \otimes e_{ji}^\alpha.$$

Notice that $\sum_i x_i g^{-1} y_i = 1$. It is straightforward to verify that $\chi' = g \rightharpoonup \tau$ coincides with χ, so χ is non-degenerate and therefore its dual left integral l has $\varepsilon_s(l) = 1$ by $l \rightharpoonup \chi = \widehat{S}^2(\widehat{1} \leftharpoonup l)$. Thus, l is a Haar integral.

(i) $\Longrightarrow$ (ii): If h is a Haar integral, then H is semisimple by Theorem 3.3.1. Let τ be a non-degenerate trace on H, then there exists a unique $i \in H$ such that $i \rightharpoonup \tau = \widehat{1} = \tau \leftharpoonup i$. One can verify that i is a two-sided non-degenerate integral in H and that $S(i) = i$ (see [BNSz], I.3.21). To prove that S^2 is inner, it is enough to construct a non-degenerate functional χ on H such that $\chi(xy) = \chi(yS^2(x))$ for all $x, y \in H$. But the proof of Lemma 7.2.1 shows that this is the case for the dual left integral to i. $\square$

REMARK 7.2.3. (a) There is a unique normalization of g from Theorem 7.2.2(ii) such that the following conditions hold ([BNSz], 4.4):

(i) $\mathrm{tr}(\pi_\alpha(g^{-1})) = \mathrm{tr}(\pi_\alpha(g))$ for all irreducible representations π_α of H (here tr is a usual trace on a matrix algebra);

(ii) $S(g) = g^{-1}$.

One can show that such a g is *group-like*, i.e.,

$$\Delta(g) = (g \otimes g)\Delta(1) = \Delta(1)(g \otimes g).$$

The element g implementing S^2 and satisfying normalization conditions (i) and (ii) is called the *canonical group-like element* of H.

(b) Since $\chi = g \rightharpoonup \tau$ is the dual left integral to h, its quasi-basis $\sum_i x_i g^{-1} \otimes y_i$ equals to $h_{(2)} \otimes S^{-1}(h_{(1)})$, which implies the formula

$$(S \otimes \mathrm{id})\Delta(h) = \sum_i x_i \otimes g^{-1} y_i = \sum_\alpha 1/\tau_\alpha \sum_{ij} e_{ij}^\alpha g^{-1/2} \otimes g^{-1/2} e_{ji}^\alpha.$$

A dual notion to that of the Haar integral is the Haar measure. Namely, a functional $\phi \in \widehat{H}$ is said to be a *Haar measure* on H if it is a normalized left and right invariant measure and $\phi \circ S = \phi$. Any of the equivalent conditions of Theorem 7.2.1 is also equivalent to existence (and uniqueness) of the Haar measure on $\widehat{H}$.

7.3. C^*-quantum groupoids. Definition 7.1.1 and the uniqueness of the unit, counit and the antipode (see Proposition 2.3.1) imply that

$$1^* = 1, \quad \varepsilon(x^*) = \overline{\varepsilon(x)}, \quad (S \circ *)^2 = \mathrm{id}$$

for all x in any $*$-quantum groupoid H. It is also easy to check the relations

$$\varepsilon_t(x)^* = \varepsilon_t(S(x)^*), \qquad \varepsilon_t(x)^* = \varepsilon_t(S(x)^*),$$

which imply that H_t and H_s are $*$-subalgebras, and to show that the dual, $\widehat{H}$, is also a $*$-quantum groupoid with respect to the $*$-operation

$$\langle x, \phi^* \rangle = \overline{\langle S(x)^*, \phi \rangle} \qquad \text{for all } \phi \in \widehat{H}, \, x \in H. \tag{7--1}$$

The $*$-operation allows to simplify the axioms of a quantum groupoid (cf. the axioms used in [NV1], [N1]). The second parts of equalities (2–1) and (2–2) of Definition 2.1.1 follow from the rest of the axioms, also $S * \mathrm{id} = \varepsilon_s$ is equivalent to $\mathrm{id} * S = \varepsilon_t$. Alternatively, under the condition that the antipode is both algebra and coalgebra anti-homomorphism, the axioms (2–1) and (2–2) and can be replaced by the identities of Proposition 2.2.1 (ii) and (v) involving the target counital map.

THEOREM 7.3.1 [BNSz, 4.5]. *In a C^*-quantum groupoid the Haar integral h exists, $h = h^*$ and*

$$(\phi, \psi) = \langle h, \phi^*\psi \rangle, \qquad \phi, \psi \in \widehat{H}$$

is a scalar product making $\widehat{H}$ a Hilbert space where the left regular representation of $\widehat{H}$ is faithful. Thus, $\widehat{H}$ is a C^*-quantum groupoid, too.

PROOF. Clearly, H and g verify all the conditions of Theorem 7.2.2, from where the existence of Haar integral follows. Since h is non-degenerate, the scalar product $(\cdot, \cdot)$ is also non-degenerate. By the equality

$$(\phi, \phi) = \langle h, \phi^*\phi \rangle = \overline{\langle S(h_{(1)})^*, \phi \rangle} \langle h_{(2)}, \phi \rangle,$$

positivity of $(\cdot, \cdot)$ follows from Remark 7.2.3(b). $\square$

We will denote by $\widehat{h}$ the Haar measure of $\widehat{H}$.

REMARK 7.3.2. ε is a positive functional, i.e., $\varepsilon(x^*x) \geq 0$ for all $x \in H$. Indeed, for all $x \in H$ we have

$$\varepsilon(x^*x) = \varepsilon(x^* 1_{(1)})\varepsilon(1_{(2)} 1'_{(2)})\varepsilon(1'_{(1)} x) = \varepsilon(\varepsilon_t(x)^* \varepsilon_t(x)) = \langle \widehat{h}, \varepsilon_t(x)^* \varepsilon_t(x) \rangle \geq 0,$$

where $\widehat{h}|_{H_t} = \varepsilon|_{H_t}$, since $\langle \widehat{h}, z \rangle = \langle \widehat{\varepsilon_t}(\widehat{h}), z \rangle = \langle \widehat{1}, z \rangle$ for all $z \in H_t$.

The Haar measure provides target and source *Haar conditional expectations* (all properties are easy to verify):

$$E_t : H \to H_t : E_t(x) = (\mathrm{id} \otimes \widehat{h})\Delta(x),$$
$$E_s : H \to H_s : E_s(x) = (\widehat{h} \otimes \mathrm{id})\Delta(x).$$

Let us introduce an element $g_t = E_t(h)^{1/2} \in H_t$.

REMARK 7.3.3 ([BNSz], 4.12). The element g_t is positive and invertible, and the canonical group-like element of H can be written as $g = g_t S(g_t^{-1})$.

PROPOSITION 7.3.4 ([NTV], 9.3). *If H is a C^*-quantum groupoid, then $D(H)$ is a quasitriangular C^*-quantum groupoid.*

PROOF. The result follows from the fact that $\widehat{D(H)}$ is a C^*-quantum groupoid, which is easy to verify using the explicit formulas from Remark 5.3.3.   □

PROPOSITION 7.3.5. *A quasitriangular C^*-quantum groupoid H is automatically ribbon with the ribbon element $\nu = u^{-1}g = gu^{-1}$, where u is the Drinfeld element from Definition 5.2.7 and g is the canonical group-like element.*

PROOF. Since u also implements S^2 (Proposition 5.2.6), $\nu = u^{-1}g$ is central, therefore $S(\nu)$ is also central. Clearly, u must commute with g. The same Proposition gives $\Delta(u^{-1}) = \mathcal{R}_{21}\mathcal{R}(u^{-1} \otimes u^{-1})$, which allows us to compute

$$\Delta(\nu) = \Delta(u^{-1})(g \otimes g) = \mathcal{R}_{21}\mathcal{R}(u^{-1}g \otimes u^{-1}g) = \mathcal{R}_{21}\mathcal{R}(\nu \otimes \nu).$$

Propositions 5.2.5 and 5.2.6 and the trace property imply that

$$\mathrm{tr}(\pi_\alpha(u^{-1})) = \mathrm{tr}(\pi_\alpha(\mathcal{R}^{(2)}S^2(\mathcal{R}^{(1)})))$$
$$= \mathrm{tr}(\pi_\alpha(S^3(\mathcal{R}^{(1)})S(\mathcal{R}^{(2)}))) = \mathrm{tr}(\pi_\alpha(S(u^{-1}))).$$

Since $u^{-1} = \nu g^{-1}$ and ν is central, the above relation means that

$$\mathrm{tr}(\pi_\alpha(\nu))\mathrm{tr}(\pi_\alpha(g^{-1})) = \mathrm{tr}(\pi_\alpha(S(\nu)))\mathrm{tr}(\pi_\alpha(g)),$$

and, therefore, $\mathrm{tr}(\pi_\alpha(\nu)) = \mathrm{tr}(\pi_\alpha(S(\nu)))$ for any irreducible representation π_α, which shows that that $\nu = S(\nu)$ (cf. [EG]).   □

REMARK 7.3.6. For a connected ribbon C^*-quantum groupoid we have:

$$\mathrm{tr}_q(f) = (\dim H_t)^{-1}\mathrm{Tr}_V(g \circ f), \ \dim_q(V) = (\dim H_t)^{-1}\mathrm{Tr}_V(g),$$

where V is an H-module and $f \in \mathrm{End}_k(V)$.

7.4. C^*-quantum groupoids and unitary modular categories. To define the (unitary) representation category $\mathrm{URep}(H)$ of a C^*-quantum groupoid H we consider *unitary* H-modules, i.e., H-modules V equipped with a scalar product

$$(\cdot, \cdot) : V \times V \to \mathbb{C} \qquad \text{such that} \qquad (h \cdot v, w) = (v, h^* \cdot w) \; \forall h \in H, v, w \in V.$$

The notion of a morphism in this category is the same as in $\mathrm{Rep}(H)$.

The monoidal product of $V, W \in \mathrm{URep}(H)$ is defined as follows. We construct a tensor product $V \otimes_{\mathbb{C}} W$ of Hilbert spaces and note that the action of $\Delta(1)$ on this left H-module is an orthogonal projection. The image of this projection is, by definition, the monoidal product of V, W in $\mathrm{URep}(H)$. Clearly, this definition is compatible with the monoidal product of morphisms in $\mathrm{Rep}(H)$.

For any $V \in \mathrm{URep}(H)$, the dual space V^* is naturally identified $(v \to \overline{v})$ with the conjugate Hilbert space, and under this identification we have $h \cdot \overline{v} = \overline{S(h)^* \cdot v}$ $(v \in V, \overline{v} \in V^*)$. In this way V^* becomes a unitary H-module with scalar product $(\overline{v}, \overline{w}) = (w, gv)$, where g is the canonical group-like element of H.

The unit object in $\mathrm{URep}(H)$ is H_t equipped with scalar product $(z, t)_{H_t} = \varepsilon(zt^*)$ (it is known [BNSz], [NV1] that the restriction of ε to H_t is a nondegenerate positive form). One can verify that the maps l_V, r_V and their inverses are isometries and that $\mathrm{URep}(H)$ is a monoidal category with duality (see also [BNSz], Section 3). In a natural way we have a notion of a *conjugation* of morphisms ([T], II.5.1).

REMARK 7.4.1. a) One can check that for a quasitriangular $*$-quantum groupoid H the braiding is an isometry in $\mathrm{URep}(H)$: $c_{V,W}^{-1} = c_{V,W}^*$.

b) For a ribbon C^*-quantum groupoid H, the twist is an isometry in $\mathrm{URep}(H)$. Indeed, the relation $\theta_V^* = \theta_V^{-1}$ is equivalent to the identity $S(u^{-1}) = u^*$, which follows from Proposition 5.2.5 and Remark 7.4.1a).

A *Hermitian ribbon category* is a ribbon category endowed with a conjugation of morphisms $f \mapsto \overline{f}$ satisfying natural conditions of ([T], II.5.2). A *unitary ribbon category* is a Hermitian ribbon category over the field $\mathbb{C}$ such that for any morphism f we have $\mathrm{tr}_q(f\overline{f}) \geq 0$. In a natural way we have a *conjugation* of morphisms in $\mathrm{URep}(H)$. Namely, for any morphism $f : V \to W$ we define $\overline{f} : W \to V$ as $\overline{f}(w) = \overline{f^*(\overline{w})}$ for any $w \in W$. Here $\overline{w} \in W^*, f^* : W^* \to V^*$ is the standard dual of f (see [T], I.1.8) and $\overline{f^*(\overline{w})} \in V$. For the proof of the following lemma see ([NTV], 9.7).

LEMMA 7.4.2. *Given a quasitriangular C^*-quantum groupoid H, $\mathrm{URep}(H)$ is a unitary ribbon Ab-category with respect to the above conjugation of morphisms.*

The next proposition extends ([EG], 1.2).

PROPOSITION 7.4.3. *If H is a connected C^*-quantum groupoid, then $\mathrm{Rep}(D(H))$ is a unitary modular category.*

PROOF. The proof follows from Lemmas 7.4.2, 5.5.1 and Propositions 5.3.4, 7.3.4. $\square$

8. C^*-Quantum Groupoids and Subfactors: The Depth 2 Case

In this section we characterize finite index type II_1 subfactors of depth 2 in terms of C^*-quantum groupoids. Recall that a II_1 factor is a $*$-algebra of operators on a Hilbert space that coincides with its second centralizer, or, equivalently, is weakly closed (i.e., is a von Neumann algebra), with the trivial center that admits a finite trace. An example of such a factor is the group von Neumann algebra of a discrete group whose every non-trivial conjugacy class is infinite, the corresponding trace given by its Haar measure. It is also known, that there exists a unique, up to an isomorphism, hyperfinite (i.e., generated by an increasing sequence of finite-dimensional C^*-algebras) II_1 factor.

There is a notion of index for subfactors, extending the notion of index of a subgroup in a group, though it can be non-integer. For the foundations of the subfactor theory see [GHJ], [JS].

8.1. Actions of C^*-quantum groupoids on von Neumann algebras. Let a von Neumann algebra M be a left H-module algebra in the sense of Definition 4.1.1 via a weakly continuous action of a C^*-quantum groupoid

$$H \otimes_{\mathbb{C}} M \ni x \otimes m \mapsto (x \triangleright m) \in M$$

such that $(x \triangleright m)^* = S(x)^* \triangleright m^*$ and $x \triangleright 1 = 0$ if and only if $\varepsilon_t(t) = 0$.

Then one can show ([NSzW], 3.4.2) that the smash product algebra (now we call it *crossed product algebra* and denote by $M \rtimes H$), equipped with an involution $[m \otimes x]^* = [(x_{(1)}^* \triangleright m^*) \otimes x_{(2)}^*]$, is a von Neumann algebra.

The collection $M^H = \{m \in M \mid x \triangleright m = \varepsilon_t(x) \triangleright m, \ \forall x \in H\}$ is a von Neumann subalgebra of M, called a *fixed point subalgebra*. The centralizer $M' \cap M \rtimes H$ always contains the source counital subalgebra H_s. Indeed, if $y \in H_s$, then $\Delta(y) = 1_{(1)} \otimes 1_{(2)} y$, therefore

$$[1 \otimes y][m \otimes 1] = [(y_{(1)} \triangleright m) \otimes y_{(2)}] = [(1_{(1)} \triangleright m) \otimes 1_{(2)} y]$$
$$= [m \otimes y] = [m \otimes 1][1 \otimes y],$$

for any $m \in M$, and $H_s \subset M' \cap M \rtimes H$. An action of H is called *minimal* if $H_s = M' \cap M \rtimes H$.

One can define the dual action of $\widehat{H}$ on the von Neumann algebra $M \rtimes H$ as in (4–1) and construct the von Neumann algebra $(M \rtimes H) \rtimes \widehat{H}$.

REMARK 8.1.1. The following results hold true if $S^2 = \mathrm{id}$, i.e., if H is involutive ([N1], 4.6, 4.7).

(a) $M \subset M \rtimes H \subset (M \rtimes H) \rtimes \widehat{H}$ is a basic construction;
(b) $M \rtimes H$ is free as a left M-module;

(c) $(M \rtimes H) \rtimes \widehat{H} \cong M \otimes_{\mathbb{C}} M_n(\mathbb{C})$ for some integer n.

EXAMPLE 8.1.2. For a trivial action of H on H_t and the corresponding dual action of $\widehat{H}$ on $H \cong H_t \rtimes H$ (see Example 4.1.3 (i) and (ii)), $H_t \subset H \subset H \rtimes \widehat{H}$ is the basic construction of finite-dimensional C^*-algebras with respect to the Haar conditional expectation E_t ([BSz2], 4.2).

If H is a connected C^*-quantum groupoid having a minimal action on a II_1 factor M then one can show ([NSzW], 4.2.5, 4.3.5) that $\widehat{H}$ is also connected and that

$$N = M^H \subset M \subset M_1 = M \rtimes H \subset M_2 = (M \rtimes H) \rtimes \widehat{H} \subset \cdots$$

is the Jones tower of factors of finite index with *the derived tower*

$$N' \cap N = \mathbb{C} \subset N' \cap M = H_t \subset N' \cap M_1 = H \subset N' \cap M_2 = H \rtimes \widehat{H} \subset \cdots$$

The fact that the last triple of finite-dimensional C^*-algebras is a basic construction, means exactly that the subfactor $N \subset M$ has depth 2. Moreover, the finite-dimensional C^*-algebras

$$
\begin{array}{ccc}
H & \subset & H \rtimes \widehat{H} \\
\cup & & \cup \\
H_s \cong \widehat{H}_t & \subset & \widehat{H},
\end{array}
\qquad (8\text{--}1)
$$

form *a canonical commuting square*, which completely determines the equivalence class of the initial subfactor. This implies that any biconnected C^*-quantum groupoid has at most one minimal action on a given II_1 factor and thus corresponds to no more than one (up to equivalence) finite index depth 2 subfactor.

REMARK 8.1.3. It was shown in ([N1], 5) that any biconnected involutive C^*-quantum groupoid has a minimal action on the hyperfinite II_1 factor. This action is constructed by iterating the basic construction for the square (8–1) in the horizontal direction, see [N1] for details.

8.2. Construction of a C^*-quantum groupoid from a depth 2 subfactor.
Let $N \subset M$ be a finite index ($[M : N] = \lambda^{-1}$) depth 2 II_1 subfactor and

$$N \subset M \subset M_1 \subset M_2 \subset \cdots \qquad (8\text{--}2)$$

the corresponding Jones tower, $M_1 = \langle M, e_1 \rangle$, $M_2 = \langle M_1, e_2 \rangle, \ldots$, where $e_1 \in N' \cap M_1$, $e_2 \in M' \cap M_2, \cdots$ are the Jones projections. The depth 2 condition means that $N' \cap M_2$ is the basic construction of the inclusion $N' \cap M \subset N' \cap M_1$. Let τ be the trace on M_2 normalized by $\tau(1) = 1$.

Let us denote

$$A = N' \cap M_1, \quad B = M' \cap M_2, \quad B_t = M' \cap M_1, \quad B_s = M_1' \cap M_2$$

and let Tr be the trace of the regular representation of B_t on itself. Since both τ and Tr are non-degenerate, there exists a positive invertible element $w \in Z(B_t)$

such that $\tau(wz) = \mathrm{Tr}\,(z)$ for all $z \in B_t$ (the index of $\tau|_{M' \cap M_1}$ in the sense of [W]).

PROPOSITION 8.2.1. *There is a canonical non-degenerate duality form between A and B defined by*

$$\langle a, b \rangle = \lambda^{-2}\tau(ae_2e_1wb) = \lambda^{-2}\mathrm{Tr}\,(E_{B_t}(bae_2e_1)), \qquad (8\text{--}3)$$

for all $a \in A$ and $b \in B$.

PROOF. If $a \in A$ is such that $\langle a, B \rangle = 0$, then, using [PP1], 1.2, one has

$$\tau(ae_2e_1 B) = \tau(ae_2e_1(N' \cap M_2)) = 0,$$

therefore, using the properties of τ and Jones projections, we get

$$\tau(aa^*) = \lambda^{-1}\tau(ae_2a^*) = \lambda^{-2}\tau(ae_2e_1(e_2a^*)) = 0,$$

so $a = 0$. Similarly for $b \in B$.   □

Using the duality $\langle a, b \rangle$, one defines a coalgebra structure on B:

$$\langle a_1 \otimes a_2, \Delta(b) \rangle = \langle a_1a_2, b \rangle, \qquad (8\text{--}4)$$
$$\varepsilon(b) = \langle 1, b \rangle, \qquad (8\text{--}5)$$

for all $a, a_1, a_2 \in A$ and $b \in B$. Similarly, one defines a coalgebra structure on A.

We define a linear endomorphism $S_B : B \to B$ by

$$E_{M_1}(be_1e_2) = E_{M_1}(e_2e_1 S_B(b)), \qquad (8\text{--}6)$$

Note that $\tau \circ S_B = \tau$. Similarly one can define a linear endomorphism $S_A : A \to A$ such that $E_{M'}(S_A(a)e_2e_1) = E_{M'}(e_1e_2a)$ and $\tau \circ S_A = \tau$.

PROPOSITION 8.2.2 [NV2, 4.5, 5.2]. *The following identities hold:*

(i) $S_B(B_s) = B_t$,
(ii) $S_B^2(b) = b$ *and* $S_B(b)^* = S_B(b^*)$,
(iii) $\Delta(b)(S_B(z) \otimes 1) = \Delta(b)(1 \otimes z), \qquad \Delta(bz) = \Delta(b)(z \otimes 1), \ (\forall z \in B_t)$,
(iv) $S_B(bc) = S_B(c)S_B(b)$,
    $\Delta(S_B(b)) = (1 \otimes w^{-1})(S_B(b_{(2)}) \otimes S_B(b_{(1)}))(S_B(w) \otimes 1)$.

Define an antipode and new involution of B by

$$S(b) = S_B(wbw^{-1}), \qquad b^\dagger = S(w)^{-1}b^*S(w). \qquad (8\text{--}7)$$

THEOREM 8.2.3. *With the above operations B becomes a biconnected C^*-quantum groupoid with the counital subalgebras B_s and B_t.*

PROOF. Since $\Delta(1) \in B_s \otimes B_t$ ([NV2], 4.6) and B_t commutes with B_s, we have

$$(\text{id} \otimes \Delta)\Delta(1) = 1_{(1)} \otimes \Delta(1)(1_{(2)} \otimes 1)$$
$$= (1 \otimes \Delta(1))(\Delta(1) \otimes 1) = (\Delta(1) \otimes 1)(1 \otimes \Delta(1))$$

which is axiom (2–1) of Definition 2.1.1. The axiom (2–2) dual to it is obtained similarly, considering the comultiplication in A.

As a consequence of the "symmetric square" relations $BM_1 = M_1B = M_2$ one obtains ([NV2], 4.12) the identity

$$bx = \lambda^{-1}E_{M_1}(b_{(1)}xe_2)b_{(2)} \quad \text{for all} \quad b \in B, \, x \in M_1, \qquad (8\text{–}8)$$

from where it follows that

$$E_{M_1}(bxye_2) = \lambda^{-1}E_{M_1}(b_{(1)}xe_2)E_{M_1}(b_{(2)}ye_2) \quad \text{for all} \quad b \in B, \, x, y \in M_1.$$
$$(8\text{–}9)$$

We use this identity to prove that Δ is a homomorphism:

$$\langle a_1a_2, bc \rangle = \langle \lambda^{-1}E_{M_1}(ca_1a_2e_2), b \rangle$$
$$= \langle \lambda^{-2}E_{M_1}(c_{(1)}a_1e_2)E_{M_1}(c_{(2)}a_2e_2), b \rangle$$
$$= \langle \lambda^{-1}E_{M_1}(c_{(1)}a_1e_2), b_{(1)} \rangle\langle \lambda^{-1}E_{M_1}(c_{(2)}a_2e_2), b_{(2)} \rangle$$
$$= \langle a_1, b_{(1)}c_{(1)} \rangle\langle a_2, b_{(2)}c_{(2)} \rangle,$$

for all $a_1, a_2 \in A$, whence $\Delta(bc) = \Delta(b)\Delta(c)$. Next,

$$\langle a, \varepsilon(1_{(1)}b)1_{(2)} \rangle = \langle 1, 1_{(1)}b \rangle\langle a, 1_{(2)} \rangle$$
$$= \langle \lambda^{-1}E_{M_1}(be_2), 1_{(1)} \rangle\langle a, 1_{(2)} \rangle$$
$$= \langle \lambda^{-1}E_{M_1}(be_2)a, 1 \rangle = \langle a, \lambda^{-1}E_{M_1}(be_2) \rangle,$$

therefore $\varepsilon_t(b) = \lambda^{-1}E_{M_1}(be_2)$ (similarly, $\varepsilon_s(b) = \lambda^{-1}E_{M_1'}(e_2b)$). Using Equation (8–9) we have

$$\langle a, b_{(1)}S(b_{(2)}) \rangle = \langle a, b_{(1)}S_B(wb_{(2)}w^{-1}) \rangle$$
$$= \lambda^{-3}\tau(E_{M_1}(S_B(wb_{(2)}w^{-1})ae_2)e_2e_1wb_{(1)})$$
$$= \lambda^{-3}\tau(E_{M_1}(e_2awb_{(2)}w^{-1})e_2e_1wb_{(1)})$$
$$= \lambda^{-3}\tau(E_{M_1}(e_2awb_{(2)}w^{-1})E_{M_1}(e_2e_1wb_{(1)}))$$
$$= \lambda^{-2}\tau(E_{M_1}(e_2ae_1wb)) = \langle a, \varepsilon_t(b) \rangle.$$

The antipode S is anti-multiplicative and anti-comultiplicative, therefore axiom (2–5) of Definition 2.1.1 follows from the other axioms. Clearly, $\varepsilon_t(B) = M' \cap M_1$ and $\varepsilon_s(B) = M_1' \cap M_2$. B is biconnected, since the inclusion $B_t = M' \cap M_1 \subset B = M' \cap M_2$ is connected ([GHJ], 4.6.3) and $B_t \cap B_s = (M' \cap M_1) \cap (M_1' \cap M_2) = \mathbb{C}$.

Finally, from the properties of Δ and S_B we have $\Delta(b^\dagger) = \Delta(b)^{\dagger \otimes \dagger}$. $\qquad \square$

REMARK 8.2.4. (i) $S^2(b) = gbg^{-1}$, where $g = S(w)^{-1}w$.

(ii) The Haar projection of B is e_2w and the normalized Haar functional is
   $\phi(b) = \tau(S(w)wb)$.

(iii) The non-degenerate duality $\langle \, , \, \rangle$ makes $A = N' \cap M_1$ the C^*-quantum groupoid dual to B.

(iv) Since the Markov trace ([GHJ], 3.2.4) of the inclusion $N \subset M$ is also the Markov trace of the finite-dimensional inclusion $A_t \subset A$, it is clear that $[M : N] = [A : A_t] = [B : B_t]$. We call last number *the index of the quantum groupoid* B.

(v) Let us mention two classification results obtained in ([NV2], 4.18, 4.19) by quantum groupoid methods:

   (a) If $N \subset M$ is a depth 2 subfactor such that $[M : N]$ is a square free integer (i.e., $[M : N]$ is an integer which has no divisors of the form n^2, $n > 1$), then $N' \cap M = \mathbb{C}$, and there is a (canonical) minimal action of a Kac algebra B on M_1 such that $M_2 \cong M_1 \rtimes B$ and $M = M_1^B$.

   (b) If $N \subset M$ is a depth 2 II_1 subfactor such that $[M : N] = p$ is prime, then $N' \cap M = \mathbb{C}$, and there is an outer action of the cyclic group $G = \mathbb{Z}/p\mathbb{Z}$ on M_1 such that $M_2 \cong M_1 \rtimes G$ and $M = M_1^G$.

8.3. Action on a subfactor. Equation (8–9) suggests the following definition of the action of B on M_1.

PROPOSITION 8.3.1. *The map* $\triangleright : B \otimes_{\mathbb{C}} M_1 \to M_1$:

$$b \triangleright x = \lambda^{-1} E_{M_1}(bxe_2) \tag{8–10}$$

defines a left action of B on M_1 (cf. [S], Proposition 17).

PROOF. The above map defines a left B-module structure on M_1, since $1 \triangleright x = x$ and

$$b \triangleright (c \triangleright x) = \lambda^{-2} E_{M_1}(b E_{M_1}(cxe_2)e_2) = \lambda^{-1} E_{M_1}(bcxe_2) = (bc) \triangleright x.$$

Next, using equation (8–9) we get

$$b \triangleright xy = \lambda^{-1} E_{M_1}(bxye_2) = \lambda^{-2} E_{M_1}(b_{(1)}xe_2)E_{M_1}(b_{(2)}ye_2)$$
$$= (b_{(1)} \triangleright x)(b_{(2)} \triangleright y).$$

By [NV2], 4.4 and properties of S_B we also get

$$S(b)^{\dagger} \triangleright x^* = \lambda^{-1} E_{M_1}(S(b)^{\dagger}x^*e_2) = \lambda^{-1} E_{M_1}(S_B(w)^{-1}S_B(wbw^{-1})^*S_B(w)x^*e_2)$$
$$= \lambda^{-1} E_{M_1}(S_B(b^*)x^*e_2) = \lambda^{-1} E_{M_1}(e_2x^*b^*)$$
$$= \lambda^{-1} E_{M_1}(bxe_2)^* = (b \triangleright x)^*.$$

Finally,

$$b \triangleright 1 = \lambda^{-1} E_{M_1}(be_2) = \lambda^{-1} E_{M_1}(\lambda^{-1} E_{M_1}(be_2)e_2) = \varepsilon_t(b) \triangleright 1,$$

and $b \triangleright 1 = 0$ if and only if $\varepsilon_t(b) = \lambda^{-1} E_{M_1}(be_2) = 0$. □

PROPOSITION 8.3.2. $M_1^B = M$, i.e. M is the fixed point subalgebra of M_1.

PROOF. If $x \in M_1$ is such that $b \triangleright x = \varepsilon_t(b) \triangleright x$ for all $b \in B$, then $E_{M_1}(bxe_2) = E_{M_1}(\varepsilon_t(b)xe_2) = E_{M_1}(be_2)x$. Taking $b = e_2$, we get $E_M(x) = x$ which means that $x \in M$. Thus, $M_1^B \subset M$.

Conversely, if $x \in M$, then x commutes with e_2 and

$$b \triangleright x = \lambda^{-1} E_{M_1}(be_2x) = \lambda^{-1} E_{M_1}(\lambda^{-1} E_{M_1}(be_2)e_2x) = \varepsilon_t(b) \triangleright x,$$

therefore $M_1^B = M$. □

PROPOSITION 8.3.3. The map $\theta : [x \otimes b] \mapsto xS(w)^{1/2}bS(w)^{-1/2}$ defines a von Neumann algebra isomorphism between $M_1 \rtimes B$ and M_2.

PROOF. By definition of the action $\triangleright$ we have:

$$\theta([x(z \triangleright 1) \otimes b]) = xS(w)^{1/2}\lambda^{-1}E_{M_1}(ze_2)bS(w)^{-1/2}$$
$$= xS(w)^{1/2}zbS(w)^{-1/2} = \theta([x \otimes zb]),$$

for all $x \in M_1$, $b \in B$, $z \in B_t$, so θ is a well defined linear map from $M_1 \rtimes B = M_1 \otimes_{B_t} B$ to M_2. It is surjective since an orthonormal basis of $B = M' \cap M_2$ over $B_t = M' \cap M_1$ is also a basis of M_2 over M_1 ([Po], 2.1.3). Finally, one can check that θ is a von Neumann algebra isomorphism ([NV2], 6.3). □

REMARK 8.3.4. (i) The action of B constructed in Proposition 8.3.1 is minimal, since we have $M_1' \cap M_1 \rtimes B = M_1' \cap M_2 = B_s$ by Proposition 8.3.3.

(ii) If $N' \cap M = \mathbb{C}$, then B is a usual Kac algebra (i.e., a Hopf C^*-algebra) and we recover the well-known result proved in [S], [L], and [D].

(iii) It is known ([Po], Section 5) that there are exactly 3 non-isomorphic subfactors of depth 2 and index 4 of the hyperfinite II_1 factor R such that $N' \cap M \neq \mathbb{C}$. One of them is $R \subset R \otimes_{\mathbb{C}} M_2(\mathbb{C})$, two others can be viewed as diagonal subfactors of the form

$$\left\{ \begin{pmatrix} x & 0 \\ 0 & \alpha(x) \end{pmatrix} \middle| x \in R \right\} \subset R \otimes M_2(\mathbb{C}),$$

where α is an outer automorphism of R such that $\alpha^2 = \mathrm{id}$, resp. α^2 is inner and $\alpha^2 \neq \mathrm{id}$.

An explicit description of the corresponding quantum groupoids can be found in ([NV1], 3.1, 3.2), ([NV2], 7). Also, in ([NV2], 7) we describe the C^*-quantum groupoid $B = M_2(\mathbb{C}) \oplus M_3(\mathbb{C})$ corresponding to the subfactor with the principal graph A_3 ([GHJ], 4.6.5). This subfactor has index $4\cos^2 \frac{\pi}{5}$ so that the index of B is equal to $16\cos^4 \frac{\pi}{5}$.

9. C^*-Quantum Groupoids and Subfactors:
The Finite Depth Case

Here we show that quantum groupoids give a description of arbitrary finite index and finite depth II_1 subfactors via a Galois correspondence and explain how to express subfactor invariants such as bimodule categories and principal graphs in quantum groupoid terms. Recall that the *depth* ([GHJ], 4.6.4) of a subfactor $N \subset M$ with finite index $\lambda^{-1} = [M : N]$ is

$$n = \inf\{k \in \mathbb{Z}^+ \mid \dim Z(N' \cap M_{k-2}) = \dim Z(N' \cap M_k)\} \in \mathbb{Z}_+ \cup \{\infty\},$$

where $N \subset M \subset M_1 \subset M_2 \subset \cdots$ is the corresponding Jones tower.

9.1. A Galois correspondence. The following simple observation gives a natural passage from arbitrary finite depth subfactors to depth 2 subfactors.

PROPOSITION 9.1.1. *Let $N \subset M$ be a subfactor of depth n then for all $k \geq 0$ the inclusion $N \subset M_k$ has depth $d + 1$, where d is the smallest positive integer $\geq \frac{n-1}{k+1}$. In particular, $N \subset M_i$ has depth 2 for all $i \geq n - 2$, i.e., any finite depth subfactor is an intermediate subfactor of some depth 2 inclusion.*

PROOF. Note that $\dim Z(N' \cap M_i) = \dim Z(N' \cap M_{i+2})$ for all $i \geq n - 2$. By [PP2], the tower of basic construction for $N \subset M_k$ is

$$N \subset M_k \subset M_{2k+1} \subset M_{3k+2} \subset \cdots,$$

therefore, the depth of this inclusion is equal to $d + 1$, where d is the smallest positive integer such that $d(k + 1) - 1 \geq n - 2$. $\qquad\square$

This result means that $N \subset M$ can be realized as an intermediate subfactor of a crossed product inclusion $N \subset N \rtimes B$ for some quantum groupoid B:

$$N \subset M \subset N \rtimes B.$$

Recall that in the case of a Kac algebra action there is a Galois correspondence between intermediate von Neumann subalgebras of $N \subset N \rtimes B$ and left coideal $*$-subalgebras of B [ILP],[E1]. Thus, it is natural to ask about a quantum groupoid analogue of this correspondence.

A unital $*$-subalgebra $I \subset B$ such that $\Delta(I) \subset B \otimes I$ (resp. $\Delta(I) \subset I \otimes B$) is said to be a left (resp. right) *coideal $*$-subalgebra*. The set $\ell(B)$ of left coideal $*$-subalgebras is a lattice under the usual operations:

$$I_1 \wedge I_2 = I_1 \cap I_2, \qquad I_1 \vee I_2 = (I_1 \cup I_2)''$$

for all $I_1, I_2 \in \ell(B)$. The smallest element of $\ell(B)$ is B_t and the greatest element is B. $I \in \ell(B)$ is said to be *connected* if $Z(I) \cap B_s = \mathbb{C}$.

To justify this definition, note that if $I = B$, then this is precisely the definition of a connected quantum groupoid, and if $I = B_t$, then this definition is equivalent to $\widehat{B}$ being connected ([N1], 3.10, 3.11, [BNSz], 2.4).

On the other hand, the set $\ell(M_1 \subset M_2)$ of intermediate von Neumann subalgebras of $M_1 \subset M_2$ also forms a lattice under the operations

$$K_1 \wedge K_2 = K_1 \cap K_2, \qquad K_1 \vee K_2 = (K_1 \cup K_2)''$$

for all $K_1, K_2 \in \ell(M_1 \subset M_2)$. The smallest element of this lattice is M_1 and the greatest element is M_2.

Given a left (resp. right) action of B on a von Neumann algebra N, we will denote (by an abuse of notation)

$$N \rtimes I = \mathrm{span}\{[x \otimes b] \mid x \in N, \, b \in I\} \subset N \rtimes B.$$

The next theorem establishes a *Galois correspondence*, i.e., a lattice isomorphism between $\ell(M_1 \subset M_2)$ and $\ell(B)$.

THEOREM 9.1.2. *Let $N \subset M \subset M_1 \subset M_2 \subset \cdots$ be the tower constructed from a depth 2 subfactor $N \subset M$, $B = M' \cap M_2$ be the corresponding quantum groupoid, and θ be the isomorphism between $M_1 \rtimes B$ and M_2 (Proposition 8.3.3). Then*

$$\phi : \ell(M_1 \subset M_2) \to \ell(B) : K \mapsto \theta^{-1}(M' \cap K) \subset B,$$
$$\psi : \ell(B) \to \ell(M_1 \subset M_2) : I \mapsto \theta(M_1 \rtimes I) \subset M_2$$

define isomorphisms between $\ell(M_1 \subset M_2)$ and $\ell(B)$ inverse to each other.

PROOF. First, let us check that ϕ and ψ are indeed maps between the specified lattices. It follows from the definition of the crossed product that $M_1 \rtimes I$ is a von Neumann subalgebra of $M_1 \rtimes B$, therefore $\theta(M_1 \rtimes I)$ is a von Neumann subalgebra of $M_2 = \theta(M_1 \rtimes B)$, so ψ is a map to $\ell(M_1 \subset M_2)$. To show that ϕ maps to $\ell(B)$, let us show that the annihilator $(M' \cap K)^0 \subset \hat{B}$ is a left ideal in $\hat{B}$.

For all $x \in A$, $y \in (M' \cap K)^0$, and $b \in M' \cap K$ we have

$$\langle xy, b \rangle = \lambda^{-2} \tau(xy e_2 e_1 w b) = \lambda^{-2} \tau(y e_2 e_1 w b x)$$
$$= \lambda^{-3} \tau(y e_2 e_1 E_{M'}(e_1 w b x)) = \langle y, \lambda^{-1} E_{M'}(e_1 w b x) \rangle,$$

and it remains to show that $E_{M'}(e_1 w b x) \in M' \cap K$. By ([GHJ], 4.2.7), the square

$$
\begin{array}{ccc}
K & \subset & M_2 \\
\cup & & \cup \\
M' \cap K & \subset & M' \cap M_2
\end{array}
$$

is commuting, so $E_{M'}(K) \subset M' \cap K$. Since $e_1 w b x \in K$, then $xy \in (M' \cap K)^0$, i.e., $(M' \cap K)^0$ is a left ideal and $\phi(K) = \theta^{-1}(M' \cap K)$ is a left coideal *-subalgebra.

Clearly, ϕ and ψ preserve $\wedge$ and $\vee$, moreover $\phi(M_1) = B_t$, $\phi(M_2) = B$ and $\psi(B_t) = M_1$, $\psi(B) = M_2$, therefore they are morphisms of lattices.

To see that they are inverses for each other, we first observe that the condition $\psi \circ \phi = \mathrm{id}$ is equivalent to $M_1(M' \cap K) = K$, and the latter follows from applying the conditional expectation E_K to $M_1(M' \cap M_2) = M_1 B = M_2$. The condition $\phi \circ \psi = \mathrm{id}$ translates into $\theta(I) = M' \cap \theta(M_1 \rtimes I)$. If $b \in I$, $x \in M = M_1^B$, then

$$\theta(b)x = \theta([1 \otimes b][x \otimes 1]) = \theta([(b_{(1)} \rhd x) \otimes b_{(2)}])$$
$$= \theta([x(1_{(2)} \rhd 1) \otimes \varepsilon(1_{(1)} b_{(1)}) b_{(2)}]) = \theta([x \otimes b]) = x\theta(b),$$

i.e., $\theta(I)$ commutes with M. Conversely, if $x \in M' \cap \theta(M_1 \rtimes I) \subset B$, then $x = \theta(y)$ for some $y \in (M_1 \rtimes I) \cap B = I$, therefore $x \in \theta(I)$. $\qquad\square$

COROLLARY 9.1.3. ([NV3], 4.5)

(i) $K = M \rtimes I$ is a factor if and only if $Z(I) \cap B_s = \mathbb{C}$.

(ii) The inclusion $M_1 \subset K = M_1 \rtimes I$ is irreducible if and only if $B_s \cap I = \mathbb{C}$.

9.2. Bimodule categories. Here we establish an equivalence between the tensor category $\mathrm{Bimod}_{N-N}(N \subset M)$ of $N - N$ bimodules of a subfactor $N \subset M$ (which is, by definition, the tensor category generated by simple subobjects of $_N L_2(M_n)_N$, $n \geq -1$, where $M_0 = M$ and $M_{-1} = N$) and the co-representation category of the C^*-quantum groupoid B canonically associated with it as in Theorem 9.1.2.

First introduce several useful categories associated to B. Recall that a left (resp., right) B-comodule V (with the structure map denoted by $v \mapsto v^{(1)} \otimes v^{(2)}$, $v \in V$) is said to be *unitary*, if

$$(v_2^{(1)})^*(v_1, v_2^{(2)}) = S(v_1^{(1)})g(v_1^{(2)}, v_2)$$
$$(\text{resp., } (v_1^{(2)})(v_1^{(1)}, v_2) = g^{-1}S((v_1^{(1)})^*)(v_1, v_2^{(1)})),$$

where $v_1, v_2 \in V$, and g is the canonical group-like element of B. This definition for Hopf $*$-algebra case can be found, e.g., in ([KS], 1.3.2).

Given left coideal $*$-subalgebras H and K of B, we consider a category $\mathcal{C}_{H-K}$ of left relative $(B, H - K)$ Hopf bimodules (cf. [Ta]), whose objects are Hilbert spaces which are both $H - K$-bimodules and left unitary B-comodules such that the bimodule action commutes with the coaction of B, i.e., for any object V of $\mathcal{C}_{H-K}$ and $v \in V$ one has

$$(h \rhd v \lhd k)^{(1)} \otimes (h \rhd v \lhd k)^{(2)} = h_{(1)} v^{(1)} k_{(1)} \otimes (h_{(2)} \rhd v^{(2)} \lhd k_{(2)}),$$

where $v \mapsto v^{(1)} \otimes v^{(2)}$ denotes the coaction of B on v, $h \in H$, $k \in K$, and morphisms are intertwining maps.

Similarly one can define a category of right relative $(B, H - K)$ Hopf bimodules.

REMARK 9.2.1. Any left B-comodule V is automatically a $B_t - B_t$-bimodule via $z_1 \cdot v \cdot z_2 = \varepsilon(z_1 v^{(1)} z_2) v^{(2)}$, $v \in V, z_1, z_2 \in B_t$. For any object of $\mathcal{C}_{H-K}$, this $B_t - B_t$-bimodule structure is a restriction of the given $H - K$-bimodule structure; it is easily seen by applying $(\varepsilon \otimes \mathrm{id})$ to both sides of the above relation of commutation and taking $h, k \in B_t$. Thus, if $H = B_t$ (resp. $K = B_t$), we can speak about right (resp. left) relative Hopf modules, a special case of weak Doi–Hopf modules [Bo].

PROPOSITION 9.2.2. *If B is a group Hopf C^*-algebra and H, K are subgroups, then there is a bijection between simple objects of $\mathcal{C}_{H-K}$ and double cosets of $H \backslash B / K$.*

PROOF. If V is an object of $\mathcal{C}_{H-K}$, then every simple subcomodule of V is 1-dimensional. Let $U = \mathbb{C}u$ ($u \mapsto g \otimes u$, $g \in B$) be one of these comodules, then all other simple subcomodules of V are of the form $h \triangleright U \triangleleft k$, where $h \in H$, $k \in K$, and

$$V = \oplus_{h,k} (h \triangleright U \triangleleft k) = \mathrm{span}\{HgK\}.$$

Vice versa, $\mathrm{span}\{HgK\}$ with natural $H-K$ bimodule and B-comodule structures is a simple object of $\mathcal{C}_{H-K}$.   □

EXAMPLE 9.2.3. If H, V, K are left coideal $*$-subalgebras of B, $H \subset V, K \subset V$, then V is an object of $\mathcal{C}_{H-K}$ with the structure maps given by $h \triangleright v \triangleleft k = hvk$ and Δ, where $v^{(1)} \otimes v^{(2)} = \Delta(v)$, $v \in V$, $h \in H$, $k \in K$. The scalar product is defined by the restriction on V of the *Markov trace* of the connected inclusion $B_t \subset B$ ([JS], 3.2). Similarly, right coideal $*$-subalgebras of B give examples of right relative $(B, H - K)$ Hopf bimodules.

Given an object V of $\mathcal{C}_{H-K}$, it is straightforward to show that the conjugate Hilbert space $\overline{V}$ is an object of $\mathcal{C}_{K-H}$ with the bimodule action

$$k \triangleright \overline{v} \triangleleft h = \overline{h^* \triangleright v \triangleleft k^*} \quad (\forall h \in H, \, k \in K)$$

(here $\overline{v}$ denotes the vector $v \in V$ considered as an element of $\overline{V}$) and the coaction $\overline{v} \mapsto \overline{v}^{(1)} \otimes \overline{v}^{(2)} = (v^{(1)})^* \otimes \overline{v}^{(2)}$. Define V^*, the dual object of V, to be $\overline{V}$ with the above structures. One can directly check that $V^{**} \cong V$ for any object V. In Example 9.2.3 the dual object can be obtained by putting $\overline{v} = v^*$ for all $v \in V$.

DEFINITION 9.2.4. Let L be another coideal $*$-subalgebra of B. For any objects $V \in \mathcal{C}_{H-L}$ and $W \in \mathcal{C}_{L-K}$, we define an object $V \otimes_L W$ from $\mathcal{C}_{H-K}$ as a tensor product of bimodules V and W [JS], 4.1 equipped with a comodule structure

$$(v \otimes_L w)^{(1)} \otimes (v \otimes_L w)^{(2)} = v^{(1)} w^{(1)} \otimes (v^{(2)} \otimes_L w^{(2)}).$$

One can verify that we have indeed an object from $\mathcal{C}_{H-K}$ and that the operation of tensor product is (i) associative, i.e., $V \otimes_L (W \otimes_P U) \cong (V \otimes_L W) \otimes_P U$; (ii) compatible with duality, i.e., $(V \otimes_L W)^* \cong W^* \otimes_L V^*$; (iii) distributive, i.e., $(V \oplus V') \otimes_L W = (V \otimes_L W) \oplus (V' \otimes_L W)$ ([NV3], 5.5). The tensor product of morphisms $T \in \mathrm{Hom}(V, V')$ and $S \in \mathrm{Hom}(W, W')$ is defined as usual:

$$(T \otimes_L S)(v \otimes_L w) = T(v) \otimes_L S(w).$$

From now on let us suppose that B is biconnected and acts outerly on the left on a II_1 factor N. Given an object V of $\mathcal{C}_{H-K}$, we construct an $N{\rtimes}H - N{\rtimes}K$-bimodule $\widehat{V}$ as follows. We put

$$\widehat{V} = \text{span}\{\Delta(1) \triangleright (\xi \otimes v) \mid \xi \otimes v \in L^2(N) \otimes_{\mathbb{C}} V\} = L^2(N) \otimes_{B_t} V$$

and denote $[\xi \otimes v] = \Delta(1) \triangleright (\xi \otimes v)$. Let us equip $\widehat{V}$ with the scalar product

$$([\xi \otimes v], [\eta \otimes w])_{\widehat{V}} = (\xi, \eta)_{\widetilde{L^2(N)}} (v, w)_V$$

and define the actions of N, H, K on $\widehat{V}$ by

$$a[\xi \otimes v] = [a\xi \otimes v], \qquad [\xi \otimes v]a = [\xi(v^{(1)} \triangleright a) \otimes v^{(2)}],$$

$$h[\xi \otimes v] = [(h_{(1)} \triangleright \xi) \otimes (h_{(2)} \triangleright v)], \qquad [\xi \otimes v]k = [\xi \otimes (v \triangleleft k)],$$

for all $a \in N, h \in H, k \in K$. One can check that these actions define the structure of an $(N{\rtimes}H) - (N{\rtimes}K)$ bimodule on $\widehat{V}$ in the algebraic sense and that this bimodule is unitary.

For any morphism $T \in \text{Hom}(V, W)$, define $\widehat{T} \in \text{Hom}(\widehat{V}, \widehat{W})$ by

$$\widehat{T}([\xi \otimes v]) = [\xi \otimes T(v)].$$

EXAMPLE 9.2.5. ([NV3], 5.6) For $V \in \mathcal{C}_{H-K}$ from Example 9.2.3, we have $\widehat{V} = N{\rtimes}V$ as $N{\rtimes}H - N{\rtimes}K$-bimodules.

The proof of the following theorem is purely technical (see [NV3], 5.7):

THEOREM 9.2.6. *The above assignments $V \mapsto \widehat{V}$ and $T \mapsto \widehat{T}$ define a functor from $\mathcal{C}_{H-K}$ to the category of $N{\rtimes}H - N{\rtimes}K$ bimodules. This functor preserves direct sums and is compatible with operations of taking tensor products and adjoints in the sense that if W is an object of $\mathcal{C}_{K-L}$, then*

$$\widehat{V \otimes_K W} \cong \widehat{V} \otimes_{N{\rtimes}K} \widehat{W}, \quad \text{and} \quad \widehat{V^*} \cong (\widehat{V})^*.$$

According to Remark 9.2.1, $\mathcal{C}_{B_t - B_t}$ is nothing but the category of B-comodules, $\text{Corep}(B)$. Let us show that this category is equivalent to $\text{Bimod}_{N-N}(N \subset M)$.

THEOREM 9.2.7. *Let $N \subset M$ be a finite depth subfactor with finite index, k be a number such that $N \subset M_k$ has depth ≤ 2, and let B be a canonical quantum groupoid such that $(N \subset M_k) \cong (N \subset N{\rtimes}B)$. Then $\text{Bimod}_{N-N}(N \subset M)$ and $\text{Rep}(B^*)$ are equivalent as tensor categories.*

PROOF. First, we observe that

$$\text{Bimod}_{N-N}(N \subset M) = \text{Bimod}_{N-N}(N \subset M_l)$$

for any $l \geq 0$. Indeed, since both categories are semisimple, it is enough to check that they have the same set of simple objects. All objects of

$$\text{Bimod}_{N-N}(N \subset M_l)$$

are also objects of $\mathrm{Bimod}_{N-N}(N \subset M)$. Conversely, since irreducible $N - N$ sub-bimodules of ${}_N L^2(M_i)_N$ are contained in the decomposition of ${}_N L^2(M_{i+1})_N$ for all $i \geq 0$, we see that objects of $\mathrm{Bimod}_{N-N}(N \subset M)$ belong to $\mathrm{Bimod}_{N-N}(N \subset M_l)$.

Hence, by Proposition 9.1.1, it suffices to consider the problem in the case when $N \subset M$ has depth 2 ($M = N \rtimes B$), i.e., to prove that

$$\mathrm{Bimod}_{N-N}(N \subset N \rtimes B)$$

is equivalent to $\mathrm{Corep}(B)$. Theorem 9.2.6 gives a functor from $\mathrm{Corep}(B) = \mathrm{Rep}(B^*)$ to $\mathrm{Bimod}_{N-N}(N \subset N \rtimes B)$. To prove that this functor is an equivalence, let us check that it yields a bijection between classes of simple objects of these categories.

Observe that B itself is an object of $\mathrm{Corep}(B)$ via $\Delta : B \to B \otimes_{\mathbb{C}} B$ and $\widehat{B} = {}_N L^2(M)_N$. Since the inclusion $N \subset M$ has depth 2, the simple objects of $\mathrm{Bimod}_{N-N}(N \subset M)$ are precisely irreducible subbimodules of ${}_N L^2(M)_N$. We have $\widehat{B} = {}_N L^2(M)_N = \oplus_i {}_N(p_i L^2(M))_N$, where $\{p_i\}$ is a family of mutually orthogonal minimal projections in $N' \cap M_1$ so every bimodule $p_i L^2(M)$ is irreducible. On the other hand, B is cosemisimple, hence $B = \oplus_i V_i$, where each V_i is an irreducible subcomodule. Note that $N' \cap M_1 = \widehat{B} = \sum p_i \widehat{B}$ and every $p_i B$ is a simple submodule of B (= simple subcomodule of B^*). Thus, there is a bijection between the sets of simple objects of $\mathrm{Corep}(B)$ and $\mathrm{Bimod}_{N-N}(N \subset M)$, so the categories are equivalent. $\qquad\qquad\square$

9.3. Principal graphs. The principal graph of a subfactor $N \subset M$ is defined as follows ([JS], 4.2, [GHJ], 4.1). Let $X = {}_N L^2(M)_M$ and consider the following sequence of $N - N$ and $N - M$ bimodules:

$${}_N L^2(N)_N, \ X, \ X \otimes_M X^*, \ X \otimes_M X^* \otimes_N X, \ldots$$

obtained by right tensoring with X^* and X. The vertex set of the principal graph is indexed by the classes of simple bimodules appearing as summands in the above sequence. Let us connect vertices corresponding to bimodules ${}_N Y_N$ and ${}_N Z_M$ by l edges if ${}_N Y_N$ is contained in the decomposition of ${}_N Z_N$, the restriction of ${}_N Z_M$, with multiplicity l.

We apply Theorem 9.2.6 to express the principal graph of a finite depth subfactor $N \subset M$ in terms of the quantum groupoid associated with it.

Let B and K be a quantum groupoid and its left coideal $*$-subalgebra such that B acts on N and $(N \subset M) \cong (N \subset N \rtimes K)$. Then $\widehat{K} = {}_N L^2(M)_M$, where we view K as a relative (B, K) Hopf module as in Example 9.2.3

By Theorem 9.2.6 we can identify irreducible $N - N$ (resp. $N - M$) bimodules with simple B-comodules (resp. relative right (B, K) Hopf modules). Consider a bipartite graph with vertex set given by the union of (classes of) simple B-comodules and simple relative right (B, K) Hopf modules and the number of edges between the vertices U and V representing B-comodule and relative right

(B, K) Hopf module respectively being equal to the multiplicity of U in the decomposition of V (when the latter is viewed as a B-comodule):

simple B-comodules

$$\cdots \mid \cdots \mid \cdots$$

simple relative right (B, K) Hopf modules

The principal graph of $N \subset M$ is the connected part of the above graph containing the trivial B-comodule.

It was shown in ([NV3], 3.3, 4.10) that:

(a) the map $\delta : I \mapsto [g^{-1/2}S(I)g^{1/2}]' \cap \widehat{B} \subset \widehat{B} \rtimes B$ defines a lattice anti-isomorphism between $\ell(B)$ and $\ell(\widehat{B})$;

(b) the triple $\delta(I) \subset \widehat{B} \subset \widehat{B} \rtimes I$ is a basic construction.

PROPOSITION 9.3.1. *If K is a coideal $*$-subalgebra of B then the principal graph of the subfactor $N \subset N \rtimes K$ is given by the connected component of the Bratteli diagram of the inclusion $\delta(K) \subset \widehat{B}$ containing the trivial representation of $\widehat{B}$.*

PROOF. First, let us show that there is a bijective correspondence between right relative (B, K) Hopf modules and $(\widehat{B} \rtimes K)$-modules. Indeed, every right (B, K) Hopf module V carries a right action of K. If we define a right action of $\widehat{B}$ by

$$v \triangleleft x = \langle v^{(1)}, x \rangle v^{(2)}, \quad v \in V, x \in \widehat{B},$$

then we have

$$(v \triangleleft k) \triangleleft x = \langle v^{(1)} k_{(1)}, x \rangle (v^{(2)} \triangleleft k_{(2)}) = \langle v^{(1)}, (k_{(1)} \triangleright x) \rangle (v^{(2)} \triangleleft k_{(2)})$$
$$= (v \triangleleft (k_{(1)} \triangleright x)) \triangleleft k_{(2)},$$

for all $x \in \widehat{B}$ and $k \in K$ which shows that kx and $(k_{(1)} \triangleright x)k_{(2)}$ act on V exactly in the same way, therefore V is a right $(\widehat{B} \rtimes K)$-module.

Conversely, given an action of $(\widehat{B} \rtimes K)$ on V, we automatically have a B-comodule structure such that

$$\langle v^{(1)} k_{(1)}, x \rangle (v^{(2)} \triangleright k_{(2)}) = \langle v^{(1)}, x_{(1)} \rangle \langle k_{(1)}, x_{(2)} \rangle (v^{(2)} \triangleleft k_{(2)})$$
$$= (v \triangleleft (k_{(1)} \triangleright x)) \triangleleft k_{(2)} = (v \triangleleft k) \triangleleft x$$
$$= \langle x, (v \triangleleft k)^{(1)} \rangle (v \triangleleft k)^{(2)},$$

which shows that $v^{(1)} k_{(1)} \otimes (v^{(2)} \triangleleft k_{(2)}) = (v \triangleleft k)^{(1)} \otimes (v \triangleleft k)^{(2)}$, i.e., that V is a right relative (B, K)-module.

Thus, we see that the principal graph is given by the connected component of the Bratteli diagram of the inclusion $\widehat{B} \subset \widehat{B} \rtimes K$ containing the trivial representation of $\widehat{B}$. Since $\widehat{B} \rtimes K$ is the basic construction for the inclusion $\delta(K) \subset \widehat{B}$, therefore the Bratteli diagrams of the above two inclusions are the same. □

COROLLARY 9.3.2. *If $N \subset N \rtimes B$ is a depth 2 inclusion corresponding to the quantum groupoid B, then its principal graph is given by the Bratteli diagram of the inclusion $\widehat{B}_t \subset \widehat{B}$.*

PROOF. In this case $K = B$ and inclusion $\widehat{B}_t \subset \widehat{B}$ is connected, so that $\delta(K) = \widehat{B}_t$ (note that $\widehat{B}$ is biconnected). $\qquad\square$

References

[BS] S. Baaj and G. Skandalis, *Unitaires multiplicatifs et dualité pour les produits croisés de C^*-algébres*, Ann. Sci. ENS., **26** (1993), 425–488.

[Ba] O. Babelon, *Universal exchange algebra for Bloch waves and Liouville theory*, Comm. Math. Phys., **139** (1991), 619–643.

[ABRR] D. Arnaudon, E. Buffenoir, E. Ragoucy, and Ph. Roche, *Universal Solutions of Quantum Dynamical Yang–Baxter Equations*, Lett.Math.Phys., **44** (1998), 201-214.

[BBB] O. Babelon, D. Bernard, and E. Billey, *A quasi-Hopf algebra interpretation of quantum 3-j and 6-j symbols and difference equations*, Phys. Lett. B, **375** (1996), no. 1-4, 89–97.

[BM] R. Blattner, S. Montgomery, *A duality theorem for Hopf module algebras*, J. Algebra, **95** (1985), 153–172.

[Bo] G. Böhm, *Doi–Hopf modules over weak Hopf algebras*, Comm. Algebra, **28**:10 (2000), 4687–4698.

[BNSz] G. Böhm, F. Nill, and K. Szlachányi, *Weak Hopf algebras I. Integral theory and C^*-structure*, J. Algebra, **221** (1999), 385–438.

[BSz1] G. Böhm, K. Szlachányi, *A coassociative C^*-quantum group with nonintegral dimensions*, Lett. in Math. Phys, **35** (1996), 437–456.

[BSz2] G. Böhm, K. Szlachányi, *Weak Hopf algebras, II: Representation theory, dimensions, and the Markov trace*, J. Algebra, **233**:1 (2000), 156–212.

[BSz3] G. Böhm, K. Szlachányi, *Weak C^*-Hopf algebras and multiplicative isometries*, J. Operator Theory, **45**:2 (2001), 357–376.

[D] M.-C. David, *Paragroupe d'Adrian Ocneanu et algebre de Kac*, Pacif. J. of Math., **172** (1996), 331–363.

[E1] M. Enock, *Sous-facteurs intermédiaires et groupes quantiques mesurés*, J. Operator Theory, **42** (1999), 305–330

[E2] M. Enock, *Inclusions of von Neumann algebras and quantum groupoids*, Inst. Math. de Jussieu, Prépublication, **231** (1999).

[EVal] M. Enock and J.-M. Vallin, *Inclusions of von Neumann algebras and quantum groupoids*, Inst. Math. de Jussieu, Prépublication, **156** (1998), to appear in J. Funct. Anal.

[EG] P. Etingof and S. Gelaki. *Some properties of finite dimensional semisimple Hopf algebras*, Math. Res. Letters **5** (1998), 191-197.

[EN] P. Etingof, D. Nikshych, *Dynamical quantum groups at roots of 1*, Duke Math. J., **108**:1 (2001), 135–168.

[ES] P. Etingof, O. Schiffmann, *Lectures on the dynamical Yang–Baxter equations*, preprint, math.QA/9908064 (1999).

[ESS] P. Etingof, T. Schedler, and O. Schiffmann, *Explicit quantization of dynamical r-matrices for finite dimensional semisimple Lie algebras*, J. Amer. Math. Soc. **13** (2000), no. 3, 595–609.

[EV] P. Etingof, A. Varchenko, *Exchange dynamical quantum groups*, Comm. Math. Phys., **205** (1999), 19–52.

[Fe] G. Felder, *Elliptic quantum groups*, XIth International Congress of Mathematical Physics (1994), 211–218, Internat. Press, Cambridge, (1995).

[GHJ] F. Goodman, P. de la Harpe, and V.F.R. Jones, *Coxeter Graphs and Towers of Algebras*, M.S.R.I. Publ. **14**, Springer, Heidelberg, 1989.

[H1] T. Hayashi, *Face algebras. I. A generalization of quantum group theory*, J. Math. Soc. Japan, **50** (1998), 293–315.

[H2] T. Hayashi, *Face algebras and their Drinfel'd doubles*, Algebraic groups and their generalizations: quantum and infinite-dimensional methods (University Park, PA, 1991), 49–61, Proc. Sympos. Pure Math., **56**, Providence, RI, 1994.

[H3] T. Hayashi, *Galois quantum groups of II_1 subfactors*, Tohoku Math. J., **51** (1999), 365–389.

[ILP] M. Izumi, R. Longo, and S. Popa, *A Galois correspondence for compact groups of automorphisms of von Neumann algebras with a generalization to Kac algebras*, J. Funct. Anal., **155** (1998), 25–63.

[KS] A. Klimyk and K. Schmüdgen, *Quantum groups and their representations*, Springer Verlag (1997).

[KY] H. Kosaki and S. Yamagami, *Irreducible bimodules associated with crossed product algebras*, Intern. J. Math., **3** (1992), 661–676.

[J] V. Jones, *Index for subfactors*, Invent. math., **72** (1983), 1-25.

[JS] V. Jones and V.S. Sunder, *Introduction to Subfactors*, London Math. Soc. Lecture Notes **234**, Cambridge University Press, 1997.

[Lu] J.-H. Lu, *Hopf algebroids and quantum groupoids*, Internat. J. Math., **7** (1996), 47–70.

[L] G. Lusztig, *On quantum groups at roots of 1*, Geom. Dedicata, **35** (1990), 89–113.

[M] S. Montgomery, *Hopf Algebras and Their Actions on Rings*, CBMS Regional Conference Series in Mathematics, **82**, AMS, 1993.

[Ma] S. Majid, *Foundations of quantum group theory*, Cambridge University Press, Cambridge, (1995).

[Mal1] G. Maltsiniotis, *Groupoïdes quantiques*, C. R. Acad. Sci. Paris, **314**, Serie I (1992), 249-252.

[Mal2] G. Maltsiniotis, *Groupoïdes quantiques*, preprint, (1999).

[N] F. Nill, *Axioms of Weak Bialgebras*, math.QA/9805104 (1998).

[N1] D. Nikshych, *Duality for actions of weak Kac algebras and crossed product inclusions of II_1 factors*, to appear in the Journal of Operator Theory, math.QA/9810049 (1998).

[N2] D. Nikshych, *A duality theorem for quantum groupoids*, in "New Trends in Hopf Algebra Theory," Contemporary Mathematics, **267** (2000), 237-243.

[NTV] D. Nikshych, V. Turaev, and L. Vainerman, *Quantum groupoids and invariants of knots and 3-manifolds*, math.QA/0006078 (2000); to appear in Topology and its Applications.

[NV1] D. Nikshych, L. Vainerman, *Algebraic versions of a finite dimensional quantum groupoid*, Lecture Notes in Pure and Appl. Math., **209** (2000), 189-221.

[NV2] D. Nikshych, L. Vainerman, *A characterization of depth 2 subfactors of II_1 factors*, J. Func. Analysis, **171** (2000), 278-307.

[NV3] D. Nikshych, L. Vainerman, *A Galois correspondence for II_1 factors and quantum groupoids*, J. Funct. Anal., **178**:1 (2000), 113-142.

[NSzW] F. Nill, K. Szlachányi, H.-W. Wiesbrock, *Weak Hopf algebras and reducible Jones inclusions of depth 2*, I: From crossed products to Jones towers, preprint math.QA/9806130 (1998).

[O1] A. Ocneanu, *Quantized groups, string algebras and Galois theory for algebras*, Operator Algebras and Applications, Vol. 2, London Math. Soc. Lecture Notes Series **135**, Cambridge Univ. Press, Cambridge, U.K., (1988).

[O2] A. Ocneanu, *Quantum Cohomology, Quantum groupoids and Subfactors* unpublished talk given at the First Caribbean School of Mathematics and Theoretical Physics, Guadeloupe, (1993).

[P] R. Pierce, *Associative algebras*, Graduate Texts in Mathematics, **88**, Springer Verlag, (1982).

[Po] S. Popa, *Classification of amenable subfactors of type II*, Acta Math., **172** (1994), 163-255.

[PP1] M. Pimsner and S. Popa, *Entropy and index for subfactors*, Ann. Sci. Ecole Norm. Sup., **19** (1986), no. 3, 57-106.

[PP2] M. Pimsner and S. Popa, *Iterating the basic construction*, Trans. Amer. Math. Soc., **310** (1988), no. 1, 127-133.

[R] J. Renault, *A groupoid approach to C^*-algebras*, Lecture Notes in Math. **793**, Springer-Verlag, 1980.

[RT] N. Reshetikhin, V. Turaev, *Invariants of 3-manifolds via link polynomials and quantum groups*, Invent. Math., **103** (1991), 547-597.

[S] W. Szymański, *Finite index subfactors and Hopf algebra crossed products*, Proc. Amer. Math. Soc., **120** (1994), no. 2, 519-528.

[Ta] M. Takeuchi, *Relative Hopf modules – equivalencies and freeness criteria*, J. Algebra, **60** (1979), 452-471.

[T] V. Turaev, *Quantum invariants of knots and 3-manifolds*, de Gruyter Studies in Mathematics, **18**, Walter de Gruyter & Co., Berlin, (1994).

[V] L. Vainerman, *A note on quantum groupoids*, C. R. Acad. Sci. Paris, **315**, Serie I (1992), 1125-1130.

[V1] J.-M. Vallin, *Bimodules de Hopf et poids opératoriels de Haar*, J. of Operator Theory, **35** (1996), 39-65.

[V2] J.-M. Vallin, *Groupoïdes quantiques finis*, Université d'Orléans, Prépublication, **6** (2000).

[W] Y. Watatani. *Index for C^*-subalgebras*, Memoirs AMS, **424** (1990).

[Xu] P. Xu, *Quantum groupoids*, preprint, math.QA/9905192 (1999).

[Y] T. Yamanouchi, *Duality for generalized Kac algebras and a characterization of finite groupoid algebras*, J. Algebra, **163** (1994), 9–50.

DMITRI NIKSHYCH
UNIVERSITY OF NEW HAMPSHIRE
DEPARTMENT OF MATHEMATICS
KINGSBURY HALL
DURHAM, NH 03824
UNITED STATES
   nikshych@hypatia.unh.edu

LEONID VAINERMAN
UNIVERSITÉ DE STRASBOURG
DÉPARTEMENT DE MATHÉMATIQUES
7, RUE RENÉ DESCARTES
F-67084 STRASBOURG FRANCE
   wain@agrosys.kiev.ua, vaynerma@math.u-strasbg.fr

New Directions in Hopf Algebras
MSRI Publications
Volume 43, 2002

On Quantum Algebras and Coalgebras, Oriented Quantum Algebras and Coalgebras, Invariants of 1–1 Tangles, Knots and Links

DAVID E. RADFORD

ABSTRACT. We outline a theory of quantum algebras and coalgebras and their resulting invariants of unoriented 1–1 tangles, knots and links, we outline a theory of oriented quantum algebras and coalgebras and their resulting invariants of oriented 1–1 tangles, knots and links, and we show how these algebras and coalgebras are related. Quasitriangular Hopf algebras are examples of quantum algebras and oriented quantum algebras; likewise coquasitriangular Hopf algebras are examples of quantum coalgebras and oriented quantum coalgebras.

Introduction

Since the advent of quantum groups [4] many algebraic structures have been described which are related to invariants of 1–1 tangles, knots, links or 3-manifolds. The purpose of this paper is to outline a theory for several of these structures, which are defined over a field k, and to discuss relationships among them. The structures we are interested in are: quantum algebras, quantum coalgebras, oriented quantum algebras, oriented quantum coalgebras and their "twist" specializations.

Quantum algebras and coalgebras account for regular isotopy invariants of unoriented 1–1 tangles. Twist quantum algebras and coalgebras, which are quantum algebras and coalgebras with certain additional structure, account for regular isotopy invariants of unoriented knots and links. As the terminology suggests,

Research supported in part by NSF Grant DMS 9802178. The author wishes to thank MSRI for providing him the opportunity and partial support to talk about joint work with Kauffman, which is the origin of this paper, at the Hopf Algebra Workshop, held at MSRI, Berkeley, CA, October 25–29, 1999. Work on this paper was completed while the author was visiting the Mathematisches Institut der Ludwig-Maximilians-Universität München during June of 2000 with the generous support of the Graduiertenkolleg for which he is very grateful. It has been the author's great pleasure to collaborate with Louis H. Kauffman.

oriented quantum algebras and coalgebras account for regular isotopy invariants of oriented 1–1 tangles; twist oriented quantum algebras and coalgebras account for regular isotopy invariants of oriented knots and links.

The notion of quantum algebra was described some time ago [10] whereas the notion of quantum coalgebra (which is more general than dual quantum algebra) was described somewhat later [19]. The notions of oriented quantum algebra and oriented quantum coalgebra have been formulated very recently and initial papers about them [16; 17] have just been circulated. This paper is basically a rough overview of a rather extensive body of joint work by the author and Kauffman [13; 15; 16; 17; 19; 20] on these and related structures. At the time of this writing oriented quantum algebras account for most known regular isotopy invariants of oriented links [20]; thus oriented quantum algebras are important for that reason. Oriented quantum algebras and quantum algebras are related in an interesting way. Every quantum algebra has an oriented quantum algebra structure which is called the associated oriented quantum algebra structure. Not every oriented quantum algebra is an associated oriented quantum algebra structure. However, a quantum algebra can always be constructed from an oriented quantum algebra in a natural way.

A quasitriangular Hopf algebra has a quantum algebra structure and a ribbon Hopf algebra has a twist quantum algebra structure. Hence there are close connections between Hopf algebras, quantum algebras and oriented quantum algebras. Quantum coalgebras and oriented quantum coalgebras are related in the same ways that quantum algebras and oriented quantum algebras are. Coquasitriangular Hopf algebras have a quantum coalgebra structure. There are coquasitriangular Hopf algebras associated with a wide class of quantum coalgebras; therefore there are close connections between Hopf algebras, quantum coalgebras and oriented quantum coalgebras.

In this paper we focus on the algebraic theory of quantum algebras and the other structures listed above and we focus on the algebraic theory of their associated invariants. For a topological perspective on these theories, in particular for a topological motivation of the definitions of quantum algebra and oriented quantum algebra, the reader is encouraged to consult [17; 21]. Here we do not calculate invariants, except to provide a few simple illustrations, nor do we classify them. For more extensive calculations see [12; 13; 14; 15; 18; 19] and [27]. We do, however, describe in detail how the Jones polynomial fits into the context of oriented quantum algebras (and thus quantum coalgebras). A major goal of this paper is to describe in sufficient detail an algebraic context which accounts for many known regular isotopy invariants of knots and links and which may prove to be fertile ground for the discovery of new invariants.

The paper is organized as follows. In Section 1 basic notations are discussed; bilinear forms, the quantum Yang–Baxter and braid equations are reviewed. We assume that the reader has a basic knowledge of coalgebras and Hopf algebras. Good references are [1; 24; 31]. Section 2 deals with Yang–Baxter algebras and the dual concept of Yang–Baxter coalgebras. Yang–Baxter algebras or coalgebras are integral parts of the algebraic structures we study in this paper. Yang–Baxter coalgebras are defined in [2] and are referred to as coquasitriangular coalgebras in [22].

The basic theory of quantum algebras is laid out in Section 3. We discuss the connection between quantum algebras and quasitriangular Hopf algebras and show how ribbon Hopf algebras give rise to oriented quantum algebras. Likewise the basic theory of oriented quantum algebras is outlined in Section 4; in particular we describe the oriented quantum algebra associated to a quantum algebra.

Section 5 gives a rather detailed description of the construction of a quantum algebra from an oriented quantum algebra. Sections 6 and 7 are about the invariants associated to quantum algebras, oriented quantum algebras and their twist specializations. These invariants are described in terms of a very natural and intuitive *bead sliding formalism*. Section 8 makes the important connection between the invariants computed by bead sliding with invariants computed by the well-established categorical method [28; 29]. The reader is encouraged to consult [13; 17; 28; 29] as background material for Sections 6–8.

The material of Section 8 motivates the notion of inner oriented quantum algebra which is discussed in Section 9. The Hennings invariant, which a 3-manifold invariant defined for certain finite-dimensional ribbon Hopf algebras, can be explained in terms of the bead sliding formalism. We comment on aspects of computation of this invariant in Section 10.

Sections 11–13 are the coalgebras versions of Sections 3–5. In Section 14 our paper ends with a discussion of invariants constructed from quantum coalgebras, oriented quantum coalgebras and their twist specializations. There may be a practical advantage to computing invariants using coalgebra structures instead of algebra structures.

Throughout k is a field and all algebras, coalgebras and vector spaces are over k. Frequently we denote algebras, coalgebras and the like by their underlying vector spaces and we denote the set of non-zero elements of k by k^*. Finally, the author would like to thank the referee for his or her very thoughtful suggestions and comments. Some of the comments led to additions to this paper, namely Propositions 1, 4, Corollary 2 and Section 9.

1. Preliminaries

For vector spaces U and V over k we denote the tensor product $U \otimes_k V$ by $U \otimes V$, the identity map of V by 1_V and the linear dual $\mathrm{Hom}_k(V, k)$ of V by V^*. If T is a linear endomorphism of V then an element $v \in V$ is T-*invariant* if $T(v) = v$. The *twist map* $\tau_{U,V} : U \otimes V \longrightarrow V \otimes U$ is defined by $\tau_{U,V}(u \otimes v) = v \otimes u$ for all $u \in U$ and $v \in V$. If V is an algebra over k we let 1_V also denote the unit of k. The meaning 1_V should always be clear from context.

By definition of the tensor product $U \otimes V$ of U and V over k there is a bijective correspondence between the set of bilinear forms $b : U \times V \longrightarrow k$ and the vector space $(U \otimes V)^*$ given by $b \mapsto b_{lin}$, where $b_{lin}(u \otimes v) = b(u, v)$ for all $u \in U$ and $v \in V$. We refer to bilinear forms $b : U \times U \longrightarrow k$ as *bilinear forms on U*.

Let $b : U \times V \longrightarrow k$ be a bilinear form and regard $U^* \otimes V^*$ as a subspace of $(U \otimes V)^*$ in the usual way. Then b is of *finite type* if $b_{lin} \in U^* \otimes V^*$ in which case we write ρ_b for b_{lin}. Thus when b is of finite type $\rho_b(u \otimes v) = b(u, v)$ for all $u \in U$ and $v \in V$. Observe that b is of finite type if and only if one of $b_\ell : U \longrightarrow V^*$ and $b_r : V \longrightarrow U^*$ has finite rank, where $b_\ell(u)(v) = b(u, v) = b_r(v)(u)$ for all $u \in U$ and $v \in V$. Consequently if one of U or V is finite-dimensional b is of finite type. The bilinear form b is *left non-singular* if b_ℓ is one-one, is *right non-singular* if b_r is one-one and is *non-singular* if b is both left and right non-singular.

Suppose that ρ is an endomorphism of $U \otimes U$ and consider the endomorphisms $\rho_{(i,j)}$ of $U \otimes U \otimes U$ for $1 \leq i < j \leq 3$ defined by

$$\rho_{(1,2)} = \rho \otimes 1_U, \quad \rho_{(2,3)} = 1_U \otimes \rho \text{ and } \rho_{(1,3)} = (1_U \otimes \tau_{U,U}) \circ (\rho \otimes 1_U) \circ (1_U \otimes \tau_{U,U}).$$

The *quantum Yang–Baxter equation* is

$$\rho_{(1,2)} \circ \rho_{(1,3)} \circ \rho_{(2,3)} = \rho_{(2,3)} \circ \rho_{(1,3)} \circ \rho_{(1,2)} \tag{1-1}$$

and the *braid equation* is

$$\rho_{(1,2)} \circ \rho_{(2,3)} \circ \rho_{(1,2)} = \rho_{(2,3)} \circ \rho_{(1,2)} \circ \rho_{(2,3)}. \tag{1-2}$$

Observe that ρ satisfies (1–1) if and only if $\rho \circ \tau_{U,U}$ satisfies (1–2) or equivalently $\tau_{U,U} \circ \rho$ satisfies (1–2). If ρ is invertible then ρ satisfies (1–1) if and only if ρ^{-1} does and ρ satisfies (1–2) if and only if ρ^{-1} does.

Let (C, Δ, ε) be a coalgebra over k. We use the notation $\Delta(c) = c_{(1)} \otimes c_{(2)}$ for $c \in C$, a variation of the Heyneman–Sweedler notation, to denote the co-product. Generally we write $\Delta^{(n-1)}(c) = c_{(1)} \otimes \cdots \otimes c_{(n)}$, where $\Delta^{(1)} = \Delta$ and $\Delta^{(n)} = (\Delta \otimes 1_C \otimes \cdots \otimes 1_C) \circ \Delta^{(n-1)}$ for $n \geq 2$. We let C^{cop} denote the coalgebra $(C, \Delta^{\mathrm{cop}}, \varepsilon)$ whose coproduct is given by $\Delta^{\mathrm{cop}}(c) = c_{(2)} \otimes c_{(1)}$ for all $c \in C$. If (M, ρ) is a right C-comodule we write $\rho(m) = m^{<1>} \otimes m^{(2)}$ for $m \in M$.

Let the set of bilinear forms on C have the algebra structure determined by its identification with the dual algebra $(C \otimes C)^*$ given by $b \mapsto b_{lin}$. Observe that $b, b' : C \times C \longrightarrow k$ are inverses if and only if

$$b(c_{(1)}, d_{(1)}) b'(c_{(2)}, d_{(2)}) = \varepsilon(c)\varepsilon(d) = b'(c_{(1)}, d_{(1)}) b(c_{(2)}, d_{(2)})$$

for all $c, d \in C$. In this case we write $b^{-1} = b'$.

For an algebra A over k we let A^{op} denote the algebra whose ambient vector space is A and whose multiplication is given by $a \cdot b = ba$ for all $a, b \in A$. An element tr of A^* is *tracelike* if $\mathrm{tr}(ab) = \mathrm{tr}(ba)$ for all $a, b \in A$. The trace function on the algebra $M_n(k)$ of $n \times n$ matrices over k is a primary example of a tracelike element.

For a vector space V over k and $\rho \in V \otimes V$ we define

$$V_{(\rho)} = \{(u^* \otimes 1_V)(\rho) + (1_V \otimes v^*)(\rho) \mid u^*, v^* \in V^*\}.$$

For an algebra A over k and invertible $\rho \in A \otimes A$ we let A_ρ be the subalgebra of A generated by $A_{(\rho)} + A_{(\rho^{-1})}$. The following lemma will be useful in our discussion of minimal quantum algebras.

LEMMA 1. *Let V be a vector space over k and $\rho \in V \otimes V$.*

(a) $\rho \in V_{(\rho)} \otimes V_{(\rho)}$ *and $V_{(\rho)}$ is the smallest subspace U of V such that $\rho \in U \otimes U$.*
(b) *Suppose that t is a linear endomorphism of V which satisfies $(t \otimes t)(\rho) = \rho$. Then $t(V_{(\rho)}) = V_{(\rho)}$.*

PROOF. Part (a) follows by definition of $V_{(\rho)}$. For part (b), we may assume that $\rho \neq 0$ and write $\rho = \sum_{i=1}^{r} u_i \otimes v_i$, where r is as small as possible. Then $\{u_1, \ldots, u_r\}$, $\{v_1, \ldots, v_r\}$ are linearly independent and the u_i's together with the v_i's span $V_{(\rho)}$. Since $(t \otimes t)(\rho) = \rho$, or equivalently $\sum_{i=1}^{r} t(u_i) \otimes t(v_i) = \sum_{i=1}^{r} u_i \otimes v_i$, it follows that the sets $\{t(u_1), \ldots, t(u_r)\}$ and $\{t(v_1), \ldots, t(v_r)\}$ are also linearly independent, that $\{u_1, \ldots, u_r\}$, $\{t(u_1), \ldots, t(u_r)\}$ have the same span and that $\{v_1, \ldots, v_r\}$, $\{t(v_1), \ldots, t(v_r)\}$ have the same span. Thus $t(V_{(\rho)}) = V_{(\rho)}$. $\square$

2. Yang–Baxter Algebras and Yang–Baxter Coalgebras

Let A be an algebra over the field k and let $\rho = \sum_{i=1}^{r} a_i \otimes b_i \in A \otimes A$. We set

$$\rho_{12} = \sum_{i=1}^{r} a_i \otimes b_i \otimes 1, \quad \rho_{13} = \sum_{i=1}^{r} a_i \otimes 1 \otimes b_i \quad \text{and} \quad \rho_{23} = \sum_{i=1}^{r} 1 \otimes a_i \otimes b_i.$$

The quantum Yang–Baxter equation for ρ is

$$\rho_{12}\rho_{13}\rho_{23} = \rho_{23}\rho_{13}\rho_{12}, \tag{2-1}$$

or equivalently

$$\sum_{i,j,\ell=1}^{r} a_i a_j \otimes b_i a_\ell \otimes b_j b_\ell = \sum_{j,i,\ell=1}^{r} a_j a_i \otimes a_\ell b_i \otimes b_\ell b_j. \tag{2-2}$$

Observe that (2–1) is satisfied when A is commutative. The pair (A, ρ) is called a *Yang–Baxter algebra over k* if ρ is invertible and satisfies (2–1).

Suppose that (A, ρ) and (A', ρ') are Yang–Baxter algebras over k. Then $(A \otimes A', \rho'')$ is a Yang–Baxter algebra over k, called the *tensor product* of (A, ρ) and (A', ρ'), where $\rho'' = (1_A \otimes \tau_{A,A'} \otimes 1_{A'})(\rho \otimes \rho')$. A *morphism* $f : (A, \rho) \longrightarrow (A', \rho')$ of Yang–Baxter algebras is an algebra map $f : A \longrightarrow A'$ which satisfies $\rho' = (f \otimes f)(\rho)$. Note that $(k, 1 \otimes 1)$ is a Yang–Baxter algebra over k. The category of Yang–Baxter algebras over k with their morphisms under composition has a natural monoidal structure.

Since ρ is invertible and satisfies (2–1) then ρ^{-1} does as well as $(\rho^{-1})_{ij} = (\rho_{ij})^{-1}$ for all $1 \leq i < j \leq 3$. Thus (A, ρ^{-1}) is a Yang–Baxter algebra over k. Note that (A^{op}, ρ) and (A, ρ^{op}) are Yang–Baxter algebras over k also, where $\rho^{\mathrm{op}} = \sum_{i=1}^{r} b_i \otimes a_i$.

An interesting example of a Yang–Baxter algebra for us [16, Example 1] is the following where $A = \mathrm{M}_n(k)$ is the algebra of all $n \times n$ matrices over k. For $1 \leq i, j \leq n$ let $E_{ij} \in \mathrm{M}_n(k)$ be the $n \times n$ matrix which has a single non-zero entry which is 1 and is located in the ith row and j^{th} column. Then $\{E_{ij}\}_{1 \leq i, j \leq n}$ is the standard basis for $\mathrm{M}_n(k)$ and $E_{ij} E_{\ell m} = \delta_{j\ell} E_{im}$ for all $1 \leq i, j, \ell, m \leq n$.

EXAMPLE 1. Let $n \geq 2$, $a, bc \in k^*$ satisfy $a^2 \neq bc, 1$ and let

$$\mathsf{B} = \{b_{ij} \,|\, 1 \leq i < j \leq n\}, \qquad \mathsf{C} = \{c_{ji} \,|\, 1 \leq i < j \leq n\}$$

be indexed subsets of k^* such that $b_{ij} c_{ji} = bc$ for all $1 \leq i < j \leq n$. Then $(\mathrm{M}_n(k), \rho_{a,\mathsf{B},\mathsf{C}})$ is a Yang–Baxter algebra over k, where

$$\rho_{a,\mathsf{B},\mathsf{C}} = \sum_{1 \leq i < j \leq n} \left(\left(a - \frac{bc}{a} \right) E_{ij} \otimes E_{ji} + b_{ij} E_{ii} \otimes E_{jj} + c_{ji} E_{jj} \otimes E_{ii} \right) + \sum_{i=1}^{n} a E_{ii} \otimes E_{ii}.$$

That $\rho_{a,\mathsf{B},\mathsf{C}}$ satisfies (2–1) follows by [27, Lemma 4 and (37)]. The notation for the scalar bc is meant to suggest a product. We point out that $\rho_{a,\mathsf{B},\mathsf{C}}$ can be derived from $\beta_{q,P}(A_\ell)$ of [6, Section 5]. See [8] also.

Representations of Yang–Baxter algebras determine solutions to the quantum Yang–Baxter equation. Suppose that A is an algebra over k and $\rho = \sum_{i=1}^{r} a_i \otimes b_i \in A \otimes A$. For a left A-module M let ρ_M be the endomorphism of $M \otimes M$ defined by $\rho_M(m \otimes n) = \sum_{i=1}^{r} a_i \cdot m \otimes b_i \cdot n$ for all $m, n \in M$. Then (A, ρ) is a Yang–Baxter algebra over k if and only if ρ_M is an invertible solution to the quantum Yang–Baxter equation (1–1) for all left A-modules M.

The notions of minimal quantum algebra and minimal oriented quantum algebra are important for theoretical reasons. These notions are based on the notion of minimal Yang–Baxter algebra. A *minimal Yang–Baxter algebra* is a Yang–Baxter algebra (A, ρ) over k such that $A = A_\rho$. Observe that only A_ρ is involved in the definition of ρ_M of the preceding paragraph.

Let (A, ρ) be a Yang–Baxter algebra over k. A *Yang–Baxter subalgebra of* (A, ρ) is a pair (B, ρ) where B is a subalgebra of A such that $\rho, \rho^{-1} \in B \otimes B$; thus $A_\rho \subseteq B$. Consequently (A, ρ) has a unique minimal Yang–Baxter subalgebra which is (A_ρ, ρ). Observe that a Yang–Baxter algebra (B, ρ') is a Yang–Baxter subalgebra of (A, ρ) if and only if B is a subalgebra of A and the inclusion $i : B \longrightarrow A$ induces a morphism of Yang–Baxter algebras $i : (B, \rho') \longrightarrow (A, \rho)$.

Let I be an ideal of A. Then there is a (unique) Yang–Baxter algebra structure $(A/I, \bar{\rho})$ on the quotient A/I such that the projection $\pi : A \longrightarrow A/I$ determines a morphism $\pi : (A, \rho) \longrightarrow (A/I, \bar{\rho})$ of Yang–Baxter algebras. If K is a field extension of k then $(A \otimes K, \rho \otimes 1_K)$ is a Yang–Baxter algebra over K where we make the identification $\rho \otimes 1_K = \sum_{i=1}^{r} (a_i \otimes 1_K) \otimes (b_i \otimes 1_K)$.

We now turn to Yang–Baxter coalgebras. A *Yang–Baxter coalgebra over k* is a pair (C, b), where C is a coalgebra over k and $b : C{\times}C \longrightarrow k$ is an invertible bilinear form, such that

$$b(c_{(1)}, d_{(1)})b(c_{(2)}, e_{(1)})b(d_{(2)}, e_{(2)}) = b(c_{(2)}, d_{(2)})b(c_{(1)}, e_{(2)})b(d_{(1)}, e_{(1)})$$

for all $c, d, e \in C$. Observe that this equation is satisfied when C is cocommutative.

Let (C, b) and (C', b') be Yang–Baxter coalgebras over k. Then $(C{\otimes}C', b'')$ is a Yang–Baxter coalgebra over k, where $b''(c{\otimes}c', d{\otimes}d') = b(c, d)b'(c', d')$ for all $c, d \in C$ and $c', d' \in C'$. A *morphism* $f : (C, b) \longrightarrow (C', b')$ of Yang–Baxter coalgebras over k is a coalgebra map $f : C \longrightarrow C'$ which satisfies $b(c, d) = b'(f(c), f(d))$ for all $c, d \in C$. Observe that (k, b) is a Yang–Baxter coalgebra over k where $b(1, 1) = 1$. The category of all Yang–Baxter coalgebras over k with their morphisms under composition has a natural monoidal structure.

Since (C, b) is a Yang–Baxter coalgebra over k it follows that (C^{cop}, b), (C, b^{-1}) and (C, b^{op}) are as well, where $b^{\mathrm{op}}(c, d) = b(d, c)$ for all $c, d \in C$.

The notions of Yang–Baxter algebra and Yang–Baxter coalgebra are dual as one might suspect. Let (A, ρ) be a Yang–Baxter algebra over k. Then (A^o, b_ρ) is a Yang–Baxter coalgebra over k, where $b_\rho(a^o, b^o) = (a^o{\otimes}b^o)(\rho)$ for all $a^o, b^o \in A^o$, and the bilinear form b_ρ is of finite type. Suppose that C is a coalgebra over k and that $b : C{\times}C \longrightarrow k$ is a bilinear form of finite type, which is the case if C is finite-dimensional. Then (C, b) is a Yang–Baxter coalgebra over k if and only if (C^*, ρ_b) is a Yang–Baxter algebra over k.

Let I be a coideal of C which satisfies $b(I, C) = (0) = b(C, I)$. Then there is a (unique) Yang–Baxter coalgebra structure $(C/I, \bar{b})$ on the quotient C/I such that the projection $\pi : C \longrightarrow C/I$ defines a morphism $\pi : (C, b) \longrightarrow (C/I, \bar{b})$ of Yang–Baxter coalgebras. Now let I the sum of all coideals J of C which satisfy $b(J, C) = (0) = b(C, J)$. Set $C_r = C/I$ and define $b_r : C_r{\times}C_r \longrightarrow k$ by $b_r(c{+}I, d{+}I) = b(c, d)$ for all $c, d \in C$. Then (C_r, b_r) is a Yang–Baxter coalgebra over k. This construction is dual to the construction (A_ρ, ρ) for Yang–Baxter algebras (A, ρ) over k.

Rational representations of Yang–Baxter coalgebras over k determine solutions to the quantum Yang–Baxter equation just as representations of Yang–Baxter algebras do. Suppose that C is a coalgebra over k and that $b : C{\times}C \longrightarrow k$ is a bilinear form. For a right C-comodule M let τ_M be the endomorphism of $M{\otimes}M$ defined by $\tau_M(m{\otimes}n) = m^{<1>}{\otimes}n^{<1>}b(m^{(2)}, n^{(2)})$ for all $m, n \in M$. Then (C, b) is a Yang–Baxter coalgebra over k if and only if τ_M is an invertible solution to the quantum Yang–Baxter equation (1–1) for all right C-comodules M.

Note that τ_M for a Yang–Baxter coalgebra (C, b) can be defined in terms (C_r, b_r). For let (M, ρ) be a right C-comodule and let $\pi : C \longrightarrow C_r$ be the projection. Then (M, ρ_r) is a right C_r-comodule, where $\rho_r = (1_M{\otimes}\pi){\circ}\rho$, and τ_M defined for (M, ρ) is the same as τ_M defined for (M, ρ_r).

Let D be a subcoalgebra of C. Then $(D, b|_{D \times D})$ is a Yang–Baxter coalgebra which we call a *Yang–Baxter subcoalgebra of* (C, b). Observe that Yang–Baxter subcoalgebras of (C, b) are those Yang–Baxter coalgebras (D, b') such that D is a subcoalgebra of C and the inclusion $\imath : D \longrightarrow C$ induces a morphism $\imath : (D, b') \longrightarrow (C, b)$ of Yang–Baxter coalgebras over k. Let K be a field extension of k. Then $(C \otimes K, b_K)$ is a Yang–Baxter coalgebra over K, where $b_K(c \otimes \alpha, d \otimes \beta) = \alpha \beta b(c, d)$ for all $c, d \in C$ and $\alpha, \beta \in K$.

3. Quantum Algebras and Quasitriangular Hopf Algebras

Quantum algebras determine regular isotopy invariants of 1–1 tangles and twist quantum algebras determine regular isotopy invariants of knots and links. The notion of quantum algebra arises in the consideration of the algebra of unoriented knot and link diagrams; see Section 6 for an indication of how the axioms for quantum algebras are related to the diagrams.

In this section we recall the definitions of quantum algebra and twist quantum algebra, discuss basic examples and outline some fundamental results about them. The reader is referred to [19].

A *quantum algebra over* k is a triple (A, ρ, s), where (A, ρ) is a Yang–Baxter algebra over k and $s : A \longrightarrow A^{\mathrm{op}}$ is an algebra isomorphism, such that

(QA.1) $\rho^{-1} = (s \otimes 1_A)(\rho)$ and

(QA.2) $\rho = (s \otimes s)(\rho)$.

Observe that (QA.1) and (QA.2) imply

(QA.3) $\rho^{-1} = (1_A \otimes s^{-1})(\rho)$;

indeed any two of (QA.1)–(QA.3) imply the third.

Quasitriangular Hopf algebras are a basic source of quantum algebras. A *quasitriangular Hopf algebra over* k is a pair (A, ρ), where A is a Hopf algebra with bijective antipode s over k and $\rho = \sum_{i=1}^{r} a_i \otimes b_i \in A \otimes A$, such that:

(QT.1) $\sum_{i=1}^{r} \Delta(a_i) \otimes b_i = \sum_{i,j=1}^{r} a_i \otimes a_j \otimes b_i b_j$,

(QT.2) $\sum_{i=1}^{r} \varepsilon(a_i) b_i = 1$,

(QT.3) $\sum_{i=1}^{r} a_i \otimes \Delta^{\mathrm{cop}}(b_i) = \sum_{i,j=1}^{r} a_i a_j \otimes b_i \otimes b_j$,

(QT.4) $\sum_{i=1}^{r} a_i \varepsilon(b_i) = 1$ and

(QT.5) $(\Delta^{\mathrm{cop}}(a))\rho = \rho(\Delta(a))$ for all $a \in A$.

Our definition of quasitriangular Hopf algebra is another formulation of the definition of quasitriangular Hopf algebra [4, page 811] in the category of finite-dimensional vector spaces over a field. Observe that (QT.1) and (QT.2) imply that ρ is invertible and $\rho^{-1} = (s \otimes 1_A)(\rho)$; apply $(m \otimes 1_A) \circ (s \otimes 1_A \otimes 1_A)$ and $(m \otimes 1_A) \circ (1_A \otimes s \otimes 1_A)$ to both sides of the equation of (QT.1). Using (QT.3) and (QT.4) it follows by a similar argument that $\rho^{-1} = (1_A \otimes s^{-1})(\rho)$. See [23, page 13].

Observe that ρ satisfies (2–1) by (QT.1) and (QT.5), or equivalently by (QT.3) and (QT.5). Thus:

EXAMPLE 2. If (A, ρ) is a quasitriangular Hopf algebra over k then (A, ρ, s) is a quantum algebra over k, where s is the antipode of A.

A fundamental example of a quasitriangular Hopf algebra is the quantum double $(D(A), \rho)$ of a finite-dimensional Hopf algebra A with antipode s over k defined in [4]. As a coalgebra $D(A) = A^{* \, cop} \otimes A$. The multiplicative identity for the algebra structure on $D(A)$ is $\varepsilon \otimes 1$ and multiplication is determined by

$$(p \otimes a)(q \otimes b) = p(a_{(1)} \rightharpoonup q \leftharpoonup s^{-1}(a_{(3)})) \otimes a_{(2)} b$$

for all $p, q \in A^*$ and $a, b \in A$; the functional $a \rightharpoonup q \leftharpoonup b \in A^*$ is defined by $(a \rightharpoonup q \leftharpoonup b)(c) = q(bca)$ for all $c \in A$. We follow [26] for the description of the quantum double.

Let $\{a_1, \ldots, a_r\}$ be a linear basis for A and let $\{a^1, \ldots, a^r\}$ be the dual basis for A^*. Then $(D(A), \rho)$ is a minimal quasitriangular Hopf algebra, where $\rho = \sum_{i=1}^{r} (\varepsilon \otimes a_i) \otimes (a^i \otimes 1)$. The definition of ρ does not depend on the choice of basis for A.

Coassociativity is not needed for Example 2. A structure which satisfies the axioms for a Hopf algebra over k with the possible exception of the coassociative axiom is called a *not necessarily coassociative Hopf algebra*.

A very important example of a quantum algebra, which accounts for the Jones polynomial when $k = \mathbb{C}$ is the field of complex numbers, is one defined on the algebra $A = M_2(k)$ of 2×2 matrices over k. The Jones polynomial and its connection with this quantum algebra is discussed in detail in Section 8.

EXAMPLE 3. Let k be a field and $q \in k^*$. Then $(M_2(k), \rho, s)$ is a quantum algebra over k, where

$$\rho = q^{-1}(E_{11} \otimes E_{11} + E_{22} \otimes E_{22}) + q(E_{11} \otimes E_{22} + E_{22} \otimes E_{11}) + (q^{-1} - q^3)E_{12} \otimes E_{21}$$

and

$$s(E_{11}) = E_{22}, \quad s(E_{22}) = E_{11}, \quad s(E_{12}) = -q^{-2}E_{12}, \quad s(E_{21}) = -q^2 E_{21}.$$

Let (A, ρ, s) and (A', ρ', s') be quantum algebras over k. The *tensor product of* (A, ρ, s) *and* (A', ρ', s') is the quantum algebra $(A \otimes A', \rho'', s \otimes s')$, where $(A \otimes A', \rho'')$ is the tensor product of the Yang–Baxter algebras (A, ρ) and (A', ρ'). A *morphism* $f : (A, \rho, s) \longrightarrow (A', \rho', s')$ *of quantum algebras* is a morphism $f : (A, \rho) \longrightarrow (A', \rho')$ of Yang–Baxter algebras which satisfies $f \circ s = s' \circ f$. Observe that $(k, 1 \otimes 1, 1_k)$, is a quantum algebra over k. The category of quantum algebras over k and their morphisms under composition has a natural monoidal structure.

Let (A, ρ, s) be a quantum algebra over k. Then (A, ρ) is a Yang–Baxter algebra over k and thus $(A^{\mathrm{op}}, \rho), (A, \rho^{-1})$ and (A, ρ^{op}) are also as we have noted.

It is not hard to see that (A^{op}, ρ, s), (A, ρ^{-1}, s^{-1}) and (A, ρ^{op}, s^{-1}) are quantum algebras over k.

Certain quotients of A have a quantum algebra structure. Let I be an ideal of A which satisfies $s(I) = I$. Then there exists a (unique) quantum algebra structure $(A/I, \bar{\rho}, \bar{s})$ on the quotient algebra A/I such that the projection $\pi : A \longrightarrow A/I$ determines a morphism $\pi : (A, \rho, s) \longrightarrow (A/I, \bar{\rho}, \bar{s})$ of quantum algebras over k. If $f : (A, \rho, s) \longrightarrow (A'\rho', s')$ is a morphism of quantum algebras, and f is onto, then $(A/I, \bar{\rho}, \bar{s})$ and (A', ρ', s') are isomorphic. We let the reader formulate and prove fundamental homomorphism theorems for quantum algebras.

We say that (A, ρ, s) is a *minimal quantum algebra over* k if $A = A_\rho$. Let B be a subalgebra of A such that $\rho \in B \otimes B$ and $s(B) = B$. Then $(B, \rho, s|_B)$ is a quantum algebra over k which is called a *quantum subalgebra of* (A, ρ, s). Since $\rho = (s \otimes s)(\rho)$ it follows that $s(A_\rho) = A_\rho$ by Lemma 1. Thus $(A_\rho, \rho, s|_{A_\rho})$ is a quantum subalgebra of (A, ρ, s). Observe that A_ρ is generated as an algebra by $A_{(\rho)} + A_{(\rho^{-1})} = A_{(\rho)}$ since $\rho^{-1} = (s \otimes 1_A)(\rho)$.

If $(B, \rho, s|_B)$ is a quantum subalgebra of (A, ρ, s) then $A_\rho \subseteq B$; thus (A, ρ, s) has a unique minimal quantum subalgebra which is $(A_\rho, \rho, s|_{A_\rho})$. Notice that the quantum subalgebras of (A, ρ, s) are the quantum algebras of the form (B, ρ', s'), where B is a subalgebra of A and the inclusion $\imath : B \longrightarrow A$ induces a morphism $\imath : (B, \rho', s') \longrightarrow (A, \rho, s)$.

Quantum algebras account for regular isotopy invariants of (unoriented) 1–1 tangles and twist quantum algebras account for regular isotopy invariants of knots and links. See Section 6.2 for details. A *twist quantum algebra over* k is a quadruple (A, ρ, s, G), where (A, ρ, s) is a quantum algebra over k and $G \in A$ is invertible, such that

$$s(G) = G^{-1} \quad \text{and} \quad s^2(a) = GaG^{-1}$$

for all $a \in A$. Twist quantum algebras arise from ribbon Hopf algebras and from quantum algebras defined on $A = M_n(k)$. It is important to note that an essential ingredient for the construction of a knot or link invariant from a twist quantum algebra is an s^*-invariant tracelike functional $\text{tr} \in A^*$.

As the referee has observed, every quantum algebra can be embedded in a twist quantum algebra. Specifically, given a quantum algebra (A, ρ, s) over k there is a twist quantum algebra $(\boldsymbol{A}, \boldsymbol{\rho}, \boldsymbol{s}, \boldsymbol{G})$ over k and an algebra embedding $\imath : A \longrightarrow \boldsymbol{A}$ which induces a morphism of quantum algebras $\imath : (A, \rho, s) \longrightarrow (\boldsymbol{A}, \boldsymbol{\rho}, \boldsymbol{s})$.

We construct $(\boldsymbol{A}, \boldsymbol{\rho}, \boldsymbol{s}, \boldsymbol{G})$ as follows. As a vector space over $\boldsymbol{A} = \oplus_{\ell \in \mathbb{Z}} A_\ell$, where $A_\ell = \{(\ell, a) \mid a \in A\}$ is endowed with the vector space structure which makes the bijection $A \longrightarrow A_\ell$ defined by $a \mapsto (\ell, a)$ a linear isomorphism. The rule $(\ell, a) \cdot (m, b) = (\ell + m, s^{-2m}(a)b)$ for all $\ell, m \in \mathbb{Z}$ and $a, b \in A$ gives $\boldsymbol{A}$ an associative algebra structure. The linear endomorphism $\boldsymbol{S}$ of $\boldsymbol{A}$ determined by $\boldsymbol{S}((\ell, a)) = (-\ell, s^{2\ell+1}(a))$ is an algebra isomorphism $\boldsymbol{S} : \boldsymbol{A} \longrightarrow \boldsymbol{A}^{op}$. The reader can easily check that $(\boldsymbol{A}, \boldsymbol{\rho}, \boldsymbol{s}, \boldsymbol{G})$ is the desired twist quantum algebra, where

$\imath : A \longrightarrow A$ is defined by $\imath(a) = (0, a)$, $\rho = (\imath \otimes \imath)(\rho)$ and $\mathbf{G} = (1, 1)$. The reader is left to supply the few remaining details of the proof of the following result which describes the pair $(\imath, (\mathbf{A}, \boldsymbol{\rho}, \boldsymbol{s}, \mathbf{G}))$.

PROPOSITION 1. *Let (A, ρ, s) be a quantum algebra over the field k. Then the pair $(\imath, (\mathbf{A}, \boldsymbol{\rho}, \boldsymbol{s}, \mathbf{G}))$ satisfies the following:*

(a) $(\mathbf{A}, \boldsymbol{\rho}, \boldsymbol{s}, \mathbf{G})$ *is a twist quantum algebra over k and $\imath : (A, \rho, s) \longrightarrow (\mathbf{A}, \boldsymbol{\rho}, \boldsymbol{s})$ is a morphism of quantum algebras.*
(b) *If (A', ρ', s', G') is a twist quantum algebra over k and $f : (A, \rho, s) \longrightarrow (A', \rho', s')$ is a morphism of quantum algebras then there exists a morphism $F : (\mathbf{A}, \boldsymbol{\rho}, \boldsymbol{s}, \mathbf{G}) \longrightarrow (A', \rho', s', G')$ of twist quantum algebras uniquely determined by $F \circ \imath = f$.* □

Recall that a ribbon Hopf algebra over k is a triple (A, ρ, v), where (A, ρ) is a quasitriangular Hopf algebra with antipode s over k and $v \in A$, such that

(R.0) v is in the center of A,

(R.1) $v^2 = us(u)$,

(R.2) $s(v) = v$,

(R.3) $\varepsilon(v) = 1$ and

(R.4) $\Delta(v) = (v \otimes v)(\rho^{\mathrm{op}} \rho)^{-1} = (\rho^{\mathrm{op}} \rho)^{-1}(v \otimes v)$.

Ribbon Hopf algebras were introduced and studied by Reshetikhin and Turaev in [29]. The element v is referred to as a special element or ribbon element in the literature. See [11] also.

Let (A, ρ, v) be a ribbon Hopf algebra over k, write $\rho = \sum_{i=1}^{r} a_i \otimes b_i$ and let $u = \sum_{i=1}^{r} s(b_i) a_i$ be the Drinfel'd element of the quasitriangular Hopf algebra (A, ρ). Then u is invertible, $s^2(a) = uau^{-1}$ for all $a \in A$ and $\Delta(u) = (u \otimes u)(\rho^{\mathrm{op}} \rho)^{-1} = (\rho^{\mathrm{op}} \rho)^{-1}(u \otimes u)$ by the results of [3]. Since u is invertible it follows by (R.1) that v is invertible. Thus $G = uv^{-1}$ is invertible, is a grouplike element of A and $s^2(a) = GaG^{-1}$ for all $a \in A$. Since G is a grouplike element of A it follows that $s(G) = G^{-1}$. Collecting results:

EXAMPLE 4. Let (A, ρ, v) be a ribbon Hopf algebra with antipode s over k. Then (A, ρ, s, uv^{-1}) is a twist quantum algebra over k, where u is the Drinfel'd element of the quasitriangular Hopf algebra (A, ρ).

For a detailed explanation of the relationship between ribbon and grouplike elements the reader is referred to [11].

Any algebra automorphism t of $M_n(k)$ has the form $t(a) = GaG^{-1}$ for all $a \in M_n(k)$, where $G \in M_n(k)$ is invertible, by the Norther–Skolem Theorem. See the corollary to [7, Theorem 4.3.1]. Such a G is unique up to scalar multiple. Let a^t be the transpose of $a \in M_n(k)$.

LEMMA 2. *Let $(M_n(k), \rho, s)$ be a quantum algebra over k.*

(a) *There exists an invertible* $M \in \mathrm{M}_n(k)$ *such that* $s(a) = Ma^t M^{-1}$ *for all* $a \in \mathrm{M}_n(k)$.

(b) $(\mathrm{M}_n(k), \rho, s, M(M^t)^{-1})$ *is a twist quantum algebra over* k.

PROOF. First observe that $t(a) = s(a^t)$ for all $a \in \mathrm{M}_n(k)$ defines an algebra automorphism t of $\mathrm{M}_n(k)$. Thus $t(a) = MaM^{-1}$ for all $a \in \mathrm{M}_n(k)$ for some invertible $M \in \mathrm{M}_n(k)$ by the preceding remarks. Since $s(a) = t(a^t)$ for all $a \in \mathrm{M}_n(k)$ part (a) follows. Part (b) is the result of a straightforward calculation. $\square$

For the quantum algebra of Example 3 we may take $M = q^{-1}E_{12} - qE_{21}$ and thus $G = -(q^{-2}E_{11} + q^2 E_{22})$.

4. Oriented Quantum Algebras

Just as the notion of quantum algebra arises in the consideration of algebra associated with diagrams of unoriented knots and links, the notion of oriented quantum algebra arises in connection with diagrams of oriented knots and links [17]. In Section 7 the reader will begin to see the relationship between the axioms for an oriented quantum algebra and oriented diagrams. For a detailed explanation of the topological motivation for the concept of oriented quantum algebra the reader is referred to [17]. An expanded version of most of what follows is found in [16].

An *oriented quantum algebra* over the field k is a quadruple $(A, \rho, t_{\mathsf{d}}, t_{\mathsf{u}})$, where (A, ρ) is a Yang–Baxter algebra over k and $t_{\mathsf{d}}, t_{\mathsf{u}}$ are commuting algebra automorphisms of A, such that

(qa.1) $(1_A \otimes t_{\mathsf{u}})(\rho)$ and $(t_{\mathsf{d}} \otimes 1_A)(\rho^{-1})$ are inverses in $A \otimes A^{\mathrm{op}}$, and

(qa.2) $\rho = (t_{\mathsf{d}} \otimes t_{\mathsf{d}})(\rho) = (t_{\mathsf{u}} \otimes t_{\mathsf{u}})(\rho)$.

Suppose that $(A, \rho, t_{\mathsf{d}}, t_{\mathsf{u}})$ and $(A', \rho', t'_{\mathsf{d}}, t'_{\mathsf{u}})$ are oriented quantum algebras over k. Then $(A \otimes A', \rho'', t_{\mathsf{d}} \otimes t'_{\mathsf{d}}, t_{\mathsf{u}} \otimes t'_{\mathsf{u}})$ is an oriented quantum algebra over k, which we refer to as the *tensor product of* $(A, \rho, t_{\mathsf{d}}, t_{\mathsf{u}})$ *and* $(A', \rho', t'_{\mathsf{d}}, t'_{\mathsf{u}})$, where $(A \otimes A', \rho'')$ is the tensor product of the Yang–Baxter algebras (A, ρ) and (A', ρ'). A *morphism* $f : (A, \rho, t_{\mathsf{d}}, t_{\mathsf{u}}) \longrightarrow (A', \rho', t'_{\mathsf{d}}, t'_{\mathsf{u}})$ *of oriented quantum algebras* is a morphism $f : (A, \rho) \longrightarrow (A', \rho')$ of Yang–Baxter algebras over k which satisfies $t'_{\mathsf{d}} \circ f = f \circ t_{\mathsf{d}}$ and $t'_{\mathsf{u}} \circ f = f \circ t_{\mathsf{u}}$. Note that $(k, 1 \otimes 1, 1_k, 1_k)$ is an oriented quantum algebra over k. The category of oriented quantum algebras over k together with their morphisms under composition has a natural monoidal structure.

An oriented quantum algebra $(A, \rho, t_{\mathsf{d}}, t_{\mathsf{u}})$ over k is *standard* if $t_{\mathsf{d}} = 1_A$ and is *balanced* if $t_{\mathsf{d}} = t_{\mathsf{u}}$, in which case we write (A, ρ, t) for $(A, \rho, t_{\mathsf{d}}, t_{\mathsf{u}})$, where $t = t_{\mathsf{d}} = t_{\mathsf{u}}$.

Standard oriented quantum algebras play an important role in the theory of oriented quantum algebras. There is always a standard oriented quantum algebra associated with an oriented quantum algebra.

PROPOSITION 2. *If* (A, ρ, t_d, t_u) *is an oriented quantum algebra over* k *then* $(A, \rho, t_u \circ t_d, 1_A)$ *and* $(A, \rho, 1_A, t_d \circ t_u)$ *are also.*

PROOF. Apply the algebra automorphisms $t_u \otimes 1_A$ and $1_A \otimes t_d$ of $A \otimes A^{op}$ to both sides of the equations of (qa.1). □

The oriented quantum algebra $(A, \rho, 1_A, t_d \circ t_u)$ is the *standard oriented quantum algebra associated with* (A, ρ, t_d, t_u).

The Yang–Baxter algebra of Example 1 has a balanced oriented quantum algebra structure.

EXAMPLE 5. Let $n \geq 2$, $a, bc \in k^*$ satisfy $a^2 \neq bc, 1$ and suppose $\omega_1, \ldots, \omega_n \in k^*$ satisfy

$$\omega_i^2 = \left(\frac{a^2}{bc} \right)^{i-1} \omega_1^2$$

for all $1 \leq i \leq n$. Then $(M_n(k), \rho_{a,B,C}, t)$ is a balanced oriented quantum algebra, where

$$t(E_{ij}) = \left(\frac{\omega_i}{\omega_j} \right) E_{ij}$$

for all $1 \leq i, j \leq n$.

Example 5 is considered in more generality in [16, Theorem 2].

Suppose that (A, ρ, t_d, t_u) is an oriented quantum algebra over k and write $\rho = \sum_{i=1}^r a_i \otimes b_i, \rho^{-1} = \sum_{j=1}^s \alpha_j \otimes \beta_j \in A \otimes A$. Then axioms (qa.1) and (qa.2) can be formulated

$$\sum_{i=1}^r \sum_{j=1}^s a_i t_d(\alpha_j) \otimes \beta_j t_u(b_i) = 1 \otimes 1 = \sum_{j=1}^s \sum_{i=1}^r t_d(\alpha_j) a_i \otimes t_u(b_i) \beta_j \qquad (4\text{--}1)$$

and

$$\sum_{i=1}^r a_i \otimes b_i = \sum_{i=1}^r t_d(a_i) \otimes t_d(b_i) = \sum_{i=1}^r t_u(a_i) \otimes t_u(b_i). \qquad (4\text{--}2)$$

respectively. Alterations to the structure of A determine other oriented quantum algebras.

Observe that (A^{op}, ρ, t_d, t_u) is an oriented quantum algebra over k in light of (2–2) and (4–1), which we denote by A^{op} as well. We have noted that (A, ρ^{-1}) is a quantum Yang–Baxter algebra over k. Let $t = t_d$ or $t = t_u$. Since $t \otimes t$ is an algebra automorphism of $A \otimes A$ and $\rho = (t \otimes t)(\rho)$ we have $\rho^{-1} = (t \otimes t)(\rho^{-1})$. By applying $t_d^{-1} \otimes t_u^{-1}$ to both sides of the equations of (4–1) we see that $(A, \rho^{-1}, t_d^{-1}, t_u^{-1})$ is an oriented quantum algebra over k. Notice that $(A, \rho^{op}, t_u^{-1}, t_d^{-1})$ is an oriented quantum algebra over k.

Let K be a field extension of k and $A \otimes K$ be the algebra over K obtained by extension of scalars. Then $(A \otimes K, \rho \otimes 1_K, t_d \otimes 1_K, t_u \otimes 1_K)$ is a unoriented quantum algebra structure over K, where $(A \otimes K, \rho \otimes 1_K)$ is the Yang–Baxter algebra described in Section 2.

Certain quotients of A have an oriented quantum algebra structure. Let I be an ideal of A and suppose that $t_d(I) = t_u(I) = I$. Then there is a unique oriented quantum algebra structure $(A/I, \overline{\rho}, \overline{t}_d, \overline{t}_u)$ on the quotient algebra A/I such that $\pi : (A, \rho, t_d, t_u) \longrightarrow (A/I, \overline{\rho}, \overline{t}_d, \overline{t}_u)$ is a morphism, where $\pi : A \longrightarrow A/I$ is the projection. Furthermore, if $f : (A, \rho, t_d, t_u) \longrightarrow (A', \rho', t'_d, t'_u)$ is a morphism of oriented quantum algebras and $f : A \longrightarrow A'$ is onto, then $(A/\ker f, \overline{\rho}, \overline{t}_d, \overline{t}_u)$ and (A', ρ', t'_d, t'_u) are isomorphic oriented quantum algebras. The reader is left to formulate and prove fundamental homomorphism theorems for oriented quantum algebras.

The oriented quantum algebra (A, ρ, t_d, t_u) is a *minimal oriented quantum algebra over* k if $A = A_\rho$. As in the case of quantum algebras, the notion of minimal oriented quantum algebra is theoretically important.

Let B be a subalgebra of A which satisfies $\rho, \rho^{-1} \in B \otimes B$ and $t_d(B) = t_u(B) = B$. Then $(B, \rho, t_d|_B, t_u|_B)$ is an oriented quantum algebra over k which is called an *oriented quantum subalgebra of* (A, ρ, t_d, t_u). Since $\rho = (t_d \otimes t_d)(\rho) = (t_u \otimes t_u)(\rho)$ it follows that $t_d(A_\rho) = t_u(A_\rho) = A_\rho$ by Lemma 1. Therefore $(A_\rho, \rho, t_d|_{A_\rho}, t_u|_{A_\rho})$ is an oriented quantum subalgebra of (A, ρ, t_d, t_u).

If $(B, \rho, t_d|_B, t_u|_B)$ is an oriented quantum subalgebra of (A, ρ, t_d, t_u) then $A_\rho \subseteq B$; thus (A, ρ, t_d, t_u) has a unique minimal oriented quantum subalgebra which is $(A_\rho, t_d|_{A_\rho}, t_u|_{A_\rho})$. Oriented quantum subalgebras of (A, ρ, t_d, t_u) are those oriented quantum algebras (B, ρ', t'_d, t'_u) over k, where B is a subalgebra of A and the inclusion $\imath : B \longrightarrow A$ determines a morphism $\imath : (B, \rho', t'_d, t'_u) \longrightarrow (A, \rho, t_d, t_u)$ of oriented quantum algebras.

Minimal Yang–Baxter algebras over k support at most one standard oriented quantum algebra structure. Observe that if (A, ρ) is a Yang–Baxter algebra over k and A is commutative, then $(A, \rho, 1_A, 1_A)$ is an oriented quantum algebra over k.

PROPOSITION 3. *Let (A, ρ) be a minimal Yang–Baxter quantum algebra over k.*

(a) *There is at most one algebra automorphism t of A such that $(A, \rho, 1_A, t)$ is a standard oriented quantum algebra over k.*

(b) *Suppose that (A, ρ, t_d, t_u) and (A, ρ, t'_d, t'_u) are oriented quantum algebras over k. Then $t_d \circ t_u = t'_d \circ t'_u$.*

(c) *Suppose that A is commutative and $(A, \rho, 1_A, t)$ is a standard oriented quantum algebra over k. Then $t = 1_A$.*

PROOF. Part (b) follows from part (a) and Proposition 2. We have noted that $(A, \rho, 1_A, 1_A)$ is an oriented quantum algebra over k when A is commutative. Thus part (c) follows from part (a) also.

To show part (a), suppose that $(A, \rho, 1_A, t)$ and $(A, \rho, 1_A, t')$ are standard oriented quantum algebras over k. Then $(1_A \otimes t)(\rho) = (1_A \otimes t')(\rho)$ since both sides of the equation are left inverses of ρ^{-1} in $A \otimes A^{\mathrm{op}}$. This equation together with (qa.2) implies $(t \otimes 1_A)(\rho) = (t' \otimes 1_A)(\rho)$ as well. These two equations imply $t|_{A_{(\rho)}} = t'|_{A_{(\rho)}}$. Now the first two equations also imply $(1_A \otimes t)(\rho^{-1}) =$

$(1_A \otimes t')(\rho^{-1})$ and $(t \otimes 1_A)(\rho^{-1}) = (t' \otimes 1_A)(\rho^{-1})$ since the maps involved are algebra automorphisms of $A \otimes A^{\mathrm{op}}$. Therefore $t|_{A_{(\rho^{-1})}} = t'|_{A_{(\rho^{-1})}}$. Since $A_{(\rho)} + A_{(\rho^{-1})}$ generates A as an algebra it now follows that $t = t'$. $\qquad\square$

Every quantum algebra accounts for a standard oriented quantum algebra.

THEOREM 1. *Suppose that* (A, ρ, s) *is a quantum algebra over the field* k. *Then* $(A, \rho, 1_A, s^{-2})$ *is an oriented quantum algebra over* k.

PROOF. Write $\rho = \sum_{i=1}^{r} a_i \otimes b_i$. Now (A, ρ) is a Yang–Baxter algebra over k since (A, ρ, s) is a quantum algebra over k. Since (QA.2) holds for s and ρ it follows that (qa.2) holds for s^{-2} and ρ. The fact that $\rho^{-1} = \sum_{i=1}^{r} s(a_i) \otimes b_i$ translates to

$$\sum_{i,j=1}^{r} s(a_i) a_j \otimes b_i b_j = 1 \otimes 1 = \sum_{j,i=1}^{r} a_j s(a_i) \otimes b_j b_i.$$

Applying $s \otimes 1_A$ to both sides of these equations yields

$$\sum_{j,i=1}^{r} s(a_j) s^2(a_i) \otimes b_i b_j = 1 \otimes 1 = \sum_{i,j=1}^{r} s^2(a_i) s(a_j) \otimes b_j b_i.$$

Since $\rho = (s^2 \otimes s^2)(\rho)$ it follows that

$$\sum_{j,i=1}^{r} s(a_j) a_i \otimes s^{-2}(b_i) b_j = 1 \otimes 1 = \sum_{i,j=1}^{r} a_i s(a_j) \otimes b_j s^{-2}(b_i);$$

that is ρ^{-1} and $(1_A \otimes s^{-2})(\rho)$ are inverses in $A \otimes A^{\mathrm{op}}$. $\qquad\square$

As a consequence of part (c) of Proposition 3 and the preceding theorem:

COROLLARY 1. *If* (A, ρ, s) *is a minimal quantum algebra over* k *and* A *is commutative then* $s^2 = 1_A$. $\qquad\square$

Let (A, ρ, s) be a quantum algebra over k. Then $(A, \rho, 1_A, s^{-2})$ and $(A, \rho, s^{-2}, 1_A)$ are oriented quantum algebras over k by Theorem 1 and Proposition 2. It may very well be the case that these are the only oriented quantum algebra structures of the form $(A, \rho, t_{\mathrm{d}}, t_{\mathrm{u}})$. Sweedler's 4-dimensional Hopf algebra A when the characteristic of k is not 2 illustrates the point. We recall that A is generated as a k-algebra by a, x subject to the relations $a^2 = 1, x^2 = 0, xa = -ax$ and the coalgebra structure of A is determined by $\Delta(a) = a \otimes a$, $\Delta(x) = x \otimes a + 1 \otimes x$.

EXAMPLE 6. Let A be Sweedler's 4-dimensional Hopf algebra with antipode s over k, suppose that the characteristic of k is not 2 and for $\alpha \in k^*$ let

$$\rho_\alpha = \frac{1}{2}(1 \otimes 1 + 1 \otimes a + a \otimes 1 - a \otimes a) + \frac{\alpha}{2}(x \otimes x + x \otimes ax + ax \otimes ax - ax \otimes x).$$

Then (A, ρ_α, s) is a minimal quantum algebra over k; moreover, $(A, \rho_\alpha, 1_A, s^{-2})$ and $(A, \rho_\alpha, s^{-2}, 1_A)$ are the only oriented quantum algebra structures of the form $(A, \rho_\alpha, t_{\mathrm{d}}, t_{\mathrm{u}})$.

If $(A, \rho, 1_A, t)$ is a standard oriented quantum algebra over k then there may be no quantum algebra of the form (A, ρ, s).

EXAMPLE 7. There is no quantum algebra of the form $(M_n(k), \rho_{a,B,C}, s)$, where $n > 2$ and $(M_n(k), \rho_{a,B,C}, t)$ is the balanced oriented quantum algebra of Example 5.

However, when $n = 2$ we have the important connection:

EXAMPLE 8. Let $(M_2(k), \rho_{q^{-1}, \{q\}, \{q\}}, t)$ be the balanced quantum algebra of Example 5 where $n = 2$ and $\omega_1 = q, \omega_2 = q^{-1}$. Then the associated standard oriented quantum algebra

$$(M_2(k), \rho_{q^{-1}, \{q\}, \{q\}}, 1_{M_2(k)}, t^2) = (M_2(k), \rho, 1_{M_2(k)}, s^{-2}),$$

where $(M_2(k), \rho, s)$ is the quantum algebra over k of Example 3.

Let A be a finite-dimensional Hopf algebra with antipode s over k. The quasitriangular Hopf algebra $(D(A), \rho)$ admits an oriented quantum algebra structure $(D(A), \rho, 1_{D(A)}, s^{*2} \otimes s^{-2})$ by Theorem 1. If t is a Hopf algebra automorphism of A which satisfies $t^2 = s^{-2}$ then $(D(A), \rho, t^{*-1} \otimes t)$ is a balanced oriented quantum algebra over k.

An oriented quantum algebra defines a regular isotopy invariant of oriented 1–1 tangles as well shall see in Section 7. An oriented quantum algebra (A, ρ, t_d, t_u) with the additional structure of an invertible $G \in A$ which satisfies

$$t_d(G) = t_u(G) = G \quad \text{and} \quad t_d \circ t_u(a) = GaG^{-1}$$

for all $a \in A$ accounts for regular isotopy invariants of knots and links. The quintuple (A, ρ, t_d, t_u, G) is called a *twist oriented quantum algebra over k*. We note that an important ingredient for the definition of these invariants is a t_d, t_u-invariant tracelike element $\mathrm{tr} \in A^*$.

Let (A, ρ, t_d, t_u, G) and $(A', \rho', t'_d, t'_u, G')$ be twist oriented quantum algebras over k. Then $(A \otimes A', \rho'', t_d \otimes t'_d, t_u \otimes t'_u, G \otimes G')$ is a twist oriented quantum algebra over k, called the *tensor product of (A, ρ, t_d, t_u, G) and $(A', \rho', t'_d, t'_u, G')$*, where $(A \otimes A', \rho'', t_d \otimes t'_d, t_u \otimes t'_u)$ is the tensor product of (A, ρ, t_d, t_u) and (A', ρ', t'_d, t'_u). A *morphism* $f : (A, \rho, t_d, t_u, G) \longrightarrow (A', \rho', t'_d, t'_u, G')$ *of twist quantum oriented algebras* over k is a morphism $f : (A, \rho, t_d, t_u) \longrightarrow (A', \rho', t'_d, t'_u)$ of oriented quantum algebras which satisfies $f(G) = G'$. The category of twist oriented quantum algebras over k and their morphisms under composition has a natural monoidal structure.

Suppose that (A, ρ, t_d, t_u, G) is a twist oriented quantum algebra over k. Then $(A, \rho, 1_A, t_d \circ t_u, G)$ is as well. There are two important cases in which a standard oriented quantum algebra over k has a twist structure.

Let (A, ρ) be a quasitriangular Hopf algebra with antipode s over k, write $\rho = \sum_{i=1}^r a_i \otimes b_i$ and let $u = \sum_{i=1}^r s(b_i) a_i$ be the Drinfel'd element of A. Then $(A, \rho, 1_A, s^{-2}, u^{-1})$ is a twist oriented quantum algebra over k.

Now suppose $A = M_n(k)$ and that (A, ρ, t_d, t_u) is a standard, or balanced, oriented quantum algebra over k. We have noticed in the remarks preceding Lemma 2 that any algebra automorphism t of A is described by $t(a) = GaG^{-1}$ for all $a \in A$, where $G \in A$ is invertible and is unique up to scalar multiple. Thus (A, ρ, t_d, t_u, G) is a twist oriented quantum algebra over k for some invertible $G \in A$. For instance the balanced oriented quantum algebra $(M_n(k), \rho_{a,B,C}, t)$ of Example 5 has a twist balanced oriented quantum algebra structure $(M_n(k), \rho_{a,B,C}, t, t, G)$, where $G = \sum_{i=1}^{n} \omega_i^2 E_{ii}$.

Just as a quantum algebra can be embedded into a twist quantum algebra, an oriented standard quantum algebra can be embedded into a twist oriented standard quantum algebra. The construction of the latter is a slight modification of the construction of (A, ρ, s, G) which precedes the statement of Proposition 1. Here $(\ell, a) \cdot (m, b) = (\ell + m, t^{-m}(a)b)$ and s is replaced by t defined by $t((\ell, a)) = (\ell, t(a))$.

PROPOSITION 4.   *Let $(A, \rho, 1_A, t)$ be a standard oriented quantum algebra over the field k. The pair $(\imath, (A, \rho, 1_A, t, G))$ satisfies the following:*

(a) $(A, \rho, 1_A, t, G)$ *is a standard twist oriented quantum algebra over k and $\imath$:*
$(A, \rho, 1_A, t) \longrightarrow (A, \rho, 1_A, t)$ *is a morphism of oriented quantum algebras.*

(b) *If $(A', \rho', 1_{A'}, t', G')$ is a twist standard oriented quantum algebra over k and $f : (A, \rho, 1_A, t) \longrightarrow (A', \rho', 1_{A'}, t')$ is a morphism of quantum algebras there exists a morphism $F : (A, \rho, 1_A, t, G) \longrightarrow (A', \rho', 1_{A'}, t', G')$ of twist standard oriented quantum algebras uniquely determined by $F \circ \imath = f$.* $\square$

We end this section with a necessary and sufficient condition for a Yang–Baxter algebra over k to have a twist oriented quantum algebra structure.

PROPOSITION 5.   *Let (A, ρ) be a Yang–Baxter algebra over the field k, write $\rho = \sum_{i=1}^{r} a_i \otimes b_i$ and $\rho^{-1} = \sum_{j=1}^{s} \alpha_j \otimes \beta_j$, suppose $G \in A$ is invertible and let t be the algebra automorphism of A defined by $t(a) = GaG^{-1}$ for all $a \in A$. Then $(A, \rho, 1_A, t, G)$ is a twist oriented quantum algebra over k if and only if*

(a) $G \otimes G$ *and ρ are commuting elements of the algebra $A \otimes A$,*
(b) $\sum_{i=1}^{r} \sum_{j=1}^{s} a_i G \alpha_j \otimes \beta_j b_i = G \otimes 1$ *and*
(c) $\sum_{j=1}^{s} \sum_{i=1}^{r} \alpha_j G^{-1} a_i \otimes b_i \beta_j = G^{-1} \otimes 1$. $\square$

When A is finite-dimensional the conditions of parts (b) and (c) of the proposition are equivalent.

5. Quantum Algebras Constructed from Standard Oriented Quantum Algebras

By Theorem 1 a quantum algebra (A, ρ, s) accounts for a standard oriented quantum algebra $(A, \rho, 1_A, s^{-2})$. By virtue of Example 7 if $(A, \rho, 1_A, t)$ is a standard oriented quantum algebra over k there may be no quantum algebra of the form (A, ρ, s).

Let $(A, \rho, 1_A, t)$ be a standard oriented quantum algebra over k. In this section we show that there is a quantum algebra structure $(\mathcal{A}, \boldsymbol{\rho}, \boldsymbol{s})$ on the direct product $\mathcal{A} = A{\oplus}A^{\mathrm{op}}$ such that the projection $\pi : \mathcal{A} \longrightarrow A$ onto the first factor determines a morphism of standard oriented quantum algebras $\pi : (\mathcal{A}, \boldsymbol{\rho}, 1_A, \boldsymbol{s}^{-2}) \longrightarrow (A, \rho, 1_A, t)$. Our result follows from a construction which starts with an oriented quantum algebra $(A, \rho, t_{\mathbf{d}}, t_{\mathbf{d}})$ and produces a quantum algebra $(\mathcal{A}, \boldsymbol{\rho}, \boldsymbol{s})$ and oriented quantum algebra $(\mathcal{A}, \boldsymbol{\rho}, t_{\mathbf{d}}, t_{\mathbf{u}})$ which are related by $t_{\mathbf{d}}{\circ}t_{\mathbf{u}} = \boldsymbol{s}^{-2}$. We follow [15, Section 3].

Let A be an algebra over k and let $\mathcal{A} = A{\oplus}A^{\mathrm{op}}$ be the direct product of the algebras A and A^{op}. Denote the linear involution of $\mathcal{A}$ which exchanges the direct summands of $\mathcal{A}$ by $\overline{(\ )}$. Thus $\overline{a{\oplus}b} = b{\oplus}a$ for all $a, b \in A$. Think of A as a subspace of $\mathcal{A}$ by the identification $a = a{\oplus}0$ for all $a \in A$. Thus $\bar{a} = 0{\oplus}a$ and every element of $\mathcal{A}$ has a unique decomposition of the form $a + \bar{b}$ for some $a, b \in A$. Note that

$$\overline{(\bar{a})} = a, \quad \overline{ab} = \bar{b}\bar{a} \quad \text{and} \quad a\bar{b} = 0 = \bar{a}b \qquad (5\text{--}1)$$

for all $a, b \in A$.

LEMMA 3. *Let* $(A, \rho, 1_A, t)$ *be a standard oriented quantum algebra over* k, *let* $\mathcal{A} = A{\oplus}A^{\mathrm{op}}$ *be the direct product of* A *and* A^{op} *and write* $\rho = \sum_{i=1}^{r} a_i{\otimes}b_i$, $\rho^{-1} = \sum_{j=1}^{s} \alpha_j{\otimes}\beta_j$. *Then* $(\mathcal{A}, \boldsymbol{\rho}, \boldsymbol{s})$ *is a quantum algebra over* k, *where*

$$\boldsymbol{\rho} = \sum_{i=1}^{r}(a_i{\otimes}b_i + \bar{a}_i{\otimes}\bar{b}_i) + \sum_{j=1}^{s}(\overline{\alpha_j}{\otimes}\beta_j + \alpha_j{\otimes}\overline{t^{-1}(\beta_j)})$$

and $\boldsymbol{s}(a{\oplus}b) = b{\oplus}t^{-1}(a)$ *for all* $a, b \in A$.

PROOF. Since t is an algebra automorphism of A it follows that t^{-1} is also. Thus $\boldsymbol{s} : \mathcal{A} \longrightarrow \mathcal{A}^{\mathrm{op}}$ is an algebra isomorphism. By definition $\boldsymbol{s}(a) = \overline{t^{-1}(a)}$ and $\boldsymbol{s}(\bar{a}) = a$ for all $a \in A$. We have noted in the discussion following (4–2) that $\rho^{-1} = (t^{-1}{\otimes}t^{-1})(\rho^{-1})$. At this point it is easy to see that $\boldsymbol{\rho} = (\boldsymbol{s}{\otimes}\boldsymbol{s})(\boldsymbol{\rho})$, or (QA.2) is satisfied for $\boldsymbol{\rho}$ and $\boldsymbol{s}$. Using the equation $\rho^{-1} = (t^{-1}{\otimes}t^{-1})(\rho^{-1})$ we calculate

$$(\boldsymbol{s}{\otimes}1_A)(\boldsymbol{\rho}) = \sum_{i=1}^{r}\left(\overline{t^{-1}(a_i)}{\otimes}b_i + a_i{\otimes}\bar{b}_i\right) + \sum_{j=1}^{s}\left(\alpha_j{\otimes}\beta_j + \overline{\alpha_j}{\otimes}\overline{\beta_j}\right).$$

Using (5–1), the equation $(t^{-1}{\otimes}1_A)(\rho) = (1_A{\otimes}t)(\rho)$, which follows by (qa.2),

$$\boldsymbol{\rho}((\boldsymbol{s}{\otimes}1_A)(\boldsymbol{\rho})) = 1{\otimes}1 + \bar{1}{\otimes}\bar{1} + \bar{1}{\otimes}1 + 1{\otimes}\bar{1} = 1_A{\otimes}1_A = ((\boldsymbol{s}{\otimes}1_A)(\boldsymbol{\rho}))\boldsymbol{\rho}.$$

Therefore $\boldsymbol{\rho}$ is invertible and $\boldsymbol{\rho}^{-1} = (\boldsymbol{s}{\otimes}1_A)(\boldsymbol{\rho})$. We have shown that (QA.1) holds for $\boldsymbol{\rho}$ and $\boldsymbol{s}$.

The fact that $\boldsymbol{\rho}$ satisfies (2–1) is a rather lengthy and interesting calculation. Using the formulation (2–2) of (2–1) one sees that (2–1) for $\boldsymbol{\rho}$ is equivalent to a set of eight equations. With the notation convention $(\rho^{-1})_{ij} = \rho_{ij}^{-1}$ for

$1 \leq \imath < \jmath \leq 3$, this set of eight equations can be rewritten as set of six equations which are:

$$\rho_{12}\rho_{13}\rho_{23} = \rho_{23}\rho_{13}\rho_{12}, \tag{5-2}$$

$$\rho_{12}\rho_{23}^{-1}\rho_{13}^{-1} = \rho_{13}^{-1}\rho_{23}^{-1}\rho_{12}, \tag{5-3}$$

$$\rho_{13}^{-1}\rho_{12}^{-1}\rho_{23} = \rho_{23}\rho_{12}^{-1}\rho_{13}^{-1}, \tag{5-4}$$

$$\sum_{\ell=1}^{r} \sum_{\jmath,m=1}^{s} a_\ell \alpha_\jmath \otimes \beta_\jmath \alpha_m \otimes t^{-1}(\beta_m) b_\ell = \sum_{\jmath,m=1}^{s} \sum_{\ell=1}^{r} \alpha_\jmath a_\ell \otimes \alpha_m \beta_\jmath \otimes b_\ell t^{-1}(\beta_m), \tag{5-5}$$

$$\sum_{\jmath,m=1}^{s} \sum_{\ell=1}^{r} \alpha_\jmath a_\ell \otimes \alpha_m t^{-1}(\beta_\jmath) \otimes b_\ell \beta_m = \sum_{\ell=1}^{r} \sum_{\jmath,m=1}^{s} a_\ell \alpha_\jmath \otimes t^{-1}(\beta_\jmath) \alpha_m \otimes \beta_m b_\ell \tag{5-6}$$

and

$$\sum_{\jmath,\ell=1}^{s} \sum_{\ell=1}^{r} \alpha_\jmath \alpha_\ell \otimes \alpha_m t^{-1}(\beta_\jmath) \otimes b_m t^{-1}(\beta_\ell) = \sum_{\ell,\jmath=1}^{s} \sum_{m=1}^{r} \alpha_\ell \alpha_\jmath \otimes t^{-1}(\beta_\jmath) a_m \otimes t^{-1}(\beta_\ell) b_m. \tag{5-7}$$

By assumption (5-2) holds. Since $\rho_{\imath\jmath}$ is invertible and $(\rho_{\imath\jmath})^{-1} = (\rho^{-1})_{\imath\jmath} = \rho_{\imath\jmath}^{-1}$, equations (5-3)–(5-4) hold by virtue of (5-2).

Now t^{-1} is an algebra automorphism of A and $\rho^{-1} = (t^{-1} \otimes t^{-1})(\rho^{-1})$. Thus applying $1_A \otimes t^{-1} \otimes 1_A$ to both sides of the equation of (5-5) we see that (5-5) and (5-6) are equivalent; applying $t^{-1} \otimes 1_A \otimes 1$ to both sides of (5-7) we see that (5-7) is equivalent to $\rho_{23}\rho_{12}^{-1}\rho_{13}^{-1} = \rho_{13}^{-1}\rho_{12}^{-1}\rho_{23}$, a consequence of (5-2). Therefore to complete the proof of the lemma we need only show that (5-5) holds.

By assumption $(1_A \otimes t)(\rho)$ and ρ^{-1} are inverses in $A \otimes A^{\mathrm{op}}$. Consequently ρ and $(1_A \otimes t^{-1})(\rho)$ are inverses in $A \otimes A^{\mathrm{op}}$ since $1_A \otimes t^{-1}$ is an algebra endomorphism of $A \otimes A^{\mathrm{op}}$. Recall that ρ^{-1} satisfies (2-1). Thus

$$\sum_{\jmath,m=1}^{s} \sum_{\ell=1}^{r} \alpha_\jmath a_\ell \otimes \alpha_m \beta_\jmath \otimes b_\ell t^{-1}(\beta_m)$$

$$= \sum_{v,\ell=1}^{r} \sum_{u,\jmath,m=1}^{s} (a_v \alpha_u) \alpha_\jmath a_\ell \otimes \alpha_m \beta_\jmath \otimes b_\ell t^{-1}(\beta_m)(t^{-1}(\beta_u) b_v)$$

$$= \sum_{v,\ell=1}^{r} \sum_{u,\jmath,m=1}^{s} a_v (\alpha_u \alpha_\jmath) a_\ell \otimes \alpha_m \beta_\jmath \otimes b_\ell t^{-1}(\beta_m \beta_u) b_v$$

$$= \sum_{v,\ell=1}^{r} \sum_{u,\jmath,m=1}^{s} a_v (\alpha_\jmath \alpha_u) a_\ell \otimes \beta_\jmath \alpha_m \otimes b_\ell t^{-1}(\beta_u \beta_m) b_v$$

$$= \sum_{v,\ell=1}^{r} \sum_{u,\jmath,m=1}^{s} a_v \alpha_\jmath (\alpha_u a_\ell) \otimes \beta_\jmath \alpha_m \otimes (b_\ell t^{-1}(\beta_u)) t^{-1}(\beta_m) b_v$$

$$= \sum_{v=1}^{r} \sum_{\jmath,m=1}^{s} a_v \alpha_\jmath \otimes \beta_\jmath \alpha_m \otimes t^{-1}(\beta_m) b_v.$$

which establishes (5-5). $\square$

THEOREM 2. *Let $(A, \rho, t_\mathbf{d}, t_\mathbf{u})$ be an oriented quantum algebra over the field k, let $\mathcal{A} = A \oplus A^{op}$ be the direct product of A and A^{op} and write $\rho = \sum_{i=1}^{r} a_i \otimes b_i$, $\rho^{-1} = \sum_{j=1}^{s} \alpha_j \otimes \beta_j$. Then:*

(a) *$(\mathcal{A}, \boldsymbol{\rho}, \boldsymbol{s})$ is a quantum algebra over k, where*

$$\boldsymbol{\rho} = \sum_{i=1}^{r} (a_i \otimes b_i + \overline{a_i} \otimes \overline{b_i}) + \sum_{j=1}^{s} (\overline{\alpha_j} \otimes \beta_j + \alpha_j \otimes \overline{t_\mathbf{d}^{-1} \circ t_\mathbf{u}^{-1}(\beta_j)})$$

and $\boldsymbol{s}(a \oplus b) = b \oplus t_\mathbf{d}^{-1} \circ t_\mathbf{u}^{-1}(a)$ for all $a, b \in A$.

(b) *$(\mathcal{A}, \boldsymbol{\rho}, t_\mathbf{d}, t_\mathbf{u})$ is an oriented quantum algebra over k, $t_\mathbf{d}, t_\mathbf{u}$ commute with $\boldsymbol{s}$ and $t_\mathbf{d} \circ t_\mathbf{u} = \boldsymbol{s}^{-2}$, where $t_\mathbf{d}(a \oplus b) = t_\mathbf{d}(a) \oplus t_\mathbf{d}(b)$ and $t_\mathbf{u}(a \oplus b) = t_\mathbf{u}(a) \oplus t_\mathbf{u}(b)$ for all $a, b \in A$.*

(c) *The projection $\pi : \mathcal{A} \longrightarrow A$ onto the first factor determines a morphism $\pi : (\mathcal{A}, \boldsymbol{\rho}, t_\mathbf{d}, t_\mathbf{u}) \longrightarrow (A, \rho, t_\mathbf{d}, t_\mathbf{u})$ of oriented quantum algebras.*

PROOF. Since $(A, \rho, 1_A, t_\mathbf{d} \circ t_\mathbf{u})$ is a standard quantum algebra over k by Proposition 2, part (a) follows by Lemma 3. Part (b) is a straightforward calculation which is left to the reader and part (c) follows by definitions. $\qquad \square$

Let $\mathcal{C}_q$ be the category whose objects are quintuples $(A, \rho, s, t_\mathbf{d}, t_\mathbf{u})$, where (A, ρ, s) is a quantum algebra over k and $(A, \rho, t_\mathbf{d}, t_\mathbf{u})$ is an oriented quantum algebra over k such that $t_\mathbf{d}, t_\mathbf{u}$ commute with s and $t_\mathbf{d} \circ t_\mathbf{u} = s^{-2}$, and whose morphisms $f : (A, \rho, s, t_\mathbf{d}, t_\mathbf{u}) \longrightarrow (A', \rho', s', t'_\mathbf{d}, t'_\mathbf{u})$ are algebra maps $f : A \longrightarrow A'$ which determine morphisms $f : (A, \rho, s) \longrightarrow (A', \rho', s')$ and $f : (A, \rho, t_\mathbf{d}, t_\mathbf{u}) \longrightarrow (A', \rho', t'_\mathbf{d}, t'_\mathbf{u})$. The construction $(\mathcal{A}, \boldsymbol{\rho}, \boldsymbol{s}, t_\mathbf{d}, t_\mathbf{d})$ of Theorem 2 is a cofree object of $\mathcal{C}_q$. Let $\pi : \mathcal{A} \longrightarrow A$ be the projection onto the first factor.

PROPOSITION 6. *Let $(A, \rho, t_\mathbf{d}, t_\mathbf{u})$ be an oriented quantum algebra over the field k. Then the pair $((\mathcal{A}, \boldsymbol{\rho}, \boldsymbol{s}, t_\mathbf{d}, t_\mathbf{u}), \pi)$ satisfies the following properties:*

(a) *$(\mathcal{A}, \boldsymbol{\rho}, \boldsymbol{s}, t_\mathbf{d}, t_\mathbf{u})$ is an object of $\mathcal{C}_q$ and $\pi : (\mathcal{A}, \boldsymbol{\rho}, t_\mathbf{d}, t_\mathbf{u}) \longrightarrow (A, \rho, t_\mathbf{d}, t_\mathbf{u})$ is a morphism of oriented quantum algebras.*

(b) *Suppose that $(A', \rho', s', t'_\mathbf{d}, t'_\mathbf{u})$ is an object of $\mathcal{C}_q$ and that $f : (A', \rho', t'_\mathbf{d}, t'_\mathbf{u}) \longrightarrow (A, \rho, t_\mathbf{d}, t_\mathbf{u})$ is a morphism of oriented quantum algebras. Then there is a morphism $F : (A', \rho', s', t'_\mathbf{d}, t'_\mathbf{u}) \longrightarrow (\mathcal{A}, \boldsymbol{\rho}, \boldsymbol{s}, t_\mathbf{d}, t_\mathbf{u})$ uniquely determined by $\pi \circ F = f$.* $\qquad \square$

6. Invariants Constructed from Quantum Algebras Via Bead Sliding

We describe a regular isotopy invariant of unoriented 1–1 tangle diagrams determined by a quantum algebra and show how the construction of this tangle invariant is modified to give a regular isotopy invariant of unoriented knot and link diagrams when the quantum algebra is replaced by a twist quantum algebra. Our discussion is readily adapted to handle the oriented case in Section 7; there quantum algebras and twist quantum algebras are replaced by oriented quantum algebras and twist oriented quantum algebras respectively. In both cases the invariants of diagrams determine invariants of 1–1 tangles, knots and links.

What follows is based on [19, Sections 6, 8]. Throughout this section all diagrams are unoriented.

6.1. Invariants of 1–1 tangles arising from quantum algebras. We represent 1–1 tangles as diagrams in the plane situated with respect to a fixed vertical. On the left below is a very simple 1–1 tangle diagram

which we refer to as T_{curl}. We require that 1–1 tangle diagrams can be drawn in a box except for two protruding line segments as indicated by the figure on the right above. Let *Tang* be the set of all 1–1 tangle diagrams in the plane situated with respect to the given vertical.

All 1–1 tangle diagrams consist of some or all of the following components:

- crossings;

    over crossings   *under crossings*

- local extrema;

    local maxima   *local minima*

and

- "vertical" lines.

There is a natural product decomposition of 1–1 tangle diagrams in certain situations. When a 1–1 tangle diagram T can be written as the union of two 1–1 tangle diagrams T_1 and T_2 such that the top point of T_1 is the base point of T_2, and the line passing through this common point perpendicular to the vertical otherwise separates T_1 and T_2, then T is called the *product of* T_1 and T_2 and this relationship is expressed by $T = T_1 \star T_2$. For example,

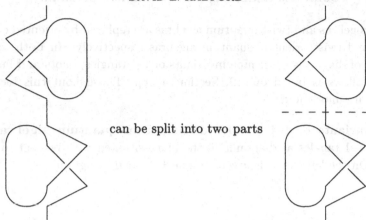

can be split into two parts

and thus

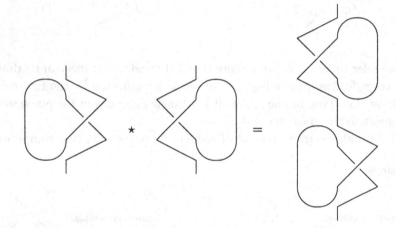

Multiplication is an associative operation. Those $T \in$ *Tang* which consist only of a vertical line can be viewed as local neutral elements with respect multiplication in *Tang*.

Let (A, ρ, s) be a quantum algebra over k. We will construct a function $\mathsf{Inv}_A : Tang \longrightarrow A$ which satisfies the following axioms:

(T.1) If $T, T' \in$ *Tang* are regularly isotopic then $\mathsf{Inv}_A(T) = \mathsf{Inv}_A(T')$,

(T.2) If $T \in$ *Tang* has no crossings then $\mathsf{Inv}_A(T) = 1$, and

(T.3) $\mathsf{Inv}_A(T \star T') = \mathsf{Inv}_A(T)\mathsf{Inv}_A(T')$ whenever $T, T' \in$ *Tang* and $T \star T'$ is defined.

The first axiom implies that Inv_A defines a regular isotopy invariant of 1–1 tangles.

Regular isotopy describes a certain topological equivalence of 1–1 tangle diagrams (and of knot and link diagrams). For the purpose of defining invariants we may view regular isotopy in rather simplistically: $T, T' \in$ *Tang* are *regularly isotopic* if T can be transformed to T′ by a finite number of local substitutions described in (M.1)–(M.4) below and (M.2rev)–(M.4rev). The symbolism $A \approx B$ in the figures below means that configuration A can be substituted for configuration B and vice versa.

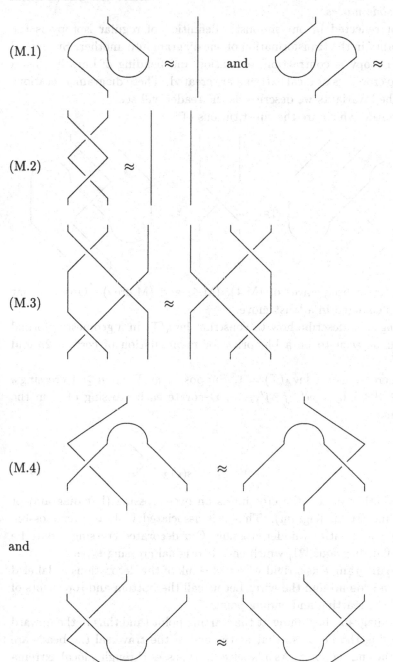

(M.1)

(M.2)

(M.3)

(M.4)

and

(M.2rev)–(M.4rev) are (M.2)–(M.4) respectively with over crossing lines replaced by under crossing lines and vice versa.

The substitutions of (M.1), (M.2) and (M.3) are known as the Reidemeister moves 0, 2 and 3 respectively. The substitutions described in (M.4) and (M.4rev) are called the *slide moves*.

What is not reflected in our simplistic definition of regular isotopy is the topological fluidity in the transformation of one diagram into another. Subsumed under regular isotopy is contraction, expansion and bending of lines in such a manner that no crossings or local extrema are created. These diagram alterations do not affect the invariants we describe as the reader will see.

The *twist moves*, which are the substitutions of

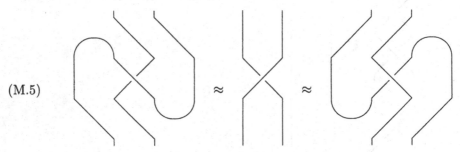

(M.5)

and (M.5rev), are consequences of (M.1), (M.4) and (M.4rev). Observe that crossing type is changed in a twist move.

Let $T \in$ *Tang*. We describe how to construct $\mathsf{Inv}_A(T)$ in a geometric, formal way which will be seen to be a blueprint for manipulation of certain $2n$-fold tensors.

If T has no crossings set $\mathsf{Inv}_A(T) = 1$. Suppose that T has $n \geq 1$ crossings. Represent $\rho \in A \otimes A$ by $e \otimes e'$, $f \otimes f'$, ... Decorate each crossing of T in the following manner

$$e \overset{\displaystyle}{\underset{\displaystyle}{\times}} e' \qquad \text{or} \qquad s(e) \overset{\displaystyle}{\underset{\displaystyle}{\times}} e' \qquad (6\text{--}1)$$

according to whether or not the crossing is an over crossing (left diagram) or an under crossing (right diagram). Thus ρ is associated with an over crossing and ρ^{-1} is associated with an under crossing. Our decorated crossings are to be interpreted as flat diagrams [21] which encode original crossing type.

Think of the diagram T as a rigid wire and think of the decorations as labeled beads which can slide around the wire. Let us call the bottom and top points of the diagram T the starting and ending points.

Traverse the diagram, beginning at the starting point (and thus in the upward direction), pushing the beads so that at the end of the traversal the beads are juxtaposed at the ending point. As a labeled bead passes through a local extrema its label x is altered: if the local extremum is traversed in the counter clockwise direction then x is changed to $s(x)$; if the local extremum is traversed in the clockwise direction then x is changed to $s^{-1}(x)$.

The juxtaposition of the beads with modified labels is interpreted as a formal product $W_A(T)$ which is read bottom to top. Write $\rho = \sum_{i=1}^r a_i \otimes b_i \in A \otimes A$. Substitution of a_i and b_i for e and e', a_j and b_j for f and $f', \ldots$ results in an element $w_A(T) \in A$. To illustrate the calculation of $W_A(T)$ and $w_A(T)$ we use the 1–1 tangle diagram T_{trefoil} depicted on the left below.

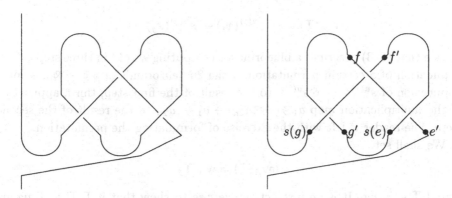

The crossing decorations are given in the diagram on the right above. Traversal of T_{trefoil} results in the formal word

$$W_A(T_{\text{trefoil}}) = s^2(e')s^2(f)s(g')s(e)s^{-1}(f')g$$

and thus

$$w_A(T_{\text{trefoil}}) = \sum_{i,j,k=1}^r s^2(b_i)s^2(a_j)s(b_k)s(a_i)s^{-1}(b_j)a_k.$$

The preceding expression can be reformulated in several ways. Since $\rho = (s \otimes s)(\rho)$ there is no harm in introducing the rule

$$W_A(T) = \cdots s^p(x)\cdots s^q(y)\cdots = \cdots s^{p+\ell}(x)\cdots s^{q+\ell}(y)\cdots \qquad (6\text{--}2)$$

where ℓ is any integer and $x \otimes y$ or $y \otimes x$ represents ρ. Under this rule we have

$$W_A(T_{\text{trefoil}}) = s(e')s^3(f)s(g')ef'g$$

and thus

$$w_A(T_{\text{trefoil}}) = \sum_{i,j,k=1}^r s(b_i)s^3(a_j)s(b_k)a_ib_ja_k$$

as well.

We now give a slightly more detailed description of $W_A(T)$ which will be useful in our discussion of knots and links. Label the crossing lines of the diagram T by $1, 2, \ldots, 2n$ in the order encountered on the traversal of T. For $1 \le i \le 2n$ let $u(i)$ be the number of local extrema traversed in the counter clockwise direction minus the number of local extrema traversed in the clockwise direction during the portion of the traversal of the diagram from crossing line i to the ending point. Let x_i be the decoration on crossing line i. Then

$$\mathsf{W}_A(\mathsf{T}) = s^{u(1)}(x_1) \cdots s^{u(2n)}(x_{2n}). \qquad (6\text{--}3)$$

We emphasize that each crossing contributes two factors to the formal product $\mathsf{W}_A(\mathsf{T})$. Let χ be a crossing of T and $i < j$ be the labels of the crossing lines of χ. Then χ contributes the ith and j^{th} factors to $\mathsf{W}_A(\mathsf{T})$ according to

$$\mathsf{W}_A(\mathsf{T}) = \cdots s^{u(i)}(x_i) \cdots s^{u(j)}(x_j) \cdots .$$

Notice that (6–3) describes a blueprint for computing $\mathsf{w}_A(\mathsf{T})$ in three steps: first, application of a certain permutation of the $2n$ tensorands of $\rho \otimes \cdots \otimes \rho$; second, application of $s^{u(1)} \otimes \cdots \otimes s^{u(2n)}$ to the result of the first step; third, application of the multiplication map $a_1 \otimes \cdots \otimes a_{2n} \mapsto a_1 \cdots a_{2n}$ to the result of the second step. The reader is left with the exercise of formulating the permutation.

We shall set

$$\mathsf{Inv}_A(\mathsf{T}) = \mathsf{w}_A(\mathsf{T})$$

for all $\mathsf{T} \in Tang$. It is an instructive exercise to show that if $\mathsf{T}, \mathsf{T}' \in Tang$ are regularly isotopic then $\mathsf{Inv}_A(\mathsf{T}) = \mathsf{Inv}_A(\mathsf{T}')$; that is (T.1) holds. By definition (T.2) holds, and in light of (6–2) it is clear that (T.3) holds. Observe that $\mathsf{Inv}_A(\mathsf{T})$ is s^2-invariant.

We end this section with a result on the relationship between Inv_A and morphisms. To do this we introduce some general terminology for comparing invariants.

Suppose that $f : X \longrightarrow Y$ and $g : X \longrightarrow Z$ are functions with the same domain. Then f *dominates* g if $x, x' \in X$ and $f(x) = f(x')$ implies $g(x) = g(x')$. If f dominates g and g dominates f then f and g are *equivalent*.

PROPOSITION 7. *Let* $f : A \longrightarrow A'$ *be a morphism of quantum algebras over* k. *Then* $f(\mathsf{Inv}_A(\mathsf{T})) = \mathsf{Inv}_{A'}(\mathsf{T})$ *for all* $\mathsf{T} \in Tang$. $\qquad\qquad\square$

Thus when $f : A \longrightarrow A'$ is a morphism of quantum algebras over k it follows that Inv_A dominates $\mathsf{Inv}_{A'}$.

6.2. Invariants of knots and links arising from twist quantum algebras.

We now turn to knots and links. In this section (A, ρ, s, G) is a twist quantum algebra over k and tr is a tracelike s^*-invariant element of A^*.

Let *Link* be the set of (unoriented) link diagrams situated with respect to our fixed vertical and let *Knot* be the set of knot diagrams in *Link*, that is the set of one component link diagrams in *Link*. If $\mathsf{L} \in Link$ is the union of two link diagrams $\mathsf{L}_1, \mathsf{L}_2 \in Link$ such that the components of L_1 and L_2 do not intersect we write $\mathsf{L} = \mathsf{L}_1 \star \mathsf{L}_2$. We shall construct a scalar valued function $\mathsf{Inv}_{A, \mathrm{tr}} : Link \longrightarrow k$ which satisfies the following axioms:

(L.1) If $L, L' \in Link$ are regularly isotopic then $Inv_{A, tr}(L) = Inv_{A, tr}(L')$,

(L.2) If $L \in Link$ is a knot with no crossings then $Inv_{A, tr}(L) = tr(G)$, and

(L.3) $Inv_{A, tr}(L \star L') = Inv_{A, tr}(L)Inv_{A, tr}(L')$ whenever $L, L' \in Link$ and $L \star L'$ is defined.

The first axiom implies that $Inv_{A, tr}$ defines a regular isotopy invariant of unoriented links.

We first define $Inv_{A, tr}(K)$ for knot diagrams $K \in Knot$. One reason we do this is to highlight the close connection between the function Inv_A and the restriction $Inv_{A, tr}|_{Knot}$.

Let $K \in Knot$. We first define an element $w(K) \in A$. If K has no crossings then $w(K) = 1$.

Suppose that K has $n \geq 1$ crossings. Decorate the crossings of K according to (6–1) and choose a point P on a vertical line of K. We refer to P as the starting and ending point. (There is no harm, under regular isotopy considerations, in assuming that K has a vertical line. One may be inserted at an end of a crossing line or at an end of a local extrema.) Traverse the diagram K, beginning at the starting point P in the upward direction and concluding at the ending point which is P again. Label the crossing lines $1, 2, \ldots, 2n$ in the order encountered on the traversal. For $1 \leq \imath \leq 2n$ let $u(\imath)$ be defined as in the case of 1–1 tangle diagrams and let $x_\imath$ be the decoration on crossing line $\imath$. Let $W(K) = s^{u(1)}(x_1) \cdots s^{u(2n)}(x_{2n})$ and let $w(K) \in A$ be computed from $W(K)$ in the same manner that $w_A(T)$ is computed from $W_A(T)$ in Section 6.1. Then

$$Inv_{A, tr}(K) = tr(G^d w(K)), \qquad (6\text{–}4)$$

where d is the Whitney degree of K with orientation determined by traversal beginning at the starting point in the upward direction. In terms of local extrema, $2d$ is the number of local extrema traversed in the clockwise direction minus the number of local extrema traversed in the counter clockwise direction.

Using a different starting point P' results in the same value for $Inv_{A, tr}(K)$. This boils down to two cases: P and P' separated by $m \geq 1$ crossing lines and no local extrema; P and P' separated by one local extrema and no crossing lines. The fact that tr is tracelike is used in the first case and the s^*-invariance of tr is used in the second.

The function $Inv_{A, tr}|_{Knot}$ can be computed in terms of Inv_A. For $T \in Tang$ let $K(T) \in Knot$ be given by

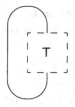

Every $\mathsf{K} \in Knot$ is regularly isotopic to $\mathsf{K}(\mathsf{T})$ for some $\mathsf{T} \in Tang$.

PROPOSITION 8. *Let (A, ρ, s, G) be a twist quantum algebra over the field k and suppose that* tr *is a tracelike s^*-invariant element of A^*. Then*

$$\mathsf{Inv}_{A, \mathrm{tr}}(\mathsf{K}(\mathsf{T})) = \mathrm{tr}(G^d \mathsf{Inv}_A(\mathsf{T}))$$

for all $\mathsf{T} \in Tang$, where d is the Whitney degree of $\mathsf{K}(\mathsf{T})$ with the orientation determined by traversal beginning at the base of T in the upward direction. □

For example, with $\rho = \sum_{i=1}^{r}$, observe that

$$\mathsf{Inv}_{A, \mathrm{tr}}(\mathsf{K}(\mathsf{T}_{\mathrm{trefoil}})) = \sum_{i,j,k=1}^{r} \mathrm{tr}(G^{-2} s(b_i) s^3(a_j) s(b_k) a_i b_j a_k)$$

and

$$\mathsf{Inv}_{A, \mathrm{tr}}(\mathsf{K}(\mathsf{T}_{\mathrm{curl}})) = \sum_{i=1}^{r} \mathrm{tr}(G^0 a_i s(b_i)) = \sum_{i=1}^{r} \mathrm{tr}(a_i s(b_i)).$$

We now define $\mathsf{Inv}_{A, \mathrm{tr}}(\mathsf{L})$ for $\mathsf{L} \in Link$ with components $\mathsf{L}_1, \ldots, \mathsf{L}_r$. Decorate the crossings of L according to (6–1). Fix $1 \le \ell \le r$. We construct a formal word $W(\mathsf{L}_\ell)$ in the following manner. If L_ℓ does not contain a crossing line then $W(\mathsf{L}_\ell) = 1$.

Suppose that L_ℓ contains $m_\ell \ge 1$ crossing lines. Choose a point P_ℓ on a vertical line of L_ℓ. We shall refer to P_ℓ as the starting point and the ending point. As in the case of knot diagrams we may assume that L_ℓ has a vertical line. Traverse the link component L_ℓ beginning at the starting point P_ℓ in the upward direction and concluding at the ending point which is also P_ℓ. Label the crossing lines contained in L_ℓ by $(\ell{:}1), \ldots, (\ell{:}m_\ell)$ in the order encountered. Let $u(\ell{:}i)$ be the counterpart of $u(i)$ for $Knot$ and let $x_{(\ell{:}i)}$ be the decoration on the crossing line $(\ell{:}i)$. Then we set

$$W(\mathsf{L}_\ell) = s^{u(\ell{:}1)}(x_{(\ell{:}1)}) \cdots s^{u(\ell{:}m_\ell)}(x_{(\ell{:}m_\ell)})$$

and

$$W(\mathsf{L}) = W(\mathsf{L}_1) \otimes \cdots \otimes W(\mathsf{L}_r).$$

Replacing the formal copies of ρ in $W(\mathsf{L})$ with ρ as was done in the case of 1–1 tangle and knot diagrams, we obtain an element $w(\mathsf{L}) = w(\mathsf{L}_1) \otimes \cdots \otimes w(\mathsf{L}_r) \in A \otimes \cdots \otimes A$. The scalar we want is

$$\mathsf{Inv}_{A, \mathrm{tr}}(\mathsf{L}) = \mathrm{tr}(G^{d_1} w(\mathsf{L}_1)) \cdots \mathrm{tr}(G^{d_r} w(\mathsf{L}_r)) \qquad (6\text{--}5)$$

which is the evaluation of $\mathrm{tr} \otimes \cdots \otimes \mathrm{tr}$ on $G^{d_1} w(\mathsf{L}_1) \otimes \cdots \otimes G^{d_r} w(\mathsf{L}_r)$, where d_ℓ is the Whitney degree of L_ℓ with orientation determined by the traversal which starts at P_ℓ in the upward direction.

The argument that $\mathsf{Inv}_{A, \mathrm{tr}}(\mathsf{K})$ does not depend on the starting point P for $\mathsf{K} \in Knot$ is easily modified to show that $\mathsf{Inv}_{A, \mathrm{tr}}(\mathsf{L})$ does not depend on the starting points $P_1, \ldots, P_r$. The argument that (T.1) holds for Inv_A shows that

(L.1) holds for $\mathsf{Inv}_{A,\,\mathrm{tr}}$ since we may assume that the starting points are not in the local part of L under consideration in (M.1)–(M.4) and (M.2rev)–(M.4rev). That (L.2) and (L.3) hold for $\mathsf{Inv}_{A,\,\mathrm{tr}}$ is a straightforward exercise. Note that (6–5) generalizes (6–4).

For example, consider the Hopf link $\mathsf{L}_{\mathrm{Hopf}}$ depicted below left. The components of L are L_1 and L_2, reading left to right. The symbol ∘ designates a starting point.

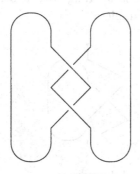 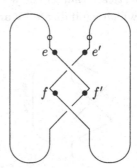

Observe that $W(\mathsf{L}_{\mathrm{Hopf}}) = f'e \otimes fe'$ and

$$\mathsf{Inv}_{A,\,\mathrm{tr}}(\mathsf{L}_{\mathrm{Hopf}}) = \sum_{i,j=1}^{r} \mathrm{tr}(G^{-1}b_i a_j)\mathrm{tr}(Ga_i b_j).$$

We end this section by noting how $\mathsf{Inv}_{A,\,\mathrm{tr}}$ and morphisms are related and revisiting the construction of Proposition 1.

PROPOSITION 9. *Let $f : (A, \rho, s, G) \longrightarrow (A', \rho', s', G')$ be a morphism of twist quantum algebras over k and suppose that tr' is a tracelike s'^*-invariant element of A'^*. Then $\mathrm{tr} = \mathrm{tr}' \circ f$ is a tracelike s^*-invariant element of A^* and*

$$\mathsf{Inv}_{A,\,\mathrm{tr}}(\mathsf{L}) = \mathsf{Inv}_{A',\,\mathrm{tr}'}(\mathsf{L})$$

for all $\mathsf{L} \in \text{Link}.$ □

Let (A, ρ, s) be a quantum algebra over k and let $(\boldsymbol{A}, \boldsymbol{\rho}, \boldsymbol{s}, \boldsymbol{G})$ be the twist quantum algebra of Proposition 1 associated with (A, ρ, s). In light of the preceding proposition it would be of interest to know what the s^*-invariant tracelike elements Tr of $\boldsymbol{A}^*$ are.

First note that $\boldsymbol{A}$ is a free right A-module with basis $\{G^n\}_{n \in Z}$. Let $\{\mathrm{tr}_n\}_{n \in Z}$ be a family of functionals of A^* which satisfies $\mathrm{tr}_n \circ s = \mathrm{tr}_{-n}$ and $\mathrm{tr}_n(ba) = \mathrm{tr}_n(as^{2n}(b))$ for all $n \in Z$ and $a, b \in A$. Then the functional $\mathrm{Tr} \in \boldsymbol{A}^*$ determined by $\mathrm{Tr}(G^n a) = \mathrm{tr}_n(a)$ for all $n \in Z$ and $a \in A$ is an s^*-invariant and tracelike. All s^*-invariant tracelike functionals of $\boldsymbol{A}^*$ are described in this manner.

7. Invariants Constructed from Oriented Quantum Algebras Via Bead Sliding

This section draws heavily from the material of the preceding section. The reader is directed to [16; 17] for a fuller presentation of the ideas found in this section.

Let **Tang** be the set of all oriented 1–1 tangle diagrams situated with respect to a fixed vertical; that is the set of all diagrams in *Tang* with a designated orientation which we indicate by arrows. For example

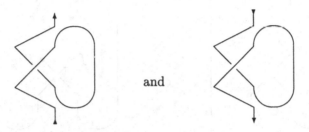

and

are elements of **Tang** whose underlying unoriented diagram is

We let $u : \textbf{Tang} \longrightarrow Tang$ be the function which associates to each $\textbf{T} \in \textbf{Tang}$ its underlying unoriented diagram $u(\textbf{T})$. For $\textbf{T} \in \textbf{Tang}$ we let $\textbf{T}^{\text{op}}$ be $\textbf{T}$ with its orientation reversed.

Likewise we let **Link** be the set of all oriented link diagrams situated with respect to the fixed vertical; that is the set of all diagrams of *Link* whose components have a designated orientation. By slight abuse of notation we also let $u : \textbf{Link} \longrightarrow Link$ be the function which associates to each $\textbf{L} \in \textbf{Link}$ its underlying unoriented diagram $u(\textbf{L})$. For $\textbf{L} \in \textbf{Link}$ we let $\textbf{L}^{\text{op}}$ be $\textbf{L}$ with the orientation on its components reversed.

In this section we redo Section 6 by making minor adjustments which result in a regular isotopy invariant $\textbf{Inv}_A : \textbf{Tang} \longrightarrow A$ of oriented 1-1 tangle diagrams, when (A, ρ, t_d, t_u) is an oriented quantum algebra over k, and in a regular isotopy invariant $\textbf{Inv}_{A,\text{tr}} : \textbf{Link} \longrightarrow k$ of oriented link diagrams, when (A, ρ, t_d, t_u, G) is a twist oriented quantum algebra over k and tr is a tracelike t_d^*, t_u^*-invariant element of A^*.

Regular isotopy in the oriented case is regular isotopy in the unoriented case with all possible orientations taken into account. For example, the one diagram of (M.2) in the unoriented case is replaced by four in the oriented case, two of which are

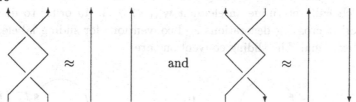

In Section 7.3 we relate invariants of a quantum algebra and invariants of its associated oriented quantum algebra.

7.1. Invariants of oriented 1–1 tangles arising from oriented quantum algebras. Oriented 1–1 tangle diagrams consist of some or all of the following components:

- oriented crossings;

  under crossings

  over crossings

- oriented local extrema;

  local maxima

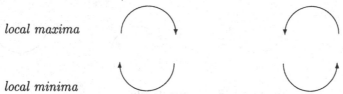

  local minima

and

- oriented "vertical" lines.

If an oriented 1–1 tangle diagram $\boxed{\begin{smallmatrix}\uparrow\\ \mathbf{T}\\ \uparrow\end{smallmatrix}}$ (respectively $\boxed{\begin{smallmatrix}\downarrow\\ \mathbf{T}\\ \downarrow\end{smallmatrix}}$) can be decomposed

into two oriented 1–1 tangle diagrams $\boxed{\begin{smallmatrix}\uparrow\\ \mathbf{T_2}\\ \uparrow\\ \mathbf{T_1}\\ \uparrow\end{smallmatrix}}$ (respectively $\boxed{\begin{smallmatrix}\downarrow\\ \mathbf{T_1}\\ \downarrow\\ \mathbf{T_2}\\ \downarrow\end{smallmatrix}}$) then we write $\mathbf{T} = \mathbf{T_1 \star T_2}$.

Now suppose that $(A, \rho, t_{\mathsf{d}}, t_{\mathsf{u}})$ is an oriented quantum algebra over k and let $\mathbf{T} \in$ *Tang*. To define $\mathbf{Inv}_A :$ **Tang** $\longrightarrow A$ we will construct formal product $\mathbf{W}_A(\mathbf{T})$ which will determine an element $\mathbf{w}_A(\mathbf{T}) \in A$. In order to do this we need to describe crossing decorations and conventions for sliding labeled beads across local extrema. The sliding conventions are:

and

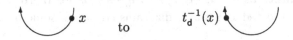

for clockwise motion;

and

for counterclockwise motion. We refer to the oriented local extrema

as having type (u_-), (u_+), (d_+) and (d_-) respectively.

There are two crossing decorations

$$(7\text{--}1)$$

from which all other crossing decorations are derived. Here $E \otimes E'$ and $e \otimes e'$ represent ρ^{-1} and ρ respectively. Compare (7–1) with (6–1). Starting with (7–1), using the above conventions for passing labeled beads across local extrema and requiring invariance under (M.4), crossings are decorated as follows:

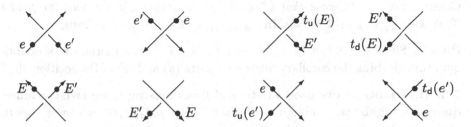

Now to define $\mathbf{W}_A(\mathbf{T})$. If $\mathbf{T}$ has no crossings then $\mathbf{W}_A(\mathbf{T}) = 1$. Suppose that $\mathbf{T}$ has $n \geq 1$ crossings. Traverse $\mathbf{T}$ in the direction of orientation and label the crossing lines $1, 2, \ldots, 2n$ in the order in which they are encountered. For $1 \leq \imath \leq 2n$ let $u_d(\imath)$ be the number of local extrema of type (d_+) minus the number of type (d_-) encountered on the portion of the traversal from line $\imath$ to the end of the traversal of $\mathbf{T}$. We define $u_u(\imath)$ in the same way where (u_+) and (u_-) replace (d_+) and (d_-) respectively. Then

$$\mathbf{W}_A(\mathbf{T}) = t_d^{u_d(1)} \circ t_u^{u_u(1)}(x_1) \cdots t_d^{u_d(2n)} \circ t_u^{u_u(2n)}(x_{2n}),$$

where $x_\imath$ is the decoration on the crossing line $\imath$. Replacing the formal representations of ρ and ρ^{-1} in $\mathbf{W}_A(\mathbf{T})$ by ρ and ρ^{-1} respectively we obtain an element $\mathbf{w}_A(\mathbf{T}) \in A$.

Set

$$\mathbf{Inv}_A(\mathbf{T}) = \mathbf{w}_A(\mathbf{T})$$

for all $\mathbf{T} \in \mathbf{Tang}$. Observe that $\mathbf{Inv}_A(\mathbf{T})$ is invariant under t_d and t_u. It can be shown that the oriented counterparts of (T.1)–(T.3) hold for $\mathbf{Inv}_A$; in particular $\mathbf{Inv}_A$ defines a regular isotopy invariant of oriented 1–1 tangles. We have noted that $(A^{\mathrm{op}}, \rho, t_d, t_u)$ is an oriented quantum algebra over k which we simply refer to as A^{op}. By Theorem 1 that $(A, \rho, 1_A, t_d \circ t_u)$ is a standard oriented quantum algebra over k which we denote by A_s. We collect some basic results on the invariant of this section.

PROPOSITION 10. *Let A be an oriented quantum algebra over k. Then*

(a) $\mathbf{Inv}_A(\mathbf{T}^{\mathrm{op}}) = \mathbf{Inv}_{A^{\mathrm{op}}}(\mathbf{T})$ *and*
(b) $\mathbf{Inv}_A(\mathbf{T}) = \mathbf{Inv}_{A_s}(\mathbf{T})$

for all $\mathbf{T} \in \mathbf{Tang}$.

(c) *Suppose that $f : A \longrightarrow A'$ is a morphism of oriented quantum algebras. Then $f(\mathbf{Inv}_A(\mathbf{T})) = \mathbf{Inv}_{A'}(\mathbf{T})$ for all $\mathbf{T} \in \mathbf{Tang}$.*

□

Part (b) of the proposition shows that standard oriented quantum algebras account for the invariants of this section. Part (c) shows that $\mathbf{Inv}_A$ dominates $\mathbf{Inv}_{A'}$ whenever there is a morphism of oriented quantum algebras $f : A \longrightarrow A'$.

COROLLARY 2. *Suppose that (A, ρ, s) is a quantum algebra over the field k. Then* $\mathbf{Inv}_{(A, \rho, 1_A, s^{-2})}(\mathbf{T}^{op}) = s(\mathbf{Inv}_{(A, \rho, 1_A, s^{-2})}(\mathbf{T}))$ *for all* $\mathbf{T} \in \mathbf{Tang}$.

PROOF. Since $s : (A^{op}, \rho, 1_A, s^{-2}) \longrightarrow (A, \rho, 1_A, s^{-2})$ is a morphism of oriented quantum algebras the corollary follows by parts (a) and (c) of Proposition 10. $\square$

7.2. Invariants of oriented knots and links arising from twist oriented quantum algebras.

Throughout this section $(A, \rho, t_{\mathsf{d}}, t_{\mathsf{u}}, G)$ is a twist oriented quantum algebra over k and tr is a tracelike t_{d}^*, t_{u}^*-invariant element of A^*. The scalar $\mathbf{Inv}_{A, \mathrm{tr}}(\mathbf{L})$ for $\mathbf{L} \in \mathbf{Link}$ is defined much in the same manner that $\mathsf{Inv}_{A, \mathrm{tr}}(\mathsf{L})$ is defined for $\mathsf{L} \in Link$ in Section 6.2.

Let $\mathbf{L} \in \mathbf{Link}$ have components $\mathbf{L}_1, \dots, \mathbf{L}_r$. Decorate the crossings of $\mathbf{L}$ according to the conventions of Section 7.1. For each $1 \leq \ell \leq r$ let d_ℓ be the Whitney degree of the link component $\mathbf{L}_\ell$. We define a formal product $\mathbf{W}(\mathbf{L}_\ell)$ as follows. If $\mathbf{L}_\ell$ contains no crossing lines then $\mathbf{W}(\mathbf{L}_\ell) = 1$. Suppose that $\mathbf{L}_\ell$ contains $m_\ell \geq 1$ crossing lines. Choose a point P_ℓ on a vertical line of $\mathbf{L}_\ell$. We may assume that $\mathbf{L}_\ell$ has a vertical line for the reasons cited in Section 6.2.

Traverse $\mathbf{L}_\ell$ in the direction of orientation beginning and ending at P_ℓ. Label the crossing lines $(\ell{:}1), \dots, (\ell{:}m_\ell)$ in the order encountered on the traversal. For $1 \leq \imath \leq m$ let $u_{\mathsf{d}}(\ell{:}\imath)$ denote the number of local extrema of type (d_+) minus the number of type (d_-) which are encountered during the portion of the traversal of $\mathbf{L}_\ell$ from the line labeled $\imath$ to its conclusion. Define $u_{\mathsf{u}}(\ell{:}\imath)$ in the same manner, where (u_+) and (u_-) replace (d_+) and (d_-) respectively. Let $x_{(\ell{:}\imath)}$ be the decoration on the line $(\ell{:}\imath)$. Set

$$\mathbf{W}(\mathbf{L}_\ell) = t_{\mathsf{d}}^{u_{\mathsf{d}}(\ell{:}1)} \mathrm{o} t_{\mathsf{u}}^{u_{\mathsf{u}}(\ell{:}1)}(x_{(\ell{:}1)}) \cdots t_{\mathsf{d}}^{u_{\mathsf{d}}(\ell{:}m)} \mathrm{o} t_{\mathsf{u}}^{u_{\mathsf{u}}(\ell{:}m)}(x_{(\ell{:}m)}),$$

set $\mathbf{W}(\mathbf{L}) = \mathbf{W}(\mathbf{L}_1) \otimes \cdots \otimes \mathbf{W}(\mathbf{L}_r)$ and replace formal copies of ρ and ρ^{-1} in $\mathbf{W}(\mathbf{L})$ to obtain an element $\mathbf{w}(\mathbf{L}) = \mathbf{w}(\mathbf{L}_1) \otimes \cdots \otimes \mathbf{w}(\mathbf{L}_r) \in A \otimes \cdots \otimes A$. We define

$$\mathbf{Inv}_{A, \mathrm{tr}}(\mathbf{L}) = \mathrm{tr}(G^{d_1} \mathbf{w}(\mathbf{L}_1)) \cdots \mathrm{tr}(G^{d_r} \mathbf{w}(\mathbf{L}_r)).$$

One can show that the oriented counterparts of (L.1)–(L.3) hold for $\mathbf{Inv}_{A, \mathrm{tr}}$; in particular $\mathbf{Inv}_{A, \mathrm{tr}}$ defines a regular isotopy invariant of oriented links.

There is an analog of Proposition 8 which we do not state here and there is an analog of part (b) of Proposition 10 and of Proposition 9 which we do record. Let A_s denote the standard twist oriented quantum algebra $(A, \rho, 1_A, t_{\mathsf{d}} \mathrm{o} t_{\mathsf{u}}, G)$ associated with $(A, \rho, t_{\mathsf{d}}, t_{\mathsf{u}}, G)$ which we denote by A.

PROPOSITION 11. *Let $(A, \rho, t_{\mathsf{d}}, t_{\mathsf{u}}, G)$ be a twist oriented quantum algebra over the field k and let* tr *be a tracelike t_{d}^*, t_{u}^*-invariant element of A^*. Then*

$$\mathbf{Inv}_{A, \mathrm{tr}}(\mathbf{L}) = \mathbf{Inv}_{A_s, \mathrm{tr}}(\mathbf{L}) \quad \text{for all} \quad \mathbf{L} \in \mathbf{Link}. \qquad \square$$

Thus the invariants of oriented links described in this section are accounted for by standard twist oriented quantum algebras.

PROPOSITION 12. *Suppose that* $f : (A, \rho, t_{\mathsf{d}}, t_{\mathsf{u}}, G) \longrightarrow (A', R', t'_{\mathsf{d}}, t'_{\mathsf{u}}, G')$ *is a morphism of twist oriented quantum algebras over* k *and that* $\mathrm{tr}' \in {A'}^*$ *is a* $t'^*_{\mathsf{d}}, t'^*_{\mathsf{u}}$-*invariant tracelike element. Then* $\mathrm{tr} \in A^*$ *defined by* $\mathrm{tr} = \mathrm{tr}' \circ f$ *is a* $t^*_{\mathsf{d}}, t^*_{\mathsf{u}}$-*invariant tracelike element and* $\mathbf{Inv}_{A, \mathrm{tr}}(\mathbf{L}) = \mathbf{Inv}_{A', \mathrm{tr}'}(\mathbf{L})$ *for all* $\mathbf{L} \in \mathbf{Link}$. $\square$

Let $(A, \rho, 1_A, t)$ be a standard oriented quantum algebra over k and let

$$(A, \rho, 1_A, t, G)$$

be the twist standard quantum algebra of Proposition 4 associated with (A, ρ, s). We describe the tracelike elements Tr of A^*.

Observe $\mathbf{A}$ is a free right A-module with basis $\{G^n\}_{n \in Z}$. Let $\{\mathrm{tr}_n\}_{n \in Z}$ be a family of functionals of A^* which satisfies $\mathrm{tr}_n \circ t = \mathrm{tr}_n$ and $\mathrm{tr}_n(ba) = \mathrm{tr}_n(at^n(b))$ for all $n \in Z$ and $a, b \in A$. Then the functional $\mathrm{Tr} \in \mathbf{A}^*$ determined by $\mathrm{Tr}(G^n a) = \mathrm{tr}_n(a)$ for all $n \in Z$ and $a \in A$ is an t^*-invariant and tracelike; all t^*-invariant tracelike functionals of $\mathbf{A}^*$ are described in this manner.

7.3. Comparison of invariants arising from quantum algebras and their associated oriented quantum algebras.
Let (A, ρ, s) be a quantum algebra over the field k and let $(A, \rho, 1_A, s^{-2})$ be the associated standard oriented quantum algebra. In this brief section we compare the invariants defined for each of them.

THEOREM 3. *Let* (A, ρ, s) *be a quantum algebra over* k. *Then:*

(a) *The equations*

$$\mathbf{Inv}_{(A, \rho, 1_A, s^{-2})}(\mathbf{T}) = \mathsf{Inv}_{(A, \rho^{-1}, s^{-1})}(u(\mathbf{T}))$$

and

$$\mathsf{Inv}_{(A, \rho, s)}(u(\mathbf{T})) = \mathbf{Inv}_{(A, \rho^{-1}, 1_A, s^2)}(\mathbf{T})$$

hold for all $\mathbf{T} \in \mathbf{Tang}$ *whose initial vertical line is oriented upward.*

(b) *Suppose further* (A, ρ, s, G^{-1}) *is a twist quantum algebra and* tr *is a tracelike* s^*-*invariant element of* A^*. *Then*

$$\mathbf{Inv}_{(A, \rho, 1_A, s^{-2}, G), \, \mathrm{tr}}(\mathbf{L}) = \mathsf{Inv}_{(A, \rho^{-1}, s^{-1}, G), \, \mathrm{tr}}(u(\mathbf{L}))$$

and

$$\mathsf{Inv}_{(A, \rho, s, G^{-1}), \, \mathrm{tr}}(u(\mathbf{L})) = \mathbf{Inv}_{(A, \rho^{-1}, 1_A, s^2, G^{-1}), \, \mathrm{tr}}(\mathbf{L})$$

for all $\mathbf{L} \in \mathbf{Link}$.

PROOF. We need only establish the first equations in parts (a) and (b). Generally to calculate a regular isotopy invariant of oriented 1–1 tangle, knot or link diagrams we may assume that all crossing lines are directed upward by virtue of the twist moves, and we may assume that diagrams have vertical lines oriented in the upward direction. We can assume that traversals begin on such lines.

As usual represent ρ by $e \otimes e'$ and ρ^{-1} by $E \otimes E'$. Since $\rho^{-1} = (s \otimes 1_A)(\rho)$ it follows that $e \otimes e' = s^{-1}(E) \otimes E'$. Thus the over crossing and under crossing labels $E \otimes E'$ and $e \otimes e'$ of (7–1) are the over crossing and under crossing labels $E \otimes E'$ and $s^{-1}(E) \otimes E'$ of (6–1). These are the decorations associated with the quantum algebra (A, ρ^{-1}, s^{-1}).

For a crossing decoration representing $x \otimes y$, traversal of the oriented 1–1 tangle or link diagram results in the modification

$$t_{\mathsf{d}}^{u_{\mathsf{d}}} \circ t_{\mathsf{u}}^{u_{\mathsf{u}}}(x) \otimes t_{\mathsf{d}}^{u'_{\mathsf{d}}} \circ t_{\mathsf{u}}^{u'_{\mathsf{u}}}(y) = s^{-2u_{\mathsf{u}}}(x) \otimes s^{-2u'_{\mathsf{u}}}(y)$$

with $(A, \rho, 1_A, s^{-2})$ and traversal in the underlying unoriented diagram results in the modification

$$s^{-(u_{\mathsf{d}} + u_{\mathsf{u}})}(x) \otimes s^{-(u'_{\mathsf{d}} + u'_{\mathsf{u}})}(y) = s^{-2u_{\mathsf{u}}}(x) \otimes s^{-2u'_{\mathsf{u}}}(y)$$

with (A, ρ^{-1}, s^{-1}); the last equation holds by our assumptions on the crossings and traversal.   □

8. Bead Sliding Versus the Classical Construction of Quantum Link Invariants

We relate the method of Section 7 for computing invariants of oriented knot and link diagrams to the method for computing invariants by composition of certain tensor products of (linear) morphisms associated with oriented crossings, local extrema and vertical lines. We will see how invariants produced by the composition method are related to those which arise from representations of twist oriented quantum algebras. For a more general discussion of the composition method, which has a categorical setting, the reader is referred to [28; 29]. This section is a reworking of [17, Sections 3, 6] which explicates its algebraic details.

In this section we confine ourselves to the category Vec_k of all vector spaces over k and their linear transformations under composition. We begin with a description of the composition method in this special case.

Let V be a finite-dimensional vector space over the field k, let $\{v_1, \ldots, v_n\}$ be a basis for V and suppose that $\{v^1, \ldots, v^n\}$ is the dual basis for V^*. Observe that

$$\sum_{\imath=1}^{n} v^{\imath}(v) v_{\imath} = v \quad \text{and} \quad \sum_{\imath=1}^{n} v^*(v_{\imath}) v^{\imath} = v^* \tag{8–1}$$

for all $v \in V$ and $v^* \in V^*$. Let $\mathcal{C}_V$ be the full subcategory of Vec_k whose objects are k and tensor powers $U_1 \otimes \cdots \otimes U_m$, where $m \geq 1$ and $U_{\imath} = V$ or $U_{\imath} = V^*$ for all $1 \leq \imath \leq m$.

The first step of the composition method is to arrange an oriented link (or knot) diagram $\mathbf{L} \in \mathbf{Link}$ so that all crossing lines are directed upward, which can be done by the twist moves, and so that $\mathbf{L}$ is stratified in such a manner that each stratum consists of a juxtaposition of oriented crossings, local extrema and vertical lines, which we refer to as components of the stratum. One such example is $\mathbf{K}_{\text{trefoil}}$ depicted below.

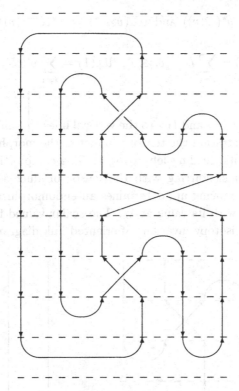

The broken lines indicate the stratification and are not part of the diagram.

The next step is to associate certain morphisms of $\mathcal{C}_V$ to the components of each stratum. Let $R : V{\otimes}V \longrightarrow V{\otimes}V$ be an invertible solution to the braid equation and let D, U be commuting linear automorphisms of V. We associate

$$R : V{\otimes}V \longrightarrow V{\otimes}V \text{ to}$$

$$R^{-1} : V{\otimes}V \longrightarrow V{\otimes}V \text{ to}$$

$$\mathcal{D}_+ : V^*{\otimes}V \longrightarrow k \text{ to}$$

$$\mathcal{D}_- : k \longrightarrow V{\otimes}V^* \text{ to}$$

$$\mathcal{U}_- : V{\otimes}V^* \longrightarrow k \text{ to}$$

$$\mathcal{U}_+ : k \longrightarrow V^*{\otimes}V \text{ to}$$

where $\mathcal{D}_+(v^* \otimes v) = v^*(D(v))$ and $\mathcal{U}_-(v \otimes v^*) = v^*(U^{-1}(v))$ for all $v^* \in V^*$, $v \in V$ and

$$\mathcal{D}_-(1) = \sum_{i=1}^{n} D^{-1}(v_i) \otimes v^i, \quad \mathcal{U}_+(1) = \sum_{i=1}^{n} v^i \otimes U(v_i),$$

and finally we associate 1_V and 1_{V^*} to the vertical lines $\uparrow$ and $\downarrow$ respectively.

Associate to each stratum the tensor product of the morphisms associated to each of its components, reading left to right. The composition of these tensor products of morphisms, starting with the tensor product associated with the bottom stratum and moving up, determines an endomorphism $\mathrm{INV}_{R,D,U}(\mathbf{L})$ of k, which we identify with its value at 1_k. The scalar valued function $\mathrm{INV}_{R,D,U}$ determines a regular isotopy invariant of oriented link diagrams if and only if

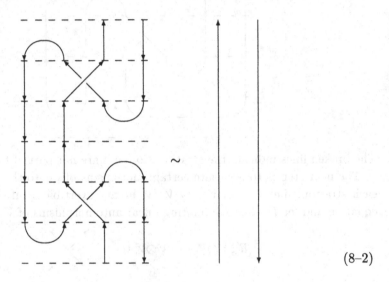

$$(8\text{–}2)$$

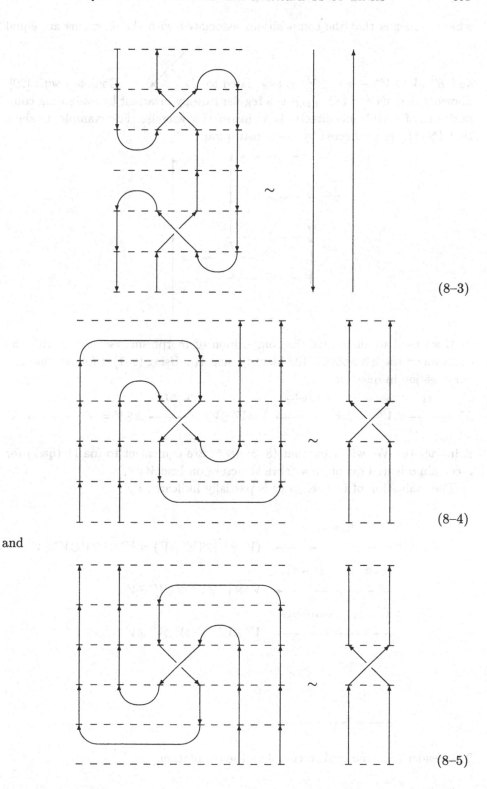

$$(8\text{--}3)$$

$$(8\text{--}4)$$

and

$$(8\text{--}5)$$

where $\sim$ means that the compositions associated with the diagrams are equal

and $R^* : V^* \otimes V^* \longrightarrow V^* \otimes V^*$ is associated to $\qquad$. Compare with [29].
Showing that $\mathrm{INV} = \mathrm{INV}_{R,D,U}$ is a regular isotopy invariant by evaluating compositions of morphisms directly is an instructive exercise. For example, to show that $\mathrm{INV}(\mathbf{L})$ is unaffected by the substitution

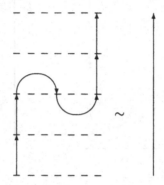

in $\mathbf{L}$ we need to show that the composition of morphisms associated with the diagram on the left above is the identity map 1_V. Using (8–1) it follows that the composition in question

$$V \xrightarrow{\;1_V\;} V = V \otimes k \xrightarrow{\;1_V \otimes u_+\;} V \otimes V^* \otimes V \xrightarrow{\;u_- \otimes 1_V\;} k \otimes V = V \xrightarrow{\;1_V\;} V$$

is in fact 1_V. We will show that (8–2)–(8–5) are equivalent to (qa.1)–(qa.2) for a certain oriented quantum algebra structure on $\mathrm{End}(V)^{\mathrm{op}}$.

The evaluation of $\mathrm{INV}(\mathbf{K}_{\mathrm{trefoil}})$ is partially indicated by

$$k = k \otimes k \xrightarrow{\;u_+ \otimes u_+\;} (V^* \otimes V) \otimes (V^* \otimes V) = V^* \otimes k \otimes V \otimes V^* \otimes V$$

$$\xrightarrow{\;1_{V^*} \otimes u_+ \otimes 1_V \otimes 1_{V^*} \otimes 1_V\;} V^* \otimes V^* \otimes V \otimes V \otimes V^* \otimes V$$

$$\xrightarrow{\;1_{V^*} \otimes 1_{V^*} \otimes R \otimes 1_{V^*} \otimes 1_V\;} V^* \otimes V^* \otimes V \otimes V \otimes V^* \otimes V$$

$$\vdots$$

$$\xrightarrow{\;\mathcal{D}_+\;} k.$$

The reader is encouraged to complete the calculation.

When the minimal polynomial of R has degree 1 or 2 there is a way of relating

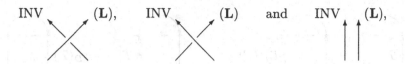

INV (L),   INV (L)   and   INV (L),

for $\mathbf{L} \in \mathbf{Link}$, where the three preceding expressions are $\mathrm{INV}(\mathbf{L})$, $\mathrm{INV}(\mathbf{L}')$ and

$\mathrm{INV}(\mathbf{L}'')$ respectively, where $\mathbf{L}'$ is $\mathbf{L}$ with a crossing of the type ⤫ replaced

by ⤫ and $\mathbf{L}''$ is $\mathbf{L}$ with the same crossing of replaced by ↑↑ .

Suppose that $\alpha R^2 - \gamma R - \beta 1_{V \otimes V} = 0$, where $\alpha, \beta, \gamma \in k$. Then $\alpha R - \beta R^{-1} = \gamma 1_{V \otimes V}$ which implies that

$$\alpha\mathrm{INV} \text{ (L)} \ - \ \beta\mathrm{INV} \text{ (L)} \ = \ \gamma\mathrm{INV} \text{ (L)} . \qquad (8\text{--}6)$$

The preceding equation is called a *skein identity* and is the basis of a recursive evaluation of INV when $\alpha, \beta \neq 0$. As we shall see the link invariant associated to Example 1 satisfies a skein identity.

In order to relate the composition method to the bead sliding method of Section 7 we introduce vertical lines decorated by endomorphisms of V. Let $T \in \mathrm{End}\,(V)$. We associate

T to $\quad \uparrow\; T \qquad$ and $\qquad T^*$ to $\quad \downarrow\; T$

There is a natural way of multiplying decorated vertical lines. For $S, T \in \mathrm{End}\,(V)$ set $S{\cdot}T = T{\circ}S$. We define

$$\begin{array}{c}\uparrow T \\ \uparrow S\end{array} \ = \ \uparrow S{\cdot}T \qquad \text{and} \qquad \begin{array}{c}\downarrow S \\ \downarrow T\end{array} \ = \ \downarrow S{\cdot}T$$

which is consistent with the composition rule for stratified diagrams.

Let t_D, t_U be the algebra automorphisms of $\mathrm{End}(V)$ defined by $t_D(T) = D{\circ}T{\circ}D^{-1}$ and $t_U(T) = U{\circ}T{\circ}U^{-1}$ for all $T \in \mathrm{End}(V)$. Then we have the analogs for sliding labeled beads across local extrema:

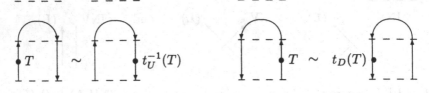

or in terms of composition formulas:

$$\mathcal{U}_-\circ(T\otimes 1_{V^*}) = \mathcal{U}_-\circ(1_V\otimes t_U^{-1}(T))^*),\quad \mathcal{D}_+\circ(1_{V^*}\otimes T) = \mathcal{D}_+\circ(t_D(T)^*\otimes 1_V)$$

and

$$(1_V\otimes T^*)\circ\mathcal{D}_- = (t_D^{-1}(T)\otimes 1_{V^*})\circ\mathcal{D}_-,\quad (T^*\otimes 1_V)\circ\mathcal{U}_+ = (1_{V^*}\otimes t_U(T))\circ\mathcal{U}_+ .$$

Regard $\mathrm{End}(V\otimes V)$ as $\mathrm{End}(V)\otimes\mathrm{End}(V)$ in the usual way, let $R = \sum_{i=1}^r e_i\otimes e_i'$ and $R^{-1} = \sum_{j=1}^s E_j\otimes E_j'$. Set $\rho = \tau_{V,V}\circ R$. Then ρ and ρ^{-1} satisfy the quantum Yang–Baxter equation (1–1). Observe that

$$\rho(u\otimes v) = \sum_{i=1}^r e_i'(v)\otimes e_i(u)\quad\text{and}\quad \rho^{-1}(u\otimes v) = \sum_{j=1}^s E_j(v)\otimes E_j'(u)$$

for all $u,v \in V$. By the multilinearity of the tensor product, to compute $\mathrm{INV}(\mathbf{L})$ we may replace $\times$ and $\times$ in the diagram $\mathbf{L}$ by

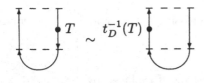 and 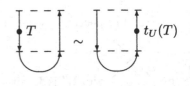 respectively, where the crossing lines of the last two configurations are regarded as vertical lines decorated by the indicated morphisms. The result of this procedure is indicated by $\times$ and $\times$ The diagram $\mathbf{L}$ with crossings replaced by these configurations is referred to as a *flat diagram*; see [17, Section 2.1] for example. Usually e_i and e_i' are denoted by e and e'; likewise E_j and E_j' are denoted by E and E'.

From this point until the end of the section oriented lines, crossings and local extrema will be components of strata. To evaluate $\mathrm{INV}(\mathbf{L})$, first choose an upward directed vertical line (not a crossing line) in each of the components

$\mathbf{L}_1, \ldots, \mathbf{L}_r$ of $\mathbf{L}$ and apply the analogs of the rules for sliding beads across local extrema described above so that the only decorated lines are the chosen verticals and each of these has a single decoration $w(\mathbf{L}_i)$. This can be done by bead sliding. Again, we may assume without loss of generality that each component has such a vertical line. Next label the intersections of the diagram $\mathbf{L}$ and the stratification lines by indices (which run over the values $1, \ldots, n$).

For a linear endomorphism T of V we write $T(v_j) = \sum_{i=1}^{n} T_j^i v_i$, where $T_j^i \in k$. Thus for endomorphisms S, T of V we have $(T \circ S)_j^i = \sum_{\ell=1}^{n} T_\ell^i S_j^\ell$. Observe that

$$\mathcal{D}_+(v^i \otimes v_j) = D_j^i \quad \text{and} \quad \mathcal{D}_-(1) = \sum_{i,j=1}^{n} (D^{-1})_j^i v_i \otimes v^j,$$

$$\mathcal{U}_+(1) = \sum_{i,j=1}^{n} v^j \otimes U_j^i v_i \quad \text{and} \quad \mathcal{U}_-(v_j \otimes v^i) = (U^{-1})_j^i.$$

We associate matrix elements to the vertical lines, the crossing lines and to the local extrema of $\mathbf{L}$ whose endpoints are now labeled by indices as follows:

δ_j^i to

that is to vertical lines with no line label or crossing lines;

T_j^i to

D_j^i to   and $(D^{-1})_j^i$ to   ;

U_j^i to   and $(U^{-1})_j^i$ to   .

The scalar INV($\mathbf{L}$) is obtained by multiplying all of the matrix elements and summing over the indices. It is a product of contributions of the components of $\mathbf{L}$. Each component $\mathbf{L}_i$ contributes a factor which is described as follows. Start at the base of the only decorated vertical line in $\mathbf{L}_i$ and let $j_1, \ldots, j_m$ be the indices of $\mathbf{L}_i$ associated to the intersections of $\mathbf{L}_i$ and the stratum lines of $\mathbf{L}$ in the order encountered on a traversal of $\mathbf{L}_i$ in the direction of orientation. Observe that j_1 and j_2, j_2 and j_3, $\ldots$, j_{m-1} and j_m, j_m and j_1 label the endpoints of stratum components whose associated matrix elements we will call $(T_1)_{j_1}^{j_2}, (T_2)_{j_2}^{j_3},$

$\ldots, (T_{m-1})^{j_m}_{j_{m-1}}, (T_m)^{j_1}_{j_m}$ respectively. The contribution which the component $\mathbf{L}_i$ makes to $\mathrm{INV}(\mathbf{L})$ is

$$\sum_{j_1,\ldots,j_m=1}^{n} (T_1)^{j_2}_{j_1}(T_2)^{j_3}_{j_2}\cdots(T_{m-1})^{j_m}_{j_{m-1}}(T_m)^{j_1}_{j_m} = \mathrm{tr}(T_m\circ\cdots\circ T_1) = \mathrm{tr}(G^{-d}w(\mathbf{L}_1)),$$

where $G = D\circ U$ and d is the Whitney degree of $\mathbf{L}_i$. If $(\mathrm{End}(V), t_D, t_U)$ is an oriented quantum algebra over k we note $(\mathrm{End}(V), t_D, t_U, D\circ U)$ is a twist oriented quantum algebra over k.

THEOREM 4. *Let V be a finite-dimensional vector space over k. Suppose that D, U are commuting linear automorphisms of V and R is a linear automorphism of $V\otimes V$ which is a solution to the braid equation (1–2). Let $\rho = \tau_{V,V}\circ R$ and let t_D, t_U be the algebra automorphisms of $\mathrm{End}(V)$ which are defined by $t_D(X) = D\circ X\circ D^{-1}$ and $t_U(X) = U\circ X\circ U^{-1}$ for all $X \in \mathrm{End}(V)$. Then:*

(a) *$(\mathrm{End}(V)^{\mathrm{op}}, \rho, t_D, t_U)$ is an oriented quantum algebra over k if and only if (8–2)–(8–5) hold for R, D and U.*

(b) *Suppose that (8–2)–(8–5) hold for R, D and U. Then $\mathrm{INV} = \mathrm{INV}_{R,D,U}$ is a regular isotopy invariant of oriented link diagrams and*

$$\mathbf{Inv}_{(\mathrm{End}(V)^{\mathrm{op}},\rho,t_D,t_U,(D\circ U)^{-1}),\mathrm{tr}}(\mathbf{L}) = \mathrm{INV}(\mathbf{L})$$

for all $\mathbf{L} \in \mathbf{Link}$.

PROOF. Using bead sliding to evaluate compositions, we note that (8–4) is equivalent to the equation $(t_U\otimes t_U)(\rho) = \rho$ and (8–5) is equivalent to the equation $(t_D^{-1}\otimes t_D^{-1})(\rho) = \rho$. Thus (8–4) and (8–5) are collectively equivalent to (qa.2) for ρ, t_D and t_U.

Assume that (8–4) and (8–5) hold for ρ, t_D and t_U. Let $A = \mathrm{End}(V)^{\mathrm{op}}$. Then (8–2) is equivalent to

$$((t_U\otimes 1)(\rho^{-1}))((1\otimes t_D)(\rho)) = 1\otimes 1$$

in $A\otimes A^{\mathrm{op}}$ and (8–3) is equivalent to

$$((1\otimes t_U)(\rho))((t_D\otimes 1)(\rho^{-1})) = 1\otimes 1$$

in $A\otimes A^{\mathrm{op}}$. Applying the algebra automorphism $t_D\otimes t_U$ of $A\otimes A^{\mathrm{op}}$ to both sides of the first equation we see that (8–2) is equivalent to

$$((t_D\otimes 1)(\rho^{-1}))((1\otimes t_U)(\rho)) = 1\otimes 1$$

in $A\otimes A^{\mathrm{op}}$. We have shown part (a). Note that (8–2) and (8–3) are equivalent since V is finite-dimensional. Part (b) follows from part (a) and the calculations preceding the statement of the theorem.   □

Let $(\mathrm{End}(V), \rho, t_D, t_U, G)$ be a twist oriented quantum algebra over k, where $G = D \circ U$. We relate the resulting invariant of oriented links with the one described in Theorem 4.

First we identify $\mathrm{End}(V)$ and $M_n(k)$ via $T \mapsto (T_j^i)$ and use this isomorphism to identify $\mathrm{End}(V \otimes V) \simeq \mathrm{End}(V) \otimes \mathrm{End}(V)$ with $M_n(k) \otimes M_n(k)$. Write $\rho = \sum_{i,j,k,\ell=1}^{n} \rho_{k\ell}^{ij} E_k^i \otimes E_\ell^j$ and let $R_{(\rho)} = \tau_{V,V} \circ \rho$ be the solution to the braid equation (1-2) associated to ρ. Then $R_{(\rho)} = \sum_{j,i,k,\ell=1}^{n} \rho_{k\ell}^{ji} E_k^i \otimes E_\ell^j$ and $R_{(\rho)}^{-1} = \sum_{j,i,k,\ell=1}^{n} (\rho^{-1})_{\ell k}^{ij} E_k^i \otimes E_\ell^j$.

Let $f : M_n(k)^{\mathrm{op}} \longrightarrow M_n(k)$ be the algebra isomorphism defined by $f(x) = x^\tau$, where x^τ is the transpose of $x \in M_n(k)$. Then $(M_n(k)^{\mathrm{op}}, \rho_f, t_{Df}, t_{Uf}, f(G))$ is a twist oriented quantum algebra over k, where $\rho_f = (f \otimes f)(\rho)$, $t_{Df} = f \circ t_D \circ f^{-1} = t_{(D^{-1})^\tau}$, $t_{Uf} = f \circ t_U \circ f^{-1} = t_{(U^{-1})^\tau}$ and $G = D \circ U$. Furthermore $f : (M_n(k)^{\mathrm{op}}, \rho, t_D, t_U, G) \longrightarrow (M_n(k)^{\mathrm{op}}, \rho_f, t_{Df}, t_{Uf}, f(G))$ is an isomorphism of twist oriented quantum algebras over k. Since $\mathrm{tr} = \mathrm{tr} \circ f$ we can use Proposition 12 to conclude that

$$\mathbf{Inv}_{(M_n(k),\rho,t_D,t_U,G),\mathrm{tr}} = \mathbf{Inv}_{(M_n(k)^{\mathrm{op}},\rho_f,t_{(D^{-1})^\tau},t_{(U^{-1})^\tau},G^\tau),\mathrm{tr}}. \qquad (8\text{--}7)$$

We note that $R_{(\rho)}$ and $R_{(\rho_\tau)}$ have the same minimal polynomial. The algebra automorphism of $M_n(k) \otimes M_n(k)$ defined by $F = \tau_{M_n(k),M_n(k)} \circ (f \otimes f)$ satisfies $R_{(\rho_\tau)} = F(R_{(\rho)})$ from which our assertion follows.

Consider the oriented quantum algebra of Example 5 which has a twist structure given by $G = \sum_{i=1}^{n} \omega_i^2 E_i^i$. The resulting invariant of oriented links satisfies a skein identity whenever bc has a square root in k. In this case

$$\frac{1}{\sqrt{\mathit{bc}}} R - \sqrt{\mathit{bc}} R^{-1} = \left(\frac{a}{\sqrt{\mathit{bc}}} - \frac{\sqrt{\mathit{bc}}}{a} \right) 1 \otimes 1;$$

thus the invariant $\mathbf{Inv} = \mathbf{Inv}_{(M_n(k),\rho,t_D,t_U,G),\mathrm{tr}}$ satisfies the skein identity (8–6) with $\alpha = \sqrt{\mathit{bc}}$, $\beta = 1/\sqrt{\mathit{bc}}$ and $\gamma = a/\sqrt{\mathit{bc}} - \sqrt{\mathit{bc}}/a$. See [16, Section 6] for an analysis of this invariant.

Suppose that further that $k = \mathbb{C}$, $t \in \mathbb{C}$ is transcendental, $q \in \mathbb{C}$ satisfies $t = q^4$, $n = 2$, $a = q^{-1}$, $b_{12} = q$ and $\sqrt{\mathit{bc}} = q$. Take $\omega_1 = q$ and $\omega_2 = q^{-1}$. In this case the skein identity is the skein identity for the bracket polynomial [10, page 50]. Thus by [10, Theorem 5.2] the Jones polynomial $V_L(t)$ in is given by

$$V_{\mathbf{L}}(t) = V_{\mathbf{L}}(q^4) = \frac{(-q^3)^{\mathrm{writhe} \mathbf{L}}}{\mathrm{tr}(G)} \mathbf{Inv}(\mathbf{L}) = \left(\frac{(-q^3)^{\mathrm{writhe} \mathbf{L}}}{q^2 + q^{-2}} \right) \mathbf{Inv}(\mathbf{L}) \qquad (8\text{--}8)$$

for all $\mathbf{L} \in \mathbf{Link}$.

By virtue of the preceding formula, Example 8 and part (b) of Theorem 3 the Jones polynomial can be computed in terms of the quantum algebra of Example 3. The calculation of the Jones polynomial in terms of Example 3 was done by Kauffman much earlier; see [10, page 580] for example.

9. Inner Oriented Quantum Algebras

The material of the last section suggests designation of a special class of oriented quantum algebras which we will refer to as inner. Formally an *inner oriented quantum algebra* is a tuple (A, ρ, D, U), where $D, U \in A$ are commuting invertible elements such that $(A, \rho, t_\mathsf{d}, t_\mathsf{u})$ is an oriented quantum algebra with $t_\mathsf{d}(a) = DaD^{-1}$ and $t_\mathsf{u}(a) = UaU^{-1}$ for all $a \in A$. Observe that if (A, ρ, D, U) is an inner oriented quantum algebra then $(A, \rho, t_\mathsf{d}, t_\mathsf{u}, DU)$ is a twist oriented quantum algebra which we denote by (A, ρ, D, U, DU).

An inner oriented quantum algebra (A, ρ, D, U) is *standard* if $D = 1_A$ and is *balanced* if $D = U$. Observe that any standard oriented quantum algebra structure on $A = \mathrm{M}_n(k)$ arises from an inner oriented quantum algebra structure.

If (A, ρ, D, U) is an inner oriented quantum algebra $(A^{\mathrm{op}}, \rho, D^{-1}, U^{-1})$, $(A, \rho^{-1}, D^{-1}, U^{-1})$ and $(A, \rho^{\mathrm{op}}, U^{-1}, D^{-1})$ are as well. See the discussion which follows Example 5.

A *morphism* $f : (A, \rho, D, U) \longrightarrow (A', \rho', D', U')$ *of inner oriented quantum algebras* is an algebra map $f : A \longrightarrow A'$ which satisfies $\rho' = (f \otimes f)(\rho)$, $f(D) = D'$ and $f(U) = U'$; that is $f : (A, \rho, t_\mathsf{d}, t_\mathsf{u}) \longrightarrow (A', \rho', t'_\mathsf{d}, t'_\mathsf{u})$ is a morphism of the associated oriented quantum algebras. For inner oriented quantum algebras (A, ρ, D, U) and (A', ρ', D', U') over k we note that $(A \otimes A', \rho'', D \otimes D', U \otimes U')$ is an inner oriented quantum algebra over k, called the *tensor product of* (A, ρ, D, U) *and* (A', ρ', D', U'), where $(A \otimes A', \rho'')$ is the tensor product of the quantum algebras (A, ρ) and (A', ρ'). Observe that $(k, 1 \otimes 1, 1, 1)$ is an inner oriented quantum algebra. Inner oriented quantum algebras together with their morphisms under composition form a monoidal category.

Let $f : A \longrightarrow A'$ be an algebra map. If (A, ρ, D, U) is an inner oriented quantum algebra over k and then $(A', (f \otimes f)(\rho), f(D), f(U))$ is as well and $f : (A, \rho, D, U) \longrightarrow (A', (f \otimes f)(\rho), f(D), f(U))$ is a morphism. In particular if V is a finite-dimensional left A-module and $f : A \longrightarrow \mathrm{End}(V)$ is the associated representation

$$f : (A^{\mathrm{op}}, \rho, D^{-1}, U^{-1}, (DU)^{-1}) \longrightarrow (\mathrm{End}(V)^{\mathrm{op}}, (f \otimes f)(\rho), f(D)^{-1}, f(U)^{-1}, f(DU)^{-1})$$

is a morphism of twist oriented quantum algebras; the latter occurs in part (b) of Theorem 4.

10. The Hennings Invariant

Let (A, ρ, v) be a finite-dimensional unimodular ribbon Hopf algebra with antipode s over the field k and let (A, ρ, s, G) be the twist quantum algebra of Example 4, where $G = uv^{-1}$. Let $\lambda \in A^*$ be a non-zero right integral for A^*. Then $\mathrm{tr} \in A^*$ defined by $\mathrm{tr}(a) = \lambda(Ga)$ for all $a \in A$ is a tracelike s^*-invariant element of A^*.

Suppose further that $\lambda(v), \lambda(v^{-1}) \neq 0$. Then the Hennings invariant for 3-manifolds is defined. It is equal to a normalization of $\mathsf{Inv}_{A, \mathrm{tr}}(\mathsf{L})$ by certain powers of $\lambda(v)$ and $\lambda(v^{-1})$, where $\mathsf{L} \in Link$ is associated with the 3-manifold.

In addition to providing a very rough description of the Hennings invariant, we would like to comment here that Proposition 9 is of very little help in computing this invariant. For suppose that $f : (A, \rho, s, G) \longrightarrow (A', \rho', s', G')$ is a morphism of twist quantum algebras over k, $\mathrm{tr}' \in A'$ is a tracelike s'^*-invariant element of A'^* and $\mathrm{tr} = \mathrm{tr}' \circ f$. Then $(0) = \lambda(G\ker f)$. The only left or right ideal of A on which λ vanishes is (0). Therefore $\ker f = (0)$, or equivalently f is one-one. Thus the right hand side of $\mathsf{Inv}_{A, \mathrm{tr}}(\mathsf{L}) = \mathsf{Inv}_{A', \mathrm{tr}'}(\mathsf{L})$ is no easier to compute than the left hand side.

The Hennings invariant [5] was defined originally using oriented links and was reformulated and conceptually simplified using unoriented links [13]. Calculations of the Hennings invariant for two specific Hopf algebras are made in [13], and calculations which are relevant to the evaluation of the Hennings invariant are made in [25].

11. Quantum Coalgebras and Coquasitriangular Hopf Algebras

To define quantum coalgebra we need the notion of coalgebra map with respect to a set of bilinear forms. Let C, D be coalgebras over k and suppose that $\mathcal{B}$ is a set of bilinear forms on D. Then a linear map $T : C \longrightarrow D$ is a *coalgebra map with respect to* $\mathcal{B}$ if $\varepsilon \circ T = \varepsilon$,

$$b(T(c_{(1)}), d)b'(T(c_{(2)}), e) = b(T(c)_{(1)}, d)b'(T(c)_{(2)}, e)$$

and

$$b(d, T(c_{(1)}))b'(e, T(c_{(2)})) = b(d, T(c)_{(1)})b'(e, T(c)_{(2)})$$

for all $c \in C$ and $d, e \in D$. A *quantum coalgebra over* k is a triple (C, b, S), where (C, b) is a Yang–Baxter coalgebra over k and $S : C \longrightarrow C^{\mathrm{cop}}$ is a coalgebra isomorphism with respect to $\{b\}$, such that

(QC.1) $b^{-1}(c, d) = b(S(c), d)$ and

(QC.2) $b(c, d) = b(S(c), S(d))$

for all $c, d \in C$. Observe that (QC.1) and (QC.2) imply

(QC.3) $b^{-1}(c, d) = b(c, S^{-1}(d))$

for all $c, d \in C$; indeed any two of (QC.1)–(QC.3) are equivalent to (QC.1) and (QC.2).

A quantum coalgebra (C, b, S) is *strict* if $S : C \longrightarrow C^{\mathrm{cop}}$ is a coalgebra isomorphism. Not every quantum coalgebra is strict. Let C be any coalgebra over k and suppose that S is a linear automorphism of C which satisfies $\varepsilon = \varepsilon \circ S$. Then (C, b, S) is a quantum coalgebra over k, where $b(c, d) = \varepsilon(c)\varepsilon(d)$ for all $c, d \in C$. There are many quantum coalgebras of this type which are not strict.

Just as quasitriangular Hopf algebras provide examples of quantum algebras, coquasitriangular Hopf algebras provide examples of quantum coalgebras. Recall that a *coquasitriangular Hopf algebra over* k is a pair (A, β), where A is a Hopf algebra over k and $\beta : A \times A \longrightarrow k$ is a bilinear form, such that

(CQT.1) $\beta(ab, c) = \beta(a, c_{(1)})\beta(b, c_{(2)})$,

(CQT.2) $\beta(1, a) = \varepsilon(a)$,

(CQT.3) $\beta(a, bc) = \beta(a_{(2)}, b)\beta(a_{(1)}, c)$,

(CQT.4) $\beta(a, 1) = \varepsilon(a)$ and

(CQT.5) $\beta(a_{(1)}, b_{(1)})a_{(2)}b_{(2)} = b_{(1)}a_{(1)}\beta(a_{(2)}, b_{(2)})$

for all $a, b, c \in A$. Let (A, β) be a coquasitriangular Hopf algebra over k. The antipode S of A is bijective [30]. By virtue of (CQT.1)–(CQT.4) it follows that β is invertible and $\beta(S(a), b) = \beta^{-1}(a, b) = \beta(a, S^{-1}(b))$ for all $a, b \in A$. Consequently $\beta(a, b) = \beta(S(a), S(b))$ for all $a, b \in A$. (CQT.1) and (CQT.5) imply that (qc.2) for A and β; apply $\beta(\ , c)$ to both sides of the equation of (CQT.5). We have shown:

EXAMPLE 9. If (A, β) is a coquasitriangular Hopf algebra over k then (A, β, S) is a quantum coalgebra over k, where S is the antipode of A.

Associativity is not necessary for the preceding example. A structure which satisfies the axioms for a Hopf algebra over k with the possible exception of the associative axiom is called a *not necessarily associative Hopf algebra over* k.

The notions of strict quantum coalgebra and quantum algebra are dual. Let (A, ρ, s) be a quantum algebra over k. Then (A^o, b_ρ, s^o) is a strict quantum coalgebra over k, and the bilinear form b_ρ is of finite type. Suppose that C is a coalgebra over k, that $b : C \times C \longrightarrow k$ is a bilinear form of finite type and that S is a linear automorphism of C. Then (C, b, S) is a strict quantum coalgebra over k if and only if (C^*, ρ_b, S^*) is a quantum algebra over k.

The dual of the quantum algebra described in Example 3 has a very simple description. For $n \geq 1$ let $C_n(k) = M_n(k)^*$ and let $\{e_{ij}\}_{1 \leq i,j \leq n}$ be the basis dual to the standard basis $\{E_{ij}\}_{1 \leq i,j \leq n}$ for $M_n(k)$. Recall that

$$\Delta(e_{ij}) = \sum_{\ell=1}^{n} e_{i\ell} \otimes e_{\ell j} \quad \text{and} \quad \varepsilon(e_{ij}) = \delta_{i,j}$$

for all $1 \leq i, j \leq n$.

EXAMPLE 10. Let k be a field and $q \in k^\star$. Then $(C_2(k), b, S)$ is a quantum coalgebra over k, where

$$b(e_{11}, e_{11}) = q^{-1} = b(e_{22}, e_{22}), \quad b(e_{11}, e_{22}) = q = b(e_{22}, e_{11}),$$

$$b(e_{12}, e_{21}) = q^{-1} - q^3$$

and

$$S(e_{11}) = e_{22}, \quad S(e_{22}) = e_{11}, \quad S(e_{12}) = -q^{-2}e_{12}, \quad S(e_{21}) = -q^2 e_{21}.$$

Let (C, b, S) and (C', b', S') be quantum coalgebras over k. The *tensor product of* (C, b, S) *and* (C', b', S') is the quantum coalgebra $(C \otimes C', b'', S \otimes S')$ over k, where $(C \otimes C', b'')$ is the tensor product of the Yang–Baxter coalgebras (C, b) and (C', b'). Observe that $(k, b, 1_k)$ is a quantum coalgebra over k, where (k, b) is the Yang–Baxter coalgebra described in Section 2. The category of quantum coalgebras over k with their morphisms under composition has a natural monoidal structure.

Suppose that (C, b, S) is a quantum coalgebra over k. Then (C, b) is a Yang–Baxter coalgebra and thus (C^{cop}, b), (C, b^{-1}) and (C, b^{op}) are as well as noted in Section 2. It is easy to see that (C^{cop}, b, S), (C, b^{-1}, S^{-1}) and $(C, b^{\mathrm{op}}, S^{-1})$ are quantum coalgebras over k.

Suppose that D is a subcoalgebra of C which satisfies $S(D) = D$. Then $(D, b|_{D \times D}, S|_D)$ is a quantum coalgebra over k which is called a *quantum subcoalgebra of* (C, b, S). Observe that the quantum subcoalgebras of (C, b, S) are the quantum coalgebras (D, b', S') over k, where D is a subcoalgebra of C and the inclusion $\imath : D \longrightarrow C$ induces a morphism of quantum coalgebras $\imath : (D, b', S') \longrightarrow (C, b, S)$.

If (C, b) is a Yang–Baxter coalgebra over k then $(D, b|_{D \times D})$ is also where D is any subcoalgebra of C. Thus (C, b) is the sum of its finite-dimensional Yang–Baxter subcoalgebras. There are infinite-dimensional quantum coalgebras (C, b, S) over k such that the only non-zero subcoalgebra D of C which satisfies $S(D) = D$ is C itself; in this case (C, b, S) is not the union of its finite-dimensional quantum subcoalgebras.

EXAMPLE 11. Let C be the grouplike coalgebra on the set of all integers and let $q \in k^*$. Then (C, b, S) is a quantum coalgebra over k, where

$$b(m, n) = \begin{cases} q & \text{:if } m + n \text{ is even} \\ q^{-1} & \text{:if } m + n \text{ is odd} \end{cases}$$

and $S(m) = m + 1$ for all integers m, n, and if D is a non-zero subcoalgebra of C such that $S(D) = D$ then $D = C$.

Suppose that I is a coideal of C which satisfies $S(I) = I$ and $b(I, C) = (0) = b(C, I)$. Then the Yang–Baxter coalgebra structure $(C/I, \bar{b})$ over k on C/I extends to is a quantum coalgebra structure on $(C/I, \bar{b}, \bar{S})$ over k such that the projection $\pi : C \longrightarrow C/I$ determines a morphism of quantum coalgebras $\pi : (C, b, S) \longrightarrow (C/I, \bar{b}, \bar{S})$. Now suppose that I is the sum of all coideals J of C such that $S(J) = J$ and $b(J, C) = (0) = b(C, J)$. Then the Yang–Baxter coalgebra (C_r, b_r) extends to a quantum coalgebra structure (C_r, b_r, S_r) such that the projection $\pi : C \longrightarrow C_r$ determines a morphism $\pi : (C, b, S) \longrightarrow (C_r, b_r, S_r)$ of quantum coalgebras over k. The quantum coalgebra construction (C_r, b_r, S_r)

is the dual of the construction of the minimal quantum algebra $(A_\rho, \rho, s|_{A_\rho})$ associated to a quantum algebra (A, ρ, s) over k.

Let K be a field extension of k. Then $(C \otimes K, b_K, S \otimes 1_K)$ is a quantum coalgebra over K, where $(C \otimes K, b_K)$ is the Yang–Baxter coalgebra described in Section 2.

We have noted that quantum coalgebras are not necessarily strict. There are natural conditions under which strictness is assured.

LEMMA 4. *A quantum coalgebra (C, b, S) over the field k is strict if b is left or right non-singular.* $\square$

Quantum coalgebra structures pull back under coalgebra maps which are onto.

PROPOSITION 13. *Let (C', b', S') be a quantum coalgebra over the field k and suppose that $f : C \longrightarrow C'$ is an onto coalgebra map. Then there is a quantum coalgebra structure (C, b, S) on C such that $f : (C, b, S) \longrightarrow (C', b', S')$ is a morphism of quantum coalgebras.* $\square$

Every finite-dimensional coalgebra over k is the homomorphic image of $C_n(k)$ for some $n \geq 1$. As a consequence of the proposition:

COROLLARY 3. *Let (C', b', S') be a quantum coalgebra over k, where C' is finite-dimensional. Then for some $n \geq 1$ there is a quantum coalgebra structure $(C_n(k), b, S)$ on $C_n(k)$ and a morphism $\pi : (C_n(k), b, S) \longrightarrow (C', b', S')$ of quantum coalgebras such that $\pi : C_n(k) \longrightarrow C'$ is onto.* $\square$

We have noted that coquasitriangular Hopf algebras have a quantum coalgebra structure. In the other direction one can always associate a coquasitriangular Hopf algebra to a quantum coalgebra (C, b, S) over k through the *free coquasitriangular Hopf algebra* $(\imath, H(C, b, S), \beta)$ on (C, b, S) whose defining mapping property is described in the theorem below. See [22, Exercise 7.4.4]. A map $f : (A, \beta) \longrightarrow (A', \beta')$ of coquasitriangular Hopf algebras is a Hopf algebra map $f : A \longrightarrow A'$ which satisfies $\beta(a, b) = \beta'(f(a), f(b))$ for all $a, b \in A$.

THEOREM 5. *Let (C, b, S) be a quantum coalgebra over the field k. Then the triple $(\imath, H(C, b, S), \beta)$ satisfies the following properties:*

(a) *The pair $(H(C, b, S), \beta)$ is a coquasitriangular Hopf algebra over k and $\imath : (C, b, S) \longrightarrow (H(C, b, S), \beta, S)$ is a morphism of quantum coalgebras, where S is the antipode of $H(C, b, S)$.*

(b) *If (A', β') is a coquasitriangular Hopf algebra over k and $f : (C, b, S) \longrightarrow (A', \beta', S')$ is a morphism of quantum coalgebras, where S' is the antipode of A', then there exists a map $F : (H(C, b, S), \beta) \longrightarrow (A', \beta')$ of coquasitriangular Hopf algebras over k uniquely determined by $F \circ \imath = f$.* $\square$

Also see [2] in connection with the preceding theorem. The quantum coalgebra (C_r, b_r, S_r) associated with a quantum coalgebra (C, b, S) plays an important role in the theory of invariants associated with (C, b, S). We end this section with a result on (C_r, b_r, S_r) in the strict case. Note that $(C_r)_r = C_r$. Compare part (a) of the following with Corollary 1.

PROPOSITION 14. *Let (C, b, S) be a quantum coalgebra over the field k and suppose that (C_r, b_r, S_r) is strict.*

(a) *If C_r is cocommutative then $S_r^2 = 1_{C_r}$.*
(b) *$S^2(g) = g$ for all grouplike elements g of C_r.*

PROOF. We may assume that $C = C_r$. To show part (a), suppose that C is cocommutative and let b' be the bilinear form on C defined by $b'(c, d) = b(S^2(c), d)$ for all $c, d \in C$. Then b' and b^{-1} are inverses since S is a coalgebra automorphism of C. Thus $b' = b$, or $b(S^2(c), d) = b(c, d)$ for all $c, d \in C$. This equation and (QC.2) imply that $b(c, S^2(d)) = b(c, d)$ for all $c, d \in C$ also. Thus the coideal $I = \mathrm{Im}\,(S^2 - 1_C)$ of C satisfies $b(I, C) = (0) = b(C, I)$. Since S is onto it follows that $S(I) = I$. Thus since $C = C_r$ we conclude $I = (0)$; that is $S^2 = 1_C$. We have established part (a). The preceding argument can be modified to give a proof of part (b). □

With the exception of Example 11 and Theorem 5 the material of this section is a very slight expansion of material found in [19].

12. Oriented Quantum Coalgebras

In this very brief section we define oriented quantum coalgebra and related concepts and discuss a few results about their structure. Most of the material of Section 11 on quantum coalgebras have analogs for oriented quantum coalgebras. The notions of oriented quantum algebra and twist oriented quantum algebra are introduced in [17].

An *oriented quantum coalgebra over k* is a quadruple (C, b, T_d, T_u), where (C, b) is a Yang–Baxter coalgebra over k and T_d, T_u are commuting coalgebra automorphisms with respect to $\{b, b^{-1}\}$, such that

(qc.1) $b(c_{(1)}, T_u(d_{(2)}))b^{-1}(T_d(c_{(2)}), d_{(1)}) = \varepsilon(c)\varepsilon(d)$,

$\qquad b^{-1}(T_d(c_{(1)}), d_{(2)})b(c_{(2)}, T_u(d_{(1)})) = \varepsilon(c)\varepsilon(d)$ and

(qc.2) $b(c, d) = b(T_d(c), T_d(d)) = b(T_u(c), T_u(d))$

for all $c, d \in C$. An oriented quantum coalgebra (C, b, T_d, T_u) over k is *strict* if T_d, T_u are coalgebra automorphisms of C, is *balanced* if $T_d = T_u$ and is *standard* if $T_d = 1_C$. We make the important observation that the T-form structures of [12] are the standard oriented quantum coalgebras over k; more precisely (C, b, T) is a T-form structure over k if and only if $(C, b, 1_C, T^{-1})$ is an oriented quantum coalgebra over k.

Let (C, b, T_d, T_u) and (C', b', T_d', T_u') be oriented quantum coalgebras over k. Then $(C \otimes C', b'', T_d \otimes T_d', T_u \otimes T_u')$ is an oriented quantum coalgebra over k, called the *tensor product of* (C, b, T_d, T_u) and (C', b', T'_d, T'_u), where $(C \otimes C', b'')$ is the tensor product of the Yang–Baxter coalgebras (C, b) and (C', b'). A *morphism* $f : (C, b, T_d, T_u) \longrightarrow (C', b', T'_d, T'_u)$ *of oriented quantum coalgebras* is a morphism $f : (C, b) \longrightarrow (C', b')$ of Yang–Baxter coalgebras which satisfies $f \circ T_d = T'_d \circ f$ and $f \circ T_u = T'_u \circ f$. Observe that $(k, b, 1_k, 1_k)$ is an oriented quantum coalgebra over k where (k, b) is the Yang–Baxter coalgebra of Section 2. The category of oriented quantum coalgebras over k and their morphisms under composition has a natural monoidal structure.

The notions of oriented quantum algebra and strict oriented quantum coalgebra are dual as the reader may very well suspect at this point. Suppose that (A, ρ, t_d, t_u) is an oriented quantum algebra over k. Then $(A^o, b_\rho, t^o_d, t^o_u)$ is a strict oriented quantum coalgebra over k and the bilinear form b_ρ is of finite type. Suppose that C is a coalgebra over k, that b is a bilinear form on C of finite type and that T_d, T_u are commuting linear automorphisms of C. Then (C, b, T_d, T_u) is a strict oriented quantum coalgebra over k if and only if $(C^*, \rho_b, T^*_d, T^*_u)$ is an oriented quantum algebra over k.

There is an analog of Proposition 2 for oriented quantum coalgebras.

PROPOSITION 15. *If* (C, b, T_d, T_u) *is an oriented quantum coalgebra over the field* k *then* $(C, b, T_d \circ T_u, 1_C)$ *and* $(C, b, 1_C, T_d \circ T_u)$ *are oriented quantum coalgebras over* k. $\qquad\qquad\square$

The oriented quantum coalgebra $(C, b, 1_C, T_d \circ T_u)$ of the proposition is called the *standard oriented quantum coalgebra associated with* (C, b, T_d, T_u). It may very well be the case that the only oriented quantum coalgebra structures (C, b, T_d, T_u) which a Yang–Baxter coalgebra (C, b) supports satisfy $T_d = 1_C$ or $T_u = 1_C$; take the dual of the oriented quantum algebra of Example 6.

As in the case of quantum algebras:

THEOREM 6.   *Let* (C, b, S) *be a quantum coalgebra over the field* k. *Then* $(C, b, 1_C, S^{-2})$ *is an oriented quantum coalgebra over the field* k. $\qquad\square$

It may very well be the case that for a standard oriented quantum coalgebra $(C, b, 1_C, T)$ there is no quantum coalgebra structure of the form (C, b, S); consider the dual of the oriented quantum algebra of Example 7.

To construct knot and link invariants from oriented quantum coalgebras we need a bit more structure. A *twist oriented quantum coalgebra over* k is a quintuple (C, b, T_d, T_u, G), where (C, b, T_d, T_u) is a strict oriented quantum coalgebra over k and $G \in C^*$ is invertible, such that

$$T^*_d(G) = T^*_u(G) = G \quad \text{and} \quad T_d \circ T_u(c) = G^{-1} \rightharpoonup c \leftharpoonup G$$

for all $c \in C$, where $c^* \rightharpoonup c = c_{(1)} c^*(c_{(2)})$ and $c \leftharpoonup c^* = c^*(c_{(1)}) c_{(2)}$ for all $c^* \in C^*$ and $c \in C$. We let the reader work out the duality between twist oriented quantum algebras and twist oriented quantum coalgebras.

The notion of strict (and twist) oriented quantum coalgebra was introduced in [16] and the general notion of quantum coalgebra was introduced in [17]. The results of this section are found in [15, Sections 2, 4].

13. Quantum Coalgebras Constructed from Standard Oriented Quantum Coalgebras

Let (C, b, T_d, T_u) be an oriented quantum coalgebra over the field k. We consider the dual of the construction discussed in Section 5. The essential part of the construction we describe in this section was made in [12] before the concepts of oriented quantum algebra and coalgebra were formulated. What follows is taken from [15, Section 4].

Let $(C, b, 1_C, T_d \circ T_u)$ be the standard oriented quantum coalgebra associated with (C, b, T_d, T_u). Let $\mathcal{C} = C \oplus C^{\mathrm{cop}}$ be the direct sum of the coalgebras C and C^{cop} and think of C as a subcoalgebra of $\mathcal{C}$ by the identification $c = c \oplus 0$ for all $c \in C$. For $c, d \in C$ define $\overline{c \oplus d} = d \oplus c$, let $\beta : \mathcal{C} \otimes \mathcal{C} \longrightarrow k$ be the bilinear form determined by

$$\beta(c, d) = \beta(\overline{c}, \overline{d}), \quad \beta(\overline{c}, d) = b^{-1}(c, d) \ \text{ and } \ \beta(c, \overline{d}) = b^{-1}(c, (T_d \circ T_u)^{-1}(d))$$

and let $\mathbf{S}$ be the linear automorphism of $\mathcal{C}$ defined by

$$\mathbf{S}(c \oplus d) = (T_d \circ T_u)^{-1}(d) \oplus c.$$

Since $(C, b, (T_d \circ T_u)^{-1})$ is a $(T_d \circ T_u)^{-1}$-form structure, it follows by [12, Theorem 1] that $(\mathcal{C}, \beta, \mathbf{S})$ is a quantum coalgebra over k. It is easy to see that $(\mathcal{C}, \beta, \mathbf{T}_d, \mathbf{T}_u)$ is an oriented quantum coalgebra over k, where

$$\mathbf{T}_d(c \oplus d) = T_d(c) \oplus T_d(d) \ \text{ and } \ \mathbf{T}_u(c \oplus d) = T_u(c) \oplus T_u(d)$$

for all $c, d \in C$. Observe that the inclusion $\imath : C \longrightarrow \mathcal{C}$ determines a morphism of oriented quantum coalgebras $\imath : (C, b, T_d, T_u) \longrightarrow (\mathcal{C}, \beta, \mathbf{T}_d, \mathbf{T}_u)$. Also observe that $\mathbf{T}_d, \mathbf{T}_u$ commute with $\mathbf{S}$.

Let $\mathcal{C}_{cq}$ be the category whose objects are quintuples (C, b, S, T_d, T_u), where (C, b, S) is a quantum coalgebra over k, (C, b, T_d, T_u) is an oriented quantum coalgebra over k and T_d, T_u commute with S, and whose morphisms

$$f : (C, b, S, T_d, T_u) \longrightarrow (C', b', S', T'_d, T'_u)$$

are morphisms of quantum coalgebras $f : (C, b, S) \longrightarrow (C', b', S')$ and morphisms of oriented quantum coalgebras $f : (C, b, T_d, T_u) \longrightarrow (C', b', T'_d, T'_u)$. Our construction gives rise to a free object of $\mathcal{C}_{cq}$.

PROPOSITION 16. *Let (C, b, T_d, T_u) be an oriented quantum coalgebra over the field k. Then the pair $(\imath, (\mathcal{C}, \beta, \mathbf{S}, \mathbf{T}_d, \mathbf{T}_u))$ satisfies the following properties:*

(a) *$(\mathcal{C}, \beta, \mathbf{S}, \mathbf{T}_d, \mathbf{T}_u)$ is an object of $\mathcal{C}_{cq}$ and $\imath : (C, b, T_d, T_u) \longrightarrow (\mathcal{C}, \beta, \mathbf{T}_d, \mathbf{T}_u)$ is a morphism of oriented quantum coalgebras over k.*

(b) *Suppose that* (C', b', S', T'_d, T'_u) *is an object of* $\mathcal{C}_{cq}$ *and* $f : (C, b, T_d, T_u) \longrightarrow$ (C', b', T'_d, T'_u) *is a morphism of oriented quantum coalgebras over* k. *There is a morphism* $F : (\mathcal{C}, \beta, S, T_d, T_u) \longrightarrow (C', b', S', T'_d, T'_u)$ *uniquely determined by* $F \circ \imath = f$. $\square$

14. Invariants Constructed from Quantum Coalgebras and Oriented Quantum Coalgebras

The invariants of Section 6 associated with finite-dimensional quantum algebras and twist quantum algebras can be reformulated in terms of dual structures in a way which is meaningful for all quantum coalgebras and twist quantum coalgebras. In this manner regular isotopy invariants of unoriented 1–1 tangles can be constructed from quantum coalgebras and regular isotopy invariants of unoriented knots and links can be constructed from twist quantum coalgebras. In the same way reformulation of the invariants described in Section 7 leads to the construction of regular isotopy invariants of oriented 1–1 tangles from oriented quantum coalgebras and to the construction of regular isotopy invariants of oriented knots and links from oriented twist quantum coalgebras.

14.1. Invariants constructed from quantum coalgebras and twist quantum coalgebras. We assume the notation and conventions of Section 6 in the following discussion. Detailed discussions of the functions Inv_C and $\mathsf{Inv}_{C,tr}$ are found in [19, Sections 6, 8] respectively. For specific calculations of these functions we refer the reader to [16; 15; 19].

Let (A, ρ, s) be a finite-dimensional quantum algebra over the field k. Then $(C, b, S) = (A^*, b_\rho, s^*)$ is a finite-dimensional strict quantum coalgebra over k. We can regard $\mathsf{Inv}_A(\mathsf{T}) \in A = C^*$ as a functional on C and we can describe $\mathsf{Inv}_A(\mathsf{T})$ in terms of $b = b_\rho$ and $S = s^*$. The resulting description is meaningful for any quantum coalgebra over k.

Let (C, b, S) be a quantum coalgebra over k. We define a function $\mathsf{Inv}_C :$ $Tang \longrightarrow C^*$ which determines a regular isotopy invariant of unoriented 1–1 tangles. If $\mathsf{T} \in Tang$ has no crossings then $\mathsf{Inv}_C(\mathsf{T}) = \varepsilon$.

Suppose that $\mathsf{T} \in Tang$ has $n \geq 1$ crossings. Traverse T in the manner described in Section 6.1 and label the crossing lines $1, 2, \ldots, 2n$ in the order encountered. For $c \in C$ the scalar $\mathsf{Inv}_C(\mathsf{T})(c)$ is the sum of products, where each crossing χ contributes a factor according to

$$\mathsf{Inv}_C(\mathsf{T})(c) =$$

$$\begin{cases} \cdots b(S^{u(\imath)}(c_{(\imath)}), S^{u(\imath')}(c_{(\imath')})) \cdots & :\chi \text{ an over crossing} \\ \cdots b^{-1}(S^{u(\imath)}(c_{(\imath)}), S^{u(\imath')}(c_{(\imath')})) \cdots & :\chi \text{ an under crossing} \end{cases} \qquad (14\text{–}1)$$

where $\imath$ (respectively $\imath'$) labels the over (respectively under) crossing line of χ.

For example

$$\mathsf{Inv}_C(\mathsf{T}_{\text{trefoil}})(c) = b^{-1}(c_{(4)}, S^2(c_{(1)}))b(S^2(c_{(2)}), c_{(5)})b^{-1}(c_{(6)}, S^2(c_{(3)}))$$

and

$$\mathsf{Inv}_C(\mathsf{T}_{\text{curl}})(c) = b(c_{(1)}, S(c_{(2)})).$$

When $(C, b, S) = (A^*, b_\rho, s^*)$ it follows that $\mathsf{Inv}_A(\mathsf{T}) = \mathsf{Inv}_C(\mathsf{T})$ for all $\mathsf{T} \in$ *Tang*. There is a difference, however, in the way in which $\mathsf{Inv}_A(\mathsf{T})$ and $\mathsf{Inv}_C(\mathsf{T})$ are computed which may have practical implications. Computation of $\mathsf{Inv}_A(\mathsf{T})$ according to the instructions of Section 6.1 usually involves non-commutative algebra calculations in tensor powers of A. Computation of $\mathsf{Inv}_C(\mathsf{T})$ based on (14–1) involves arithmetic calculations in the commutative algebra k. There is an analog of Proposition 7 for quantum coalgebras. Thus by virtue of Corollary 3 we can further assume that $C = C_n(k)$ for some $n \geq 1$.

Final comments about the 1–1 tangle invariant described in (14–1). Even the simplest quantum coalgebras, specifically ones which are pointed and have small dimension, can introduce interesting combinatorics into the study of 1–1 tangle invariants. See [12] for details. Generally quantum coalgebras, and twist quantum coalgebras, seem to provide a very useful perspective for the study of invariants of 1–1 tangles, knots and links.

We now turn to knots and links. Let (C, b, S, G) be a twist quantum coalgebra over k and suppose that c is a cocommutative S-invariant element of C. We construct a scalar valued function $\mathsf{Inv}_{C,c} : Link \longrightarrow k$ which determines a regular isotopy invariant of unoriented links.

Let $\mathsf{L} \in Link$ be a link diagram with components $\mathsf{L}_1, \ldots, \mathsf{L}_r$. Let $d_1, \ldots, d_r$ be the associated Whitney degrees and set $c(\ell) = c {\llcorner} G^{d_\ell}$ for all $1 \leq \ell \leq r$. Let $\omega \in k$ be the product of the $G^{d_\ell}(c(\ell))$'s such that L_ℓ contains no crossing lines; if there are no such components set $\omega = 1$. The scalar $\mathsf{Inv}_{C,c}(\mathsf{L})$ is ω times a sum of products, where each crossing contributes a factor according to

$$\mathsf{Inv}_C(\mathsf{T})(c) =$$

$$\omega \begin{cases} \cdots b(S^{u(\ell:\imath)}(c(\ell)_{(\imath)}), S^{u(\ell':\imath')}(c(\ell')_{(\imath')})) \cdots & : \chi \text{ an over crossing} \\ \cdots b^{-1}(S^{u(\ell:\imath)}(c(\ell)_{(\imath)}), S^{u(\ell':\imath')}(c(\ell')_{(\imath')})) \cdots : & \chi \text{ an under crossing} \end{cases} \quad (14\text{–}2)$$

where $(\ell:\imath)$ (respectively $(\ell':\imath')$) labels the over (respectively under) crossing line of χ. We note that $\mathsf{Inv}_{A,\text{tr}}(\mathsf{L}) = \mathsf{Inv}_{C,\text{tr}}(\mathsf{L})$ when (A, ρ, s, G) is a finite-dimensional twist quantum algebra over k and $(C, b, S, G) = (A^*, b_\rho, s^*, G)$.

14.2. Invariants constructed from oriented quantum coalgebras and twist quantum coalgebras.

There are analogs $\mathbf{Inv}_C : \mathbf{Tang} \longrightarrow C^*$ and $\mathbf{Inv}_{C,c} : \mathbf{Link} \longrightarrow k$ of the invariants Inv_C and $\mathsf{Inv}_{C,c}$ of Section 14.1 for oriented quantum coalgebras and twist oriented quantum coalgebras respectively. There is a version of Theorem 3 for the coalgebra construction of Section 5; see [15, Section 8.3].

The descriptions of $\mathbf{Inv}_C$ and $\mathbf{Inv}_{C,c}$ are fairly complicated since there are eight possibilities for oriented crossings situated with respect to a vertical. If all crossings of $\mathbf{T} \in \mathbf{Tang}$ and $\mathbf{L} \in \mathbf{Link}$ are oriented in the upward direction, which we may assume by virtue of the twist moves, then the descriptions of $\mathbf{Inv}_C$ and $\mathbf{Inv}_{C,c}$ take on the character of (14–1) and (14–2) respectively. See [15, Section 8.2] for details.

References

[1]   E. Abe, "Hopf Algebras," **74**, *Cambridge Tracts in Mathematics*, Cambridge University Press, Cambridge, UK, 1980.

[2] Yukio Doi, Braided bialgebras and quadratic algebras, *Comm. in Algebra* **21**, (1993), 1731–1749.

[3]   V. G. Drinfel'd, On almost cocommutative Hopf algebras, *Leningrad Math J.* (translation) **1** (1990), 321–342.

[4]   V. G. Drinfel'd, Quantum Groups, *Proceedings of the International Congress of Mathematicians*, Berkeley, California, USA (1987), 798–820.

[5]   Mark Hennings, Invariants of links and 3-manifolds obtained from Hopf algebras, *J. London Math. Soc.* **54** (1996), 594–624.

[6]   Takahiro Hayashi, Quantum groups and quantum determinants, *J. Algebra* **152** (1992), 146–165.

[7]   I. N. Herstein, "Noncommutative rings." The Carus Mathematical Monographs, No. 15 Published by The Mathematical Association of America; distributed by John Wiley & Sons, Inc., New York 1968.

[8] M. Jimbo, Quantum R matrix related to the generalized Toda system: An algebraic approach, *in* "Lecture Notes in Physics", **246** (1986), 335–361.

[9] Louis Kauffman, Gauss Codes, quantum groups and ribbon Hopf algebras, *Reviews in Math. Physics* **5** (1993), 735–773.

[10]   Louis Kauffman, "Knots and Physics," World Scientific, Singapore/New Jersey/London/Hong Kong, 1991, 1994.

[11]   Louis H. Kauffman and David E. Radford, A necessary and sufficient condition for a finite-dimensional Drinfel'd double to be a ribbon Hopf algebra, *J. of Algebra* **159** (1993), 98–114.

[12]   Louis H. Kauffman and David E. Radford, A separation result for quantum coalgebras with an application to pointed quantum coalgebras of small dimension, *J. of Algebra* **225** (2000), 162–200.

[13] Louis H. Kauffman and David E. Radford, Invariants of 3-manifolds derived from finite-dimensional Hopf algebras, *J. Knot Theory Ramifications* **4** (1995), 131–162.

[14] Louis H. Kauffman and David E. Radford, On invariants of knots and links which arise from quantum coalgebras, *in preparation*.

[15]   Louis H. Kauffman and David E. Radford, Oriented quantum algebras and coalgebras, invariants of knots and links, *in preparation*.

[16]   Louis H. Kauffman and David E. Radford, Oriented quantum algebras and invariants of knots and links, *J. of Algebra* (**to appear**), 45 LATEXpages.

[17] Louis H. Kauffman and David E. Radford, Oriented quantum algebras, categories, knots and links, *J. Knot Theory Ramifications* (**to appear**), 47 LaTeXpages.

[18] Louis H. Kauffman and David E. Radford, Quantum algebra structures on $n \times n$ matrices, *J. of Algebra* **213** (1999), 405–436.

[19] Louis H. Kauffman and David E. Radford, Quantum algebras, quantum coalgebras, invariants of 1–1 tangles and knots, *Comm. Algebra* **28** (2000), 5101–5156.
    Quantum algebras, quantum coalgebras, invariants of 1–1 tangles and knots, *Comm. in Alg.*, to appear (2000), 62 LaTeXpages.

[20] Louis H. Kauffman and David E. Radford, State sums and oriented quantum algebras, *in preparation.*

[21] Louis H. Kauffman, David E. Radford and Steve Sawin, Centrality and the KRH invariant. *J. Knot Theory Ramifications* **7** (1998), 571–624.

[22] Larry A. Lambe and David E. Radford, "Introduction to the Quantum Yang–Baxter Equation and Quantum Groups: An algebraic Approach," Kluwer Academic Publishers, Boston/Dordrecht/London, 1997.

[23] Shahn Majid, Quasitriangular Hopf algebras and Yang–Baxter equations, *Int. J. Mod. Physics A* **5** (1990), 1–91.

[24] S. Montgomery, "Hopf Algebras and their actions on rings," **82**, *Regional Conference Series in Mathematics*, AMS, Providence, RI, 1993.

[25] D. E. Radford, On Kauffman's knot invariants arising from finite-dimensional Hopf algebras. In *Advances in Hopf algebras (Chicago, Il 1992)*, **148** of *Lecture Notes in Pure and Applied Math. Series*, 205–266. Dekker, New York.

[26] D. E. Radford, Minimal quasitriangular Hopf algebras. *J. Algebra* **157** (1993), 285–315.

[27] D. E. Radford, On a parameterized family of twist quantum coalgebras and the bracket polynomial, *J. of Algebra* **225** (2000), 93–123.

[28] N. Yu. Reshetikhin and V. G. Turaev, Invariants of 3-manifolds via link polynomials and quantum groups, *Invent. Math.* **103** (1991), 547–597.

[29] N. Yu. Reshetikhin and V. G. Turaev, Ribbon graphs and their invariants derived from quantum groups, *Comm. Math. Physics* **127** (1990), 1–26.

[30] P. Schauenburg, On coquasitriangular Hopf algebras and the quantum Yang–Baxter equation, *Algebra Berichte* **67** (1992), 1–76.

[31] M. E. Sweedler, "Hopf Algebras," Benjamin, New York, 1969.

DAVID E. RADFORD
DEPARTMENT OF MATHEMATICS, STATISTICS AND COMPUTER SCIENCE (M/C 249)
851 SOUTH MORGAN STREET
UNIVERSITY OF ILLINOIS AT CHICAGO
CHICAGO, ILLINOIS 60607-7045
UNITED STATES
    radford@uic.edu

New Directions in Hopf Algebras
MSRI Publications
Volume 43, 2002

Hopf Algebra Extensions and Monoidal Categories

PETER SCHAUENBURG

ABSTRACT. Tannaka reconstruction provides a close link between monoidal
categories and (quasi-)Hopf algebras. We discuss some applications of the
ideas of Tannaka reconstruction to the theory of Hopf algebra extensions,
based on the following construction: For certain inclusions of a Hopf algebra into a coquasibialgebra one can consider a natural monoidal category
consisting of Hopf modules, and one can reconstruct a new coquasibialgebra
from that monoidal category.

CONTENTS

1. Introduction

In most applications of Hopf algebras, it is a key property that one can form tensor products of representations of Hopf algebras. A most fruitful — if incorrect — truism in quantum group and Hopf algebra theory says that the converse holds: For every nice category with abstract tensor product there has to be some kind of bialgebra or Hopf algebra whose representations (modules or comodules) are (or are related to) that category.

In reality, additional data and conditions are needed to transform this rather vague statement into precise mathematical facts — however, we would like to maintain the general idea that once we find a monoidal category in nature, there is, with some luck, a Hopf algebra behind it, and we only have to compute or "reconstruct" what it looks like. An early example for this "philosophy" is in [37].

The correspondence between Hopf algebras and their representation categories is often a good way to understand Hopf algebraic constructions. A good example is Drinfeld's construction of the quantum double of a Hopf algebra, which has a very conceptual explanation in the construction of the center of a monoidal category. It is the author's firm (though easily refutable) belief that for every reasonable construction in Hopf algebra theory there ought to be a natural construction on the level of categories with tensor product that explains the Hopf algebra construction.

This paper presents some applications of that general principle to the theory of Hopf algebra extensions. In the broadest sense of the term, an extension of Hopf algebras is perhaps just some injective Hopf algebra map $K \to H$ (or, dually, a Hopf algebra surjection, which is the viewpoint taken in [35]). We will be much stricter in our terminology by calling an extension only a short exact sequence $K \to H \to Q$ of Hopf algebra maps, which is in addition cleft. However, we will also be interested in more general inclusions $K \subset H$ of Hopf algebras, which we shall always require to be (co)cleft. The central result in our considerations, taken from [47], is a construction that assigns to each such inclusion another inclusion $K^* \subset \tilde{H}$, provided K is finite. Actually it will turn out that $\tilde{H}$ is often just a coquasibialgebra (the dual notion to Drinfeld's quasibialgebras), and it will be natural to also allow coquasibialgebras for H — in this setting, the construction is often an involution, and it can be made into a functor, which we denote by $H \mapsto \mathfrak{F}(H) := \tilde{H}$.

The description of short exact sequences of Hopf algebras is quite complicated: In any such sequence $K \to H \to Q$ the middle term is a bicrossproduct $H = K^\tau\#_\sigma Q$ described in terms of a weak action of Q on K, a weak coaction of K on Q, a two-cocycle $\sigma: Q \otimes Q \to K$, and a two-cycle $\tau: Q \to K \otimes K$. These data have to fulfill a rather large list of axioms to ensure that the multiplication and comultiplication on $K \otimes Q$ built from them will result in an honest bialgebra. Worse than the size of the list is its lack of a conceptual interpretation. If

K is commutative and Q is cocommutative the axioms have a cohomological description, although the responsible cohomology theory is in its turn hard to handle; if K and Q are arbitrary, even the cohomological interpretation fails, and the equations become merely a rather complicated combinatorical description of a bialgebra structure on $K \otimes Q$. Matters can only get worse if we allow more general inclusions $K \subset H$, or even such inclusions in which H is only a coquasibialgebra. But many prominent constructions in the theory of Hopf algebras and quantum groups do yield cocleft Hopf algebra inclusions: besides bicrossproducts, these are double crossproducts, including the Drinfeld double as an example, and Radford biproducts of a Hopf algebra with a Yetter–Drinfeld Hopf algebra. The combinatorics of these, and of more general cases, have been analyzed in the literature.

We would like to emphasize that our general construction does not require the knowledge of most of the combinatorial data involved in an extension or inclusion. Rather, it is quite intrinsic, and proceeds according to the general creed formulated at the beginning. To any inclusion $K \subset H$, there is naturally associated a monoidal category, namely ${}^H_K\mathcal{M}_K$, the category of K-K-bimodules within the monoidal category of H-comodules. It only remains to reconstruct the bialgebra, or, in this case, the coquasibialgebra, responsible for this monoidal category structure. This turns out to be possible if H is cocleft over K; some additional conditions have to be met if the coquasibialgebra H is not an ordinary bialgebra.

In several instances the construction has well-known special cases. For example, one can apply it to a bismash product Hopf algebra $K\#Q$, and obtains a double crossproduct Hopf algebra $Q \bowtie K^*$. Equally well, one can apply the construction to an inclusion $L \subset Q \bowtie L$ of a Hopf algebra L into a double crossproduct, and obtains a bismash product $L^*\#Q$. On one hand, this is nothing new: The combinatorial data (termed a Singer pair by Masuoka) needed for the construction of a bismash product have long been known to be in bijection with the combinatorial data (called a matched pair) needed for the construction of a double crossproduct. In fact, one of the primary sources for Singer pairs is matched pairs arising naturally from groups that are the product of two subgroups. On the other hand, some aspects are new after all: We have found an intrinsic connection between a bismash product Hopf algebra and the double crossproduct Hopf algebra built from the same data. This gives an explanation for the bijection between Singer pairs and matched pairs which refers to the actual Hopf algebras in consideration, rather than the combinatorial data used to construct them. It is clear that one may hope to also gain insights through categorical considerations into properties of the two Hopf algebras thus related, although we can at the moment only offer a rather strange application to their Drinfeld doubles.

For most Hopf algebra inclusions $K \subset H$, the resulting inclusion $K^* \subset \tilde{H} = \mathfrak{F}(H)$ has $\tilde{H}$ a coquasibialgebra rather than an ordinary bialgebra. However, this turns out to have as a special case a result on ordinary cocommutative Hopf algebras and their cohomology: Let F, G be finite groups, $L = k[G]$ and $Q = k[F]$ the group algebras, and $K = k^G$ the dual of L. We assume given a group $F \bowtie G$ having F and G as subgroups satisfying $FG = F \bowtie G$. This also gives rise to a double crossproduct $Q \bowtie L \cong k[F \bowtie G]$, hence to a Singer pair (K, Q). We denote by $\mathcal{H}^n(L)$ the Sweedler cohomology group with coefficients in the base field k, which is the same as group cohomology of G with coefficients in the multiplicative group of k. The long exact sequence

$$0 \to \mathcal{H}^1(Q \bowtie L) \overset{\text{res}}{\to} \mathcal{H}^1(Q) \oplus \mathcal{H}^1(L) \to \text{Aut}(K \# Q) \to \mathcal{H}^2(Q \bowtie L) \overset{\text{res}}{\to}$$

$$\overset{\text{res}}{\to} \mathcal{H}^2(Q) \oplus \mathcal{H}^2(L) \to \text{Opext}(Q, K) \to \mathcal{H}^3(Q \bowtie L) \overset{\text{res}}{\to} \mathcal{H}^3(Q) \oplus \mathcal{H}^3(L)$$

connecting a certain group $\text{Opext}(Q, K)$ of Hopf algebra extensions $K \to H \to Q$ and the automorphism group of the special extension $K \# Q$ with the group cohomologies of F, G and $F \bowtie G$ was discovered by Kac [22] (in a slightly different setting, cf. [32]). The Kac sequence is the long exact sequence arising from a short exact sequence of double complexes: One of these is the intricate cohomological description of bicrossproducts that we alluded to above, while the other two turn out to compute the three group cohomologies involved in the sequence. We note that the notations chosen above for the Kac sequence are deceivingly Hopf algebraic: the sequence does not at first make sense when we consider general cocommutative Hopf algebras L, Q in place of group algebras. In Kac' work the cohomology groups are group cohomology with coefficients in the multiplicative group of the field (which is naturally isomorphic to Sweedler cohomology), and the techniques leading to the sequence are specific to the case of group algebras. In [47], we have replaced these techniques by arguments using the functor $\mathfrak{F}$: One of the maps in Kac' sequence, the one from $\text{Opext}(Q, K)$ to $\mathcal{H}^3(Q \bowtie L)$, assigns to a Hopf algebra (one of the extensions of Q by K) another Hopf algebra (the double crossproduct $Q \bowtie L$) and a three-cocycle. As it turns out, this map is a special case of the functor $\mathfrak{F}$, when we interpret $Q \bowtie L$ together with the cocycle as a coquasibialgebra. Exactness of Kac' sequence means in particular that the map from $\text{Opext}(Q, K)$ to $\mathcal{H}^3(Q \bowtie L)$ has a partial inverse (back from a subgroup of its codomain to a quotient of its domain). As it turns out, this partial inverse is also a special case of the functor $\mathfrak{F}$. Besides a new proof, these facts give a new interpretation to Kac' result: The correspondence between certain Hopf algebras, and certain Hopf algebras with cocycles, which originally arose from cohomological data and their interpretation as a combinatorical description of Hopf algebras, can now be described more intrinsically. Moreover, the new explanation generalizes immediately to the case where Q and L are cocommutative Hopf algebras, L is finite, and $K = L^*$. It generalizes even further to the case of arbitrary Hopf algebras Q and L with L finite, if we replace the cohomology groups by suitable sets of classes of coquasibialgebras. Unfortunately, it does not generalize to cover the analog of the Kac sequence proved by Masuoka for Lie algebras and their enveloping algebras in place of groups and their group algebras, although some things can be said about this case as well.

The plan of the paper is as follows: In the first section we recall the basic machinery of reconstruction that allows us to find Hopf algebras associated to monoidal categories, and we review the relevant notions of Hopf modules that will furnish the necessary supply of examples underlying our application.

In the second section we recall various notions of extensions of Hopf algebras. Only very little emphasis is placed on the cohomological descriptions, since these are treated in more detail in Akira Masuoka's report [31] in this volume. We will pay equally little attention to the precise combinatorical description of the various constructions, in accordance with the fact that our constructions need hardly any of this information.

In the third section we first review the basic construction of the functor $\mathfrak{F}$ for the case of inclusions $K \subset H$ with finite K, as found in [47]. After discussing some examples and properties of $\mathfrak{F}$, we turn to two extensions of the theory: First, we discuss what remains of the construction if we drop finiteness of K (but work over a field). Second, we get into the question when $\mathfrak{F}(H)$ has an antipode, provided H is an ordinary Hopf algebra; this is closely related to the question when the category $_K^H\mathcal{M}_K$ has duals. Further, we discuss how the functor $\mathfrak{F}$ specializes to a construction treated earlier by Yongchang Zhu [61], when applied to an inclusion of groups. Zhu's results inspired the considerations on antipodes in the present paper, and a byproduct is Example 4.5.1, a counterexample to the following (very desirable) statement: If the category of finite dimensional comodules over a coquasibialgebra H is rigid (that is, all objects have dual objects), then H is a coquasi-Hopf algebra.

In the final section we treat Kac' sequence and its variations. We review the results in [47], adding two features: One is a short discussion of some of Masuoka's results (cf. [31]). These are not covered by the general constructions in subsection 4.1, but we shall investigate in some detail where this really fails. The other new feature concerns a result of Kreimer [25] on Galois objects over tensor products. We show how this derives easily from the generalized Kac sequence.

Preliminaries and notation. Throughout, we work over a commutative base ring k, all algebras, coalgebras, etc. are over k, and most maps are tacitly supposed to be k-linear maps. We write $\otimes$ and Hom for tensor product over k and the set of k-linear maps. We use Sweedler notation without summation symbol for comultiplication, $\Delta(c) = c_{(1)} \otimes c_{(2)}$, and the versions $\delta(v) = v_{(-1)} \otimes v_{(0)}$ and $\delta(v) = v_{(0)} \otimes v_{(1)}$ for left and right comodules, respectively. There will be much need for variations in the shape of the parentheses to distinguish various structures. We denote the opposite algebra of an algebra A by A^{op}, the opposite coalgebra of a coalgebra C by C^{cop}, and the opposite algebra and opposite coalgebra of a bialgebra B by B^{bop}. We write $\triangleright$ for the usual left action of an algebra A on its dual space A^*, and use $\triangleleft$ for the right action. For an algebra A over a field k, we denote its finite dual by A°. We denote the category of left A-modules by $_A\mathcal{M}$, with similar meanings of $\mathcal{M}_A$, $_A\mathcal{M}_B$. We denote the category of right C-comodules by $\mathcal{M}^C$, with the obvious variations $^C\mathcal{M}$, $^C\mathcal{M}^D$. The notations $_f\mathcal{M}_A$ and $\mathcal{M}^C_f$ (with the obvious variations) denote the full subcategories of those modules or comodules that are finitely generated projective k-modules. Our general references for Hopf algebra theory are [55; 1; 34].

2. Hopf Algebras and Monoidal Categories

The topic of this section is mainly the translation procedure between Hopf algebras and monoidal categories. Its basis is the simple observation that there is a tensor product of representations (modules or comodules) of a bialgebra B. In this way $_B\mathcal{M}$ and $\mathcal{M}^B$ become monoidal categories. In the reverse direction, one hopes to associate to any monoidal category a bialgebra having the objects of that category as its representations. Specifically, if the category of representations of an algebra or coalgebra has a monoidal category structure, one wants to reconstruct a bialgebra that is responsible for the tensor product in the category. While this last step seems almost trivial, one should not forget that rather special circumstances are necessary to make it work: The tensor product of two representations (be it modules or comodules) of a bialgebra is formed as the tensor product over k, carrying a representation defined using the bialgebra structure. Stated in a fancy manner, the underlying functor $_A\mathcal{M} \to {}_k\mathcal{M}$ is a monoidal functor. But we would like to point out that even when this rather strict requirement is not fulfilled, the general scheme by which every monoidal category should come from a bialgebra is still valid to some extent: When the underlying functor $_A\mathcal{M} \to {}_k\mathcal{M}$ is not quite as nice, it often turns out that A fulfills a weakened set of axioms compared to those defining a bialgebra. Drinfeld's notion of a (co)quasibialgebra [14] is central to this paper: it occurs when the underlying functor does preserve tensor products, but in an incoherent fashion; thus the representations of a (co)quasibialgebra form a monoidal category, with the ordinary tensor product over k, but the associativity of tensor products is modified. The representations of a Hopf face algebra in the sense of Hayashi [18] form a monoidal category with respect to a "truncated" tensor product. The representations of a weak Hopf algebra in the sense of Böhm, Nill and Szlachányi [6] form a monoidal category with tensor product a certain submodule of the tensor product over k. The representations of a $\times_R$-bialgebra in the sense of Takeuchi [56] form a monoidal category with the tensor product over the k-algebra R [43].

2.1. Monoidal categories.

The relevant notion of a category with a nice tensor product is called a tensor category or monoidal category. A monoidal category consists of a category $\mathcal{C}$, a bifunctor $\otimes\colon \mathcal{C} \times \mathcal{C} \to \mathcal{C}$, a neutral object $I \in \mathcal{C}$ and isomorphisms $\alpha\colon (X \otimes Y) \otimes Z \to X \otimes (Y \otimes Z)$, $\lambda\colon I \otimes X \to X$ and $\rho\colon X \otimes I \to X$ which are natural in $X, Y, Z \in \mathcal{C}$ and required to be *coherent*. The latter means by definition that the pentagonal diagrams

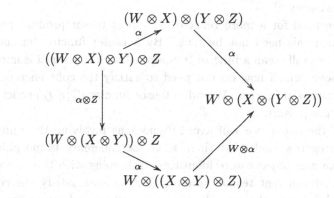

and the diagrams

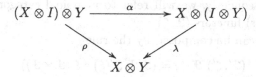

commute for all $W, X, Y, Z \in \mathcal{C}$. We say that a monoidal category $\mathcal{C}$ is strict if α, λ, and ρ are identity morphisms.

The meaning of coherence is in Mac Lane's coherence theorem [26], which says that every diagram formally composed from instances of α, λ, and ρ (like the two in the definition) commutes. Informally, one can then 'identify' any multiple tensor products (like the source and the sink of the pentagon) that only differ in the way that parentheses are set — just like one usually writes multiple tensor products of vector spaces without ever using parentheses, or just like in a strict monoidal category.

With the notion of a monoidal category comes the notion of a monoidal functor between them. A monoidal functor $(\mathcal{F}, \xi, \xi_0)\colon \mathcal{C} \to \mathcal{D}$ consists of a functor $\mathcal{F}\colon \mathcal{C} \to \mathcal{D}$, an isomorphism $\xi\colon \mathcal{F}(X) \otimes \mathcal{F}(Y) \to \mathcal{F}(X \otimes Y)$, natural in $X, Y \in \mathcal{C}$, and an isomorphism $\xi_0\colon I \to \mathcal{F}(I)$, required to fulfill the coherence conditions

$$
\begin{array}{ccccc}
(\mathcal{F}(X) \otimes \mathcal{F}(Y)) \otimes \mathcal{F}(Z) & \xrightarrow{\xi \otimes 1} & \mathcal{F}(X \otimes Y) \otimes \mathcal{F}(Z) & \xrightarrow{\xi} & \mathcal{F}((X \otimes Y) \otimes Z) \\
\downarrow{\alpha} & & & & \downarrow{\mathcal{F}(\alpha)} \\
\mathcal{F}(X) \otimes (\mathcal{F}(Y) \otimes \mathcal{F}(Z)) & \xrightarrow[1 \otimes \xi]{} & \mathcal{F}(X) \otimes \mathcal{F}(Y \otimes Z) & \xrightarrow[\xi]{} & \mathcal{F}(X \otimes (Y \otimes Z))
\end{array}
$$

along with the two coherence conditions

$$\mathcal{F}(\lambda)\xi(\xi_0 \otimes \mathrm{id}) = \lambda \colon I \otimes \mathcal{F}(X) \to \mathcal{F}(X)$$
$$\mathcal{F}(\rho)\xi(\mathrm{id} \otimes \xi_0) = \rho \colon \mathcal{F}(X) \otimes I \to \mathcal{F}(X)$$

involving the neutral objects. Considering the case where $\mathcal{F}$ is a forgetful functor, this means that the coherence morphisms 'up' in the category $\mathcal{C}$ are the same as 'down' in the category $\mathcal{D}$.

We will have need for a more relaxed version of tensor product preserving functor, in which this need not be true. By a tensor functor (or incoherent tensor functor) we will mean a functor $\mathcal{F}$ equipped with functorial isomorphisms ξ and ξ_0 as above, which now do not need to satisfy the coherence conditions imposed for a monoidal functor. We call a tensor functor $(\mathcal{F}, \xi, \xi_0)$ strict if ξ and ξ_0 are identity morphisms.

Throughout the paper, we will avoid discussions involving the unit objects of monoidal categories, and treat them as if the coherence isomorphisms λ, ρ dealing with the unit objects were identities. In keeping with this, we will only be dealing with incoherent tensor functors that at least satisfy the coherence conditions involving unit objects, which we will call neutral tensor functors; we will never deal with unit objects explicitly, and thus suppress all considerations involving ξ_0. Consequently, we will refer to monoidal categories $(\mathcal{C}, \otimes, \alpha)$ and monoidal (or tensor) functors $(\mathcal{F}, \xi)$.

Tensor functors can be composed by the rule

$$(\mathcal{F}', \xi')(\mathcal{F}, \xi) = (\mathcal{F}'\mathcal{F}, \mathcal{F}'(\xi) \circ \xi'(\mathcal{F} \times \mathcal{F}))$$

i. e. by giving $\mathcal{F}'\mathcal{F}$ the tensor functor structure defined by

$$\mathcal{F}'\mathcal{F}(X) \otimes \mathcal{F}'\mathcal{F}(Y) \xrightarrow{\xi'} \mathcal{F}'(\mathcal{F}(X) \otimes \mathcal{F}(Y)) \xrightarrow{\mathcal{F}'(\xi)} \mathcal{F}'\mathcal{F}(X \otimes Y).$$

By a morphism $\varphi \colon (\mathcal{F}, \xi) \to (\mathcal{F}', \xi')$ between tensor functors we mean a natural morphism $\varphi \colon \mathcal{F} \to \mathcal{F}'$ compatible with the tensor functor structure in the sense that all diagrams

$$
\begin{array}{ccc}
\mathcal{F}(X) \otimes \mathcal{F}(Y) & \xrightarrow{\ \xi\ } & \mathcal{F}(X \otimes Y) \\
{\scriptstyle \varphi_X \otimes \varphi_Y}\big\downarrow & & \big\downarrow{\scriptstyle \varphi_{X \otimes Y}} \\
\mathcal{F}'(X) \otimes \mathcal{F}'(Y) & \xrightarrow{\ \xi'\ } & \mathcal{F}'(X \otimes Y)
\end{array}
$$

commute. A monoidal equivalence between monoidal categories $\mathcal{C}$ and $\mathcal{D}$ is a monoidal functor $(\mathcal{F}, \xi) \colon \mathcal{C} \to \mathcal{D}$ such that the functor $\mathcal{F}$ is a category equivalence. One can show that there is then a monoidal functor $(\mathcal{G}, \zeta) \colon \mathcal{D} \to \mathcal{C}$ such that $(\mathcal{G}, \zeta)(\mathcal{F}, \xi) \cong (\mathcal{I}d, \mathrm{id})$ and $(\mathcal{F}, \xi)(\mathcal{G}, \zeta) \cong (\mathcal{I}d, \mathrm{id})$ as monoidal functors. Thus a monoidal equivalence roughly speaking establishes a one-to-one correspondence between any statements or concepts expressed in terms of the monoidal category structure of $\mathcal{C}$, and the same statements or concepts in $\mathcal{D}$. Mac Lane's coherence theorem can be expressed nicely as saying that every monoidal category is monoidally equivalent to a strictly monoidal category [23, XI.5]. This gives support to the following informal statement used widely in the literature: To prove a general claim about monoidal categories, it will always suffice to treat the case of a strict monoidal category.

We will already follow this principle when we now define the notion of a dual object of an object X in a monoidal category $\mathcal{C}$. A left dual $X^\vee$ is by definition a triple $(X^\vee, \mathrm{db}, \mathrm{ev})$ in which $\mathrm{db}: I \to X \otimes X^\vee$ and $\mathrm{ev}: X^\vee \otimes X \to I$ are morphisms satisfying

$$\left(X \xrightarrow{\mathrm{db} \otimes X} X \otimes X^\vee \otimes X \xrightarrow{X \otimes \mathrm{ev}} X \right) = \mathrm{id}_X,$$

$$\left(X^\vee \xrightarrow{X^\vee \otimes \mathrm{db}} X^\vee \otimes X \otimes X^\vee \xrightarrow{\mathrm{ev} \otimes X^\vee} X^\vee \right) = \mathrm{id}_{X^\vee}.$$

A right dual $^\vee X$ is defined similarly with morphisms $\mathrm{db}: k \to {}^\vee X \otimes X$ and $\mathrm{ev}: X \otimes {}^\vee X \to k$. For a k-module to have a dual in $_k\mathcal{M}$ means to be finitely generated projective. More generally $M \in {}_R\mathcal{M}_R$ has a left dual iff it is finitely generated projective as right R-module, and the same holds with left and right interchanged. We say that a category $\mathcal{C}$ is left (or right) rigid if every object in $\mathcal{C}$ has a left (or right) dual. By definition, a right inner hom-functor $\underline{\mathrm{hom}}(X, -)$ is a right adjoint to the functor $\mathcal{C} \ni Y \mapsto Y \otimes X \in \mathcal{C}$; that is, it is defined by an adjunction

$$\mathcal{C}(Y \otimes X, Z) \cong \mathcal{C}(Y, \underline{\mathrm{hom}}(X, Z))$$

natural in $Y, Z \in \mathcal{C}$. A left inner hom-functor is a right adjoint for $X \otimes -$. If $\mathcal{C}$ has inner hom-functors $\underline{\mathrm{hom}}(X, -)$ for all X, then we say that $\mathcal{C}$ is right closed. If X has a dual, then an inner hom-functor $\underline{\mathrm{hom}}(X, -)$ exists: one can take $\underline{\mathrm{hom}}(X, Z) = Z \otimes X^\vee$. In particular, left rigid categories are right closed. Conversely, assume that a right inner hom-functor $\underline{\mathrm{hom}}(X, -)$ exists. Then from the adjointness defining $\underline{\mathrm{hom}}(X, -)$ one can construct natural morphisms $Y \otimes \underline{\mathrm{hom}}(X, Z) \to \underline{\mathrm{hom}}(X, Y \otimes Z)$ for all $Y, Z \in \mathcal{C}$. If all these are isomorphisms, then X has a dual, namely $\underline{\mathrm{hom}}(X, I)$.

It is important to note that monoidal functors automatically preserve duals in the following sense: If $X \in \mathcal{C}$ has a left dual $X^\vee$, and $(\mathcal{F}, \xi): \mathcal{C} \to \mathcal{D}$ is a monoidal functor, then $\mathcal{F}(X^\vee)$ is a dual to $\mathcal{F}(X)$ endowed with the maps

$$\mathcal{F}(X^\vee) \otimes \mathcal{F}(X) \xrightarrow{\xi} \mathcal{F}(X^\vee \otimes X) \xrightarrow{\mathcal{F}(\mathrm{ev})} \mathcal{F}(I) = I$$

$$I = \mathcal{F}(I) \xrightarrow{\mathcal{F}(\mathrm{db})} \mathcal{F}(X \otimes X^\vee) \xrightarrow{\xi^{-1}} \mathcal{F}(X) \otimes \mathcal{F}(X^\vee)$$

If both $\mathcal{C}$ and $\mathcal{D}$ are rigid, this results in isomorphisms $\mathcal{F}(X^\vee) \cong \mathcal{F}(X)^*$ (where $(-)^*$ denotes the dual in $\mathcal{D}$), which can be chosen to be functorial in X. In general, monoidal functors need not preserve inner hom-functors, and it will turn out that incoherent tensor functors need not preserve duals.

2.2. Bialgebras, quasibialgebras, and their representations. Key examples of monoidal categories arise from representations of bialgebras. For simplicity we shall pretend throughout that the category $_k\mathcal{M}$ of k-modules with the tensor product over k is strictly monoidal. When B is a bialgebra, then the category $_B\mathcal{M}$ of left B-modules is (strictly) monoidal: the tensor product of two B-modules V, W is their tensor product over k with the diagonal module structure $b(v \otimes w) = b_{(1)}v \otimes b_{(2)}w$. Dually, the category $\mathcal{M}^B$ of right B-comodules is monoidal: the tensor product of two B-comodules V, W is their tensor product over k, endowed with the codiagonal comodule structure $V \otimes W \ni v \otimes w \mapsto v_{(0)} \otimes w_{(0)} \otimes v_{(1)}w_{(1)}$.

In all of these examples the obvious underlying functors to the category of k-modules are strictly monoidal. If we are given an algebra A, and a structure of monoidal category on $_A\mathcal{M}$ such that the underlying functor to $_k\mathcal{M}$ is a strict monoidal functor, then we can actually reconstruct a unique bialgebra structure on A inducing the given monoidal category structure on $_A\mathcal{M}$. A similar statement holds for the category $^C\mathcal{M}$ of comodules over a coalgebra C: If $^C\mathcal{M}$ is monoidal, and the underlying functor to $_k\mathcal{M}$ is strictly monoidal, then there is a unique map $\nabla \colon C \otimes C \to C$ such that $(v \otimes w)_{(-1)} \otimes (v \otimes w)_{(0)} = v_{(-1)}w_{(-1)} \otimes v_{(0)} \otimes w_{(0)}$ for all $V, W \in {}^C\mathcal{M}$, $v \in V$, and $w \in W$, and ∇ makes C into a bialgebra. Throughout this section we will state 'reconstruction' results like this without a hint of a proof, before sketching some of the techniques in the background in the next section.

The situation is *not* essentially worse if we assume that the underlying functor $_A\mathcal{M} \to {}_k\mathcal{M}$ (or $^C\mathcal{M} \to {}_k\mathcal{M}$) is a non-strict monoidal functor. In that case we can also find a monoidal category structure on $_A\mathcal{M}$ such that the identity is a (non-strict) monoidal equivalence between the two monoidal category structures on $_A\mathcal{M}$, and the underlying functor $_A\mathcal{M} \to {}_k\mathcal{M}$ is strictly monoidal for the new structure.

An essential change of the situation occurs if we assume the underlying functor $_A\mathcal{M} \to {}_k\mathcal{M}$ to be a (neutral) tensor functor instead of a monoidal one.

Drinfeld [14] has defined a quasibialgebra A to be an algebra equipped with a not necessarily coassociative comultiplication $\Delta \colon A \to A \otimes A$ and a counit $\varepsilon \colon A \to k$ for Δ, both of which are algebra maps, and an invertible element $\phi \in A \otimes A \otimes A$, the associator, satisfying $(A \otimes \varepsilon \otimes A)(\phi) = 1_A \otimes 1_A \in A \otimes A$, $(A \otimes \varepsilon)\Delta = \mathrm{id}_A = (\varepsilon \otimes A)\Delta$,

$$(A \otimes \Delta)\Delta(a) \cdot \phi = \phi \cdot (\Delta \otimes A)\Delta(a) \in A \otimes A \otimes A$$

for all $a \in A$, and

$$(A \otimes A \otimes \Delta)(\phi) \cdot (\Delta \otimes A \otimes A)(\phi) = (1 \otimes \phi) \cdot (A \otimes \Delta \otimes A)(\phi) \cdot (\phi \otimes 1)$$

in $A \otimes A \otimes A \otimes A$.

The meaning of the definition is that the category $_A\mathcal{M}$ of A-modules over a quasibialgebra (A, ϕ) is a monoidal category. The tensor product of $V, W \in {}_A\mathcal{M}$

is formed just as in the case of ordinary bialgebras, by setting $a(v \otimes w) = a_{(1)}v \otimes a_{(2)}w$. The essential change compared to the bialgebra case is that the underlying functor $_A\mathcal{M} \to {}_k\mathcal{M}$ is no longer monoidal, but only a strict neutral tensor functor. This means that the associator morphism in $_A\mathcal{M}$ is not the same as the ordinary one for k-modules. Instead, one defines $\alpha\colon (U \otimes V) \otimes W \to U \otimes (V \otimes W)$ for $U, V, W \in {}_A\mathcal{M}$ as left multiplication by $\phi \in A \otimes A \otimes A$, that is $\alpha(u \otimes v \otimes w) = \phi^{(1)}u \otimes \phi^{(2)}v \otimes \phi^{(3)}w$, if we write formally $\phi = \phi^{(1)} \otimes \phi^{(2)} \otimes \phi^{(3)}$. One can check that the converse holds: Every structure of monoidal category for $_A\mathcal{M}$ for which the underlying functor is a strict neutral tensor functor arises from a quasibialgebra structure on A.

The dual notion was used first by Majid [28]: A coquasibialgebra H is a coalgebra equipped with a not necessarily associative multiplication $\nabla\colon H \otimes H \to H$, which is a coalgebra map, a grouplike element $1_H \in H$ which is a unit for ∇, and a convolution invertible trilinear form $\phi\colon H \otimes H \otimes H \to k$, the coassociator, satisfying $\phi(g \otimes 1 \otimes h) = \varepsilon(g)\varepsilon(h)$,

$$(f_{(1)}g_{(1)})h_{(1)}\phi(f_{(2)} \otimes g_{(2)} \otimes h_{(2)}) = \phi(f_{(1)} \otimes g_{(1)} \otimes h_{(1)})f_{(2)}(g_{(2)}h_{(2)}),$$

and

$$\phi(d_{(1)}f_{(1)} \otimes g_{(1)} \otimes h_{(1)})\phi(d_{(2)} \otimes f_{(2)} \otimes g_{(2)}h_{(2)})$$
$$= \phi(d_{(1)} \otimes f_{(1)} \otimes g_{(1)})\phi(d_{(2)} \otimes f_{(2)}g_{(2)} \otimes h_{(1)})\phi(f_{(3)} \otimes g_{(3)} \otimes h_{(2)})$$

for $d, f, g, h \in H$. For any coquasibialgebra (H, ϕ) the equations $\phi(1 \otimes g \otimes h) = \phi(g \otimes h \otimes 1) = \varepsilon(gh)$ hold for all $g, h \in H$.

If (H, ϕ) is a coquasibialgebra, then the category $^H\mathcal{M}$ has the following structure of a monoidal category: The tensor product of $V, W \in {}^H\mathcal{M}$ is $V \otimes W$ with the codiagonal comodule structure $v \otimes w \mapsto v_{(-1)}w_{(-1)} \otimes v_{(0)} \otimes w_{(0)}$ as in the case of an ordinary bialgebra, and the associator isomorphisms $\alpha\colon (U \otimes V) \otimes W \to U \otimes (V \otimes W)$ for $U, V, W \in {}^H\mathcal{M}$ are given by $\alpha(u \otimes v \otimes w) = \phi(u_{(-1)} \otimes v_{(-1)} \otimes w_{(-1)})u_{(0)} \otimes v_{(0)} \otimes w_{(0)}$. Conversely, if $^H\mathcal{M}$ is a monoidal category, and the underlying functor to $_k\mathcal{M}$ is a strict neutral tensor functor, then there are unique maps $\nabla\colon H \otimes H \to H$ and $\phi \in (H \otimes H \otimes H)^*$ such that $(v \otimes w)_{(-1)} \otimes (v \otimes w)_{(0)} = v_{(-1)}w_{(-1)} \otimes v_{(0)} \otimes w_{(0)}$ and $\alpha(u \otimes v \otimes w) = \phi(u_{(-1)} \otimes v_{(-1)} \otimes w_{(-1)})u_{(0)} \otimes v_{(0)} \otimes w_{(0)}$ hold for all $U, V, W \in {}^H\mathcal{M}$, $u \in U$, $v \in V$, and $w \in W$. With the structures ∇ and ϕ, H is a coquasibialgebra.

We note that for any coquasibialgebra $H = (H, \phi)$ there are opposite, coopposite, and biopposite coquasibialgebras $H^{\mathrm{op}}, H^{\mathrm{cop}}, H^{\mathrm{bop}}$ in which multiplication, comultiplication, or both are opposite, and the coassociators are given by $\phi^{\mathrm{op}}(f \otimes g \otimes h) = \phi^{-1}(h \otimes g \otimes f)$, $\phi^{\mathrm{cop}} = \phi^{-1}$, and $\phi^{\mathrm{bop}}(f \otimes g \otimes h) = \phi(h \otimes g \otimes f)$.

It goes without saying that coquasibialgebras form a category (which, incidentally, is even monoidal): There is an obvious notion of morphism between two coquasibialgebras (H, ϕ) and (H', ϕ'), namely, this should be a multiplicative, unit-preserving coalgebra map $f\colon H \to H'$ satisfying $\phi'(f \otimes f \otimes f) =$

$\phi\colon H^{\otimes 3} \to k$. If $f\colon H \to H'$ is a coquasibialgebra map, then the functor $^{f}\mathcal{M}\colon {}^{H}\mathcal{M} \to {}^{H'}\mathcal{M}$ is a strictly monoidal functor. Conversely, any strict monoidal functor $\mathcal{F}\colon {}^{H}\mathcal{M} \to {}^{H'}\mathcal{M}$ that commutes with the underlying functors to $_{k}\mathcal{M}$ is of this form. Thus maps of coquasibialgebras classify strict monoidal functors commuting with underlying functors, and this statement makes the need for a more relaxed notion obvious: There should be morphisms between coquasibialgebras that classify non-strict monoidal functors. The following definition of such morphisms is clearly folklore, although only the cases $F = \mathrm{id}$ or $\theta = \varepsilon$ appear to be in the literature:

DEFINITION 2.2.1. Let (H, ϕ) and (H', ϕ') be two coquasibialgebras. A coquasimorphism $(F, \theta)\colon (H, \phi) \to (H', \phi')$ consists of a unital coalgebra map $F\colon H \to H'$ and a convolution invertible $\theta\colon H \otimes H \to k$ satisfying

$$\theta(g_{(1)} \otimes h_{(1)})F(g_{(2)}h_{(2)}) = F(g_{(1)})F(h_{(1)})\theta(g_{(2)} \otimes h_{(2)})$$

and

$$\theta(f_{(1)} \otimes g_{(1)})\theta(f_{(2)}g_{(2)} \otimes h_{(1)})\phi(f_{(3)} \otimes g_{(3)} \otimes h_{(2)})$$
$$= \phi'(F(f_{(1)}) \otimes F(g_{(1)}) \otimes F(h_{(1)}))\theta(g_{(2)} \otimes h_{(2)})\theta(f_{(2)} \otimes g_{(3)}h_{(3)})$$

for all $f, g, h \in H$.

When $F\colon H \to H'$ is a coalgebra map between two coquasibialgebras (H, ϕ) and (H', ϕ'), then a bijection between forms $\theta\colon H \otimes H \to k$ making (F, θ) a coquasimorphism, and monoidal functor structures on $^{F}\mathcal{M}\colon {}^{H}\mathcal{M} \to {}^{H'}\mathcal{M}$ is given as follows: When (F, θ) is a coquasimorphism, then $^{(F,\theta)}\mathcal{M} := (^{F}\mathcal{M}, \xi)\colon {}^{H}\mathcal{M} \to {}^{H'}\mathcal{M}$ is a monoidal functor with the isomorphism $\xi\colon {}^{F}\mathcal{M}(V) \otimes {}^{F}\mathcal{M}(W) \to {}^{F}\mathcal{M}(V \otimes W)$ given, for $V, W \in {}^{H}\mathcal{M}$, by

$$V \otimes W \ni v \otimes w \mapsto \theta(v_{(-1)} \otimes w_{(-1)})v_{(0)} \otimes w_{(0)} \in V \otimes W.$$

If we define the composition of two coquasimorphisms $(F, \theta)\colon H \to H'$ and $(F', \theta')\colon H' \to H''$ by $(F', \theta')(F, \theta) = (F'F, \theta'(F \otimes F) * \theta)$, then we have

$$\left(^{(F',\theta')}\mathcal{M}\right)\left(^{(F,\theta)}\mathcal{M}\right) = {}^{(F',\theta')(F,\theta)}\mathcal{M}$$

as monoidal functors.

A special case of coquasimorphisms are cotwists dual to the twists defined in [14]: These can be considered as coquasimorphisms in which the underlying coalgebra map is the identity. Let (H, ∇, ϕ) be a coquasibialgebra, and $\theta\colon H \otimes H \to k$ a convolution invertible map satisfying $\theta(1 \otimes h) = \theta(h \otimes 1) = \varepsilon(h)$ for all $h \in H$. Then there is a unique coquasibialgebra structure $(H, \nabla, \phi)^{\theta} = (H^{\theta}, \nabla^{\theta}, \phi^{\theta})$ on the coalgebra $H^{\theta} := H$ such that $(\mathrm{id}_{H}, \theta)\colon H \to H^{\theta}$ is a coquasimorphism: One can easily solve the equations defining a coquasimorphism for ∇^{θ} and ϕ^{θ}. By the above, a cotwist $(\mathrm{id}_{H}, \theta)$ gives rise to a monoidal equivalence $\mathcal{I}d\colon {}^{H}\mathcal{M} \to {}^{H^{\theta}}\mathcal{M}$, whose monoidal functor structure is induced by θ.

Note that every coquasimorphism $(F, \theta): H \to H'$ in which θ is invertible factors into a cotwist and a coquasibialgebra map:

$$(F, \theta) = \left(H \xrightarrow{(\mathrm{id}_H, \theta)} H^\theta \xrightarrow{(F, \varepsilon)} H' \right).$$

If H is a bialgebra and $\theta: H \otimes H \to k$ is a cotwist, then H^θ is a bialgebra (with trivial coassociator) if and only if $\theta(f_{(1)} \otimes g_{(1)})\theta(f_{(2)}g_{(2)} \otimes h) = \theta(g_{(1)} \otimes h_{(1)})\theta(f \otimes g_{(2)}h_{(2)})$ holds for all $f, g, h \in H$. We say that θ is a (normalized) two-cocycle on H. If H is a Hopf algebra, then any two-cocycle (co)twist H^θ is also a Hopf algebra. Cotwists by two-cocycles appear in [12].

The dual (and older, cf. [14]) notion to a cotwist is the twist of a quasibialgebra H by an invertible element $t \in H \otimes H$. We will only need the case where H is an ordinary bialgebra, and t is a two-cycle, that is, satisfies $t_{23}(\mathrm{id} \otimes \Delta)(t) = t_{12}(\Delta \otimes \mathrm{id})(t) \in H \otimes H \otimes H$, where $t_{23} = 1 \otimes t$ and $t_{12} = t \otimes 1$, and $(\varepsilon \otimes \mathrm{id})(t) = (\mathrm{id} \otimes \varepsilon)(t) = 1$. Then H_t is a bialgebra, with underlying algebra the same as H, and comultiplication $\Delta_t(h) = t\Delta(h)t^{-1}$. If H is a Hopf algebra, then so is H_t. Note that if H is finite, then t is a two-cycle if and only if t, considered as a map $H^* \otimes H^* \to k$, is a two-cocycle on H^*.

There is also a way of (co)twisting coquasimorphisms: If H and H' are coquasibialgebras, $(F, \theta): H \to H'$ is a coquasimorphism, and $t: H \to k$ satisfies $t(1) = 1$, then $(F, \theta)^t := (F^t, \theta^t): H \to H'$, defined by

$$\theta^t(g \otimes h) = t(g_{(1)})t(h_{(1)})\theta(g_{(2)} \otimes h_{(2)})t^{-1}(g_{(3)}h_{(3)}),$$
$$F^t(h) = t(h_{(1)})F(h_{(2)})t^{-1}(h_{(3)})$$

for $g, h \in H$, is also a coquasimorphism, called the cotwist of (F, θ) by t. This procedure has its interpretation in terms of the comodule category as well: Two coquasomorphisms are each other's cotwists if and only if the corresponding monoidal functors are isomorphic as monoidal functors.

A coquasibialgebra is called a coquasi-Hopf algebra if it has a coquasiantipode. By definition, this is a triple (S, β, γ) consisting of a coalgebra antiautomorphism $S: H \to H$ and linear forms $\beta, \gamma \in H^*$ such that

$$h_{(1)}\beta(h_{(2)})S(h_{(3)}) = \beta(h)1_H,$$
$$S(h_{(1)})\gamma(h_{(2)})h_{(3)} = \gamma(h)1_H,$$
$$\phi(S(h_{(1)}) \otimes \gamma(h_{(2)})h_{(3)}\beta(h_{(4)}) \otimes S(h_{(5)})) = \varepsilon(h),$$
$$\phi^{-1}(h_{(1)} \otimes \beta(h_{(2)})S(h_{(3)})\gamma(h_{(4)}) \otimes h_{(5)}) = \varepsilon(h)$$

hold for all $h \in H$. The meaning of the definition (whose dual, a quasi-Hopf algebra, appears in [14]) is that — just as for ordinary Hopf algebras — the category of k-finitely generated projective H-comodules is left and right rigid. The right dual of $V \in {}^H\mathcal{M}_f$ is V^* equipped with the comodule structure defined by

$\varphi_{(-1)}\varphi_{(0)}(v) = S(v_{(-1)})\varphi(v_{(0)})$ for all $v \in V$ and $\varphi \in V^*$, and the maps

$$\mathrm{ev} \colon V \otimes V^* \ni v \otimes \varphi \mapsto \beta(v_{(-1)})\varphi(v_{(0)}) \in k,$$
$$\mathrm{db} \colon k \ni 1 \mapsto v^i \otimes \gamma(v_{i(-1)})v_{i(0)} \in V^* \otimes V.$$

The first two equations in the definition of a quasi-antipode express colinearity of the maps ev and db, while the other two say that the maps fulfill the axioms for a dual object in the comodule category.

Conversely, if we are given a coquasibialgebra H such that the category of finite projective H-comodules is left and right rigid, then we would like to be able to reconstruct a coquasi-antipode for H; we can only hope for this to work if k is a field or H itself is finite; otherwise finite comodules give us too little information on H. This reconstruction result for coquasi-Hopf algebras is indeed claimed to hold, at least over a field, in [29, Sec. 9.4.1]. In fact the same belief has been held (independently, as it were) by the author for considerable time, but we shall present a counterexample in subsection 4.5. The analogous statement for ordinary bialgebras was proved by Ulbrich [59]. The key difference between the cases is that monoidal functors (such as the underlying functor from the comodule category over a bialgebra to the category of vector spaces) preserve duals, while incoherent tensor functors need not. Majid's proof relies on an (explicitly given, [29, (9.37)]) isomorphism between the dual object of an H-comodule V in the monoidal category of H-comodules (which is assumed to exist) and the dual vector space of V. However, we shall see that the dual object of V needs not even have the same dimension as V. Nevertheless, we shall without giving further details refer to [29] for the rest of the proof of the following fact: If H is a coquasibialgebra over a field k, if the category $^H\mathcal{M}_f$ is left and right rigid, and if the underlying functor $^H\mathcal{M}_f \to {}_k\mathcal{M}$ preserves duals, then H has a coquasiantipode.

2.3. Coendomorphism coalgebras. In the preceding section we have discussed how to get from a (coquasi)bialgebra to a monoidal category, and stated without proof that one can go the other way. In this section we shall discuss in a little more detail how the reconstruction process works. This is largely an elaboration of work of Ulbrich [59; 58]; see also [29; 21; 41; 40].

Assume first that we are given a coalgebra C, and consider the underlying functor $\mathcal{U} \colon \mathcal{M}^C \to {}_k\mathcal{M}$. When we want to reconstruct properties of C (like being a bialgebra), the key fact is that the coaction of C on each of its comodules defines a natural transformation $\delta \colon \mathcal{U} \to \mathcal{U} \otimes C$ which has a universal property: For every k-module T, every natural transformation $\phi \colon \mathcal{U} \to \mathcal{U} \otimes T$ factors as $\phi = (\mathcal{U} \otimes f)\delta$ for a unique k-linear map $f \colon C \to T$. This describes an isomorphism

$$\mathrm{Hom}(C, T) \ni f \mapsto (\mathcal{U} \otimes f)\delta \in \mathrm{Nat}(\mathcal{U}, \mathcal{U} \otimes T),$$

whose inverse maps a natural transformation ψ to $(\varepsilon \otimes T)\psi_C$.

When we are given an abstract monoidal category $\mathcal{C}$ and want to reconstruct a Hopf algebra H such that $\mathcal{C}$ is the category of modules (or comodules) over H, the least additional data we need is a specified functor $\omega\colon \mathcal{C} \to {}_k\mathcal{M}$ to the category of k-modules, which will play the role of the underlying functor above. Then a coendomorphism coalgebra of ω is by definition a k-module $\operatorname{coend}(\omega)$ together with a universal natural transformation $\delta\colon \omega \to \omega \otimes \operatorname{coend}(\omega)$ giving an isomorphism

$$\operatorname{Hom}(\operatorname{coend}(\omega), T) \ni f \mapsto (\omega \otimes f)\delta \in \operatorname{Nat}(\omega, \omega \otimes T)$$

for every k-module T. We have seen that any coalgebra C can be recovered as the coendomorphism coalgebra of the underlying functor $\mathcal{M}^C \to {}_k\mathcal{M}$. If k is a field, then C can equally well be recovered from the underlying functor $\mathcal{M}_f^C \to {}_k\mathcal{M}$, due to the finiteness theorem for comodules. This is one of the reasons why coendomorphism coalgebras are often to be preferred over just taking the endomorphism algebra of ω, which we can always do if we are not too worried about set theoretical complications, and relating the category $\mathcal{C}$ to the modules over $\operatorname{End}(\omega)$. Another reason is that if ω has a coendomorphism coalgebra, then the natural morphism

$$\operatorname{Hom}(\operatorname{coend}(\omega) \otimes M, T) \cong \operatorname{Nat}(\omega \otimes M, \omega \otimes T) \qquad (2\text{--}1)$$
$$f \mapsto (\omega \otimes f)(\delta \otimes M)$$

is an isomorphism for all $M \in {}_k\mathcal{M}$, by the slightly sketchy calculation

$$\operatorname{Hom}(\operatorname{coend}(\omega) \otimes M, T) \cong \operatorname{Hom}(M, \operatorname{Hom}(\operatorname{coend}(\omega), T))$$
$$\cong \operatorname{Hom}(M, \operatorname{Nat}(\omega, \omega \otimes T)) \cong \operatorname{Nat}(\omega \otimes M, \omega \otimes T).$$

This generalization of the defining property of coend is important when we try to reconstruct additional properties of $\operatorname{coend}(\omega)$ from properties of $\mathcal{C}$ and ω, such as being monoidal.

The basis for such reconstructions is the universal property of $\operatorname{coend}(\omega)$ and the natural transformation δ. The universal property first defines a unique coalgebra structure Δ for $\operatorname{coend}(\omega)$ such that the natural transformation δ endows every $\omega(X)$ with a $\operatorname{coend}(\omega)$-comodule structure. This defines a functor $\hat{\omega}\colon \mathcal{C} \to \mathcal{M}^{\operatorname{coend}(\omega)}$ that lifts ω as in the following diagram

$$
\begin{array}{ccc}
\mathcal{C} & \xrightarrow{\;\hat{\omega}\;} & \mathcal{M}^{\operatorname{coend}(\omega)} \\
& {}_\omega\searrow \quad \swarrow{}_{\mathcal{U}} & \\
& {}_k\mathcal{M} &
\end{array}
\qquad (2\text{--}2)
$$

in which $\mathcal{U}$ denotes the underlying functor. The functor $\hat{\omega}$ in its turn has a universal property: Every other functor $\mathcal{F}\colon \mathcal{C} \to \mathcal{M}^C$ compatible with the underlying functor as in the outer triangle of the following diagram

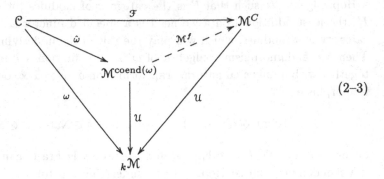

$$(2\text{-}3)$$

factors through a dashed arrow induced by a coalgebra map $f\colon \mathrm{coend}(\omega) \to C$, which is the unique k-linear map making all the diagrams

$$
\begin{array}{ccc}
\omega(X) & \xrightarrow{\ \delta\ } & \omega(X) \otimes \mathrm{coend}(\omega) \\
\Vert & & \downarrow{\mathrm{id}\otimes f} \\
U\mathcal{F}(X) & \xrightarrow{\ \delta\ } & U\mathcal{F}(X) \otimes C
\end{array}
$$

commute. In particular, if $\omega\colon \mathcal{C} \to {}_k\mathcal{M}$ as well as $\nu\colon \mathcal{D} \to {}_k\mathcal{M}$ have coendomorphism coalgebras, then every functor $\mathcal{F}\colon \mathcal{C} \to \mathcal{D}$ satisfying $\nu\mathcal{F} = \omega$ induces a coalgebra map $\mathrm{coend}(\mathcal{F})\colon \mathrm{coend}(\omega) \to \mathrm{coend}(\nu)$ making the diagram

$$
\begin{array}{ccc}
\mathcal{C} & \xrightarrow{\ \mathcal{F}\ } & \mathcal{D} \\
\hat{\omega}\downarrow & & \downarrow\hat{\nu} \\
\mathcal{M}^{\mathrm{coend}(\omega)} & \xrightarrow{\ \mathcal{M}^f\ } & \mathcal{M}^{\mathrm{coend}(\nu)}
\end{array}
$$

commute, and f is the unique k-linear map making all the diagrams

$$
\begin{array}{ccc}
\omega(X) & \xrightarrow{\ \delta\ } & \omega(X) \otimes \mathrm{coend}(\omega) \\
\Vert & & \downarrow{\mathrm{id}\otimes f} \\
\nu\mathcal{F}(X) & \xrightarrow{\ \delta\ } & \nu\mathcal{F}(X) \otimes \mathrm{coend}(\nu)
\end{array}
$$

commute

The reconstruction process for morphisms needs a slight refinement, since one usually finds the outer triangle in (2-3) to commute only up to a specified isomorphism $\zeta\colon \omega \to U$. However, it is sufficiently general as stated to distill a bialgebra structure on $\mathrm{coend}(\omega)$ out of a monoidal category $\mathcal{C}$ and *strict* monoidal underlying functor ω: We have to apply the procedure to the functor $\otimes\colon \mathcal{C} \times \mathcal{C} \to \mathcal{C}$, with respect to the functors

$$\omega \otimes \omega\colon \mathcal{C} \times \mathcal{C} \ni (X, Y) \mapsto \omega(X) \otimes \omega(Y) \in {}_k\mathcal{M},$$

and $\omega\colon \mathcal{C} \to {}_k\mathcal{M}$, which should work since ω being strictly monoidal means a commutative triangle

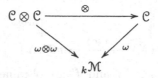

We only have to know what coalgebra is attached to the functor $\omega \otimes \omega$. More generally, for $\omega \otimes \nu\colon \mathcal{C} \otimes \mathcal{D} \to {}_k\mathcal{M}$ the stronger property (2–1) of the coendomorphism coalgebra allows us to calculate

$$\mathrm{Nat}(\omega \otimes \nu, \omega \otimes \nu \otimes T) \cong \mathrm{Nat}(\omega, \mathrm{Nat}(\nu, \omega \otimes \nu \otimes T))$$
$$\cong \mathrm{Nat}(\omega, \mathrm{Hom}(\mathrm{coend}(\nu), \omega \otimes T))$$
$$\cong \mathrm{Nat}(\omega \otimes \mathrm{coend}(\nu), \omega \otimes T)$$
$$\cong \mathrm{Hom}(\mathrm{coend}(\omega) \otimes \mathrm{coend}(\nu), T),$$

so that in particular $\mathrm{coend}(\omega \otimes \omega) = \mathrm{coend}(\omega) \otimes \mathrm{coend}(\omega)$. Thus a strict monoidal $\omega\colon \mathcal{C} \to {}_k\mathcal{M}$ gives rise to a coalgebra map $\nabla\colon \mathrm{coend}(\omega) \otimes \mathrm{coend}(\omega) \to \mathrm{coend}(\omega)$, and ∇ makes $\mathrm{coend}(\omega)$ a bialgebra such that (2–2) is a commutative diagram of monoidal functors. To see that ∇ is associative, we only need to look at the diagram

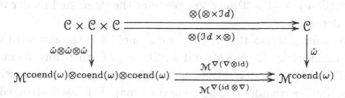

which commutes both with the upper and the lower horizontal arrows. But the two arrows at the top are identical, so ∇ is associative.

The reconstruction of a multiplication ∇ on $\mathrm{coend}(\omega)$ can be carried over directly to the case where ω is a strict incoherent tensor functor, but the last step fails in this situation. But it still follows, as first observed by Majid [28], that there is a unique coquasibialgebra structure on $\mathrm{coend}(\omega)$ such that ω lifts to a monoidal functor $\hat{\omega}$ making (2–2) a commutative diagram of incoherent tensor functors. To see this, one has to reconstruct a trilinear form $\phi\colon \mathrm{coend}(\omega)^{\otimes 3} \to k$ from the associativity isomorphism in $\mathcal{C}$. More generally, let $\mathcal{F}, \mathcal{F}'\colon \mathcal{C} \to \mathcal{M}^C$ be two functors with $\mathcal{UF} = \omega = \mathcal{UF}'$ and let $f, f'\colon \mathrm{coend}(\omega) \to C$ be the corresponding coalgebra maps. Assume further that there is some isomorphism $\alpha\colon \mathcal{F} \to \mathcal{F}'$. Then $\mathcal{U}(\alpha)$ is an endomorphism of ω, which corresponds to a linear form $\phi\colon \mathrm{coend}(\omega) \to k$ by the universal property of $\mathrm{coend}(\omega)$. Note that if $\mathcal{U}(\alpha) = \mathrm{id}$, then $\phi = \varepsilon$. One can show that $f * \phi = \phi * f'$. In particular, if $\mathcal{F}, \mathcal{F}'\colon \mathcal{C} \to \mathcal{D}$ are functors with $\nu\mathcal{F} = \omega = \nu\mathcal{F}'$, and $\alpha\colon \mathcal{F} \to \mathcal{F}'$ is a natural isomorphism, then there is a unique (invertible) linear form ϕ on $\mathrm{coend}(\omega)$ making the diagrams

$$\omega(X) \xrightarrow{\delta} \omega(X) \otimes \text{coend}(\omega)$$

$$\Big\| \qquad\qquad\qquad \downarrow \text{id} \otimes \phi$$

$$\nu\mathcal{F}(X) \xrightarrow{\nu(\alpha)} \nu\mathcal{F}'(X)$$

commute, and $f * \phi = \phi * f'$, if f and f' are obtained from $(\mathcal{F}, \zeta)$ and $(\mathcal{F}', \zeta')$, respectively. A coquasibialgebra structure can be reconstructed by applying this to the two functors $\mathcal{F}, \mathcal{F}' \colon \mathcal{C}^3 \to \mathcal{C}$ composed from tensor product, and the associativity isomorphism α between them. The defining equations of a coquasibialgebra can be deduced from the coherence axioms for α. In fact there is a rather general principle why axioms for categories, functors, and natural transformations give rise to 'analogous' axioms for coalgebras, coalgebra maps, and linear forms. The reconstruction is a two-functor assigning, on the object level, coalgebras to categories with functors to $_k\mathcal{M}$; on the morphism level, coalgebra maps to functors commuting with underlying functors; on the two-morphism level, linear forms to natural transformations. The two-functor laws say that composition of functors corresponds to compositon of coalgebra maps, and that composition of natural transformations corresponds to convolution of linear forms. The reconstruction two-functor is even monoidal: we have seen that it assigns the tensor product of $\text{coend}(\omega)$ and $\text{coend}(\nu)$ to a 'tensor product' $(\mathcal{C}, \omega) \otimes (\mathcal{D}, \nu) := (\mathcal{C} \times \mathcal{D}, \omega \otimes \nu)$. We refer the interested reader to [40, Sec. 2] for more details.

Similar reasoning shows that if $\mathcal{C}, \mathcal{D}$ are monoidal categories with strict neutral tensor functors $\omega \colon \mathcal{C} \to {}_k\mathcal{M}$ and $\nu \colon \mathcal{D} \to {}_k\mathcal{M}$, then any monoidal functor $(\mathcal{F}, \xi) \colon \mathcal{C} \to \mathcal{D}$ such that $\nu\mathcal{F} = \omega$ as functors, yields a coquasimorphism $(F, \theta) \colon \text{coend}(\omega) \to \text{coend}(\nu)$: The coalgebra map F is reconstructed from the functor $\mathcal{F}$, and the bilinear form θ is reconstructed from the isomorphism

$$\xi \colon \otimes \circ (\mathcal{F} \times \mathcal{F}) \to \mathcal{F} \circ \otimes \colon \mathcal{C} \otimes \mathcal{C} \to \mathcal{D}.$$

As a particular case, we see how changing the tensor functor structure of a functor $\omega \colon \mathcal{C} \to {}_k\mathcal{M}$ affects the coquasibialgebra attached to it: If ξ, ξ' are two tensor functor structures for ω, and if H, H' are the coquasibialgebra structures on $\text{coend}(\omega)$ obtained from them, then one can regard the identity functor with the trivial monoidal functor structure as the functor $(\mathcal{F}, \beta)$ giving rise to a coquasimorphism $(F, \theta) \colon H \to H'$, in which F is the identity; in short, changing the tensor functor structure of ω amounts to changing $\text{coend}(\omega)$ by a cotwist.

It remains to discuss some cases where coendomorphism coalgebras exist. First of all, any coalgebra C is the coendomorphism coalgebra of the underlying functor $\mathcal{M}^C \to {}_k\mathcal{M}$. If k is a field, then C is also the coendomorphism coalgebra of the underlying functor from the category of finite dimensional comodules over C. Generally, if $\omega \colon \mathcal{C} \to {}_k\mathcal{M}$ is a functor that takes values in finitely generated projective k-modules, then ω has a coendomorphism coalgebra. If k is a field, $\mathcal{C}$

is a k-linear abelian category, and ω is an exact faithful k-linear functor taking values in finite dimensional vector spaces, then the functor $\hat{\omega}\colon \mathcal{C} \to \mathcal{M}^{\mathrm{coend}(\omega)}$ in (2–2) induces an equivalence with the category of finite dimensional coend(ω)-comodules.

Even when general reasons guarantee the existence of a coendomorphism coalgebra, it may be hard to figure out its structure. The following observation will prove useful to justify an educated guess. The situation we have in mind is that we can guess the coalgebra coend(ω) and even show that we have a category equivalence $\mathcal{C} \cong \mathcal{M}^{\mathrm{coend}(\omega)}$ — except we don't know that our candidate for coend(ω) is really a coassociative coalgebra. The rather simle idea is that, dually, if we have a nonassociative algebra that has a faithful associative representation, then the algebra is associative after all.

LEMMA 2.3.1. *Let C be a k-module endowed with maps $\Delta\colon C \to C \otimes C$ and $\varepsilon\colon C \to k$. We define the category $\mathcal{M}^C$ of C-comodules to have objects all pairs (V, ρ) with $V \in {}_k\mathcal{M}$ and $\rho\colon V \to V \otimes C$ satisfying $(V \otimes \Delta)\rho = (\rho \otimes C)\rho\colon V \to V \otimes C \otimes C$.*

If for every $c \in C$ there exist $V \in \mathcal{M}^C$, elements $v_i \in V$, and linear forms $\varphi_i \in V^*$ such that $c = \sum_i (\varphi_i \otimes C)\delta(v_i)$, then C is a coalgebra.

PROOF. For any $V \in \mathcal{M}^C$ with "comodule" structure map δ the diagram

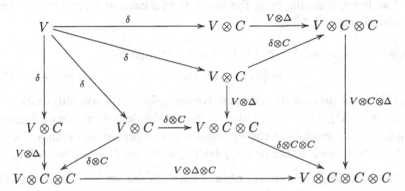

shows that for $f = (\Delta \otimes C)\Delta - (C \otimes \Delta)\Delta\colon C \to C \otimes C \otimes C$ at least the equation $(V \otimes f)\delta = 0$ holds. But this entails, for $c \in C$ and V, v_i, φ_i as above, that $f(c) = f(\sum(\varphi_i \otimes \mathrm{id})\delta(v_i)) = \sum(\varphi_i \otimes \mathrm{id})(V \otimes f)\delta(v_i) = 0$. Hence C is a coassociative coalgebra, and ε is a counit by a similar reasoning. $\square$

The following Lemma is useful when we try to reconstruct a coalgebra over a field k from its finite dimensional comodules. One should think of the case where the finite dual A° of an algebra A is found as a coalgebra which is a subset of A^* with comultiplication the restriction of the map $A^* \to (A \otimes A)^*$ dual to multiplication. The subset A° consists of all the entries of the finite matrix representations of A, or of all the elements in A^* that "occur" in the maps $V \to V \otimes A^*$ associated to finite dimensional A-modules V.

LEMMA 2.3.2. *Let k be a field, C^1 and C^2 two vector spaces, $\iota: C^1 \otimes C^1 \to C^2$ an injection, $\Delta^1: C^1 \to C^2$ and $\varepsilon^1: C^1 \to k$ linear maps.*

Define the category $\mathcal{M}^{C^1}$ of right C^1-comodules to have as objects vector spaces V equipped with a map $\rho^1: V \to C^1$ such that the diagrams

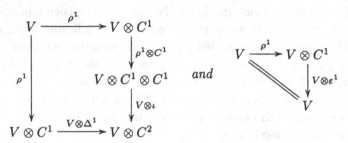

commute.

For a C^1-comodule V, element $v \in V$, and linear form $\varphi \in V^*$ put $[\varphi|v] := \langle \varphi, v_{(0)} \rangle v_{(1)} \in C^1$, where $v_{(0)} \otimes v_{(1)} = \rho^1(v)$.

Put $C := k\text{-}\mathrm{span}\{[\varphi|v] | V \in \mathcal{M}_f^{C^1}, v \in V, \varphi \in V^*\}$.

Then C is a coalgebra, and a category equivalence $\mathcal{M}_f^C \to \mathcal{M}_f^{C^1}$ is induced by the inclusion $C \subset C^1$.

PROOF. We have $\Delta^1(C) \subset \iota(C \otimes C)$, since for every $V \in \mathcal{M}_f^{C^1}$, $v \in V$, and $\varphi \in V^*$ we have, denoting by v_i the elements of a basis of V, and v^i the elements of the dual basis of V^*:

$$\Delta^1([\varphi|v]) = \langle \varphi, v_{(0)} \rangle \Delta^1(v_{(1)}) = \langle \varphi, v_{(0)(0)} \rangle \iota(v_{(0)(1)} \otimes v_{(1)})$$
$$= \langle \varphi, v_{i(0)} \rangle \langle v^i, v_{(0)} \rangle \iota(v_{i(1)} \otimes v_{(1)}) = \iota([\phi|v_i] \otimes [v^i|v])$$

Thus, C is a not necessarily coassociative coalgebra with comultiplication Δ defined by $\iota\Delta = \Delta^1|_C$, and counit $\varepsilon|_C$. Moreover, by definition every C^1-comodule is actually a C-comodule in the sense used in the preceding Lemma. By definition of C, the preceding Lemma applies to show that C is coassociative. $\square$

2.4. Coquasibialgebras and cohomology.

So far we have dealt with the interpretation of coquasibialgebras in terms of monoidal categories of comodules. There is another, cohomological interpretation coming from the cocommutative case. A cocommutative coquasibialgebra is the same as an ordinary bialgebra, equipped in addition with a three-cocycle of Sweedler cohomology [54]: The modified associativity in a coquasibialgebra specializes to ordinary associativity in the cocommutative case, and the equation corresponding to Mac Lane's pentagonal axiom is precisely the condition for $\phi: H \otimes H \otimes H \to k$ to be a Sweedler three-cocycle with coefficients in the trivial H-module algebra k. A coquasimorphism $(F, \theta): H \to H'$ between cocommutative coquasibialgebras (H, ϕ) and (H', ϕ') is just an ordinary bialgebra homomorphism F, with, in addition, a normalized invertible $\theta: H \otimes H \to k$ such that its Sweedler coboundary is the quotient $\phi'(F \otimes F \otimes F) * \phi^{-1}$. In particular, cotwisting a cocommutative coquasibialgebra does not affect its multiplication, but only its coassociator, and two coassociators for a cocommutative bialgebra H are cotwists of each other if and only if they are cohomologous Sweedler cocycles. Finally, if $(F, \theta): H \to H'$ is a coquasi-morphism between cocommutative coquasibialgebras, and $t \in H^*$ is normalized and convolution invertible, then $(F, \theta)^t = (F^t, \theta^t)$ has $F^t = F$, and θ^t differs from θ by the Sweedler coboundary of t. Motivated by this, we shall, for non-cocommutative coquasibialgebras as well, say that two coquasimorphisms (F, θ) and (F', θ') are cohomologous if there exists t with $(F', \theta') = (F, \theta)^t$. As a special case, we have a notion of when two cotwists on a coquasibialgebra H are coho-mologous — but usually multiplication of cotwists, while trivially well-defined (it is just convolution), loses its meaning in the non-cocommutative case, since a cotwist will usually have a codomain different from its domain. However, we can consider the group of cohomology classes of "self-twists" $(\mathrm{id}, \theta): (H, \phi) \to (H, \phi)$, that is, of twists that do not affect the coquasibialgebra (H, ϕ) they act on. The conditions for this to happen are easily worked out: If the cocycle ϕ is trivial, they mean that θ is a two-cocycle (this is the condition for ϕ^θ to be trivial if ϕ is trivial), but also that

$$\theta(g_{(1)} \otimes h_{(1)})g_{(2)}h_{(2)} = g_{(1)}h_{(1)}\theta(g_{(2)} \otimes h_{(2)})$$

holds for all $g, h \in H$ (this is the condition that twisted multiplication is the same as the original one). We will call such two-cocycles *central*, and denote the group of cotwisting classes of central two-cocycles by $\mathcal{H}^2_c(H)$. We will use the same notation for the group of self-twists of a coquasibialgebra H with nontrivial coassociator.

2.5. Hopf modules.
The most conceptual definition of a Hopf module is that it is a module (or comodule) over a certain algebra (or coalgebra) within a certain monoidal category. Although all Hopf modules in this paper will be of this type, it should be noted that there are meaningful Hopf modules that do not fit easily into this scheme, namely Doi–Koppinen Hopf modules [11; 24], and more generally entwined modules [7]. An entwining structure consists of an algebra A, a coalgebra C, and a map $\psi: C \otimes A \to A \otimes C$ satisfying the identities $\psi(C \otimes \nabla) = (\nabla \otimes C)(A \otimes \psi)(\psi \otimes A)$ and $(A \otimes \Delta)\psi = (\psi \otimes C)(C \otimes \psi)(\Delta \otimes A)$ along with some identities involving (co)units. An entwined module is an A-module and C-comodule, with the module structure μ and comodule structure ρ satisfying a compatibility condition governed by ψ, namely $\rho\mu = (\mu \otimes C)(M \otimes \psi)(\rho \otimes A): M \otimes A \to M \otimes C$. Doi–Koppinen Hopf modules can be seen as special cases: When C is an H-module coalgebra and A is an H-comodule algebra, then an entwining structure can be defined by $\psi(c \otimes a) = a_{(0)} \otimes c \leftharpoonup a_{(1)}$; a Doi–Koppinen Hopf module is by definition a right A-module as well as right C-comodule satisfying $(ma)_{(0)} \otimes (ma)_{(1)} = m_{(0)}a_{(0)} \otimes m_{(1)} \leftharpoonup a_{(1)} \in M \otimes C$ for all $m \in M$ and $a \in A$.

We will now turn to the more conceptual versions of Hopf modules defined in terms of monoidal categories. This is based on the theory of algebras within monoidal categories, as developed in [36].

Suppose that $\mathcal{C}$ is a monoidal category. There is a natural notion of associative algebra in $\mathcal{C}$. This is most obvious if $\mathcal{C}$ is strictly monoidal. Then an algebra (A, ∇) within $\mathcal{C}$ consists of an object A in $\mathcal{C}$, equipped with a multiplication $\nabla \colon A \otimes A \to A$ satisfying associativity $\nabla(\nabla \otimes A) = \nabla(A \otimes \nabla) \colon A \otimes A \otimes A \to A$, and admitting a unit $\eta \colon I \to A$ satisfying $\nabla(A \otimes \eta) = \mathrm{id}_A = \nabla(\eta \otimes A) \colon A \to A$.

When $(\mathcal{C}, \otimes, \alpha)$ is a no longer strict monoidal category, an algebra (A, ∇, η) in $\mathcal{C}$ is an object A together with a multiplication ∇ and unit η as above, now satisfying associativity in the form

$$\nabla(\nabla \otimes A) = \nabla(A \otimes \nabla)\alpha \colon (A \otimes A) \otimes A \to A$$

and the unit conditions as above. Obviously the instance of α was introduced into the associativity condition as the only conceivable way of making it make sense; one cannot compare $\nabla(A \otimes \nabla)$ and $\nabla(\nabla \otimes A)$ directly as they have distinct domains, but the domains are isomorphic via the morphism α making part of the definition of the monoidal category $\mathcal{C}$.

There are more instances where one has an obvious notion generalizing a well-known concept of ring and module theory to ring and module theory within a *strict* monoidal category, and a perhaps slightly less obvious version in a non-strict monoidal category, obtained by adorning the strict version with suitable copies of α. We will tacitly maintain each time that it is clear how this has to be done. Although this is quite informal, at least a hint may be given as to how one might try to formalize the idea: Each monoidal category $\mathcal{C}$ is equivalent as a monoidal category to a strict monoidal category $\hat{\mathcal{C}}$. Thus we may define an algebra in $\mathcal{C}$ to be a pair (A, ∇) which is mapped by that monoidal equivalence to an algebra in $\hat{\mathcal{C}}$. However, we should stress once again that there appears to be no rigorous metatheorem in the literature that would allow us to work only in the strict case as long as a certain type of notions and statements is concerned.

Given an algebra A in a monoidal category $\mathcal{C}$, a right A-module is an object M of $\mathcal{C}$ with a morphism $\mu \colon M \otimes A \to M$ satisfying $\mu(M \otimes \nabla) = \mu(\mu \otimes A) \colon M \otimes A \otimes A \to M$. The notion of a left A-module is analogous. Given two algebras A, B, an A-B-bimodule is a triple (M, μ_ℓ, μ_r) such that (M, μ_ℓ) is a left A-module, (M, μ_r) is a right B-module, and $\mu_r(\mu_\ell \otimes B) = \mu_\ell(A \otimes \mu_r) \colon A \otimes M \otimes B \to M$. We denote the categories of left (right, bi-) modules by ${}_A\mathcal{C}, \mathcal{C}_A, {}_A\mathcal{C}_B$.

For any algebra A in $\mathcal{C}$, the underlying functor $\mathcal{C}_A \to \mathcal{C}$ has a left adjoint; that is, there exists a free right A-module over any object V of $\mathcal{C}$. This can be constructed as $V \otimes A$, equipped with the right A-module structure $V \otimes \nabla \colon V \otimes A \otimes A \to V \otimes A$. More generally, if B is another algebra, then the underlying functor ${}_A\mathcal{C}_B \to {}_A\mathcal{C}$ has a left adjoint assigning to a left A-module N the free A-B-bimodule over N, which can be constructed as $N \otimes B$ with right B-module structure $N \otimes \nabla$ and left A-module structure $\mu \otimes B$, where μ is the A-module structure of N. If the category $\mathcal{C}$ has coequalizers, then one can define the tensor product of $M \in \mathcal{C}_A$ and $N \in {}_A\mathcal{C}$ by the coequalizer

$$M \otimes A \otimes N \underset{M \otimes \mu}{\overset{\mu \otimes N}{\rightrightarrows}} M \otimes N \longrightarrow M \underset{A}{\otimes} N$$

Provided that the tensor product in $\mathcal{C}$ preserves coequalizers, one can endow the tensor product $M \underset{A}{\otimes} N$ of $M \in {}_B\mathcal{C}_A$ and $N \in {}_A\mathcal{C}_R$ with an A-R-bimodule structure. In particular, the category ${}_A\mathcal{C}_A$ of A-A-bimodules in $\mathcal{C}$ is a monoidal category. We note [36, sec. 3] that if $\mathcal{C}$ is closed and has equalizers, then the category ${}_A\mathcal{C}_A$ is closed as well. If $M = V \otimes A$ is the free right A-module generated by $V \in \mathcal{C}$, then for any $N \in {}_A\mathcal{C}$ we have a canonical isomorphism $M \underset{A}{\otimes} N \cong V \otimes N$.

If the category $\mathcal{C}$ is the category of representations of a bialgebra H, there is standard terminology for the concepts above (see e. g. [34]): An algebra A in $\mathcal{M}^H$ is called an H-comodule algebra, A right A-module M in the category $\mathcal{M}^H$ is called a (relative) Hopf module; we'll denote the category of these by $\mathcal{M}^H_A$, with obvious variants ${}_A\mathcal{M}^H$, ${}_A\mathcal{M}^H_A$, etc. A coalgebra in the monoidal category of right H-modules, which is supposed to mean an algebra in the opposite category, is called an H-module coalgebra, and a comodule over C in $\mathcal{M}_H$ is also called a relative Hopf module, belonging to the category $\mathcal{M}^C_H$. The first Hopf modules in the literature are examples of both situations at the same time: If H is a bialgebra, then the regular right coaction of H on itself makes H an H-comodule algebra, and the regular right action of H on itself makes it a right H-module coalgebra. Both definitions of a Hopf module in $\mathcal{M}^H_H$ as given above agree in this situation. More generally, if $\iota \colon K \to H$ is a bialgebra map, which we consider an inclusion for simplicity, then H is a K-module coalgebra, and K an H-comodule algebra, via ι. Both notions of a Hopf module in $\mathcal{M}^H_K$ defined in these situations agree: Such an object is a right K-module as well as right H-comodule M satisfying $(mx)_{(0)} \otimes (mx)_{(1)} \in M \otimes H$ for all $m \in M$ and $x \in K$.

The conceptual definition of Hopf module allows the effortless generalization to cases involving (co)quasibialgebras in place of bialgebras. This means that we have to introduce instances of the associator maps in various places in the definitions and constructions we wrote down above for algebras and modules in strict categories; this being done, we can specialize to the case of comodules over coquasibialgebras. Thus if (H, ϕ) is a coquasibialgebra, then a left H-comodule algebra A is defined to be an algebra in the category of left H-comodules, which is by definition a left H-comodule A, equipped with a multiplication $A \otimes A \to A$ and unit $1 \in A$. The multiplication is required to be a map in ${}^H\mathcal{M}$, that is, to satisfy $(ab)_{(-1)} \otimes (ab)_{(0)} = a_{(-1)}b_{(-1)} \otimes a_{(0)}b_{(0)}$ for all $a, b \in A$, and to be associative in the sense that $(xy)z = \phi(x_{(-1)} \otimes y_{(-1)} \otimes z_{(-1)})x_{(0)}(y_{(0)}z_{(0)})$ holds. The unit has to satisfy $1_{(-1)} \otimes 1_{(0)} = 1 \otimes 1 \in H \otimes A$ and is an ordinary unit for multiplication. A left A-module in ${}^H\mathcal{M}$ is an H-comodule M with a map $A \otimes M \ni a \otimes m \mapsto am \in M$ of H-comodules satisfying $(xy)m = \phi(x_{(-1)} \otimes$

$y_{(-1)} \otimes m_{(-1)})x_{(0)}(y_{(0)}m_{(0)})$; similarly for right modules and bimodules. The free right A-module generated by $V \in {}^H\mathcal{M}$ is $V \otimes A$ with the right A-module structure defined by $(v \otimes x)y = \phi(v_{(-1)} \otimes x_{(-1)} \otimes y_{(-1)})v_{(0)} \otimes x_{(0)}y_{(0)}$. If V was a left B-module, then $V \otimes A$ is a B-A-bimodule with $b(v \otimes x) = \phi^{-1}(b_{(-1)} \otimes v_{(-1)} \otimes x_{(-1)})b_{(0)}v_{(0)} \otimes x_{(0)}$. The tensor product of a right A-module M and a left A-module N in ${}^H\mathcal{M}$ is by definition the quotient of $M \otimes N$ by the relations $mx \otimes n = \phi(m_{(-1)} \otimes x_{(-1)} \otimes n_{(-1)})m_{(0)} \otimes x_{(0)}n_{(0)}$ for all $m \in M, n \in N$, and $x \in A$.

3. Hopf Algebra Extensions

3.1. Cleft extensions and crossed products. Let H be a bialgebra. An H-comodule algebra A is said to be cleft if there exists a convolution invertible H-colinear map $j: H \to A$ (a cleaving map). One can show that such a j can always be chosen to satisfy $j(1) = 1$ (otherwise replace it with $\tilde{j}$ defined by $\tilde{j}(h) = j^{-1}(1)j(h)$). Assume given a cleaving map j for the H-comodule algebra A; put $R := A^{\mathrm{co}\,H} := \{x \in A | \rho(x) = x \otimes 1\}$. One can show that $\pi: A \ni x \mapsto x_{(0)}j^{-1}(x_{(1)}) \in B$ is well-defined (i. e. takes values in the coinvariants R), and one has an isomorphism $\psi: R \otimes H \to A$ of left R-modules and right H-comodules, given by $\psi(r \otimes h) = rj(h)$ and $\psi^{-1}(x) = \pi(x_{(0)}) \otimes x_{(1)}$. Hence classifying cleft H-comodule algebras with coinvariant subalgebra R amounts to classifying algebra structures on $R \otimes H$ such that $R \ni r \mapsto r \otimes 1 \in R \otimes H$ is an algebra map, and $R \otimes H$ is an H-comodule algebra with the comodule structure induced by that of H. One can show that any such multiplication is given by a formula

$$(r \otimes g)(s \otimes h) = r(g_{(1)} \rightharpoonup s)\sigma(g_{(2)}|h_{(1)}) \otimes g_{(3)}h_{(2)},$$

in which $\rightharpoonup: H \otimes R \to R$ and $\sigma: H \otimes H \to R$ are linear maps. Certain conditions have to be met by $\rightharpoonup$ and σ to ensure that multiplication is associative with unit $1 \otimes 1$. We make a point of not recalling these conditions, since we will never need them explicitly. We do mention, however, that in the case where H is cocommutative, and R is commutative, there is a reasonable cohomological interpretation of the axioms: They say that R is an H-module algebra, and σ is a two-cocycle in the Sweedler cohomology [54] of H with coefficients in R. Hence the term cocycle for σ, which is also used in the general situation where there is no cohomology theory behind it.

We will call an algebra $A := R\#_\sigma H := R \otimes H$ with the multiplication according to the above formula a crossed product of R and H, not quite in accordance with the literature, where crossed products are usually required to have σ invertible. If H is a Hopf algebra, this is equivalent to the H-comodule algebra $R\#_\sigma H$ being cleft; the map $j: H \ni h \mapsto 1 \otimes h \in R \otimes H$ can serve as a cleaving map.

If H is a Hopf algebra, then a crossed product $R\#_\sigma H$ can also be characterized as a Hopf–Galois extension with normal basis. Here a Hopf–Galois extension

A/R is an H-comodule algebra A with coinvariants $A^{\mathrm{co}\,H} = R$, such that the Galois map $\beta\colon A \underset{B}{\otimes} A \to A \otimes H$ given by $\beta(x \otimes y) = xy_{(0)} \otimes y_{(1)}$ is a bijection. A normal basis for an H-Galois extension A/B is an isomorphism $A \cong B \otimes H$ of left B-modules and right H-comodules. Any cleft extension with cleaving map j is a Galois extension (obviously with a normal basis) with $\beta^{-1}(x \otimes h) = xj^{-1}(h_{(1)}) \otimes j(h_{(2)})$. Conversely, Hopf–Galois extensions with normal basis are cleft.

The theory of cleft extensions, crossed products, and Hopf–Galois extensions with normal basis is developed in [13; 4; 5], see also [34]. There is a dual counterpart, which was worked out in detail in [9]. A crossed coproduct of a coalgebra C and a bialgebra H is a coalgebra structure on $H \otimes C$ making $H \otimes C$ an H-module coalgebra with the obvious left H-module structure, such that the map $\varepsilon \otimes C\colon H \otimes C \to C$ is a coalgebra map. One can show that any such comultiplication is necessarily of the form

$$\Delta(h \otimes c) = h_{(1)}\tau^{(1)}(c_{(1)}) \otimes c_{(2)[0]} \otimes h_{(2)}\tau^{(2)}(c_{(2)})c_{(2)[1]} \otimes c_{(3)},$$

involving maps $\rho\colon C \ni c \mapsto c_{[0]} \otimes c_{[1]} \in C \otimes H$ and $\tau\colon C \ni c \mapsto \tau^{(1)}(c) \otimes \tau^{(2)}(c) \in H \otimes H$, subject to a list of axioms that we'll skip once more. If the "two-cycle" τ is invertible (if C is cocommutative and H is commutative it is then an honest two-cycle in a suitable homology theory), then $C \otimes H$ is a cocleft H-module coalgebra with cocleaving map $\pi = (\varepsilon \otimes H)\colon C \otimes H \to H$. Here, a right H-module coalgebra D is called cocleft if there exists a convolution invertible right H-module map (a cocleaving) $\pi\colon D \to H$. Such a cocleaving may always be chosen to be counital, i. e. to satisfy $\varepsilon\pi = \varepsilon$. A cocleft H-module coalgebra is always an H-Galois coextension; this notion appears (without a name) in [48]. By definition, an H-Galois coextension is a surjection $\nu\colon D \to C$ of an H-module coalgebra D onto the factor coalgebra $C := D/DH^+$, such that the Galois map

$$\beta\colon D \otimes H \ni d \otimes h \mapsto d_{(1)} \otimes d_{(2)}h \in D\underset{C}{\square}D$$

is a bijection. In the case where D is cocleft, one may write $\beta^{-1}(\sum d_i \otimes d_i') = d_i\pi^{-1}(d_{i(1)}) \otimes \pi(d_{i(2)})$. Any H-Galois coextension D with a normal basis, that is, such that $D \cong C \otimes H$ as left C-comodules and right H-comodules, is cocleft.

A very important tool in our constructions will be Schneider's structure theorem for Hopf modules, in the version for Galois coextensions [48]. It states that when D an H-Galois coextension of C such that H is flat and D is faithfully coflat as C-comodule (for example, this is fulfilled if D has a normal basis), then the functor

$$\mathcal{M}^C \ni V \mapsto V\underset{C}{\square}D \in \mathcal{M}^D_H$$

is a category equivalence with inverse mapping $M \in \mathcal{M}_H^D$ to M/MH^+. Here $V \square_C D \subset V \otimes D$ is the cotensor product

$$V \underset{C}{\square} D := \left\{ \sum v_i \otimes d_i \,\middle|\, \sum v_{i(0)} \otimes v_{i(1)} \otimes d_i = \sum v_i \otimes \nu(d_{i(1)}) \otimes d_{i(2)} \right\}$$

(incidentally, a special case of the tensor product in monoidal categories, namely the tensor product over the algebra C in the opposite of the category of k-modules.) The Hopf module structures of $V \square_C D$ are induced by the structures of the right tensor factor, and the comodule structure of M/MH^+ is that of a factor comodule of M. We are most interested in the case that D is cocleft. Then $V \underset{C}{\square} D \cong V \otimes H$ as right H-modules; the isomorphism $M \cong M/MH^+ \otimes H$ for $M \in \mathcal{M}_H^D$ that makes part of Schneider's category equivalence is then given by $m \mapsto \overline{m_{(0)}} \otimes m_{(1)}$, and its inverse maps $\overline{m} \otimes h$ to $m_{(0)} \pi^{-1}(m_{(1)}) h$.

3.2. Short exact sequences. One of the main objects of interest in this paper will be extensions of Hopf algebras, or short exact sequences. We refer to [50] for a detailed analysis of this notion.

For the purpose of this paper, a sequence of Hopf algebras and Hopf algebra maps $K \xrightarrow{\iota} H \xrightarrow{\nu} Q$ is exact if it fulfills the conditions stated to be equivalent in the following Proposition.

PROPOSITION 3.2.1. *Let K, H, Q be Hopf algebras with bijective antipodes, $\iota \colon K \to H$ and $\nu \colon H \to Q$ Hopf algebra maps. The following are equivalent:*

(1) (i) *ν is surjective and conormal,*

    (ii) *ι is the kernel of ν, and*

    (iii) *H is cleft as a right Q-comodule algebra.*

(2) (i) *ι is injective and normal,*

    (ii) *ν is the cokernel of ι, and*

    (iii) *H is cocleft as a left K-module coalgebra.*

(3) (i) *ν is surjective, $\nu\iota = \eta\varepsilon$, and*

    (ii) *there is a unital and counital right Q-comodule map $j \colon Q \to H$ such that $\nabla(\iota \otimes j) \colon K \otimes Q \to H$ is an isomorphism.*

(4) (i) *ι is injective, $\nu\iota = \eta\varepsilon$, and*

    (ii) *there is a unital and counital left K-module map $\pi \colon H \to K$ such that $(\pi \otimes \nu)\Delta \colon H \to K \otimes Q$ is an isomorphism.*

Even if we don't know the precise definitions involved, it seems obvious that (1) and (2) as well as (3) and (4) are mutually dual. It is also obvious that (1) contains a naïve definition of short exact sequence, which would state that ν is surjective and ι the kernel of ν; the meaning of the remaining conditions in (1) is perhaps less obvious, and we shall also be very short in even explaining the ingredients.

From [50] we recall that the (Hopf) kernel of a Hopf algebra map ν (i. e. the equalizer of ν and the trivial morphism $\eta\varepsilon$ in the category of Hopf algebras) can be computed by $\mathrm{HKer}(\nu) = H^{co\,Q} = {}^{co\,Q}H$ if ν is conormal, and $\mathrm{HKer}(\nu)$ will then be a Hopf subalgebra. Dually, the Hopf cokernel of a normal morphism ι is $\mathrm{HCoker}(\iota) = H/K^+H = H/HK^+$.

The equivalence of (1) and (2) is essentially contained in [50]; however, only faithful (co)flatness conditions are required there in place of the (co)cleftness conditions, giving two weaker, mutually dual and equivalent sets of conditions (by the way, bijectivity of the antipodes is not needed for the equivalence of (1) and (2)). Conditions (c) should be seen as analogs of the fact that, in an extension $N \hookrightarrow G \to G/N$ of groups one can always choose a section for the surjection by just choosing coset representatives. For the equivalence of conditions (c) we just note that, given a cleaving j, a cocleaving π can be obtained just like the map π for general cleft extensions at the beginning of the previous section, and a cleaving can be obtained from a cocleaving in the dual fashion. We will not show the equivalence with (3) and (4), but point out that they state that H has normal basis as Q-comodule algebra and K-module coalgebra.

By a result of Schneider [49], the (co)cleftness conditions on a short exact sequence are automatic whenever the Hopf algebras involved are finite dimensional over a base field. More generally, whenever H is a finite dimensional Hopf algebra over the field k, and $K \subset H$ is a Hopf subalgebra H is cocleft as a left K-module coalgebra, and, dually, when $I \subset H$ is a Hopf ideal, then H is cleft as right comodule algebra over H/I. As an immediate consequence we have:

COROLLARY 3.2.2. *Let K, H, Q be finite dimensional Hopf algebras, $\iota\colon K \to H$ an injective, and $\nu\colon H \to Q$ a surjective Hopf algebra map with $\nu\iota = \eta\varepsilon$.*
The sequence $K \overset{\iota}{\hookrightarrow} H \overset{\nu}{\twoheadrightarrow} Q$ is exact if and only if $\dim H = \dim K \cdot \dim Q$.

Although this result in finite dimensions makes the notion of short exact sequence look very natural, we should note that in the infinite dimensional case cleftness as we required it does rule out important examples of short exact sequences, for instance of affine algebraic groups.

Given two extensions $K \overset{\iota}{\hookrightarrow} H \overset{\nu}{\twoheadrightarrow} Q$, and $K \overset{\iota'}{\hookrightarrow} H' \overset{\nu'}{\twoheadrightarrow} Q$ an isomorphism of extensions is an isomorphism $f\colon H \to H'$ with $\nu'f = \nu$ and $f\iota = \iota'$.

We note that the following situation arises from the conditions in (3) and (4) in Proposition 3.2.1, which are equivalent even if we are dealing with bialgebras in place of Hopf algebras: We are given a mapping system

$$K \underset{\pi}{\overset{\iota}{\rightleftarrows}} H \underset{j}{\overset{\nu}{\rightleftarrows}} Q$$

in which K, H, Q are bialgebras, ι and ν are bialgebra maps, π is a unital counital left K-module map, j is a unital and counital Q-comodule map, and the conditions $\pi\iota = \mathrm{id}_K$, $\nu j = \mathrm{id}_Q$, $\pi j = \eta\varepsilon$, $\nu\iota = \eta\varepsilon$, and $\mathrm{id}_H = \iota\pi * j\nu$ are satisfied.

Assume given such a situation (where we can now drop the assumption that k is a field, but assume that all the bialgebras are k-flat). We can identify $H = K \otimes Q$ with $\pi = K \otimes \varepsilon$, $\nu = \varepsilon \otimes Q$, $j = \eta \otimes Q$ and $\iota = K \otimes \eta$.

Then $H = K \#_\sigma Q$ is a crossed product algebra with respect to $\rightharpoonup: Q \otimes K \to K$ and $\sigma: Q \otimes Q \to K$, and simultaneously a crossed coproduct coalgebra $H = K^\tau \# Q$ with $\rho: Q \ni q \mapsto q_{[0]} \otimes q_{[1]} \in Q \otimes K$ and $\tau: Q \to K \otimes K$. Such a bialgebra is called a bicrossproduct, and was introduced by Majid, see [29], generalizing the previously known cases where K is commutative and Q is cocommutative [53; 20]. An important special case are bismash products, in which σ and τ are trivial. This case can be characterized by saying that the map j is an algebra map, and π is a coalgebra map. Masuoka has called a collection $(K, Q) = (K, Q, \rightharpoonup, \rho)$ with maps $\rightharpoonup: Q \otimes K \to K$ and $\rho: Q \to Q \otimes K$ a Singer pair, if the data fulfill the necessary conditions to build such a bismash product. If K is commutative and Q cocommutative, we will speak of an abelian Singer pair. If $K^\tau \#_\sigma Q$ is a bicrossproduct with K commutative and Q cocommutative, then the maps $\rightharpoonup$ and ρ in the bicrossproduct form a Singer pair, and they only depend on the isomorphism class of the extension $K \to K^\tau \#_\sigma Q \to Q$. The set of extensions giving rise to a fixed abelian Singer pair is denoted $\text{Opext}(Q, K)$. It has a group structure (given by a kind of Baer product), and it has a cohomological description [53; 20] through a double complex built from the Singer pair.

3.3. More general inclusions.
Assume that $\iota: K \to H$ is a map of flat bialgebras such that H is a cocleft left K-module algebra (for example, that H and K are finite dimensional Hopf algebras over a field). If we put $Q := H/K^+ H$, then Q is a quotient coalgebra and right H-module of H, and we have an isomorphism $(\pi \otimes \nu)\Delta: H \to K \otimes Q$, where ν denotes the canonical surjection and $\pi: H \to K$ is a cocleaving map, which we may assume unital and counital. If we define $j: Q \to H$ by $j(\nu(h)) = h_{(1)} \pi^{-1}(h_{(2)})$, then the inverse isomorphism is given by $\nabla(\iota \otimes j): K \otimes Q \to H$. Hence we have a mapping system of the following type:

$$K \xrightarrow[\pi]{\iota} H \xrightarrow[j]{\nu} Q , \qquad (3\text{-}1)$$

where K and H are bialgebras, ι is a bialgebra map, π is a unital counital left K-module map, Q is a right H-module coalgebra, ν is a right H-module coalgebra map, $j: Q \to H$ is a unital counital right Q-comodule map, and the equations $\pi\iota = \text{id}_K$, $\nu j = \text{id}_Q$, $\pi j = \eta\varepsilon$, $\nu\iota = \eta\varepsilon$, and $\text{id}_H = \iota\pi * j\nu$ are satisfied. In any such system we can again identify $H = K \otimes Q$ with $\nu = \varepsilon \otimes Q$, $\iota = K \otimes \eta$, $\pi = K \otimes \varepsilon$, and $j = \eta \otimes Q$.

Several well-known constructions in quantum group theory lead to mapping systems of this type.

One of these is the bicrossproduct construction discussed above; here Q is a bialgebra and ν is a bialgebra map. A special case of this is the bismash product construction, in which both of the "cocycles" σ and τ are trivial; this is characterized by π being a coalgebra map, and j being an algebra map. The even more special case of a tensor product Hopf algebra is characterized by all four maps ι, π, ν, and j being bialgebra maps.

Another type of mapping system is furnished by the double crossproduct construction [27], where ι and j are bialgebra maps (and Q a bialgebra), whereas π and ν are coalgebra maps. The coalgebra structure of a double crossproduct $H = K \bowtie Q := K \otimes Q$ is the tensor product coalgebra, whereas multiplication is given by

$$(x \otimes p)(y \otimes q) = x(p_{(1)} \rightharpoonup y_{(1)}) \otimes (p_{(2)} \leftharpoonup y_{(2)})q,$$

in terms of certain maps $\rightharpoonup : Q \otimes K \to K$ and $\leftharpoonup : Q \otimes K \to Q$. A collection $(K, Q) = (K, Q, \rightharpoonup, \leftharpoonup)$ of data fulfilling the necessary conditions to build a double crossproduct is called a matched pair. If K and Q are cocommutative, we speak of an abelian matched pair; this case is treated in [57]. One can characterize double crossproducts as the general form of a bialgebra H equipped with bialgebra maps $\iota : K \to H$ and $j : Q \to H$ such that $\nabla(\iota \otimes j): K \otimes Q \to H$ is a bijection. From this description it should be obvious that a matched pair of group algebras arises whenever two groups F, G are subgroups of a third group $F \bowtie G$ such that multiplication induces a bijection $F \times G \to F \bowtie G$.

Another class of mapping systems of type (3–1) arises from Radford's biproduct construction [38]. This is the case in which both ι and π are bialgebra maps. In a biproduct $K \star Q$, comultiplication and multiplication are given by

$$\Delta(x \otimes q) = x_{(1)} \otimes q_{(1)[0]} \otimes x_{(2)} q_{(1)[1]} \otimes q_{(2)}$$
$$(x \otimes p)(y \otimes q) = x y_{(1)} \otimes (p \leftharpoonup y_{(2)})q,$$

that is, $K \star Q$ is a cosmash product coalgebra, and a smash product algebra, but on the other side compared to a bismash product. The combinatorics of the action and coaction have, for once, a conceptual interpretation in this case: In more modern language they say that Q is a Hopf algebra within the braided monoidal category of Yetter–Drinfeld modules over K.

Biproducts, bismash products, and double crossproducts were unified in the construction of trivalent products in [2], and independently but later in [8] and [44]. In particular, the papers [2; 8] describe how one can construct bialgebra structures on a tensor product $K \otimes Q$ in which both K and Q are algebras and coalgebras (but not necessarily bialgebras). This is precisely the case of a mapping system (3–1) in which π is a coalgebra map and j an algebra map. The coalgebra structure in such a trivalent product is a cosmash product as in all the examples before, and the multiplication has the same form as for a double crossproduct.

The construction in [44] is less general in that the first factor is always a bialgebra, but more general in that it allows the second factor Q to be a nonassociative algebra. This will perhaps seem more natural after subsection 4.2 below. Nonassociative algebras are unavoidable in the construction, which completely classifies mapping systems of the type discussed above, with ι a bialgebra map and π a left K-module coalgebra map.

A still more (and perhaps most) general version can be found in [3]. Here tensor products $K \otimes Q$ are covered in which K is an algebra and non-coassociative coalgebra, and Q is a coalgebra and nonassociative algebra; moreover, everything is set within a braided monoidal category.

4. Coquasibialgebras Reconstructed from Hopf Modules

In this section we will use the concepts of section 2 to give a general construction of coquasibialgebras from cocleft inclusions of Hopf algebras (and in fact more general inclusions involving coquasibialgebras). The construction will have a conceptual description that, at the utmost level of compression, fits in a few lines; the outcome can be computed quite explicitly, and we will discuss several interesting special cases.

4.1. The general construction. Assume $K \subset H$ is an inclusion of Hopf algebras such that H is cocleft as a left K-module coalgebra with cocleaving π. We consider the monoidal category $^H_K\mathcal{M}_K = {}_K(^H\mathcal{M})_K$ of K-bimodules in the category of H-comodules (which we can do since K is an H-comodule algebra via the inclusion). Since H is cocleft, we have $M \cong K \otimes M/K^+M$ for each $M \in {}^H_K\mathcal{M}$, hence the functor $^H_K\mathcal{M}_K \ni M \mapsto M/K^+M \in {}_k\mathcal{M}$ preserves tensor products (by the calculation $M \underset{K}{\otimes} N = M \underset{K}{\otimes} (K \otimes N/K^+N) \cong M \otimes N/K^+N$, hence $(M \underset{K}{\otimes} N)/K^+(M \underset{K}{\otimes} N) \cong M/K^+M \otimes N/K^+N$). Thus if there is a coendomorphism coalgebra of the functor ω, then it is naturally a coquasibialgebra, which we denote $\tilde{H}$ (it depends, of course, on H, K, the inclusion, and the cleaving map).

Directly from the abstract description of $\tilde{H}$ we can see some things about its structure; we assume that K is finite. Every Hopf module in $^K_K\mathcal{M}_K$ is trivially a Hopf module in $^H_K\mathcal{M}_K$. Thus we get a functor $^K_K\mathcal{M}_K \to {}^H_K\mathcal{M}_K$. Now the former category is equivalent to $\mathcal{M}_K \cong {}^{K^*}\mathcal{M}$ by sending a right K-module V to $K \otimes V$ with the obvious structure of a Hopf module in $^K_K\mathcal{M}$, and the diagonal right K-module structure [42]. Observe that the composition $\mathcal{M}_K \to {}^K_K\mathcal{M}_K \to {}^H_K\mathcal{M}_K \overset{\omega}{\to} {}_k\mathcal{M}$ is just the underlying functor. Thus we have a natural map $K^* \to \tilde{H}$; upon closer inspection, this turns out to be a coquasibialgebra map, and to turn $\tilde{H}$ into a right K^*-module coalgebra. Next, we also have a natural map $\tilde{H} \to K^*$: By the universal property of $\tilde{H}$, this amounts to giving a natural transformation $\omega \to K^* \otimes \omega$, or $\omega \otimes K \to \omega$. Such a transformation can easily be assembled from the right K-module structures of the objects in $^H_K\mathcal{M}_K$, that is, we can define $\omega \otimes K \to \omega$ by $\overline{m} \otimes x \mapsto \overline{mx}$ for $m \in M$ and $x \in K$; upon closer inspection, the resulting map $\tilde{H} \to K^*$ will turn out to be a cocleaving map for the right K^*-module coalgebra $\tilde{H}$.

All in all, we have (up to existence of the necessary coendomorphism coalgebras) a natural construction that requires an inclusion of a Hopf algebra K

into a bialgebra H, such that H is left K-cocleft. The construction yields as its result an inclusion of K^* into a coquasibialgebra $\tilde{H}$, such that $\tilde{H}$ is a right K^*-module coalgebra, and cocleft as such. Stated rather sloppily, the output of the construction is similar to, but not quite as good as, its input.

We would like to have a generalization that can do with input of the same type as its output.

So assume that H is only a coquasibialgebra, and $K \subset H$ a sub-coquasibi-algebra, and let us investigate what additional hypotheses we need to arrive at a similar construction $\tilde{H}$ as above. We will endeavor to make Definition 4.1.1 below look inevitable, though the discussion will not be sufficiently rigorous to yield necessary conditions; in fact, we will sketch in subsection 4.6 a situation where it appears one can get a result from weaker conditions.

To be able to consider $_K(^H\mathcal{M})_K$, we needed to have K an algebra in the category $^H\mathcal{M}$, with respect to the comodule structure induced by the inclusion. This means that for x, y, z the associativity in the coquasibialgebra H has to coincide with the modified associativity

$$(xy)z = \phi(x_{(1)} \otimes y_{(1)} \otimes z_{(1)})x_{(2)}(y_{(2)}z_{(2)})$$

for an algebra in $^H\mathcal{M}$. Upon applying ε, this entails that $\phi|_{K \otimes K \otimes K} = \varepsilon$, and K is hence an ordinary bialgebra. To apply the structure theorem for Hopf modules, we used that K is a Hopf algebra, and H a K-module coalgebra with cocleaving map π. To have H a K-module means that

$$x(yh) = (xy)h = \phi(x_{(1)} \otimes y_{(1)} \otimes h_{(1)})x_{(2)}(y_{(2)}h_{(2)})$$

for all $x, y \in K$ and $h \in H$, which is true if and only if $\phi|_{K \otimes K \otimes H} = \varepsilon$. As above, we have a natural functor $_K^K\mathcal{M}_K \to {}_K^H\mathcal{M}_K$ that yields a natural map $K^* \to \tilde{H}$. Assume that we know it is a bialgebra map, and that it makes $\tilde{H}$ a right K^*-module coalgebra. Then a cocleaving map $\tilde{\pi} \colon \tilde{H} \to K^*$ was constructed above by using the natural transformation $\omega \otimes K \to \omega$ given by $\overline{m} \otimes y \mapsto \overline{my}$ for each $m \in M \in {}_K^H\mathcal{M}_K$ and $y \in K$. If H is a coquasibialgebra, this is no longer obviously well-defined, since objects in $_K^H\mathcal{M}_K$ are not ordinary bimodules; it is well-defined, however, if we assume that right multiplication by $y \in K$ on $M \in {}_K^H\mathcal{M}_K$ is a left K-module map; this means $x(my) = (xm)y = \phi(x_{(1)} \otimes m_{(-1)} \otimes y_{(1)})x_{(2)}(m_{(0)}y_{(2)})$ for all $x, y \in K$ and $m \in M$. Assuming there are enough objects in $_K^H\mathcal{M}_K$, this means that $\phi|_{K \otimes H \otimes K}$ should be trivial. Another vital part of the construction was using the structure theorem for Hopf modules to assemble the isomorphism

$$\theta \colon M \underset{K}{\otimes} N \cong M \underset{K}{\otimes} (K \otimes N/K^+N) \cong M \otimes N/K^+N$$

with $\theta(m \otimes n) = m\pi(n_{(-1)}) \otimes \overline{n_{(0)}}$. To maintain this in the case where H is a coquasibialgebra, we need to remember that the tensor product $M \underset{K}{\otimes} N$ is now defined in the category $^H\mathcal{M}$. We need to have, abusing notation

$$\theta(mx \otimes n) = \theta(\phi(m_{(-1)} \otimes x_{(1)} \otimes n_{(-1)})m_{(0)} \otimes x_{(2)}n_{(0)})$$

for $m \in M$, $x \in K$, and $n \in N$. Now

$$\theta(mx \otimes n) = (mx)\pi(n_{(-1)}) \otimes \overline{n_{(0)}}$$
$$= \phi(m_{(-1)} \otimes x_{(1)} \otimes \pi(n_{(-1)})_{(1)})m_{(0)}(x_{(2)}\pi(n_{(-1)})_{(2)}) \otimes \overline{n_{(0)}}$$

and

$$\theta(\phi(m_{(-1)} \otimes x_{(1)} \otimes n_{(-1)})m_{(0)} \otimes x_{(2)}n_{(0)})$$
$$= \phi(m_{(-1)} \otimes x_{(1)} \otimes n_{(-2)})m_{(0)}\pi(x_{(2)}n_{(-1)}) \otimes \overline{x_{(3)}n_{(0)}}$$
$$= \phi(m_{(-1)} \otimes x_{(1)} \otimes n_{(-2)})m_{(0)}(x_{(2)}\pi(n_{(-1)})) \otimes \overline{n_{(0)}}.$$

We can only expect these to be equal for all objects M, N, if

$$\phi(g \otimes x \otimes h_{(1)})\pi(h_{(2)}) = \phi(g \otimes x \otimes \pi(h)_{(1)})\pi(h)_{(2)}$$

for all $g, h \in H$ and $x \in K$. Next, θ was used above to endow the functor $\omega\colon {}^H_K\mathcal{M}_K \to {}_k\mathcal{M}$ with an incoherent tensor functor structure, which was easy since θ was a left K-module map. In our generalized situation, we have $\theta(x(m \otimes n)) = \theta(\phi^{-1}(x_{(1)} \otimes m_{(1)} \otimes n_{(1)})x_{(2)}m_{(0)} \otimes n_{(0)}) = \theta^{-1}(x_{(1)} \otimes m_{(-1)} \otimes n_{(-2)})x_{(2)}m_{(0)}\pi(n_{(-1)}) \otimes \overline{n_{(0)}}$ and $x\theta(m \otimes n) = xm\pi(n_{(1)}) \otimes \overline{n_{(0)}}$. To have these equal for all $x \in K$, $m \in M \in {}^H_K\mathcal{M}_K$ and $n \in N \in {}^H_K\mathcal{M}_K$ should imply $\phi|_{K \otimes H \otimes H}$ is trivial.

We have arrived at the set of conditions for objects in the following definition [47, Def. 3.3.3]:

DEFINITION 4.1.1. Let K be a k-flat Hopf algebra. The category $\mathfrak{E}_\ell(K)$ is defined as follows:

An object of $\mathfrak{E}_\ell(K)$ is

(1) a k-flat coquasibialgebra H together with
(2) a coquasibialgebra map $\iota\colon K \to H$ such that $\phi(\iota \otimes H \otimes H) = \varepsilon$, and
(3) a convolution invertible, unit and counit preserving left K-module map $\pi\colon H \to K$ such that $\phi(g \otimes x \otimes h_{(1)})\pi(h_{(2)}) = \phi(g \otimes x \otimes \pi(h)_{(1)})\pi(h)_{(2)}$ holds for all $g, h \in H$ and $x \in K$.

A morphism in $\mathfrak{E}_\ell(K)$ is a coquasimorphism $(F, \theta)\colon H \to H'$ such that $F\iota = \iota'$ and $\theta(\iota \otimes H) = \varepsilon$

We note once more that condition (3) makes sense since condition (2) implies that H is a K-module coalgebra. We also note that the cocleftness assumption implies that ι is injective; we shall treat it as an inclusion.

There is an obvious left-right switched version of $\mathfrak{E}_\ell(K)$ which we denote by $\mathfrak{E}_r(K)$. It is defined so that $H \in \mathfrak{E}_\ell(K)$ means the same as $H^{\mathrm{bop}} \in \mathfrak{E}_r(K^{\mathrm{bop}})$.

An object $H \in \mathfrak{E}_\ell(K)$ is really a sextuple of data $(H, \Delta, \nabla, \phi, \iota, \pi)$; we shall always use these notations, and similar ones for other objects of $\mathfrak{E}_\ell(K)$, say $H' = (H', \Delta', \nabla', \phi', \iota', \pi')$. Also, we will always use the following conventions: We define the functor $\omega\colon {}^H_K\mathcal{M}_K \to {}_k\mathcal{M}$ by $\omega(M) := \overline{M} := M/K^+M$, writing

$\nu\colon M \to \overline{M}$ for the canonical surjection. For any $M \in {}^{H}_{K}\mathcal{M}$ we define $j\colon \overline{M} \to M$ by $j(\overline{m}) = \pi^{-1}(m_{(-1)})m_{(0)}$. We note that $M \ni m \mapsto \pi(m_{(-1)}) \otimes \overline{m_{(0)}} \in K \otimes \overline{M}$ is an isomorphism of left K-modules for every $M \in {}^{H}_{K}\mathcal{M}$. We write $Q = H/K^{+}H$, a quotient coalgebra of H, endowed with a canonical right action of K defined by $\overline{h} \leftharpoonup x := \overline{hx} := \overline{hx}$. We endow the functor ω with the tensor functor structure $\xi\colon \overline{M} \otimes \overline{N} \to \overline{M \underset{K}{\otimes} N}$ given by $\xi(\overline{m} \otimes v) = \overline{m \otimes j(v)}$, with $\xi^{-1}(\overline{m \otimes n}) = \overline{m\pi(n_{(-1)})} \otimes \overline{n_{(0)}}$.

Our task in the rest of the section is to see that $\mathrm{coend}(\omega)$ exists, and to find its coquasibialgebra structure. To describe the coquasibialgebra structure, it is convenient to introduce an intermediate step in the general procedure described in Sections 2.2 and 2.3. We define [47, Def. 3.2.4] the category $\left({}^{Q}\mathcal{M}\right)_{K}$ to consist of left Q-comodules V equipped with a map $V \otimes K \ni v \otimes x \mapsto vx \in V$ satisfying $(vx)_{(-1)} \otimes (vx)_{(0)} = v_{(-1)}x_{(1)} \otimes v_{(0)}x_{(2)}$ and $(vx)y = \phi(v_{(-1)} \otimes x_{(1)} \otimes y_{(1)})v_{(0)}(x_{(2)}y_{(2)})$, along with $v1_{K} = v$, for all $v \in V$ and $x,y \in K$; the modified associativity condition makes sense since ϕ induces a well-defined map $\phi\colon Q \otimes H \otimes H \to k$ by [47, Lem.3.2.1]. Schneider's structure theorem for Hopf modules says that ω induces a category equivalence ${}^{H}_{K}\mathcal{M} \cong {}^{Q}\mathcal{M}$, and this extends [47, Prop. 3.2.5] to a category equivalence ${}^{H}_{K}\mathcal{M}_{K} \cong \left({}^{Q}\mathcal{M}\right)_{K}$. This splits the problem of lifting ω to an equivalence $\hat{\omega}\colon {}^{H}_{K}\mathcal{M}_{K} \cong {}^{\mathrm{coend}(\omega)}\mathcal{M}$ of monoidal categories into two parts:

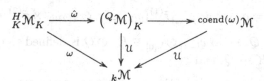

where the two underlying functors, denoted $\mathcal{U}$, can be made into strict neutral tensor functors in such a way that the two triangles are commutative triangles of tensor functors: The monoidal category structure of $\left({}^{Q}\mathcal{M}\right)_{K}$ achieving this for the left hand triangle is obtained by transporting the structures of ${}^{H}_{K}\mathcal{M}_{K}$ through the equivalence $\hat{\omega}$. For example, we have to determine the action and coaction on the tensor product of $V, W \in \left({}^{Q}\mathcal{M}\right)_{K}$ in such a way that the isomorphism $\xi\colon \omega(M \underset{K}{\otimes} N) \to \omega(M) \otimes \omega(N)$ are compatible with these actions and coactions. This yields

$$(v \otimes w)x = \phi(j(v_{(-1)}) \otimes j(w_{(-1)})_{(1)} \otimes x_{(1)})v_{(0)}\pi(j(w_{(-1)})_{(2)}x_{(2)}) \otimes w_{(0)}x_{(3)}$$
$$(v \otimes w)_{(-1)} \otimes (v \otimes w)_{(0)} = \overline{j(v_{(-1)})j(w_{(-1)})_{(1)}} \otimes v_{(0)}\pi(j(w_{(-1)})_{(2)}) \otimes w_{(0)}$$

both of which can be determined from the equation

$$(v \otimes w)_{(-1)} \otimes (v \otimes w)_{(0)}x = \overline{j(v_{(-2)})j(w_{(-2)})_{(1)}}\phi(v_{(-1)} \otimes j(w_{(-2)})_{(2)} \otimes x_{(1)})$$
$$\otimes v_{(0)}\pi(j(v_{(-2)})_{(3)})(v_{(-1)} \leftharpoonup x_{(2)}) \otimes w_{(0)}x_{(3)}$$

in which we have written $q \rightharpoonup x = \pi(j(q)x)$ for $q \in Q$ and $x \in K$. The transport of structures also means that, since $\hat{\omega}$ has to be a monoidal functor, the diagrams

$$
\begin{array}{ccc}
(\omega(L) \otimes \omega(M)) \otimes \omega(N) \xrightarrow{\xi \otimes \mathrm{id}} \omega(L \underset{K}{\otimes} M) \otimes \omega(N) \xrightarrow{\xi} \omega((L \underset{K}{\otimes} M) \underset{K}{\otimes} N) \\
\Big\downarrow{\tilde{\alpha}} \qquad\qquad\qquad\qquad\qquad\qquad\qquad\qquad \Big\downarrow{\omega(\alpha)} \\
\omega(L) \otimes (\omega(M) \otimes \omega(N)) \xrightarrow{\mathrm{id} \otimes \xi} \omega(L) \otimes \omega(M \underset{K}{\otimes} N) \xrightarrow{\xi} \omega(L \underset{K}{\otimes} (M \underset{K}{\otimes} N))
\end{array}
$$

have to commute, which leads to the formula

$$
\tilde{\alpha}(u \otimes v \otimes w) = \phi(j(u_{(-1)}) \otimes j(v_{(-1)})_{(1)} \otimes j(w_{(-1)})_{(1)})
$$
$$
u_{(0)} \pi(j(v_{(-1)})_{(2)} j(w_{(-1)})_{(2)}) \otimes v_{(0)} \pi(j(w_{(-1)})_{(3)}) \otimes w_{(0)}
$$

The category equivalence ${}^{H}_{K}\mathcal{M}_{K} \cong \left({}^{Q}\mathcal{M}\right)_{K}$ and the monoidal category structure of the latter category did not depend as yet on finiteness of K. The next step is to assume K finite, and to find a coalgebra $\tilde{H}$ with ${}^{\tilde{H}}\mathcal{M} \cong \left({}^{Q}\mathcal{M}\right)_{K}$. It is easy to understand how we may be led to the idea that $\tilde{H}$ can be modelled on $Q \otimes K^{*}$. After all, an object in $\left({}^{Q}\mathcal{M}\right)_{K}$ is determined by maps $V \to Q \otimes V$ and $V \to K^{*} \otimes V$, the latter corresponding to the action $V \otimes K \to V$ — but of course this is not a sufficient explanation; we shall sketch one now.

Define $\tilde{H} := Q \otimes K^{*}$, and define $\Delta \colon \tilde{H} \to \tilde{H} \otimes \tilde{H}$ by

$$
\Delta(q \bowtie \varphi) = q_{(1)} \bowtie q_{(2)[-\tilde{1}]} \tilde{\tau}^{(1)}(q_{(3)}) \varphi_{(1)} \otimes q_{(2)[\tilde{0}]} \bowtie \tilde{\tau}^{(2)}(q_{(3)}) \varphi_{(2)},
$$

where the map $\tilde{\rho} \colon Q \ni q \mapsto q_{[-\tilde{1}]} \otimes q_{[\tilde{0}]} \in K^{*} \otimes Q$ is defined to correspond to the right action $Q \otimes K \to Q$, and

$$
\tilde{\tau} \colon Q \ni q \mapsto \tilde{\tau}^{(1)}(q) \otimes \tilde{\tau}^{(2)}(q) \in K^{*} \otimes K^{*}
$$

is defined by $\tilde{\tau}(q)(x \otimes y) = \phi(j(q) \otimes x \otimes y)$. Define $\varepsilon \colon \tilde{H} \to k$ by $\varepsilon(q \bowtie \varphi) = \varepsilon(q)\varphi(1)$, $\tilde{\nu} \colon \tilde{H} \to Q$ by $\tilde{\nu}(q \bowtie \varphi) = q\varphi(1)$, and $\tilde{\pi} := \varepsilon \otimes K^{*} \colon \tilde{H} \to Q$.

We claim that $\tilde{H}$ is the object we are looking for, and a category equivalence between ${}^{\tilde{H}}\mathcal{M}$ and $\left({}^{Q}\mathcal{M}\right)_{K}$ is given in the 'obvious' way, that is, by the fact that a map $V \to \tilde{H} \otimes V$ induces maps $V \to Q \otimes V$ and $V \to K^{*} \otimes V$, the latter corresponding to a map $V \otimes K \to V$.

For the time being, we do not know whether $\tilde{H}$ is a coalgebra, but it is obvious that $\tilde{\nu}$ is a morphism of coalgebras, one of which is not necessarily coassociative. In [47] we have shown explicitly that $\tilde{H}$ is coassociative, and we will repeat essentially the same calculations in subsection 4.3 for a more general case. However, it seems worthwhile to indicate a trick that lets us get away without any explicit calculations with the cocycle ϕ. After all, the construction of $\tilde{H}$ has been intrinsic so far in only referring to the representation categories coming from H; we already used the defining axioms for a coquasibialgebra H by saying that ${}^{H}\mathcal{M}$ is a monoidal category, and since this characterizes coquasibialgebras, nothing else 'should' be needed. We will apply Lemma 2.3.1: $\tilde{\nu}$ being comultiplicative and counital, any $\tilde{H}$-comodule V (in the sense of Lemma 2.3.1) with comodule structure Λ is a Q-comodule via $\lambda = (\tilde{\nu} \otimes V)\Lambda$. In addition, the obvious left Q-comodule structure on $\tilde{H}$ coincides with $(\tilde{\nu} \otimes \tilde{H})\tilde{\Delta}$. Thus, applying $\tilde{\nu} \otimes \tilde{H} \otimes V$ to the coassociativity condition of a $\tilde{H}$-comodule V yields that the comodule structure map $\Lambda \colon V \to \tilde{H} \otimes V$ is left Q-colinear, hence can be written in the form $\Lambda(v) = v_{(-1)} \otimes \mu'(v_{(-0)})$ for some map $\mu' \colon V \to K^* \otimes V$, which is the same as a map $\mu \colon V \otimes K \to V$. Conversely, given a left Q-comodule V and a map $\mu \colon V \otimes K \to V$, let μ' denote the corresponding map $V \to K^* \otimes V$. Then we can define

$$\Lambda \colon V \ni v \mapsto v_{(-1)} \otimes \mu'(v_{(-0)}) \in \tilde{H} \otimes V.$$

To have a category equivalence $({}^Q\mathcal{M})_K \cong {}^{\tilde{H}}\mathcal{M}$ of the claimed form, we need to show that Λ is a comodule structure if and only if μ makes V an object of $({}^Q\mathcal{M})_K$. The condition $(\tilde{H} \otimes \Lambda)\Lambda(v) = (\tilde{\Delta} \otimes V)\Lambda(v) \in \tilde{H} \otimes \tilde{H} \otimes V$ needs only be verified after applying the maps $Q \otimes x^* \otimes Q \otimes y^* \otimes V \colon Q \otimes K^* \otimes Q \otimes K^* \otimes V \to Q \otimes Q \otimes V$ for all $x, y \in K$ (where $x^* \colon K^* \ni \varphi \mapsto \varphi(x) \in k$.) Now $(Q \otimes x^* \otimes Q \otimes y^* \otimes V)(\tilde{H} \otimes \Lambda)\Lambda(v) = v_{(-1)} \otimes (v_{(0)}x)_{(-1)} \otimes (v_{(0)}x)_{(0)}y$ and $(Q \otimes x^* \otimes Q \otimes y^*)\Delta(q \otimes \varphi) = q_{(1)} \otimes q_{(2)}x_{(1)}\phi(q_{(3)} \otimes x_{(2)} \otimes y_{(1)})\varphi(x_{(3)}y_{(2)})$ and hence $(Q \otimes x^* \otimes Q \otimes y^* \otimes V)(\tilde{\Delta} \otimes V)\Lambda(v) = v_{(-3)} \otimes v_{(-2)}x_{(1)} \otimes \phi(v_{(-1)} \otimes x_{(2)} \otimes y_{(1)})v_{(0)}(x_{(3)}y_{(2)})$ and thus V is a Λ-comodule if and only if $V \in ({}^Q\mathcal{M})_K$ (for the "only if" part apply once $\varepsilon \otimes \varepsilon \otimes V$, and once apply $\varepsilon \otimes Q \otimes V$ after specializing $y = 1$). This establishes the category equivalence $({}^Q\mathcal{M})_K \cong {}^{\tilde{H}}\mathcal{M}$, where the latter is still in the sense of Lemma 2.3.1; we will use the notation $v \mapsto v_{(-\tilde{1})} \otimes v_{(\tilde{0})}$ for $\tilde{H}$-comodule structures below. To show that $\tilde{H}$ is, after all, coassociative, it remains to apply Lemma 2.3.1. We consider $H \in {}^H_K\mathcal{M}$, and the free K-K-bimodule $H \otimes K$ generated by the left K-module H in ${}^H\mathcal{M}$. The corresponding object in $({}^Q\mathcal{M})_K$ is $\overline{H \otimes K} \cong Q \otimes K$ with left comodule structure $(q \otimes x)_{(-1)} \otimes (q \otimes x)_{(0)} = q_{(1)}x_{(1)} \otimes q_{(2)} \otimes x_{(2)}$ and K-action $(q \otimes x)y = \phi(q_{(1)} \otimes x_{(1)} \otimes y_{(1)})q_{(2)} \otimes x_{(2)}y_{(2)}$. We claim that simply $q \otimes \varphi = (q \otimes 1)_{(-\tilde{1})}(\varepsilon \otimes \varphi)((q \otimes 1)_{(\tilde{0})})$, for which it suffices to check that

$$(Q \otimes x^*)(\tilde{H} \otimes \varepsilon \otimes \varphi)\Lambda(q \otimes 1) = (Q \otimes \varepsilon \otimes \varphi)((q \otimes 1)_{(-1)} \otimes (q \otimes 1)_{(0)}x)$$

$$= (Q \otimes \varepsilon \otimes \varphi)(q_{(1)} \otimes q_{(2)} \otimes x) = q\varphi(x) = (Q \otimes x^*)(q \otimes \varphi)$$

for all $x \in K$.

This finishes the proof that $\tilde{H}$ is a coalgebra and we have a category equivalence ${}^{\tilde{H}}\mathcal{M} \cong ({}^Q\mathcal{M})_K$. Since we know the monoidal category structure on $({}^Q\mathcal{M})_K$, we know it on ${}^{\tilde{H}}\mathcal{M}$, and can assemble the corresponding coquasibialgebra structure: Multiplication is given by

$$(p \otimes \varphi)(q \otimes \psi) = \tilde{\nabla}(p \otimes \varphi \otimes q \otimes \psi)$$

$$:= \overline{j(p_{(1)})j(q_{(1)})_{(1)}} \otimes \kappa(j(p_{(2)}) \otimes j(q_{(1)})_{(2)})((\varphi \triangleleft \pi(j(q_{(1)})_{(3)}))\stackrel{\leftharpoonup}{-}q_{(2)})\psi$$

where $\kappa \colon H \otimes H \to K^*$ is given by $\kappa(g \otimes h)(x) = \phi(g \otimes h \otimes x)$, and $\stackrel{\leftharpoonup}{-} \colon K^* \otimes Q \to K^*$ is defined by $\langle \varphi \stackrel{\leftharpoonup}{-} q, x \rangle = \langle \varphi, q \rightharpoonup x \rangle$; for if we define $\tilde{\nabla} \colon \tilde{H} \otimes \tilde{H} \to \tilde{H}$ by this formula, then

$$(Q \otimes x^*)\tilde{\nabla}(p \otimes \varphi \otimes q \otimes \psi) = \overline{j(p_{(1)})j(q_{(1)})_{(1)}}\phi(j(p_{(2)}) \otimes j(q_{(1)})_{(2)} \otimes x_{(1)})$$
$$\langle \varphi, \pi(j(q_{(1)})_{(3)})(q_{(2)} \rightharpoonup x_{(2)}) \rangle \langle \psi, x_{(3)} \rangle$$

for $x \in K$, and we see immediately that

$$(Q \otimes x^*)\tilde{\nabla}(v_{(-\tilde{1})} \otimes w_{(-\tilde{1})}) \otimes v_{(\tilde{0})} \otimes w_{(\tilde{0})} = (v \otimes w)_{(-1)} \otimes (v \otimes w)_{(0)}x$$
$$= (Q \otimes x^*)((v \otimes w)_{(-\tilde{1})}) \otimes (v \otimes w)_{(\tilde{0})}$$

holds for all $v \in V \in {}^{\tilde{H}}\mathcal{M}$ and $w \in W \in {}^{\tilde{H}}\mathcal{M}$.

The coassociator $\tilde{\phi}$ is read off from the associator in $\left({}^{Q}\mathcal{M}\right)_K$ to be

$$\tilde{\phi}(p \otimes \varphi \otimes q \otimes \psi \otimes r \otimes \vartheta) =$$
$$\phi(j(p) \otimes j(q)_{(1)} \otimes j(r)_{(1)})\langle \varphi | \pi(j(q)_{(2)}j(r)_{(2)}) \rangle \langle \psi | \pi(j(r)_{(3)}) \rangle \varepsilon(\vartheta)$$

for all $p, q, r \in Q$, $\varphi, \psi, \vartheta \in K^*$.

It turns out that $\tilde{H} := \mathfrak{F}(H)$ is naturally an object in $\mathfrak{E}_r(K^*)$. We will skip the details showing that $\mathfrak{F}$ can be made into a functor from $\mathfrak{E}_\ell(K) \to \mathfrak{E}_r(K^*)$. The idea is that a morphism $H \to H'$ in $\mathfrak{E}_\ell(K)$ leads to a functor ${}^{H}_{K}\mathcal{M}_K \to {}^{H'}_{K}\mathcal{M}_K$, which induces a coquasimorphism $\tilde{H} \to \tilde{H}'$ by reconstruction. We will denote the version of the functor $\mathfrak{F} \colon \mathfrak{E}_\ell(K) \to \mathfrak{E}_r(K^*)$ obtained by switching sides by $\mathfrak{F}^{\mathrm{bop}} \colon \mathfrak{E}_r(K) \to \mathfrak{E}_\ell(K^*)$.

4.2. First examples. The least complicated examples of the functor $\mathfrak{F}$ occur when we require both $H \in \mathfrak{E}_\ell(K)$ and $\tilde{H} \in \mathfrak{E}_r(K^*)$ to be ordinary bialgebras with trivial coassociators ϕ and $\tilde{\phi}$. When we inspect the formula for the coassociator of $\tilde{H}$ in subsection 4.1, we find (by specializing $q = 1$) that we have to have $\pi(j(r)_{(1)}) \otimes \pi(j(r)_{(2)}) = \varepsilon(r)1 \otimes 1$ for all $r \in Q$, from which it is easy to see that π is a coalgebra map. By specializing $\psi = \varepsilon$, we find $\pi(j(q)j(r)) = \varepsilon(q)\varepsilon(r)$, from which one can deduce that j is multiplicative. This means that H is a trivalent product of K and Q as recalled in subsection 3.3, that is, comultiplication and multiplication in $H \cong K \otimes Q$ have to be given by

$$(x \otimes p)(y \otimes q) = x(p_{(1)} \rightharpoonup y_{(1)}) \otimes (p_{(2)} \leftharpoonup y_{(2)})q,$$
$$\Delta(x \otimes q) = x_{(1)} \otimes q_{(1)[0]} \otimes x_{(2)}q_{(1)[1]} \otimes q_{(3)}$$

for some maps $\rightharpoonup \colon Q \otimes K \to K$, $\leftharpoonup \colon Q \otimes K \to Q$, and $\rho \colon Q \ni q \mapsto q_{[0]} \otimes q_{[1]} \in Q \otimes K$, which are necessarily given by $q \rightharpoonup x = \pi(j(q)x)$, $q \leftharpoonup x = \nu(j(q)x)$, and $\rho(q) = \nu(j(q)_{(1)}) \otimes \pi(j(q)_{(2)})$. When we now inspect multiplication and comultiplication in $\tilde{H}$, we find that they are given by

$$(p \otimes \varphi)(q \otimes \psi) = p(\varphi \xrightarrow{\sim} q_{(1)}) \otimes (\varphi_{(2)} \xleftarrow{\sim} q_{(2)})\psi,$$
$$\Delta(q \otimes \varphi) = q_{(1)} \otimes q_{(1)[-\tilde{1}]}\varphi_{(1)} \otimes q_{(2)[\tilde{0}]} \otimes \varphi_{(2)},$$

where $\overset{\sim}{\rightharpoonup}\colon K^* \otimes Q \to Q$, $\overset{\sim}{\leftharpoonup}\colon K^* \otimes Q \to Q$ and $\tilde{\rho}\colon Q \ni q \mapsto q_{[-\tilde{1}]} \otimes q_{[\tilde{0}]} \in K^* \otimes Q$ are induced in the obvious way by ρ, $\rightharpoonup$, and $\leftharpoonup$, respectively. In particular, $\tilde{H}$ is a trivalent product as well (on the other side), with the necessary actions and coactions obtained in essence by permuting the data that led to H. Note that we never had to check explicitly that the new data fulfill the axioms that they have to in order that a trivalent product can be constructed. This is not particularly hard to do, but we are still happy to have a new insight into why it works. Obviously applying $\mathfrak{F}^{\mathrm{bop}}$ to $\tilde{H}$ gives us H back, so we have established a bijection between trivalent products $K \otimes Q$ and trivalent products $Q \otimes K^*$. As a particular case we have recovered the well-known bijection between Singer pairs $(K, Q, \rho, \rightharpoonup)$ and matched pairs $(Q, K^*, \overset{\sim}{\rightharpoonup}, \overset{\sim}{\leftharpoonup})$. This bijection is a major source for Singer pairs, since — as we already recalled briefly in subsection 3.3 — matched pairs arise naturally from groups composed from two subgroups; see [31] for examples. With a view towards our next application, we note a rare example where the bijection can be used in the other direction: If K is a finite Hopf algebra and $Q = K^{\mathrm{op}}$, then the tensor product Hopf algebra $K \otimes Q$ is obviously a bismash product extension of Q by K. We can endow it with the not so obvious cocleaving π defined by $\pi(x \otimes q) = xq$. This leads to $j(q) = S(q_{(1)}) \otimes q_{(2)}$, which is multiplicative, while π is a coalgebra map. Hence we have written $K \otimes K^{\mathrm{op}}$ as a bismash product in a nontrivial way. As it turns out, the associated double crossproduct $K^{\mathrm{op}} \bowtie K^*$ is (a version of) the Drinfeld double of K.

Free with our approach to the bijection between Singer and matched pairs, we get an intrinsic connection between the bismash product $K \# Q$ constructed from a Singer pair (K, Q), and the double crossproduct $Q \bowtie K^*$ constructed from the associated matched pair (Q, K^*) in the shape of category equivalences

$$ {}^{K \# Q}_{K}\mathcal{M}_K \cong {}^{Q \bowtie K^*}\mathcal{M} \quad \text{and} \quad {}_{K^*}\mathcal{M}^{Q \bowtie K^*}_{K^*} \cong \mathcal{M}^{K \# Q}. $$

Admittedly, it is not obvious what kind of information on the Hopf algebras might be drawn from this connection, but at least we can give a rather strange application to Drinfeld doubles via the center construction. Recall that the category of modules over the Drinfeld double $D(H)$ of a finite dimensional Hopf algebra H is equivalent to the category ${}^{H}_{H}\mathcal{YD}$ of (left-left) Yetter–Drinfeld modules, which is in turn equivalent to the center $\mathcal{Z}({}^{H}\mathcal{M})$ of the monoidal category ${}^{H}\mathcal{M}$, as well as to the center of the monoidal category ${}_{H}\mathcal{M}$. Since our application will not appear again elsewhere in this paper, we just refer to [23] for more details on these facts (with slightly different conventions), and proceed to do the following calculation, with the assumption that we are given a Singer pair (K, Q) with K finite:

$$ {}^{Q \bowtie K^*}_{Q \bowtie K^*}\mathcal{YD} \cong \mathcal{Z}\left({}^{Q \bowtie K^*}\mathcal{M}\right) \cong \mathcal{Z}\left({}^{K \# Q}_{K}\mathcal{M}_K\right) \cong \mathcal{Z}\left({}^{K \# Q}\mathcal{M}\right) \cong {}^{K \# Q}_{K \# Q}\mathcal{YD}. $$

The only nontrivial part is the third category equivalence in the chain. This is the fact that (under additional hypotheses of exactness of certain tensor products) for any algebra K in any monoidal category $\mathcal{C}$, the center of ${}_K\mathcal{C}_K$ is equivalent to the center of $\mathcal{C}$; the proof is quite complicated [46], but purely categorical and not at all specific to the Hopf algebra situation. As a consequence we get, if Q is also finite, an equivalence ${}_{D(Q\bowtie K^*)}\mathcal{M} \cong {}_{D(K\#Q)}\mathcal{M}$ of braided monoidal categories, and we conclude that the Drinfeld doubles $D(K\#Q)$ and $D(Q \bowtie K^*)$ are isomorphic up to a Drinfeld twist. Note that they should not be expected to be isomorphic, since Q is not (at least not in the obvious way) a subcoalgebra of $D(K\#Q)$, but Q is a Hopf subalgebra of $D(Q \bowtie K^*)$. A special case of the above occurs if the double crossproduct $A = Q \bowtie K^*$ is itself a Drinfeld double, which can be arranged with the associated bismash product $K \otimes K^{\mathrm{op}}$. Thus $D(A) \cong D(K \otimes K^{\mathrm{op}}) \cong D(K) \otimes D(K^{\mathrm{op}}) \cong A \otimes A$ up to a twist. This was stated more generally for factorizable quasitriangular A by Reshetikhin and Semenov–Tian–Shansky [39], see [51].

Trivalent products exhaust the examples where both H and $\mathfrak{F}(H)$ are ordinary bialgebras *with trivial coassociators* (while it may occur that $\mathfrak{F}(H)$ is a nontrivial coquasibialgebra, but also happens to be a bialgebra with the same multiplication, typically when $\mathfrak{F}(H)$ is cocommutative; we shall look at this case in more detail in section 5 below).

But of course one can apply $\mathfrak{F}$ to any inclusion $K \subset H$ of a Hopf algebra into a bialgebra with cocleaving $\pi: H \to K$. This situation yields a mapping system (3–1) in which Q is a coalgebra and nonassociative algebra, satisfying axioms that are very unpleasant even in the case [44] where π is a coalgebra map. One can characterize precisely [47, Sec.5.1] which mapping systems

$$K^* \underset{\tilde{\pi}}{\overset{\tilde{\iota}}{\rightleftarrows}} \tilde{H} \underset{j}{\overset{\tilde{\nu}}{\rightleftarrows}} Q$$

arise under $\mathfrak{F}$ from bialgebras H. What's more, one can show that the functor $\mathfrak{F}^{\mathrm{bop}}$ is an inverse mapping for $\mathfrak{F}$ as a map between the class of bialgebras in $\mathfrak{E}_\ell(K)$, and the corresponding subclass of $\mathfrak{E}_r(K^*)$.

In general the functor $\mathfrak{F}$ has a vague tendency to be an involution in this sense: When one lists all the combinatorial data involving only K and Q rather than the whole object $H \in \mathfrak{E}_\ell(K)$ (such data are the restrictions of ϕ to combinations of copies of K and Q in each argument, the mutual actions and coactions between K and Q, or the cocycles in a bicrossproduct), then it turns out that $\mathfrak{F}$ will look like an involution on all these data, acting on them essentially by reinterpreting maps involving K as maps involving K^* through duality. However, there is no reason to expect that the combinatorial data determines H completely (as opposed to the ordinary bialgebra case), and we know of no intrinsic arguments leading to a general isomorphism $\mathfrak{F}^{\mathrm{bop}}(\mathfrak{F}(H)) \cong H$. See [47, Sec. 4].

4.3. Remedies for infinite dimensional cases.
The general construction reviewed in subsection 4.1 applies only when the subobject $K \subset H$ is finitely generated projective. In this section we shall investigate how much can be done

in the case where K is infinite. We assume throughout this section that the base ring k is a field.

THEOREM 4.3.1. *Let K be a Hopf algebra, and $H \in \mathfrak{E}_\ell(K)$. Let ${}_K^H\mathcal{M}_K$ be the full subcategory of ${}_K^H\mathcal{M}_K$ whose objects are finitely generated left K-modules. ${}_K^H\mathcal{M}_K$ is a monoidal subcategory of ${}_K^H\mathcal{M}_K$. The functor $\omega: {}_K^H\mathcal{M}_K \to {}_k\mathcal{M}$ with $\omega(M) = \overline{M} = M/K^+M$ factors over an equivalence $\hat{\omega}: {}_K^H\mathcal{M}_K \to {}^{\tilde{H}}\mathcal{M}_f$, where $\tilde{H}$ is a coquasibialgebra that can be realized as a subspace of $Q \otimes K^*$ with coquasibialgebra structure determined by*

$$(Q \otimes x^* \otimes Q \otimes y^*)\Delta(q \otimes \varphi) = q_{(1)} \otimes q_{(2)} \leftharpoonup x_{(1)}\phi(q_{(3)} \otimes x_{(2)} \otimes y_{(1)})\langle \varphi, x_{(3)}y_{(2)}\rangle,$$

$$(p \otimes \varphi)(q \otimes \psi) = \overline{j(p_{(1)})j(q_{(1)})_{(1)}} \otimes \kappa(j(p_{(2)}) \otimes j(q_{(1)})_{(2)})((\varphi \triangleleft \pi(j(q_{(1)})_{(3)})) \leftharpoonup q_{(2)})\psi,$$

and

$$\tilde{\phi}(p \otimes \varphi \otimes q \otimes \psi \otimes r \otimes \vartheta) =$$
$$\phi(j(p) \otimes j(q)_{(1)} \otimes j(r)_{(1)})\langle \varphi | \pi(j(q)_{(2)}j(r)_{(2)})\rangle\langle \psi | \pi(j(r)_{(3)})\rangle\varepsilon(\vartheta)$$

for all $p, q, r \in Q$, $\varphi, \psi, \vartheta \in K^*$, and $x, y \in K$, where $x^*, y^*: K^* \to k$ correspond to $x, y \in K$.

PROOF. Although the tensor product in ${}_K^H\mathcal{M}_K$ is not an ordinary tensor product, it is still true that if $M, N \in {}_K^H\mathcal{M}_K$ are finitely generated as left K-modules, then so is $M \underset{K}{\otimes} N$: Clearly the image of $M \otimes V$ in $M \underset{K}{\otimes} N$ is finitely generated whenever V is a finite dimensional subspace of N. Now since N is finitely generated, we can choose a finite dimensional H-subcomodule of N generating N as left K-module. Then $M \underset{K}{\otimes} N$ is the image of $M \otimes V$, since for $m \in M$, $x \in K$ and $v \in V$ we have

$$m \otimes xv = \phi(m_{(-1)} \otimes x_{(1)} \otimes v_{(-1)})m_{(0)}x_{(2)} \otimes v_{(0)}$$

and V is a subcomodule.

With the aim of applying Lemma 2.3.2, we define $\tilde{H}^1 := Q \otimes K^*$, $\tilde{H}^2 :=$ $\mathrm{Hom}(K \otimes K, Q \otimes Q)$, and $\iota: Q \otimes K^* \otimes Q \otimes K^* \to \mathrm{Hom}(K \otimes K, Q \otimes Q)$ by

$$\iota(p \otimes \varphi \otimes q \otimes \psi)(x \otimes y) = p \otimes q\langle \varphi, x\rangle\langle \psi, y\rangle,$$

that is $\iota(F)(x \otimes y) = (Q \otimes x^* \otimes Q \otimes y^*)(F)$. Further, we define $\Delta^1: \tilde{H}^1 \to \tilde{H}^2$ by

$$\Delta^1(q \otimes \varphi)(x \otimes y) = q_{(1)} \otimes q_{(2)} \leftharpoonup x_{(1)}\phi(q_{(3)} \otimes x_{(2)} \otimes y_{(1)})\langle \varphi, x_{(3)}y_{(2)}\rangle,$$

and $\varepsilon^1(q \otimes \varphi) = \varepsilon(q)\varphi(1)$.

Now ${}_K^H\mathcal{M}_K$ is equivalent to the full subcategory $({}^Q\mathcal{M}_f)_K$ of $({}^Q\mathcal{M})_K$ consisting of finite-dimensional objects. We have an equivalence $({}^Q\mathcal{M}_f)_K \cong {}^{\tilde{H}^1}\mathcal{M}_f$, where the latter category is to be understood in the sense of Lemma 2.3.2: This can be proved in essentially the same manner as in subsection 4.1. In fact the form of comultiplication we have given here was already used in the proof there. The only difference is that this time the action $V \otimes K \to V$ on $V \in \left({}^{Q}\mathcal{M}\right)_K$ gives rise to $V \to K^* \otimes V$ since V is finite-dimensional. Now we can apply Lemma 2.3.2. The formulas for multiplication and coassociator are proved as in subsection 4.1.

$\square$

The inclusion functor ${}^{K}_{K}\mathcal{M}_K \to {}^{H}_{K}\mathcal{M}_K$ restricts to those Hopf modules that are finitely generated left K-modules to give an inclusion ${}_f\mathcal{M}_K \to {}^{\tilde{H}}\mathcal{M}_f$ from finite dimensional K-modules to finite dimensional $\tilde{H}$-comodules. Hence we have a natural map $K^\circ \to \tilde{H}$, which is easily seen to have the obvious form when we consider $\tilde{H} \subset Q \otimes K^*$. We also have an obvious coalgebra map $\tilde{\nu}\colon \tilde{H} \to Q$, and a natural map $\tilde{\pi}\colon H \to K^*$ coming from the natural transformations $\omega \to \omega$ that arise from right multiplication with an element of K.

However, we do not know if $\tilde{H}$ contains all of $Q \otimes K^\circ$, nor if it is contained in $Q \otimes K^\circ$, nor if $\tilde{\pi}$ takes values in K°. When H is an ordinary bialgebra with trivial coquasibialgebra structure ϕ, then all of the $\omega(M)$ for $M \in {}^{H}_{K}\underline{\mathcal{M}}_K$ are ordinary right K-modules, of finite dimension, so all elements of $\tilde{H}$ are contained in $Q \otimes K^\circ$, and $\tilde{\pi}$ takes values in K°. The same conclusion can be obtained under the weaker assumption that the map $\tilde{\tau}\colon H \to (K \otimes K)^*$ induced by ϕ takes values in $K^\circ \otimes K^\circ$. We then write

$$\tilde{\tau}(q) := \tilde{\tau}^{(1)}(q) \otimes \tilde{\tau}^{(2)}(q) \in K^\circ \otimes K^\circ$$

and denote the convolution inverse of $\tilde{\tau}$ by

$$\tilde{\tau}^{-1}(q) := \tilde{\tau}^{-(1)}(q) \otimes \tilde{\tau}^{-(2)}(q).$$

To show that $v_{[-\tilde{1}]} \otimes v_{[\tilde{0}]} \in K^\circ \otimes V$ for all $v \in V \in \left({}^{Q}\mathcal{M}_f\right)_K$ we calculate

$$
\begin{aligned}
\langle v_{[-\tilde{1}]}, xy \rangle v_{[\tilde{0}]} &= v(xy) = \phi^{-1}(v_{(-1)} \otimes x_{(1)} \otimes y_{(1)})(v_{(0)}x_{(2)})y_{(2)} \\
&= \phi^{-1}(v_{(-1)} \otimes x_{(1)} \otimes y_{(1)})\langle v_{(0)[\tilde{0}][-\tilde{1}]}, x_{(2)} \rangle \langle v_{(0)[\tilde{0}][-\tilde{1}]}, y_{(2)} \rangle v_{(0)[\tilde{0}][\tilde{0}]} \\
&= \langle \tilde{\tau}^{-(1)}(v_{(-1)})v_{(0)[-\tilde{1}]}, x \rangle \langle \tilde{\tau}^{-(2)}(v_{(-1)})v_{(0)[\tilde{0}][-\tilde{1}]}, y \rangle v_{(0)[\tilde{0}][\tilde{0}]}
\end{aligned}
$$

for $x, y \in K$, showing $\Delta(v_{[-\tilde{1}]}) \otimes v_{[\tilde{0}]} \in K^* \otimes K^* \otimes V$.

To show that, under additional hypotheses, $\tilde{H}$ contains all of $Q \otimes K^\circ$, we need a sufficient supply of objects in $\left({}^{Q}\mathcal{M}_f\right)_K$. This time, unfortunately, we cannot get away without a calculation involving cocycle identities. It is unfortunate that there appears to be no intrinsic reason to expect the object $H \otimes K^*$ to be defined below to be in ${}^{H}_{K}\mathcal{M}_K$ (as opposed to the object $H \otimes K$ used in subsection 4.1.) If K is finite, the object $\overline{H \otimes K^*}$ is just the regular left $\tilde{H}$-comodule $\tilde{H}$.

Define $\beta\colon H \otimes K \to K^\circ$ by $\beta(h \otimes x)(y) = \phi(h \otimes x \otimes y)$ for $h \in H$ and $x, y \in K$. We note that $\beta(xh \otimes y) = \varepsilon(x)\beta(h \otimes y)$, a reformulation of [47, Lem. 3.2.1]. Moreover

$$\beta(h_{(1)}x_{(1)} \otimes y_{(1)})(\beta(h_{(2)} \otimes x_{(2)}) \triangleleft y_{(2)}) = \phi(h_{(1)} \otimes x_{(1)} \otimes y_{(1)})\beta(h_{(2)} \otimes x_{(2)}y_{(2)})$$

as a special case of the cocycle identity of ϕ, taking $\phi|_{K \otimes K \otimes K} = \varepsilon$ into account. It follows that $H \otimes K^* \in {}^H_K \mathcal{M}_K$ with the left H-comodule and K-module structures induced by those of the left tensor factor, and the right K-action defined by $(h \otimes \varphi)x = h_{(1)}x_{(1)} \otimes \beta(h_{(2)} \otimes x_{(2)})(\varphi \triangleleft x_{(3)})$. In fact

$$\phi(h_{(1)} \otimes x_{(1)} \otimes y_{(1)})(h_{(2)} \otimes \varphi)(x_{(2)}y_{(2)})$$
$$= \phi(h_{(1)} \otimes x_{(1)} \otimes y_{(1)})h_{(2)}(x_{(2)}y_{(2)}) \otimes \beta(h_{(3)} \otimes x_{(3)}y_{(3)})(\varphi \triangleleft x_{(4)}y_{(4)})$$
$$= (h_{(1)}x_{(1)})y_{(1)} \otimes \phi(h_{(2)} \otimes x_{(2)} \otimes y_{(2)})\beta(h_{(3)} \otimes x_{(3)}y_{(3)})(\varphi \triangleleft x_{(4)}y_{(4)})$$
$$= (h_{(1)}x_{(1)})y_{(1)} \otimes \beta(h_{(2)}x_{(2)} \otimes y_{(2)})(\beta(h_{(3)} \otimes x_{(3)}) \triangleleft y_{(3)})(\varphi \triangleleft x_{(4)}y_{(4)})$$
$$= (h_{(1)}x_{(1)})y_{(1)} \otimes \beta(h_{(2)}x_{(2)} \otimes y_{(2)})((\beta(h_{(3)} \otimes x_{(3)})(\varphi \triangleleft x_{(4)})) \triangleleft y_{(3)})$$
$$= (h_{(1)}x_{(1)} \otimes \beta(h_{(2)} \otimes x_{(2)})(\varphi \triangleleft x_{(3)}))y = ((h \otimes \varphi)x)y$$

and

$$(xh \otimes \varphi)y = x_{(1)}h_{(1)}y_{(1)} \otimes \beta(x_{(2)}h_{(2)} \otimes y_{(2)})(\varphi \triangleleft y_{(3)})$$
$$= xh_{(1)}y_{(1)} \otimes \beta(h_{(2)} \otimes y_{(2)})(\varphi \triangleleft y_{(3)}) = x((h \otimes \varphi)y)$$

for all $x, y \in K$, $h \in H$, and $\varphi \in K^*$, while $H \otimes K^* \in {}^H_K \mathcal{M}$ holds trivially.

THEOREM 4.3.2. *Let k be a field, K a Hopf algebra, and $H \in \mathfrak{E}_\ell(K)$. Assume that H is an ordinary right K-module (for example, ϕ is trivial or H is cocommutative), and that the action of K on Q is locally finite. Assume that the map $\tilde{\tau}: Q \to (K \otimes K)^*$ defined by ϕ takes values in $K^\circ \otimes K^\circ$.*
Then $\tilde{H} = Q \otimes K^\circ \in \mathfrak{E}_r(K^\circ)$.

PROOF. We have already shown $\tilde{H} \subset Q \otimes K^\circ$. Since $\beta: Q \otimes K \to K^*$ takes values in K°, and $K^\circ \subset K^*$ is stable under the right action of K, we see that $H \otimes K^\circ \subset H \otimes K^*$ is a subobject in ${}^H_K \mathcal{M}_K$. We consider the object $\omega(H \otimes K^*) \cong Q \otimes K^* \in ({}^Q \mathcal{M})_K$. We claim that $Q \otimes K^\circ$ is the union of its finite dimensional subobjects. In fact by local finiteness of the K-module Q, any $q \in Q$ is contained in a finite dimensional subcoalgebra $C \subset Q$ that is in addition stable under the action of K. The image of $\beta|_{C \otimes K}$ is contained in some finite dimensional subspace of K°, and for any $\varphi \in K^\circ$ we know that $\varphi \triangleleft K$ is finite dimensional. We conclude that $(q \otimes \varphi)K$ is finite dimensional, and of course a subobject of $Q \otimes K^\circ$ in $({}^Q \mathcal{M})_K$. We conclude the proof by observing that $q \otimes \varphi = (q \otimes \varphi)_{(-1)}(\varepsilon \otimes \eta^*)((q \otimes \varphi)_{(0)})$. $\square$

4.4. Hopf algebra inclusions and antipodes. So far, we have never treated the question when the coquasibialgebra associated to an inclusion $K \subset H$ is a coquasi-Hopf algebra. We shall turn to this question now, but only derive some criteria for the case of an inclusion of Hopf algebras.

If $K \subset H$ is an inclusion of Hopf algebras with bijective antipodes, and $\pi: H \to K$ is a cocleaving map for the left K-module coalgebra H, then it is straightforward to check that $\pi' = S_K \pi S_H^{-1}: H \to K$ is a cocleaving for the right

K-module coalgebra H, so we can simply say below that H is K-cocleft in this case.

Obviously a cocleaving map for the left K-module coalgebra H is also a cocleaving map for the right K^{bop}-module coalgebra H^{bop}, so that H is K-cocleft if and only if H^{bop} is K^{bop}-cocleft, and H^{op} is K^{op}-cocleft if and only if H^{cop} is K^{cop}-cocleft.

For the rest of the section we assume that $K \subset H$ is an inclusion of Hopf algebras with bijective antipodes and H is K-cocleft, that is, $H \in \mathfrak{E}_\ell(K)$.

To see whether $\tilde{H}$ has a coquasi-antipode, we shall first investigate dual objects in the Hopf module category $_K^H\mathcal{M}_K$.

If $M \in {}_K^H\mathcal{M}_K$ has a right (resp. left) dual, then, since monoidal functors preserve duals, M is necessary finitely generated projective as left (resp. right) K-module, and the underlying K-K-bimodule of the right (resp. left) dual of M is ${}^\vee M = \mathrm{Hom}_{K-}(M, K)$ (resp. $M^\vee = \mathrm{Hom}_{-K}(M, K)$.) On the other hand, when M is finitely generated projective as left (resp. right) K-module, Ulbrich [60] has given a comodule structure on $\mathrm{Hom}_{K-}(M, N)$ (resp. $\mathrm{Hom}_{-K}(M, N)$) for every Hopf module $N \in {}_K^H\mathcal{M}_K$ that defines an inner hom-functor in the category $_K^H\mathcal{M}_K$. The comodule structure is defined by

$$f_{(-1)} \otimes f_{(0)}(m) = S(m_{(-1)}) f(m_{(0)})_{(-1)} \otimes f(m_{(0)})_{(0)}$$

for $f \in \mathrm{Hom}_{K-}(M, N)$, and

$$f_{(-1)} \otimes f_{(0)}(m) = f(m_{(0)})_{(-1)} S^{-1}(m_{(-1)}) \otimes f(m_{(0)})_{(0)}$$

for $f \in \mathrm{Hom}_{-K}(M, N)$. (More generally, if k is a field, then a certain submodule of $\mathrm{Hom}_{K-}(M, N)$ is is an inner hom-functor if M is not finitely generated. In fact, from the results in [36] cited in subsection 2.5, we can expect $_K^H\mathcal{M}_K$ to be closed even if H is only a coquasi-Hopf algebra, since then $^H\mathcal{M}$ is closed; however, the inner hom-functors have a complicated description which we have not been able to use.) Since the canonical maps $^\vee M \otimes N \to \mathrm{Hom}_{K-}(M, N)$ are bijective if $_K M$ is finitely generated projective, it follows that $^\vee M$ is a right dual of M in the category $_K^H\mathcal{M}_K$. Similarly $M^\vee$ is a left dual of M if M is finitely generated projective as right K-module. All these considerations do not give us any left or right rigid subcategory of $_K^H\mathcal{M}_K$, since, for example, when M is left finitely generated projective, the same needs not hold for $^\vee M$. We do get two subcategories of $_K^H\mathcal{M}_K$ that are in full duality with each other by taking left, resp. right duals.

Thus, we are interested in cases where the two finiteness conditions on a Hopf module in $_K^H\mathcal{M}_K$ coincide. In fact only finite generation will be an issue: If k is a field, then by our general cocleftness assumption every Hopf module in $_K^H\mathcal{M}$ is a free K-module. In addition every Hopf module in $^H\mathcal{M}_K$ is a projective K-module by [48, Rem. 4.3]. In Theorem 4.4.3 we shall assume moreover that H^{cop} is K^{cop}-cocleft, so that every Hopf module in $^H\mathcal{M}_K \cong {}_{K^{\mathrm{op}}}^{H^{\mathrm{op}}}\mathcal{M}$ will even be a free K-module by more elementary reasons. The key to dealing with finite generation is the following simple observation, which follows immediately from the fact that Hopf modules in $_K^H \mathcal{M}$ are free K-modules:

REMARK 4.4.1. Let k be a field and K a Hopf algebra, and H a cocleft left K-module coalgebra. Then $M \in {}_K^H \mathcal{M}$ is a finitely generated K-module if and only if $M/K^+ M$ is finite dimensional. If this is the case, then every Hopf submodule of M is also a finitely generated K-module.

For the proof of Theorem 4.4.3 it is convenient to provide

LEMMA 4.4.2. Let k be a field and $H \in \mathfrak{E}_\ell(K)$, where H is a bialgebra, and K is a Hopf algebra. If $Q := H/K^+ H$ is a locally finite right K-module, then H is the union of those of its K-K-bimodule subcoalgebras that are finitely generated as left K-modules.

PROOF. Let $h \in H$, and choose a finite dimensional subcoalgebra $E \subset H$ with $h \in E$. Put $D := KEK \ni h$, which is by construction a K-K-bimodule subcoalgebra of H, and in particular a subobject of H in the Hopf module category $_K^H \mathcal{M}_K$. By local finiteness $\nu(D) = \nu(E)K \subset Q$ is finite dimensional, hence D is a finitely generated left K-module by the preceding remark. □

THEOREM 4.4.3. Let k be a field, and $H \in \mathfrak{E}_\ell(K)$, where H and K are Hopf algebras with bijective antipodes.

Assume that H^{cop} is cocleft as a K^{cop}-module coalgebra, and $Q := H/K^+ H$ is a locally finite right K-module.

Then the category $_K^H \underline{\mathcal{M}}_K$ is left and right rigid.

PROOF. The cocleftness assumption on H^{cop} is equivalent to $H^{\mathrm{op}} \in \mathfrak{E}_\ell(K^{\mathrm{op}})$ (we have used $H^{\mathrm{cop}} \in \mathfrak{E}_\ell(K^{\mathrm{cop}})$ for the statement of the theorem since H will be cocommutative in our application in subsection 4.5). Using the bijective antipodes one sees that H/HK^+ is a locally finite left K-module, or stated differently $H^{\mathrm{op}}/(K^{\mathrm{op}})^+ H^{\mathrm{op}}$ is a locally finite right K^{op}-module. In other words, the inclusion $K^{\mathrm{op}} \subset H^{\mathrm{op}}$ satisfies the same hypotheses as $K \subset H$.

We will show that these hypotheses imply that $M \in {}_K^H \mathcal{M}_K$ is a finitely generated left K-module if and only if it is a finitely generated right K-module, which is all we need by the discussion preceding Remark 4.4.1.

Due to the identification $_K^H \mathcal{M}_K \cong {}_{K^{\mathrm{op}}}^{H^{\mathrm{op}}} \mathcal{M}_{K^{\mathrm{op}}}$ and the remarks at the beginning of the proof we need only show one of the implications. So let $M \in {}_K^H \mathcal{M}_K$ be finitely generated as left K-module. We can apply Remark 4.4.1 to the left K^{op}-module structure of M and need only show that M can be embedded in a larger Hopf module $N \in {}^H \mathcal{M}_K$ such that N is a finitely generated right K-module. To construct N, let C be a finite dimensional subcoalgebra of Q such that the Q-comodule $\overline{M}$ is a C-comodule. Then $m_{(-1)} \otimes \overline{m_{(0)}} = \pi(m_{(-2)}) j(\overline{m_{(-1)}}) \otimes \overline{m_{(0)}} \in Kj(C) \otimes \overline{M}$ for all $m \in M$. We can apply Lemma 4.4.2 to the left K^{op}-module coalgebra H^{op} to conclude that $j(C)$ and hence $Kj(C)$ is contained in a K-K-bimodule subcoalgebra $D \subset H$ that is finitely generated as right K-module, say by a finite-dimensional subspace $D' \subset D$. We have $M \cong H \square_Q \overline{M} \cong D \square_Q \overline{M} \subset$ $D. \otimes \overline{M}.$ as Hopf modules in $^H\mathcal{M}_K$, and $N := D \otimes \overline{M}$ is generated as right K-module by $D' \otimes \overline{M}$, since for all $d \in D'$, $x \in K$, and $v \in \overline{M}$ we have $dx \otimes v = dx_{(1)} \otimes vS^{-1}(x_{(3)})x_{(2)} = (d \otimes vS^{-1}(x_{(2)}))x_{(1)} \in (D' \otimes \overline{M})K.$ □

REMARKS 4.4.4. (1) If H is finite, and a Hopf algebra, then the conditions in Theorem 4.4.3 are always satisfied.

(2) Under the assumptions of Theorem 4.4.3, $\tilde{H} \cong Q \otimes K^\circ$ is a cosmash product as a coalgebra.

(3) If k is a commutative base ring, but we assume that K is finitely generated projective, and H^{cop} is cocleft as K^{cop}-module coalgebra, then we can also conclude that $^H_K\mathcal{M}_K$ is left and right rigid, because now every Hopf module $M \in {}^H_K\mathcal{M}_K$ can be written as $K \otimes (M/K^+M)$ as well as $(M/MK^+) \otimes K$; thus if M is finite projective as left K-module, then M/K^+M is finitely generated projective over k, hence so is M, hence M/MK^+ is finitely generated projective over k, and M finitely generated projective as right K-module.

Now for $\tilde{H}$ to have a coquasi-antipode, we not only need $^H_K\mathcal{M}_K$ to have duals, but we also need the underlying functor $^H\mathcal{M} \to {}_k\mathcal{M}$, hence the functor $\omega : {}^H_K\mathcal{M}_K \to {}_k\mathcal{M}$, to preserve these duals.

THEOREM 4.4.5. *Let* $H \in \mathfrak{E}_\ell(K)$, *where* H *and* K *are Hopf algebras with bijective antipodes. Assume that* K *and* H *are finitely generated projective, or that* k *is a field.*

If there is an isomorphism $H \cong K. \otimes Q.$ *of right* Q-*comodules and* K-*modules, then* $\tilde{H}$ *is a quasi-Hopf algebra.*

PROOF. The isomorphism $H \cong K. \otimes Q.$ yields an isomorphism of right K-modules, natural in $M \in {}^H_K\mathcal{M}_K$,

$$M \cong H \square_Q \overline{M} \cong (K. \otimes Q.) \square_Q \overline{M}. \cong K. \otimes \overline{M}. \cong \overline{M} \otimes K.$$

(where the last isomorphism maps $x \otimes v \mapsto vS^{-1}(x_{(2)}) \otimes x_{(1)}$). In particular, $M \in {}^H_K\mathcal{M}_K$ is a finitely generated left K-module, iff it is a finitely generated projective right K-module, iff $\overline{M}$ is a finitely generated projective k-module. In particular, $^H_K\mathcal{M}_K \cong {}^{\tilde{H}}\mathcal{M}_f$ is left and right rigid. Moreover, one has a natural isomorphism $\mathrm{Hom}_{K-}(M, K) \cong \mathrm{Hom}_{K-}(K \otimes \overline{M}, K) \cong \overline{M}^* \otimes K$ of right K-modules, and thus a natural isomorphism $\omega(\mathrm{Hom}_{K-}(M, K)) \cong \mathrm{Hom}_{K-}(M, K)/\mathrm{Hom}_{K-}(M, K)K^+ \cong \overline{M}^*$ of k-modules. Thus, the underlying functor $^{\tilde{H}}\mathcal{M}_f \to {}_k\mathcal{M}$ preserves duals, and $\tilde{H}$ is a quasi-Hopf algebra. □

The criterion given in Theorem 4.4.5 is sharp whenever K is finite:

THEOREM 4.4.6. *Let* k *be a field,* K *a finite dimensional Hopf algebra, and* $H \in \mathfrak{E}_\ell(K)$ *a Hopf algebra with bijective antipode. If* $\tilde{H}$ *is a quasi-Hopf algebra, then there is an isomorphism* $H \cong K. \otimes Q.$ *of right* Q-*comodules and* K-*modules.*

PROOF. By assumption the functor $\omega\colon {}^H_K\mathcal{M}_K \to {}_k\mathcal{M}$ preserves duals. Now for any finite dimensional $M \in {}^H_K\mathcal{M}_K$ the right dual in ${}^H_K\mathcal{M}_K$ is

$$^\vee M = \mathrm{Hom}_{K-}(M, K) \cong \overline{M}^* \otimes K$$

as right K-modules, and hence we have a natural isomorphism $\omega(^\vee M) \cong \overline{M}^* \cong$ $^\vee M/(^\vee M)K^+$ for $M \in {}^H_K\mathcal{M}_K$. Since $^\vee(-)$ is a full duality, we have a natural isomorphism $M/K^+M \cong M/MK^+$ for $M \in {}^H_K\mathcal{M}_K$, which specializes for $M = H \otimes K$ to an isomorphism $f\colon Q \otimes K \to H$. For $x \in K$ an automorphism of $H \otimes K$ in ${}^H_K\mathcal{M}_K$ is given by $h \otimes y \mapsto hx_{(2)} \otimes S^{-1}(x_{(1)})y$. Applying naturality of f to this automorphism yields $f(qx_{(2)} \otimes S^{-1}(x_{(1)})y) = f(q \otimes y)x$. For any $\gamma \in Q^*$, an automorphism of $H \otimes K$ in ${}^H_K\mathcal{M}_K$ is given by $h \otimes x \mapsto h_{(1)}\gamma(\overline{h_{(2)}}) \otimes x$. Applying naturality of f yields $f(q_{(1)}\gamma(q_{(2)}) \otimes y) = f(q \otimes y)_{(1)} \otimes \overline{f(q \otimes y)_{(2)}}$. From this we conclude that $K \otimes Q \ni y \otimes q \mapsto f(q \otimes S(y)) \in H$ is a Q-comodule and K-module map as required. $\qquad\square$

The following Lemma gives a sufficient condition to have an isomorphism as needed in Theorem 4.4.5. It is designed to apply to the examples in subsection 4.5 below, and asks essentially for H to be right K-cocleft in a very particular way.

LEMMA 4.4.7. *Let $K \subset H$ be an inclusion of Hopf algebras with bijective antipodes, such that H is cleft as a left K-module coalgebra.*

Denote by $\underline{h}$ the class in H/HK^+ of $h \in H$. Assume that there exists a coalgebra section $\gamma\colon H/HK^+ \to H$ for the canonical surjection, such that

$$\vartheta\colon H/HK^+ \otimes K \ni \underline{h} \otimes x \mapsto \gamma(\underline{h})x \in H$$

and $\nu\gamma\colon H/HK^+ \to Q$ are bijections.

Then there is an isomorphism $H \cong K \otimes Q$ of right K-modules and right Q-comodules.

PROOF. Being a coalgebra map, γ is convolution invertible with inverse $\gamma^{-1} = S\gamma$.

Since γ is a coalgebra map, ϑ and hence its inverse are left H/HK^+-comodule maps, and thus ϑ^{-1} can be written in the form $\vartheta^{-1}(h) = \underline{h_{(1)}} \otimes \kappa(h_{(2)})$ for some right K-module map $\kappa\colon H \to K$. By definition $\vartheta(h_{(1)})\kappa(\overline{h_{(2)}}) = h$ for all $h \in H$, hence $\kappa(h) = \vartheta^{-1}(h_{(1)})h_{(2)}$. Since $\Delta\gamma^{-1}(h) = \gamma^{-1}(h_{(2)}) \otimes \gamma^{-1}(h_{(1)})$, we have $\Delta(\kappa(h)) = \vartheta^{-1}(h_{(2)})h_{(3)} \otimes \vartheta^{-1}(h_{(1)})h_{(4)} = \kappa(h_{(2)}) \otimes \vartheta^{-1}(h_{(1)})h_{(3)}$. Now we claim that the (bijective) map

$$\ell\colon H \ni h \mapsto \nu(\gamma(\underline{h_{(1)}})) \otimes \kappa(h_{(2)}) \in Q \otimes K$$

is a map of right Q-comodules and K-modules as indicated. From this, the claim follows since

$$Q \otimes K \ni q \otimes x \mapsto x_{(1)} \otimes qx_{(2)} \in K \otimes Q.$$

is an isomorphism of right Q-comodules and K-modules. Now ℓ is obviously bijective (since $\nu\gamma$ is), and right K-linear. Moreover, the right comodule structure on $Q \otimes K$ gives

$$\ell(h)_{(0)} \otimes \ell(h)_{(1)} = (\nu\gamma(h_{(1)}) \otimes \kappa(h_{(2)}))_{(0)} \otimes (\nu\gamma(h_{(1)}) \otimes \kappa(h_{(2)}))_{(1)}$$

$$= \nu(\gamma(h_{(1)}))_{(1)} \otimes \kappa(h_{(2)})_{(1)} \otimes \nu(\gamma(h_{(1)}))_{(2)}\kappa(h_{(2)})_{(2)}$$

$$= \nu(\gamma(h_{(1)})) \otimes \kappa(h_{(4)}) \otimes \nu(\gamma(h_{(2)})\gamma(h_{(3)})h_{(5)})$$

$$= \nu(\gamma(h_{(1)})) \otimes \kappa(h_{(2)}) \otimes \nu(h_{(3)}) = \ell(h_{(1)}) \otimes \overline{h_{(2)}}$$

so that ℓ is Q-colinear. $\qquad\qquad\qquad\qquad\qquad\qquad\qquad\qquad\qquad\qquad\square$

4.5. Yongchang Zhu's example. A special case of our construction assigning a coquasibialgebra to any inclusion of Hopf algebras was previously obtained by Yongchang Zhu [61] who constructed a quasi-Hopf algebra $A(G, B)$ associated to any inclusion $B \subset G$ of finite groups. Some indications are also given for the case of an inclusion of infinite groups. The motivation in [61] is a close relation between the category of representations of $A(G, B)$ and the Hecke algebra $H(G//B)$ associated to the inclusion. In this section we shall give a brief overview how [61] relates to the construction in [47] cited above.

Plainly, if $B \subset G$ is an inclusion of finite groups, we can consider $k[G]$ as an object in $\mathfrak{E}_\ell(k[B])$ once we choose a cocleaving map $\pi : k[G] \to k[B]$. To have a result that compares easily to [61], we do this in the following rather roundabout manner: Choose a set R of representatives for the set G/B of right cosets. Then every $g \in G$ can be written uniquely in the form $g = [g]\{g\}^{-1}$ for $[g] \in R$ and $\{g\} \in B$. We choose the cocleaving $\pi : k[G] \to k[B]$ with $\pi(g) = \{g^{-1}\}$ and find $g = \{g^{-1}\}[g^{-1}]^{-1}$, hence $j(\bar{g}) = [g^{-1}]^{-1}$ for the associated map $j : Q = k[G]/k[B]^+k[G] \to k[G]$. We identify Q with $k[R]$ by $\bar{g} = [g^{-1}]$, so that we have $j(r) = r^{-1}$ for $r \in R$. We denote by $\hat{b}$ for $b \in B$ the elements of the basis of k^B dual to the basis B of $k[B]$. The coquasibialgebra $\tilde{H}$ from subsection 4.1 identifies with $k[R] \otimes k^B$, with canonical basis consisting of the $r \otimes \hat{b}$ with $r \in R$ and $b \in B$, and it remains to specialize the formulas for the coquasibialgebra structure. The right action of $k[B]$ on Q translates as $r \leftharpoonup b = \overline{r^{-1}b} = [b^{-1}r]$. The action $Q \otimes k[B] \to k[B]$ translates as $r \rightharpoonup b = \pi(j(r)b) = \{b^{-1}r\}$. Then the coaction of k^B on $k[R]$ is given by $r_{[-\bar{1}]} \otimes r_{[\bar{0}]} = \sum_{b \in B} \hat{b} \otimes [b^{-1}r]$, and the dualized action of $k[R]$ on k^B satisfies $(\hat{b} \leftharpoonup r)\hat{c} = (\hat{b} \leftharpoonup r)(c)\hat{c} = \hat{b}(\{c^{-1}r\})\hat{c} = \delta_{b,\{c^{-1}r\}}\hat{c}$. Finally we note that j is a coalgebra map, $j(r)j(s) = r^{-1}s^{-1}$ and hence $\overline{j(r)j(s)} = [sr]$ and $\pi(j(r)j(s)) = \{sr\}$. Now we can calculate

$$\Delta(r \otimes \hat{b}) = \sum_{c,d \in B} r \otimes \hat{c}\hat{d} \otimes [c^{-1}r] \otimes \widehat{d^{-1}b} = \sum_{c \in B} r \otimes \hat{c} \otimes [c^{-1}r] \otimes \widehat{c^{-1}b},$$

$$(r \otimes \hat{b})(s \otimes \hat{c}) = \overline{j(r)j(s)} \otimes (\hat{b} \leftharpoonup s)\hat{c} = \delta_{b,\{c^{-1}s\}}[sr] \otimes \hat{c},$$

and

$$\phi(r \otimes \hat{a} \otimes s \otimes \hat{b} \otimes t \otimes \hat{c}) = \langle \hat{a}, \pi(j(s)j(t)) \rangle \delta_{b,e} \delta_{c,e} = \langle \hat{a}, \{ts\} \rangle \delta_{b,e} \delta_{c,e}.$$

We can dualize these formulas to get a quasibialgebra structure on $\tilde{H}^* \cong k^R \otimes k[B]$ with its canonical basis of elements $\hat{r} \otimes b$ for $r \in R$ and $b \in B$, which is also the dual basis to the basis $r \otimes \hat{b}$ of $\tilde{H}$. We compute

$$(\hat{r} \otimes b)(\hat{s} \otimes c) = \sum_{\substack{t \in R \\ a,d \in B}} \delta_{rt} \delta_{bd} \delta_{s,[d^{-1}t]} \delta_{c,d^{-1}a} \hat{t} \otimes a$$

$$= \delta_{s,[b^{-1}r]} \hat{r} \otimes bc = \delta_{r,[bs]} \hat{r} \otimes bc,$$

$$\Delta(\hat{r} \otimes b) = \sum_{\substack{s,t \in R \\ c,d \in B}} \delta_{c,\{d^{-1}t\}} \delta_{r,[ts]} \delta_{bd} \hat{s} \otimes c \otimes \hat{t} \otimes d$$

$$= \sum_{s,t \in R} \delta_{[t^{-1}r],s} \hat{s} \otimes \{b^{-1}t\} \otimes \hat{t} \otimes b = \sum_{t \in R} \widehat{[t^{-1}r]} \otimes \{b^{-1}t\} \otimes \hat{t} \otimes b,$$

$$\phi = \sum_{r,s,t \in R} \hat{r} \otimes \{ts\} \otimes \hat{e} \otimes s \otimes \hat{e} \otimes t.$$

Comparing with [61] we see that $\tilde{H}^* \cong A(G, B)^{\mathrm{cop}}$ as quasibialgebras (note that between the conventions in [61] and the present paper ϕ is replaced by its inverse).

It goes without saying that the category $\mathcal{C}(G, B)$ constructed by Zhu is equivalent to the category $_{k[B]}^{k[G]}\mathcal{M}_{k[B]}$ of finite-dimensional of G-graded vector spaces with a compatible two-sided B-action; after all, the former is shown to be equivalent to the category of finite dimensional $A(G, B)$-modules in [61], while the latter is, by definition of $\tilde{H}$, the category of finite-dimensional $\tilde{H}$-comodules. We will sketch a direct proof: Let $D \in B \backslash G / B$ be a double coset in G. Then for every $M \in {}_{k[B]}^{k[G]}\mathcal{M}_{k[B]} = {}_{k[B]}\mathcal{M}_{k[B]}^{k[G]}$ the subspace $M_D := \bigoplus_{g \in D} M_g = \{m \in M | m_{(0)} \otimes m_{(-1)} \in M \otimes k[D]\}$ is a $k[B]$-subbimodule, hence a subobject in $_{k[B]}^{k[G]}\mathcal{M}_{k[B]}$, and $M = \bigoplus_{D \in B \backslash G / B} M_D$. It remains to describe the objects in $_{k[B]}^{k[D]}\mathcal{M}_{k[B]}$ when D is a double coset. We know that this subcategory of $_{k[B]}\mathcal{M}_{k[B]}^{k[G]}$ is equivalent to the category $\mathcal{M}_{k[B]}^{k[B \backslash D]}$. If $S \subset B$ is the stabilizer of an element of $B \backslash D$, then $B \backslash D \cong S \backslash B$ as right B-sets. But a special case of Schneider's structure theorem [48, Thm. 3.7] for Hopf modules gives a category equivalence $\mathcal{M}_{k[S]} \cong \mathcal{M}_{k[B]}^{k[S \backslash B]}$ mapping V to $V \underset{k[S]}{\otimes} k[B]$. To wrap up, objects of $_{k[B]}\mathcal{M}_{k[B]}^{k[G]}$ decompose as direct sums of objects of $_{k[B]}\mathcal{M}_{k[B]}^{k[D]}$ for each double coset D, and the latter category is equivalent to the category of representations of the stabilizer S of some element in $B \backslash D$, matching the description of the category $\mathcal{C}(G, B)$ by Zhu [61].

The motivation for $\mathcal{C}(G, B)$ in [61] is that its Grothendieck ring $\mathcal{G}(\mathcal{C}(G, B))$ maps surjectively onto the Hecke ring $H(G//B)$ associated to the inclusion $B \subset G$. We shall reproduce this now for the category ${}^{k[G]}_{k[B]}\underline{\mathcal{M}}_{k[B]}$. Let $H(G//B)$ be the free $\mathbb{Z}$-module over the set $G//B = B\backslash G/B$ of double cosets. We choose a set W of representatives for the double cosets and write D_w for the double coset containing $w \in W$. The universal property of the Grothendieck ring implies immediately that we have a well-defined group homomorphism

$$T \colon \mathcal{G}\left({}^{k[G]}_{k[B]}\underline{\mathcal{M}}_{k[B]}\right) \ni M \mapsto \sum_{w \in W} \dim(M_w) D_w \in H(G//B),$$

which is onto since $T(k[D]) = D$. We wish to show that T is a homomorphism to the Hecke ring; as a byproduct we will obtain a proof that the usual multiplication on the Hecke ring really is a ring structure.

For $M, N \in {}^{k[G]}_{k[B]}\underline{\mathcal{M}}_{k[B]}$, double cosets D, D', and $w'' \in W$ we have

$$(M_D \otimes N_{D'})_{w''} = \bigoplus_{\substack{d \in D, d' \in D' \\ dd' = w''}} M_d \otimes N_{d'}.$$

We can write $M \underset{k[B]}{\otimes} N$ as the quotient of $M \otimes N$ by the right action of B through the automorphisms t_b defined by $t_b(m \otimes n) = mb \otimes b^{-1}n$ for $b \in B$. Since t_b preserves the grading and maps $M_x \otimes N_y$ bijectively onto $M_{xb} \otimes N_{b^{-1}y}$, we have

$$(M_D \underset{k[B]}{\otimes} N_{D'})_{w''} \cong \bigoplus_{\substack{r \in D \cap R, d' \in D' \\ rd' = w''}} M_r \otimes N_{d'} \cong \bigoplus_{\substack{r \in D \cap R, r' \in D' \cap R \\ rr' \in w''B}} M_r \otimes N_{r'}$$

and thus, for $D = D_w$ and $D' = D_{w'}$:

$$\dim((M_D \underset{k[B]}{\otimes} N_{D'})_{w''})$$

$$= \left|\{(r, r') \in D \cap R \times D' \cap R \mid rr' \in w''B\}\right| \cdot \dim(M_w)\dim(N_{w'})$$

and finally $T(M \underset{k[B]}{\otimes} N) = T(M)T(N)$, if we define multiplication in $H(G//B)$ by

$$DD' = \sum_{w'' \in W} \left|\{(r, r') \in D \cap R \times D' \cap R \mid rr' \in w''B\}\right| D_{w''}$$

In particular, it follows that this formula defines the structure of an associative unital ring on $H(G//B)$; it is the usual multiplication in the Hecke ring as given in [52]. It also follows that T is a ring homomorphism.

Thus far we have seen that Zhu's construction $A(G, B)$ for a finite group G and subgroup B is, as a quasibialgebra, contained as a special case in [47]. We postpone discussing the quasi-Hopf structure, and turn first to a generalization of $A(G, B)$ to the case of infinite groups G and B. This is sketched slightly informally in [61] by saying that all formulas in the construction of $A(G, B)$ make sense for infinite groups, provided one restricts the attention to finite dimensional representations of $A(G, B)$, while otherwise the formulas involve infinite sums. We believe that the generalized coquasibialgebra construction $\tilde{H}$ from subsection 4.3 for $H = k[G] \in \mathfrak{E}_\ell(k[B])$ is a more formal version of this. Recall that finite comodules of $\tilde{H}$ correspond to Hopf modules in $^{k[G]}_{k[B]}\mathcal{M}_{k[B]}$ that are finitely generated left B-modules. In view of the description of $^{k[G]}_{k[B]}\mathcal{M}_{k[B]}$ obtained above (which holds just as well for infinite groups), we see that no objects can occur that involve degrees in G whose double cosets have infinitely many orbits under the left (or right) action of B. Thus we assume for the rest of this section that we are given a group G and a subgroup $B \subset G$ such that every double coset $D \in B\backslash G/B$ contains only finitely many left (or, equivalently, finitely many right) cosets. In this case, the right $k[B]$-module $Q = k[G]/k[B]^+k[G]$ is locally finite as right $k[B]$-module, so that Theorem 4.3.2 applies, and we know that $\tilde{H}$ can be modelled on the k-space $k[R] \otimes k[B]^\circ$. We also note that the above construction of a surjective homomorphism T from the Grothendieck ring of $^{k[G]}_{k[B]}\underline{\mathcal{M}}_{k[B]}$ to the Hecke ring $H(G//B)$ works just as well in this situation.

The question remains how the antipode given by Zhu for $A(G, B)$ fits in our picture. (Co)quasiantipodes are not considered at all in [47]. In fact the author learned about [61] when visiting MSRI, and the construction of a quasiantipode in [61] motivated the constructions in subsection 4.4.

Zhu gives an antipode for $A(G, B)$ under the following conditions: There should exist a choice R of right coset representatives that is, at the same time, a set of left coset representatives. This is clearly true if G is finite. The criterion in Lemma 4.4.7 is designed to be applicable to this case, so that we can use Theorem 4.4.5 to arrive at the desired conclusion: Whenever there is a common set of representatives for the left and right cosets of B in G, then $\tilde{H} = k[R] \otimes k[B]^\circ$ is a coquasi-Hopf algebra.

If $B \subset G$ is an inclusion of infinite groups, then one can show that there is a common set of representatives for the left and right cosets of B in G if and only if every double coset contains as many left as right cosets. It is certainly known that this may fail to be true; we shall nevertheless give an example: We consider the subgroup $2^{\mathbb{Q}}$ of the multiplicative group $\mathbb{R}^+$. We let an infinite cyclic group $\langle\sigma\rangle$ act on $2^{\mathbb{Q}}$ by $\sigma \rightharpoonup 2^x = 2^{2x}$, and let $G = 2^{\mathbb{Q}} \rtimes \langle\sigma\rangle$ be the semidirect product. We let B be the subgroup $2^{\mathbb{Z}} \subset 2^{\mathbb{Q}} \subset G$. The left coset of $2^x\sigma^k$ consists of all elements $2^{x+m}\sigma^k$ for $m \in \mathbb{Z}$, while the right coset consists of all $2^{x+2^k m}\sigma^k$ for $m \in \mathbb{Z}$. It follows that the double coset D of $2^x\sigma^k$ is a left coset containing 2^k right cosets when $k \geq 0$, and D is a right coset containing 2^{-k} left cosets when $k < 0$.

At any rate, if every double coset in G is the union of finitely many left cosets, then the quotient $k[G]/k[B]^+k[G]$ is a locally finite right $k[B]$-module, and since $k[G]$ is cocommutative, the inclusion $k[B] \subset k[G]$ satisfies the conditions of Theorem 4.4.3. It follows that $^{\tilde{H}}\mathcal{M}$ is left and right rigid. If D is a double coset, and $V = k[D]/k[B]^+k[D]$ the $\tilde{H}$-comodule corresponding to $k[D] \in {}^{k[G]}_{k[B]}\mathcal{M}_{k[B]}$, then the discussion preceding Remark 4.4.1 shows that $\dim V = |B \backslash D|$ is the number of left cosets in D, while $\dim({}^\vee V) = |D/B|$ is the number of right cosets in D, where ${}^\vee V$ is the right dual of V in ${}^H\mathcal{M}$. As we have seen, the two numbers, hence the dimensions of V and its dual can differ. Such a phenomenon cannot occur if $\tilde{H}$ is a coquasi-Hopf algebra, so we have seen that the sufficient condition considered by Zhu to construct a quasiantipode is in fact necessary to construct a coquasiantipode for $\tilde{H}$, and we have found the counterexample announced in subsection 2.2:

EXAMPLE 4.5.1. There is a coquasibialgebra H such that the category ${}^H\mathcal{M}_f$ of finite dimensional left H-comodules is left and right rigid, but the dimension of the dual ${}^\vee V$ of an object $V \in {}^H\mathcal{M}_f$ is in general different from $\dim V$. In particular, H is not a coquasi-Hopf algebra

We will show in a separate paper (now available as a preprint [45]) that when a coquasibialgebra H is finite dimensional, and $V \in {}^H\mathcal{M}$ has a dual object ${}^\vee V$, then $\dim({}^\vee V) = \dim V$. One can deduce from this that if H is finite dimensional and cosemisimple, and ${}^H\mathcal{M}$ is left and right rigid, then H is a coquasi-Hopf algebra.

4.6. Not an example: Quantum doubles. Contrary to this section's title, we already explained in subsection 4.2 how the quantum double of a finite Hopf algebra K is contained as a special case in the constructions of subsection 4.1, namely, as the double crossproduct associated to $K \otimes K^{\mathrm{op}}$ and the cocleaving $\pi = \nabla \colon K \otimes K^{\mathrm{op}} \to K$. This means that we reconstruct $D(K)$ from its comodule category, which turns out to be ${}^{K \otimes K^{\mathrm{op}}}_K \mathcal{M}_K$. But the real meaning of $D(K)$ lies in its module category, which is braided due to the quasitriangular structure of $D(K)$.

If K is cocommutative, there is a 'better' way: We let $H = K \otimes K$, and consider $H \in \mathfrak{E}_\ell(K)$ with respect to the inclusion $\iota := \Delta \colon K \to H$, and the cocleaving $\pi := K \otimes \varepsilon \colon H \to K$. It is straightforward to check that ${}^H_K\mathcal{M}_K$ identifies canonically with ${}^K_K\mathcal{M}^K_K$, which is equivalent the category of Yetter–Drinfeld modules over K by [42], and hence equivalent to the category of modules over the Drinfeld double. Thus we expect that $\tilde{H}$ will be dual to the Drinfeld double of K. In detail we can identify Q with K according to the surjection $\nu \colon H \ni x \otimes y \mapsto S(x)y \in K$. We then find $j = \eta \otimes K \colon K \to H$. In particular, j is a bialgebra map, and the action of Q on K defined by $q \rightharpoonup x = \pi(j(q)\iota(x))$ turns out to be trivial. Hence the algebra $\tilde{H}$ is just $K \otimes K^*$. The coalgebra structure of $\tilde{H}$ is that of a cosmash product with respect to the coaction of K on K^* dual to the coadjoint coaction of K on itself.

As it turns out, the second way admits a generalization beyond the case of arbitrary Hopf algebras to the construction of the Drinfeld double of a quasi-Hopf algebra. The construction does not fit into the framework of subsection 4.1, but it is rather close in spirit, so I would like to sketch the approach, although it is unfinished work at present. (A variant of the constructions sketched below is now contained in [45].)

The origin of Drinfeld doubles of quasi-Hopf algebras is the construction by Dijkgraaf, Pasquier and Roche [10] of a variation $D^\omega(G)$ of the Drinfeld double $D(G) := D(k[G])$ of a finite group G, depending on a three-cocycle of the group G with values in the multiplicative group of the base field k. The result is a quasi-Hopf algebra, which was interpreted by Majid [30] as the Drinfeld double quasi-Hopf algebra of the quasi-Hopf algebra (k^G, ω), with the coassociator $\omega \in k^G \otimes k^G \otimes k^G$. Majid also announces the construction of a Drinfeld double for general quasi-Hopf algebras. It is surprising at first sight that a double quasi-Hopf algebra of a quasi-Hopf algebra H should exist: After all, the double of a Hopf algebra H is modelled on $H \otimes H^*$, with H and H^* as subalgebras. But if H is just a quasi-Hopf algebra, then H^* is not an associative algebra, so one is at a loss looking for an associative algebra structure for $H \otimes H^*$. However, as explained in [30], there is a good reason to expect that the construction works anyway: The module category over the Drinfeld double $D(H)$ of an ordinary Hopf algebra H is equivalent to the center of the category of H-modules. The center construction is a purely categorical procedure assigning a braided monoidal category to any monoidal category; of course it can be applied to the category of H-modules also when H is just a quasi-Hopf algebra. The result should be the module category for another quasi-Hopf algebra $D(H)$, the Drinfeld double of H — by reconstruction principles. While he gives some indications, Majid falls (in the author's opinion) short of showing that $D(H)$ can be realized on the vector space $H \otimes H^*$. This was achieved by Hausser and Nill in [15; 16] through explicit computations.

We shall sketch very briefly how one can derive the construction from ideas similar to those in subsection 4.1. To stay close to the formalism used so far, we dualize the problem and look for a coquasi-Hopf algebra analog $D^*(K)$ of the dual Drinfeld double of a coquasi-Hopf algebra (K, ϕ). First, we note that K, though not an associative algebra, is an algebra in the monoidal category $^K \mathcal{M}^K$ of K-bicomodules, due to the modified associativity of K as a coquasibialgebra. Thus, we can consider the "Hopf module" categories $^K_K \mathcal{M}^K = {}_K \left(^K \mathcal{M}^K \right)$ and $^K_K \mathcal{M}^K_K = {}_K \left(^K \mathcal{M}^K \right)_K$. The category $^K_K \mathcal{M}^K_K$ has a natural structure of monoidal category with a suitably modified tensor product over K. By comparing with [42] we expect $^K_K \mathcal{M}^K_K$ to be equivalent to the category of comodules over $D^*(K)$. To make the setup look more like that in subsection 4.1, we put $H = K \otimes K^{\text{cop}}$, so that $^K_K \mathcal{M}^K_K \cong {}^H_K \mathcal{M}_K$, and we want to find $D^*(K) = \tilde{H}$ with $^H_K \mathcal{M}_K \cong {}^{\tilde{H}} \mathcal{M}$. The solution in subsection 4.1 was based on Schneider's category equivalence $^H_K \mathcal{M} \cong {}^Q \mathcal{M}$. For the present situation, another paper of Hausser and Nill [17] provides a category equivalence $^H_K \mathcal{M} \cong {}^K_K \mathcal{M}^K \cong \mathcal{M}^K \cong {}^{K^{\text{cop}}} \mathcal{M}$ mapping $V \in \mathcal{M}^K$ to $K \otimes V \in {}^K_K \mathcal{M}^K$, the free left K-module in $^K \mathcal{M}^K$ generated by V. One can infer that objects of $^K_K \mathcal{M}^K_K$, which are objects of $^K_K \mathcal{M}^K$ with an additional right action of K, can be classified by objects of $^{K^{\mathrm{cop}}}\mathcal{M}$ with an additional right action of K (classically, the latter objects would just be Yetter–Drinfeld modules by [42]). From the necessary compatibility condition between the comodule structure and the action, one derives a (not a priori coassociative) comultiplication on $\tilde{H} = K \otimes K^*$ such that $^{\tilde{H}}\mathcal{M} \cong {}^H_K\mathcal{M}_K$. To show that it is coassociative after all, one applies Lemma 2.3.1 after furnishing 'enough' objects in $^K_K\mathcal{M}^K_K$, and for this last step one may use the construction of free modules within the monoidal category $^K\mathcal{M}^K$, which provides the free right K-module over the free left K-module generated by the object $K \in {}^K\mathcal{M} \subset {}^K\mathcal{M}^K$.

While this approach to constructing the (dual) Drinfeld double is a spitting image of the procedure in subsection 4.1, we should stress again that it is not at all a special case. Rather, it poses the question for a common generalization.

5. Kac' Exact Sequence

In [22], Kac describes the following exact sequence (we have adopted notations from [32])

$$0 \to \mathcal{H}^1(F \bowtie G, k^\times) \overset{\mathrm{res}}{\to} \mathcal{H}^1(F, k^\times) \oplus \mathcal{H}^1(G, k^\times) \to \mathrm{Aut}(k^G \# kF) \to$$

$$\to \mathcal{H}^2(F \bowtie G, k^\times) \overset{\mathrm{res}}{\to} \mathcal{H}^2(F, k^\times) \oplus \mathcal{H}^2(G, k^\times) \to \mathrm{Opext}(kF, k^G) \to$$

$$\to \mathcal{H}^3(F \bowtie G, k^\times) \overset{\mathrm{res}}{\to} \mathcal{H}^3(F, k^\times) \oplus \mathcal{H}^3(G, k^\times) \to \cdots$$

Here $\mathcal{H}^\bullet(G, A)$ stands for the cohomology of a group G with coefficients in a G-module A, and $k^\times$ is the multiplicative group of a fixed base field (in Kac' work $k = \mathbb{C}$, but see [32]), with the trivial group action. The group $F \bowtie G$ is assumed to be a group containing F and G as subgroups such that the map $F \times G \to F \bowtie G$ given by multiplication is a bijection. By the remarks in subsection 4.2 this leads to a Singer pair between the Hopf algebras k^G and kF, hence to a bismash product Hopf algebra $k^G \# kF$. By $\mathrm{Aut}(k^G \# kF)$ we mean the group of automorphisms of the extension $k^G \# kF$ of kF by k^G. By $\mathrm{Opext}(kF, k^G)$ we denote the group of extensions of kF by k^G giving rise to the same Singer pair.

We refer to Akira Masuoka's paper [31] in this volume for more information on Kac' sequence beyond the following remarks: There are in essence three steps leading to the sequence. First, one has to translate the Opext group (and the automorphism group) in the sequence into cohomological data. This part of Kac' work considerably predates the more general cohomological description of extensions of cocommutative by commutative Hopf algebras as carried out by Singer [53] and Hofstetter [19; 20], who do not seem to have been aware of [22]. The description involves a certain double complex. Second, one exhibits this double complex as the middle term of a short exact sequence of double complexes, and obtains a long exact cohomology sequence. Third, one interprets the cohomology groups of the end terms of the short exact sequence of double complexes. One of them is trivially related to the bar resolutions computing the group cohomology of F and G, while it is more intricate to show that the third term is a non-standard resolution of $\mathbb{Z}$ that can be used to compute the cohomology of $F \bowtie G$. It is particularly this last step which is specific to the group case.

5.1. A Kac sequence for general Hopf algebras. Of course we can pass from the groups F, G, and $F \bowtie G$ in the Kac sequence to their group algebras $Q := kF, L := kG$, and $k[F \bowtie G] \cong kF \bowtie kG$. From [54] we know that group cohomology with coefficients in the multiplicative group of the field k is the same as Sweedler cohomology of the group algebra with coefficients in the trivial module algebra k. Abbreviating the latter by $\mathcal{H}^\bullet(H) = \mathcal{H}^\bullet(H, k)$, and writing $K := L^*$, we obtain the form

$$0 \to \mathcal{H}^1(Q \bowtie L) \overset{\mathrm{res}}{\to} \mathcal{H}^1(Q) \oplus \mathcal{H}^1(L) \to \mathrm{Aut}(K \# Q) \to \mathcal{H}^2(Q \bowtie L) \overset{\mathrm{res}}{\to}$$
$$\overset{\mathrm{res}}{\to} \mathcal{H}^2(Q) \oplus \mathcal{H}^2(L) \to \mathrm{Opext}(Q, K) \to \mathcal{H}^3(Q \bowtie L) \overset{\mathrm{res}}{\to} \mathcal{H}^3(Q) \oplus \mathcal{H}^3(L)$$

of Kac' sequence. It shows no explicit reference any more to the groups involved, so one may write this down equally well for cocommutative Hopf algebras Q and L, with L finite, and $K = L^*$. However, from [22] or [32] we cannot draw a definition of all the maps in the sequence, much less infer its exactness; specifically the maps from the automorphism group to the second cohomology, and from the Opext group to the third cohomology, will need a new explanation. We will not transfer Kac' techniques to the situation with general cocommutative Hopf algebras, but rather give an explanation through the functor $\mathfrak{F}$; this generalizes beyond the cohomological picture to an analog of the Kac sequence for non-cocommutative Hopf algebras. However, to formulate this 'quantum' analog, we have to alter the appearance of the sequence quite completely. If Q and K are noncommutative and non-cocommutative, then the congruence class of an extension $K \to H \to Q$ no longer determines the action of Q on K and the coaction of K on Q uniquely, and in the description of H as a bicrossproduct $K {}^\tau \#_\sigma Q$ with cocycle σ and cycle τ it is no longer possible to obtain a bismash product from the same action and coaction with trivial (co)cycles. The only solution appears to be to consider, instead of the Opext groups for each Singer pair separately, the set $\mathrm{Ext}(Q, K)$ of congruence classes of all extensions of Q by K; in the abelian case, this is the disjoint union of all the Opext groups for various Singer pairs. Next, we have to consider what happens to the map from the Opext group to degree three cohomology in the general case. We have seen that a good non-cocommutative replacement for a three-cocycle is a coquasibi-algebra structure. If we are given a double crossproduct of cocommutative Q and L, then for a three-cocycle ϕ on it to be in the kernel of the homomorphism $\mathcal{H}^3(Q \bowtie L) \to \mathcal{H}^3(Q) \oplus \mathcal{H}^3(L)$ induced by the restrictions means that the restrictions of ϕ to $Q \otimes Q \otimes Q$ as well as to $L \otimes L \otimes L$ are trivial, and this in turn means that Q and L, ordinary bialgebras with trivial coassociators, are subcoquasibibialgebras of $(Q \bowtie L, \phi)$. In the non-cocommutative case, we cannot fix our attention on a Singer pair at a time, and thus it also makes no sense to consider only one double crossproduct at a time. Thus, we define the set $\mathfrak{P}(Q, L)$ to consist of the cohomology classes of generalized product coquasibialgebras of Q and L. Here a generalized product coquasibialgebra is by definition a co-quasibialgebra (P, ϕ) with injective coquasibialgebra maps $Q \to P$ and $L \to P$, such that multiplication in P induces an isomorphism $Q \otimes L \to P$. We call two generalized product coquasibialgebras P, P' cohomologous if there exists a coquasiisomorphism $(F, \theta) \colon P \to P'$ whose underlying map F commutes with the respective inclusion maps from L and Q. One can show that in the case where both Q and L are cocommutative, $\mathfrak{P}(Q, L)$ is in natural bijection with the sum of the kernels of all the homomorphism $\mathcal{H}^3(Q \bowtie L) \to \mathcal{H}^3(Q) \oplus \mathcal{H}^3(L)$ induced by the restrictions, for all possible matched pairs between Q and L: the bijection is obtained by interpreting a three-cocycle ϕ on a double crossed product $Q \bowtie L$ as a coquasibialgebra $(Q \bowtie L, \phi)$. Thus, a good replacement for the exact sequences

$$\mathrm{Opext}(Q, K) \to \mathcal{H}^3(Q \bowtie L) \to \mathcal{H}^3(Q) \oplus \mathcal{H}^3(L)$$

is a surjection

$$\mathrm{Ext}(Q, K) \to \mathfrak{P}(Q, L).$$

We can describe such a surjection in terms of the functor $\mathfrak{F}$: By choosing a co-cleaving, an extension $K \to H \to Q$ can be regarded as an element of $\mathfrak{E}_\ell(K)$, and gives rise to $\tilde{H} := \mathfrak{F}(H) \in \mathfrak{E}_r(L)$. One can specialize the formulas in subsection 4.1 to find that Q is a subcoquasibialgebra of $\tilde{H}$, with $\tilde{\phi}|_{Q \otimes \tilde{H} \otimes \tilde{H}} = \varepsilon$. The obvious isomorphism $Q \otimes L \to \tilde{H}$ is given by multiplication. Finally, modifying the cleaving in H leads to a different result $\tilde{H}$, but one that only differs by a twist. Thus, the functor $\mathfrak{F}$ induces a well-defined map $[\mathfrak{F}] \colon \mathrm{Ext}(Q, K) \to \mathfrak{P}(Q, L)$. It appears to be far from surjective at first sight, however, since the coassociator $\tilde{\phi}$ on $\tilde{H}$ satisfies stronger conditions than just $\tilde{\phi}|_{Q \otimes Q \otimes Q} = \varepsilon$ and $\tilde{\phi}|_{L \otimes L \otimes L} = \varepsilon$. But one can show [47, Prop. 6.2.3] that any element of $\mathfrak{P}(Q, L)$ does contain a representative (P, ϕ) that fulfills the stronger conditions $\phi|_{Q \otimes P \otimes P} = \varepsilon$ and $\phi|_{P \otimes P \otimes L}$. This not only removes the obvious obstacle to surjectivity of $[\mathfrak{F}]$, but also allows to prove it: The special representative (P, ϕ) is, by the second condition, an element of $\mathfrak{E}_r(L)$, and it turns out that the functor $\mathfrak{F}^{\mathrm{bop}} \mathfrak{E}_r(L) \to \mathfrak{E}_\ell(K)$, applied to such representatives, defines a section for $[\mathfrak{F}]$. This is a special case of the fact, mentioned in subsection 4.2, that $\mathfrak{F}$ is an involution on certain classes of objects of $\mathfrak{E}_\ell(K)$.

We turn now to the part

$$\mathcal{H}^2(Q) \oplus \mathcal{H}^2(L) \to \mathrm{Opext}(Q, K) \to \mathcal{H}^3(Q \bowtie L)$$

of the cocommutative Kac sequence. Taking the union over all possible Singer pairs, we obtain an exact sequence

$$\mathcal{H}^2(Q) \oplus \mathcal{H}^2(L) \to \text{Ext}(Q, K) \to \mathfrak{P}(Q, L),$$

with a certain sloppiness: since neither $\text{Ext}(Q, K)$ nor $\mathfrak{P}(Q, L)$ are groups, we have to say what exact means. Now via the group homomorphisms $\mathcal{H}^2(Q) \oplus \mathcal{H}^2(L) \to \text{Opext}(Q, K)$, the group $\mathcal{H}^2(Q) \oplus \mathcal{H}^2(L)$ acts on each Opext group, hence on the set $\text{Ext}(Q, K)$, and exactness means that the fibers of the map $\text{Ext}(Q, K) \to \mathfrak{P}(Q, L)$ are the orbits of this action. For the non-cocommutative case, we now have to find a replacement for this action on $\text{Ext}(Q, K)$. In fact this is quite easy: Recall that any two-cocycle $\theta \colon Q \otimes Q \to k$ can be used to cotwist Q to give a different Hopf algebra Q^θ. Of course we can pull θ back to give a two-cocycle on any extension H in $\text{Ext}(Q, K)$, and thus twist H to give an extension H^θ in $\text{Ext}(Q^\theta, K)$. Similarly, if $t \colon L \otimes L \to k$ is a two-cocycle, we can consider it as a cycle $t \in K \otimes K$, and twist K to obtain a new Hopf algebra K_t. Surely t is also a two-cycle in any extension H in $\text{Ext}(Q, K)$, and we can twist H by it to get a new extension $H_t \in \text{Ext}(Q, K_t)$. Now for our purposes we are not interested in modifying the end terms of an extension, but only the middle term. Thus, we need to restrict our attention to those cocycles θ and t that do not affect Q and K, or, in the terminology introduced in subsection 2.4, we have to consider central cocycles. It turns out [47, Lem. 6.3.1] that one has indeed a well-defined action of $\mathcal{H}_c^2(Q) \oplus \mathcal{H}_c^2(L)$ on $\text{Ext}(Q, K)$ as just described, and the orbits of this action are [47, Thm. 6.3.6] precisely the fibers of $[\mathfrak{F}] \colon \text{Ext}(Q, K) \to \mathfrak{P}(Q, L)$.

Next, we consider the portion

$$\mathcal{H}^2(Q \bowtie L) \to \mathcal{H}^2(Q) \oplus \mathcal{H}^2(L) \to \text{Opext}(Q, K)$$

of the cocommutative Kac sequence. For the non-cocommutative version, we have already replaced the right hand map by an action of $\mathcal{H}_c^2(Q) \times \mathcal{H}_c^2(L)$ on $\text{Ext}(Q, K)$, so now exactness amounts to determining the stabilizers of the action. In the cocommutative case, the stabilizer of an extension H in $\text{Ext}(Q, K)$ is the image of the second cohomology of the associated double crossproduct $Q \bowtie L$. In our situation it turns out that the stabilizer is the image of the self-twist group $\mathcal{H}_c^2(\mathfrak{F}(H))$ of the associated coquasibialgebra [47, Thm. 6.3.5].

We have thus discussed the non-cocommutative replacements for exactness of the portion

$$\mathcal{H}^2(Q \bowtie L) \to \mathcal{H}^2(Q) \oplus \mathcal{H}^2(L) \to \text{Opext}(Q, K) \to$$
$$\to \mathcal{H}^3(Q \bowtie L) \to \mathcal{H}^3(Q) \oplus \mathcal{H}^3(L)$$

of Kac' exact sequence. We will skip discussing the lower order terms [47, Sec. 6.5], although we will present an application in the next section. It is worthwhile noting that the cocommutative Kac sequence that we obtain as a special case is in fact essentially the same as the original Kac sequence for groups [47, Sec. 6.4]

5.2. Masuoka's sequence for Lie algebras.
In [33], Masuoka gives variants of Kac' exact sequence that apply to Lie algebras and their enveloping algebras instead of groups and their group algebras.

The purpose of this section is not to show how these derive from the author's sequence for cocommutative Hopf algebras. The results from subsection 5.1 simply do not apply because the enveloping algebras are always infinite-dimensional. Even though we showed in subsection 4.3 how to modify the construction underlying the generalized Kac sequence to the case of infinite Hopf subalgebras, we will find that this does not explain Masuoka's version of Kac' sequence. We shall indicate here which parts appear to generalize smoothly, and where the real difficulties particular to the Lie case come in to obstruct the way. Throughout the section we assume the base ring k is a field.

We refer the reader to [31], in this volume, for details on Masuoka's sequences. We will discuss here only the variant in [33, Cor. 4.12.], which is an exact sequence

$$0 \to \mathcal{H}^1(\mathfrak{f} \bowtie \mathfrak{g}) \to \mathcal{H}^1(\mathfrak{f}) \oplus \mathcal{H}^1(\mathfrak{g}) \to \mathrm{Aut}((U\mathfrak{g})^\circ \# U(\mathfrak{f})) \to \mathcal{H}^2(\mathfrak{f} \bowtie \mathfrak{g})$$

$$\to \mathcal{H}^2(\mathfrak{f}) \oplus \mathcal{H}^2(\mathfrak{g}) \to \mathrm{Opext}(U(\mathfrak{f}), (U\mathfrak{g})^\circ) \to \mathcal{H}^3(\mathfrak{f} \bowtie \mathfrak{g}) \to \mathcal{H}^3(\mathfrak{f}) \oplus \mathcal{H}^3(\mathfrak{g})$$

in which $\mathfrak{f}$ and $\mathfrak{g}$ are Lie algebras, with $\mathfrak{g}$ finite dimensional, $\mathfrak{f} \bowtie \mathfrak{g}$ is a double crossproduct Lie algebra, and the cohomology groups are Lie algebra cohomology with coefficients in the trivial module k. The universal enveloping algebra of $\mathfrak{f} \bowtie \mathfrak{g}$ is a double crossproduct Hopf algebra $U(\mathfrak{f} \bowtie \mathfrak{g}) \cong U(\mathfrak{f}) \bowtie U(\mathfrak{g})$ with respect to a matched pair of Hopf algebras in which the action of $U(\mathfrak{g})$ on $U(\mathfrak{f})$ is locally finite. This matched pair gives rise to a Singer pair $(U(\mathfrak{g})^\circ, U(\mathfrak{f}))$ which features in the smash product and the Opext group that appear in the sequence.

By [54] we can replace these cohomology groups by Sweedler cohomology groups of the associated universal enveloping algebras. If we then put $Q = U\mathfrak{f}$, $L = U\mathfrak{g}$, and $K = L^\circ$, then Masuoka's sequence looks just the same as the generalized cocommutative Kac sequence in subsection 5.1. However, since L has infinite dimension, the sequence in subsection 5.1 does not contain Masuoka's sequence as a special case. We will now discuss how far the results from the previous sections apply to Masuoka's situation.

Assume first that we are given a matched pair $(Q, L, \to, \leftarrow)$ of Hopf algebras. Assume that the action $\to$ of L on Q is locally finite. Then by the left-right switched version of Theorem 4.3.2 we obtain $\mathfrak{F}^{\mathrm{bop}}(Q \bowtie L) \cong L^\circ \otimes Q$. The formulas obtained for the multiplication and comultiplication on $\mathfrak{F}^{\mathrm{bop}}(Q \bowtie L)$ specialize to say that $L^\circ \otimes Q = L^\circ \# Q$ is a bismash product with respect to the coaction of L° on Q dual to $\to$, and the action of Q on L° dual to $\leftarrow$. In particular, it follows that the action of Q on L^* stabilizes L°, and that we have a Singer pair (Q, K) with $K := L^\circ$ (cf. [33, Lem. 4.1]). On the other hand, let K be any Hopf algebra and assume given a bicrossproduct extension $K^\tau \#_\sigma Q$.

Since the action of K on Q is trivial, Theorem 4.3.2 applies again to show that $\mathfrak{F}(K \, ^\tau \#_\sigma Q) = Q \otimes K^\circ \in \mathfrak{P}(Q, K^\circ)$ is a generalized product coquasibialgebra.

Assume further that L is cocommutative, $K = L^\circ$, and the abelian Singer pair underlying the bicrossproduct $K \, ^\tau \#_\sigma Q$ is the same as that arising from the matched pair in the double crossproduct $Q \bowtie L$. Then the natural map $L \to K^\circ$ induces a bialgebra map $Q \bowtie L \to Q \bowtie K^\circ$, and we can restrict the coquasibialgebra structure on $Q \bowtie K^\circ$ arising from $\mathfrak{F}(K \, ^\tau \#_\sigma Q)$ to give a three-cocycle on $Q \bowtie L$. All in all, we have defined a map $[\mathfrak{F}]$: $\mathrm{Opext}(Q, K) \to \mathcal{H}^3(Q \bowtie L)$ which takes values in the kernel of the map $\mathcal{H}^3(Q \bowtie L) \to \mathcal{H}^3(Q) \oplus \mathcal{H}^3(L)$ induced by the restrictions. Precisely as in [47], one can also show that any element of the latter kernel contains a representative ϕ that fulfills $\phi|_{(Q\bowtie L)\otimes(Q\bowtie L)\otimes L} = \varepsilon$ and $\phi|_{Q\otimes(Q\bowtie L)\otimes(Q\bowtie L)} = \varepsilon$, so that $(Q \bowtie L, \phi) \in \mathfrak{C}_r(L)$. We can apply the functor $\mathfrak{F}^{\mathrm{bop}}$ to obtain some bialgebra $H = \widetilde{Q \bowtie L}$ equipped with maps $K \to H \to Q$ of bialgebras. However, we do not know that $H = K \otimes Q$ is a bicrossproduct in $\mathrm{Opext}(Q, K)$ unless the map $Q \to (L \otimes L)^*$ induced by ϕ takes values in $K \otimes K$; this is what we would need to apply Theorem 4.3.2 once more.

Of course we can specialize the results above to the case where Q and L are enveloping algebras as in Masuoka's sequence. We have not checked that the map $\mathrm{Opext}(U(\mathfrak{f}), (U\mathfrak{g})^\circ) \to \mathcal{H}^3(\mathfrak{f} \bowtie \mathfrak{g})$ we obtain is the same as that obtained by Masuoka. If this is the case, then Masuoka's sequence seems to indicate that every cohomology class in the kernel of the map

$$\mathcal{H}^3(U\mathfrak{f} \bowtie U\mathfrak{g}) \to \mathcal{H}^3(U\mathfrak{f}) \oplus \mathcal{H}^3(U\mathfrak{g})$$

has a representative ϕ which is trivial on $U(\mathfrak{f} \bowtie \mathfrak{g}) \otimes U(\mathfrak{f} \bowtie \mathfrak{g}) \otimes U(\mathfrak{g})$ as well as $U(\mathfrak{f}) \otimes U(\mathfrak{f} \bowtie \mathfrak{g}) \otimes U(\mathfrak{f} \bowtie \mathfrak{g})$, and in addition induces a map $U(\mathfrak{f} \bowtie \mathfrak{g}) \to U(\mathfrak{g})^\circ \otimes U(\mathfrak{g})^\circ$.

5.3. Galois extensions over tensor products.

In this final section we sketch how the generalized Kac sequence contains a result of Kreimer [25, Thm. 3.7] on Galois objects over tensor products of Hopf algebras. It states that when Q and L are finitely generated projective cocommutative Hopf algebras, then the group of Galois objects $\mathrm{Gal}(Q \otimes L)$ can be computed as

$$\mathrm{Gal}(Q \otimes L) \cong \mathrm{Gal}(Q) \oplus \mathrm{Gal}(L) \oplus \mathrm{Hopf}(Q, L^*),$$

the last term being the group of Hopf algebra homomorphisms under convolution.

We will prove this result from the low order terms of the generalized Kac sequence; one should note that this cannot adequately be called a short proof of Kreimer's result, since we just shifted the complications to the proof of the Kac sequence. However, we believe that the connection between the two results is of some interest.

We first recall some background: If H is a cocommutative finitely generated projective Hopf algebra, then the H-Galois extensions A/k that are faithfully flat k-modules form an abelian group $\mathrm{Gal}(H)$ under cotensor product over H. The

abelian group $\mathrm{Gal}(H)$ is a contravariant functor of H, again by cotensor product: For a k-split Hopf algebra map $f\colon H \to F$, the associated group homomorphism maps $A \in \mathrm{Gal}(F)$ to $A \square_F H$; in case f is an injection, we can identify this with
$$A(H) := \{a \in A | a_{(0)} \otimes a_{(1)} \in A \otimes H\}.$$

If $H = Q \otimes L$ is a tensor product of cocommutative Hopf algebras, it follows that the map $\mathrm{Gal}(H) \to \mathrm{Gal}(Q) \oplus \mathrm{Gal}(L)$ induced by the projections of H to Q and L is a split surjection with splitting induced by the injections of Q and L into H. By the same reason the homomorphisms $\mathcal{H}^n(H) \to \mathcal{H}^n(Q) \oplus \mathcal{H}^n(L)$ are split epimorphisms. Thus the Kac sequence for the trivial double crossproduct $Q \otimes L$ reduces in low dimensions to a split short exact sequence

$$0 \to \mathrm{Aut}(L^* \otimes Q) \to \mathcal{H}^2(Q \bowtie L) \to \mathcal{H}^2(Q) \oplus \mathcal{H}^2(L) \to 0.$$

The first term is the group of such automorphisms of α of $L^* \otimes Q$ that fix L^* and induce the identity on the quotient Q. These are easily seen to be in bijection with Hopf algebra maps $f\colon Q \to L^*$ by the formula $\alpha(x \otimes q) = xf(q_{(1)}) \otimes q_{(2)}$.

To obtain Kreimer's result, we would like to replace the right half of the sequence by the split epimorphism $\mathrm{Gal}(Q \otimes L) \to \mathrm{Gal}(Q) \oplus \mathrm{Gal}(L)$, that is, we have to show that the latter has the same kernel. If we recall that the second Sweedler cohomology groups describe precisely Galois objects which are cleft, then we just have to see that if $A \in \mathrm{Gal}(Q \otimes L)$ becomes trivial under both maps to $\mathrm{Gal}(Q)$ and $\mathrm{Gal}(L)$, then A is cleft. But even when we only assume that $A(Q)$ is Q-cleft with cleaving map $j\colon Q \to A(Q)$, and that $A(L)$ is L-cleft with cleaving map $\gamma\colon L \to A(L)$, we arrive immediately at the desired conclusion that A is cleft with cleaving map $Q \otimes L \ni q \otimes \ell \mapsto j(q)\gamma(\ell) \in A$.

References

[1] ABE, E. *Hopf Algebras*. Cambridge University Press, Cambridge etc., 1977.

[2] BESPALOV, Y., AND DRABANT, B. Cross product bialgebras, part I. *J. Algebra* (1999), 466–505.

[3] BESPALOV, Y., AND DRABANT, B. Cross product bialgebras, part II. *preprint* (math.QA/9904142).

[4] BLATTNER, R. J., COHEN, M., AND MONTGOMERY, S. Crossed products and inner actions of Hopf algebras. *Trans. AMS 298* (1986), 671–711.

[5] BLATTNER, R. J., AND MONTGOMERY, S. Crossed products and Galois extensions of Hopf algebras. *Pacific J. of Math. 137* (1989), 37–54.

[6] BÖHM, G., NILL, F., AND SZLACHÁNYI, K. Weak Hopf algebras I: Integral theory and C^*-structure. *J. Algebra 221* (1999), 385–438.

[7] BRZEZIŃSKI, T. On modules associated to coalgebra Galois extensions. *J. Algebra 215* (1999), 290–317.

[8] CAENEPEEL, S., ION, B., MILITARU, G., AND ZHU, S. The factorization problem and the smash biproduct of algebras and coalgebras. *Algebras and Representation Theory 3* (2000), 19–42.

[9] DASCALESCU, S., MILITARU, G., AND RAIANU, S. Crossed coproducts and cleft coextensions. *Comm. Algebra 24* (1996), 1229–1243.

[10] DIJKGRAAF, R., PASQUIER, V., AND ROCHE, P. QuasiHopf algebras, group cohomology and orbifold models. *Nuclear Phys. B Proc. Suppl. 18B* (1990), 60–72.

[11] DOI, Y. Unifying Hopf modules. *J. Algebra 153* (1992), 373–385.

[12] DOI, Y. Braided bialgebras and quadratic bialgebras. *Comm. in Alg. 21* (1993), 1731–1749.

[13] DOI, Y., AND TAKEUCHI, M. Cleft comodule algebras for a bialgebra. *Comm. in Alg. 14* (1986), 801–817.

[14] DRINFEL'D, V. G. Quasi-Hopf algebras. *Leningrad Math. J. 1* (1990), 1419–1457.

[15] HAUSSER, F., AND NILL, F. Diagonal crossed products by duals of quasi-quantum groups. *Rev. Math. Phys. 11* (1999), 553–629.

[16] HAUSSER, F., AND NILL, F. Doubles of quasi-quantum groups. *Comm. Math. Phys. 199* (1999), 547–589.

[17] HAUSSER, F., AND NILL, F. Integral theory for quasi-Hopf algebras. *preprint* (math.QA/9904164).

[18] HAYASHI, T. Face algebras I–A generalization of quantum group theory. *J. Math. Soc. Japan 50* (1998), 293–315.

[19] HOFSTETTER, I. *Erweiterungen von Hopf-Algebren und ihre kohomologischc Beschreibung.* PhD thesis, Universität München, 1990.

[20] HOFSTETTER, I. Extensions of Hopf algebras and their cohomological description. *J. Algebra 164* (1994), 264–298.

[21] JOYAL, A., AND STREET, R. An introduction to Tannaka duality and quantum groups. In *Category Theory (Proceedings, Como 1990)* (1991), A. Carboni, M. C. Pedicchio, and G. Rosolini, Eds., vol. 1488 of *Lec. Notes in Math.*, Springer, pp. 412–492.

[22] KAC, G. I. Extensions of groups to ring groups. *Math. USSR Sbornik 5* (1968), 451–474.

[23] KASSEL, C. *Quantum Groups*, vol. 155 of *GTM*. Springer, 1995.

[24] KOPPINEN, M. Variations on the smash product with applications to group-graded rings. *J. Pure Appl. Algebra 104* (1995), 61–80.

[25] KREIMER, H. F. Hopf–Galois theory and tensor products of Hopf algebras. *Comm. in Alg. 23* (1995), 4009–4030.

[26] MAC LANE, S. Natural associativity and commutativity. *Rice Univ. Studies 49* (1963), 28–46.

[27] MAJID, S. Physics for algebraists: Non-commutative and non-cocommutative Hopf algebras by a bicrossproduct construction. *J. Algebra 130* (1990), 17–64.

[28] MAJID, S. Tannaka–Krein theorem for quasi-Hopf algebras. In *Deformation theory and quantum groups with applications to mathematical physics (Amherst, MA, 1990)* (1992), M. Gerstenhaber and J. Stasheff, Eds., vol. 134 of *Contemp. Math.*, AMS, pp. 219–232.

[29] MAJID, S. *Foundations of quantum group theory.* Cambridge Univ. Press, 1995.

[30] MAJID, S. Quantum double for quasi-Hopf algebras. *Lett. Math. Phys.* *45* (1998), 1–9.

[31] MASUOKA, A. Hopf algebra extensions and cohomology. *in this volume.*

[32] MASUOKA, A. Calculations of some groups of Hopf algebra extensions. *J. Algebra* *191* (1997), 568–588.

[33] MASUOKA, A. Extensions of Hopf algebras and Lie bialgebras. *Trans. AMS 352* (2000), 3837–3879.

[34] MONTGOMERY, S. *Hopf algebras and their actions on rings*, vol. 82 of *CBMS Regional Conference Series in Mathematics.* AMS, Providence, Rhode Island, 1993.

[35] MONTGOMERY, S. Classifying finite-dimensional semisimple Hopf algebras. In *Trends in the representation theory of finite dimensional algebras* (1998), E. L. Green and B. Huisgen-Zimmermann, Eds., vol. 229 of *Contemp. Math.*, AMS.

[36] PAREIGIS, B. Non-additive ring and module theory I. General theory of monoids. *Publ. Math. Debrecen 24* (1977), 189–204.

[37] PAREIGIS, B. A non-commutative non-cocommutative Hopf algebra in "nature". *J. Algebra 70* (1981), 356–374.

[38] RADFORD, D. E. The structure of Hopf algebras with a projection. *J. Algebra 92* (1985), 322–347.

[39] RESHETIKHIN, N. Y., AND SEMENOV-TIAN-SHANSKY, M. A. Quantum *R*-matrices and factorization problems. *J. Geom. Physics 5* (1988), 205–266.

[40] SCHAUENBURG, P. On coquasitriangular Hopf algebras and the quantum Yang–Baxter equation. *Algebra Berichte 67* (1992).

[41] SCHAUENBURG, P. Tannaka duality for arbitrary Hopf algebras. *Algebra Berichte 66* (1992).

[42] SCHAUENBURG, P. Hopf modules and Yetter–Drinfel'd modules. *J. Algebra 169* (1994), 874–890.

[43] SCHAUENBURG, P. Bialgebras over noncommutative rings and a structure theorem for Hopf bimodules. *Appl. Categorical Structures 6* (1998), 193–222.

[44] SCHAUENBURG, P. The structure of Hopf algebras with a weak projection. *Algebras and Representation Theory 3* (2000), 187–211.

[45] SCHAUENBURG, P. Hopf modules and the double of a quasi-Hopf algebra, *preprint* (2001). To appear in *Trans. Amer. Math. Soc.*

[46] SCHAUENBURG, P. The monoidal center construction and bimodules. *J. Pure Appl. Algebra 158* (2001), 325–346.

[47] SCHAUENBURG, P. Hopf bimodules, coquasibialgebras, and an exact sequence of Kac. *Adv. Math.* (to appear).

[48] SCHNEIDER, H.-J. Principal homogeneous spaces for arbitrary Hopf algebras. *Israel J. of Math. 72* (1990), 167–195.

[49] SCHNEIDER, H.-J. Normal basis and transitivity of crossed products for Hopf algebras. *J. Algebra 152* (1992), 289–312.

[50] SCHNEIDER, H.-J. Some remarks on exact sequences of quantum groups. *Comm. in Alg. 21* (1993), 3337–3357.

[51] SCHNEIDER, H.-J. Some properties of factorizable Hopf algebras. *Proc. AMS 129*, 7 (2001), 1891–1898.

[52] SHIMURA, G. *Introduction to the Arithmetic Theory of Automorphic Functions.* Princeton Univ. Press, 1971.

[53] SINGER, W. Extension theory for connected Hopf algebras. *J. Algebra 21* (1972), 1–16.

[54] SWEEDLER, M. E. Cohomology of algebras over Hopf algebras. *Trans. AMS 133* (1968), 205–239.

[55] SWEEDLER, M. E. *Hopf Algebras.* Benjamin, New York, 1969.

[56] TAKEUCHI, M. Groups of algebras over $A \otimes \overline{A}$. *J. Math. Soc. Japan 29* (1977), 459–492.

[57] TAKEUCHI, M. Matched pairs of groups and bismash products of Hopf algebras. *Comm. Algebra 9* (1981), 841–882.

[58] ULBRICH, K.-H. Fiber functors of finite dimensional comodules. *manuscripta math. 65* (1989), 39–46.

[59] ULBRICH, K.-H. On Hopf algebras and rigid monoidal categories. *Israel J. Math. 72* (1990), 252–256.

[60] ULBRICH, K.-H. Smash products and comodules of linear maps. *Tsukuba J. Math. 14* (1990), 371–378.

[61] ZHU, Y. Hecke algebras and representation ring of Hopf algebras. In *First International Congress of Chinese Mathematicians (Beijing, 1998)*. Amer. Math. Soc., Providence, RI, 2001, pp. 219–227.

PETER SCHAUENBURG
MATHEMATISCHES INSTITUT DER UNIVERSITÄT MÜNCHEN
THERESIENSTRASSE 39
D-80333 MÜNCHEN
GERMANY
schauen@rz.mathematik.uni-muenchen.de

New Directions in Hopf Algebras
MSRI Publications
Volume 43, 2002

A Short Course on Quantum Matrices

MITSUHIRO TAKEUCHI

NOTES TAKEN BY BERND STRÜBER

CONTENTS

Introduction

The notion of quantum matrices has interesting and important connections with various topics in mathematics such as quantum group theory, Hopf algebra theory, braided tensor categories, knot and link invariants, the Yang–Baxter equation, representation theory and so on.

This course is an introduction to quantum groups. I have made an effort to have the exposition as elementary as possible. I intended to talk in some informal way, and did not intend to give detailed proofs. The present notes have somewhat more formal and rigorous flavor than the actual lectures, in

These notes are based on a course given at University of Munich in Summer, 1997 with arrangements by H.-J. Schneider and financial support from the Humboldt Foundation and the Graduiertenkolleg of University of Munich.

which I tried to visualize and illustrate connections with knot theory by using transparency sheets.

I begin with introducing 2×2 q-matrices in Section 1. They are characterized in two ways, one by the q-adjoint matrix (Proposition 1.3) and the other by the R-matrix R_q (Proposition 1.6). Each of them admits an interesting interpretation in the theory of knot invariants, and this is illustrated in Section 2 by talking about a basic knot invariant, called Kauffman's bracket polynomial.

The 2×2 q-matrices have the remarkable property that their n-th powers are q^n-matrices. This fact was found around 1991 by several physicists. I reproduce Umeda and Wakayama's elegant proof in Section 3. The first characterization (Proposition 1.3) plays a role in the proof.

General q-matrices of degree n are introduced in Section 4 as well as a brief exposition of fundamental facts involving the R-matrix R_q, the q-symmetric and q-exterior algebras, the q-determinant, the q-adjoint matrix and so on. The q-matrix bialgebra $\mathcal{O}_q(\mathrm{M}(n))$ and the coordinate Hopf algebras $\mathcal{O}_q(\mathrm{GL}(n))$ and $\mathcal{O}_q(\mathrm{SL}(n))$ of quantum GL and SL are also considered in this section.

The material of Section 4 leads to a q-analogue of linear algebra. In Section 5, I reproduce J. Zhang's result on a q-analogue of the Cayley–Hamilton theorem which will be one of the most interesting topics in this area.

The R-matrix R_q introduced in Section 4 plays a remarkable role in the construction of the so-called Homfly polynomial which is a two-variable invariant of oriented links. This is illustrated in Section 6 following Kauffman's idea.

In Sections 7 to 9, we talk about some Hopf algebraic properties of $\mathcal{O}_q(\mathrm{GL}(n))$ and $\mathcal{O}_q(\mathrm{SL}(n))$. In Section 7, we talk about the duality of two quantum Hopf algebras $\mathcal{O}_q(\mathrm{SL}(n))$ and $\mathrm{U}_q(\mathbf{sl_n})$. Details are exposed in case $n = 2$ and q is not a root of unity.

We show how to determine all group-like and skew-primitive elements of $\mathcal{O}_q(\mathrm{GL}(n))^\circ$, the dual Hopf algebra of $\mathcal{O}_q(\mathrm{GL}(n))$, in Section 8. This technical result is used to describe all quantum group homomorphisms $\mathrm{SL}_q(n) \to \mathrm{GL}_q(m)$ in Section 9. We explain the main result (Theorem 9.13) when q is not a root of unity, since this case is very easy to handle. However, we note that the result is also valid at roots of unity.

These nine sections were delivered in six consecutive lectures in Munich, while the remaining part came out from my rough draft written in Japanese with translation into English by A. Masuoka.

In Section 10, we introduce 2-parameter quantum matrices and construct the bialgebra $\mathcal{O}_{\alpha,\beta}(\mathrm{M}(n))$ and the Hopf algebra $\mathcal{O}_{\alpha,\beta}(\mathrm{GL}(n))$ as generalizations of $\mathcal{O}_q(\mathrm{M}(n))$ and $\mathcal{O}_q(\mathrm{GL}(n))$. If we take $(\alpha, \beta) = (1, q)$, the Hopf algebra $\mathcal{O}_{1,q}(\mathrm{GL}(n))$ defines the Dipper–Donkin quantum GL whose polynomial representations become equivalent with comodules for $\mathcal{O}_{1,q}(\mathrm{M}(n))$. This idea leads to a q-analogue of the Schur algebra which is discussed in Section 11 as well as the Hecke algebra.

Y. Doi has introduced the idea of cocycle deformations of a bialgebra through his study of braided and quadratic bialgebras. The notion of a braided bialgebra is dual to Drinfeld's quasi-triangular bialgebras. These items are explained briefly in Section 12 and I reproduce my own result on determining all cocycle deformations of the bialgebra $\mathcal{O}_{\alpha,\beta}(M(n))$.

This course ends with a short remark on 2×2 R-matrices in Section 13 including my student Suzuki's new results.

These lecture notes were written up by B. Strüber who has made an excellent job, especially when he made my rough drafts and transparency sheets into a formal and rigorous exposition. I thank him most warmly for his efforts. I would also like to thank H.-J. Schneider who invited me to Munich and arranged my lectures.

Section 14 was added after acceptance of this exposition in this publication in order to update the contents. I introduce E. Müller's and E. Letzter's current results concerning the quantum Frobenius map in this addendum. Finally, Section 15 is a short annotated bibliography.

1. (2×2) q-Matrices

Throughout this paper, let k be a fixed base field, $k^{\times} := k \setminus \{0\}$ and $q \in k^{\times}$.

DEFINITION 1.1 (q-MATRICES). Let $A = \left(\begin{smallmatrix} a & b \\ c & d \end{smallmatrix} \right)$ be a (2×2)-matrix over some k-algebra. We call A a q-matrix if its entries satisfy the following relations:

$$
\begin{aligned}
ba &= qab, & dc &= qcd, \\
ca &= qac, & db &= qbd, \\
cb &= bc, & da - ad &= (q - q^{-1})bc.
\end{aligned}
$$

(The last relation implies $ad - q^{-1}bc = da - qbc$.) We call this expression the q-determinant of A and denote it by $|A|_q$ (for $q = 1$, this is just the usual determinant).

DEFINITION 1.2. The q-adjoint matrix of any matrix A is defined as

$$
\tilde{A} := \begin{pmatrix} d & -qb \\ -q^{-1}c & a \end{pmatrix}.
$$

It is easily verified that $\tilde{A}$ is a q^{-1}-matrix, if A is a q-matrix.

PROPOSITION 1.3. (a) For any q-matrix A, we have:

$$
|\tilde{A}|_{q^{-1}} = |A|_q, \tilde{A} = \begin{pmatrix} & q \\ -1 & \end{pmatrix} A^t \begin{pmatrix} & -1 \\ q^{-1} & \end{pmatrix}.
$$

(b) A matrix A is a q-matrix, with $\delta = |A|_q$, if and only if $A\tilde{A} = \delta I = \tilde{A}A$.

PROOF. Part (a) is immediate. Part (b) follows by comparing the matrix coefficients of

$$
A\tilde{A} = \begin{pmatrix} ad - q^{-1}bc & ba - qab \\ cd - q^{-1}dc & da - qcb \end{pmatrix}, \quad \tilde{A}A = \begin{pmatrix} da - qbc & db - qbd \\ -q^{-1}ca + ac & ad - q^{-1}cb \end{pmatrix}.
$$

□

COROLLARY 1.4. *Let A, B be q-matrices over the same algebra.*

(a) *The q-determinant δ of A commutes with the entries of A.*

(b) *If δ is invertible then A is invertible and $A^{-1} = \delta^{-1}\tilde{A}$, which is a q^{-1}-matrix, with determinant*

$$|A^{-1}|_{q^{-1}} = \delta^{-2}|\tilde{A}|_{q^{-1}} = \delta^{-1} = |A|_q^{-1}.$$

(c) *If every entry of A commutes with every entry of B, then AB is a q-matrix, with $|AB|_q = |A|_q|B|_q$.*

PROOF. Part (a) follows from $\delta A = A\tilde{A}A = A\delta$. Part (b) is easily obtained from Proposition 1.3. To show (c), observe that

$$(\widetilde{AB}) = \begin{pmatrix} & q \\ -1 & \end{pmatrix}(AB)^t \begin{pmatrix} & -1 \\ q^{-1} & \end{pmatrix} = \begin{pmatrix} & q \\ -1 & \end{pmatrix}B^tA^t \begin{pmatrix} & -1 \\ q^{-1} & \end{pmatrix} = \tilde{B}\tilde{A}.$$

Hence, writing $\delta := |A|_q$ and $\delta' := |B|_q$, we get:

$$AB\widetilde{AB} = AB\tilde{B}\tilde{A} = A\delta'\tilde{A} = A\tilde{A}\delta' = \delta\delta'I.$$

Similarly, $\widetilde{AB}AB = \delta\delta'I$. This proves the claim. □

We give an equivalent description of q-matrices, in terms of "R-matrices".

DEFINITION 1.5. In the following, we identify $M_4(k)$ with $M_2(k) \otimes M_2(k)$, by

$$\mathbb{E}_{(ik),(j\ell)} = \mathbb{E}_{ij} \otimes \mathbb{E}_{k\ell}.$$

For a (2×2)-matrix $A = (a_{ij})$ over some k-algebra, we write

$$A^{(2)} := (a_{ij}a_{k\ell})_{(ik),(j\ell)},$$

which is a (4×4)-matrix with rows and columns indexed by (11), (12), (21), (22).

We write:

$$R_q := \begin{pmatrix} q & 0 & 0 & 0 \\ 0 & 0 & 1 & 0 \\ 0 & 1 & q - q^{-1} & 0 \\ 0 & 0 & 0 & q \end{pmatrix}.$$

PROPOSITION 1.6 (THE R-MATRIX R_q AND q-MATRICES). (a) *R_q is invertible and satisfies the following equation in $M_2(k) \otimes M_2(k) \otimes M_2(k)$ (called the braid relation or the Yang–Baxter equation):*

$$(R_q \otimes I)(I \otimes R_q)(R_q \otimes I) = (I \otimes R_q)(R_q \otimes I)(I \otimes R_q). \tag{1-1}$$

(b) *We have $(R_q - qI)(R_q + q^{-1}I) = 0$.*

(c) *A (2×2)-matrix A is a q-matrix if and only if $A^{(2)}R_q = R_qA^{(2)}$.*

The proof is easy and straightforward, but tedious, so we omit it.

2. q-Matrices and Kauffman's Bracket Polynomial

There is a close relation between R-matrices and polynomials associated to diagrams of links.

It is not easy to give a formal definition of a *link diagram*, so we illustrate this concept with some examples:

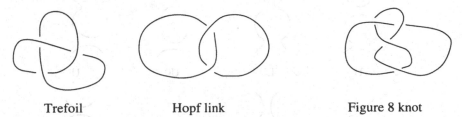

Trefoil   Hopf link   Figure 8 knot

Figure 1. Some link diagrams

DEFINITION 2.1 (KAUFFMAN'S BRACKET POLYNOMIAL). Let L be a link diagram. Kauffman's bracket polynomial $\langle L \rangle \in \mathbb{Z}[t, t^{-1}]$ is defined by the following rules:

(a) $\langle \, \slash\!\!\!\diagdown \, \rangle = t \langle \,)(\, \rangle + t^{-1} \langle \asymp \rangle$;

(b) $\langle \bigcirc \ldots \bigcirc \rangle = d^n$, for $d := -t^2 - t^{-2}$ (for n circles).

It is an easy exercise to determine the bracket polynomial for a given link diagram, using these rules. We have, for example

$$\langle \text{Hopf link} \rangle = (-t^4 - t^{-4})d, \quad \langle \text{Trefoil} \rangle = (-t^5 - t^{-3} + t^{-9})d.$$

Kauffman's polynomial is invariant under operations which transform a link diagram in another diagram of "the same" link.

The operations shown in Figure 2 are called *Reidemeister moves* of type II and III, respectively.

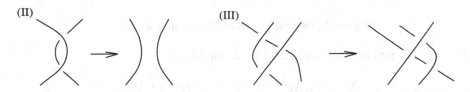

Figure 2. Reidemeister moves of type II and III

PROPOSITION 2.2. *Kauffman's bracket polynomial is invariant under Reidemeister moves of type II and III, i.e., under regular isotopy.*

PROOF. We give the proof for type II, the same method applies to type III. By replacing successively the crossings $\langle \, \slash\!\!\!\diagdown \, \rangle$ by either $\langle)(\rangle$ or $\langle \asymp \rangle$, we get the *skein tree* shown in Figure 3.

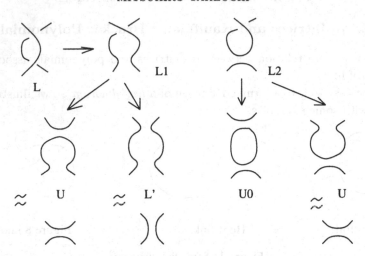

Figure 3. Skein tree for Reidemeister move II

Now we can apply the rules of Definition 2.1 to get the desired result:

$$\langle L \rangle = t\langle L1 \rangle + t^{-1}\langle L2 \rangle$$
$$= t(t\langle U \rangle + t^{-1}\langle L' \rangle) + t^{-1}(t\langle U0 \rangle + t^{-1}\langle U \rangle)$$
$$= (t^2 + t^{-2})\langle U \rangle + \langle L' \rangle + d\langle U \rangle = \langle L' \rangle. \qquad \square$$

There is another method to compute $\langle L \rangle$, closely related to the R-matrix R_q.

Any link diagram L can be decomposed into *elements* of the form $\wr$ (arc), $\cap$ (cap), $\cup$ (cup), $\diagdown\!\!\!\diagup$, $\diagup\!\!\!\diagdown$ (crossings). Consider the following example:

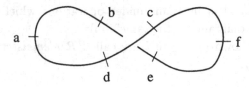

Figure 4. Decomposition of the link diagram of ∞

The bracket polynomial $\langle \infty \rangle$ can be calculated as

$$\sum \langle_a \cap_b \rangle \langle^a \cup^d \rangle \langle^b_d \diagdown^c_e \rangle \langle_c \cap_f \rangle \langle^e \cup^f \rangle,$$

where a, b, c, d, e, f range over $\{1, 2\}$, if one associates suitable values to the factors $\langle C \rangle$, where C denotes a labelled element of the link diagram.

THEOREM 2.3 (KAUFFMAN). *The bracket polynomial $\langle L \rangle$ of a link diagram L is*

$$\langle L \rangle = \sum \left(\prod (\langle C \rangle \mid C \text{ is a labelled element of } L) \mid \text{all labels in } \{1, 2\} \right),$$

where $\langle C \rangle$, for a labelled element C of L, is defined as

$$\langle {}^a \,\rceil_b \rangle := \delta_{ab},$$

$$\langle {}_a \cap_b \rangle := \langle {}^a \cup^b \rangle := M_{ab},$$

$$\langle {}^a_c \diagup\hspace{-6pt}\diagdown {}^b_d \rangle := t\delta_{ac}\delta_{bd} + t^{-1}M_{ab}M_{cd} =: R^{ab}_{cd},$$

$$\langle {}^a_c \diagdown\hspace{-6pt}\diagup {}^b_d \rangle := t^{-1}\delta_{ac}\delta_{bd} + tM_{ab}M_{cd} =: \overline{R}^{ab}_{cd},$$

with $M := \begin{pmatrix} 0 & it \\ -it^{-1} & 0 \end{pmatrix}$ and $i^2 = -1$.

DEFINITION 2.4. Let $R := (R^{ab}_{cd})_{(ab),(cd)}$ denote the (4×4)-matrix, with entries in $\mathbb{Z}[t, t^{-1}]$, defined by

$$R^{ab}_{cd} := t\delta_{ac}\delta_{bd} + t^{-1}M_{ab}M_{cd} = \langle {}^a_c \diagup\hspace{-6pt}\diagdown {}^b_d \rangle.$$

Hence,

$$R = \begin{pmatrix} t & 0 & 0 & 0 \\ 0 & 0 & t^{-1} & 0 \\ 0 & t^{-1} & t - t^{-3} & 0 \\ 0 & 0 & 0 & t \end{pmatrix},$$

which means that for $q := t^2$, we have (cf. Definition 1.5)

$$R = t^{-1}R_q. \tag{2-1}$$

Recall that a matrix A is a q-matrix if and only if $A^{(2)}$ commutes with R_q (Proposition 1.6(c)). This can be interpreted as a certain compatibility condition with link diagrams. In the following, we allow link diagrams to contain "nodes", labelled with matrices in some algebra.

DEFINITION 2.5. Let $A = (A_{ij})$ be a (2×2)-matrix with entries in some algebra K. By defining

$$< \!\boxed{A}\; \boxed{A}\! > \;=\; \sum_{i,j,k,l \in \{1,2\}} \left\langle {}^i_j \;\; {}^k_l \right\rangle A_{ij}A_{kl}, \tag{2-2}$$

we associate a bracket polynomial (in $K[t, t^{-1}]$) to link diagrams which may contain nodes with the matrix A.

We examine the invariance conditions given on Figure 5.

PROPOSITION 2.6. Let $q := t^2$ and let A be any matrix with entries in some k-algebra.

(a) Conditions (1) and (2) are satisfied if and only if A is a q-matrix, with $|A|_q = 1$.

(b) Condition (3), condition (4) and the condition that A is a q-matrix are equivalent.

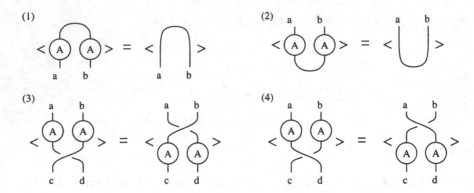

Figure 5. Invariance conditions for link diagrams with nodes

PROOF. (a) Using (2–2) and Kauffman's theorem, the left hand side of condition (1) is

$$\sum_{i,j\in\{1,2\}} A_{ia}M_{ij}A_{jb} = (A^t MA)_{ab}.$$

Hence, (1) is satisfied if and only if $A^t MA = M$. Now M^2 is the identity matrix and $\tilde{A} = MA^t M$. Therefore, condition (1) is equivalent to $\tilde{A}A = I$.

It is shown similarly that (2) is equivalent to $A\tilde{A} = I$. Now part (a) follows from Proposition 1.3 (b).

(b) It easy to check that (3) is equivalent to the condition that $A^{(2)}$ commutes with R, and that (4) is equivalent to $\overline{R} = R^{-1}$. Recall that $R = t^{-1}R_q$, by (2–1). Now (b) follows from Proposition 1.6(c). □

Proposition 2.6 means that the bracket polynomial is invariant under the action of the quantum group $SL_q(2)$, with $q := t^2$.

3. Powers of (2×2) q-Matrices

THEOREM 3.1 (POWERS OF (2×2) q-MATRICES). *If A is a (2×2) q-matrix then A^n is a q^n-matrix, and its q^n-determinant is:*

$$|A^n|_{q^n} = |A|_q^n.$$

PROOF (Umeda, Wakayama, 1993). Let $A = \begin{pmatrix} a & b \\ c & d \end{pmatrix}$ be a (2×2) q-matrix and $\jmath := \begin{pmatrix} 1 & \\ & q \end{pmatrix}$. Writing $\tau := a + q^{-1}d$ and $\delta := |A|_q$, it is easy to check that

$$A^2 = \tau\jmath A - q^{-1}\delta\jmath^2. \tag{3-1}$$

Since the transpose A^t is a q-matrix as well, with the same q-determinant, it follows that

$$(A^t)^2 = \tau\jmath A^t - q^{-1}\delta\jmath^2. \tag{3-2}$$

We show by induction that for all $n \geq 0$:

$$(A^t)^n = \jmath^{n-1}(A^n)^t\jmath^{1-n}. \tag{3-3}$$

The cases $n = 0, 1$ are obvious. Assume (3–3) is true for $n, n + 1$. Then

$$
\begin{aligned}
(A^t)^{n+2} \;&\overset{(3\text{–}2)}{=}\; \tau \mathcal{J}(A^t)^{n+1} - q^{-1}\delta\mathcal{J}^2(A^t)^n \\
&\overset{(3\text{–}3)}{=}\; \tau \mathcal{J}^{n+1}(A^{n+1})^t \mathcal{J}^{-n} - q^{-1}\delta\mathcal{J}^{n+1}(A^n)^t \mathcal{J}^{1-n} \\
&=\; \mathcal{J}^{n+1}\underbrace{\left(\tau(A^{n+1})^t \mathcal{J} - q^{-1}\delta(A^n)^t \mathcal{J}^2 \right)}_{(*)} \mathcal{J}^{-1-n}. \qquad (3\text{–}4)
\end{aligned}
$$

From (3–1), it follows that

$$
A^{n+2} = \tau \mathcal{J} A^{n+1} - q^{-1}\delta\mathcal{J}^2 A^n.
$$

Hence, the expression $(*)$ is equal to $(A^{n+2})^t$ and (3–4) implies

$$
(A^t)^{n+2} = \mathcal{J}^{n+1}(A^{n+2})^t \mathcal{J}^{-1-n},
$$

which proves the induction step.

By Proposition 1.3, we have $A\tilde{A} = \tilde{A}A = \delta I$. Since δ commutes with the entries of A, we get

$$
A^n \tilde{A}^n = \tilde{A}^n A^n = \delta^n I. \qquad (3\text{–}5)
$$

It follows that

$$
\begin{aligned}
\tilde{A}^n \;&=\; \begin{pmatrix} & q \\ -1 & \end{pmatrix} (A^t)^n \begin{pmatrix} & -1 \\ q^{-1} & \end{pmatrix} \\
&\overset{(3\text{–}3)}{=}\; \begin{pmatrix} & q \\ -1 & \end{pmatrix} \mathcal{J}^{n-1}(A^n)^t \mathcal{J}^{1-n} \begin{pmatrix} & -1 \\ q^{-1} & \end{pmatrix} \\
&=\; \begin{pmatrix} & q^n \\ -1 & \end{pmatrix} (A^n)^t \begin{pmatrix} & -1 \\ q^{-n} & \end{pmatrix} = \widetilde{(A^n)}.
\end{aligned}
$$

Using (3–5), this implies

$$
A^n \widetilde{(A^n)} = \widetilde{(A^n)} A^n = \delta^n I,
$$

which means A^n is a q^n-matrix, with q^n-determinant δ^n (cf. Proposition 1.3(b)). $\qquad\square$

4. The Quantum Linear Groups $\mathrm{GL}_q(n)$ and $\mathrm{SL}_q(n)$

DEFINITION 4.1 (($n \times n$) q-MATRICES). Let $A = (a_{ij})$ be an $(n \times n)$-matrix, with entries in some k-algebra. We call A a q-matrix, if every (2×2)-minor of A is a q-matrix.

A (2×2)-minor of A is a (2×2)-matrix obtained from A by removing some rows and columns.

Similar to Proposition 1.6, we can describe $(n \times n)$ q-matrices using an R-matrix. The next definition generalizes Definition 1.5.

DEFINITION 4.2. We identify $M_{n^2}(k)$ with $M_n(k) \otimes M_n(k)$, by

$$\mathbb{E}_{(ik),(j\ell)} = \mathbb{E}_{ij} \otimes \mathbb{E}_{k\ell}.$$

For an $(n \times n)$-matrix $A = (a_{ij})$ over some k-algebra, we write

$$A^{(2)} := (a_{ij}a_{k\ell})_{(ik),(j\ell)},$$

which is an $(n^2 \times n^2)$-matrix.

The matrix $R_q \in M_n(k) \otimes M_n(k)$ is given by

$$R_q := q \sum_i \mathbb{E}_{ii} \otimes \mathbb{E}_{ii} + \sum_{i \neq j} \mathbb{E}_{ij} \otimes \mathbb{E}_{ji} + (q - q^{-1}) \sum_{i<j} \mathbb{E}_{jj} \otimes \mathbb{E}_{ii}.$$

PROPOSITION 4.3 (THE R-MATRIX R_q AND q-MATRICES). (a) R_q *is invertible and satisfies the Yang–Baxter equation* (1–1).
(b) *We have* $(R_q - qI)(R_q + q^{-1}I) = 0$.
(c) *An* $(n \times n)$-*matrix* A *is a* q-*matrix if and only if* $A^{(2)}R_q = R_q A^{(2)}$.

We omit the proof since it is easy but tedious.

COROLLARY 4.4. *Let* A, B *be* $(n \times n)$ q-*matrices (over the same algebra), such that every entry of* A *commutes with every entry of* B. *Then* AB *is again a* q-*matrix*.

PROOF. Since $(AB)^{(2)} = A^{(2)}B^{(2)}$, the claim follows from Proposition 4.3. ☐

DEFINITION 4.5 (THE q-MATRIX BIALGEBRA). Let $M := \mathcal{O}_q(M(n))$ denote the k-algebra generated by elements x_{ij} $(1 \leq i, j \leq n)$, subject to the relations that $X := (x_{ij})_{i,j}$ is a q-matrix. We call M the q-*matrix bialgebra*.

PROPOSITION 4.6. *The following algebra maps make* M *a bialgebra*:

$$\Delta : M \to M \otimes M, \quad x_{ij} \mapsto \sum_{s=1}^n x_{is} \otimes x_{sj},$$
$$\varepsilon : M \to k, \quad x_{ij} \mapsto \delta_{ij}.$$

PROOF. The matrices $A := (x_{ij} \otimes 1)_{i,j}$ and $B := (1 \otimes x_{ij})_{i,j}$ are both q-matrices over $M \otimes M$. Since every entry of A commutes with every entry of B, the product $AB = (\Delta(x_{ij}))_{i,j}$ is also a q-matrix. Hence, Δ is a well-defined algebra map.

Since the identity matrix is a q-matrix, ε is well-defined, too. The bialgebra axioms are easily checked on the generators. ☐

PROPOSITION 4.7 (THE ALGEBRAIC STRUCTURE OF M). *The algebra* M *is an integral domain, which is non-commutative, if* $n > 1$ *and* $q \neq 1$.
It is a polynomial algebra in $x_{11}, \ldots, x_{nn}$, *i.e. the ordered monomials in* $x_{11}, \ldots, x_{nn}$ *(with respect to any total ordering) form a basis of* M.

The proof is not easy and requires some calculations. We omit it here.

Let V_n be an n-dimensional vector space with basis $\{e_1, \ldots, e_n\}$. It becomes a right M-comodule by

$$\varrho : V_n \to V_n \otimes M, e_j \mapsto \sum_{i=1}^{n} e_i \otimes x_{ij}. \tag{4-1}$$

The tensor algebra $T(V_n)$ is a right M-comodule algebra by the algebra map extension of $\varrho : T(V_n) \to T(V_n) \otimes M$,

$$\varrho(e_{j_1} \cdot \ldots \cdot e_{j_r}) := \sum_{i_1, \ldots i_r} e_{i_1} \cdot \ldots \cdot e_{i_r} \otimes x_{i_1, j_1} \cdot \ldots \cdot x_{i_r, j_r}. \tag{4-2}$$

The appropriate restriction of ϱ makes $V_n \otimes V_n$ into an M-comodule. (For a general survey of comodules and comodule algebras, see for example Section 4 of Montgomery's book *Hopf algebras and their actions on rings*, CBMS Regional Conference Series in Mathematics 82, AMS, 1993.)

LEMMA 4.8 (THE R-MATRIX INDUCES A COLINEAR MAP). *The linear map* $R_q : V_n \otimes V_n \to V_n \otimes V_n$ *defined by* $R_q \in M_n(k) \otimes M_n(k)$ *is given as follows:*

$$e_i \otimes e_j \mapsto \begin{cases} e_j \otimes e_i, & \text{if } i < j, \\ q e_i \otimes e_i, & \text{if } i = j, \\ e_j \otimes e_i + (q - q^{-1}) e_i \otimes e_j, & \text{if } i > j. \end{cases}$$

It is an M-comodule map.

PROOF. The $((ik), (j\ell))$-component of $X^{(2)}$ is, by definition, $x_{ij} x_{k\ell}$. Since

$$\varrho(e_j \otimes e_\ell) = \sum_{i,k} (e_i \otimes e_k) \otimes x_{ij} x_{k\ell},$$

the relation $X^{(2)} R_q = R_q X^{(2)}$ implies the claim.   □

It follows that the kernel and the image of $(R_q - qI)$ are M-subcomodules of $V_n \otimes V_n$. They are given explicitly as follows:

$$\mathrm{Im}(R_q - qI) = k\{e_j \otimes e_i - q e_i \otimes e_j \mid i < j\},$$
$$\mathrm{Ker}(R_q - qI) = k\{e_i \otimes e_i, e_i \otimes e_j + q e_j \otimes e_i \mid i < j\}.$$

This motivates the following definition:

DEFINITION 4.9 (DEFORMED SYMMETRIC AND EXTERIOR ALGEBRAS). Let I_S denote the ideal of the tensor algebra $T(V_n)$ generated by the image of $(R_q - qI)$, and $I_\wedge$ the ideal generated by the kernel of $(R_q - qI)$. We call $S_q(V_n) := T(V_n)/I_S$ the *q-symmetric* algebra and $\bigwedge_q(V_n) := T(V_n)/I_\wedge)$ the *q-exterior* algebra.

It is not difficult to show that $S_q(V_n)$ has the basis

$$\{e_1^{s_1} \cdot \ldots \cdot e_n^{s_n} \mid \text{all } s_i \geq 0\},$$

and $\bigwedge_q(V_n)$ has the basis

$$\{e_{i_1} \cdot \ldots \cdot e_{i_m} \mid 1 \leq i_1 < \cdots < i_m \leq n\}.$$

They are made into graded algebras by defining the degree of all e_i to be 1.

COROLLARY 4.10 (THE COACTION OF M AND THE q-DETERMINANT). (a) *The algebras $S_q(V_n)$ and $\bigwedge_q(V_n)$ are right M-comodule algebras by the coaction induced by ϱ.*
(b) *There is a unique $g \in M$ such that $\varrho(e_1 \cdot \ldots \cdot e_n) = e_1 \cdot \ldots \cdot e_n \otimes g$, where $e_1 \cdot \ldots \cdot e_n \in \bigwedge_q(V_n)$, namely*

$$g = \sum_{\sigma \in S_n} (-q)^{-\ell(\sigma)} x_{\sigma(1),1} \cdot \ldots \cdot x_{\sigma(n),n}$$

(where $\ell(\sigma)$ denotes the number of inversions of a permutation σ).

PROOF. The ideals $I_S, I_\wedge$ considered in Definition 4.9 are M-subcomodules by Lemma 4.8. This implies (a).

All homogeneous components of $S_q(V_n)$ and $\bigwedge_q(V_n)$ are M-subcomodules. In particular, the n-th component of $\bigwedge_q(V_n)$, which is $k \cdot e_1 \cdot \ldots \cdot e_n$, is a 1-dimensional M-subcomodule. Hence, there is a group-like element $g \in M$ with the required property. Uniqueness is clear, since $e_1 \cdot \ldots \cdot e_n \neq 0$ in $\bigwedge_q(V_n)$.

To calculate g, observe that in $\bigwedge_q(V_n)$, we have

$$e_{\sigma(1)} \cdot \ldots \cdot e_{\sigma(n)} = (-q)^{-\ell(\sigma)} e_1 \cdot \ldots \cdot e_n,$$

for all $\sigma \in S_n$. From (4–2), we obtain the required result:

$$
\begin{aligned}
\varrho(e_1 \cdot \ldots \cdot e_n) &= \sum_{i_1, \ldots i_n} e_{i_1} \cdot \ldots \cdot e_{i_n} \otimes x_{i_1,1} \cdot \ldots \cdot x_{i_n,n} \\
&\overset{(!)}{=} \sum_{\sigma \in S_n} e_{\sigma(1)} \cdot \ldots \cdot e_{\sigma(n)} \otimes x_{\sigma(1),1} \cdot \ldots \cdot x_{\sigma(n),n} \\
&= \sum_{\sigma \in S_n} (-q)^{-\ell(\sigma)} e_1 \cdot \ldots \cdot e_n \otimes x_{\sigma(1),1} \cdot \ldots \cdot x_{\sigma(n),n}
\end{aligned}
$$

(for "(!)", notice that $e_{i_1} \cdot \ldots \cdot e_{i_r} = 0$ in $\bigwedge_q(V_n)$, if $i_j = i_\ell$ for some $j \neq \ell$).   □

DEFINITION 4.11 (THE QUANTUM DETERMINANT). The element $g \in M$ is called the q-*determinant* (or *quantum determinant*) of X and denoted by $|X|_q$.

Note that for $n = 2$, we obtain the q-determinant as defined in Section 1 (see Definition 1.1).

DEFINITION 4.12 (THE QUANTUM ADJOINT MATRIX). Let X_{ij} denote the $(n-1) \times (n-1)$ minor of X, obtained by removing the i-th row and the j-th column. Note that X_{ij} is again a q-matrix. The q-*adjoint matrix* of X is defined as

$$\tilde{X} := ((-q)^{j-i} |X_{ji}|_q)_{i,j}.$$

PROPOSITION 4.13 (QUANTUM ADJOINT MATRIX, QUANTUM DETERMINANT). *We have*

$$X\tilde{X} = \tilde{X}X = |X|_q I.$$

PROOF. Writing $\widehat{e_i} := e_1 \cdots e_{i-1} e_{i+1} \cdots e_n \in \bigwedge_q(V_n)$, it is easily checked that

$$\varrho(\widehat{e_j}) = \sum_{i=1}^{n} \widehat{e_i} \otimes |X_{ij}|_q. \tag{4-3}$$

It follows from the relations in $\bigwedge_q(V_n)$ that

$$\widehat{e_i} e_j = \delta_{ij}(-q)^{i-n} e_1 \cdots e_n. \tag{4-4}$$

The last two equations imply

$$\delta_{jk}(-q)^{j-n} e_1 \cdots e_n \otimes g \overset{(4-4)}{=} \varrho(\widehat{e_j} e_k) \overset{(4-3)}{=} \sum_{i=1}^{n} \widehat{e_i} e_i \otimes |X_{ij}|_q x_{ik}$$

$$\overset{(4-4)}{=} \sum_{i=1}^{n} e_1 \cdots e_n \otimes (-q)^{i-n} |X_{ij}|_q x_{ik}.$$

Comparison of the coefficients of $e_1 \cdots e_n$ yields

$$\delta_{jk}(-q)^{j-n} g = \sum_i (-q)^{i-n} |X_{ij}|_q x_{ik},$$

which means that $gI = \tilde{X}X$. Similarly, one sees $gI = X\tilde{X}$ by using the fact that $|X|_q = |X^t|_q$. $\qquad\square$

Since $gX = X\tilde{X}X = Xg$, we get:

COROLLARY 4.14. *The quantum determinant* $|X|_q$ *is central in* M.

THEOREM 4.15 (THE QUANTIZED GENERAL LINEAR GROUP). *Suppose that* $M := \mathcal{O}_q(\mathrm{M}(n))$ *and let* $H := M[g^{-1}]$. *Then* H *is a Hopf algebra.*

PROOF. We denote the images in H of the generators $x_{ij} \in M$ again by x_{ij}.
By Proposition 4.13, the matrix X is invertible in $\mathrm{M}_n(H)$, with inverse

$$X^{-1} = g^{-1}\tilde{X} = \tilde{X}g^{-1}.$$

In the opposite algebra H^{op} of H, we have

$$(X^{-1})^{(2)} = ((X^{-1})_{k\ell}(X^{-1})_{ij})_{(ik),(j\ell)} = (X^{(2)})^{-1}.$$

This matrix commutes with R_q, since so does $X^{(2)}$ (cf. Proposition 4.3). This shows that X^{-1} is a q-matrix with entries in H^{op}.
Hence, there is an algebra map S', defined by

$$S' : M \to H^{\mathrm{op}}, \quad X \mapsto X^{-1}(\text{componentwise}).$$

This induces an anti-algebra map $S : M \to H$.
We show that S factors through H. Let $C : M \to H$ denote the canonical map ($x_{ij} \mapsto x_{ij}$). It is easy to check that the relation

$$S * C = u\varepsilon = C * S \tag{4-5}$$

holds on the generators x_{ij}. (Here "$*$" denotes the convolution product and u, resp. ε the unit of H, resp. the counit of M.)

It is not difficult to deduce that (4–5) holds (everywhere). We apply (4–5) to g and obtain, since g is group-like:

$$S(g)g = 1 = gS(g).$$

Hence, $S(g) = g^{-1}$, and we can extend S to an anti-algebra map

$$S : H \to H, \quad g \mapsto g^{-1}, \quad g^{-1} \mapsto g.$$

This is the antipode of H (it suffices to verify the axiom for the antipode on generators, which is easily done). □

DEFINITION 4.16. The *category of quantum groups* is, by definition, the opposite category of the category of Hopf algebras. The Hopf algebra associated to the quantum group G is denoted by $\mathcal{O}(G)$.

DEFINITION 4.17. We write

$$\mathcal{O}_q(\mathrm{GL}(n)) := H, \quad \mathcal{O}_q(\mathrm{SL}(n)) := H/(g-1).$$

The associated quantum groups $\mathrm{GL}_q(n)$ and $\mathrm{SL}_q(n)$ are called the *quantized general* and *special linear groups*.

5. A q-Analogue of the Cayley–Hamilton Theorem

This section discusses a quantized version of the Cayley–Hamilton theorem, found by J. Zhang (1991). We start with a few notations:

DEFINITION 5.1. Let $A = (a_{ij})$ be an $(n \times n)$ q-matrix with entries in some k-algebra. We use the abbreviation

$$q_{ij} := \begin{cases} q & \text{for } i < j, \\ 1 & \text{for } i = j, \\ q^{-1} & \text{for } i > j. \end{cases}$$

For elements $i_1 < \cdots < i_m$ and $j_1 < \cdots < j_m$ of $[1,n] := \{1, 2, \ldots, n\}$, we write

$$D(i_1, \ldots, i_m \mid j_1, \ldots, j_m) := |A_{\{i_1, \ldots, i_m\}\{j_1, \ldots, j_m\}}|_q, \qquad (5\text{–}1)$$

and define, for $j, m \in [1, n]$:

$$tr_j^m := \sum_{1 \le i_1 < \cdots < i_m \le n} q_{i_1, j} \cdot \ldots \cdot q_{i_m, j} D(i_1, \ldots, i_m \mid i_1, \ldots, i_m),$$

$$\mathrm{Tr}^m := \begin{pmatrix} tr_1^m & & 0 \\ & \ddots & \\ 0 & & tr_n^m \end{pmatrix}.$$

In particular, $tr_j^1 = \sum_{i=1}^n q_{ij} a_{ii}$ and $tr_j^n = q_{1j} \cdot \ldots \cdot q_{nj} |A|_q$.

THEOREM 5.2 (THE q-CAYLEY–HAMILTON THEOREM). *For any q-matrix A,
we have*:

$$A^n - \mathrm{Tr}^1 A^{n-1} + \cdots + (-1)^{n-1}\mathrm{Tr}^{n-1}A + (-1)^n\mathrm{Tr}^n = 0.$$

PROOF. (1) Take any $i_1,\ldots,i_m,j_1,\ldots,j_m \in [1,n]$. Writing $i := (i_1,\ldots,i_m)$,
$j := (j_1,\ldots,j_m)$, and $i\sigma := (i_{\sigma(1)},\ldots,i_{\sigma(m)})$ (for a permutation $\sigma \in S_m$), we
put (in accordance with (5–1), if $i_1 < \cdots < i_m$ and $j_1 < \cdots < j_m$):

$$D(i \mid j) := \sum_{\sigma \in S_m} (-q)^{\ell(i)-\ell(i\sigma)} a_{i_{\sigma(1)},j_1} \cdot \cdots \cdot a_{i_{\sigma(m)},j_m},$$

and this term vanishes, unless $i_1,\ldots,i_m$ are distinct and $j_1,\ldots,j_m$ are distinct.
(Here $\ell(i)$ denotes the number of inversions in $(i_1,\ldots,i_m)$.)

(2) When $i_1,\ldots,i_m$ are distinct and $j_1,\ldots,j_m$ are distinct, put $\{i'_1,\ldots,i'_m\}$
$:= \{i_1,\ldots,i_m\}$, such that $i'_1 < \cdots < i'_m$, and define $j'_1 < \cdots < j'_m$ similarly.
Writing $i' := (i'_1,\ldots,i'_m)$ and $j' := (j'_1,\ldots,j'_m)$, we obtain:

$$D(i \mid j) = (-q)^{\ell(i)-\ell(j)}D(i' \mid j'). \tag{5–2}$$

For any $k \in [1,m]$, we write

$$S_m^k := \{\sigma \in S_m \mid \sigma(1) < \cdots < \sigma(k) \quad \text{and} \quad \sigma(k+1) < \cdots < \sigma(m)\}.$$

Then the following Laplace expansion formula can be shown (for all $1 \le k \le m$):

$$D(i \mid j) = \sum_{\sigma \in S_m^k} (-q)^{\ell(i)-\ell(i\sigma)} D(i_{\sigma(1)},\ldots,i_{\sigma(k)} \mid j_1,\ldots,j_k)$$
$$\cdot D(i_{\sigma(k+1)},\ldots,i_{\sigma(m)} \mid j_{k+1},\ldots,j_m).$$

For $(m+1,m)$ instead of (m,k), we obtain, in particular, for all $i_0,j_0 \in [1,n]$:

$$D(i_0,i \mid i,j_0)$$
$$= (-q)^{\ell(i_0,i)-\ell(i,i_0)} D(i \mid i)a_{i_0,j_0}$$
$$+ \sum_{t=1}^{m}(-q)^{\ell(i_0,i)-\ell(i_0,\ldots,\widehat{i_t},\ldots,i_m,i_t)} D(i_0,\ldots,\widehat{i_t},\ldots,i_m \mid i)\, a_{i_t,j_0}$$
$$= (-q_{i_1,i_0}) \cdot \cdots \cdot (-q_{i_m,i_0})D(i \mid i)\, a_{i_0,j_0}$$
$$+ \sum_{t=1}^{m} D(i_0,\ldots,\widehat{i_t},\ldots,i_m \mid i_1,\ldots,\widehat{i_t},\ldots,i_m,i_t)\, a_{i_t,j_0}. \tag{5–3}$$

(3) For any $i,j \in [1,n]$, we write

$$b_{ij}^m := \sum_{1 \le i_1 < \cdots < i_m \le n} D(i,i_1,\ldots,i_m \mid i_1,\ldots,i_m,j),$$
$$B^m := (b_{ij}^m)_{i,j}.$$

It follows from (5–3) that

$$B^m = (-1)^m\mathrm{Tr}^m A + B^{m-1}A$$

for all $1 \leq m \leq n-1$, where $B^0 = A$. It follows immediately, by induction, that

$$B^m = (-1)^m \mathrm{Tr}^m A + (-1)^{m-1} \mathrm{Tr}^{m-1} A^2 + \ldots + (-1) \mathrm{Tr}^1 A^m + A^{m+1},$$

hence

$$B^{n-1} = (-1)^{n-1} \mathrm{Tr}^{n-1} A + (-1)^{n-2} \mathrm{Tr}^{n-2} A^2 + \ldots + (-1) \mathrm{Tr}^1 A^{n-1} + A^n.$$

It remains to verify that $B^{n-1} = (-1)^{n-1} \mathrm{Tr}^n$.

Because of step (1), we have

$$b_{ij}^{n-1} = \sum_{1 \leq i_1 < \cdots < i_{n-1} \leq n} D(i, i_1 \ldots, i_{n-1} \mid i_1, \ldots, i_{n-1}, j) = 0,$$

unless $i = j$. If $i = j$, however, we get:

$$b_{ii}^{n-1} = D(i, 1, \ldots, \hat{i}, \ldots, n \mid 1, \ldots, \hat{i}, \ldots, n, i)$$
$$\overset{(5-2)}{=} (-q)^{\ell(i,1,\ldots,\hat{i},\ldots,n) - \ell(1,\ldots,\hat{i},\ldots,n,i)} |A|_q$$
$$= (-1)^{n-1} (q_{1i} \cdot \ldots \cdot q_{ni}) |A|_q$$
$$= (-1)^{n-1} tr_i^n,$$

where we have used that $\ell(i, 1, \ldots, \hat{i}, \ldots, n) = i-1$ and $\ell(1, \ldots, \hat{i}, \ldots, n, i) = n-i$.
Hence, $B^{n-1} = (-1)^{n-1} \mathrm{Tr}^n$. $\qquad \square$

6. The R-Matrix R_q and the Homfly Polynomial

The Homfly polynomial can be associated to every oriented link diagram and is invariant under regular isotopy. The name comes from the initials of Hoste, Ocneanu, Millett, Freyd, Lickorish and Yetter.

In this section, we use the following abbreviations for parts of oriented link diagrams:

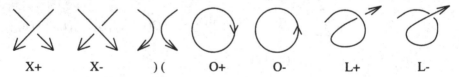

Figure 6. Types of crossings and loops

DEFINITION 6.1 (THE HOMFLY POLYNOMIAL). Let K be an oriented link diagram. The *Homfly polynomial* of K, denoted by $H_K(\alpha, z) \in \mathbb{Z}[\alpha, \alpha^{-1}, z, z^{-1}]$ is characterized by the following properties:

(a) If K is regularly isotopic to K' (denoted as $K \approx K'$), then $H_K = H_{K'}$;
(b) $H_{O+} = 1$;
(c) $H_{X+} - H_{X-} = z H_{)(}$;
(d) $H_{L+} = \alpha H_{\to}, H_{L-} = \alpha^{-1} H_{\to}$.

Using these rules, one can calculate H_K, using so-called *skein trees*. We illustrate this procedure with the trefoil.

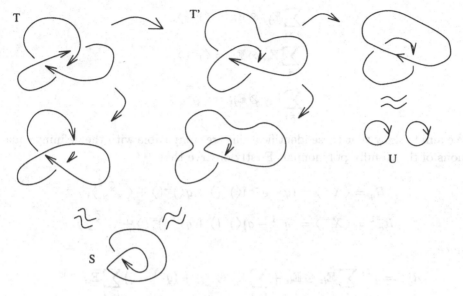

Figure 7. A skein tree for the trefoil

Using (c), (d) of Definition 6.1, it is an easy exercise to show that, for any K:

$$H_{O+K} = \delta H_K,$$

with $\delta := (\alpha - \alpha^{-1})z^{-1}$. Hence, in particular, $H_U = H_{O+O+} = \delta$.

Now we can calculate the Homfly polynomial of the trefoil T as follows:

$$H_T \overset{(c)}{=} H_S + zH_{T'} \overset{(c)}{=} H_S + z(zH_S + H_U)$$
$$= \alpha + z^2\alpha + z\delta = 2\alpha - \alpha^{-1} + z^2\alpha.$$

In general, many skein trees derive from a given K. But the computation always yields the same polynomial H_K, which is, hence, well-defined.

We prove this using the R-matrix R_q introduced in Definition 4.2.

Let q be an indeterminate and R_q have size $(n+1)^2 \times (n+1)^2$. Recall that

$$R_q = q\sum_i \mathbb{E}_{ii} \otimes \mathbb{E}_{ii} + \sum_{i \neq j} \mathbb{E}_{ij} \otimes \mathbb{E}_{ji} + (q - q^{-1})\sum_{i<j} \mathbb{E}_{jj} \otimes \mathbb{E}_{ii}.$$

Here, we let i, j range over the set of indices $\mathcal{J} := \{-n, -n+2, \ldots, n-2, n\}$.

We associate R_q (resp. R_q^{-1}) to crossings X^+ (positive crossing), resp. X^- (negative crossing).

We write

$$\mathbb{E}_{jj} \otimes \mathbb{E}_{ii} =: \langle^j\rangle(^i\rangle,$$
$$\mathbb{E}_{ij} \otimes \mathbb{E}_{ji} =: \langle^i \diagup \diagdown^j \rangle.$$

Thus the summands of R_q have the following form (we omit the summation sign in our symbolic notation)

$$\sum_{j>i} \mathbb{E}_{jj} \otimes \mathbb{E}_{ii} = \langle\rangle^>(),$$

$$\sum_{i} \mathbb{E}_{ii} \otimes \mathbb{E}_{ii} = \langle\rangle^=(),$$

$$\sum_{i\neq j} \mathbb{E}_{ij} \otimes \mathbb{E}_{ji} = \langle \overset{\neq}{\swarrow\hspace{-0.5em}\searrow} \rangle.$$

We aim to show that these identifications are compatible with the defining equations of the Homfly polynomial. Firstly observe that

$$R_q = \langle X^+ \rangle = (q - q^{-1})\langle\rangle^>() + q\langle\rangle^=() + \langle \overset{\neq}{\swarrow\hspace{-0.5em}\searrow} \rangle,$$

$$R_q^{-1} = \langle X^- \rangle = (q^{-1} - q)\langle\rangle^<() + q^{-1}\langle\rangle^=() + \langle \overset{\neq}{\swarrow\hspace{-0.5em}\searrow} \rangle,$$

since

$$R_q^{-1} = q^{-1}\sum_{i} \mathbb{E}_{ii} \otimes \mathbb{E}_{ii} + \sum_{i\neq j} \mathbb{E}_{ij} \otimes \mathbb{E}_{ji} + (q^{-1} - q)\sum_{i<j} \mathbb{E}_{ii} \otimes \mathbb{E}_{jj}.$$

The coefficients $(q - q^{-1})$, q, $(q^{-1} - q)$, q^{-1}, 1 are called *vertex weights*.

REMARK 6.2. For $z := q - q^{-1}$, the equation in Definition 6.1 (c) is satisfied:

$$\langle X^+ \rangle - \langle X^- \rangle = (q - q^{-1})\left(\langle\rangle^>() + \langle\rangle^<() + \langle\rangle^=()\right)$$
$$= (q - q^{-1})\langle\rangle().$$

We apply the following procedures (S1) and (S2) to a given oriented link diagram K.

(S1) Replace each crossing X^+, X^- in K by $\swarrow\hspace{-0.5em}\searrow$ or $)($.

For the trefoil, there are three possible results shown in Figure 8.

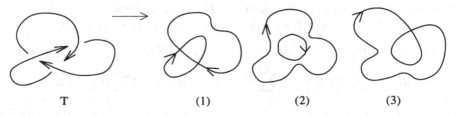

T   (1)   (2)   (3)

Figure 8. Procedure (S1) applied to the trefoil

From this step, we obtain oriented *link shadows*. We only admit link shadows that have no self-intersection. In Figure 8, this still allows (1) and (2), but rules out (3).

Let σ be any admissible link shadow obtained from K by (S1).

(S2) Label an element of $\mathcal{J}$ (called *spin*) to each component of σ, such that
a) Two components of the same spin do not intersect;
b) If $_a)(_b$ in σ comes from X^+ (resp. X^-) in K, then $a \geq b$ (resp. $a \leq b$).
In the example above, we must have $a > b$ for $_a)(_b$ in shadow (1) and $a \geq b$ for $_a)(_b$ in shadow (2).

DEFINITION 6.3 (STATES OF A LINK DIAGRAM). A *state* of a link diagram is an admissible link shadow, with spins labelled to its components according to (S2).

The trefoil has three kinds of states of type (1) and states of type (2).

DEFINITION 6.4. Let σ be a state of K. Let $\langle K \mid \sigma \rangle$ be the product of vertex weights of σ at the crossings of K, i.e. the product of the following values:

$$\langle X^+ \mid \langle\rangle^>(\rangle\rangle = q - q^{-1}, \quad \langle X^- \mid \langle\rangle^<(\rangle\rangle = q^{-1} - q,$$
$$\langle X^+ \mid \langle\rangle^=(\rangle\rangle = q, \quad \langle X^- \mid \langle\rangle^=(\rangle\rangle = q^{-1},$$
$$\langle X^+ \mid \times\!\!\!\!/ \rangle = 1, \quad \langle X^- \mid \times\!\!\!\!/ \rangle = 1$$

(the left hand side symbols the type of the crossing in K, the right hand side indicates the relation between the spins in σ).

Note that conditions (a) and (b) in (S2) imply that always (precisely) one of the cases in Definition 6.4 holds.

In the example of the trefoil, we obtain (cf. Figure 8):

$$\langle T \mid \sigma_{(1)} \rangle = q - q^{-1}, \text{since } a > b,$$
$$\langle T \mid \sigma_{(2)} \rangle = \begin{cases} (q - q^{-1})^3, & \text{if } a > b, \\ q^3, & \text{if } a = b. \end{cases}$$

DEFINITION 6.5. For a state σ of an oriented link diagram K, define

$$\|\sigma\| := \sum(\text{not}(\ell)\text{label}(\ell) \mid \ell \text{ is a component of } \sigma),$$

where $\text{not}(O^+) := 1$, $\text{not}(O^-) := -1$, and $\text{label}(\ell) \in \mathcal{J}$ denotes the spin labelled to ℓ.

DEFINITION 6.6 (THE KAUFFMAN POLYNOMIAL). For an oriented link diagram K, define *Kauffman's polynomial* $\langle K \rangle$ as

$$\langle K \rangle := \sum(\langle K \mid \sigma \rangle q^{\|\sigma\|} \mid \sigma \text{ is a state of } K),$$

which is a polynomial in $\mathbb{Z}[q, q^{-1}]$, depending on n.

REMARK 6.7. The polynomial associated to the circle can be calculated as follows:

$$\langle O^+ \rangle = \sum_{a \in \mathcal{J}} \underbrace{\langle O^+ \mid O^{+a} \rangle}_{=1} q^{\|O^{+a}\|} = \sum_{a \in \mathcal{J}} q^a$$
$$= \frac{q^{n+1} - q^{-n-1}}{q - q^{-1}} = \delta,$$

if one puts $\alpha := q^{n+1}$ (and δ and z as before; cf. the beginning of this section and Remark 6.2).

Let K denote any oriented link diagram containing a part of the form L^+ (cf. Figure 6). By (S1) and (S2), any part of K of the form L^+ is transformed into a state of the form

$$\sigma(a,b) := \overset{\rightarrow a}{\underset{O+b}{}}, \quad \text{where } a \le b.$$

Let $\langle L^+ \rangle$ denote the polynomial associated to K and let $\langle \rightarrow \rangle$ denote the polynomial of the diagram obtained from K by replacing one occurrance of L^+ by "$\rightarrow$". Then we get

$$\langle L^+ \mid \sigma(a,b) \rangle = \langle \rightarrow \mid \rightarrow^a \rangle \cdot \begin{cases} q & \text{if } a = b, \\ q - q^{-1} & \text{if } a < b, \end{cases}$$

$$\|\sigma(a,b)\| = b + \| \rightarrow^a \|,$$

which implies

$$\langle L^+ \rangle = \sum_{a \le b} \langle L^+ \mid \sigma(a,b) \rangle q^{\|\sigma(a,b)\|}$$

$$= \sum_a \underbrace{\left(qq^a + \sum_{b>a}(q-q^{-1})q^b \right)}_{=q^{n+1}} \langle \rightarrow \mid \rightarrow^a \rangle q^{\|\rightarrow^a\|}$$

$$= q^{n+1} \sum_a \langle \rightarrow \mid \rightarrow^a \rangle q^{\|\rightarrow^a\|} = q^{n+1} \langle \rightarrow \rangle.$$

REMARK 6.8. This coincides with Definition 6.1 (d), for $\alpha = q^{n+1}$ (as above). Similarly, one sees the analogue relation for $\langle L^- \rangle$.

THEOREM 6.9 (Kauffman). *The polynomial $\langle K \rangle$ is invariant under regular isotopy and satisfies*

$$\langle O^+ \rangle = \delta = \frac{q^{n+1} - q^{-n-1}}{q - q^{-1}}, \tag{6-1}$$

$$\langle X^+ \rangle - \langle X^- \rangle = (q - q^{-1})\langle\rangle\langle\rangle, \tag{6-2}$$

$$\langle L^+ \rangle = q^{n+1} \langle \rightarrow \rangle, \tag{6-3}$$

$$\langle L^- \rangle = q^{-n-1} \langle \rightarrow \rangle. \tag{6-4}$$

PROOF. We have shown Equations (6–1)–(6–1) in Remarks 6.7, 6.2 and 6.8.

We leave it to the reader to verify that $\langle K \rangle$ is invariant under Reidemeister moves (II) and (III) for oriented diagrams (Figure 2). This gives invariance under regular isotopy. □

COROLLARY 6.10. *Kauffman's polynomial and the Homfly polynomial satisfy*

$$\langle K \rangle = \langle O^+ \rangle H_K(q^{n+1}, q-q^{-1}).$$

Since the map $\mathbb{Z}[\alpha, \alpha^{-1}, z, z^{-1}] \to \mathbb{Q}(q)$ defined by $\alpha \mapsto q^{n+1}$, $z \mapsto q - q^{-1}$ is injective, the Homfly polynomial is well-defined.

It is an easy exercise to verify that Kauffman's polynomial of the trefoil T is indeed

$$\langle T \rangle = ((q^2 + q^{-2})q^{n+1} - q^{-n-1})\delta = H_T(q^{n+1}, q - q^{-1})\delta.$$

7. Duality with the Quantized Enveloping Algebra $U_q(\mathbf{sl_n})$

In this section, we examine the duality of the Hopf algebra $\mathcal{O}_q(\mathrm{SL}(n))$ and the quantized enveloping algebra of the Lie algebra $\mathbf{sl_n}$.

DEFINITION 7.1 (THE QUANTIZED ENVELOPING ALGEBRA OF $\mathbf{sl_2}$). For $q^2 \neq 1$, the algebra $U_q(\mathbf{sl_2})$ is generated by elements K, K^{-1}, E, F, subject to the following relations:

$$KK^{-1} = K^{-1}K = 1,$$

$$KEK^{-1} = q^2 E,$$

$$KFK^{-1} = q^{-2}F,$$

$$EF - FE = \frac{K - K^{-1}}{q - q^{-1}}.$$

PROPOSITION 7.2. *The algebra* $U_q(\mathbf{sl_2})$ *has a basis* $\{F^i K^s E^j \mid i, j \geq 0, s \in \mathbb{Z}\}$. *It carries a Hopf algebra structure as follows:*

$$\Delta(K) = K \otimes K, \qquad\qquad \varepsilon(K) = 1, \quad S(K) = K^{-1},$$

$$\Delta(E) = 1 \otimes E + E \otimes K, \qquad \varepsilon(E) = 0, \quad S(E) = -EK^{-1},$$

$$\Delta(F) = K^{-1} \otimes F + F \otimes 1, \quad \varepsilon(F) = 0, \quad S(F) = -KF.$$

The proof is not easy and involves tedious computations. We omit it here (see Section 15).

We examine the relationship of the dual Hopf algebra $U_q(\mathbf{sl_2})^\circ$ with $\mathcal{O}_q(\mathrm{GL}(n))$. Recall the definition of the dual of a Hopf algebra:

DEFINITION 7.3. Let H be a Hopf algebra. Let $\pi : H \to M_n(k)$ be a finite dimensional representation of H and $\pi^* : (M_n(k))^* \to H^*$ be the dual map. The *dual* of H is defined as

$$H^\circ := \sum \left(\mathrm{Im}(\pi^*) \mid \pi : H \to M_n(k), \text{ as above}, n \geq 1\right).$$

DEFINITION 7.4 (TENSOR PRODUCTS AND TRANSPOSES OF REPRESENTATIONS). Let $\pi : H \to M_n(k)$ and $\pi' : H \to M_m(k)$ be representations of a Hopf algebra H.

(a) The tensor product $\pi \otimes \pi'$ is defined as the representation

$$H \overset{\Delta}{\to} H \otimes H \overset{\pi \otimes \pi'}{\longrightarrow} M_n(k) \otimes M_m(k) \overset{\cong}{\to} M_{nm}(k).$$

(b) The transpose π^t is defined as the representation

$$H \xrightarrow{S} H \xrightarrow{\pi} M_n(k) \xrightarrow{\text{transpose}} M_n(k).$$

This definition implies that

$$\operatorname{Im}((\pi \underline{\otimes} \pi')^*) = \operatorname{Im}(\pi^*)\operatorname{Im}(\pi'^*), \quad \operatorname{Im}((\pi^t)^*) = S^*(\operatorname{Im}(\pi^t)).$$

Using the tensor product and the transpose of representations, one can show, successively:

PROPOSITION 7.5. (a) *There is precisely one coalgebra structure on H°, such that π^* is a coalgebra map, for all finite dimensional representations π of H.*
(b) H° *is a subalgebra of H^* (with the convolution product).*
(c) H° *is S^* invariant.*
(d) H° *is a Hopf algebra with antipode S^*.*
(e) *Assume all finite dimensional representations of H are completely reducible. Let $\{\pi_\lambda \mid \lambda \in \Lambda\}$ be a complete set of (pairwise non-isomorphic) finite dimensional representations of H. Then*

$$H^\circ = \bigoplus_{\lambda \in \Lambda} \operatorname{Im}(\pi_\lambda^*)$$

(cf. the Peter–Weyl theorem in the theory of Lie groups).

We assume q is not a root of unity and $\operatorname{char}(k) \neq 2$ up to 7.11.

THEOREM 7.6 (REPRESENTATIONS OF $U_q(\mathbf{sl_2})$) (ROSSO). (a) *All finite dimensional representations of $U_q(\mathbf{sl_2})$ are completely reducible.*
(b) *all finite dimensional irreducible representations of $U_q(\mathbf{sl_2})$ are exhausted by*

$$\pi_n : U_q(\mathbf{sl_2}) \to M_{n+1}(k), \pi_n' : U_q(\mathbf{sl_2}) \to M_{n+1}(k),$$

which are defined as follows:

$$\pi_n(K) := \begin{pmatrix} q^n & & & 0 \\ & q^{n-2} & & \\ & & \ddots & \\ 0 & & & q^{-n} \end{pmatrix},$$

$$[6pt]\pi_n(E) := \begin{pmatrix} 0 & [n] & & 0 \\ & \ddots & \ddots & \\ & & \ddots & [1] \\ 0 & & & 0 \end{pmatrix}, \pi_n(F) := \begin{pmatrix} 0 & & & 0 \\ [1] & \ddots & & \\ & \ddots & \ddots & \\ 0 & & [n] & 0 \end{pmatrix},$$

$$[6pt]\pi_n'(K) := -\pi_n(K), \pi_n'(E) := \pi_n(E), \pi_n'(F) := -\pi_n(F),$$

where $[i] := \dfrac{q^i - q^{-i}}{q - q^{-1}}$.

COROLLARY 7.7 (THE DUAL OF $U_q(\mathbf{sl_2})$). *The dual Hopf algebra of $U_q(\mathbf{sl_2})$ is*

$$U_q(\mathbf{sl_2})^\circ = \bigoplus_{n=0}^{\infty} (\operatorname{Im}(\pi_n^*) \oplus \operatorname{Im}((\pi_n')^*)).$$

PROPOSITION 7.8. *The following equivalences hold for representations of $U_q(\mathbf{sl_2})$:*

(a) $\pi'_n \cong \pi_n \otimes \pi'_0 \cong \pi'_0 \otimes \pi_n$;

(b) $\pi'_n \cong \pi_n$;

(c) (CLEBSCH–GORDAN RULE) $\pi_m \otimes \pi_n \cong \pi_{m+n} \oplus \pi_{m+n-2} \oplus \cdots \oplus \pi_{|m-n|}$.

In particular, $\pi'_0 : U_q(\mathbf{sl_2}) \to k$, $K \mapsto -1$, $E, F \mapsto 0$ defines an algebra map and, hence, corresponds to a group-like element $\gamma \in U_q(\mathbf{sl_2})^\circ$. This element γ has order 2.

It follows from Proposition 7.8 (a) that

$$\mathrm{Im}(\pi'^*_n) = \gamma \mathrm{Im}(\pi^*_n) = \mathrm{Im}(\pi^*_n)\gamma.$$

Hence, the following sub-coalgebra is normalized by γ:

$$A := \bigoplus_{n=0}^{\infty} \mathrm{Im}(\pi^*_n).$$

Because of (b), A is S^*-invariant. For the case $n = 1$, we have

$$\pi_1^t = \mathrm{inn}\left(\begin{smallmatrix} & -1 \\ q & \end{smallmatrix}\right) \circ \pi_1, \tag{7--1}$$

which is easily checked on the generators of $U_q(\mathbf{sl_2})$. (Here, $\mathrm{inn}(P)$ denotes the inner automorphism induced by an invertible matrix P, that is, $\mathrm{inn}(P) : M_n(k) \to M_n(k)$, $Y \mapsto PYP^{-1}$.)

The Clebsch–Gordan rule implies that A is a subalgebra (hence a Hopf subalgebra), which is generated, as algebra, by $\mathrm{Im}(\pi^*_1)$.

THEOREM 7.9 ($\mathcal{O}_q(\mathrm{SL}(2))$ AS A HOPF SUBALGEBRA OF $U_q(\mathbf{sl_2})^\circ$). *Let* $\pi_1 : U_q(\mathbf{sl_2}) \to M_2(k)$ *and* $A \subset U_q(\mathbf{sl_2})$ *be as above and define* $a, b, c, d \in U_q(\mathbf{sl_2})^\circ$ *by*

$$\begin{pmatrix} a(x) & b(x) \\ c(x) & d(x) \end{pmatrix} := \pi_1(x) \quad \text{for all } x \in U_q(\mathbf{sl_2}).$$

Then there is a unique isomorphism of Hopf algebras, given by

$$\varphi : \mathcal{O}_q(\mathrm{SL}(2)) \to A, \quad X \mapsto \left(\begin{smallmatrix} a & b \\ c & d \end{smallmatrix}\right).$$

PROOF. We only prove that φ is a well-defined, surjective Hopf algebra map (injectivity can be shown by tedious computational arguments, for which we refer to the references).

By definition, (a, b, c, d) is a basis of $\mathrm{Im}(\pi^*_1)$. Since π^*_1 is a coalgebra map, it follows that

$$\begin{pmatrix} \Delta(a) & \Delta(b) \\ \Delta(c) & \Delta(d) \end{pmatrix} = \begin{pmatrix} a \otimes 1 & b \otimes 1 \\ c \otimes 1 & d \otimes 1 \end{pmatrix} \begin{pmatrix} 1 \otimes a & 1 \otimes b \\ 1 \otimes c & 1 \otimes d \end{pmatrix}. \tag{7--2}$$

From the definition of transposed representations, we get for all $x \in U_q(\mathbf{sl_2})$:

$$\pi_1^t(x) = \begin{pmatrix} a(S(x)) & b(S(x)) \\ c(S(x)) & d(S(x)) \end{pmatrix}^t = \begin{pmatrix} S(a) & S(c) \\ S(b) & S(d) \end{pmatrix}(x).$$

From (7--1), it follows that

$$\begin{pmatrix} S(a) & S(c) \\ S(b) & S(d) \end{pmatrix} = \begin{pmatrix} & -1 \\ q & \end{pmatrix} \begin{pmatrix} a & b \\ c & d \end{pmatrix} \begin{pmatrix} & q^{-1} \\ -1 & \end{pmatrix}.$$

Now we obtain, using the transpose of this equation at "(!)", that

$$\begin{pmatrix} a & b \\ c & d \end{pmatrix}^{-1} = \begin{pmatrix} S(a) & S(b) \\ S(c) & S(d) \end{pmatrix} \overset{(!)}{=} \begin{pmatrix} & -1 \\ q^{-1} & \end{pmatrix} \begin{pmatrix} a & c \\ b & d \end{pmatrix} \begin{pmatrix} & q \\ -1 & \end{pmatrix}$$

$$= \begin{pmatrix} & q \\ -1 & \end{pmatrix} \begin{pmatrix} a & c \\ b & d \end{pmatrix} \begin{pmatrix} & -1 \\ q^{-1} & \end{pmatrix} = \widetilde{\begin{pmatrix} a & b \\ c & d \end{pmatrix}}$$

(cf. Definition 1.2). It follows from Proposition 1.3 that $\begin{pmatrix} a & b \\ c & d \end{pmatrix}$ is a q-matrix with q-determinant 1. Hence, φ is a (well-defined) algebra map.

It follows from (7–2) that φ respects the comultiplication as well and is, hence, a Hopf algebra map (any bialgebra map between Hopf algebras automatically respects the antipode).

As shown before, (a, b, c, d) is a basis of $\mathrm{Im}(\pi_1^*)$, which generates A as an algebra. Therefore, φ is surjective. $\square$

In the following, we identify $\mathcal{O}_q(\mathrm{SL}(2))$ with the Hopf subalgebra A of $U_q(\mathrm{sl}_2)^\circ$.

The conjugation with γ on A can be computed as

$$\gamma \begin{pmatrix} a & b \\ c & d \end{pmatrix} \gamma^{-1} = \begin{pmatrix} a & -b \\ -c & d \end{pmatrix}. \tag{7–3}$$

COROLLARY 7.10 (THE HOPF ALGEBRA STRUCTURE OF $U_q(\mathrm{sl}_2)^\circ$). *If q is not a root of unity and $\mathrm{char}(k) \neq 2$, the dual of $U_q(\mathrm{sl}_2)$ can be considered as semi-direct product*

$$U_q(\mathrm{sl}_2)^\circ = \mathcal{O}_q(\mathrm{SL}(2)) \rtimes \mathbb{Z}_2$$

with respect to the action of $\gamma \in \mathbb{Z}_2$ on $\mathcal{O}_q(\mathrm{SL}(2))$ given in (7–3).

These results may be generalized to $\mathcal{O}_q(\mathrm{SL}(n))$.

DEFINITION 7.11 (THE ALGEBRA $U_q(\mathrm{sl}_n)$). For $q^2 \neq 1$, the algebra $U_q(\mathrm{sl}_n)$ is generated by elements K_i, K_i^{-1}, E_i, F_i, ($i \in \{1, \ldots, n-1\}$), subject to the following relations:

$$K_i K_i^{-1} = K_i^{-1} K_i = 1,$$
$$K_i K_j = K_j K_i,$$
$$K_i E_j K_i^{-1} = q^{\alpha_{ij}} E_j,$$
$$K_i F_j K_i^{-1} = q^{-\alpha_{ij}} F_j,$$
$$E_i F_j - F_j E_i = \delta_{ij} \frac{K_i - K_i^{-1}}{q - q^{-1}},$$
$$E_i E_j = E_j E_i, \quad \text{if } |i-j| \geq 2,$$
$$F_i F_j = F_j F_i, \quad \text{if } |i-j| \geq 2,$$
$$E_i^2 E_j - (q + q^{-1}) E_i E_j E_i + E_j E_i^2 = 0, \quad \text{if } |i-j| = 1,$$
$$F_i^2 F_j - (q + q^{-1}) F_i F_j F_i + F_j F_i^2 = 0, \quad \text{if } |i-j| = 1.$$

Here, $\alpha = (\alpha_{ij})_{i,j}$ denotes the Cartan matrix of the Lie algebra sl_n:

$$\alpha_{ij} = 2, -1, 0 \quad \text{if } |i-j| = 0, 1, \geq 2 \text{ respectively.}$$

PROPOSITION 7.12. *The algebra* $U_q(\mathbf{sl_n})$ *carries a Hopf algebra structure, such that the subalgebras* $k\langle K_i^{\pm 1}, E_i, F_i \rangle$ *are isomorphic to* $U_q(\mathbf{sl_2})$.

The proof requires several computations, which are omitted here.

THEOREM 7.13 (REPRESENTATIONS OF $U_q(\mathbf{sl_n})$) (Rosso). *If q is not a root of unity and* char$(k) \neq 2$ *then all finite dimensional representations of* $U_q(\mathbf{sl_n})$ *are completely reducible.*

The irreducible representations can be given explicitly, but this is rather complicated and we omit it here.

DEFINITION 7.14. The *basic representation* $\pi : U_q(\mathbf{sl_n}) \to M_n(k)$ of $U_q(\mathbf{sl_n})$ is defined as the algebra map given by

$$K_i \mapsto I + (q-1)\mathbb{E}_{ii} + (q^{-1}-1)\mathbb{E}_{i+1,i+1}, E_i \mapsto \mathbb{E}_{i,i+1}, F_i \mapsto \mathbb{E}_{i+1,i}.$$

It is easy, but tedious, to verify that this is indeed a well-defined algebra map. (For $n = 2$, it is the map π_1 defined in Theorem 7.6.)

Theorem 7.9 generalizes to the case $n \geq 2$.

THEOREM 7.15 ($\mathcal{O}_q(\mathrm{SL}(n))$ AS A HOPF SUBALGEBRA OF $U_q(\mathbf{sl_n})^{\circ}$). *Define* $a_{ij} \in U_q(\mathbf{sl_n})^{\circ}$, *for* $i, j \in [1, n]$ *by*

$$(a_{ij}(x))_{i,j} := \pi(x) \quad \text{for all } x \in U_q(\mathbf{sl_n}).$$

Then there is an injective Hopf algebra map given by

$$\varphi : \mathcal{O}_q(\mathrm{SL}(n)) \to U_q(\mathbf{sl_n})^{\circ}, \quad x_{ij} \mapsto a_{ij} \quad \text{for all } i, j \in [1, n].$$

PROOF. As before, we only show that φ is a (well-defined) Hopf algebra map (injectivity is not trivial and requires several computations).

As in the case $n = 2$, the family $(a_{ij})_{i,j}$ is a basis of $\mathrm{Im}(\pi^*)$, such that

$$\Delta(a_{ij}) = \sum_{\ell=1}^{n} a_{i\ell} \otimes a_{\ell j}. \tag{7-4}$$

It remains to show that $(a_{ij})_{i,j}$ is an $(n \times n)$ q-matrix of q-determinant 1. Its entries are in $U_q(\mathbf{sl_n})^{\circ}$.

Let V_n be an n-dimensional k-vector space, with basis $\{e_1, \ldots e_n\}$. In a natural way, V_n is a left $M_n(k)$-module; and it becomes a left $U_q(\mathbf{sl_n})$-module via π. Hence, $V_n \otimes V_n$ is a left $U_q(\mathbf{sl_n})$-module as well, via the comultiplication Δ.

It is easy (but tedious) to verify that

$$R_q : V_n \otimes V_n \to V_n \otimes V_n$$

is $U_q(\mathbf{sl_n})$-linear (it suffices to do the calculations for generators of $U_q(\mathbf{sl_n})$ and basis elements of $V_n \otimes V_n$).

Hence, the q-exterior algebra $\bigwedge_q(V_n) = T(V_n)/\mathrm{Ker}(R_q - qI)$ and the q-symmetric algebra $S_q(V_n)$ are $U_q(\mathbf{sl_n})$-module algebras (cf. Definition 4.9).

Now, V_n is a right $U_q(\mathbf{sl_n})^\circ$-comodule via

$$e_j \mapsto \sum_{i=1}^n e_i \otimes a_{ij},$$

and the above implies that R_q is a $U_q(\mathbf{sl_n})^\circ$-comodule map. If we put $A = (a_{ij})$, this means $A^{(2)}$ commutes with R_q (cf. Lemma 4.8). Hence, A is a q-matrix by Proposition 1.6.

The subspace $k \cdot e_1 \cdot \ldots \cdot e_n \subset \bigwedge_q(V_n)$ is a one-dimensional submodule (cf. Section 4), and it is not difficult to check that $U_q(\mathbf{sl_n})$ acts trivially on it (i.e. $u \cdot e_1 \cdot \ldots \cdot e_n = \varepsilon(u)e_1 \cdot \ldots \cdot e_n$), it suffices to verify this for generators. In other words, it is a trivial $U_q(\mathbf{sl_n})^\circ$-comodule. This implies that (a_{ij}) has q-determinant 1.

Therefore, φ defines an algebra map. By (7–4), the map φ respects the comultiplication as well and is, hence, a Hopf algebra map.   □

Theorem 7.15 allows us to identify $\mathcal{O}_q(\mathrm{SL}(n))$ with a Hopf subalgebra of $U_q(\mathbf{sl_n})^\circ$.

Also Corollary 7.10 generalizes to the case $n \geq 2$. Firstly, note that algebra maps γ_i are defined (for $i \in [1, n-1]$) by

$$\gamma_i : U_q(\mathbf{sl_n}) \to k, \quad E_j, F_j \mapsto 0, \quad K_j \mapsto \begin{cases} 1, & \text{if } i \neq j, \\ -1, & \text{if } i = j. \end{cases}$$

This means that all γ_i are group-like elements in $U_q(\mathbf{sl_n})^\circ$. It is easy to see that they generate a group isomorphic to $\mathbb{Z}_2^{n-1}$.

Conjugation with γ_i is given as follows:

$$\gamma_i a_{st} \gamma_i^{-1} = \begin{cases} a_{st}, & \text{if } i < s, t \text{ or } s, t \leq i, \\ -a_{st}, & \text{if } s \leq i < t \text{ or } t \leq i < s. \end{cases} \tag{7–5}$$

In particular, the γ_i normalize the Hopf subalgebra $\mathcal{O}_q(\mathrm{SL}(n))$ of $U_q(\mathbf{sl_n})^\circ$.

Using the explicit description of all irreducible representations of $U_q(\mathbf{sl_n})$, one can prove the following result:

THEOREM 7.16 (THE HOPF ALGEBRA STRUCTURE OF $U_q(\mathbf{sl_n})^\circ$). *If q is not a root of unity and* $\mathrm{char}(k) \neq 2$, *then*

$$U_q(\mathbf{sl_n})^\circ = \mathcal{O}_q(\mathrm{SL}(n)) \rtimes \mathbb{Z}_2^{n-1},$$

with respect to the action of $\mathbb{Z}_2^{n-1} = \langle \gamma_1, \ldots, \gamma_{n-1} \rangle$ *on* $\mathcal{O}_q(\mathrm{SL}(n))$ *given in* (7–5).

8. Skew Primitive Elements

In this section, we determine all group-like and skew primitive elements of $\mathcal{O}_q(\mathrm{GL}(n))^\circ$. We assume throughout that q is not a root of unity.

Recall from the last section, that we have an injective Hopf algebra map $\varphi : \mathcal{O}_q(\mathrm{SL}(n)) \to U_q(\mathbf{sl_n})^\circ$.

One can show that the associated pairing $\mathcal{O}_q(\mathrm{SL}(n)) \otimes U_q(\mathbf{sl_n}) \to k$ is non-degenerate, i.e., we obtain injective maps

$$U_q(\mathbf{sl_n}) \hookrightarrow \mathcal{O}_q(\mathrm{SL}(n))^\circ \overset{\subseteq}{\hookrightarrow} \mathcal{O}_q(\mathrm{GL}(n))^\circ.$$

In the sequel, we consider $U_q(\mathbf{sl_n})$ as a sub Hopf algebra of $\mathcal{O}_q(\mathrm{GL}(n))^\circ$.

The basic representation $\pi : U_q(\mathbf{sl_n}) \to M_n(k)$ extends to an algebra map

$$\tilde{\pi} : \mathcal{O}_q(\mathrm{GL}(n))^\circ \to M_n(k), \quad f \mapsto (f(x_{ij}))_{i,j} \qquad (8\text{--}1)$$

(where (x_{ij}) denotes the canonical set of generators of $\mathcal{O}_q(\mathrm{GL}(n))$).

Recall that the set of group-like elements of a Hopf algebra H is denoted by $G(H)$. Given $g, h \in G(H)$, the set of (g,h)-*skew primitive* elements is

$$P_{g,h}(H) := \{u \in H \mid \Delta(u) = g \otimes u + u \otimes h\}.$$

In particular, $P(H) := P_{1,1}(H)$ is called the set of *primitive* elements of H. It is always a Lie algebra ($x, y \in P(H)$ implies $xy - yx \in P(H)$).

Note that, in any case, $g - h \in P_{g,h}(H)$. We call $P_{g,h}(H)$ *trivial*, if it is spanned by this element.

To determine all skew primitive elements of H, it suffices to determine all $(1,g)$-skew primitive elements for all $g \in G(H)$, since for any $\gamma \in G(H)$, we have

$$\gamma P_{g,h}(H) = P_{\gamma g, \gamma h}(H), \quad P_{g,h}(H)\gamma = P_{g\gamma, h\gamma}(H).$$

In the following, we put $H := \mathcal{O}_q(\mathrm{GL}(n))^\circ$. We start with some examples of group-like and skew primitive elements of H.

REMARK 8.1. The elements $K_1, \dots K_{n-1}$ are group-like in H and $1 - K_i$, E_i, $K_i F_i$ are linearly independent $(1, K_i)$-skew primitive elements of H.

Note that

$$\mathcal{O}(T) := \mathcal{O}_q(\mathrm{GL}(n))/(x_{ij} \mid i \neq j)$$

is a factor Hopf algebra of $\mathcal{O}_q(\mathrm{GL}(n))$, which is naturally isomorphic to the group Hopf algebra $k\langle \mathbb{Z}^n \rangle$.

The corresponding quantum subgroup $T \subset \mathrm{GL}_q(n)$ is called *canonical maximal torus* (cf. Definition 4.16).

The dual Hopf algebra $\mathcal{O}(T)^\circ$ is isomorphic to the function algebra $(k\langle \mathbb{Z}^n \rangle)^\circ$ on the group $\mathbb{Z}^n$, hence has an n-dimensional space of primitive elements.

Since $\mathcal{O}(T)^\circ \subset \mathcal{O}_q(\mathrm{GL}(n))^\circ = H$, the space $P(H)$ is at least n-dimensional. We will show that it has exactly this dimension.

The first aim is to determine all group-like elements of H.

LEMMA 8.2 (q-MATRICES WITH SCALAR ENTRIES). *An invertible $(n \times n)$-matrix $c = (c_{ij})$ with entries in k is a q-matrix if and only if it is diagonal.*

PROOF. Assume that $c_{11} = 0$. Since c is invertible, there are $i, j > 1$, such that $c_{1j}, c_{i1} \neq 0$. By the q-relations for c, we get $(q - q^{-1})c_{1j}c_{i1} = c_{ij}0 - 0c_{ij} = 0$, hence $q^2 = 1$, contradicting our asumption on q.

Therefore, $c_{11} \neq 0$. Since $q \neq 1$, the relation $c_{1j}c_{11} = qc_{11}c_{1j}$ implies that $c_{1j} = 0$, for all $j > 1$. Similarly, we obtain $c_{i1} = 0$, for all $i > 1$.

Since $(c_{ij})_{i,j \geq 2}$ is a q-matrix as well, it follows by induction that c is diagonal. Conversely, a diagonal matrix over k is obviously a q-matrix. $\square$

PROPOSITION 8.3 (THE GROUP-LIKE ELEMENTS OF H). *The algebra map* $\tilde{\pi}$: $H \to M_n(k)$ *defined in* (8–1) *induces an isomorphism of groups* $\pi : G(H) \to T(k)$, *where* $T(k)$ *denotes the group of invertible diagonal matrices in* $M_n(k)$.

PROOF. The group-like elements in $H = \mathcal{O}_q(\mathrm{GL}(n))^\circ$ are exactly the algebra maps

$$g : \mathcal{O}_q(\mathrm{GL}(n)) \to k,$$

and these are in one-to-one correspondence with the invertible $(n \times n)$ q-matrices with entries in k. According to Lemma 8.2, these are exactly the invertible diagonal matrices, which proves the claim. $\square$

Now we determine all skew primitive elements of H.

Let C_3 denote the 3-dimensional coalgebra, with basis (β, γ, δ), such that γ, δ are group-like and β is (γ, δ)-skew primitive.

Fix two group-like elements $g, h \in G(H)$ and let

$$\mathrm{diag}(g_1, \ldots, g_n) := \pi(g), \quad \mathrm{diag}(h_1, \ldots, h_n) := \pi(h) \qquad (8\text{–}2)$$

denote the corresponding matrices in $T(k)$.

The elements $u \in P_{g,h}(H)$ are in one-to-one correspondence with the coalgebra maps $\varphi : C_3 \to H$, such that $(\beta, \gamma, \delta) \mapsto (u, g, h)$.

Note that the dual algebra C_3^* is isomorphic to the algebra $M_2^+(k)$ of upper triangular matrices in $M_2(k)$, via

$$C_3^* \to M_2^+(k), \quad f \mapsto \begin{pmatrix} f(\gamma) & f(\beta) \\ & f(\delta) \end{pmatrix}.$$

Since $H = \mathcal{O}_q(\mathrm{GL}(n))^\circ$, the coalgebra maps φ considered before give rise to algebra maps of the following form

$$\psi : \mathcal{O}_q(\mathrm{GL}(n)) \to C_3^* \cong M_2^+(k), \quad x_{ij} \mapsto \begin{pmatrix} g(x_{ij}) & u(x_{ij}) \\ & h(x_{ij}) \end{pmatrix}. \qquad (8\text{–}3)$$

An algebra map is given by (8–3) if and only if $(\psi(x_{ij}))_{i,j}$ is an invertible q-matrix (with entries in $M_2^+(k)$). Note that, by definition, we have $g(x_{ij}) = \delta_{ij}g_i$ and $h(x_{ij}) = \delta_{ij}h_i$; compare (8–2) and (8–1). The correspondence $\varphi \leftrightarrow \psi$ is one-to-one.

Take any $u \in H$, write $c_{ij} := u(x_{ij})$ and

$$\tilde{C} := (\tilde{c}_{ij}), \quad \tilde{c}_{ij} := \begin{pmatrix} \delta_{ij}g_i & c_{ij} \\ & \delta_{ij}h_i \end{pmatrix}.$$

By the observations before, $u \in P_{g,h}(H)$ implies that $\tilde{C}$ is an invertible q-matrix.

LEMMA 8.4. *Assume $\tilde{C}$ is a q-matrix.*

(a) *For all $i < j$, we have $(h_i - qg_i)c_{ij} = (h_i - qg_i)c_{ji} = (g_j - qh_j)c_{ij} = (g_j - qh_j)c_{ji} = 0$.*
(b) *For all $i < j$, we have $(g_j - h_j)c_{ii} = (g_i - h_i)c_{jj}$.*
(c) *If $|i - j| \geq 2$ then $c_{ij} = 0$.*
(d) *If $c_{i,i+1} \neq 0$ or $c_{i+1,i} \neq 0$ then $g^{-1}h = K_i$.*

PROOF. Suppose $i < j$. By the q-relations for $\tilde{C}$,

$$\begin{pmatrix} 0 & c_{ij} \\ 0 & 0 \end{pmatrix} \begin{pmatrix} g_i & c_{ii} \\ 0 & h_i \end{pmatrix} = \tilde{c}_{ij}\tilde{c}_{ii} = q\tilde{c}_{ii}\tilde{c}_{ij} = q \begin{pmatrix} g_i & c_{ii} \\ 0 & h_i \end{pmatrix} \begin{pmatrix} 0 & c_{ij} \\ 0 & 0 \end{pmatrix}.$$

Comparing the matrix entries yields $c_{ij}h_i = qg_ic_{ij}$, hence $(h_i - qg_i)c_{ij} = 0$. The other relations in (a) are shown similarly.

Moreover, the q-relations for $\tilde{C}$ give

$$\tilde{c}_{jj}\tilde{c}_{ii} - \tilde{c}_{ii}\tilde{c}_{jj} = (q - q^{-1})\tilde{c}_{ij}\tilde{c}_{ji} = \begin{pmatrix} 0 & * \\ 0 & 0 \end{pmatrix} \begin{pmatrix} 0 & * \\ 0 & 0 \end{pmatrix} = 0,$$

hence, $\tilde{c}_{ii}$ and $\tilde{c}_{jj}$ commute. Since

$$\tilde{c}_{jj}\tilde{c}_{ii} = \begin{pmatrix} g_j & c_{jj} \\ 0 & h_j \end{pmatrix} \begin{pmatrix} g_i & c_{ii} \\ 0 & h_i \end{pmatrix} = \begin{pmatrix} g_ig_j & g_jc_{ii}+h_ic_{jj} \\ 0 & h_ih_j \end{pmatrix},$$

it follows that $g_jc_{ii} + c_{jj}h_i = g_ic_{jj} + c_{ii}h_j$, which implies (b).

To show (c), suppose $i < j < k$. We claim $c_{ik} = 0$. Using the q-relations for $\tilde{C}$, we get

$$(q - q^{-1})\tilde{c}_{jj}\tilde{c}_{ik} = [\tilde{c}_{jk}, \tilde{c}_{ij}] = [\begin{pmatrix} 0 & * \\ 0 & 0 \end{pmatrix}, \begin{pmatrix} 0 & * \\ 0 & 0 \end{pmatrix}] = 0.$$

Since $q^2 \neq 1$ and $\tilde{c}_{jj}$ is (by definition) invertible, it follows that $\tilde{c}_{ik} = 0$, hence $c_{ik} = 0$. It is checked similarly that $c_{ki} = 0$, so (c) is proved.

Now assume $c_{i,i+1} \neq 0$. The relations in (a) imply $h_i = qg_i$ and $g_{i+1} = qh_{i+1}$. It remains to show $g_j = h_j$ for all $j \notin \{i, i+1\}$ (then $\pi(g^{-1}h) = \pi(K_i)$ and the claim follows from Proposition 8.3).

By (c), we have $\tilde{c}_{j,i+1} = 0$ if $j < i$, and $\tilde{c}_{ij} = 0$ if $i+1 < j$. In both cases, the q-relations for $\tilde{C}$ imply that $\tilde{c}_{jj}$ and $\tilde{c}_{i,i+1}$ commute. One calculates

$$\tilde{c}_{jj}\tilde{c}_{i,i+1} = \begin{pmatrix} 0 & g_jc_{i,i+1} \\ 0 & 0 \end{pmatrix}, \quad \tilde{c}_{i,i+1}\tilde{c}_{jj} = \begin{pmatrix} 0 & c_{i,i+1}h_j \\ 0 & 0 \end{pmatrix}.$$

Hence, $g_jc_{i,i+1} = c_{i,i+1}h_j$, which implies $g_j = h_j$ (since $c_{i,i+1} \neq 0$).
The case $c_{i+1,i} \neq 0$ is treated similarly. $\square$

THEOREM 8.5 (THE SKEW PRIMITIVE ELEMENTS OF $\mathcal{O}_q(\mathrm{GL}(n))^\circ$). *Let $H := \mathcal{O}_q(\mathrm{GL}(n))^\circ$ and $g, h \in G(H)$.*

(a) *$P_{g,h}(H)$ is trivial if and only if $g^{-1}h \notin \{1, K_1, \ldots K_{n-1}\}$.*
(b) *If $g^{-1}h = K_i$, then $P_{g,h}(H)$ is 3-dimensional and spanned by $g - h, gE_i, hF_i$.*
(c) *$P(H)$ is n-dimensional and equal to $P(\mathcal{O}(T)^\circ)$, considering $\mathcal{O}(T)^\circ \subset H$.*

PROOF. The restriction of $\tilde{\pi} : H \to M_n(k)$, $f \mapsto (f(x_{ij}))$ to the set $P_{g,h}(H)$ is injective (since the algebra map ψ (8–3) is determined by its values on algebra generators — but note that $\tilde{\pi}$ itself need not be injective).

Let $u \in P_{g,h}(H)$ and $c_{ij} := u(x_{ij})$, which means $c := (c_{ij}) = \tilde{\pi}(u)$.

(1) Suppose $g^{-1}h \notin \{1, K_1, \ldots K_{n-1}\}$. Then, by (c) and (d) of Lemma 8.4, the matrix c is diagonal. Moreover, (b) implies

$$(c_{11}, \ldots, c_{nn}) = \lambda(g_1 - h_1, \ldots, g_n - h_n),$$

for some scalar λ, which means $c = \tilde{\pi}(\lambda(g - h))$. Since $\tilde{\pi}$ is injective on skew primitive elements, $u = \lambda(g - h)$, so $P_{g,h}(H)$ is trivial.

(2) Suppose $g^{-1}h = K_i$, for some $1 \le i < n$. By (c) and (d) of Lemma 8.4, the non-zero off-diagonal entries of c are at most $c_{i,i+1}$ and $c_{i+1,i}$. As before, (b) implies that

$$(c_{11}, \ldots, c_{nn}) = \lambda(g_1 - h_1, \ldots, g_n - h_n),$$

for some scalar λ.

It is easily checked that $g - h, gE_i, hF_i$ are (g, h)-skew primitive. One calculates

$$\tilde{\pi}(gE_i) = g_i \mathbb{E}_{i,i+1},$$
$$\tilde{\pi}(hF_i) = h_{i+1} \mathbb{E}_{i+1,i},$$
$$\tilde{\pi}(g - h) = \mathrm{diag}(g_1 - h_1, \ldots g_n - h_n).$$

By injectivity, it follows that $P_{g,h}(H)$ is spanned by $(g - h, gE_i, hF_i)$.

(3) Let $g = h$. Then (c) and (d) of Lemma 8.4 imply that the matrix c is diagonal. Hence, $\tilde{\pi}$ maps $P_{g,g}(H) = gP(H)$ injectively to the set of diagonal matrices in $M_n(k)$. Therefore, $P_{g,g}(H)$ is at most n-dimensional.

The claim now follows from the observations after Remark 8.1.   □

The results of Theorem 8.5 still hold if we only assume $q^2 \ne 1$. However, the considered map $U_q(\mathbf{sl_n}) \to \mathcal{O}_q(GL(n))^\circ$ is injective only if q is not a root of unity.

9. Group Homomorphisms $SL_q(n) \to GL_q(m)$

Recall that the category of quantum groups is the opposite category of the category of Hopf algebras (Definition 4.16).

A morphism $\varrho : SL_q(n) \to GL_q(m)$ of quantum groups is thus a Hopf algebra map $\mathcal{O}(\varrho) : \mathcal{O}_q(GL(m)) \to \mathcal{O}_q(SL(n))$. In this section, we determine all such morphisms.

We assume that q is not a root of unity.

DEFINITION 9.1 (THE DERIVED HOMOMORPHISM). Let $\varrho : SL_q(n) \to GL_q(m)$ be a morphism of quantum groups. The dual map of

$$\mathcal{O}_q(GL(m)) \xrightarrow{\mathcal{O}(\varrho)} \mathcal{O}_q(SL(n)) \subset U_q(\mathbf{sl_n})^\circ$$

is called the derived morphism and denoted as $\partial\varrho : U_q(\mathbf{sl_n}) \to \mathcal{O}_q(GL(m))^\circ$.

REMARK 9.2. *The map $\partial\varrho$ is determined by its values on all K_i, E_i, F_i; more-over, $\partial\varrho_1 = \partial\varrho_2$ implies $\varrho_1 = \varrho_2$.*

PROOF. The first property is obvious. Now suppose $\varrho : \mathrm{SL}_q(n) \to \mathrm{GL}_q(m)$. It is straightforward to check that the following diagram commutes:

$$
\begin{array}{ccc}
\mathcal{O}_q(\mathrm{GL}(m)) & \xrightarrow{\ \ \mathrm{can}\ \ } & \mathcal{O}_q(\mathrm{GL}(m))^{\circ\circ} \\[4pt]
{\scriptstyle \mathcal{O}(\varrho)} \downarrow & & \downarrow {\scriptstyle (\partial\varrho)^{\circ}} \\[4pt]
\mathcal{O}_q(\mathrm{SL}(n)) & \xrightarrow{\ \ \varphi\ \ } & \mathrm{U}_q(\mathbf{sl_n})^{\circ}
\end{array}
$$

where φ is as in Theorem 7.15 and can denotes the canonical map. Since φ is injective, $\mathcal{O}(\varrho)$ and, hence, ϱ are uniquely determined by $\partial\varrho$. $\square$

A quantum group G can be considered, equivalently, as the functor

$$
G : \mathrm{Alg}_k \to \mathrm{Set}, R \mapsto G(R) := \mathrm{Alg}_k(\mathcal{O}(G), R),
$$

where Alg_k denotes the category of k-algebras and Set the category of sets.
We adopt this point of view in this section.

LEMMA 9.3. *There is no nontrivial morphism $\mathrm{SL}_q(2) \to G_m(:= \mathrm{GL}_q(1))$.*

PROOF. The quantum group G_m corresponds to the group Hopf algebra $k[\mathbb{Z}]$. The algebra maps from $k[\mathbb{Z}]$ to an algebra R are in 1-to-1 correspondence with the invertible elements of R (this explains the name "G_m": multiplicative group of units).

Suppose $f : \mathrm{SL}_q(2) \to G_m$ is a morphism (of quantum groups). There is an embedding $G_m \hookrightarrow \mathrm{SL}_q(2)$, given by

$$
G_m(R) \to \mathrm{SL}_q(2)(R), a \mapsto \left(\begin{smallmatrix} a & \\ & a^{-1} \end{smallmatrix} \right).
$$

Consider $G_m \hookrightarrow \mathrm{SL}_q(2) \xrightarrow{f} G_m$. There is some $N \in \mathbb{Z}$, such that this morphism is given by

$$
G_m(R) \to G_m(R), a \mapsto a^N.
$$

It follows that $\partial f : \mathrm{U}_q(\mathbf{sl_2}) \to \mathcal{O}(G_m)^{\circ}$ maps K — which corresponds to $\left(\begin{smallmatrix} q & \\ & q^{-1} \end{smallmatrix} \right)$ — to q^N. (We identify group-like elements in $\mathcal{O}(G_m)^{\circ}$ with non-zero elements in k; see Proposition 8.3).

Since $\mathcal{O}(G_m)^{\circ}$ is commutative and $KEK^{-1} = q^2 E$, it follows that $\partial f(E) = 0$, similarly $\partial f(F) = 0$. We obtain

$$
q^N - q^{-N} = \partial f(K - K^{-1}) = (q - q^{-1})\partial f(EF - FE) = 0,
$$

which implies $N = 0$, since q is not a root of unity. Hence, ∂f is trivial, so f is trivial as well. $\square$

PROPOSITION 9.4. *Let* $\varrho : SL_q(2) \to GL_q(m)$ *be a nontrivial morphism of quantum groups* $(m > 1)$. *Then there is some* $i < m$ *and* $c \in k^\times$, *such that*

$$\partial\varrho(K) = K_i, \quad \partial\varrho(E) = cE_i, \quad \partial\varrho(F) = c^{-1}F_i.$$

PROOF. Let $T \subset GL_q(m)$ denote the canonical maximal torus (Section 8). By Proposition 8.3, $\mathcal{O}(T)^\circ$ and $\mathcal{O}_q(GL(m))^\circ$ have the same group-like elements. Since $\partial\varrho(K)$ is group-like, $\partial\varrho(K) \in \mathcal{O}(T)^\circ$.

Assume $\partial\varrho(E) = \partial\varrho(F) = 0$. Then the image of $\partial\varrho$ is contained in $\mathcal{O}(T)^\circ$, which means that $\mathrm{Im}(\varrho) \subset T$. But T is isomorphic to the direct product of n copies of G_m. Hence, by Lemma 9.3, the morphism ϱ is trivial, contradicting the hypothesis.

Therefore, $\partial\varrho(E), \partial\varrho(F)$ are not both zero. Suppose $\partial\varrho(E) \neq 0$. Then $\partial\varrho(E)$ is a nontrivial $(1, \partial\varrho(K))$-skew primitive element. It follows from Theorem 8.5 that $\partial\varrho(K) = K_i$, for some $i < m$ (it is impossible that $\partial\varrho(K) = 1$, since conjugation with $\partial\varrho(K)$ is not the identity on $\partial\varrho(E)$).

The space of $(1, K_i)$-primitive elements is spanned by $(1 - K_i, E_i, K_iF_i)$ (cf. Theorem 8.5). Since

$$K_i\partial\varrho(E)K_i^{-1} = \partial\varrho(KEK^{-1}) = q^2\partial\varrho(E),$$

we get $\partial\varrho(E) = cE_i$, for some $c \in k^\times$.

The relation

$$0 \neq K_i - K_i^{-1} = (q - q^{-1})[\partial\varrho(E), \partial\varrho(F)] \tag{9-1}$$

implies $\partial\varrho(F) \neq 0$. Similarly as above, we get $\partial\varrho(F) = c'F_i$ for some $c' \in k^\times$, and from (9–1) again, it follows that $cc' = 1$. ☐

Conversely, there actually *exist* morphisms as described in Proposition 9.4:

DEFINITION 9.5. Suppose $m > 1$. For $0 \le s < m - 1$, the morphism $\eta^{(s)} : SL_q(2) \to SL_q(m)$ is defined as follows:

$$\eta^{(s)}(R) : SL_q(2)(R) \to SL_q(m)(R), \quad A \mapsto \begin{pmatrix} I_s & & \\ & A & \\ & & I_{m-2-s} \end{pmatrix}$$

(for any algebra R, where I_n denotes the $(n \times n)$-identity matrix).

For $a, b \in k^\times$, the morphism $\mathrm{inn}(a, b) : SL_q(2) \to SL_q(2)$ is defined by

$$\mathrm{inn}(a, b)(R) : SL_q(2)(R) \to SL_q(2)(R), \quad A \mapsto \begin{pmatrix} a & \\ & b \end{pmatrix} A \begin{pmatrix} a & \\ & b \end{pmatrix}^{-1}$$

(for any algebra R).

REMARK 9.6. The derived maps are given as follows:

$$\partial\eta^{(s)} : \quad K, E, F \mapsto K_{s+1}, E_{s+1}, F_{s+1},$$
$$\partial\mathrm{inn}(a, b) : \quad K, E, F \mapsto K, ab^{-1}E, a^{-1}bF.$$

We summarize what has been proved so far:

PROPOSITION 9.7 (MORPHISMS $\mathrm{SL}_q(2) \to \mathrm{GL}_q(m)$). *Nontrivial morphisms of quantum groups* $\mathrm{SL}_q(2) \to \mathrm{GL}_q(m)$ *exist only if* $m > 1$. *In this case, all of them are exhausted by the compositions* $\mathrm{inn}(a,b)\eta^{(s)}$ *for* $0 \leq s < m-1$ *and* $a, b \in k^{\times}$.

We now turn to the general case. Let $\varrho : \mathrm{SL}_q(n) \to \mathrm{GL}_q(m)$ be a nontrivial morphism.

By applying Proposition 9.7 to $\varrho\eta^{(s)}$, for $0 \leq s < m-1$, we obtain the following result:

There is a non-empty set $I \subset [1, n-1]$, a map $\sigma : I \to [1, m-1]$, and $c_i \in k^{\times}$, for $i \in I$, such that

$$\partial\varrho : K_i,\ E_i,\ F_i \mapsto \begin{cases} K_{\sigma(i)},\ c_i E_{\sigma(i)},\ c_i^{-1} F_{\sigma(i)} & \text{if } i \in I, \\ 1,\ 0,\ 0 & \text{if } i \notin I. \end{cases}$$

LEMMA 9.8. *Writing* (α_{ij}) *for the Cartan matrix of* $\mathbf{sl_n}$ *(Definition 7.11), we have:*

(a) $I = [1, n-1]$;
(b) $\alpha_{ij} = \alpha_{\sigma(i),\sigma(j)}$;
(c) σ *is injective, in particular* $n \leq m$;
(d) *if* $|i - j| = 1$ *then* $|\sigma(i) - \sigma(j)| = 1$.

PROOF. (a) Suppose $j \in I$ and $|i-j| = 1$. Then $K_i E_j K_i^{-1} = q^{-1} E_j$. Application of $\partial\varrho$ yields

$$\partial\varrho(K_i) E_{\sigma(j)} \partial\varrho(K_i^{-1}) = q^{-1} E_{\sigma(j)},$$

in particular, $\partial\varrho(K_i) \neq 1$, which means $i \in I$. This proves $I = [1, n-1]$.

(b) By the relations of $\mathrm{U}_q(\mathbf{sl_n})$, we have $K_i E_j K_i^{-1} = q^{\alpha_{ij}} E_j$. We apply $\partial\varrho$ and get

$$K_{\sigma(i)} E_{\sigma(j)} K_{\sigma(i)}^{-1} = q^{\alpha_{ij}} E_{\sigma(j)}.$$

Since the left hand side is equal to $q^{\alpha_{\sigma(i),\sigma(j)}} E_{\sigma(j)}$ and q is not a root of unity, we get $\alpha_{\sigma(i),\sigma(j)} = \alpha_{ij}$.

Parts (c) and (d) follow from (b), since $i = j$ if and only if $\alpha_{ij} = 2$, and $|i - j| = 1$ if and only if $\alpha_{ij} = -1$. $\qquad\square$

It is easily checked that the maps $\sigma : [1,\ n-1] \to [1,\ m-1]$ of the form described in Lemma 9.8 are precisely the maps of the form

$$\sigma_s, \sigma_s' : [1,\ n-1] \to [1,\ m-1], \quad \sigma_s(i) := s + i, \quad \sigma_s'(i) := s + n - i,$$

for $0 \leq s \leq m - n$. We have proved the following:

PROPOSITION 9.9. *Let* $\varrho : \mathrm{SL}_q(n) \to \mathrm{GL}_q(m)$ *be a nontrivial morphism of quantum groups. Then* $m \geq n$, *and there are* $0 \leq s \leq m - n$ *and* $c_i \in k^{\times}$, *for* $1 \leq i < n$, *such that*

$$\partial\varrho : K_i,\ E_i,\ F_i \ \mapsto \ K_{s+i},\ c_i E_{s+i},\ c_i^{-1} F_{s+i} \qquad\qquad \text{for } 1 \leq i < n,$$

or $\quad \partial\varrho : K_i,\ E_i,\ F_i \ \mapsto \ K_{s+n-i},\ c_i E_{s+n-i},\ c_i^{-1} F_{s+n-i} \quad \text{for } 1 \leq i < n.$

We show that morphisms of the form descibed above do exist.

DEFINITION 9.10. Suppose $m \geq n$. For $0 \leq s \leq m - n$, the morphism $\eta^{(s)}$: $\mathrm{SL}_q(n) \to \mathrm{SL}_q(m)$ is defined as follows:

$$\eta^{(s)}(R) : \mathrm{SL}_q(n)(R) \to \mathrm{SL}_q(m)(R), \quad A \mapsto \begin{pmatrix} I_s \\ & A \\ & & I_{m-n-s} \end{pmatrix}.$$

For $a_1, \ldots, a_n \in k^\times$, the morphism $\mathrm{inn}(a_1, \ldots, a_n) : \mathrm{SL}_q(n) \to \mathrm{SL}_q(n)$ is defined by

$$\mathrm{inn}(a_1, \ldots, a_n)(R) : \mathrm{SL}_q(n)(R) \to \mathrm{SL}_q(n)(R),$$
$$A \mapsto \mathrm{diag}(a_1, \ldots, a_n) A \, \mathrm{diag}(a_1, \ldots, a_n)^{-1}$$

(for any algebra R).

REMARK 9.11. The derived maps are given by

$$\partial \eta^{(s)} : \qquad\qquad K_i, E_i, F_i \;\; \mapsto \;\; K_{s+i}, E_{s+i}, F_{s+i},$$
$$\partial \mathrm{inn}(a_1, \ldots, a_n) : \;\; K_i, E_i, F_i \;\; \mapsto \;\; K_i, a_i a_{i+1}^{-1} E_i, a_i^{-1} a_{i+1} F_i.$$

LEMMA 9.12. *There is exactly one automorphism Φ of the quantum group* $\mathrm{SL}_q(n)$, *such that*

$$\partial \Phi : K_i, \; E_i, \; F_i \mapsto K_{n-i}, \; E_{n-i}, \; F_{n-i},$$

for all $i \in [1, n-1]$. If $n = 2$ then Φ is the identity, otherwise, Φ has order 2.

PROOF. Let (x_{ij}) denote, as usual, the canonical generators of $\mathcal{O}_q(\mathrm{GL}(n))$. There is an automorphism

$$I := \mathrm{inn}\,\mathrm{diag}(-q, (-q)^2, \ldots, (-q)^n) : \mathcal{O}_q(\mathrm{GL}(n)) \to \mathcal{O}_q(\mathrm{GL}(n)),$$
$$x_{ij} \mapsto (-q)^{i-j} x_{ij},$$

an anti-automorphism

$$\Gamma : \mathcal{O}_q(\mathrm{GL}(n)) \to \mathcal{O}_q(\mathrm{GL}(n)), \quad x_{ij} \mapsto x_{n+1-j, n+1-i},$$

and the antipode (cf. Definition 4.12 and the proof of Theorem 4.15)

$$S : \mathcal{O}_q(\mathrm{GL}(n)) \to \mathcal{O}_q(\mathrm{GL}(n)), \quad x_{ij} \mapsto (-q)^{j-i} |X_{ji}|_q |X|_q^{-1}.$$

It can be shown by checking on generators that Γ, I, S commute with one another. Moreover, $\Gamma^2 = \mathrm{id}$, $S^2 = I^{-2}$ and $\Gamma(|X|_q) = I(|X|_q) = |X|_q$, $S(|X|_q) = |X|_q^{-1}$.

It follows that the composite $\mathcal{O}(\Phi) := S\Gamma I$ is an automorphism of $\mathcal{O}_q(\mathrm{GL}(n))$, given on generators by

$$\mathcal{O}(\Phi)(x_{ij}) = |X_{n+1-i, n+1-j}|_q |X|_q^{-1}.$$

Since $\mathcal{O}(\Phi)$ maps the quantum determinant to its inverse, $\mathcal{O}(\Phi)$ induces an automorphism of $\mathcal{O}_q(\mathrm{SL}(n))$. Direct calculation shows that $\partial \Phi$ has the described form. $\qquad\qquad\qquad\qquad\qquad\qquad\qquad\qquad\qquad\qquad\qquad\qquad\qquad\qquad\qquad\;\;\square$

Summarizing these results, we get the following main theorem:

THEOREM 9.13 (ALL MORPHISMS $\mathrm{SL}_q(n) \to \mathrm{GL}_q(m)$). *Nontrivial morphisms of quantum groups* $\mathrm{SL}_q(n) \to \mathrm{GL}_q(m)$ *exist only if* $n \le m$. *If this is the case, all of them are exhausted by*

$$\mathrm{inn}(a_1, \ldots, a_n)\eta^{(s)}, \quad \mathrm{inn}(a_1, \ldots, a_n)\Phi\eta^{(s)},$$

for $0 \le s \le m - n$ *and* $a_1, \ldots, a_n \in k^\times$.

In particular:

COROLLARY 9.14 (ENDOMORPHISMS AND AUTOMORPHISMS OF $\mathrm{SL}_q(n)$).
(a) *Every nontrivial endomorphism of the quantum group* $\mathrm{SL}_q(n)$ *is an automorphism.*
(b) *Every automorphism of* $\mathrm{SL}_q(n)$ *is inner (by a diagonal matrix in* k*) or the composite with* Φ.

Theorem 9.13 is valid (more generally) if $q^2 \ne 1$ or $q = -1$, char$k = 0$.

The endomorphism theorem (Corollary 9.14(a)) is valid if $q^2 \ne 1$ or char$k = 0$.

The automorphism theorem (Corollary 9.14(b)) is valid for $\mathrm{GL}_q(n)$ and $\mathrm{SL}_q(n)$ if $q^2 \ne 1$. It is valid for any $q \in k^\times$, if one replaces "diagonal matrix in k" by "q-matrix with entries in k".

10. The 2-Parameter Quantization

We now take invertible scalars α, β instead of the parameter q, and extend the results for q-matrices to those for (α, β)-matrices.

DEFINITION 10.1 (TWO-PARAMETER QUANTUM MATRICES). A (2×2) matrix $A = \begin{pmatrix} a & b \\ c & d \end{pmatrix}$ is called an (α, β)-*matrix* if the following relations hold:

$$ba = \alpha ab, \qquad\qquad dc = \alpha cd,$$
$$ca = \beta ac, \qquad\qquad db = \beta bd,$$
$$cb = \beta\alpha^{-1}bc, \quad da - ad = (\beta - \alpha^{-1})bc.$$

The *quantum determinant* of A is defined by

$$\delta = ad - \alpha^{-1}bc = da - \beta bc,$$

and is denoted by $|A|_{\alpha,\beta}$ or simply by $|A|$.

Many of the results on q-matrices stated so far will be extended to (α, β)-matrices.

There are two analogues of the matrix $\tilde{A}$ as follows:

$$A \begin{pmatrix} d & -\alpha b \\ -\alpha^{-1}c & a \end{pmatrix} = \begin{pmatrix} d & -\beta b \\ -\beta^{-1}c & a \end{pmatrix} A = \begin{pmatrix} \delta & 0 \\ 0 & \delta \end{pmatrix}.$$

This implies

$$\begin{pmatrix} d & -\beta b \\ -\beta^{-1}c & a \end{pmatrix} \delta = \begin{pmatrix} d & -\beta b \\ -\beta^{-1}c & a \end{pmatrix} A \begin{pmatrix} d & -\alpha b \\ -\alpha^{-1}c & a \end{pmatrix} = \delta \begin{pmatrix} d & -\alpha b \\ -\alpha^{-1}c & a \end{pmatrix},$$

so that δ commutes with a, d, but not with b, c. However, as in the one-parameter case, A is invertible if δ is invertible.

DEFINITION 10.2 (GENERAL TWO-PARAMETER QUANTUM MATRICES). An $(n \times n)$ matrix A is called an (α, β)-*matrix*, if any (2×2) minor in A is an (α, β)-matrix.

The matrix R_q (cf. Definition 4.2) is extended as follows:

$$R_{\alpha,\beta} := \alpha\beta \sum_{i=1}^{n} \mathbb{E}_{ii} \otimes \mathbb{E}_{ii} + \sum_{i<j}(\alpha\mathbb{E}_{ij} \otimes \mathbb{E}_{ji} + \beta\mathbb{E}_{ji} \otimes \mathbb{E}_{ij} + (\alpha\beta - 1)\mathbb{E}_{jj} \otimes \mathbb{E}_{ii}).$$

Note that $R_q = q^{-1}R_{q,q}$.

We may regard $R_{\alpha,\beta}$ as a linear transformation $V_n \otimes V_n \to V_n \otimes V_n$ defined by

$$R_{\alpha,\beta}(e_i \otimes e_j) = \begin{cases} \beta e_j \otimes e_i & \text{if } i < j, \\ \alpha\beta e_i \otimes e_i & \text{if } i = j, \\ \alpha e_j \otimes e_i + (\alpha\beta - 1)e_i \otimes e_j & \text{if } i > j. \end{cases}$$

The next proposition generalizes Proposition 4.3.

PROPOSITION 10.3 (THE R-MATRIX $R_{\alpha,\beta}$ AND (α, β)-MATRICES).
(a) $R_{\alpha,\beta}$ *is invertible and satisfies the braid condition* (1–1).
(b) *We have* $(R_{\alpha,\beta} - \alpha\beta I)(R_{\alpha,\beta} + I) = 0$.
(c) *An* $(n \times n)$ *matrix* A *is an* (α, β)-*matrix if and only if* $A^{(2)}$ *commutes with* $R_{\alpha,\beta}$.

As in the case of $\mathcal{O}_q(\mathrm{M}(n))$, part (c) of this proposition ensures that the algebra $\mathcal{O}_{\alpha,\beta}(\mathrm{M}(n))$ defined by n^2 generators $x_{11}, x_{12}, \ldots, x_{nn}$ and the relation that $X = (x_{ij})$ is an (α, β)-matrix forms in a natural way a bialgebra over which V_n is a right comodule (cf. Proposition 4.6 and Equation (4–1)).

Since $R_{\alpha,\beta} : V_n \otimes V_n \to V_n \otimes V_n$ is a right $\mathcal{O}_{\alpha,\beta}(\mathrm{M}(n))$ comodule isomorphism, it follows, by considering the images $\mathrm{Im}(R_{\alpha,\beta} - \alpha\beta I)$ and $\mathrm{Im}(R_{\alpha,\beta} + I)$, that $S_\alpha(V_n)$ and $\bigwedge_\beta(V_n)$ are right $\mathcal{O}_{\alpha,\beta}(\mathrm{M}(n))$ comodule algebras in a natural way (cf. Definition 4.9). Similarly, $S_\beta(V_n)$ and $\bigwedge_\alpha(V_n)$ are left $\mathcal{O}_{\alpha,\beta}(\mathrm{M}(n))$ comodule algebras.

The group-likes arising from the $\mathcal{O}_{\alpha,\beta}(\mathrm{M}(n))$ coaction on each n-th component of $\bigwedge_\alpha(V_n)$ and of $\bigwedge_\beta(V_n)$ coincide with each other, and are equal to

$$g := \sum_{\sigma \in S_n}(-\beta)^{-\ell(\sigma)}x_{\sigma(1),1} \cdot \ldots \cdot x_{\sigma(n),n}$$

$$= \sum_{\sigma \in S_n}(-\alpha)^{-\ell(\sigma)}x_{1,\sigma(1)} \cdot \ldots \cdot x_{n,\sigma(n)}.$$

This is called the *quantum determinant* and is denoted by $|X|_{\alpha,\beta}$ or simply by $|X|$.

We have

$$X\left((-\alpha)^{j-i}|X_{ji}|\right)_{i,j} = \left((-\beta)^{j-i}|X_{ji}|\right)_{i,j} X = gI.$$

Since this implies that

$$X \operatorname{diag}((-\alpha)^{-1},\ldots,(-\alpha)^{-n})(|X_{ji}|)_{i,j}\operatorname{diag}(-\beta,\ldots,(-\beta)^n)X$$
$$= g\operatorname{diag}(\alpha^{-1}\beta,\ldots,(\alpha^{-1}\beta)^n)X$$
$$= X\operatorname{diag}(\alpha^{-1}\beta,\ldots,(\alpha^{-1}\beta)^n)g,$$

we have $x_{ij}g = (\beta\alpha^{-1})^{i-j}gx_{ij}$.

This allows us to define $\mathcal{O}_{\alpha,\beta}(\mathrm{GL}(n))$ to be the localization $\mathcal{O}_{\alpha,\beta}(\mathrm{M}(n))[g^{-1}]$. If we let g^{-1} be a group-like element, then $\mathcal{O}_{\alpha,\beta}(\mathrm{GL}(n))$ forms a Hopf algebra including $\mathcal{O}_{\alpha,\beta}(\mathrm{M}(n))$ as a sub-bialgebra.

The antipode S of $\mathcal{O}_{\alpha,\beta}(\mathrm{GL}(n))$ satisfies:

$$S(x_{ij}) = (-\beta)^{j-i}g^{-1}|X_{ji}| = (-\alpha)^{j-i}|X_{ji}|g^{-1},$$
$$S^2(x_{ij}) = (\alpha\beta)^{j-i}x_{ij}.$$

The Hopf algebra $\mathcal{O}_{\alpha,\beta}(\mathrm{GL}(n))$ defines the 2-parameter quantization $\mathrm{GL}_{\alpha,\beta}(n)$ of $\mathrm{GL}(n)$.

11. The q-Schur Algebra and the Hecke Algebra

Fix a non-zero element q in k and a non-negative integer n.

DEFINITION 11.1 (THE HECKE ALGEBRA). The Hecke algebra $\mathcal{H}$ is the algebra generated by $n-1$ elements $T_1,\ldots,T_{n-1}$ with the relations

$$(T_i - q)(T_i + 1) = 0, \tag{11-1}$$
$$T_iT_{i+1}T_i = T_{i+1}T_iT_{i+1}, \tag{11-2}$$
$$T_iT_j = T_jT_i \quad \text{if } |i-j| > 1. \tag{11-3}$$

PROPOSITION 11.2. *Let $\pi \in S_n$ and suppose that $\pi = s_{i_1} \cdot \ldots \cdot s_{i_\ell}$ is a reduced expression with the transpositions $s_a = (a, a+1)$. (Thus $\ell = \ell(\pi)$, the length of π.) Then*

$$T_\pi := T_{i_1} \cdot \ldots \cdot T_{i_\ell}$$

is independent of the choice of the reduced expression for π. Moreover, $\{T_\pi \mid \pi \in S_n\}$ is a basis of $\mathcal{H}$.

If $q = 1$, then $\mathcal{H} = kS_n$, the group algebra of the symmetric group S_n.

If $q = p^r$, a power of a prime p, then $\mathcal{H} = H_k(\mathrm{GL}_n(q), B)$, the Iwahori–Hecke algebra, with B the Borel subgroup.

DEFINITION 11.3 (THE q-SCHUR ALGEBRA). Suppose that $q = \alpha\beta$, where $\alpha, \beta \in k^{\times}$. The n-th component of $\mathcal{O}_{\alpha,\beta}(\mathrm{M}(d))$ is denoted by $A(d,n)$; it is a subcoalgebra. The dual algebra

$$S(d,n) := A(d,n)^{*}$$

is determined by the product q, as we will see soon, so it is denoted by $S_q(d,n)$ and called the q-Schur algebra.

The vector space $(V_d)^{\otimes n}$ is a right $A(d,n)$-comodule, with respect to the diagonal coaction by $\mathcal{O}_{\alpha,\beta}(\mathrm{M}(d))$, which is given by

$$(V_d)^{\otimes n} \to (V_d)^{\otimes n} \otimes A(d,n),$$

$$e_{i_1} \otimes \ldots \otimes e_{i_n} \mapsto \sum_{j_1,\ldots,j_n} e_{j_1} \otimes \ldots \otimes e_{j_n} \otimes x_{j_1,i_1} \cdot \ldots \cdot x_{j_n,i_n}.$$

Right $A(d,n)$ comodules are interpreted as polynomial representations of $\mathrm{GL}_{\alpha,\beta}(d)$ of degree n.

PROPOSITION 11.4. *The algebra* $S(d,n)$ *(with* d, n *fixed) is determined, up to isomorphism, by* q *(rather than* α,β*).*

This allows us to write $S_q(d,n) = S(d,n)$ and justifies the name "q"-Schur algebra.

PROOF. To see this, we first make $(V_d)^{\otimes n}$ into a right $\mathcal{H}$ module by identifying

$$T_i = \mathrm{id}_{(V_d)^{\otimes(i-1)}} \otimes R_{\alpha,\beta} \otimes \mathrm{id}_{(V_d)^{\otimes(n-i-1)}},$$

a linear endomorphism of $(V_d)^{\otimes n}$, where $R_{\alpha,\beta}$ acts on $V_d \otimes V_d$ by left multiplication (this action is well-defined, cf. Proposition 10.3 (b)).

By the construction of $\mathcal{O}_{\alpha,\beta}(\mathrm{M}(d))$, the coalgebra $A(d,n)$ is the "cocentralizer" of $T_1, \ldots, T_{n-1}$, or in other words the largest quotient coalgebra of $\mathrm{End}((V_d)^{\otimes n})^{*}$, over which $T_1, \ldots, T_{n-1}$ are all comodule endomorphisms.

This means that $A(d,n)^{*}$ is the centralizer of $T_1, \ldots, T_{n-1}$. Thus we have a natural isomorphism

$$S(d,n) \xrightarrow{\cong} \mathrm{End}_{\mathcal{H}}((V_d)^{\otimes n}).$$

Hence, it is enough to show that the right $\mathcal{H}$ module $(V_d)^{\otimes n}$ is determined by q.

Let $i = (i_1, \ldots, i_n)$ be an n-tuple of integers $1 \le i_k \le d$. Write $e_i = e_{i_1} \otimes \ldots \otimes e_{i_n}$. All the e_i's form a basis of $(V_d)^{\otimes n}$. For the transposition $s = (a, a+1)$, it follows from the definition of T_s ($= T_a$) that

$$e_i T_s = \begin{cases} q e_i & \text{if } i_a = i_{a+1}, \\ \beta e_{is} & \text{if } i_a < i_{a+1}, \\ (q-1)e_i + \alpha e_{is} & \text{if } i_a > i_{a+1}. \end{cases}$$

where S_n acts naturally on the set of the n-tuples i from the right.

Define an equivalence relation among the n-tuples by

$$i \sim j :\Leftrightarrow \exists \pi \in S_n : j = i\pi.$$

The equivalence classes are in one-to-one correspondence with the set $\Lambda(d, n)$ of the compositions $\lambda = (\lambda_1, \ldots, \lambda_d)$ of n into d parts (i.e. $\lambda_1 + \cdots + \lambda_d = n$, where all $\lambda_k \geq 0$). Here, i belongs to λ if and only if $i \sim i_\lambda$, where

$$i_\lambda := (\underbrace{1, \ldots, 1}_{\lambda_1}, \underbrace{2, \ldots, 2}_{\lambda_2}, \ldots, \underbrace{d, \ldots, d}_{\lambda_d}).$$

Clearly, the right $\mathcal{H}$ module $(V_d)^{\otimes n}$ decomposes as

$$(V_d)^{\otimes n} = \bigoplus_{\lambda \in \Lambda(d,n)} \left(\bigoplus_{i \sim i_\lambda} ke_i \right),$$

a direct sum of the $\mathcal{H}$ submodules $\bigoplus_{i \sim i_\lambda} ke_i$.

Fix some $\lambda = (\lambda_1, \ldots, \lambda_d)$ in $\Lambda(d, n)$, and let $Y_\lambda (\subset S_n)$ denote the stabilizer of the d subsets $\{1, \ldots, \lambda_1\}, \{\lambda_1 + 1, \ldots, \lambda_1 + \lambda_2\}, \ldots, \{\lambda_1 + \cdots + \lambda_{d-1} + 1, \ldots, n\}$. Write $x_\lambda := \sum_{\pi \in Y_\lambda} T_\pi$.

PROPOSITION 11.5 (Dipper and James). *There is a right $\mathcal{H}$ module isomorphism*

$$\bigoplus_{i \sim i_\lambda} ke_i \cong x_\lambda \mathcal{H}, \text{ given by } e_{i_\lambda} \mapsto x_\lambda.$$

Hence $(V_d)^{\otimes n} \cong \bigoplus_\lambda x_\lambda \mathcal{H}$, which implies that $(V_d)^{\otimes n}$, hence the algebra $S(d, n)$ also, is determined by q. $\square$

COROLLARY 11.6. *If $\alpha\beta = \alpha'\beta'$, then $\mathcal{O}_{\alpha,\beta}(M(n)) \cong \mathcal{O}_{\alpha',\beta'}(M(n))$, as coalgebras.*

The q-Schur algebra was introduced by Dipper and James, and its representations have been investigated in detail.

REMARK 11.7 (Du, Parshall, Wang). The isomorphism mentioned in the last corollary is given explicitly as follows: Suppose $\alpha\beta = \alpha'\beta'$ and set $\xi := \alpha'/\alpha = \beta/\beta'$. Then there is a coalgebra isomorphism

$$\varphi_\xi : \mathcal{O}_{\alpha,\beta}(M(n)) \xrightarrow{\cong} \mathcal{O}_{\alpha',\beta'}(M(n)), x_{ij} \mapsto \xi^{\ell(i)-\ell(j)} x'_{ij},$$

where $x_{ij} := x_{i_1, j_1} \cdot \ldots \cdot x_{i_r, j_r}$, $x'_{ij} := x'_{i_1, j_1} \cdot \ldots \cdot x'_{i_r, j_r}$, denote the monomials which span $\mathcal{O}_{\alpha,\beta}(M(n))$ and $\mathcal{O}_{\alpha',\beta'}(M(n))$, respectively, and $\ell(i)$ is the number of inversions in i.

Furthermore, this isomorphism is extended uniquely to a coalgebra isomorphism

$$\mathcal{O}_{\alpha,\beta}(GL(n)) \xrightarrow{\cong} \mathcal{O}_{\alpha',\beta'}(GL(n)).$$

12. Cocycle Deformations

DEFINITION 12.1 (2-COCYCLES FOR A GROUP). A *2-cocycle for a group G* (with coefficients in the trivial G-module $k^\times$) is a map $\sigma : G \times G \to k^\times$, which satisfies

$$\sigma(x,y)\sigma(xy,z) = \sigma(y,z)\sigma(x,yz), \quad x,y,z \in G.$$

Let us generalize this notion to a bialgebra A:

DEFINITION 12.2 (2-COCYCLES FOR A BIALGEBRA). A *2-cocycle for a bialgebra A* is a bilinear form $\sigma : A \times A \to k$, that is invertible (in the algebra $(A \otimes A)^*$) and that satisfies

$$\sum \sigma(x_1, y_1)\sigma(x_2 y_2, z) = \sum \sigma(y_1, z_1)\sigma(x, y_2 z_2), \quad x,y,z \in A.$$

A 2-cocycle σ for A is said to be *normal*, if it satisfies

$$\sigma(1,x) = \varepsilon(x) = \sigma(x,1), \quad x \in A.$$

For any 2-cocycle σ, the map $\sigma^{-1}(1,1)\sigma$ is a normal 2-cocycle. In the following, we assume that all 2-cocycles are normal.

PROPOSITION 12.3 (DEFORMATION OF BIALGEBRAS BY 2-COCYCLES) (Doi).
(a) *Using a 2-cocycle σ, define a new multiplication on A as follows*

$$x \bullet y := \sum \sigma(x_1, y_1) x_2 y_2 \sigma^{-1}(x_3, y_3), \quad x,y \in A.$$

   This makes A into an algebra with the same unit element.
(b) *With this new algebra structure and the original coalgebra structure, A forms a bialgebra, which is denoted by A^σ. It is called the* deformation *of A by cocycle σ.*
(c) *If A is a Hopf algebra, A^σ is also a Hopf algebra, with the antipode S^σ, defined by*

$$S^\sigma(x) = \sum \sigma(x_1, S(x_2)) S(x_3) \sigma^{-1}(S(x_4), x_5), \quad x \in A.$$

EXAMPLE 12.4 (THE QUANTUM DOUBLE). Let H be a finite-dimensional Hopf algebra and define $A := H^{*\mathrm{cop}} \otimes H$. The bilinear form $\sigma : A \times A \to k$ determined by

$$\sigma(p \otimes x, q \otimes y) = \langle p, 1 \rangle \langle q, x \rangle \langle \varepsilon, y \rangle$$

is a 2-cocycle for A. The algebra A^σ is a bicrossed product of H with $H^{*\mathrm{cop}}$ determined by the following relations:

$$(p \otimes 1) \bullet (1 \otimes x) = p \otimes x,$$

$$(1 \otimes x) \bullet (p \otimes 1) = \sum \langle p_3, x_1 \rangle p_2 \otimes x_2 \langle p_1, S(x_3) \rangle,$$

where $p \in H^*$, $x \in H$. Hence the Hopf algebra A^σ coincides with the *quantum double* $D(H)$ of H, which is due to Drinfeld.

In the remainder of this section, we shall consider mainly cocycle deformations of $\mathcal{O}_{\alpha,\beta}(\mathrm{M}(n))$ and $\mathcal{O}_q(\mathrm{GL}(n))$.

We first generalize the construction of $\mathcal{O}_{\alpha,\beta}(\mathrm{M}(n))$, following Doi's method.

DEFINITION 12.5 (THE BIALGEBRA $M(C,\sigma)$). Let C be a coalgebra, and let $\sigma : C \times C \to k$ be an invertible bilinear form. Let $M(C,\sigma)$ denote the quotient algebra of the tensor algebra $T(C)$ by the relation

$$\sum \sigma(x_1, y_1) x_2 \otimes y_2 = \sum \sigma(x_2, y_2) y_1 \otimes x_1, \quad x, y \in C.$$

In fact, $M(C,\sigma)$ is a quotient bialgebra of $T(C)$, where $T(C)$ has the unique bialgebra structure making C a subcoalgebra.

EXAMPLE 12.6 (THE BIALGEBRA $\mathcal{O}_{\alpha,\beta}(\mathrm{M}(n))$). Define $C_n = M_n(k)^*$, and let x_{ij} ($\in C_n$) be the dual basis of the matrix units $\mathbb{E}_{ij}$ ($\in M_n(k)$). Define an invertible bilinear form $\sigma_{\alpha,\beta} : C_n \times C_n \to k$, where $\alpha, \beta \in k^\times$, by

$$\sigma_{\alpha,\beta}(x_{ii}, x_{jj}) = \begin{cases} \beta, & \text{if } i < j, \\ \alpha\beta, & \text{if } i = j, \\ \alpha, & \text{if } i > j, \end{cases}$$

$$\sigma_{\alpha,\beta}(x_{ij}, x_{ji}) = \alpha\beta - 1, \text{ if } i < j,$$

$$\sigma_{\alpha,\beta}(x_{ij}, x_{k\ell}) = 0, \text{ otherwise.}$$

This bilinear form is related with the linear transformation $R_{\alpha,\beta}$ (introduced below Definition 10.2) as follows:

$$R_{\alpha,\beta}(e_k \otimes e_\ell) = \sum_{i,j} \sigma_{\alpha,\beta}(x_{jk}, x_{i\ell}) e_i \otimes e_j.$$

One sees the defining relations for $M(C_n, \sigma_{\alpha,\beta})$ are interpreted as $X^{(2)} R_{\alpha,\beta} = R_{\alpha,\beta} X^{(2)}$, so that we have (cf. Proposition 10.3 (c)):

$$M(C_n, \sigma_{\alpha,\beta}) = \mathcal{O}_{\alpha,\beta}(\mathrm{M}(n)).$$

LEMMA 12.7. *Let τ be a 2-cocycle for $M(C,\sigma)$. Then there is a bialgebra isomorphism*

$$M(C, \sigma^\tau) \xrightarrow{\cong} M(C, \sigma)^\tau,$$

which is the identity on C, where $\sigma^\tau : C \times C \to k$ is the invertible bilinear form defined by

$$\sigma^\tau(x, y) = \sum \tau(y_1, x_1) \sigma(x_2, y_2) \tau^{-1}(x_3, y_3), (x, y, z \in C). \tag{12-1}$$

DEFINITION 12.8. **Braided bialgebras**

A *braiding* on a bialgebra A is an invertible bilinear form $\sigma : A \times A \to k$ such that for all $x, y, z \in A$, we have:

$$\sigma(xy, z) = \sum \sigma(x, z_1)\sigma(y, z_2),$$

$$\sigma(x, yz) = \sum \sigma(x_1, z)\sigma(x_2, y),$$

$$\sigma(x_1, y_1)x_2 y_2 = \sum y_1 x_1 \sigma(x_2, y_2).$$

A braiding on A is a 2-cocycle for A. The last equation means that $A^\sigma = A^{\mathrm{op}}$. The first and the last equations imply the following Yang–Baxter condition (for all $x, y, z \in A$):

$$\sum \sigma(x_1, y_1)\sigma(x_2, z_1)\sigma(y_2, z_2) = \sum \sigma(y_1, z_1)\sigma(x_1, z_2)\sigma(x_2, y_2). \qquad (12\text{-}2)$$

If we regard σ as an element in $(A \otimes A)^*$, then the last equation is rewritten as

$$\sigma_{12}\sigma_{13}\sigma_{23} = \sigma_{23}\sigma_{13}\sigma_{12}$$

in the algebra $(A \otimes A \otimes A)^*$.

If σ is a braiding on A, then for any right A comodules V, W, it follows that

$$R_\sigma : V \otimes W \to W \otimes V, \quad v \otimes w \mapsto \sum \sigma(v_1, w_1)w_0 \otimes v_0$$

is a right A comodule isomorphism. The monoidal category of right A comodules becomes a braided category with the structure R_σ.

PROPOSITION 12.9 (A BRAIDING ON THE BIALGEBRA $\mathcal{O}_{\alpha,\beta}(\mathrm{M}(n))$). *If an invertible bilinear form $\sigma : C \times C \to k$ satisfies the Yang–Baxter condition, then it is extended uniquely to a braiding on $M(C, \sigma)$.*

In particular, $\mathcal{O}_{\alpha,\beta}(\mathrm{M}(n))$ has a natural braiding $\sigma_{\alpha,\beta}$ (the extension of $\sigma_{\alpha,\beta}$ in Example 12.6).

REMARK 12.10. If τ is a 2-cocycle for a bialgebra A and if σ is a braiding on A, then σ^τ as defined in (12-1) is a braiding on A^τ.

PROPOSITION 12.11. *Let $\alpha, \beta, \alpha', \beta' \in k^\times$. If $\alpha'\beta' = \alpha\beta$ or $(\alpha\beta)^{-1}$, then $\mathcal{O}_{\alpha',\beta'}(\mathrm{M}(n))$ is a cocycle deformation of $\mathcal{O}_{\alpha,\beta}(\mathrm{M}(n))$.*

PROOF. Let $T \subset \mathrm{M}_{\alpha,\beta}(n)$ be the canonical maximal torus with the corresponding bialgebra projection

$$\mathcal{O}_{\alpha,\beta}(\mathrm{M}(n)) \to \mathcal{O}(T) = k[t_1, t_1^{-1}, \ldots, t_n, t_n^{-1}],$$

defined by $x_{ij} \mapsto \delta_{ij}t_i$. For $q \in k^\times$, set

$$\tau_q(t_1^{e(1)} \cdot \ldots \cdot t_n^{e(n)}, t_1^{f(1)} \cdot \ldots \cdot t_n^{f(n)}) = \prod_{i<j} q^{e(i)f(j)}.$$

Then τ_q gives a 2-cocycle for $\mathcal{O}(T)$, which may be regarded as a 2-cocycle for $\mathcal{O}_{\alpha,\beta}(\mathrm{M}(n))$ through the projection. Since one computes

$$(\sigma_{\alpha,\beta})^{\tau_q} = \sigma_{q\alpha, q^{-1}\beta},$$

we get, using Lemma 12.7:

$$\mathcal{O}_{\alpha,\beta}(M(n))^{\tau_q} \cong \mathcal{O}_{q\alpha,q^{-1}\beta}(M(n)),$$

which yields the conclusion in the case where $\alpha'\beta' = \alpha\beta$.

For the assertion in the other case, it is enough to note that there is a bialgebra isomorphism

$$\mathcal{O}_{\alpha,\beta}(M(n)) \cong \mathcal{O}_{\alpha^{-1},\beta^{-1}}(M(n)), x_{ij} \mapsto x_{n+1-i,n+1-j}. \qquad \square$$

We claim that the converse of Proposition 12.11 holds true.

THEOREM 12.12 (COCYCLE DEFORMATIONS OF $\mathcal{O}_{\alpha,\beta}(M(n))$). *The bialgebra $\mathcal{O}_{\alpha',\beta'}(M(n))$ is a cocycle deformation of $\mathcal{O}_{\alpha,\beta}(M(n))$ if and only if $\alpha'\beta' = \alpha\beta$ or $(\alpha\beta)^{-1}$.*

PROOF. It remains to show the "only if" part. Suppose that τ is a 2-cocycle for $\mathcal{O}_{\alpha,\beta}(M(n))$ such that there exists a bialgebra isomorphism

$$\varphi : \mathcal{O}_{\alpha',\beta'}(M(n)) \xrightarrow{\cong} \mathcal{O}_{\alpha,\beta}(M(n))^{\tau},$$

which may be regarded as

$$\varphi : M(C_n, \sigma_{\alpha',\beta'}) \xrightarrow{\cong} M(C_n, (\sigma_{\alpha,\beta})^{\tau}).$$

From a simple observation, it follows that the restriction of φ to C_n gives an automorphism of C_n.

By the Noether–Skolem theorem, there exists a linear isomorphism $\psi : V_n \xrightarrow{\cong} V_n$ which is "semi-colinear" with respect to φ in the sense that the diagram

$$
\begin{array}{ccc}
V_n & \xrightarrow{\psi} & V_n \\
\varrho \downarrow & & \downarrow \varrho \\
V_n \otimes C_n & \xrightarrow{\psi \otimes \varphi} & V_n \otimes C_n
\end{array}
$$

commutes, where ϱ denotes the canonical comodule structure, i.e. $\varrho(e_j) = \sum_i e_i \otimes x_{ij}$. Let σ denote the braiding on $M(C_n, \sigma_{\alpha',\beta'})$ which is the pull-back of $(\sigma_{\alpha,\beta})^{\tau}$ through φ. Then the last commutative diagram makes the following commute.

$$
\begin{array}{ccc}
V_n \otimes V_n & \xrightarrow{\psi \otimes \psi} & V_n \otimes V_n \\
R_\sigma \downarrow & & \downarrow R_{(\sigma_{\alpha,\beta})^{\tau}} \\
V_n \otimes V_n & \xrightarrow{\psi \otimes \psi} & V_n \otimes V_n
\end{array}
$$

Note further that $R_{(\sigma_{\alpha,\beta})^{\tau}} = P_\tau R_{\alpha,\beta} P_\tau^{-1}$, where

$$P_\tau(e_k \otimes e_\ell) = \sum_{i,j} \tau(x_{ik}, x_{j\ell}) e_i \otimes e_j.$$

Then we see that R_σ satisfies the following conditions:

(a) R_σ is a right $\mathcal{O}_{\alpha',\beta'}(M(n))$ comodule automorphism;

(b) R_σ satisfies the braid condition;

(c) $R_\sigma^2 = (\alpha\beta - 1)R_\sigma + \alpha\beta$.

Condition (a) is equivalent to

$$R_\sigma \in \mathrm{End}_{S(n,2)}(V_n^{\otimes 2}),$$

where $S(n,2)$ is taken with respect to α', β' (Section 11). This means that R_σ is in the double centralizer of $R_{\alpha',\beta'}$. It is known that the double centralizer of a linear transformation of a finite dimensional vector space consists of all polynomials of the linear transformation.

Since $R_{\alpha',\beta'}$ satisfies a quadratic equation, this implies that

(a') R_σ is a linear combination of 1, $R_{\alpha',\beta'}$.

If we look for a linear map R_σ satisfying (a'), (b) and which can be extended to a braiding on $\mathcal{O}_{\alpha',\beta'}(M(n))$, we see that R_σ should be equal, up to non-zero scalar multiplication, to $R_{\alpha',\beta'}$ or $R_{\alpha',\beta'} + 1 - \alpha'\beta'$.

Suppose first $R_\sigma = cR_{\alpha',\beta'}$, with $c \in k^\times$. Then, by Proposition 10.3 (b),

$$R_\sigma^2 = c^2 R_{\alpha',\beta'}^2 = c^2(\alpha'\beta' - 1)R_{\alpha',\beta'} + c^2\alpha'\beta'.$$

Since it follows from (c) that

$$R_\sigma^2 = (\alpha\beta - 1)R_\sigma + \alpha\beta = c(\alpha\beta - 1)R_{\alpha',\beta'} + \alpha\beta,$$

we have $c^2(\alpha'\beta' - 1) = c(\alpha\beta - 1)$ and $c^2\alpha'\beta' = \alpha\beta$, whence, by eliminating c,

$$\frac{(\alpha\beta - 1)^2}{\alpha\beta} = \frac{(\alpha'\beta' - 1)^2}{\alpha'\beta'}$$

or $(\alpha\beta - \alpha'\beta')(\alpha\beta\alpha'\beta' - 1) = 0$.

The same equation is obtained in the other case.   □

REMARK 12.13. The proof shows that the braidings on $\mathcal{O}_{\alpha,\beta}(M(n))$ are exhausted essentially by the two of $\sigma_{\alpha,\beta}$ and the pull-back $\sigma'_{\alpha,\beta}$ of $\sigma_{\alpha^{-1},\beta^{-1}}$ through the isomorphism $\mathcal{O}_{\alpha,\beta}(M(n)) \cong \mathcal{O}_{\alpha^{-1},\beta^{-1}}(M(n))$ (cf. the proof of Proposition 12.11).

Note that $R_{\sigma'_{\alpha,\beta}} = R_{\alpha,\beta} + 1 - \alpha\beta$.

In a similar way:

COROLLARY 12.14. If $q^2 \neq 1$, then $\mathcal{O}_q(GL(n))$ cannot be a cocycle deformation of any commutative Hopf algebra.

13. (2×2) R-Matrices

By a (2×2) R-*matrix*, we mean an invertible matrix R in $M_2(k) \otimes M_2(k)$ which satisfies the braid condition. We ask:

How many (2×2) R-matrices exist?

For such a matrix R, we define the k-bialgebra $\mathcal{O}_R(M(2))$ generated by the entries $x_{11}, x_{12}, x_{21}, x_{22}$ in $X = (x_{ij})_{i,j \in \{1,2\}}$ with the relation $X^{(2)} R = RX^{(2)}$.

Let $C_2 := M_2(k)^*$, as in Example 12.6. The (2×2) R-matrices R are in one-to-one correspondence with the invertible bilinear forms $\sigma : C_2 \times C_2 \to k$, which satisfy the Yang–Baxter condition (12–2), in such a way that

$$R_{(ij),(k\ell)} = \sigma(x_{jk}, x_{i\ell}). \tag{13–1}$$

Furthermore, we have $\mathcal{O}_R(M(2)) = M(C_2, \sigma)$. Hence, $\mathcal{O}_R(M(2))$ is a braided bialgebra.

Kauffman classifies the R-matrices of the form:

$$\begin{pmatrix} n & & & \\ & r & d & \\ & s & \ell & \\ & & & p \end{pmatrix} \tag{13–2}$$

(where rows and columns are indexed by (11), (12), (21), (22)). This is invertible if and only if $p, n \neq 0$ and $r\ell - ds \neq 0$.

REMARK 13.1. (a) A matrix of the form (13–2) satisfies the braid condition if and only if the following relations hold:

$$r\ell d = r\ell s = r\ell(r - \ell) = 0,$$
$$p^2 \ell = p\ell^2 + \ell ds,$$
$$n^2 \ell = n\ell^2 + \ell ds,$$
$$p^2 r = pr^2 + rds,$$
$$n^2 r = nr^2 + rds.$$

(b) The following are examples of R-matrices of this form:

$$\begin{pmatrix} n & & & \\ & & d & \\ & s & & \\ & & & p \end{pmatrix}, \quad \begin{pmatrix} \gamma & & & \\ & & \alpha & \\ & \beta & \alpha\beta - 1 & \\ & & & \delta \end{pmatrix}, \tag{13–3}$$

where $\gamma, \delta \in \{\alpha\beta, -1\}$ and $\alpha, \beta, n, p, d, s \in k^\times$ are arbitrary. For $\gamma := \delta := \alpha\beta$, we get $R_{\alpha,\beta}$.

The bialgebras $\mathcal{O}_R(M(2))$ for these R-matrices are not yet investigated except for $R_{\alpha,\beta}$. However, for the following two examples of R-matrices, the bialgebras $\mathcal{O}_R(M(2))$ have been investigated.

EXAMPLE 13.2 (Takeuchi–Tambara). Assume $\text{char}(k) \neq 2$. For $q \in k^\times$, the matrix

$$R := 1/2 \begin{pmatrix} 2-(q-1)^2 & & & (q-1)^2 \\ & 1-q^2 & 1+q^2 & \\ & 1+q^2 & 1-q^2 & \\ (q+1)^2 & & & 2-(q+1)^2 \end{pmatrix}$$

is invertible and satisfies the braid condition and the relation

$$(R-I)(R+q^2I) = 0.$$

If $q^2 \neq -1$, then R is diagonalizable to $\text{diag}(1,1,-q^2,-q^2)$.

EXAMPLE 13.3 (Suzuki). For $\alpha, \beta \in k^\times$, write

$$R := \begin{pmatrix} & & & \alpha \\ & \beta & & \\ & & \beta & \\ \alpha & & & \end{pmatrix}.$$

The corresponding invertible form $\tau_{\alpha,\beta} : C_2 \times C_2 \to k$ according to (13–1) is given by

$$\tau_{\alpha,\beta}(x_{ij}, x_{k\ell}) = \begin{cases} \alpha & \text{if } (ij,k\ell) \in \{(12,12),(21,21)\}, \\ \beta & \text{if } (ij,k\ell) \in \{(12,21),(21,12)\}, \\ 0 & \text{otherwise.} \end{cases}$$

Suppose $\alpha^2 \neq \beta^2$. Then $\mathcal{O}_R(M(2)) = M(C_2, \tau_{\alpha,\beta})$ is independent of the choice of α, β. Denote by B this bialgebra. Then

(a) B is generated by $x_{11}, x_{12}, x_{21}, x_{22}$ with relations $x_{11}^2 = x_{22}^2$, $x_{12}^2 = x_{21}^2$, and $x_{ij}x_{\ell m} = 0$, for $i - j \not\equiv \ell - m \bmod 2$;
(b) B is cosemisimple;
(c) the maps $\tau_{\alpha,\beta}$, for $\alpha, \beta \in k^\times$, exhaust the braidings on B.

14. The Quantum Frobenius Map and Related Topics

The quantum Frobenius map for $\mathrm{GL}_q(n)$ was introduced by Parshall–Wang (1991) and independently by myself (1992). Assume q is a root of unity, let ℓ be the order of q^2 and put $\varepsilon = q^{\ell^2}$. We have

	ℓ odd	ℓ even
$q^\ell = 1$	$\varepsilon = 1$	$\times$
$q^\ell = -1$	$\varepsilon = -1$	$\varepsilon = 1$

PROPOSITION 14.1. If $X = (x_{ij})$ is a q-matrix, then $X^{(\ell)} = (x_{ij}{}^\ell)$ is an ε-matrix.

By associating $X^{(\ell)}$ to X, we get a homomorphism of quantum groups

$$\mathcal{F}\colon \mathrm{GL}_q(n) \to \mathrm{GL}_\varepsilon(n)$$

which is called the quantum Frobenius map. The corresponding Hopf algebra map

$$\mathcal{O}(\mathcal{F})\colon \mathcal{O}_\varepsilon(\mathrm{GL}(n)) \to \mathcal{O}_q(\mathrm{GL}(n))$$

is injective and free of rank ℓ^{n^2}. If $q^\ell = 1$ (hence ℓ odd), then the image of $\mathcal{O}(\mathcal{F})$ is contained in the center. Let $\mathrm{GL}_q'(n)$ be the quantum subgroup of $\mathrm{GL}_q(n)$ represented by the quotient Hopf algebra $\mathcal{O}_q(\mathrm{GL}(n))/\mathcal{O}_q(\mathrm{GL}(n))\mathcal{O}(\mathrm{GL}(n))^+$ which is ℓ^{n^2}-dimensional. We may think of it as the kernel of $\mathcal{F}$, and we obtain an exact sequence of quantum groups

$$1 \to \mathrm{GL}_q'(n) \to \mathrm{GL}_q(n) \xrightarrow{\mathcal{F}} \mathrm{GL}(n) \to 1. \tag{14-1}$$

A finite quantum subgroup of $\mathrm{GL}_q(n)$ means a finite dimensional quotient Hopf algebra of $\mathcal{O}_q(\mathrm{GL}(n))$. If q has an odd order, $\mathrm{GL}_q'(n)$ is such an example. Recently, E. Müller has determined all finite subgroups of $\mathrm{GL}_q(n)$. We describe his results in the following.

If q is not a root of unity, all finite subgroups of $\mathrm{GL}_q(n)$ are contained in the canonical torus T. We assume q is a root of unity of odd order ℓ and k is an algebraically closed field of characteristic 0.

Let I be a subset of $I_0 = \{(i, i+1), (i+1, i) \mid i = 1, 2, \ldots, n-1\}$. A quantum subgroup $P_{q,I}$ of $\mathrm{GL}_q(n)$ is determined by the following condition: If $i \le a, a+1 \le j$ for some $(a, a+1) \notin I$ of if $j \le b, b+1 \le i$ for some $(b+1, b) \notin I$, then the (i, j) component is zero.

EXAMPLE 14.2. $n = 5$, $I = \{(1, 2), (3, 4), (2, 1), (5, 4)\}$.

$$P_{q,I} = \begin{pmatrix} * & * & 0 & 0 & 0 \\ * & * & 0 & 0 & 0 \\ 0 & 0 & * & * & 0 \\ 0 & 0 & 0 & * & 0 \\ 0 & 0 & 0 & * & * \end{pmatrix}.$$

Let s be the number of i such that $(i, i+1) \notin I$ and $(i+1, i) \notin I$. In the above example, $s = 2$. Then $P_{q,I}$ factors as the direct product of s blocks. By associating the q-determinant with each block, we get a homomorphism of quantum groups

$$D_q\colon P_{q,I} \to (G_m)^s.$$

When $q = 1$, we write $P_I = P_{1,I}$ and $D = D_1$. Thus

$$D\colon P_I \to (G_m)^s.$$

We have the following commutative diagram of quantum groups with exact rows:

$$
\begin{array}{ccccccccc}
1 & \longrightarrow & P'_{q,I} & \longrightarrow & P_{q,I} & \overset{\mathcal{F}}{\longrightarrow} & P_I & \longrightarrow & 1 \\
& & \downarrow{\scriptstyle D'_q} & & \downarrow{\scriptstyle D_q} & & \downarrow{\scriptstyle D} & & \\
1 & \longrightarrow & (_\ell G_m)^s & \longrightarrow & (G_m)^s & \overset{\ell\text{-th power}}{\longrightarrow} & (G_m)^s & \longrightarrow & 1
\end{array}
$$

where $P'_{q,I} = P_{q,I} \cap \mathrm{GL}'_q(n)$ and D'_q denotes the map induced by D_q.

Let G be a finite quantum subgroup of $\mathrm{GL}_q(n)$. The image $\mathcal{F}(G)$ which is a finite algebraic subgroup of $\mathrm{GL}(n)$ is identified with a finite (abstract) subgroup Γ of $\mathrm{GL}_n(k)$, since k is algebraically closed of characteristic 0. The exact sequence (14–1) induces an exact sequence of finite quantum groups:

$$1 \to G' \to G \overset{\mathcal{F}}{\to} \Gamma \to 1 \tag{14–2}$$

The following are key results of Müller.

PROPOSITION 14.3. *If G' is a quantum subgroup of $\mathrm{GL}'_q(n)$, there are a subset I of I_0 and a quotient group Q' of $(_\ell G_m)^s$ such that*

$$G' = \mathrm{Ker}\Big(P'_{q,I} \overset{D'_q}{\longrightarrow} Q'\Big).$$

PROPOSITION 14.4. *Let G be a finite quantum subgroup of $\mathrm{GL}_q(n)$ and let I be a subset of I_0. Assume $\ell > n^2/4$. If $G' \subset P'_{q,I}$, then $G \subset P_{q,I}$. In particular, we have $\Gamma \subset P_I(k)$.*

If we are in this situation, we have the following commutative diagram of quantum groups with exact rows:

$$
\begin{array}{ccccccccc}
1 & \longrightarrow & G' & \longrightarrow & G & \overset{\mathcal{F}}{\longrightarrow} & \Gamma & \longrightarrow & 1 \\
& & \cap & & \cap & & \cap & & \\
1 & \longrightarrow & P'_{q,I} & \longrightarrow & P_{q,I} & \overset{\mathcal{F}}{\longrightarrow} & P_I & \longrightarrow & 1 \\
& & \downarrow{\scriptstyle D'_q} & & \downarrow{\scriptstyle D_q} & & \downarrow{\scriptstyle D} & & \\
1 & \longrightarrow & (_\ell G_m)^s & \longrightarrow & (G_m)^s & \overset{\ell\text{-th power}}{\longrightarrow} & (G_m)^s & \longrightarrow & 1 \\
& & \downarrow & & \downarrow{\scriptstyle \text{push out}} & & \| & & \\
1 & \longrightarrow & Q' & \longrightarrow & Q & \longrightarrow & (G_m)^s & \longrightarrow & 1
\end{array}
\tag{14–3}
$$

By an easy diagram-chasing, we conclude there is a homomorphism of (abstract) groups $\alpha_G \colon \Gamma \to Q(k)$ such that the diagram

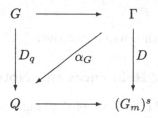

commutes.

Conversely, consider a set of data as follows:

$I \subset I_0$
Q' a quotient group of $(_\ell G_m)^s$,
G' as in Proposition 14.3
$\Gamma \subset P_I(k)$ a subgroup,
Q as in (14–3),
$\alpha \colon \Gamma \to Q(k)$ a group homomorphism such that

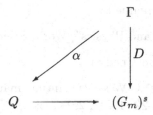

commutes.

We have the following main result.

THEOREM 14.5 (E. Müller). *With the set of data above, let*

$$G = \mathrm{Ker}\left(\mathcal{F}^{-1}(\Gamma) \cap P_{q,I} \underset{D_q}{\overset{\alpha\mathcal{F}}{\rightrightarrows}} Q\right).$$

Then G is a finite quantum subgroup of $\mathrm{GL}_q(n)$ which fits the exact sequence (14–2). If $\ell > n^2/4$, then these G for all possible previous sets of data exhaust all finite quantum subgroups of $\mathrm{GL}_q(n)$.

Finally, we mention the following result of E. Letzter concerning $\mathrm{Spec}\,\mathcal{O}_q(\mathrm{GL}(n))$, the set of prime ideals of the non-commutative ring $\mathcal{O}_q(\mathrm{GL}(n))$. We assume q is a root of unity of odd order ℓ. Multiplication of a row or a column of a q-matrix by a constant yields a q-matrix. Considering multiplication of all rows and columns of the generating q-matrix by q, one obtains a group action of $(\mathbb{Z}/\ell)^{2n-1}$ on $\mathcal{O}_q(\mathrm{GL}(n))$ as ring automorphisms. The image of $\mathcal{O}(\mathcal{F})$ is contained in the invariants by this action. Hence the Frobenius map $\mathcal{F}$ induces a map

$$(\mathrm{Spec}\,\mathcal{O}_q(\mathrm{GL}(n)))/(\mathbb{Z}/\ell)^{2n-1} \;\to\; \mathrm{Spec}\,\mathcal{O}(\mathrm{GL}(n))$$
$$P \qquad\qquad \mapsto \qquad \mathcal{O}(\mathcal{F})^{-1}(P).$$

E. Letzter has shown that this map is bijective.

15. References and Notes

For the general concepts of the theory of Hopf algebras, see:

- S. Montgomery, "Hopf algebras and their actions on rings", CBMS Reg. Conf. Ser. in Math. 82 (1993), AMS.
- M. Sweedler, "Hopf algebras", Benjamin, New York, 1969.

Section 1. Proposition 1.3 is due to Umeda and Wakayama (see below under Section 3). An elementary approach (slightly different from ours) to (2×2) q-matrices is given in

- Yu. I. Manin, "Quantum Groups and Noncommutative Geometry," CRM, Montreal, 1988.

Section 2. These results are due to

- L. H. Kauffman, "Knots and Physics", World Scientific, 1991, pp. 124–136.

Section 3. The result is contained in

- H. Ewen, O. Ogievetsky, J. Wess, "Quantum matrices in two dimensions", Letters in Math. Physics 22 (1991), 297–305;
- S. P. Vokos, B. Zumino, J. Wess, "Analysis of the basic matrix representation of $\mathrm{GL}_q(2,\mathbb{C})$", Z. Phys. C-Particles and Fields 48 (1990), 65–74.

The proof given here is due to

- T. Umeda, M. Wakayama, "Powers of (2×2) quantum matrices", Comm. Alg. 21 (1993), 4461–4465.

Section 4. The notion of $(n \times n)$ q-matrices was first introduced by Drinfeld in the following paper, although he did not use this terminology.

- V. G. Drinfeld, "Quantum groups", Proc. ICM Berkeley, 1986, pp. 798–820.

In the following paper, the celebrated construction of $A(R)$ was given and applied to define the q-function algebras $\mathrm{Func}_q(\mathrm{GL}(n,\mathbb{C}))$, $\mathrm{Func}_q(\mathrm{SL}(n,\mathbb{C}))$.

- L. D. Faddeev, N. Yu. Reshetikhin, L. A. Takhtajan, "Quantization of Lie groups and Lie algebras", Leningrad Math. J. 1 (1990), 193–225.

Around 1987, Woronowicz reached the same notion from the viewpoint of operator algebras. See

- S. L. Woronowicz, "Twisted SU(2)-group", Publ. RIMS, Kyoto Univ. 23 (1987), 117–181.

Afterwards, it has become clear gradually that the "linear algebra" can be quantized by using q-matrices. In the following paper, the FRT construction of $A(R)$ was reformulated from the coalgebraic viewpoint by using the notion of "cocentralizers" and "conormalizers", and it was applied to clarify the relations between R_q, $\bigwedge_q(V_n)$, $S_q(V_n)$ and $O_q(M(n))$.

- M. Takeuchi, "Matric bialgebras and quantum groups", Israel J. Math. 72 (1990), 232–251.

The q-linear algebra including the Laplace expansion is discussed thoroughly in the following two articles.

- E. Taft, J. Towber, "Quantum deformations of flag schemes and Grassman schemes I", J. Alg. 142 (1991), 1–36;
- B. Parshall, J.-P. Wang, "Quantum linear groups", Memoirs of the AMS 439 (1991).

Section 5. The result is due to

- J. Zhang, "The quantum Cayley–Hamilton theorem", preprint, 1991.

Section 6. The results of this section are from

- L. H. Kauffman, "Knots and Physics", World Scientific, 1991, pp. 161–173.

Section 7. These results are from

- M. Takeuchi, "Hopf algebra techniques applied to the quantum group $U_q(\mathbf{sl_2})$" Contemp. Math. 134 (1992), 309–323.

Section 8. For Proposition 8.3 refer to § 5, especially to Prop. 5.3, Thm. 5.4 of

- M. Takeuchi, "q-Representations of quantum groups", Canad. Math. Soc. Conf. Proc. 16 (1995), 347–385.

Section 9. The paper just noted, in which a morphism $\varrho : G \to \mathrm{GL}_q(n)$ is called a q-representation of the quantum group G, determines the q-representations, the one-representations of $\mathrm{SL}_q(n)$, and also the automorphisms of $\mathrm{GL}_q(n)$ and of $\mathrm{SL}_q(n)$.

What is stated in the text is part of these results, namely the results in the case where q is not a root of unity. To discuss the general case, we have to extend the duality of $U_q(\mathbf{sl_n})$ stated in Section 7 to that of the Lusztig form $\tilde{U}_q(\mathbf{sl_n})$. This technical device is investigated thoroughly in the next papers.

- M. Takeuchi, "Some topics on $\mathrm{GL}_q(n)$", J. Alg. 147 (1992), 379–410;
- M. Takeuchi, "The quantum hyperalgebra of $\mathrm{SL}_q(2)$", Proc. Symp. Pure Math. 56(2) (1994), 121–134.

Section 10. The concept of (α, β)-matrices is due to

- M. Takeuchi, "A two-parameter quantization of GL(n)", Proc. Japan Acad. 66, Ser. A (1990), 112–114.

Quantization with more parameters is also investigated by many mathematicians.

Section 11. The q-Schur algebra was introduced in

- R. Dipper, G. James, "The q-Schur algebra", Proc. London Math. Soc. (3) 59 (1989), 23–50.

Its relations with quantum groups have been clarified in

- R. Dipper, S. Donkin, "Quantum GL$_n$", Proc. London Math. Soc. (3) 63 (1991), 165–211.

The last remark of this section is due to

- Du, B. Parshall, J.-P. Wang, "Two-parameter quantum linear groups and the hyperbolic invariance of q-Schur algebras", J. London Math. Soc. (2) 44 (1991), 420–436.

Dipper and James have shown that there are interesting relations between the representation of the q-Schur algebra, where q is a power of a prime, and those of GL$_n(\mathbb{F}_q)$. Although their arguments are very complicated, they can be simplified considerably if we discuss only the unipotent representations.

For this result, refer to

- M. Takeuchi, "Relations of representations of quantum groups and finite groups", Advances in Hopf algebras, L.N. pure and appl. math. Vol. 158, pp. 319–326, Marcel Dekker, 1994;
- M. Takeuchi, "The group ring of GL$_n(\mathbb{F}_q)$ and the q-Schur algebra", J. Math. Soc. Japan 48 (1996), 259–274.

Section 12. The cocycle deformation A^σ and the construction of $M(C, \sigma)$ were introduced in

- Y. Doi, "Braided bialgebras and quadratic bialgebras", Com. Alg. 21 (1993), 1731–1749.

As the dual notion of quasi-triangular bialgebras due to Drinfeld, the braided bialgebra was introduced, for example, in

- T. Hayashi, "Quantum groups and quantum determinants", J. Alg. 152 (1992), 146–165;
- R. G. Larson, J. Towber, "Two dual classes of bialgebras related to the concepts of "quantum groups" and "quantum Liealgebra"", Com. Alg. 19 (1991), 3295–3345.

The results stated in this section are taken from

• M. Takeuchi, "Cocycle deformations of coordinate rings of quantum matrices",
  J. Alg. 189 (1997), 23–33.

Section 13. The results of this section are taken from

• L. H. Kauffman, "Knots and Physics", World Scientific, 1991, pp. 316;
• M. Takeuchi and D. Tambara, "A new one-parameter family of (2×2) matrix
  bialgebras", Hokkaido Math. J. 21 (1992), 405–419;
• Satoshi Suzuki, "A family of braided cosemisimple Hopf algebras of finite
  dimension", Tsukuba J. Math. 22 (1998), 1–30

Section 14. The results of this section come from

• E. Müller, "Finite subgroups of the quantum general linear group", Proc.
  London Math. Soc. (3) 81 (2000), 190–210;
• E. S. Letzter, "On the quantum Frobenius map for general linear groups", J.
  Algebra 179 (1996), 115–126.

MITSUHIRO TAKEUCHI
NOTES TAKEN BY BERND STRÜBER
INSTITUTE OF MATHEMATICS
UNIVERSITY OF TSUKUBA
TSUKUBA-SHI IBARAKI, 305-8571 JAPAN

New Directions in Hopf Algebras
MSRI Publications
Volume 43, 2002

The Brauer Group of a Hopf Algebra

FREDDY VAN OYSTAEYEN AND YINHUO ZHANG

ABSTRACT. Let H be a Hopf algebra with a bijective antipode over a
commutative ring k with unit. The Brauer group of H is defined as the
Brauer group of Yetter–Drinfel'd H-module algebras, which generalizes the
Brauer–Long group of a commutative and cocommutative Hopf algebra and
those known Brauer groups of structured algebras.

Introduction

The Brauer group is something like a mathematical chameleon, it assumes the
characteristics of its environment. For example, if you look at it from the point of
view of representation theory you seem to be dealing with classes of noncommutative algebras appearing in the representation theory of finite groups, a purely
group theoretical point of view presents it as the second Galois-cohomology
group, over number fields it becomes an arithmetical tool related to the local
theory via complete fields, over an algebraic function field or some coordinate
rings it gets a distinctive geometric meaning and category theoretical aspects
are put in evidence when relating the Brauer group to K-theory, in particular
the K_2-group. When looking at the vast body of theory existing for the Brauer
group one cannot escape to note the central role very often played by group
actions and group gradings. This is most evident for example in the appearance
of crossed products or generalizations of these.

Another typical case is presented by Clifford algebras and the $\mathbb{Z}_2$ (i.e., $\mathbb{Z}/2\mathbb{Z}$)
graded theory contained in the study of the well-known Brauer–Wall group [57],
as well as the generalized Clifford algebras in the Brauer–Long group for an
abelian group [29]. At that point the theory was ripe for an approach via Hopf
algebras where certain actions and co-actions (like the grading by a group) may
be adequately combined in one unifying theory [30], but for commutative cocommutative Hopf algebras only. However, the cohomological interpretation for such

This work was supported by UIA and the URC of USP.

Brauer–Long groups presented some technical problems that probably slowed down the development of a general theory. The cohomological description was obtained years later by S. Caenepeel a.o. [6; 7; 8; 9] prompted by new interest in the matter stemming from earlier work of Van Oystaeyen and Caenepeel, Van Oystaeyen on another type of graded Brauer group. The problem of considering noncommutative noncocommutative Hopf algebras remained and became more fascinating because of the growing interest in quantum groups. The present authors then defined and studied the Brauer group of a quantum group first in terms of the category of Yetter–Drinfel'd modules, but quickly generalized it to the Brauer group of a braided category [53], thus arriving at the final generality one would hope for after [38]. The Brauer group of a quantum group or even of a general Hopf algebra presents us with an interesting new invariant but a warning is in place. Not only is this group non-abelian, it is even non-torsion in general! Even restriction to cohomology describable or split parts does not reduce the complexity much. On the other hand, at least for finite dimensional Hopf algebras explicit calculations should be possible. Note that even the case of the Brauer group of the group ring of a non-abelian group is a very new and interesting object. Recently concrete calculations have been finalized for Sweedler's four dimensional Hopf algebra, group rings of dihedral groups and a few more low dimensional examples [16; 54; 56].

The arrangement of this paper is as follows:

(i)   Basic notions and conventions
(ii) Quaternion algebras
(iii) The definition of the Brauer group
(iv) An exact sequence for the Brauer group $BC(k, H, R)$
(v)   The Hopf automorphism group
(vi) The second Brauer group

We do not repeat here a survey of main results because the paper is itself an expository paper albeit somewhat enriched by new results at places. We have adopted a very constructive approach starting with a concrete treatment of actions and coactions on quaternion algebras (Section 2), so that the abstractness of the definition in Section 3 is well-motivated and is made look natural. We shall not include the Brauer–Long group theory in this paper as the reader may find a comprehensive introduction in the book [6].

1. Basic Notions and Conventions

Throughout k is a commutative ring with unit unless it is specified and $(H, \Delta, \varepsilon, S)$ or simply H is a Hopf algebra over k where (H, Δ, ε) is the underlying coalgebra and S is a bijective antipode. Since the antipode S is bijective, the opposite H^{op} and the co-opposite H^{cop} are again Hopf algebras with antipode

S^{-1}. We will often use Sweedler's sigma notation; for example, we will write, for $h \in H$,

i. $\qquad \Delta h = \sum h_{(1)} \otimes h_{(2)},$

ii. $\qquad (1 \otimes \Delta)\Delta h = (\Delta \otimes 1)\Delta h = \sum h_{(1)} \otimes h_{(2)} \otimes h_{(3)}$

etc. For more detail concerning the theory of Hopf algebras we refer to [1; 35; 47].

1.1. Dimodules and Yetter–Drinfel'd modules. A k-module is said to be of *finite type* if it is finitely generated projective. If a k-module M of finite type is faithful, then M is said to be *faithfully projective*

Let H be a Hopf algebra. We will take from [47] the theory of H-modules and H-comodules for granted. Sigma notations such as $\sum m_{(0)} \otimes m_{(1)}$ for the comodule structure $\chi(m)$ of an element m of a left H-comodule M will be adapted from [47]. In this paper we will use χ for comodule structures over Hopf algebras and use ρ for comodule structures over coalgebras in order to distinguish two comodule structures when they happen to be together.

Write $_H\mathbf{M}$ for the category of left H-modules and H-module morphisms. If M and N are H-modules the diagonal H-module structure on $M \otimes N$ and the adjoint H-module structure on $\text{Hom}(M, N)$ are given by:

i. $\qquad h \cdot (m \otimes n) = \sum h_{(1)} \cdot m \otimes h_{(2)} \cdot n,$

ii. $\qquad (h \cdot f)(m) = \sum h_{(1)} \cdot f(S(h_{(2)}) \cdot m),$

for $h \in H, n \in N, f : M \longrightarrow N$. The category $_H\mathbf{M}$ together with the tensor product and the trivial H-module k forms a *monoidal category* (see [31]). If M is left H-module, we have a k-module of invariants

$$M^H = \{m \in M \mid h \cdot m = \varepsilon(h)m.\}$$

In a dual way, we have a monoidal category of right H-comodules, denoted $(\mathbf{M}^H, \otimes, k)$ or simply $\mathbf{M}^H$. For instance, if M and N are two right H-comodules, the codiagonal H-comodule structure on $M \otimes N$ is given by

$$\chi(m \otimes n) = \sum m_{(0)} \otimes n_{(0)} \otimes m_{(1)}n_{(1)}$$

for $m \in M$ and $n \in N$. If an H-comodule M is of finite type, then $\text{Hom}(M, N) \cong N \otimes M^*$ has a comodule structure:

$$\chi(f)(m) = \sum f(m_{(0)})_{(0)} \otimes f(m_{(0)})_{(1)}S(m_{(1)})$$

for $f \in \text{Hom}(M, N)$ and $m \in M$. For a right H-comodule M the k-module

$$M^{coH} = \{m \in M \mid \chi(m) = m \otimes 1\}$$

is called the coinvariant submodule of M.

A k-module M which is both an H-module and an H-comodule is called an H-*dimodule* if the action and the coaction of H commute, that is, for all $m \in M, h \in H$,

$$\sum (h \cdot m)_{(0)} \otimes (h \cdot m)_{(1)} = \sum h \cdot m_{(0)} \otimes m_{(1)}.$$

Write $\mathbf{D}^H$ for the category of H-dimodules. When the Hopf algebra H is finite, we obtain equivalences of categories

$$\mathbf{M}^{H \otimes H^*} \cong \mathbf{D}^H \cong_{H^* \otimes H} \mathbf{M};$$

here $H \otimes H^*$ and $H^* \otimes H$ are tensor Hopf algebras. For more details on dimodules, we refer to [29; 30].

Recall that a Yetter–Drinfel'd H-module (simply a YD H-module) M is a left crossed H-bimodule [58]. That is, M is a k-module which is a left H-module and a right H-comodule satisfying the following equivalent compatibility conditions [27, 5.1.1]:

i. $\qquad \sum h_{(1)} \cdot m_{(0)} \otimes h_{(2)} m_{(1)} = \sum (h_{(2)} \cdot m)_{(0)} \otimes (h_{(2)} \cdot m)_{(1)} h_{(1)}$

ii. $\qquad \chi(h \cdot m) = \sum (h_{(2)} \cdot m_{(0)}) \otimes h_{(3)} m_{(1)} S^{-1}(h_{(1)})$.

Denote by Q^H ($_H Y D^H$ in several references) the category of YD H-modules and YD H-module morphisms. For two YD H-modules M and N, the diagonal H-module structure and the codiagonal H^{op}-comodule structure on tensor product $M \otimes N$ satisfy the compatibility conditions of a YD H-module. So $M \otimes N$ is a YD H-module, denoted $M \tilde{\otimes} N$. It is easy to see that the natural map

$$\Gamma : (X \tilde{\otimes} Y) \tilde{\otimes} Z \longrightarrow X \tilde{\otimes} (Y \tilde{\otimes} Z)$$

is a YD H-module isomorphism, and the trivial YD H-module k is a unit with respect to $\tilde{\otimes}$. Therefore $(Q^H, \tilde{\otimes}, k)$ forms a monoidal category (for details concerning monoidal categories we refer to [31; 58]).

Let M and N be YD H-modules. Then there exists a YD H-module isomorphism Ψ between $M \tilde{\otimes} N$ and $N \tilde{\otimes} M$:

$$\Psi : M \tilde{\otimes} N \longrightarrow N \tilde{\otimes} M, m \tilde{\otimes} n \mapsto \sum n_{(0)} \tilde{\otimes} n_{(1)} \cdot m$$

with inverse $\Psi^{-1}(n \tilde{\otimes} m) = \sum S(n_{(1)}) \cdot m \tilde{\otimes} n_{(0)}$. It is not hard to check that $(Q^H, \tilde{\otimes}, \Gamma, \Psi, k)$ is a braided monoidal (or quasitensor) category (see [31; 58]). If in addition, H is a finite Hopf algebra, then there is a category equivalence:

$$_{D(H)} \mathbf{M} \sim Q^H$$

where $D(H)$ is the Drinfel'd double $(H^{\mathrm{op}})^* \bowtie H$ which is a finite quasitriangular Hopf algebra over k as described in [23; 32; 40].

1.2. H-dimodule and YD H-module algebras. An algebra A is a (left) H-module algebra if there is a measuring action of H on A, i.e., for $h \in H, a, b \in A$,

i. $\qquad A$ is a left H-module,

ii. $\qquad h \cdot (ab) = \sum (h_{(1)} \cdot a)(h_{(2)} \cdot b)$,

iii. $\qquad h \cdot 1 = \varepsilon(h)1$.

Similarly, an algebra is called a (right) H-comodule algebra if A is a right H-comodule with the comodule structure $\chi : A \longrightarrow A \otimes H$ being an algebra map, i.e., for $a, b \in A$,

i. $\quad \chi(ab) = \sum a_{(0)} b_{(0)} \otimes a_{(1)} b_{(1)},$

ii. $\quad \chi(1) = 1 \otimes 1.$

An *H-dimodule algebra* A is an H-dimodule and a k-algebra which is both an H-module algebra and an H-comodule algebra. Suppose that H is both commutative and cocommutative. Let A and B be two H-dimodule algebras. The smash product $A\#B$ is defined as follows: $A\#B = A \otimes B$ as a k-module and the multiplication is given by

$$(a\#b)(c\#d) = \sum a(b_{(1)} \cdot c)\#b_{(0)}d.$$

Then $A\#B$ furnished with the diagonal H-module structure and codiagonal comodule structure $A \otimes B$ is again an H-dimodule algebra.

The *H-opposite* $\bar{A}$ of an H-dimodule algebra A is equal to A as an H-dimodule, but with multiplication given by

$$\bar{a} \cdot \bar{b} = \overline{\sum (a_{(1)} \cdot b)a_{(0)}}$$

which is again an H-dimodule algebra.

A *Yetter–Drinfel'd H-module algebra* A is a YD H-module and a k-algebra which is a left H-module algebra and a right H^{op}-comodule algebra. Note that here we replace H by H^{op} when we deal with *comodule algebra structures*.

As examples pointed out in [11], $(H^{op}, \Delta, \mathrm{ad}')$ and (H, χ, ad) are regular Yetter–Drinfel'd H-module algebras with H-structures defined as follows:

$$h \, \mathrm{ad}' \, x = \sum h_{(2)} x S^{-1}(h_{(1)})$$

$$\Delta(x) = \sum x_{(1)} \otimes x_{(2)} \, .$$

$$h \, \mathrm{ad} \, x = \sum h_{(1)} x S(h_{(2)})$$

$$\chi(x) = \sum x_{(2)} \otimes S^{-1}(x_{(1)}).$$

Let A and B be two YD H-module algebras. We may define a *braided product*, still denoted $\#$, on the YD H-module $A \tilde{\otimes} B$:

$$(a\#b)(c\#d) = \sum ac_{(0)}\#(c_{(1)} \cdot b)d \qquad (1\text{--}1)$$

for $a, c \in A$ and $b, d \in B$. The braided product $\#$ makes $A\#B$ a left H-module algebra and a right H^{op}-comodule algebra so that $A\#B$ is a YD H-module algebra. Note that the braided product $\#$ is associative.

Now let A be a YD H-module algebra. The *H-opposite algebra* $\bar{A}$ of A is the YD H-module algebra defined as follows: $\bar{A}$ equals A as a YD H-module, with multiplication given by the formula

$$\bar{a} \circ \bar{b} = \sum \overline{b_{(0)}(b_{(1)} \cdot a)}$$

for all $\bar{a}, \bar{b} \in \bar{A}$. In case the antipode of H is of order two, $\bar{A}$ is equal to A as a YD H-module algebra.

Let M be a YD H-module such that M is of finite type. The endomorphism algebra $\mathrm{End}_k(M)$ is a YD H-module algebra with the H-structures induced by those of M, i.e., for $h \in H$, $f \in \mathrm{End}_k(M)$ and $m \in M$,

$$
\begin{aligned}
(h \cdot f)(m) &= \sum h_{(1)} \cdot f(S(h_{(2)}) \cdot m), \\
\chi(f)(m) &= \sum f(m_{(0)})_{(0)} \otimes S^{-1}(m_{(1)}) f(m_{(0)})_{(1)}.
\end{aligned}
\tag{1-2}
$$

Recall from [11, 4.2] that the H-opposite of $\mathrm{End}_k(M)$ is isomorphic as an YD H-algebra to $\mathrm{End}_k(M)^{\mathrm{op}}$, where the latter has YD H-module structure given by

$$
\begin{aligned}
(h \cdot f)(m) &= \sum h_{(2)} \cdot f(S^{-1}(h_{(1)}) \cdot m), \\
\chi(f)(m) &= \sum f(m_{(0)})_{(0)} \otimes f(m_{(0)})_{(1)} S(m_{(0)})
\end{aligned}
\tag{1-3}
$$

for $m \in M$, $h \in H$ and $f \in \mathrm{End}_k(M)$.

1.3. Quasitriangular and coquasitriangular Hopf algebras. A *quasitriangular Hopf algebra* is a pair $(H, \mathcal{R})$, where H is a Hopf algebra with an invertible element $\mathcal{R} = \sum R^{(1)} \otimes R^{(2)} \in H \otimes H$ satisfying the following axioms $(r = \mathcal{R})$:

(QT1) $\sum \Delta(R^{(1)}) \otimes R^{(2)} = \sum R^{(1)} \otimes r^{(1)} \otimes R^{(2)} r^{(2)}$,
(QT2) $\sum \varepsilon(R^{(1)}) R^{(2)} = 1$,
(QT3) $\sum R^{(1)} \otimes \Delta(R^{(2)}) = \sum R^{(1)} r^{(1)} \otimes r^{(2)} \otimes R^{(2)}$,
(QT4) $\sum R^{(1)} \varepsilon(R^{(2)}) = 1$,
(QT5) $\Delta^{\mathrm{cop}}(h) \mathcal{R} = \mathcal{R} \Delta(h)$,

where $\Delta^{\mathrm{cop}} = \tau \Delta$ is the comultiplication of the Hopf algebra H^{cop} and τ is the switch map.

Now let M be a left H-module. It is well-known that there is an induced H-comodule structure on M as follows:

$$
\chi(m) = \sum R^{(2)} \cdot m \otimes R^{(1)}
\tag{1-4}
$$

for $m \in M$ such that the left H-module M together with (1-4) is a YD H-module. When M is a left H-module algebra, then (1-4) makes M into a right H^{op}-comodule algebra and hence a YD H-module algebra. It is easy to see that $\mathrm{Hom}_H(M, N) = \mathrm{Hom}_H^H(M, N)$ for any two YD H-modules M, N with comodule structures (1-4) stemming from the left module structures. Thus the category $_H\mathbf{M}$ of left H-modules and H-morphisms can be embedded into the category $\mathcal{Q}^H$ as a full subcategory, which we denote by $_H\mathbf{M}^{\mathcal{R}}$. Moreover $_H\mathbf{M}^{\mathcal{R}}$ is a braided monoidal subcategory of $\mathcal{Q}^H$ since the tensor product is closed in $_H\mathbf{M}$ and the braiding Ψ of $\mathcal{Q}^H$ restricts to the braiding of $_H\mathbf{M}^{\mathcal{R}}$ which is nothing but $\Psi_{\mathcal{R}}$ induced by the $\mathcal{R}$-matrix:

$$
\Psi_{\mathcal{R}}(m \otimes n) = \sum R^{(2)} \cdot n \otimes R^{(1)} \cdot m
$$

where $m \in M, n \in N$ and $M, N \in {}_H\mathbf{M}^{\mathcal{R}}$.

A *coquasitriangular Hopf algebra* is a pair (H, R), where H is a Hopf algebra and $R \in (H \otimes H)^*$ is a convolution invertible element and satisfies the following axioms:

(CQT1) $R(h \otimes 1) = R(1 \otimes h) = \varepsilon(h)1_H$,
(CQT2) $R(ab \otimes c) = \sum R(a \otimes c_{(1)})R(b \otimes c_{(2)})$,
(CQT3) $R(a \otimes bc) = \sum R(a_{(1)} \otimes c)R(a_{(2)} \otimes b)$,
(CQT4) $\sum b_{(1)}a_{(1)}R(a_{(2)} \otimes b_{(2)}) = \sum R(a_{(1)} \otimes b_{(1)})a_{(2)}b_{(2)}$.

Let M be a right H-comodule. There is an induced left H-module structure on M given by

$$h \triangleright_1 a = \sum a_{(0)} R(h \otimes a_{(1)}) \tag{1-5}$$

for all $a \in A, h \in H$, such that M is a YD H-module. The right H-comodule category $\mathbf{M}^H$ can be embedded into $\mathcal{Q}^H$ as a full braided monoidal subcategory. We denote by $\mathbf{M}_R^H$ the braided subcategory of $\mathcal{Q}^H$.

It is easy to check that an H^{op}-comodule algebra A with the H-module structure described in (1-5) is a YD H-module algebra.

2. Quaternion Algebras

Let k be a field. Quaternion algebras play a very important role in the study of the Brauer group $\mathrm{Br}(k)$ of k. On the other hand, quaternion algebras also represent elements in the Brauer–Wall group $\mathrm{BW}(k)$ of $\mathbb{Z}_2$-graded algebras. The natural $\mathbb{Z}_2$-gradings of quaternion algebras are obtained from certain involutions related to the canonical quadratic forms of quaternion algebras. However, one may find that the same quaternion algebra will represent two different elements in $\mathrm{BW}(k)$. When one turns to the Brauer–Long group $\mathrm{BD}(k, \mathbb{Z}_2)$ of $\mathbb{Z}_2$-dimodule algebras where actions of $\mathbb{Z}_2$ commute with the $\mathbb{Z}_2$-gradings, the quaternion algebras now represent four different elements of order two. Now if we add a differential on the $\mathbb{Z}_2$-graded algebras such that they become differential superalgebras, we may form the Brauer group of differential superalgebras, and the quaternion algebras are now differential superalgebras. If we mimic the process used by C.T.C. Wall, we obtain a Brauer group $\mathrm{BDS}(k)$ of differential superalgebras. A new interesting fact now shows, i.e., a quaternion algebra may represent an element of infinite order in $\mathrm{BDS}(k)$. As a consequence, the Brauer group $\mathrm{BDS}(k)$ is a non-torsion infinite group if k has characteristic zero.

Recall the definition of a quaternion algebra. For $\alpha, \beta \in k^{\bullet} = k \backslash 0$, define a 4-dimensional algebra with basis $\{1, u, v, w\}$ by the following multiplication table:

$$uv = w, \quad u^2 = \alpha 1, \quad v^2 = \beta 1, \quad vu = -w.$$

Here 1 denotes the unit. We denote this algebra by $\left(\frac{\alpha, \beta}{k}\right)$. The elements in the subspace $ku + kv + kw$ are called *pure quaternions*. The subspace of pure quaternions is independent of the choice of standard basis and is determined by the algebra structure of $\left(\frac{\alpha,\beta}{k}\right)$.

There exists a canonical linear involution given by

$$^{-} : \left(\frac{\alpha,\beta}{k}\right) \longrightarrow \left(\frac{\alpha,\beta}{k}\right), \quad \overline{x} = \overline{x_0 + x_1} = x_0 - x_1$$

where $x_0 \in k$ and $x_1 \in ku + kv + kw$. It follows that $\left(\frac{\alpha,\beta}{k}\right)$ is isomorphic to its opposite algebra $\left(\frac{\alpha,\beta}{k}\right)^{\mathrm{op}}$. One may easily calculate that the center of $\left(\frac{\alpha,\beta}{k}\right)$ is k and that $\left(\frac{\alpha,\beta}{k}\right)$ has no proper ideals except $\{0\}$. An algebra is called a *central simple algebra* if its center is canonically isomorphic to k and it has no proper non-zero ideals. Any $n \times n$-matrix algebra $M_n(k)$ is a central simple algebra. The opposite algebra of a central simple algebra is obviously a central simple algebra. The tensor product of two central simple algebras is still a central simple algebra. There are several characterizations of a central simple algebra [19; 39]:

PROPOSITION 2.1. *Let A be a finite dimensional algebra over a field k. The following are equivalent:*

(1) *A is a central simple algebra.*
(2) *A is a central separable algebra (here A is separable if mult : $A \otimes A \longrightarrow A$ splits as an A-bimodule map).*
(3) *A is isomorphic to a matrix algebra $M_n(D)$ over a skew field D where the center of D is k.*
(4) *The canonical linear algebra map can : $A \otimes A^{\mathrm{op}} \longrightarrow \mathrm{End}(A)$ given by $\mathrm{can}(a \otimes b)(c) = acb$ for $a, b, c \in A$ is an isomorphism.*

A finite dimensional algebra satisfying one of the above equivalent conditions is called an *Azumaya algebra*. Let $B(k)$ be the set of all isomorphism classes of Azumaya algebras. Then $B(k)$ is a semigroup with the multiplication induced by the tensor product and with the unit represented by the one dimensional algebra k.

Define an equivalence relation $\sim$ on $B(k)$ as follows: Two central simple algebras A and B are equivalent, denoted $A \sim B$, if there are two positive integers m and n such that

$$A \otimes M_n(k) \cong B \otimes M_m(k)$$

as algebras. Then the quotient set of $B(k)$ modulo the equivalence relation $\sim$ is a group and is called the *Brauer group* of k, denoted $\mathrm{Br}(k)$.

The Brauer group $\mathrm{Br}(k)$ can be defined more intuitively as the quotient $B(k)/M(k)$, where $M(k)$ is a sub-semigroup generated by the isomorphism classes of matrix algebras over k. If $[A]$ is an element in $\mathrm{Br}(k)$ represented by a central simple algebra A, then the inverse $[A]^{-1}$ is represented by the opposite algebra A^{op} because $A \otimes A^{\mathrm{op}}$ is isomorphic to a matrix algebra. The Brauer group $\mathrm{Br}(k)$ can be generalized to the Brauer group of a commutative ring by making use of the equivalent condition (2) or (4) of Proposition 2.1. That is, an Azumaya algebra A over a commutative ring is a faithfully projective algebra such that the condition (2) or (4) of Proposition.2.1 holds. One may refer to [2; 19] for the details on the Brauer group of a commutative ring. However, in this section we restrict our attention to the case where k is a field.

Let us return to the consideration of quaternion algebras. We know that a quaternion algebra is a central simple algebra and it is isomorphic to its opposite algebra due to the canonical involution map. Thus $[(\frac{\alpha,\beta}{k})]$ is an element of order not greater than two. Actually any element of order two in $\mathrm{Br}(k)$ can be represented by a tensor product of quaternion algebras (see [39]).

The quaternion algebra $(\frac{\alpha,\beta}{k})$ has a canonical $\mathbb{Z}_2$-grading defined as follows:

$$\left(\frac{\alpha,\beta}{k}\right) = A_0 + A_1, \quad A_0 = k + kw, \quad A_1 = ku + kv. \tag{2-1}$$

In [57], Wall introduced the notion of a $\mathbb{Z}_2$-*graded Azumaya algebra* which is a graded central and a graded separable algebra A in the following sense:

i. the graded center $Z_g(A) = \{a \in A \mid ab = ba_0 + b_0a_1 - b_1a_1, \forall b \in A\} = k$.

ii. A is a simple graded algebra, i.e., A has no proper non-zero graded ideals.

As in Proposition 2.1, we may replace 'graded simplicity' by 'graded separability' if the characteristic of k is different from 2. That is, condition ii can be replaced by

iii. $A \otimes A \longrightarrow A$ splits as a $\mathbb{Z}_2$-graded A-bimodule map, where the grading on $A \otimes A$ is the diagonal one.

Given two graded algebras A and B. The product $A\hat{\otimes}B$ of two graded algebras A and B is defined as follows:

$$(a\hat{\otimes}b)(c\hat{\otimes}d) = (-1)^{\partial(b)\partial(c)}ac\hat{\otimes}bd \tag{2-2}$$

where b and c are homogeneous elements and $\partial(b), \partial(c)$ are the graded degrees of b and c respectively. If A and B are graded Azumaya algebras, then the product $A\hat{\otimes}B$ is a graded Azumaya algebra. Now one may repeat the definition of $\mathrm{Br}(k)$ by adding the term '($\mathbb{Z}_2-$) graded' to obtain the Brauer group of graded algebras which is referred to as *the Brauer–Wall group*, denoted $\mathrm{BW}(k)$. Notice that in the definition of the equivalence $\sim$, the grading of any matrix algebra $M_n(k)$ must be 'good', namely, $M_n(k) \cong \mathrm{End}(M)$ as graded algebras for some n-dimensional graded module M. The Brauer–Wall group $\mathrm{BW}(k)$ can be completely described in terms of the usual Brauer group $\mathrm{Br}(k)$ and the group of graded quadratic extensions:

$$1 \longrightarrow \mathrm{Br}(k) \longrightarrow \mathrm{BW}(k) \longrightarrow Q_2(k) \longrightarrow 1$$

where $Q_2(k) = \mathbb{Z}_2 \times k^\bullet/k^{\bullet 2}$ with multiplication given by

$$(e,d)(e',d') = (e + e', (-1)^{ee'}dd')$$

for $e, e' \in \mathbb{Z}_2$ and $d, d' \in k^{\bullet}/k^{\bullet 2}$. One may write down the multiplication rule for the product $\text{Br}(k) \times Q_2(k)$ so that $\text{BW}(k)$ is isomorphic to $\text{Br}(k) \times Q_2(k)$ (for details see [18; 46]).

Again let us look at the quaternion algebras $\left(\frac{\alpha,\beta}{k}\right)$, $\alpha, \beta \in k^{\bullet}$. Let $k\langle\sqrt{\alpha}\rangle$ be the graded algebra $k \oplus ku$ and $k\langle\sqrt{\beta}\rangle$ be the graded algebra $k \oplus kv$. It is easy to show that $k\langle\sqrt{\alpha}\rangle$ is a graded Azumaya algebra and that $\left(\frac{\alpha,\beta}{k}\right)$ is isomorphic to the graded product $k\langle\sqrt{\alpha}\rangle\hat{\otimes}k\langle\sqrt{\beta}\rangle$. It follows that $\left(\frac{\alpha,\beta}{k}\right)$ is a graded Azumaya algebra. We denote by $\left\langle\frac{\alpha,\beta}{k}\right\rangle$ the graded Azumaya algebra $\left(\frac{\alpha,\beta}{k}\right)$ in order to make the difference between $\left\langle\frac{\alpha,\beta}{k}\right\rangle$ and $\left(\frac{\alpha,\beta}{k}\right)$. Since $[k\langle\sqrt{\alpha}\rangle] \in \text{BW}(k)$ is an element of order two (or one) if $\alpha \notin k^{\bullet 2}$ (or $\alpha \in k^{\bullet 2}$), $[\left(\frac{\alpha,\beta}{k}\right)]$ is of order equal or less than two. Though $\left(\frac{\alpha,\beta}{k}\right)$ and $\left\langle\frac{\alpha,\beta}{k}\right\rangle$ are the same algebra, they do represent two different elements of order two in $\text{BW}(k)$ when $\left(\frac{\alpha,\beta}{k}\right)$ is a division algebra.

Furthermore, the quaternion algebras are no longer the 'smallest' nontrivial graded Azumaya algebras in terms of dimension. Here the smallest ones are quadratic extensions of k. This prompts the idea that the more extra structures you put on algebras, the more classes of such structured Azumaya algebras you will get, and the richer the corresponding Brauer group will be. So we look again at quadratic extensions and quaternion algebras. For a quadratic extension $k\langle\sqrt{\alpha}\rangle$, there is a natural k-linear $\mathbb{Z}_2$-action on it:

$$\sigma(1) = 1, \quad \sigma(u) = -u \tag{2-3}$$

where σ is the generator of the group $\mathbb{Z}_2$ and u is the generator of the field $k\langle\sqrt{\alpha}\rangle$. It is easy to see that the action (2-3) commutes with the canonical grading on $k\langle\sqrt{\alpha}\rangle$. The action (2-3) extends to any quaternion algebra $\left(\frac{\alpha,\beta}{k}\right)$ in the way of diagonal group action. In fact, to any graded algebra $A = A_0 \oplus A_1$, one may associate a natural $\mathbb{Z}_2$-action on A as follows:

$$\sigma(a_i) = (-1)^i a_i \tag{2-4}$$

where $a_i \in A_i$ is a homogeneous element of A.

A $\mathbb{Z}_2$-graded algebra with a $\mathbb{Z}_2$-action that commutes with the grading is a $\mathbb{Z}_2$-dimodule algebra. The notion of a dimodule algebra for a finite abelian group was introduced by F. W. Long in 1972 [29], it is extended for a commutative and cocommutative Hopf algebra in [30]. Let A be a $\mathbb{Z}_2$-graded algebra. Having the canonical $\mathbb{Z}_2$-action (2-4), A is a $\mathbb{Z}_2$-dimodule algebra. The product (2-2) respects the action (2-4). In this case, we may forget the action (2-4). However, if we take any two $\mathbb{Z}_2$-dimodule algebras A and B, the graded product (2-2) may not respect actions of $\mathbb{Z}_2$. For instance, A is a graded Azumaya algebra with the action (2-4) and B is a graded Azumaya algebra with the trivial $\mathbb{Z}_2$-action (i.e., σ acts as the identity map). Both A and B are dimodule algebras, but $A\hat{\otimes}B$ is not a dimodule algebra.

In order to have a product for dimodule algebras, we have to modify the product (2-2) such that the action of $\mathbb{Z}_2$ is involved. This is the situation dealt with by F.W. Long. Let A and B be two dimodule algebras. Long defined a product $\#$ on $A \otimes B$ as follows:

$$(a\#b)(c\#d) = ac\#\sigma^{\partial(c)}(b)d \qquad (2\text{--}5)$$

where c is a homogeneous element. The product (2–5) preserves the dimodule structures, and restricts to the product (2–2) when the dimodule algebras have the canonical action (2–4). With this product (2–5) Long was able to define the Brauer group of dimodule algebras which is now referred to as the Brauer–Long group of $\mathbb{Z}_2$ and is denoted $BD(k, \mathbb{Z}_2)$. The definition of an Azumaya $\mathbb{Z}_2$-dimodule algebra is similar to the definition of a graded Azumaya algebra.

Suppose that the characteristic of the field k is different from two. A $\mathbb{Z}_2$-dimodule algebra is called an Azumaya dimodule algebra if A satisfies the following two conditions:

i. A is $\mathbb{Z}_2$-central, namely, $\{a \in A \mid ab = b\sigma^i(a), \forall b \in A_i\} = \{a \in A \mid ba = a_0 b + a_1 \sigma(b), \forall b \in A\} = k$.

ii. the multiplication map $A\#A \longrightarrow A$ splits as A-bimodule and $\mathbb{Z}_2$-dimodule map.

Note that the foregoing definition is not the original definition given by Long, but it is equivalent to that if the characteristic of k is different from two. The equivalence relation $\sim$ is defined as follows: for two Azumaya dimodule algebras A and B, $A \sim B$ if and only if there exists two finite dimensional dimodules M and N such that

$$A\#\mathrm{End}(M) \cong B\#\mathrm{End}(N)$$

as dimodule algebras. The Brauer–Long group $BD(k, \mathbb{Z}_2)$ contains the Brauer Wall-group $BW(k)$ as a subgroup.

Let us investigate the role played by quaternion algebras in $BD(k, \mathbb{Z}_2)$. If $\left(\frac{\alpha, \beta}{k}\right)$ is a quaternion algebra, then there are eight types of dimodule structures on $\left(\frac{\alpha, \beta}{k}\right)$:

(1) the trivial action and the trivial grading,
(2) the trivial action and the canonical grading (2–1),
(3) the canonical action (2–4) and the trivial grading,
(4) the action (2–4) and the grading (2–1), i.e., the dimodule structure of $\left\langle \frac{\alpha, \beta}{k} \right\rangle$,
(5) the action (2–4) and the grading $A_0 = k \oplus ku$, $A_1 = kv \oplus kw$,
(6) the grading (2–1) and the action given by $\sigma(u) = u, \sigma(v) = -v$.

If we switch the roles of u and v in (5) and (6), we will obtain two more dimodule structures on $\left(\frac{\alpha, \beta}{k}\right)$. One may take a while to check that the first four types of dimodule structures make $\left(\frac{\alpha, \beta}{k}\right)$ into $\mathbb{Z}_2$-Azumaya dimodule algebras. However, though $\left(\frac{\alpha, \beta}{k}\right)$ is an Azumaya algebra the dimodule algebra $\left(\frac{\alpha, \beta}{k}\right)$ of type five or six is not a $\mathbb{Z}_2$-Azumaya algebra because it is not $\mathbb{Z}_2$-central. For instance, the left center

$$\{a \in A \mid ab = b\sigma^i(a), \forall b \in A_i\} = k \oplus ku.$$

is not trivial in the case of type five.

Nevertheless, the $\left(\frac{\alpha,\beta}{k}\right)$ of type (1)–(4) represent four different elements of order two in $BD(k,\mathbb{Z}_2)$ when $\left(\frac{\alpha,\beta}{k}\right)$ is a division algebra. Let $\left(\frac{\alpha,\beta}{k}\right)_i$ be the $\mathbb{Z}_2$-Azumaya dimodule algebra of type (i), where $i = 1, 2, 3, 4$. Since the multiplication of the group is induced by the braided product (2–5), $[\left(\frac{\alpha,\beta}{k}\right)_1]$ commutes with $[\left(\frac{\alpha,\beta}{k}\right)_i]$, but in general $[\left(\frac{\alpha,\beta}{k}\right)_2][\left(\frac{\alpha,\beta}{k}\right)_3] \neq [\left(\frac{\alpha,\beta}{k}\right)_3][\left(\frac{\alpha,\beta}{k}\right)_2]$ (pending on k, see [18, Thm]). For example, when $k = \mathbb{R}$, the real number field, we have

$$[\mathbb{H}_2][\mathbb{H}_3] = (1,-1,1,-1)(1,1,-1,-1) = (1,-1,-1,1),$$

$$[\mathbb{H}_3][\mathbb{H}_2] = (1,1,-1,-1)(1,-1,1,-1) = (1,-1,-1,-1),$$

where $\mathbb{H} = \left(\frac{-1,-1}{k}\right)$ and $BD(\mathbb{R},\mathbb{Z}_2) = \mathbb{Z}_2 \times \mathbb{Z}_2 \times \mathbb{Z}_2 \times Br(\mathbb{R})$ ($Br(\mathbb{R}) = \mathbb{Z}_2$) with multiplication rules given by [18, Thm] or [7, 13.12.14]. In fact, $BD(\mathbb{R},\mathbb{Z}_2) \cong D_8$, the dihedral group of 16 elements and $BW(\mathbb{R}) \cong \mathbb{Z}_8$, the cyclic group of 8 elements (see [29] or [7, 13.12.15]). Thus $BD(k,\mathbb{Z}_2)$ may not be an abelian group though $BW(k)$ is an abelian group. Nonetheless, both $BW(k)$ and $BD(k,\mathbb{Z}_2)$ are torsion groups (see [18]). This will not be the case for the Brauer group of differential superalgebras introduced hereafter.

For convenience we call a $\mathbb{Z}_2$-graded algebra a *super* algebra. Let $A = A_0 \oplus A_1$ be a superalgebra. A linear endomorphism δ of A is called a *super-derivation* of A if δ is a degree one graded endomorphism and satisfies the following condition:

$$\delta(ab) = a\delta(b) + (-1)^{\partial(b)}\delta(a)b$$

where b is homogeneous and a is arbitrary. A super-derivation δ is called a *differential* if $\delta^2 = 0$.

A graded algebra A with a differential δ is called a *differential superalgebra* (simply DS algebra), denoted (A,δ) or just A if there is no confusion. Two DS algebras (A,δ_A) and (B,δ_B) can be multiplied by means of the graded product (2–2). So we obtain a new DS algebra $(A\hat{\otimes}B, \delta_{A\hat{\otimes}B})$, where $\delta_{A\hat{\otimes}B}$ is given by

$$\delta_{A\hat{\otimes}B}(a\hat{\otimes}b) = a\hat{\otimes}\delta_B(b) + (-1)^i\delta_A(a)\hat{\otimes}b$$

for $a \in A$ and $b \in B_i$.

Let M be a graded module. M is called a *differential graded module* if there exists a degree one graded linear endomorphism δ_M of M such that $\delta_M^2 = 0$ (in the sequel δ_M will be called a *differential on M*). The endomorphism ring $End(M)$ is a DS algebra. The grading on $End(M)$ is the induced grading and the differential δ on $End(M)$ is induced by δ_M, namely,

$$\delta(f)(m) = f(\delta_M(m)) + (-1)^{\partial(m)}\delta(f(m)) \tag{2–6}$$

for any homogeneous element $m \in M$ and $f \in End(M)$.

Now let A be the graded Azumaya algebra $\left\langle\frac{\alpha,\beta}{k}\right\rangle$. There is a natural differential on A given by Doi and Takeuchi [22]:

$$\delta(1) = \delta(u) = 0, \quad \delta(v) = 1, \quad \delta(w) = u. \tag{2–7}$$

So any quaternion algebra is a DS algebra. As mentioned before, the graded Azumaya algebra $\left\langle \frac{\alpha,\beta}{k} \right\rangle$ represents an element of order not greater than two in $BW(k)$. This means that the product graded Azumaya algebra $A\hat{\otimes}A$ is a graded 4×4-matrix algebra. So there is a 4-dimensional graded module M such that $A\hat{\otimes}A \cong \text{End}(M)$ as graded algebras. Is this the same when we add the canonical differential (2–7) to $\left\langle \frac{\alpha,\beta}{k} \right\rangle$? In other words, does there exist a graded module M with a differential δ_M such that $A\hat{\otimes}A$ is isomorphic to $\text{End}(M)$ as a DS algebra? The answer is even negative for the graded matrix algebra $\left\langle \frac{1,-1}{k} \right\rangle$. Before we answer the question let us first define the Brauer group of DS algebras.

DEFINITION 2.2. A DS algebra A is called a DS Azumaya algebra if A is a graded Azumaya algebra. Two DS Azumaya algebras A and B are said to be equivalent, denoted $A \sim B$, if there are two differential graded module M and N such that $A\hat{\otimes}\text{End}(M) \cong B\hat{\otimes}\text{End}(N)$ as DS algebras.

Let $B(k)$ be the set of isomorphism classes of DS Azumaya algebras. It is a routine verification that the quotient set of $B(k)$ modulo the equivalence relation $\sim$ is a group and is called the *Brauer group of DS algebras*, denoted $\text{BDS}(k)$. If A represents an element $[A]$ of $\text{BDS}(k)$, then the graded opposite $\bar{A}$ represents the inverse of $[A]$ in $\text{BDS}(k)$, where A and $\bar{A}$ share the same differential. The unit of the group $\text{BDS}(k)$ is represented by matrix DS algebras which are the endomorphism algebras of some finite dimensional differential graded modules. In other words, if A is a DS Azumaya algebra such that $[A] = 1$, then A is isomorphic to $\text{End}(M)$ as a DS algebra for some finite dimensional differential graded module M. This follows from the fact that the Brauer equivalence $\sim$ is the same as the differential graded Morita equivalence which can be done straightforward by adding the 'differential' to graded Morita equivalence. We would rather wait till next section to see a far more general H-Morita theory.

Following Definition 2.2 all quaternion algebras $\left\langle \frac{\alpha,\beta}{k} \right\rangle$, $\alpha,\beta \in k^\bullet$, are DS Azumaya algebras. We show first that the DS matrix algebra $\left\langle \frac{\alpha,-\alpha}{k} \right\rangle$ does not represents the unit of $\text{BDS}(k)$. In order to prove this, we need to consider the associated automorphism σ given by (2–4) of a differential graded module M. Since the differential δ of M is a degree one graded endomorphism, it follows that δ anti-commutes with σ, namely, $\sigma\delta + \delta\sigma = 0$.

LEMMA 2.3. *For any* $\alpha \in k^\bullet$, *the DS algebra* $[\left\langle \frac{\alpha,-\alpha}{k} \right\rangle] \neq 1$ *in* $\text{BDS}(k)$.

PROOF. Let A be $\left\langle \frac{\alpha,-\alpha}{k} \right\rangle$ and assume that $[\left\langle \frac{\alpha,-\alpha}{k} \right\rangle] = 1$ in $\text{BDS}(k)$. Then there exists a two dimensional differential graded module M such that $\left\langle \frac{\alpha,-\alpha}{k} \right\rangle \cong \text{End}(M)$. Since the differential δ_E and the automorphism σ_E of $\text{End}(M)$ are induced by the differential δ_M and automorphism σ_M of M respectively (see (2–6) for the differential), $A \cong \text{End}(M)$ implies that there exist two elements ν and ω in A such that the canonical differential δ given by (2–7) is induced by ν and the automorphism σ_A given by (2–4) is the inner automorphism induced by

ω, i.e.,

$$\delta(a) = a\nu - (-1)^{\partial(a)}\nu a, \quad \sigma(b) = \omega b \omega^{-1}$$

where $a, b \in A$ and a is homogeneous. Furthermore, ν and ω satisfy the relations that δ_M and σ_M obey, i.e.,

$$\nu^2 = 0, \omega^2 = 1, \quad \nu\omega + \omega\nu = 0.$$

Let u, v be the two generators of $\langle \frac{\alpha, -\alpha}{k} \rangle$. Then we have the following relations:

$$\delta(u) = u\nu + \nu u = 0,$$

$$\delta(v) = v\nu + \nu v = 1,$$

$$\sigma(u) = \omega u \omega^{-1} = -u,$$

$$\sigma(v) = \omega v \omega^{-1} = -v.$$

It follows that $\nu = -\frac{\alpha^{-1}}{2}v + suv$ for some $s \in k$ and $\omega = \alpha^{-1}uv$. Since $\nu^2 = 0$, we have

$$0 = (-\frac{\alpha^{-1}}{2}v + suv)^2 = -\frac{\alpha^{-1}}{4} + s^2\alpha^2$$

So s cannot be zero. However, the anti-commutativity of ν with ω implies that

$$0 = \nu\omega + \omega\nu$$

$$= (-\tfrac{1}{2}\alpha^{-1}v + suv)\alpha^{-1}uv + \alpha^{-1}uv(-\tfrac{1}{2}\frac{\alpha^{-1}}{v} + suv)$$

$$= 2s\alpha.$$

So s must be zero. Contradiction! Thus we have proved that it is impossible to have $\langle \frac{\alpha, -\alpha}{k} \rangle \cong \mathrm{End}(M)$ for some 2-dimensional differential graded module M, and hence $[\langle \frac{\alpha, -\alpha}{k} \rangle] \neq 1$. $\square$

From Lemma 2.3 we see that a DS matrix algebra $\langle \frac{\alpha, -\alpha}{k} \rangle$ ($\alpha \in k^\bullet$) representing the unit in $\mathrm{BW}(k)$ now represents a non-unit element in $\mathrm{BDS}(k)$. In the following we show that $\langle \frac{\alpha, -\alpha}{k} \rangle$ represents an element of infinite order in $\mathrm{BDS}(k)$ if the characteristic of k is zero. In fact:

PROPOSITION 2.4 [54, Prop. 7]. *Let* $(k, +)$ *be the additive group of* k. *Then*

$$\tau : (k, +) \longrightarrow \mathrm{BDS}(k), \quad \alpha \mapsto \left[\left\langle \frac{\alpha^{-1}, -\alpha^{-1}}{k} \right\rangle\right], \quad \alpha \neq 0, 0 \mapsto 1$$

is a group monomorphism.

PROOF. By Lemma 2.3 it is sufficient to show that τ is a group homomorphism. Consider the product $\langle \frac{\alpha^{-1}, -\alpha^{-1}, 0}{k} \rangle \hat{\otimes} \langle \frac{\beta^{-1}, -\beta^{-1}, 0}{k} \rangle$. If $\alpha + \beta = 0$, then $\langle \frac{\beta^{-1}, -\beta^{-1}, 0}{k} \rangle = \overline{\langle \frac{\alpha^{-1}, -\alpha^{-1}, 0}{k} \rangle}$ and

$$\left\langle \frac{\alpha^{-1}, -\alpha^{-1}, 0}{k} \right\rangle \hat{\otimes} \left\langle \frac{\beta^{-1}, -\beta^{-1}, 0}{k} \right\rangle \cong \mathrm{End}\left\langle \frac{\alpha^{-1}, -\alpha^{-1}, 0}{k} \right\rangle,$$

which represents the unit in $BDS(k)$.

Assume that $\alpha + \beta \neq 0$. Let $\{u, v\}$ and $\{u', v'\}$ be the generators of $\left\langle \frac{\alpha^{-1}, -\alpha^{-1}}{k} \right\rangle$ and of $\left\langle \frac{\beta^{-1}, -\beta^{-1}}{k} \right\rangle$ respectively. Let

$$U = \frac{\alpha\beta}{\alpha + \beta}(u\hat{\otimes}w' + w\hat{\otimes}u'), \quad V = \frac{\alpha}{\alpha + \beta}v\hat{\otimes}1 + \frac{\beta}{\alpha + \beta}1\hat{\otimes}v'.$$

Then

$$U^2 = (\alpha + \beta)^{-1}, \quad V^2 = -(\alpha + \beta)^{-1}, \quad UV + VU = 0.$$

Thus U and V generate the matrix algebra $\left(\frac{\sigma^{-1}, -\sigma^{-1}}{k}\right)$, where $\sigma = \alpha + \beta$. One may further check that the induced $\mathbb{Z}_2$-grading and the induced differential on $\left(\frac{\sigma^{-1}, -\sigma^{-1}}{k}\right)$ are given by (2–1) and (2–7). Thus U and V generate a DS quaternion subalgebra $\left\langle \frac{(\alpha+\beta)^{-1}, -(\alpha+\beta)^{-1}}{k} \right\rangle$ in $\left\langle \frac{\alpha^{-1}, -\alpha^{-1}}{k} \right\rangle \hat{\otimes} \left\langle \frac{\beta^{-1}, -\beta^{-1}}{k} \right\rangle$. Applying the commutator theorem for Azumaya algebras (see [19]), we obtain

$$\left\langle \frac{\alpha^{-1}, -\alpha^{-1}}{k} \right\rangle \hat{\otimes} \left\langle \frac{\beta^{-1}, -\beta^{-1}}{k} \right\rangle = \left\langle \frac{\sigma^{-1}, -\sigma^{-1}}{k} \right\rangle \otimes M_2(k)$$

as algebras, where $\sigma = \alpha + \beta$. We leave it to readers to check that they are equal as DS algebras (or see [55, Coro.2]). It follows that

$$\tau(\alpha)\tau(\beta) = \left[\left\langle \frac{\alpha^{-1}, -\alpha^{-1}}{k} \right\rangle \hat{\otimes} \left\langle \frac{\beta^{-1}, -\beta^{-1}}{k} \right\rangle \right] = \left[\left\langle \frac{(\alpha + \beta)^{-1}, -(\alpha + \beta)^{-1}}{k} \right\rangle \right]$$

$$= \tau(\alpha + \beta)$$

in the Brauer group $BDS(k)$. So we have proved that τ is a group homomorphism. □

Note that when the characteristic of k is 0, $(k, +)$ is not a torsion group. The element represented by the matrix algebra $\left\langle \frac{1, -1}{k} \right\rangle$ in $BDS(k)$ generates a subgroup which is isomorphic to $\mathbb{Z}$. In this case any quaternion algebra $\left\langle \frac{\alpha, \beta}{k} \right\rangle$ with the canonical grading and the canonical differential represents an element of infinite order in the Brauer group $BDS(k)$. If the characteristic of k is $p \neq 2$, then $\left\langle \frac{\alpha, \beta}{k} \right\rangle$ represents an element of order not greater than p in $BDS(k)$. The group $(k, +)$ indicates the substantial difference between the Brauer–Wall group $BW(k)$ and the Brauer group $BDS(k)$ of DS algebras. Actually, this subgroup comes only from extra differentials added to graded Azumaya algebras.

THEOREM 2.5 [54, Thm. 8]. $BDS(k) = BW(k) \times (k, +)$.

PROOF. By definition of a DS Azumaya algebra, we have a well-defined group homomorphism

$$\gamma : BDS(k) \longrightarrow BW(k), \quad [A] \longrightarrow [A]$$

by forgetting the differential on the latter A. It is clear that γ is a surjective map as a graded Azumaya algebra with a trivial differential is a DS Azumaya algebra. Since the graded Azumaya algebra $\left\langle \frac{\alpha^{-1}, -\alpha^{-1}}{k} \right\rangle$ represents the unit in $BW(k)$, we have $\tau(k, +) \subseteq \mathrm{Ker}(\gamma)$. To prove that $\mathrm{Ker}(\gamma) \subseteq \tau(k, +)$, we need to use the associated automorphism σ given by (2–4) of a graded algebra. Let A be a DS Azumaya algebra representing a non-trivial element in $\text{Ker}(\gamma)$. Since $[A] = 1$ in $\text{BW}(k)$, A is a graded matrix algebra. Since A is Azumaya, the associated automorphism σ is an inner automorphism induced by some invertible element $u \in A$ such that $u^2 = 1$. Similarly, the differential δ is an inner super-derivation induced by some element $v \in A$ in the sense that

$$\delta(a) = va - (-1)^{\partial(a)}va$$

for any homogeneous element $a \in A$. Note that $v^2 \neq 0$ by the proof of Lemma 2.3. Now one may apply the properties that $\delta^2 = 0$, $\sigma\delta + \delta\sigma = 0$ and $\sigma^2 = 1$ to obtain that u and v generate a quaternion subalgebra $\left(\frac{\alpha,\beta}{k}\right)$ for some $\alpha, \beta \in k^\bullet$ with α being a square number. Here u, v are not necessarily the canonical generators of $\left(\frac{\alpha,\beta}{k}\right)$ (see [54, Thm. 8] for more detail). Thus $\left(\frac{\alpha,\beta}{k}\right)$ is a matrix algebra and A is a tensor product $\left(\frac{\alpha,\beta}{k}\right) \otimes M_n(k)$ of two matrix algebras for some integer n. Since u and v generate $\left(\frac{\alpha,\beta}{k}\right)$ and $M_n(k)$ commutes with $\left(\frac{\alpha,\beta}{k}\right)$, σ and δ act on $M_n(k)$ trivially and $\left(\frac{\alpha,\beta}{k}\right)$ is a DS subalgebra of A. It follows that $A = \left(\frac{\alpha,\beta}{k}\right) \hat{\otimes} M_n(k)$. Finally one may take a while to check that there is a pair of new generators u', v' of $\left(\frac{\alpha,\beta}{k}\right)$ such that the DS algebra $\left(\frac{\alpha,\beta}{k}\right)$ can be written as $\left\langle \frac{\alpha,\beta}{k} \right\rangle$ with u', v' being the canonical generators. So $[A] = [\left\langle \frac{\alpha,\beta}{k} \right\rangle \hat{\otimes} M_n(k)] = [\left\langle \frac{\alpha,\beta}{k} \right\rangle] \in \tau(k, +)$. Finally, since γ is split by the inclusion map, the Brauer group $\text{BDS}(k)$ is a direct product of $\text{BW}(k)$ with $(k, +)$. $\square$

DS algebras may be generalized to differential $\mathbb{Z}_2$-dimodule algebras adding one differential to a dimodule algebra such that the action of the differential anticommutes with the action of the non-unit element of $\mathbb{Z}_2$. The Brauer group $\text{BDD}(k, \mathbb{Z}_2)$ of differential dimodule algebras can be defined and computed. Once again quaternion algebras play the same roles as they do in the Brauer group of DS algebras. As an exercise for readers, the Brauer group $\text{BDD}(k, \mathbb{Z}_2)$ is isomorphic to the group $(k, +) \times \text{BD}(k, \mathbb{Z}_2)$ [56]. Other exercises include adding more differentials, say n differentials $\delta_1, \cdots, \delta_n$, to graded Azumaya algebras or dimodule Azumaya algebras. For instance, one may obtain the Brauer group $\text{BDS}_n(k)$ of n-differential superalgebras which is isomorphic to the group $(k, +)^n \times \text{BW}(k)$.

From the proofs of Lemma 2.3 and Theorem 2.5 one may find that the argument there is actually involved with actions of an automorphism and a differential which satisfy the relations:

$$\sigma^2 = 1, \quad \delta^2 = 0, \quad \sigma\delta + \delta\sigma = 0$$

where σ is the non-unit element of $\mathbb{Z}_2$. In fact, the four dimensional algebra generated by σ and δ is a Hopf algebra with comultiplication given by

$$\Delta(\sigma) = \sigma \otimes \sigma, \quad \Delta(\delta) = 1 \otimes \delta + \delta \otimes \sigma$$

and counit given by $\varepsilon(\sigma) = 1$ and $\varepsilon(\delta) = 0$. This Hopf algebra is called Sweedler Hopf algebra, denoted H_4. The two generators σ and δ are usually replaced by g and h. Thus a DS algebra is nothing else but an H_4-module algebra.

Conversely if A is an H_4-module algebra, there is a natural $\mathbb{Z}_2$-grading on A given by

$$A_0 = \{x \in A \mid g(x) = x\}, \quad A_1 = \{x \in A \mid g(x) = -x\}. \qquad (2\text{--}8)$$

With respect to the grading (2–8), the action of h is a differential on A so that A is a DS algebra. Moreover, the H_4-module algebra is a YD H_4-module algebra with the coaction given by the grading (2–8).

In this way we may identify DS algebras with YD H_4-module algebras with coactions given by the grading (2–8). This is the reason why the Brauer group of DS algebras can be defined. The elements in the Brauer group of DS algebras are eventually represented by those so called YD H_4-Azumaya algebras which will be introduced in the next section. In particular, quaternion algebras are YD H_4-Azumaya algebras.

3. The Definition of the Brauer Group

Throughout this section H is a flat k-Hopf algebra with a bijective antipode S, and all k-modules (except H) are faithfully projective over k. Let A be a YD H-module algebra. The two YD H-module algebras $A\#\overline{A}$ and $\overline{A}\#A$ are called the left and right H-enveloping algebras of A (see (1–1) for definition of $\#$). We are now able to define the concept of an H-Azumaya algebra, and construct the Brauer group of the Hopf algebra H.

DEFINITION 3.1. A YD H-module algebra A is called an H-Azumaya algebra if it is faithfully projective as a k-module and if the following YD H-module algebra maps are isomorphisms:

$$F : A\#\overline{A} \longrightarrow \operatorname{End}(A), \qquad F(a\hat{\otimes}b)(x) = \sum ax_{(0)}(x_{(1)} \cdot b),$$
$$G : \overline{A}\#A \longrightarrow \operatorname{End}(A)^{\operatorname{op}}, \qquad G(\overline{a}\#b)(x) = \sum a_{(0)}(a_{(1)} \cdot x)b.$$

where the YD H-structures of $\operatorname{End}(A)$ and $\operatorname{End}(A)^{\operatorname{op}}$ are given by (1–2) and (1–3).

It follows from the definition that a usual Azumaya algebra with trivial H-structures is an H-Azumaya algebra. One may take a while to check (or see [11]) that the H-opposite algebra $\overline{A}$ of an H-Azumaya algebra A and the braided product $A\#B$ of two H-Azumaya algebras A and B are H-Azumaya algebras. In particular, the YD H-module algebra $\operatorname{End}(M)$ of any faithfully projective YD H-module M is an H-Azumaya algebra. An H-Azumaya algebra of the form $\operatorname{End}(M)$ is called an elementary H-Azumaya algebra. As usual we may define an equivalence relation on the set $B(k, H)$ of isomorphism classes of H-Azumaya algebras.

DEFINITION 3.2. Let A and B be two H-Azumaya algebras. A and B are said to be Brauer equivalent, denoted $A \sim B$, if there exist two faithfully projective YD H-modules M and N such that $A\#\text{End}(M) \cong B\#\text{End}(N)$ as YD H-module algebras.

As expected the quotient set of $B(k, H)$ modulo the Brauer equivalence is a group with multiplication induced by the braided product $\#$ and with inverse operator induced by the H-opposite $^{-}$. Denote by $\text{BQ}(k, H)$ the group $B(k, H)/\sim$ and call it the *Brauer group of the Hopf algebra* H or the *Brauer group of Yetter-Drinfel'd* H-*module algebras*. Since a usual Azumaya algebra with trivial YD H-module structures is H-Azumaya and the Brauer equivalence restricts to the usual Brauer equivalence, the classical Brauer group $\text{Br}(k)$ of k is a subgroup of $\text{BQ}(k, H)$ sitting in the center of $\text{BQ}(k, H)$.

Let E be a commutative ring with unit. Suppose that we have a ring homomorphism $f : k \longrightarrow E$. By usual base change $H_E = H \otimes_k E$ is a E-Hopf algebra. Now in a way similar to [30, 4.7, 4.8] we obtain an induced group homomorphism on the Brauer group level.

PROPOSITION 3.3. *The functor* $M \mapsto M \otimes_k E$ *induces a group homomorphism* $\text{BQ}(k, H) \longrightarrow \text{BQ}(E, H_E)$, *mapping the class of* A *to the class of* A_E.

The kernel of the foregoing homomorphism, denoted by $\text{BQ}(E/k, H)$, is called *the relative Brauer group of* H *w.r.t. the extension* E/k. Denote by $\text{BQ}^s(k, H)$ the union of relative Brauer groups $\text{BQ}(E/k, H)$ of all faithfully flat extensions E of k. $\text{BQ}^s(k, H)$ is called the *split part* of $\text{BQ}(k, H)$. In [12], $\text{BQ}^s(k, H)$ was described in a complex:

$$1 \longrightarrow \text{BQ}^s(k, H) \longrightarrow \text{BQ}(k, H) \longrightarrow \text{O}(E(H))$$

where $\text{O}(E(H))$ is a subgroup of the automorphism group $\text{Aut}(E(H))$ and $E(H)$ is the group of group-like elements of the dual Drinfel'd double $D(H)^*$ of H (see [12, 3.11-3.14] for details).

Now let H be a commutative and cocommutative Hopf algebra. In this situation, a YD H-module (algebra) is an H-dimodule (algebra). But an H-Azumaya algebra in the sense of Definition 3.1 is not an Azumaya H-dimodule algebra in the sense of Long (see [30] for detail on the Brauer group of dimodule algebras we refer to [6; 30]). The reason for this is that the braided product we choose in $\mathcal{Q}^H$ is the inverse product of $\mathbf{D}^H$ when H is commutative and cocommutative. However we have the following:

PROPOSITION 3.4 [12, Prop.5.8]. *Let* H *be a commutative and cocommutative Hopf algebra. If* A *is an* H-*Azumaya algebra, then* A^{op} *is an* H-*Azumaya dimodule algebra. Moreover,* $\text{BQ}(k, H)$ *is isomorphic to* $\text{BD}(k, H)$. *The isomorphism is given by* $[A] \mapsto [A^{\text{op}}]^{-1}$.

When H is a commutative and cocommutative Hopf algebra, the Brauer–Long group $BD(k, H)$ has two subgroups $BM(k, H)$ and $BC(k, H)$ (see [30, 1.10, 2.13]). The subgroup $BM(k, H)$ consists of isomorphism classes represented by H-Azumaya dimodule algebras with trivial H-comodule structures, and the subgroup $BC(k, H)$ consists of isomorphism classes represented by H-Azumaya dimodule algebras with trivial H-module structures. These two subgroups were calculated by M. Beattie in [3] which are completely determined by the groups of Galois objects of H and H^* respectively (e.g., see Corollary 4.3.5).

If H is not cocommutative (or not commutative) an H-module (or comodule) algebra with the trivial H^{op}-comodule (or the trivial H-module) structure does not need to be YD H-module algebra. In general, we do not have subgroups like $BM(k, H)$ or $BC(k, H)$ in the Brauer group $BQ(k, H)$ when H is non-commutative or non-cocommutative. However, when H is quasitriangular or coquasitriangular, we have subgroups similar to $BM(k, H)$ or $BC(k, H)$.

It is well known that the notion of a quasitriangular Hopf algebra is the generalization of the notion of a cocommutative Hopf algebra. If $(H, \mathcal{R})$ is a quasitriangular Hopf algebra, an H-module algebra is automatically a YD H-module algebra with the freely granted H^{op}-comodule structure (1–4). Since $_H\mathcal{M}^{\mathcal{R}}$ is a braided subcategory of $\mathcal{Q}^H$ and the braided product $\#$ given by (1–1) commutes (or is compatible) with the H^{op}-coaction (1–4), the canonical H-linear map F and G in Definition 3.1 are automatically H^{op}-colinear. Thus the subset of $BQ(k, H)$ consisting of isomorphism classes represented by H-Azumaya algebras with H^{op}-coactions of the form (1–4) stemming from H-actions is a subgroup, denoted $BM(k, H, \mathcal{R})$. We now have the following inclusions for a QT Hopf algebra.

$$Br(k) \subseteq BM(k, H, \mathcal{R}) \subseteq BQ(k, H).$$

Similarly if (H, R) is a coquasitriangular Hopf algebra, the Brauer group $BQ(k, H)$ possesses a subgroup $BC(k, H, R)$ consisting of isomorphism classes represented by H-Azumaya algebras with H-actions (1–5) stemming from the H^{op}-coactions. In this case we have the following inclusions of groups for a CQT Hopf algebra:

$$Br(k) \subseteq BC(k, H, R) \subseteq BQ(k, H)$$

When H is a finite commutative and cocommutative Hopf algebra, a CQT structure can be interpreted by a Hopf algebra map from H into H^*. As a matter of fact, there is one-to-one correspondence between the CQT structures on H and the Hopf algebra maps from H to H^* which form a group $\mathrm{Hopf}(H, H^*)$ with the convolution product. The correspondence is given by

$$\{\text{CQT structures on } H\} \longrightarrow \mathrm{Hopf}(H, H^*), \quad R \mapsto \theta_R, \ \theta_R(h)(l) = R(h \otimes l)$$

for any $h, l \in H$. In this case the Brauer group $BC(k, H, R)$ is Orzech's Brauer group $B_\theta(k, H)$ of $BD(k, H)$ consisting of classes of θ-dimodule algebras. For θ-dimodule algebras one may refer to [36] in the case that H is a group Hopf algebra with a finite abelian group and to [6, §12.4] in the general case. In a special case that $H = kG$ is a finite abelian group algebra, $\gamma : G \times G \longrightarrow k^{\bullet}$ a bilinear map, we may view γ as a coquasitriangular structure on H. Then the Brauer group $B_{\gamma}(k, G)$ of graded algebras investigated by Childs, Garfinkel and Orzech (see [14; 15]) is isomorphic to $BC(k, H, \gamma)$.

Note that a cocommutative Hopf algebra with a coquasitriangular structure is necessarily commutative. Similarly, a commutative quasitriangular Hopf algebra is cocommutative.

Now let us consider the finite case. Suppose that H is a faithfully projective Hopf algebra. Then the Drinfel'd double $D(H)$ is a quasitriangular Hopf algebra with the canonical QT structure $\mathcal{R}$ represented by a pair of dual bases of H and H^{*}, e.g., [23; 40], and there is a one-to-one correspondence between left $D(H)$-module algebras and Yetter–Drinfel'd H-module algebras, [32]. It follows that $BQ(k, H) = BM(k, D(H), \mathcal{R})$. So

$$BQ(k, H) \subseteq BQ(k, D(H)).$$

Now write $D^{n}(H)$ for $D(D^{n-1}(H))$, the n-*th* Drinfel'd double. Then we have the following chain of inclusions:

$$BQ(k, H) \subseteq BQ(k, D(H)) \subseteq BQ(k, D^{2}(H)) \subseteq \cdots \subseteq BQ(k, D^{n}(H)) \subseteq \cdots$$

A natural question arises: when is the foregoing ascending chain finite?

To end this section, let us look once again at Definition 3.1 and the definition of the Brauer equivalence. It is not surprising that these definitions are essentially categorical in nature. This means that an H-Azumaya algebra can be characterized in terms of monoidal category equivalences. The Brauer equivalence is in essence the Morita equivalence. In particular, the unit in the Brauer group $BQ(k, H)$ is only represented by elementary H-Azumaya algebras.

Let A be a YD H-module algebra. A *left A-module M in $\mathcal{Q}^{H}$* is both a left A-module and a YD H-module satisfying the compatibility conditions:

i. $h \cdot (am) = \sum (h_{(1)} \cdot a)(h_{(2)} \cdot m),$

ii. $\chi(am) = \sum a_{(0)} m_{(0)} \otimes m_{(1)} a_{(1)}.$

That is, M is a left $A\#H$-module and a right Hopf module in $_{A}\mathcal{M}^{H^{op}}$. Here $A\#H$ is the usual smash product rather than the braided product. Denote by $_{A}\mathcal{Q}^{H}$ the category of left A-modules in $\mathcal{Q}^{H}$ and A-module morphisms in $\mathcal{Q}^{H}$. Similarly, we may define a *right A-module M in $\mathcal{Q}^{H}$* as a right A-module and a YD H-module such that the following two compatibility conditions hold:

i. $h \cdot (ma) = \sum (h_{(1)} \cdot m)(h_{(2)} \cdot a),$

ii. $\chi(ma) = \sum m_{(0)} a_{(0)} \otimes a_{(1)} m_{(1)}.$

Denote by $\mathcal{Q}_{A}^{H}$ the category of right A-modules in $\mathcal{Q}^{H}$ and their morphisms. Now let A and B be two YD H-module algebras. An *(A-B)-bimodule M in $\mathcal{Q}^{H}$* is an $(A$-$B)$-bimodule which belongs to $_{A}\mathcal{Q}^{H}$ and $\mathcal{Q}_{B}^{H}$. Denote by $_{A}\mathcal{Q}_{B}^{H}$ the category of $(A$-$B)$-bimodules. View k as a trivial YD H-module algebra. Then $_{A}\mathcal{Q}^{H}$ is just the category of $(A\text{-}k)$-bimodules in $\mathcal{Q}^H$. Similarly $\mathcal{Q}_A^H$ is the category of $(k\text{-}A)$-bimodules in $\mathcal{Q}^H$.

DEFINITION 3.5. A (strict) Morita context $(A, B, P, Q, \varphi, \psi)$ is called a (strict) H-Morita context in $\mathcal{Q}^H$ if the following conditions hold:

(1) A and B are YD H-module algebras,
(2) P is an $(A\text{-}B)$-bimodule in $\mathcal{Q}^H$ and Q is an $(B\text{-}A)$-bimodule in $\mathcal{Q}^H$,
(3) φ and ψ are (surjective) YD H-module algebra maps: $\varphi : P\tilde{\otimes}_B Q \longrightarrow A$ and $\psi : Q\tilde{\otimes}_A P \longrightarrow B$.

An H-Morita context in $\mathcal{Q}^H$ is a usual Morita context if one forgets the H-structures. When one works with the base category $\mathcal{Q}^H$, the usual Morita theory applies fully and one obtains an H-Morita theory in $\mathcal{Q}^H$. Here are few basic properties of H-Morita contexts.

PROPOSITION 3.6. (1) *If P is a faithfully projective YD H-module, then*

$$(\operatorname{End}(P), k, P, P^*, \varphi, \psi)$$

is a strict H-Morita context in $\mathcal{Q}^H$. Here φ and ψ are given by $\varphi(p \otimes f)(x) = pf(x)$ and $\psi(f \otimes p) = f(p)$.
(2) *Let B be a YD H-module algebra. If $P \in \mathcal{Q}_B^H$ is a B-progenerator, then $(A = \operatorname{End}_B(P), B, P, Q = \operatorname{Hom}_B(P, B), \varphi, \psi)$ is a strict H-Morita context in $\mathcal{Q}^H$. Here φ and ψ are given by $\varphi(p \otimes f)(x) = pf(x)$ and $\psi(f \otimes p) = f(p)$,*

where $\operatorname{End}_B(P)$ is a YD H-module algebra with adjoint H-structures given in Subsection 1.1. Like usual Morita theory, if $(A, B, P, Q, \varphi, \psi)$ is a strict H-Morita context, then the pairs of functors

$$Q\tilde{\otimes}_A - : {}_A\mathcal{Q}^H \longrightarrow {}_B\mathcal{Q}^H \quad \text{and} \quad P\tilde{\otimes}_B - : {}_B\mathcal{Q}^H \longrightarrow {}_A\mathcal{Q}^H,$$
$$-\tilde{\otimes}_A P : \mathcal{Q}_A^H \longrightarrow \mathcal{Q}_B^H \quad \text{and} \quad -\tilde{\otimes}_B Q : \mathcal{Q}_B^H \longrightarrow \mathcal{Q}_A^H$$

define equivalences between the categories of bimodules in $\mathcal{Q}^H$.

Let A be a YD H-module algebra, $\overline{A}$ is the H-opposite of A. Write A^e for $A\#\overline{A}$ and ${}^eA$ for $\overline{A}\#A$. Then A may be regarded as a left A^e-module and a right ${}^eA$-module as follows:

$$(a\#\overline{b}) \cdot x = \sum ax_{(0)}(x_{(1)} \cdot b), \text{ and } x \cdot (\overline{a}\#b) = \sum a_{(0)}(a_{(1)} \cdot x)b. \qquad (3\text{--}1)$$

It is clear that A with foregoing A^e and ${}^eA$-module structures is in ${}_{A^e}\mathcal{Q}^H$ and $\mathcal{Q}_{{}^eA}^H$ respectively. Now consider the categories ${}_{A^e}\mathcal{Q}^H$ and $\mathcal{Q}_{{}^eA}^H$. To a left A^e-module M in ${}_{A^e}\mathcal{Q}^H$ we associate a YD H-submodule

$$M^A = \{m \in M \mid (a\#1)m = (1\#\overline{a})m, \forall a \in A\}.$$

This correspondence gives rise to a functor $(-)^A$ from ${}_{A^e}\mathcal{Q}^H$ to $\mathcal{Q}^H$. On the other hand, we have an *induction functor* $A\tilde{\otimes}-$ from $\mathcal{Q}^H$ to ${}_{A^e}\mathcal{Q}^H$. It is easy to see that

$$A \tilde{\otimes} - : \Omega^H \longrightarrow {}_{A^e}\Omega^H, N \mapsto A \tilde{\otimes} N,$$
$$(-)^A : {}_{A^e}\Omega^H \longrightarrow \Omega^H, \quad M \mapsto M^A. \tag{3-2}$$

is an adjoint pair of functors. Similarly we have an adjoint pair of functors between categories Ω^H and $\Omega^H_{e\,A}$:

$$- \tilde{\otimes} A : \Omega^H \longrightarrow \Omega^H_{e\,A}, \quad N \mapsto N \tilde{\otimes} A,$$
$${}^A(-) : \Omega^H_{e\,A} \longrightarrow \Omega^H, \quad M \mapsto {}^A M \tag{3-3}$$

where ${}^A M = \{m \in M \mid m(1 \# a) = m(\bar{a} \# 1), \forall a \in A\}$.

PROPOSITION 3.7 [12, Prop.2.6]. *Let A be a YD H-module algebra. Then A is H-Azumaya if and only if (3-2) and (3-3) define equivalences of categories.*

In fact, (3-2) and (3-3) define the equivalences between the braided monoidal categories if A is H-Azumaya (see [12]).

With the previous preparation one is able to show that the 'Brauer equivalence' is equivalent to the 'H-Morita equivalence' in Ω^H. We denote by $A \overset{m}{\sim} B$ that A is H-Morita equivalent to B, and denote by $A \overset{b}{\sim} B$ that A is Brauer equivalent to B, i.e., $[A] = [B] \in \mathrm{BQ}(k, H)$.

THEOREM 3.8 [12, Thm. 2.10]. *Let A, B be H-Azumaya algebras. $A \overset{b}{\sim} B$ if and only if $A \overset{m}{\sim} B$.*

As a direct consequence, we have that if $[A] = 1$ in $\mathrm{BQ}(k, H)$, then $A \cong \mathrm{End}(P)$ for some faithfully projective YD H-module P.

4. An Exact Sequence for the Brauer Group $\mathrm{BC}(k, H, R)$

As we explained in Section 3, when a Hopf algebra H is finite, the Brauer group $\mathrm{BQ}(k, H)$ of H is equal to the Brauer group $\mathrm{BC}(k, D(H)^*, R)$. So in the finite case, it is sufficient to consider the Brauer group $\mathrm{BC}(k, H, R)$ of a finite coquasitriangular Hopf algebra. We present a general approach to the calculation of the Brauer group $\mathrm{BC}(k, H, R)$. The idea of this approach is basically the one of Wall in [57] where he introduced the first Brauer group of structured algebras, i.e., the Brauer group of super (or $\mathbb{Z}_2$-graded) algebras which is now called the Brauer–Wall group, denoted $\mathrm{BW}(k)$. He proved that the Brauer group of super algebras over a field k is an extension of the Brauer group $\mathrm{Br}(k)$ by the group $Q_2(k)$ of $\mathbb{Z}_2$-graded quadratic extensions of k, i.e., there is an exact sequence of group homomorphisms:

$$1 \longrightarrow \mathrm{Br}(k) \longrightarrow \mathrm{BW}(k) \longrightarrow Q_2(k) \longrightarrow 1$$

In 1972, Childs, Garfinkel and Orzech studied the Brauer group of algebras graded by a finite abelian group [14], and Childs (in [15]) generalized Wall's sequence by constructing a non-abelian group $\mathrm{Galz}(G)$ of bigraded Galois objects replacing the graded quadratic group of k in Wall's sequence. They obtained an exact group sequence:

$$1 \longrightarrow \mathrm{Br}(k) \longrightarrow \mathrm{BC}(k, G, \phi) \longrightarrow \mathrm{Galz}(G)$$

where $\mathrm{Galz}(G)$ may be described by another exact sequence when G is a p-group.

The complexity of the group $\mathrm{Galz}(G)$ is evident. An object in $\mathrm{Galz}(G)$ involves both two-sided G-gradings and two-sided G-actions such that the actions and gradings commute. In 1992, K. Ulbrich extended the exact sequences of Childs to the case of the Brauer–Long group of a commutative and cocommutative Hopf algebra (see [48]). The technique involved is essentially the Hopf Galois theory of a finite Hopf algebra. However, when the Hopf algebra is not commutative, a similar group of Hopf Galois objects does not exists although one still obtains a group of Hopf bigalois objects with respect to the cotensor product (see [42; 55]). The idea of this section is to apply the Hopf quotient Galois theory. This requires the deformation of the Hopf algebra H. Throughout, (H, R) is a finite CQT Hopf algebra.

4.1. The algebra $\mathcal{H}_R$. We start with the definition of the new product $\star$ on the k-module H:

$$h \star l = \sum l_{(2)} h_{(2)} R(S^{-1}(l_{(3)}) l_{(1)} \otimes h_{(1)})$$

$$= \sum h_{(2)} l_{(1)} R(l_{(2)} \otimes S(h_{(1)}) h_{(3)})$$

where h and l are in H. $(H, \star)$ is an algebra with unit 1. We denote by $\mathcal{H}_R$ the algebra $(H, \star)$. It is easy to see that the counit map ε of H is still an augmentation map from $\mathcal{H}_R$ to k.

There is a double Hopf algebra for a CQT Hopf algebra (H, R) (not necessarily finite). This double Hopf algebra, denoted $D[H]$ due to Doi and Takeuchi (see [21]), is equal to $H \otimes H$ as a coalgebra with the multiplication given by

$$(h \otimes l)(h' \otimes l') = \sum hh'_{(2)} \otimes l_{(2)} l' R(h'_{(1)} \otimes l_{(1)}) R(S(h'_{(3)}) \otimes l_{(3)})$$

for h, l, h' and $l' \in H$. The antipode of $D[H]$ is given by

$$S(h \otimes l) = (1 \otimes S(l))(S(h) \otimes 1)$$

for all $h, l \in H$. The counit of $D[H]$ is $\varepsilon \otimes \varepsilon$.

Since H is finite, the canonical Hopf algebra homomorphism $\Theta_l : H \longrightarrow H^{*\mathrm{op}}$ given by $\Theta_l(h)(l) = R(h \otimes l)$ induces an Hopf algebra homomorphism from $D[H]$ to $D(H)$, the Drinfel'd quantum double $H^{*\mathrm{op}} \bowtie H$.

$$\Phi : D[H] \longrightarrow D(H), \quad \Phi(h \bowtie l) = \Theta_l(h) \bowtie l.$$

When Θ_l is an isomorphism, we may identify $D[H]$ with $D(H)$. Thus a YD H-module is automatically a left $D[H]$-module. Moreover, the following algebra monomorphism ϕ shows that $\mathcal{H}_R$ can be embedded into $D[H]$.

$$\phi : \mathcal{H}_R \longrightarrow D[H], \quad \phi(h) = \sum S^{-1}(h_{(2)}) \bowtie h_{(1)}.$$

Thus we may view $\mathcal{H}_R$ as a subalgebra of the double $D[H]$. Moreover, one may check that the image of ϕ in $D[H]$ is a left coideal of $D[H]$. We obtain the following:

PROPOSITION 4.1.1. $\mathcal{H}_R$ is a left $D[H]$-comodule algebra with the comodule structure given by

$$\chi : \mathcal{H}_R \longrightarrow D[H] \otimes \mathcal{H}_R, \quad \chi(h) = \sum (S^{-1}(h_{(3)}) \bowtie h_{(1)}) \otimes h_{(2)}.$$

The left $D[H]$-comodule structure of $\mathcal{H}_R$ in Proposition 4.1.1 demonstrates that $\mathcal{H}_R$ can be embedded into $D[H]$ as a left coideal subalgebra. In fact, $\mathcal{H}_R$ can be further embedded into $D(H)$ as a left coideal subalgebra.

CORROLLARY 4.1.2. The composite algebra map

$$\mathcal{H}_R \xrightarrow{\ \phi\ } D[H] \xrightarrow{\ \Phi\ } D(H)$$

is injective, and $\mathcal{H}_R$ is isomorphic to a left coideal subalgebra of $D(H)$.

Let us now consider Yetter–Drinfel'd H-modules and $\mathcal{H}_R$-bimodules. Let M be a Yetter–Drinfel'd module over H, or a left $D(H)$-module. The following composite map:

$$\mathcal{H}_R \otimes M \xrightarrow{\phi \otimes \iota} D[H] \otimes M \xrightarrow{\Phi \otimes \iota} D(H) \otimes M$$

makes M into a left $\mathcal{H}_R$-module. If we write $-\triangleright$ for the above left action, then we have the explicit formula:

$$h \multimap m = \sum (h_{(2)} \cdot m_{(0)}) R(S^{-1}(h_{(4)}) \otimes h_{(3)} m_{(1)} S^{-1}(h_{(1)})) \tag{4-1}$$

for $h \in \mathcal{H}_R$ and $m \in M$.

Since there is an augmentation map ε on $\mathcal{H}_R$, we may define the $\mathcal{H}_R$-invariants of a left $\mathcal{H}_R$-module M which is

$$M^{\mathcal{H}_R} = \{ m \in M \mid h \multimap m = \varepsilon(h)m, \forall h \in \mathcal{H}_R \}.$$

When a left $\mathcal{H}_R$-module comes from a YD H-module we have

$$M^{\mathcal{H}_R} = \{ m \in M \mid h \cdot m = h \triangleright_1 m = \sum m_{(0)} R(h \otimes m_{(1)}), \forall h \in H \}.$$

Now we define a right $\mathcal{H}_R$-module structure on a YD H-module M. Observe that the right H-comodule structure of M induces two left H-module structures. The first one is (1–5), and the second one is given by

$$h \triangleright_2 m = \sum m_{(0)} R(S(m_{(1)}) \otimes h) \tag{4-2}$$

for $h \in H$ and $m \in M$. With this second left H-action (4–2) on M, M can be made into a right $D[H]$-module.

LEMMA 4.1.3. *Let M be a Yetter–Drinfel'd H-module. Then M is a right $D[H]$-module defined by*

$$m \leftharpoonup (h \bowtie l) = S(l) \rhd_2 (S(h) \cdot m)$$

for $h, l \in H$ and $m \in M$. Moreover, if A is a YD H-module algebra, then A is a right $D[H]^{\mathrm{cop}}$-module algebra.

The right $D[H]$-module structure in Lemma 4.1.3 does not match the canonical left $D[H]$-module structure induced by the Hopf algebra map Φ so as to yield a $D[H]$-bimodule structure on M. However the right $\mathcal{H}_R$-module structure on M given by

$$M \otimes \mathcal{H}_R \xrightarrow{\iota \otimes \phi} M \otimes D[H] \longrightarrow M,$$

more precisely:

$$m \triangleleft\!- h = \sum (h_{(3)} \cdot m_{(0)}) R(h_{(4)} m_{(1)} S^{-1}(h_{(2)}) \otimes h_{(1)}) \qquad (4\text{-}3)$$

for $m \in M$ and $h \in \mathcal{H}_R$, together with the left $\mathcal{H}_R$-module structure (4–1), defines an $\mathcal{H}_R$-bimodule structure on M.

PROPOSITION 4.1.4. *Let M be a YD H-module. Then M is an $\mathcal{H}_R$-bimodule via (4–1) and (4–3).*

If A is a YD H-module algebra, then Proposition 4.1.4 implies that A is an $\mathcal{H}_R$-bimodule algebra in the sense that:

$$\begin{aligned} h \rightharpoonup (ab) &= \sum (h_{(-1)} \cdot a)(h_{(0)} \rightharpoonup b) \\ (ab) \triangleleft\!- h &= \sum (a \cdot h_{(0)})(b \triangleleft\!- h_{(-1)}) \end{aligned} \qquad (4\text{-}4)$$

for $a, b \in A$ and $h \in \mathcal{H}_R$, where $\chi(h) = \sum h_{(-1)} \otimes h_{(0)} \in D[H] \otimes \mathcal{H}_R$.

To end this subsection we present the dual comodule version of (4–4) which is needed in the next subsection. Observe that the dual coalgebra $\mathcal{H}_R^*$ is a left $D[H]^*$-module quotient coalgebra of the dual Hopf algebra $D[H]^*$ in the sense that the following coalgebra map is a surjective $D[H]$-comodule map:

$$\phi^* : D[H]^* \longrightarrow \mathcal{H}_R^*, \quad p \bowtie q \mapsto qS^{-1}(p).$$

Thus a left (or right) $D[H]$-comodule M is a left (or right) $\mathcal{H}_R^*$-comodule in the natural way through ϕ^*. In order to distinguish $D[H]$ or $\mathcal{H}_R^*$-comodule structures from the H-comodule structures (e.g., a YD H-module has all three comodule structures) we use different uppercase Sweedler sigma notations:

i. $\quad \sum x^{[-1]} \otimes x^{[0]}, \ \sum x^{[0]} \otimes x^{[1]}$ stand for left and right $D[H]^*$-comodule structures,

ii. $\quad \sum x^{(-1)} \otimes x^{(0)}, \ \sum x^{(0)} \otimes x^{(1)}$ stand for left and right $\mathcal{H}_R^*$-comodule structures, where x is an element in a due comodule. Now let A be a YD H-module algebra. Then A is both a left and right $D[H]$-module algebra, and therefore an $\mathcal{H}_R$-bimodule algebra in the sense of (4–4). Thus the dual comodule versions of the formulas in (4–4) read as follows:

$$\sum (ab)^{(0)} \otimes (ab)^{(1)} = \sum a^{[0]} b^{(0)} \otimes a^{[1]} \rightharpoonup b^{(1)},$$
$$\sum (ab)^{(-1)} \otimes (ab)^{(0)} = \sum b^{[-1]} \rightharpoonup a^{(-1)} \otimes a^{(0)} b^{[0]} \qquad (4\text{--}5)$$

for $a, b \in A$, where $\rightharpoonup$ is the left action of $D[H]^*$ on $\mathcal{H}_R^*$. We will call A a *right (or left) $\mathcal{H}_R^*$-comodule algebra* in the sense of (4–5).

Finally, for a YD H-module M, we will write $M_\diamond$ (or $_\diamond M$) for the right (or left) $\mathcal{H}_R^*$-coinvariants. For instance,

$$M_\diamond = \{ m \in M \mid \sum m^{(0)} \otimes m^{(1)} = m \otimes \varepsilon. \}$$

It is obvious that $M_\diamond = M^{\mathcal{H}_R}$.

4.2. The group $\mathrm{Gal}(\mathcal{H}_R)$. We are going to construct a group $\mathrm{Gal}(\mathcal{H}_R)$ of 'Galois' objects for the deformation $\mathcal{H}_R$. The group $\mathrm{Gal}(\mathcal{H}_R)$ plays the vital role in an exact sequence to be constructed.

DEFINITION 4.2.1. Let A be a right $D[H]^*$-comodule algebra. $A/A_\diamond$ is said to be a right $\mathcal{H}_R^*$-Galois extension if the linear map

$$\beta^r : A \otimes_{A_\diamond} A \longrightarrow A \otimes \mathcal{H}_R^*, \quad \beta^r(a \otimes b) = \sum a^{(0)} b \otimes a^{(1)}$$

is an isomorphism. Similarly, if A is a left $D[H]^*$-comodule algebra, then $A/_\diamond A$ is said to be left Galois if the linear map

$$\beta^l : A \otimes_{\diamond A} A \longrightarrow \mathcal{H}_R^* \otimes A, \quad \beta^l(a \otimes b) = \sum b^{(-1)} \otimes ab^{(0)}$$

is an isomorphism. If in addition the subalgebra $_\diamond A$ (or $A_\diamond$) is trivial, then A is called a *left* (or *right*) $\mathcal{H}_R^*$-*Galois object*. For more detail on Hopf quotient Galois theory, readers may refer to [33; 44; 45].

The objects we are interested in are those $\mathcal{H}_R^*$-bigalois objects which are both left and right $\mathcal{H}_R^*$-Galois such that the left and right $\mathcal{H}_R^*$-coactions commute. Denote by $\mathcal{E}(\mathcal{H}_R)$ the category of YD H-module algebras which are $\mathcal{H}_R^*$-bigalois objects. The morphisms in $\mathcal{E}(\mathcal{H}_R)$ are YD H-module algebra isomorphisms. We are going to define a product in the category $\mathcal{E}(\mathcal{H}_R)$. Let $\#_R$ be the braided product in the category $\mathbf{M}_R^H$ to differ from the braided product in $\mathcal{Q}^H$. This makes sense when a YD H-module algebra A can be treated as an algebra in $\mathbf{M}_R^H$ forgetting the H-module structure of A and endowing with the induced H-module structure (1–5).

Given two objects X and Y in $\mathcal{E}(\mathcal{H}_R)$, we define a generalized cotensor product $X \wedge Y$ (in terms of $\mathcal{H}_R^*$-bicomodules) as a subset of $X \#_R Y$:

$$\left\{ \sum x_i \# y_i \in X \#_R Y \mid \sum x_i \triangleleft\!- h \# y_i = \sum x_i \# h -\!\triangleright y_i, \forall h \in \mathcal{H}_R. \right\}$$

In the foregoing formula we may change the actions $\triangleleft-$ and $-\triangleright$ of $\mathcal{H}_R$ into the actions of H which are easier to check.

$$X \wedge Y = \{\sum x_i \# y_i \in X \#_R Y \mid \sum h_{(1)} \cdot x_i \# h_{(2)} \triangleright_1 y_i$$
$$= \sum h_{(1)} \triangleright_2 x_i \# h_{(2)} \cdot y_i, \forall h \in H\}. \tag{4-6}$$

The formula (4–6) allow us to define a left H-action on $X \wedge Y$:

$$h \cdot \sum(x_i \# y_i) = \sum h_{(1)} \cdot x_i \# h_{(2)} \triangleright_1 y_i = \sum h_{(1)} \triangleright_2 x_i \# h_{(2)} \cdot y_i \tag{4-7}$$

whenever $\sum x_i \# y_i \in X \wedge Y$ and $h \in H$. The left H-action is YD compatible with the right diagonal H-coaction, so we obtain:

PROPOSITION 4.2.2. *If X, Y are two objects of $\mathcal{E}(\mathcal{H}_R)$, then $X \wedge Y$ with the H-action (4–7) and the H-coaction inherited from $X \#_R Y$ is a YD H-module algebra. Moreover, $X \wedge Y$ is an object of $\mathcal{E}(\mathcal{H}_R)$.*

Let H^* be the convolution algebra of H. There is a canonical YD H-module structure on H^* such that H^* is a YD H-module algebra:

$$h \cdot p = \sum p_{(1)} < p_{(2)}, h >, \ H\text{-action}$$
$$h^* \cdot p = \sum h^*_{(2)} p S^{-1}(h^*_{(1)}), \ H\text{-coaction} \tag{4-8}$$

for $h^*, p \in H^*$ and $h \in H$. One may easily check that H^* with the YD H-module structure (4–8) is an object in $\mathcal{E}(\mathcal{H}_R)$, denoted I. Moreover I is the unit object of $\mathcal{E}(H)$ with respect to the product $\wedge$. It follows that the category $\mathcal{E}(\mathcal{H}_R)$ is a monoidal category.

Denote by $E(\mathcal{H}_R)$ the set of the isomorphism classes of objects in $\mathcal{E}(\mathcal{H}_R)$. The fact that $\mathcal{E}(\mathcal{H}_R)$ is a monoidal category implies that the set $E(\mathcal{H}_R)$ is a semigroup. In general, $E(\mathcal{H}_R)$ is not necessarily a group. However, it contains a subgroup of a nice type.

Recall that a YD H-module algebra A is said to be *quantum commutative* (q.c.) if

$$ab = \sum b_{(0)}(b_{(1)} \cdot a) \tag{4-9}$$

for any $a, b \in A$. That is, A is a commutative algebra in $\mathcal{Q}^H$.

Let X be a q.c. object in $\mathcal{E}(\mathcal{H}_R)$. Let $\overline{X}$ be the opposite algebra in $\mathbf{M}_R^H$. That is, $\overline{X} = X$ as a right H^{op}-comodule, but with the multiplication given by

$$\overline{x} \circ \overline{y} = \sum \overline{y_{(0)} x_{(0)}} R(y_{(1)} \otimes x_{(1)})$$

where $\overline{x}, \overline{y} \in \overline{X}$. Since the H-action on X does not define an H-module algebra structure on $\overline{X}$, we have to define a new H-action on $\overline{X}$ such that $\overline{X}$ together with the inherited H^{op}-comodule structure is a YD H-module algebra. Let H act on $\overline{X}$ as follows:

$$h \rightharpoonup \overline{x} = \sum \overline{h^u_{(3)} \cdot (h_{(2)} \triangleright_2 (h_{(5)} \triangleright_1 x))} R(S(h_{(4)}) \otimes h_{(1)}) \tag{4-10}$$

where $h \in \mathcal{H}_R$, $\overline{x} \in \overline{X}$, $h^u = \sum S(h_{(2)})u^{-1}(h_{(1)})$ and $u = \sum S(R^{(2)})R^{(1)} \in H^*$ is the Drinfel'd element of H^*.

PROPOSITION 4.2.3. *Let X be an object in $\mathcal{E}(\mathcal{H}_R)$ such that X is q.c. Then:*

(1) $\overline{X}$ *together with the H-action (4–10) is a YD H-module algebra.*

(2) $\overline{X}$ *is a q.c. object in $\mathcal{E}(\mathcal{H}_R)$ and $X \wedge \overline{X} \cong I = \overline{X} \wedge X$.*

Since the proof is lengthy, we refer reader to [59] for the complete proof. Denote by $\mathrm{Gal}(\mathcal{H}_R)$ the subset of $E(\mathcal{H}_R)$ consisting of the isomorphism classes of objects in $\mathcal{E}(\mathcal{H}_R)$ such that the objects are quantum commutative in $\mathcal{Q}^H$. We have

THEOREM 4.2.4 [59, Thm. 3.12]. *The set $\mathrm{Gal}(\mathcal{H}_R)$ is a group with product induced by $\wedge$ and inverse operator induced by H-opposite .*

4.3. The exact sequence. For convenience we will call an H-Azumaya algebra A an *R-Azumaya algebra* if the H-action on A is of form (1–5). That is, A represents an element of $\mathrm{BC}(k, H, R)$. In this subsection we investigate the R-Azumaya algebras which are Galois extensions of the coinvariants, and establish a group homomorphism from $\mathrm{BC}(k, H, R)$ to the group $\mathrm{Gal}(\mathcal{H}_R)$ constructed in the previous subsection.

In the sequel, we will write:

$$M_0 = \{m \in M \mid \sum m_{(0)} \otimes m_{(1)} = m \otimes 1\}$$

for the coinvariant k-submodule of a right H-comodule M, in order to make a difference between $\mathcal{H}_R^*$-coinvariants and H-coinvariants. We start with a special elementary R-Azumaya algebra.

LEMMA 4.3.1. *Let $M = H^{\mathrm{op}}$ be the right H^{op}-comodule, and let A be the elementary R-Azumaya algebra $\mathrm{End}(M)$. Then $A \cong H^{*\mathrm{op}} \# H^{\mathrm{op}}$, where the left H^{op}-action on $H^{*\mathrm{op}}$ is given by $h \cdot p = \sum p_{(1)} \langle p_{(2)}, S^{-1}(h) \rangle = S^{-1}(h) \rightharpoonup p$, whenever $h \in H^{\mathrm{op}}$ and $p \in H^{*\mathrm{op}}$.*

Let A be an R-Azumaya algebra. We have $[A \# \mathrm{End}(H^{\mathrm{op}})] = [A]$ since $\mathrm{End}(H^{\mathrm{op}})$ represents the unit of $\mathrm{BC}(k, H, R)$. Now the composite algebra map

$$H^{\mathrm{op}} \xrightarrow{\lambda} \mathrm{End}(H^{\mathrm{op}}) \hookrightarrow A \# \mathrm{End}(H^{\mathrm{op}})$$

is H^{op}-colinear. It follows that $A \# \mathrm{End}(H^{\mathrm{op}})$ is a smash product algebra $B \# H^{\mathrm{op}}$ where $B = (A \# \mathrm{End}(H^{\mathrm{op}}))_0$. Thus we obtain that any element of $\mathrm{BC}(k, H, R)$ can be represented by an R-Azumaya algebra which is a smash product. Since any smash product algebra is a Galois extension of its coinvariants, we have that any element of $\mathrm{BC}(k, H, R)$ can be represented by an R-Azumaya algebra which is an H^{op}-Galois extension of its coinvariants. Moreover, one may easily prove that if A is an R-Azumaya algebra such that it is an H^{op}-Galois extension of A_0, then $\overline{A}$ is a H^{op}-Galois extension of $(A_0)^{\mathrm{op}}$.

An R-Azumaya algebra A is said to be *Galois* if it is a right H^{op}-Galois extension of its coinvariant subalgebra A_0. Let A be a Galois R-Azumaya algebra. Denote by $\pi(A)$ the centraliser subalgebra $C_A(A_0)$ of A_0 in A. It is clear that $\pi(A_0)$ is an H^{op}-comodule subalgebra of A. The *Miyashita–Ulbrich–Van Oystaeyen* (MUVO) *action* (see [34; 48; 50; 51]; the last author mentioned considered it first in the situation of purely inseparable splitting rings in [50]) of H on $\pi(A)$ is given by

$$h \rightharpoonup a = \sum X_i^h a Y_i^h \qquad (4\text{--}11)$$

where $\sum X_i^h \otimes Y_i^h = \beta^{-1}(1 \otimes h)$, for $h \in H$ and β is the canonical Galois map given by $\beta(a \otimes b) = \sum ab_{(0)} \otimes b_{(1)}$. It is well-known (e.g., see [48; 11]) that $\pi(A)$ together with the MUVO action (4–11) is a new YD H-module algebra. Moreover, $\pi(A)$ is quantum commutative in the sense of (4–9) [48; 52].

Recall that when a Galois H^{op}-comodule algebra A is an Azumaya algebra, the centraliser $\pi(A)$ is a right H^*-Galois extension of k with respect to the MUVO action (4–11); compare [48]. This is not the case when A is an R-Azumaya algebra. However, $\pi(A)$ turns out to be an $\mathcal{H}_R^*$-Galois object, instead of an H^*-Galois object.

PROPOSITION 4.3.2 [59, Prop. 4.5]. *Let A be a Galois R-Azumaya algebra. Then $\pi(A)/k$ is an $\mathcal{H}_R^*$-biextension and $\pi(A)$ is an object in $\mathrm{Gal}(\mathcal{H}_R)$.*

It is natural to expect the functor π to be a monoidal functor from the monoidal category of Galois R-Azumaya algebras to the monoidal category $\mathcal{E}(\mathcal{H}_R)$. This is indeed the case.

PROPOSITION 4.3.3. *π is a monoidal functor. That is:*

(1) *If A and B are two Galois R-Azumaya algebras, then $\pi(A\#B) = \pi(A) \wedge \pi(B)$; and*

(2) *If M is a finite right H^{op}-comodule, and $A = \mathrm{End}(M)$ is the elementary R-Azumaya algebra such that A is a Galois R-Azumaya algebra, then $\pi(A) \cong I$.*

It follows that π induces a group homomorphism $\tilde{\pi}$ from the Brauer group $\mathrm{BC}(k, H, R)$ to the group $\mathrm{Gal}(\mathcal{H}_R)$ sending element $[A]$ to element $[\pi(A)]$, where A is chosen as a Galois R-Azumaya algebra.

In order to describe the kernel of $\tilde{\pi}$, one has to analyze the H-coactions on the Galois R-Azumaya algebras. We obtain that the kernel of $\tilde{\pi}$ is isomorphic to the usual Brauer group $\mathrm{Br}(k)$. Thus we obtain the following exact sequence:

THEOREM 4.3.4 [59, Thm. 4.11]. *We have an exact sequence of group homomorphisms:*

$$1 \longrightarrow \mathrm{Br}(k) \xrightarrow{\iota} \mathrm{BC}(k, H, R) \xrightarrow{\tilde{\pi}} \mathrm{Gal}(\mathcal{H}_R). \qquad (4\text{--}12)$$

Note that the exact sequence (4–12) indicates that the factor group

$$\mathrm{BC}(k, H, R)/\mathrm{Br}(k)$$

is completely determined by the $\mathcal{H}^*_R$-bigalois objects. In particular, when k is an algebraically closed field $\mathrm{BC}(k, H, R)$ is a subgroup of $\mathrm{Gal}(\mathcal{H}_R)$.

Now let us look at some special cases. First let H be a commutative Hopf algebra. H has a trivial coquasitriangular structure $R = \varepsilon \otimes \varepsilon$. In this case. $\mathcal{H}_R$ is equal to H as an algebra and $D[H] = H \otimes H$ is the tensor product algebra. An R-Azumaya algebra is an Azumaya algebra which is a right H-comodule algebra with the trivial left H-action. On the other hand, the $\mathcal{H}_R$-bimodule structures (4–1) and (4–3) of a YD H-module M coincide and are exactly the left H-module structure of M. So in this case an object in the category $\mathcal{E}(\mathcal{H}_R)$ is nothing but an H^*-Galois object which is automatically an H^*-bigalois object since H^* is cocommutative. So the group $\mathrm{Gal}(\mathcal{H}_R)$ is the group $E(H^*)$ of H^*-Galois objects with the cotensor product over H^*. So we obtain the following exact sequence due to Beattie.

CORROLLARY 4.3.5 [3]. *Let H be a finite commutative Hopf algebra. Then the following group sequence is exact and split:*

$$1 \longrightarrow \mathrm{Br}(k) \overset{\iota}{\longrightarrow} \mathrm{BC}(k, H) \overset{\widetilde{\pi}}{\longrightarrow} E(H^*) \longrightarrow 1$$

where the group map $\widetilde{\pi}$ is surjective and split because any H^*-Galois object B is equal to $\pi(B \# H)$ and the smash product $B \# H$ is a right H-comodule Azumaya algebra which represents an element in $\mathrm{BC}(k, H)$.

Secondly we let R be a non-trivial coquasitriangular structure of H, but let H be a commutative and cocommutative finite Hopf algebra over k. In this case, $\mathcal{H}_R$ is isomorphic to H as an algebra and becomes a Hopf algebra. In this case, an object in $\mathrm{Gal}(\mathcal{H}_R)$ is an H^*-bigalois object. It is not difficult to check that YD H-module (or H-dimodule) structures commute with both H^*-Galois structures.

Let θ be the Hopf algebra homomorphism corresponding to the coquasitriangular structure R, that is,

$$\theta : H \longrightarrow H^*, \quad \theta(h)(l) = R(l \otimes h)$$

for $h, l \in H$. Let $\rightharpoonup$ be the induced H-action on a right H-comodule M:

$$h \rightharpoonup m = \sum m_{(0)} \theta(h)(m_{(1)}) = \sum m_{(0)} R(m_{(1)} \otimes h)$$

for $h \in H$ and $m \in M$. In [49], Ulbrich constructed a group $D(\theta, H^*)$ consisting of isomorphism classes of H^*-bigalois objects which are also H-dimodule algebras such that all H and H^* structures commute, and satisfy the following additional conditions interpreted by means of R [49, (14), (16)]:

$$h \rightharpoonup a = \sum a_{(0)} \leftharpoonup h_{(1)} R(a_{(1)} \otimes S(h_{(2)})) R(S(h_{(3)}) \otimes a_{(2)})$$
$$\sum x_{(0)} (a \leftharpoonup x_{(1)}) = \sum (x_{(1)} \rightharpoonup a) x_{(0)}, \tag{4–13}$$

Let us check that any object A in the category $\mathcal{E}(\mathcal{H}_R)$ satisfies the conditions (4-13) so that A represents an element of $D(\theta, H^*)$. Indeed, since H is commutative and cocommutative, we have

$$h \dashrightarrow a = \sum (h_{(2)} \cdot a_{(0)}) R(S^{-1}(h_{(4)}) \otimes h_{(3)} a_{(1)} S^{-1}(h_{(1)}))$$

$$= \sum (h_{(1)} \cdot a_{(0)}) R(S(h_{(2)}) \otimes a_{(1)})$$

$$= \sum (h_{(2)} \cdot a_{(0)}) R(a_{(1)} \otimes h_{(1)}) R(a_{(2)} \otimes S(h_{(3)})) R(S(h_{(4)}) \otimes a_{(3)})$$

$$= \sum (a_{(0)} \vartriangleleft\!\!- h_{(1)}) R(a_{(1)} \otimes S(h_{(2)})) R(S(h_{(3)}) \otimes a_{(2)}),$$

and

$$\sum x_{(0)} (a \vartriangleleft\!\!- x_{(1)}) = \sum x_{(0)} (x_{(1)} \cdot a_{(0)}) R(a_{(1)} \otimes x_{(1)})$$

$$= \sum a_{(0)} x_{(0)} R(a_{(1)} \otimes x_{(1)}) \quad (by \ q.c.)$$

$$= \sum (x_{(1)} \rightharpoonup a) x_{(0)}$$

for any $a, x \in A$ and $h \in H$. It follows that the group $\mathrm{Gal}(\mathcal{H}_R)$ is contained in $D(\theta, H^*)$. As a consequence, we obtain Ulbrich's exact sequence [49, 1.10]:

$$1 \longrightarrow \mathrm{Br}(k) \longrightarrow \mathrm{BD}(\theta, H^*) \overset{\pi_\theta}{\longrightarrow} D(\theta, H^*)$$

for a commutative and cocommutative finite Hopf algebra with a Hopf algebra homomorphism θ from H to H^*.

4.4. An example. Let k be a field with characteristic different from two. Let H_4 be the Sweedler four dimensional Hopf algebra over k. That is, H_4 is generated by two elements g and h satisfying

$$g^2 = 1, h^2 = 0, gh + hg = 0.$$

The comultiplication, the counit and the antipode are as follows:

$$\Delta(g) = g \otimes g, \Delta(h) = 1 \otimes h + h \otimes g,$$
$$\varepsilon(g) = 1, \qquad \varepsilon(h) = 0,$$
$$S(g) = g, \qquad S(h) = gh.$$

There is a family of CQT structures R_t on H_4 parameterized by $t \in k$ as follows:

R_t	1	g	h	gh
1	1	1	0	0
g	1	-1	0	0
h	0	0	t	$-t$
gh	0	0	t	t

It is not hard to check that the Hopf algebra homomorphisms Θ_l and Θ_r induced by R_t are as follows:

$$\Theta_l : H_4^{\text{COP}} \longrightarrow H_4^*, \quad \Theta_l(g) =, \overline{1} - \overline{g}, \quad \Theta_l(h) = t(\overline{h} - \overline{gh}),$$
$$\Theta_r : H_4^{\text{OP}} \longrightarrow H_4^*, \quad \Theta_r(g) = \overline{1} - \overline{g}, \quad \Theta_r(h) = t(\overline{h} + \overline{gh}).$$

When t is non-zero, Θ_l and Θ_r are isomorphisms, so that H_4 is a self-dual Hopf algebra.

We have that $\mathcal{H}_{R_t}$ is a 4-dimensional algebra generated by two elements u and v such that u and v satisfy the relations:

$$u^2 = 1, uv - vu = 0, v^2 = t(1 - u),$$

which is isomorphic to the commutative algebra $k[y]/\langle y^4 - 2ty^2 \rangle$ when t is not zero.

The double algebra $D[H_4]$ with respect to R_t is generated by four elements, g_1, g_2, h_1 and h_2 such that

$$g_i^2 = 1, \ h_i^2 = 0, \ g_i h_j + h_j g_i = 0,$$
$$g_1 g_2 = g_2 g_1, \quad h_1 h_2 + h_2 h_1 = t(1 - g_1 g_2).$$

The comultiplication of $D[H_4]$ is easy because the Hopf subalgebras generated by g_i, h_i, $i = 1, 2$, are isomorphic to H_4. Thus the algebra embedding ϕ reads as follows:

$$\mathcal{H}_{R_t} \longrightarrow D[H_4], \quad \phi(u) = g_1 g_2, \phi(v) = g_1(h_2 - h_1).$$

Let us consider the triangular case (H_4, R), where $R = R_0$. In this case, an algebra A is an H_4-module algebra if and only if it a DS-algebra (see section 2), and A is R-Azumaya algebra if and only if A is a DS-Azumaya algebra. From Theorem 2.5 we know that the Brauer group $BC(k, H_4, R)$ is isomorphic to $(k, +) \times BW(k)$. Let us work out the group $\text{Gal}(\mathcal{H}_R)$ and calculate the Brauer group $BC(k, H_4, R)$ using the sequence (4–12).

First of all we have the following structure theorem of bigalois objects in $\mathcal{E}(\mathcal{H}_R)$ [59].

THEOREM 4.4.1 [59, Thm. 5.7]. *Let A be a bigalois object in $\mathcal{E}(\mathcal{H}_R)$. Then A is either of type (A) or of type (B):*
Type (A): *A is a generalized quaternion algebra $\left(\frac{\alpha, \beta}{k}\right)$, $\alpha \neq 0$, with the following YD H-module structures:*

$$g \cdot u = -u, \qquad g \cdot v = -v,$$
$$h \cdot u = 0, \qquad h \cdot v = 1,$$
$$\rho(u) = u \otimes 1 - 2uv \otimes gh, \quad \rho(v) = v \otimes g + 2\beta \otimes h.$$

Type (B): *A is a commutative algebra* $k(\sqrt{\alpha}) \otimes k(\sqrt{\beta})$ *with the following YD H-module structures*:

$$g \cdot u = u, \qquad\qquad g \cdot v = -v,$$
$$h \cdot u = 0, \qquad\qquad h \cdot v = 1,$$
$$\rho(u) = u \otimes 1 + 2uv \otimes h, \qquad \rho(v) = v \otimes g + 2\beta \otimes h,$$

where $k(\sqrt{\alpha})$ *and* $k(\sqrt{\beta})$ *are generated by elements* u *and* v *respectively, and* u, v *satisfy the relations:* $u^2 = \alpha, uv = vu$ *and* $v^2 = \beta$.

As a consequence of Theorem 4.4.1, we have the group structure of the group $\mathrm{Gal}(\mathcal{H}_R)$:

PROPOSITION 4.4.2 [59, 5.8–5.9]. *The group* $\mathrm{Gal}(\mathcal{H}_R)$ *is equal to* $k \times (k^\bullet/k^{\bullet 2}) \times \mathbb{Z}_2$ *as a set. The multiplication rule on the set is given by*

$$(\beta, \alpha, i)(\beta', \alpha', j) = (\beta + \beta', (-1)^{ij}\alpha\alpha', i + j).$$

The foregoing multiplication rule of $\mathrm{Gal}(\mathcal{H}_R)$ shows that $\mathrm{Gal}(\mathcal{H}_R)$ is a direct product of $(k, +)$ and the group $k^\bullet/k^{\bullet 2} \rtimes \mathbb{Z}_2$ which is isomorphic to the group $Q_2(k)$ of graded quadratic extensions of k (see [57]).

Notice that an object of type (A) in $\mathrm{Gal}(\mathcal{H}_R)$ is some generalized quaternion algebra $\left(\frac{\alpha,\beta}{k}\right)$ with the H_4-action and coaction given in Theorem 4.4.1, where $\alpha \in k^\bullet$ and $\beta \in k$. When $\beta \neq 0$, $\left(\frac{\alpha,\beta}{k}\right)$ is a Galois R-Azumaya algebra if we forget the left H_4-module structure. Since the coinvariant subalgebra of $\left(\frac{\alpha,\beta}{k}\right)$ is trivial, we have $\pi\left(\left(\frac{\alpha,\beta}{k}\right)\right) = \left(\frac{\alpha,\beta}{k}\right)$ if $\beta \neq 0$. To get an object $\left(\frac{\alpha,0}{k}\right)$ in $\mathrm{Gal}(\mathcal{H}_R)$, where $\alpha \in k^\bullet$, we consider the Galois R-Azumaya algebra $\left(\frac{\alpha,1}{k}\right)\#\left(\frac{1,-1}{k}\right)$. Since π is monoidal, we have

$$\pi\left(\left(\frac{\alpha,1}{k}\right)\#\left(\frac{1,-1}{k}\right)\right) = \left(\frac{\alpha,1}{k}\right) \wedge \left(\frac{1,-1}{k}\right) = \left(\frac{\alpha,0}{k}\right)$$

for any $\alpha \in k^\bullet$.

For an object $k(\sqrt{\alpha}) \otimes k(\sqrt{\beta})$ of type (B) in $\mathrm{Gal}(\mathcal{H}_R)$, we choose a Galois R-Azumaya algebra A such that $\pi(A) = \left(\frac{\alpha,\beta}{k}\right)$ (assured by the foregoing arguments). Then it is easy to check that

$$\pi(A\#k(\sqrt{1})) = k(\sqrt{\alpha}) \otimes k(\sqrt{\beta})$$

for $\alpha \in k^\bullet$ and $\beta \in k$. Thus we have shown that the homomorphism $\tilde{\pi}$ is surjective and we have an exact sequence:

$$1 \longrightarrow \mathrm{Br}(k) \longrightarrow \mathrm{BC}(k, H_4, R) \xrightarrow{\tilde{\pi}} \mathrm{Gal}(\mathcal{H}_R) \longrightarrow 1. \qquad (4\text{--}14)$$

Recall that the Brauer–Wall group $\mathrm{BW}(k)$ is $\mathrm{BC}(k, k\mathbb{Z}_2, R')$, where $k\mathbb{Z}_2$ is the sub-Hopf algebra of H_4 generated by the group-like element $g \in H_4$, and R' is the restriction of R to $k\mathbb{Z}_2$. The following well-known exact sequence is a special case of (4–12):

$$1 \longrightarrow \mathrm{Br}(k) \longrightarrow \mathrm{BW}(k) \xrightarrow{\tilde{\pi}} Q_2(k) \longrightarrow 1, \qquad (4\text{--}15)$$

where $Q_2(k)$ is nothing but $\mathrm{Gal}(\mathcal{H}_{R'})$ and $\mathcal{H}_{R'} \cong k\mathbb{Z}_2$, here $H = k\mathbb{Z}_2$.

The sequence (4–15) can be also obtained if we restrict the homomorphism $\tilde{\pi}$ in (4–14) to the subgroup $\mathrm{BW}(k)$ of $\mathrm{BC}(k, H_4, R)$. The subgroup $\tilde{\pi}(\mathrm{BW}(k))$ of $\mathrm{Gal}(\mathcal{H}_R)$ consists of all objects of forms: $\left(\frac{\alpha, 0}{k}\right)$ of type (A) and $k(\sqrt{\alpha}) \otimes k(\sqrt{0})$ of type (B), which is isomorphic to $Q_2(k)$ (see [59] for details). In fact we have the following commutative diagram:

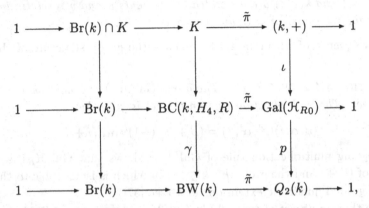

where γ is the canonical map defined in Theorem 2.5, K is the kernel of γ, ι is the inclusion map and p is the projection from $(k, +) \times Q_2(k)$ onto $Q_2(k)$. Here $\tilde{\pi}(K) = (k, +)$ because $\tilde{\pi} \circ \gamma = p \circ \tilde{\pi}$ (which can be easily checked on Galois R-Azumaya algebras $\left(\frac{\alpha, \beta}{k}\right)$ and $\left(\frac{\alpha, \beta}{k}\right)\#k(\sqrt{1})$). By definition of γ we have $\mathrm{Br}(k) \cap K = 1$. It follows that $K \cong (k, +)$. Since γ is split, we obtain that the Brauer group $\mathrm{BC}(k, H_4, R)$ is isomorphic to the direct product group $(k, +) \times \mathrm{BW}(k)$, which coincides with Theorem 2.5.

Recently, G. Carnovale proved in [13] that the Brauer group $\mathrm{BC}(k, H_4, R_t)$ is isomorphic to $\mathrm{BC}(k, H_4, R_0)$ for any $t \neq 0$ although (H_4, R_t) is not coquasitriangularly isomorphic to (H_4, R_0) when $t \neq 0$ [40].

5. The Hopf Automorphism Group

Let H be a faithfully projective Hopf algebra over a commutative ring k. As we have seen from the previous section, the Brauer group $\mathrm{BQ}(k, H)$ may be approximated by computing the group $\mathrm{Gal}(\mathcal{H}_R)$, where $\mathcal{H}_R$ is a deformation of the dual $D(H)^*$ of the quantum double $D(H)$. However, to compute explicitly the group $\mathrm{BQ}(k, H)$ is a hard task. On the other hand, there are some subgroups of $\mathrm{BQ}(k, H)$ which are (relatively) easier to calculate. For instance, when H is commutative and cocommutative, various subgroups of the Brauer–Long group could more easily be studied [3; 4; 7; 8; 10; 17]. One of these subgroups is Deegan's subgroup introduced in [17] which involves the Hopf algebra structure of H itself and in fact turns out to be isomorphic to the Hopf algebra automorphism group $\mathrm{Aut}(H)$ [17; 8]. The connection between $\mathrm{Aut}(H)$ and $\mathrm{BD}(k, H)$ for some particular commutative and cocommutative Hopf algebra H was probably first studied by M.Beattie in [3] where she established the existence of an exact sequence:

$$1 \longrightarrow \mathrm{BC}(k,G)/\mathrm{Br}(k) \times \mathrm{BM}(k,G)/\mathrm{Br}(k) \longrightarrow \mathrm{B}(k,G)/\mathrm{Br}(k) \xrightarrow{\beta} \mathrm{Aut}(G) \longrightarrow 1$$

where $\mathrm{B}(k,G)$ is the subgroup of $\mathrm{BD}(k,G)$ consisting of the classes represented by G-dimodule Azumaya algebras whose underlying algebras are Azumaya, $\mathrm{Aut}(G)$ is the automorphism group of G, G is a finite abelian group and k is a connected ring. Based on Beattie's construction of the map β, Deegan constructed his subgroup $\mathrm{BT}(k,G)$ which is then isomorphic to $\mathrm{Aut}(G)$; the resulting embedding of $\mathrm{Aut}(G)$ in the Brauer–Long group (group case) is known as Deegan's embedding theorem. In [8], S.Caenepeel looked at the Picard group of a Hopf algebra, and extended Deegan's embedding theorem from abelian groups to commutative and cocommutative Hopf algebras. But if H is a quantum group (i.e., a (co)quasitriangular Hopf algebra) or just any non-commutative non-cocommutative Hopf algebra then it seems that the method of Deegan and Caenepeel cannot be extended to obtain a group homomorphism from some subgroup of $\mathrm{BQ}(k,H)$ to the automorphism group $\mathrm{Aut}(H)$. In fact, $\mathrm{Aut}(H)$ can no longer be embedded into $\mathrm{BQ}(k,H)$. On the other hand, the idea of Deegan's construction can still be applied to our non-commutative and non-cocommutative case.

Let M be a faithfully projective Yetter–Drinfel'd H-module. Then $\mathrm{End}_k(M)$ is an H-Azumaya YD H-module algebra. However, if M is an H-bimodule, that is, a left H-module and a right H-comodule, but not a YD H-module, it may still happen that $\mathrm{End}_k(M)$ is a YD H-module algebra.

Take a non-trivial Hopf algebra isomorphism $\alpha \in \mathrm{Aut}(H)$ (for example, if the antipode S of H is not of order two, S^2 is a non-trivial Hopf automorphism). We define a left H-module and a right H-comodule H_α as follows: as a k-module $H_\alpha = H$; we equip H_α with the obvious H-comodule structure given by Δ, and an H-module structure given by

$$h \cdot x = \sum \alpha(h_{(2)})xS^{-1}(h_{(1)})$$

for $h \in H$, $x \in H_\alpha$. Since α is nontrivial H_α is not a YD H-module. Let $A_\alpha = \mathrm{End}(H_\alpha)$ with H-structures induced by the H-structures of H_α, that is,

$$(h \cdot f)(x) = \sum h_{(1)}f(S(h_{(2)}) \cdot x)$$

$$\chi(f)(x) = \sum f(x_{(0)})_{(0)} \otimes S^{-1}(x_{(1)})f(x_{(0)})_{(1)}$$

for $f \in A_\alpha$, $x \in H_\alpha$.

LEMMA 5.1 [12, 4.6, 4.7]. *If H is a faithfully projective Hopf algebra and α is a Hopf algebra automorphism of H, then A_α is an Azumaya YD-module algebra and the following map defines a group homomorphism:*

$$\omega : \mathrm{Aut}(H) \longrightarrow \mathrm{BQ}(k,H), \quad \alpha \mapsto [A_{\alpha^{-1}}].$$

In the sequel, we will compute the kernel of the map ω. Let $D(H)$ denote the Drinfel'd double of H. Let A be an H-module algebra. Recall from [5] that the H-action on A is said to be *strongly inner* if there is an algebra map $f : H \longrightarrow A$ such that

$$h \cdot a = \sum f(h_{(1)}) a f(S(h_{(2)})), \quad a \in A, h \in H.$$

LEMMA 5.2. *Let M be a faithfully projective k-module. Suppose that $\text{End}(M)$ is a $D(H)$-Azumaya algebra. Then $[\text{End}(M)] = 1$ in $\text{BM}(k, D(H))$ if and only if the $D(H)$-action on A is strongly inner.*

The proof has its own interest. Suppose that the $D(H)$-action on A is strongly inner. There is an algebra map $f : D(H) \to A$ such that $t \cdot a = \sum f(t_{(1)}) a f(S(t_{(2)}))$, $t \in D(H), a \in A$. This inner action yields a $D(H)$-module structure on M given by

$$t \to m = f(t)(m), \quad t \in D(H), m \in M.$$

Since f is an algebra representation map the above action does define a module structure. Now it is straightforward to check that the $D(H)$-module structure on A is exactly induced by the $D(H)$-module structure on M defined above. By definition $[\text{End}(M)] = 1$ in $\text{BM}(k, D(H))$.

Conversely, if $[A] = 1$, then there exists a faithfully projective $D(H)$-module N such that $A \cong \text{End}(N)$ as $D(H)$-module algebras by [12, 2.11]. Now $D(H)$ acts strongly innerly on $\text{End}(N)$. Let $u : D(H) \longrightarrow \text{End}(N)$ be the algebra representation map. Now one may easily verify that the strongly inner action induced by the composite algebra map:

$$\mu : D(H) \xrightarrow{u} \text{End}(N) \cong A$$

exactly defines the $D(H)$-module structure on A.

LEMMA 5.3. *For a faithfully projective k-module M, let $u, v : H \longrightarrow \text{End}(M)$ define H-module structures on M, call them M_u and M_v. If $\text{End}(M_u) = \text{End}(M_v)$ as left H-modules via (1–2), then $(v \circ S) * u$ is an algebra map from H to k, i.e., a grouplike element in H^*. Similarly, if M admits two H-comodule structures ρ, χ such that the induced H-comodule structures on $\text{End}(M)$ given by (1–2) coincide, then there is a grouplike element $g \in G(H)$ such that $\chi = (1 \otimes g)\rho$, i.e., $\chi(x) = \sum x_{(0)} \otimes g x_{(1)}$ if $\rho(x) = \sum x_{(0)} \otimes x_{(1)}$ for $x \in M$.*

PROOF. For any $m \in M, h \in H, \phi \in \text{End}(M_u) = \text{End}(M_v)$,

$$\sum u(h_{(1)})[\phi[u(S(h_{(2)}))(m)]] = \sum v(h_{(1)})[\phi[v(S(h_{(2)}))(m)]],$$

or equivalently,

$$\sum v(S(h_{(1)}))[u(h_{(2)})(\phi[u(S(h_{(3)}))(m)])] = \phi(v(S(h))(m)).$$

Let $\lambda = (v \circ S) * u : H \longrightarrow \text{End}(M)$ with convolution inverse $(u \circ S) * v$. Letting $m = u(h_{(4)})(x)$ for any $x \in H$ in the equation above, we obtain $\lambda(h) \in$

$Z(\text{End}(M)) = k$ for all $h \in H$. Since u, v are algebra maps, it is easy to see that λ is an algebra map from H to k.   □

Given a group-like element $g \in G(H)$, g induces an inner Hopf automorphism of H denoted $\bar{g}$, i.e., $\bar{g}(h) = g^{-1}hg$, $h \in H$. Similarly, if λ is a group-like element of H^*, then λ induces a Hopf automorphism of H, denoted by $\bar{\lambda}$ where $\bar{\lambda}(h) = \sum \lambda(h_{(1)})h_{(2)}\lambda^{-1}(h_{(3)})$, $h \in H$. Since $G(D(H)) = G(H^*) \times G(H)$ ([40, Prop.9]) and $\bar{g}$ commutes with $\bar{\lambda}$ in $\text{Aut}(H)$, we have a homomorphism θ:

$$G(D(H)) \longrightarrow \text{Aut}(H), \quad (\lambda, g) \mapsto \bar{g}\bar{\lambda}.$$

Let $K(H)$ denote the subgroup of $G(D(H))$ consisting of elements

$$\{(\lambda, g) \mid \overline{g^{-1}}(h) = \bar{\lambda}(h), \forall h \in H\}.$$

LEMMA 5.4. *Let H be a faithfully projective Hopf algebra. Then $K(H) \cong G(D(H)^*)$.*

PROOF. By [40, Prop.10], an element $g \otimes \lambda$ is in $G(D(H)^*)$ if and only if $g \in G(H), \lambda \in G(H^*)$ and g, λ satisfy the identity:

$$g(\lambda \rightharpoonup h) = (h \leftharpoonup \lambda)g, \quad \forall h \in H,$$

where, $\lambda \rightharpoonup h = \sum h_{(1)}\lambda(h_{(2)})$ and $h \leftharpoonup \lambda = \sum h_{(2)}\lambda(h_{(1)})$. Let $g \in G(H), \lambda \in G(H^*)$, for any $h \in H$, we have

$$\sum gh_{(1)}\lambda(h_{(2)}) = \sum \lambda(h_{(1)})h_{(2)}g \iff \sum h_{(1)}\lambda(h_{(2)}) = \sum \lambda(h_{(1)})g^{-1}h_{(2)}g$$

$$\iff \sum \lambda^{-1}(h_{(1)})h_{(2)}\lambda(h_{(3)}) = \sum g^{-1}hg.$$

This means $g \otimes \lambda$ is in $G(D(H)^*)$ if and only if $(\lambda, g) \in K(H)$. Therefore $K(H) = G(D(H)^*)$.   □

Applying Lemmas 5.1–5.4, one may able to show that the group homomorphism θ can be embedded into the following long exact sequence:

THEOREM 5.5 [52, Thm. 5]. *Let H be a faithfully projective Hopf algebra over k. The following sequence is exact:*

$$1 \longrightarrow G(D(H)^*) \longrightarrow G(D(H)) \xrightarrow{\theta} \text{Aut}(H) \xrightarrow{\omega} \text{BQ}(k, H), \qquad (5\text{–}1)$$

where $\theta(\lambda, g) = \bar{\lambda}\bar{g}$ and $\omega(\alpha) = A_{\alpha^{-1}} = \text{End}(H_{\alpha^{-1}})$.

As a consequence of the theorem, we rediscover the Deegan–Caenepeel's embedding theorem for a commutative and cocommutative Hopf algebra [8; 17].

COROLLARY 5.6. *Let H be a faithfully projective Hopf algebra such that $G(H)$ and $G(H^*)$ are contained in the centers of H and H^* respectively. Then the map ω in the sequence (5–1) is a monomorphism. In particular, if H is a commutative and cocommutative faithfully projective Hopf algebra over k, then $\text{Aut}(H)$ can be embedded into $\text{BQ}(k, H)$.*

Note that in this case, $G(D(H)^*) = G(D(H))$. It follows that the homomorphism θ is trivial, and hence the homomorphism ω is a monomorphism.

In the following, we present two examples of the exact sequence (5–1).

EXAMPLE 5.7. Let H be the Sweedler Hopf algebra over a field k in Subsection 4.4. H is a self-dual Hopf algebra, i.e., $H \cong H^*$ as Hopf algebras. It is straightforward to show that the Hopf automorphism group $\text{Aut}(H)$ is isomorphic to $k^{\bullet} = k \backslash 0$ via:

$$f \in \text{Aut}(H), f(g) = g, f(h) = zh, z \in k^{\bullet}.$$

Considering the group $G(D(H))$ of group-like elements, it is easy to see that

$$G(D(H)) = \{(\varepsilon, 1), (\lambda, 1), (\varepsilon, g), (\lambda, g)\} \cong \mathbb{Z}_2 \times \mathbb{Z}_2$$

where $\lambda = p_1 - p_g$, and p_1, p_g is the dual basis of $1, g$. One may calculate that the kernel of the map θ is given by:

$$K(H) = \{(\varepsilon, 1), (\lambda, g)\} \cong \mathbb{Z}_2$$

The image of θ is $\{\bar{1}, \bar{g}\}$ which corresponds to the subgroup $\{1, -1\}$ of k^*. Thus by Theorem 5.5 we have an exact sequence:

$$1 \longrightarrow \mathbb{Z}_2 \longrightarrow \mathbb{Z}_2 \times \mathbb{Z}_2 \longrightarrow k^{\bullet} \longrightarrow BQ(k, H),$$

It follows that $k^{\bullet}/\mathbb{Z}_2$ can be embedded into the Brauer group $BQ(H)$. In particular, if $k = \mathcal{R}$, the real field, then $Br(\mathcal{R}) = \mathbb{Z}_2 \subset BQ(\mathcal{R}, H)$, and $\mathcal{R}^{\bullet}/\mathbb{Z}_2$ is a non-torsion subgroup of $BQ(\mathcal{R}, H)$.

In the previous example, the subgroup $k^{\bullet}/\mathbb{Z}_2$ of the Brauer group $BQ(k, H)$ is still an abelian group. The next example shows that the general linear group $GL_n(k)$ modulo a finite group of roots of unity for any positive number n may be embedded into the Brauer group $BQ(k, H)$ of some finite dimensional Hopf algebra H.

EXAMPLE 5.8. Let $m > 2, n$ be any positive numbers. Let H be Radford's Hopf algebra of dimension $m2^{n+1}$ over $\mathbb{C}$ (complex field) generated by $g, x_i, 1 \le i \le n$ such that

$$g^{2m} = 1, {x_i}^2 = 0, gx_i = -x_i g, x_i x_j = -x_j x_i.$$

The coalgebra structure Δ and the counit ε are given by

$$\Delta g = g \otimes g, \quad \Delta x_i = x_i \otimes g + 1 \otimes x_i, \quad \varepsilon(g) = 1, \quad \varepsilon(x_i) = 0, \quad 1 \le i \le n.$$

By [40, Prop.11], the Hopf automorphism group of H is $GL_n(\mathbb{C})$. Now we compute the groups $G(D(H))$ and $G(D(H)^*)$. It is easy to see that $G(H) = (g)$ (see also [40, p353]) is a cyclic group of order $2m$. Let $\omega_i, 1 \le i \le m$ be the m-th roots of 1, and let ζ_j be the m-*th* roots of -1. Define the algebra maps η_i and λ_i from H to $\mathbb{C}$ as follows:

$$\eta_i(g) = \omega_i g, \quad \eta_i(x_j) = 0, 1 \le i, j \le m,$$

and

$$\lambda_i(g) = \zeta_i g, \quad \lambda_i(x_j) = 0, 1 \leq i, j \leq m.$$

One may check that $\{\eta_i, \lambda_i\}_{i=1}^n$ is the group $G(H^*)$. It follows that $G(D(H)) = G(H) \times G(H^*) \cong (g) \times U$, where U is the group of $2m$-th roots of 1. To compute $G(D(H)^*)$ it is enough to calculate $K(H)$. Since

$$\bar{g}^i = \begin{cases} \text{id if } i \text{ is even}, \\ \nu \text{ if } i \text{ is odd}, \end{cases}$$

where $\nu(g) = g$, $\nu(x_j) = -x_j, 1 \leq j \leq n$, and

$$\bar{\eta}_i(g) = g, \quad \bar{\eta}_i(x_j) = \omega_i x_j, \quad 1 \leq i, j \leq n,$$
$$\bar{\lambda}_i(g) = g, \quad \bar{\lambda}_i(x_j) = \zeta_i x_j, \quad 1 \leq i, j \leq n.$$

It follows that

$$K(H) = \{(\varepsilon, g^{2i}), (\psi, g^{2i-1}), 1 \leq i \leq m\},$$

where ψ is is given by:

$$\psi(g) = -g, \quad \psi(x_i) = 0.$$

Consequently $G(D(H)^*) \cong U$, Since the base field is $\mathbb{C}$, $(g) \cong U$, and we have an exact sequence

$$1 \longrightarrow U \longrightarrow U \times U \longrightarrow GL_n(\mathbb{C}) \longrightarrow \mathrm{BQ}(\mathbb{C}, H).$$

The two examples above highlight the interest of the study of the Brauer group of a Hopf algebra. In Example 5.8, even though the classical Brauer group $Br(\mathbb{C})$ is trivial, the Brauer group $\mathrm{BQ}(\mathbb{C}, H)$ is still large enough.

In the rest of this section, we consider a natural action of $\mathrm{Aut}(H)$ on $\mathrm{BQ}(k, H)$. Let A be an H-Azumaya algebra, and α a Hopf algebra automorphism of H. Consider the YD H-module algebra $A(\alpha)$, which equals A as a k-algebra, but with H-structures of $(A(\alpha), \rightharpoonup, \chi')$ given by

$$h \rightharpoonup a = \alpha(h) \cdot a \text{ and } \chi'(a) = \sum a_{(0)} \otimes \alpha^{-1}(a_{(1)}) = (1 \otimes \alpha^{-1})\chi(a)$$

for all $a \in A(\alpha)$, $h \in H$.

LEMMA 5.9. *Let A, B be H-Azumaya algebras. If α is a Hopf algebra automorphism of H, then $A(\alpha)$ is an H-Azumaya algebra, and $(A\#B)(\alpha) \cong A(\alpha)\#B(\alpha)$.*

The action of $\mathrm{Aut}(H)$ on $\mathrm{BQ}(k, H)$ is an inner action. Indeed, we have:

THEOREM 5.10 [12, Thm. 4.11]. *$\mathrm{Aut}(H)$ acts innerly on $\mathrm{BQ}(k, H)$, more precisely, for any H-Azumaya algebra B and $\alpha \in \mathrm{Aut}(H)$, we have $[B(\alpha)] = [A_\alpha][B][A_{\alpha^{-1}}]$.*

Theorem 5.10 yields the multiplication rule for two elements $[B]$ and $\omega(\alpha) = [A_\alpha]$ where B is any H-Azumaya algebra and α is in $\text{Aut}(H)$. In particular, if T is a subgroup such that T is invariant (or stable) under the action of $\text{Aut}(H)$, then the subgroup generated by T and $\omega(\text{Aut}(H))$ is $(\iota \otimes \omega)(T \rtimes \text{Aut}(H))$, where $\rtimes$ is the usual semi-direct product of groups.

EXAMPLE 5.11. Let (H_4, R_0) be the CQT Hopf algebra described in Subsection 4.4. The automorphism group $\text{Aut}(H_4)$ of H_4 is isomorphic to $k^\bullet$. If A is an R_0-Azumaya algebra and $\alpha \in \text{Aut}(H_4)$, then $A(\alpha)$ is still an R_0-Azumaya algebra as the automorphism α does not affect the induced action (1-5). Thus the subgroup $\text{BC}(k, H_4, R_0)$ is stable under the action of $\text{Aut}(H_4)$. By Example 5.7 we get a non-abelian subgroup of $\text{BQ}(k, H_4)$ (see [56]):

$$(\iota \otimes \omega)(\text{BC}(k, H_4, R_0) \rtimes k^\bullet) \cong \text{BC}(k, H_4, R_0) \rtimes (k^\bullet/\mathbb{Z}_2)$$

$$\cong \text{BW}(k) \times (k, +) \rtimes (k^\bullet/\mathbb{Z}_2).$$

6. The Second Brauer Group

In the classical Brauer group theory of a commutative ring k, an Azumaya algebra can be characterized as a central separable algebra over k. However this is not the case when we deal with the Brauer group of structured algebras. For instance, in the Brauer–Wall group of a commutative ring k, a representative (i.e., a $\mathbb{Z}_2$-graded Azumaya algebra) is not necessarily a central separable algebra, instead it is a graded central and graded separable algebra. Motivated by the example of the Brauer–Wall group, one may be inspired to try to define the Brauer group of a Hopf algebra H by using H-separable algebras in a natural way as we did for the Brauer-long group of $\mathbb{Z}_2$-dimodule algebras in section 2. In 1974, B. Pareigis for the first time defined two Brauer groups in a symmetrical category (see [38]). The first Brauer group was defined in terms of Morita equivalence whereas the second Brauer group was defined by so called 'central separable algebras' in the category. The two Brauer groups happen to be equal if the unit of the symmetrical category is a projective object. This is the case if the category is the $\mathbb{Z}_2$-graded module category with the graded product (2-2).

However the dimodule category of a finite abelian group (or a finite commutative cocommutative Hopf algebra) is not a symmetric category, instead a braided monoidal category ([32]). Therefore, the definition of the second Brauer group due to Pareigis can not be applied to the dimodule category. Nevertheless, we are able to modify Pareigis's definitions to get the proper definitions of the two Brauer groups for a braided monoidal category (see [52]) so that they allow to recover all known Brauer groups. We are not going into the details of the categorical definitions given in [52]. Instead we will focus our attention on the Yetter Drinfel'd module category of a Hopf algebra.

Like the Brauer–Wall group $\text{BW}(k)$, the Brauer–Long group $\text{BD}(k, H)$ of a finite commutative and cocommutative Hopf algebra H over a commutative ring

k can be defined by central separable algebras in the category when k is nice (e.g., k is a field with characteristic 0. However, a counter example exists when k is not so nice, e.g., $BD(k, \mathbb{Z}_2)$ when 2 is not a unit in k (see [7]).

When a Hopf algebra is not commutative and cocommutaive, even if k is a field with $\text{ch}(k) = 0$, the second Brauer group defined by H-separable algebras turns out to be smaller than the Brauer group of H-Azumaya algebras. In other words, an H-Azumaya algebra is not necessarily an H-separable algebra. Some examples of this will be presented. Let H be a Hopf algebra over k and let A^e (or $^e A$) be the H-enveloping algebra $A \# \bar{A}$ (or $\bar{A} \# A$).

DEFINITION 6.1. Let A be a YD H-module algebra. A is said to be H-*separable* if the following exact sequence splits in $_{A^e} \mathcal{Q}^H$:

$$A^e \xrightarrow{\pi_A} A \longrightarrow 0.$$

In this section π_A is the usual multiplication of A. We will often use M_0 to stand for $M^H \bigcap M^{coH}$, the intersection of the invariants and the coinvariants of YD H-module M. A H-separable algebra can be described by separability idempotent elements.

PROPOSITION 6.2. *Let A be a* YD H-module algebra. The following statements are equivalent:

(1) A is H-separable.
(2) There exists an element $e_l \in A_0^e$ such that $\pi_A(e_l) = 1$ and $(a\#1)e_l = (1\#\bar{a})e_l$ for all $a \in A$.
(3) There exists an element $c \in (A\#A)_0$ such that $\pi_A(e_l) = 1$ and $(a\#1)e = e(1\#a)$ for all $a \in A$.
(4) There exists an element $e_r \in {}^e A_0$ such that $\pi_A(e_r) = 1$ and $e_r(\bar{a}\#1) = e_r(1\#a)$ for all $a \in A$.
(5) $\pi_A : {}^e A \longrightarrow A \longrightarrow 0$ splits in $\mathcal{Q}^H_{eA}$.

The proof of these statements is straightforward. We emphasize that H-separable algebras are k-separable by the statement (3). However a separable YD H-module algebra is not necessarily H-separable. For instance, the H_4-Azumaya algebra $\langle \frac{1,-1}{k} \rangle$ is a separable algebra and $k\mathbb{Z}_2$-separable, but not a H_4-separable algebra. We also have that H_4 itself is an H_4-Azumaya algebra, but it is certainly not a separable algebra over any field.

If $e_l = \sum x_i \# \bar{y}_i$ is a separability idempotent in $A \# \bar{A}$, we may choose $e = \sum x_i \# y_i$ and $e_r = \sum \bar{x}_i \# y_i$. Thus we may write e_A for e_l and e'_A for e_r without ambiguity. Since e_A is an idempotent element in each of the above cases, it follows that if A is H-separable then $M^A = e_A \rightharpoonup M$ and $^A N = N \leftharpoonup e'_A$ for $M \in {}_{A^e}\mathcal{Q}^H$ and $N \in \mathcal{Q}^H_{eA}$ respectively. In particular, $A^A = e_A \rightharpoonup A$ and $^A A = A \leftharpoonup e'_A$. For a YD H-module algebra A we shall call A^A and $^A A$ the *left* and the *right H-center* of A respectively. In case $A^A = k$ or $^A A = k$ we shall say that A is *left* or *right central* respectively, and A is *H-central* if A is both left and right central.

PROPOSITION 6.3. (1) *Let $f : A \longrightarrow B$ be an epimorphism of* YD *H-module algebras. If A is H-separable then B is H-separable.*

(2) *Let E be a commutative k-algebra. If A is H-separable then $E \otimes_k A$ is an $E \otimes_k H$-separable E-algebra.*

(3) *If A is H-separable, then $\bar{A}$ is H-separable. If in addition, A is left (or right) central then $\bar{A}$ is right (or left) central respectively.*

(4) *If A, B are H-separable, so is $A \# B$. If in addition, A and B are left (or right) central, then $A \# B$ is left (or right) central.*

Like the classical case, the ground ring k is an H-direct summand of of a left (or right) central H-separable algebra.

LEMMA 6.4. *Let A be a left or right H-central H-separable algebra. Then the inclusion map embeds k as a direct summand of A in $\mathcal{Q}^H$.*

PROOF. Let e be an H-separability idempotent of A. Then the map $T_e : A \longrightarrow k$ given by $T_e(a) = e \rightharpoonup a$ for $a \in A$ is a YD H-module map. We have $e \rightharpoonup 1 = \pi_A(e) = 1$. □

The map T_e described above is a section for the inclusion map $\iota : k \hookrightarrow A$ in $\mathcal{Q}^H$, that is, $T_e \circ \iota = $ id. We will call a YD H-module map $T : A \longrightarrow k$ an *H-trace map* of a YD H-module algebra A if $T(1) = 1$. Notice that usually a trace map is an onto map but does not necessarily carry the unit to the unit. We will show later that an H-Azumaya algebra A is an H-central H-separable algebra if and only if A has an H-trace map. It follows that H-trace maps in a one-to-one way correspond to H-separability idempotents when A is an H-Azumaya algebra.

A YD H-module algebra A is said to be *H-simple* if A has no proper YD H-module ideal (simply *H-ideal*). This is equivalent to A being simple in $_{A^e}\mathcal{Q}^H$ or $\mathcal{Q}^H_{e A}$.

PROPOSITION 6.5. *Let A be a left (or right) H-central H-separable algebra. Then A is H-simple if and only if k is a field.*

PROOF. Suppose that A is H-simple and I is a non-zero ideal of k. IA is an H-ideal of A and $IA = A$. Let t be the H-trace map described in Lemma 6.4. Then $t(IA) = t(A)$ implies $I = k$. It follows that k is a field.

Conversely, suppose that A is an H-separable algebra over a field k. Since H-separability implies k-separability, A is semisimple artinian. Let M be an H-ideal of A, then there exists a central idempotent $c \in A$ such that $M = cA = Ac$. c must be in A_0. Now for any $a \in A$, we have

$$\sum a_{(0)}(a_{(1)} \cdot c) = ac = ca, \quad \sum c_{(0)}(c_{(1)} \cdot a) = ca = ac.$$

Thus c is in both A^A and $^A A$. Now if A is left or right H-central H-separable algebra then $c \in k$, and hence $M = cA = A$. □

LEMMA 6.6. *If A is a left or right H-central H-separable algebra, then for any maximal H-ideal I of A there exists a maximal ideal α of k such that $I = \alpha A$ and $I \cap k = \alpha$.*

In view of Proposition 6.5 and Lemma 6.6, one may use an arqument similar to the classical case [14, 2.8] to obtain that a H-central H-separable algebra is an H-Azumaya algebra with an H-trace map. In fact, we have the following:

THEOREM 6.7. *A YD H-module algebra A is an H-central H-separable algebra if and only if A is an H-Azumaya algebra with an H-trace map.*

PROOF. By the foregoing remark, it is sufficient to show that an H-Azumaya algebra with an H-trace map is H-central and H-separable. Assume that A is an H-Azumaya algebra with an H-trace map T. Since A is H-Azumaya we have the isomorphism $A\#\overline{A} \cong \text{End}(A)$. In this way we may view T as an element in $A\#\overline{A}$. In fact T is in $(A\#\overline{A})_0$ since T is H-linear and H-colinear. We now can show that T is an H-separability idempotent of A^e. Now $\pi_A(T) = T(1) = 1$, and for any $a, x \in A$,

$$(a\#1)T(x) = aT(x) = (1\#\overline{a})T(x),$$

because $T(x) \in k$. It follows from the foregoing equalities that we have $(a\#1)T = (1\#\overline{a})T$ for any $a \in A$. Therefore, A is H-separable. $\qquad\square$

In general, an H-Azumaya algebra is not necessarily H-separable, in other words, an H-Azumaya algebra need not have an H-trace map. For example, H_4 is not a separable algebra, but it is an H_4-Azumaya algebra (see [56]). For this reason, we call an H-central H-separable algebra a *strongly H-Azumaya* algebra (for short we say that it is strong).

CORROLLARY 6.8. *Let A, B be H-Azumaya algebras. If $A\#B$ is strong, so are A and B.*

PROOF. By Theorem 6.7 it is enough to show that both A and B have an H-trace map. This is the case since $A\#B$ has an H-trace map T and the restriction map $T_A(a) = T(a\#1)$ and T_B are clearly H-trace maps of A and B respectively. $\qquad\square$

This corollary indicates that even the trivial H-Azumaya algebra $\text{End}(M)$, for M a faithfully projective YD H-module, is not necessarily strong. For example, if A is non-strongly H-Azumaya, e.g., $A = H_4$, then $\overline{A}$ is not strong, and hence $\text{End}(A) \cong A\#\overline{A}$ is not strong by Corollary 6.8. So a strongly H-Azumaya algebra may be Brauer equivalent to a non-strongly H-Azumaya algebra. Now a natural question arises. What condition has to be imposed on H so that any H-Azumaya algebra is strongly H-Azumaya? We have a complete answer for a faithfully projective Hopf algebra, and a partial answer for an infinite Hopf algebra over a field with characteristic 0.

PROPOSITION 6.9. *Let H be a faithfully projective Hopf algebra over k. The following are equivalent*:

(1) *Any H-Azumaya algebra is strongly H-Azumaya.*
(2) *Any elementary H-Azumaya algebra is strongly H-Azumaya.*
(3) *There exist an integral $t \in H$ and an integral $\varphi \in H^*$ such that $\varepsilon(t) = 1$ and $\varphi(1) = 1$.*
(4) *k is a projective object in $\mathcal{Q}^H$.*

PROOF. (1) $\Longleftrightarrow$ (2) due to Corollary 6.8. To prove that (2) $\Longrightarrow$ (3), we take the faithfully projective YD H-module M which is the left regular H-module of H itself, with the H-comodule structure given by

$$\rho(h) = \sum h_{(2)} \otimes h_{(3)} S^{-1}(h_{(1)})$$

for any $h \in M$. Now let A be the elementary H-Azumaya algebra $\mathrm{End}(M)$. Since A is strong, A has an H-trace map, say, $T : A \longrightarrow k$. Since A is faithfully projective, A^* is a YD H-module. We may view T as an element in A_0^* as T is H-linear and H-colinear. Identify A^* with $M \tilde{\otimes}^* M$ as a YD H-module. One may easily check that the k-module A^{*H} of H-invariants of A^* consists of elements of form

$$\left\{ \sum t_{(1)} \otimes t_{(2)} \rightharpoonup f \mid t \in \int_l, \ f \in {}^*M \right\}$$

where $\int_l$ is the rank one k-module of left integrals of H and $(t_{(2)} \rightharpoonup f)(h) = f(S^{-1}(t_{(2)})h)$ for any $h \in M$. It follows that $T = \sum t_{(1)} \otimes t_{(2)} \rightharpoonup f$ for some left integral $t \in H$ and an element f in ${}^*M$. Let $\{m_i \otimes p_i\}$ be a dual basis of M so that $\sum m_i \otimes p_i = 1_A$. Since $T(1_A) = 1_k$, we obtain:

$$T(1_A) = \sum p_i(t_{(1)}) f(S^{-1}(t_{(2)})m_i) = \sum f(S^{-1}(t_{(2)})t_{(1)}) = \varepsilon(t)f(1) = 1.$$

So $\varepsilon(t)$ is a unit of k, and one may choose a left integral t' to replace t so that $\varepsilon(t') = 1$.

Similarly, if we choose a faithfully projective YD H-module M as follows: $M = H$ as a right H-comodule with the comultiplication as the right comodule structure and with the adjoint left H-action given by

$$h \cdot m = \sum h_{(2)} m S^{-1}(h_{(1)}),$$

then one may find a left integral $\varphi \in H^*$ such that $\varphi(1) = 1$.

(3) $\Longleftrightarrow$ (4) is the Maschke theorem. Since H is a faithfully projective Hopf algebra, the quantum double Hopf algebra $D(H)$ is faithfully projective over k, and $D(H)$ is a projective object in $\mathcal{Q}^H$. Assume that there are two left integrals $t \in H$ and $\varphi \in H^*$ such that $\varepsilon(t) = 1$ and $\varphi(1) = 1$. The counit of $D(H)$ is a YD H-module map which is split by the YD H-module map $\iota' : k \longrightarrow D(H)$ sending the unit 1 to the element $\varphi \bowtie t$. So k is a YD H-module direct summand of $D(H)$, and hence it is projective in $\mathcal{Q}^H$. The converse holds as the foregoing argument can be reversed.

Finally, we show that (3) $\Longrightarrow$ (1). Assume $\varphi \bowtie t$ is a left integral of $D(H)$ such that $\varepsilon(t) = 1 = \varphi(1)$. If A is an H-Azumaya algebra, then the multiplication map

$$\pi : A \# \bar{A} \longrightarrow A$$

splits as a left $A\#\bar{A}$-module. Let $\mu : A \longrightarrow A\#\bar{A}$ be the split map, and let $e = \mu(1)$. Then $(\varphi \bowtie t) \cdot e$ is an H-separability element of A. So A is strongly H-Azumaya. $\qquad\square$

If H is not a faithfully projective Hopf algebra, we have a sufficient condition which requires that the antipode of H be involutory.

PROPOSITION 6.10. *Let k be a field with* $\mathrm{ch}(k) = 0$ *and let H be a Hopf algebra over k. If the antipode S of H is involutory, then any H-Azumaya algebra is strongly H-Azumaya.*

PROOF. By Corollary 6.8, it is enough to show that any elementary H-Azumaya algebra is strong. Let M be a faithfully projective YD H-module, and let $A = \mathrm{End}(M)$. We show that A has an H-trace map. Identify A with $M \otimes M^*$ as YD H-modules. Let tr be the normal trace map of A which sends the element $m \otimes m^*$ to $m^*(m)$. Since $\mathrm{ch}(k) = 0$, we have that $\mathrm{tr}(1_A) = n$ for some integer is a unit. We show that tr is a YD H-module map, then the statement follows. Indeed, if $h \in H$, $m \in M$ and $m^* \in M^*$, we have

$$\mathrm{tr}(h \cdot (m \otimes m^*)) = \sum \mathrm{tr}(h_{(1)} \cdot m \otimes h_{(2)} \cdot m^*) = \sum (h_{(2)} \cdot m^*)(h_{(1)} \cdot m)$$

$$= \sum m^*(S(h_{(2)})h_{(1)} \cdot m) = \varepsilon(h) \, \mathrm{tr}(m).$$

Similarly, tr is H-colinear as well. $\qquad\square$

Note that when $\mathrm{ch}(k) = 0$ the condition $S^2 = \mathrm{id}$ is equivalent to the condition (3) in Proposition 6.9 if H is finite dimensional (see [28]). However, this is not the case when H is not finite. There is an example of a Hopf algebra that is involutory (e.g., $T(V)$, the universal enveloping Hopf algebra), but without integrals. Nevertheless, it remains open whether k is a projective object in $\mathcal{Q}^H$ if and only if $S^2 = \mathrm{id}$ in case $\mathrm{ch}(k) = 0$.

As mentioned in the title of this section, we are able to define the second Brauer group of strongly H-Azumaya algebras as a result of proposition 6.3. That is, the second Brauer group, denoted $\mathrm{BQs}(k, H)$, consists of isomorphism classes of strongly H-Azumaya algebras modulo the same Brauer equivalence, where the elementary H-Azumaya algebras $\mathrm{End}(M)$ are required to be H-separable. It is evident from Theorem 6.7 that the second Brauer group $\mathrm{BQs}(k, H)$ is a subgroup of $\mathrm{BQ}(k, H)$, which contains the usual Brauer group $\mathrm{Br}(k)$ as a normal subgroup. Let us summarize it as follows:

COROLLARY 6.11. *The subset $\mathrm{BQs}(k, H)$ represented by the strongly H-Azumaya algebras is a subgroup of $\mathrm{BQ}(k, H)$.*

PROOF. The only thing left to check is the coincidence of the two Brauer equivalence relations. Assume that A and B are two strongly H-Azumaya algebras and $[A] = [B]$ in BQ(k, H). Then $A \# \overline{B} \cong \text{End}(M)$ for some faithfully projective YD H-module M. It follows that $A \# \text{End}(B) \cong B \# \text{End}(M)$. By Proposition 6.3, End(B) and End(M) are strongly H-Azumaya algebras, and we obtain that $[A] = [B]$ in BQs(k, H). $\qquad\qquad\qquad\qquad\qquad\qquad\qquad\qquad\qquad$ $\square$

Now the question arises: when is BQs(k, H) equal to BQ(k, H)? When a Hopf algebra satisfies the assumption in Proposition 6.9 or Proposition 6.10, BQ$(k, H) = $ BQs(k, H). However, since a strongly H-Azumaya algebra may be Brauer equivalent to a non-strongly H-Azumaya algebra, there is a possibility that for some Hopf algebra H, any BQ(k, H) element can be represented by a strongly H-Azumaya algebra, but at the same time there may exist non-strongly H-Azumaya algebras.

Note that for a quasitriangular or coquasitriangular Hopf algebra H, The H-central and H-separable or strongly H-Azumaya algebras are special cases of those above. For example, If (H, R) is a cosemisimple-like coquasitriangular Hopf algebra, then BCs$(k, H, R) = $ BC(k, H, R). If G a finite abelian group, $H = kG$ with a bilinear map $\varphi : G \times G \longrightarrow k$, Then H is a cosemisimple-like coquasitriangular Hopf algebra. Thus all graded $(H\text{-})$ Azumaya algebra are strongly H-Azumaya [14; 15].

Acknowledgement. We are grateful to the referee for valuable comments and helpful suggestions that made the current version more readable.

References

[1] E. Abe, Hopf algebras, Cambridge University Press, 1977.

[2] M. Auslander and O. Goldman, The Brauer group of a commutative ring, Trans. Amer. Math. 97 (1960), 367–409.

[3] M. Beattie, A direct sum decomposition for the Brauer group of H-module algebras, J. Algebra 43 (1976), 686–693.

[4] M. Beattie, The Brauer group of central separable G-Azumaya algebras, J. Algebra 54 (1978), 516–525

[5] R. J. Blattner, M. Cohen, and S. Montgomery, Crossed products and inner actions of Hopf algebras, Trans. Amer. Math. Soc. 298 (1986), 672–711.

[6] S. Caenepeel, Brauer groups, Hopf algebras and Galois theory, K-Monographs in Mathematics 4 (1998), Kluwer Academic Publishers.

[7] S. Caenepeel, Computing the Brauer–Long group of a Hopf algebra I: the cohomological theory, Israel J. Math. 72 (1990), 38–83.

[8] S. Caenepeel, Computing the Brauer–Long group of a Hopf algebra II: the Skolem–Noether theory, J. Pure Appl. Algebra 84 (1993),107–144.

[9] S. Caenepeel, The Brauer–Long group revisited: the multiplication rules, Algebra and number theory (Fez), Lecture Notes in Pure and Appl. Math. 208 (2000), 61–86, Dekker, New York.

[10] S. Caenepeel and M. Beattie, A cohomological approach to the Brauer–Long group and the groups of Galois extensions and strongly graded rings, Trans. Amer. Math. Soc. 324 (1991), 747–775.

[11] S. Caenepeel, F. Van Oystaeyen and Y. H. Zhang, Quantum Yang–Baxter module algebras, K-Theory 8 (1994), 231–255.

[12] S. Caenepeel, F. Van Oystaeyen and Y. H. Zhang, The Brauer group of Yetter–Drinfel'd module algebras, Trans. Amer. Math. Soc. 349 (1997), 3737–3771.

[13] G. Carnovale, Some isomorphism for the Brauer groups of a Hopf algebra, preprint UIA, 2000.

[14] L. N. Childs, G. Garfinkel and M. Orzech, The Brauer group of graded Azumaya algebras, Trans. Amer. Math. Soc. 175 (1973), 299–326.

[15] L. N. Childs, The Brauer group of graded Azumaya algebras II: graded Galois extensions, Trans. Amer. Math. Soc. 204 (1975), 137–160.

[16] J. Cuadra, F. Van Oystaeyen and Y. H. Zhang, The Brauer group of Dihedral group D_3, in preparation.

[17] A. P. Deegan, A subgroup of the generalized Brauer group of Γ-Azumaya algebras, J. London Math. Soc. 2 (1981), 223–240.

[18] F. DeMeyer and T. Ford, Computing the Brauer group of $\mathbb{Z}_2$-dimodule algebras, J. Pure Appl. Algebra 54 (1988), 197–208.

[19] F. DeMeyer and E. Ingraham, Separable algebras over commutative rings, Lecture Notes in Mathematics 181 (1971), Springer-Verlag, Berlin. 151 (1971).

[20] Y. Doi and M. Takeuchi, Hopf–Galois extensions of algebras, the Miyashita–Ulbrich action, and Azumaya algebras. J. Algebra 121 (1989), 488–516.

[21] Y. Doi and M. Takeuchi, Multiplication alternation by two cocycles-the quantum version, Comm. in Algebra, 22 (1994), 5715–5732.

[22] Y. Doi and M. Takeuchi, Quaternion algebras and Hopf crossed products, Comm. in Algebra 23 (1995), 3291–3325.

[23] V. G. Drinfel'd, Quantum groups, Proc. of the Int. Congress of Math. Berkeley, Ca. (1987), 798–819.

[24] M. Koppinen, A Skolem–Noether theorem for Hopf algebra measurings, Arch. Math. 13 (1981), 353–361.

[25] M. A. Knus and M. Ojanguren, Théorie de la descente et algèbres d'Azumaya, Lecture Notes in Math. 389 (1974), Springer-Verlag, Berlin.

[26] H. K. Kreimer and M. Takeuchi, Hopf algebras and Galois extensions, Indiana Univ. Math. J. 30 (1981), 675–692.

[27] L. A. Lambe and D. E. Radford, Algebraic aspects of the quantum Yang–Baxter equation, J. Algebra 154 (1992), 228–288.

[28] R. G. Larson and D. E. Radford, Semisimple cosemisimple Hopf algebras, J. Algebra Amer. J. Math. 109 (1987), 187–195.

[29] F. W. Long A generalization of the Brauer group graded algebras, Proc. London Math. Soc. 29 (1974), 237–256.

[30] F. W. Long, The Brauer group of dimodule algebras, J. Algebra 31 (1974), 559–601.

[31] S. Majid, Foundations of quantum group theory, Cambridge University Press, 1995.

[32] S. Majid, Doubles of quasitriangular Hopf algebras, Comm. in Algebra 19 (1991), 3061–3073.

[33] A. Masuoka, Quotient theory of Hopf algebras, Advances in Hopf algebras (Chicago, IL, 1992), Lecture Notes in Pure and Appl. Math. 158 (1994), 107–133, Dekker, New York.

[34] Y. Miyashita, An exact sequence associated with a generalized crossed product, Nagoya Math. J. 49 (1973), 21–51.

[35] S. Montgomery, Hopf algebras and their actions on rings, CBMS, 82 (1992).

[36] M. Orzech, Brauer groups of graded algebras, Lecture Notes in Math. 549 (1976), 134–147, Springer-Verlag, Berlin.

[37] M. Orzech, On the Brauer group of algebras having a grading and an action, Canadian J. Math. 32 (1980), 533–552.

[38] B. Pareigis, Non-additive ring and module theory IV, the Brauer group of a symmetric monoidal category, Lecture Notes in Math. 549 (1976), 112–133, Springer-Verlag.

[39] R. S. Pierce, Associative algebras, Springer-Verlag, Berlin, 1982.

[40] D. E. Radford, Minimal quasitriangular Hopf algebras, J. Algebra 157 (1993), 285–315.

[41] D. J. Saltman, The Brauer group is torsion, Proc. Amer. Math. Soc. 81 (1981), 385–387.

[42] P. Schauenburg, Hopf bi-Galois extensions. Comm. in Algebra 24 (1996), no. 12, 3797–3825.

[43] P. Schauenburg, Bi-Galois objects over the Taft algebras. Israel J. Math. 115 (2000), 101–123.

[44] H.-J., Schneider, Principal homogeneous spaces for arbitrary Hopf algebras, Israel J. Math. 72 (1990), 167–195.

[45] H.-J., Schneider, Representation theory of Hopf Galois extensions, Israel J. Math. 72 (1990), 196–231.

[46] C. Small, The Brauer–Wall group of a commutative ring, Trans. Amer. Soc. 156 (1971), 455–491.

[47] M. E. Sweedler, Hopf algebras, Benjamin, 1969.

[48] K.-H. Ulbrich, Galoiserweiterungen von nicht-kommutativen Ringen, Comm. in Algebra 10 (1982), 655–672.

[49] K.-H. Ulbrich, A exact sequence for the Brauer group of dimodule Azumaya algebras, Math. J. Okayama Univ. 35 (1993), 63–88.

[50] F. Van Oystaeyen, Pseudo-places algebras and the symmetric part of the Brauer group, Ph.D. dissertation, March 1972, Vrije Universiteit Amsterdam.

[51] F. Van Oystaeyen, Derivations of Graded Rings and Clifford Systems, J. Algebra 98 (1986), 485–498.

[52] F. Van Oystaeyen and Y. H. Zhang, The embedding of automorphism group into the Brauer group, Canadian Math. Bull. 41 (1998), 359–367.

[53] F. Van Oystaeyen and Y. H. Zhang, The Brauer group of a braided monoidal category, J. Algebra 202 (1998), 96–128.

[54] F. Van Oystaeyen and Y. H. Zhang, The Brauer group of the Sweedler's Hopf algebra H_4, Proc. Amer. Math. Soc. 129 (2001), 371–380.

[55] F. Van Oystaeyen and Y. H. Zhang, Bi-Galois Objects form a group, preprint 1993 (unpublished).

[56] F. Van Oystaeyen and Y. H. Zhang, Computing subgroups of the Brauer group of H_4, to appear in Comm. in Algebra.

[57] C. T. C. Wall, Graded Brauer groups, J. Reine Agnew. Math. 213 (1964), 187–199.

[58] D. N. Yetter, Quantum groups and representations of monoidal categories, Math. Proc. Cambridge Philos. Soc. 108 (1990), 261–290.

[59] Y. H. Zhang, An exact sequence for the Brauer group of a finite quantum group, preprint.

FREDDY VAN OYSTAEYEN
DEPARTMENT OF MATHEMATICS
UNIVERSITY OF ANTWERP (UIA)
ANTWERP B-2610
BELGIUM
voyst@uia.ua.ac.be

YINHUO ZHANG
DEPARTMENT OF MATHEMATICS AND COMPUTING SCIENCE
THE UNIVERSITY OF THE SOUTH PACIFIC
SUVA
FIJI
zhang_y@manu.usp.ac.fj